T0181565

Lecture Notes in Computer Science 6650

Commenced Publication in 1973
Founding and Former Series Editors:
Gerhard Goos, Juris Hartmanis, and Jan van Leeuwen

Oded Goldreich et al.

Studies in Complexity and Cryptography

Miscellanea on the Interplay
between Randomness and Computation

In Collaboration with
Lidor Avigad, Mihir Bellare, Zvika Brakerski
Shafi Goldwasser, Shai Halevi, Tali Kaufman, Leonid Levin
Noam Nisan, Dana Ron, Madhu Sudan, Luca Trevisan
Salil Vadhan, Avi Wigderson, David Zuckerman

 Springer

Volume Editor

Oded Goldreich
Weizmann Institute of Science
Faculty of Mathematics and Computer Science
76100 Rehovot, Israel
E-mail: oded.goldreich@weizmann.ac.il

Cover illustration: Artwork by Harel Luz, Tel Aviv, Israel.

ISSN 0302-9743   e-ISSN 1611-3349
ISBN 978-3-642-22669-4   e-ISBN 978-3-642-22670-0
DOI 10.1007/978-3-642-22670-0
Springer Heidelberg Dordrecht London New York

Library of Congress Control Number: 2011932979

CR Subject Classification (1998): F.1, E.3, E.1, F.4.1, G.2.2, G.1

LNCS Sublibrary: SL 1 – Theoretical Computer Science and General Issues

Typesetting: Camera-ready by author, data conversion by Scientific Publishing Services, Chennai, India

Printed on acid-free paper

Springer is part of Springer Science+Business Media (www.springer.com)

Preface

This volume contains a collection of studies in the areas of complexity theory and foundations of cryptography. These studies were conducted at different times during the last couple of decades. Although many of these studies have been referred to by other works, none of them was formally published before.

Indeed, this volume is quite unusual, and it raises two opposite questions regarding the publication of the foregoing studies: (1) why were these studies not published (formally) before, and (2) why are they being published now?

Let me start with the second question. In the years that have elapsed since the completion of many of these individual studies, I have occasionally looked at them for some reason. On these occasions, I felt that it is somewhat inappropriate that these works were never published formally (although many of them were posted on forums such as ECCC). The current volume is aimed at amending this situation somewhat.

Turning to the first question, the answer varies according to the case. Regarding the surveys and the programmatic and/or reflective articles, the answer is quite straightforward: The standard publication venues for research in complexity and/or cryptography do not welcome such articles, which may reflect the unfortunate fact that our community does not hold such articles in high esteem. Regarding the articles that describe research contributions, the answer varies from the non-existence of an adequate venue (at least at the relevant time), to unjustified (in retrospect) timidness regarding the work.

The late publication of some of these articles also raises questions regarding the relation of the current versions to the original ones. These questions are addressed at the beginning of each individual article, where the original posting is stated and the nature of the revision is outlined. In general, all articles were revised (based on their last posted version), but the revision attempts to preserve the spirit of the original work. In the few cases that later developments suggest a different perspective and/or technical improvements, this is stated explicitly while comparing the original perspective and/or results with the current one.

The compilation of this volume led me to complete the writing of a couple of surveys. In addition, I decided to also include in this volume a few rather recent research contributions.

The studies in this volume are arranged in three parts. Part I contains 20 research contributions, Part II contain 12 surveys (and one overview essay on "Randombess and Computation"), and Part III contains three programmatic and/or reflective articles. Most studies in Part I (and a couple of the studies in Part II) were conducted by me in collaboration with other researchers.

The topics addressed in the various studies include average-case complexity, complexity of approximation, derandomization, expander graphs, hashing functions, locally testable codes, machines that take advice, NP-completeness, one-way functions, probabilistically checkable proofs (PCPs), proofs of knowledge, property testing, pseudorandomness, randomness extractors, sampling, trapdoor permutations, zero-knowledge and non-interative zero-knowledge (NIZK). Indeed, one may say that most of these works belong to the interplay between randomness and computation.

Part I: Research Contributions

Part II: Surveys

Part III: Programmatic and Reflective Articles

1. On Security Preserving Reductions – A Suggested Terminology
2. Contemplations on Testing Graph Properties
3. Another Motivation for Reducing the Randomness Complexity of Algorithms

I am grateful to all of my co-authors of the papers included in the current volume: Lidor Avigad, Mihir Bellare, Zvika Brakerski, Shafi Goldwasser, Shai Halevi, Tali Kaufman, Leonid Levin, Noam Nisan, Dana Ron, Madhu Sudan, Luca Trevisan, Salil Vadhan, Avi Wigderson, and David Zuckerman. In addition, I wish to thank all researchers who have contributed to the research being surveyed in this volume.

Oded Goldreich

Table of Contents

Research Contributions

Surveys

Programmatic and Reflective Articles

Finding the Shortest Move-Sequence in the Graph-Generalized 15-Puzzle Is NP-Hard

Oded Goldreich

Abstract. Following Wilson (*J. Comb. Th.* (B), 1975), Johnson (*J. of Alg.*, 1983), and Kornhauser, Miller and Spirakis (*25th FOCS*, 1984), we consider a game that consists of moving distinct pebbles along the edges of an undirected graph. At most one pebble may reside in each vertex at any time, and it is only allowed to move one pebble at a time (which means that the pebble must be moved to a previously empty vertex). We show that the problem of finding the shortest sequence of moves between two given "pebble configuations" is NP-Hard.

Keywords: NP-Completeness, Games' Complexity, Computational Group Theory.

This work was completed in July 1984, and later appeared as Technical Report No. 792 of the Computer Science Department of the Technion (Israel). The current revision is quite minimal.

1 Problem's Definition

The following generalization of the "15-Puzzle" appeared in [4,2,3]:

Board: The game is played on a finite, undirected, simple graph. The graph will be denoted by $G(V, E)$.

Legal Board Configuration: Every vertex contains at most one pebble, and one vertex is empty. That is, $BC : V \rightarrow \{0, 1, 2, \ldots, |V| - 1\}$ is a legal board configuration if it is one-to-one and onto. The board configuration is interpreted as follows: if $BC(v) \neq 0$, then vertex v contains pebble $BC(v)$, and if $BC(v) = 0$, then vertex v is empty.

Legal Moves: A legal move consists of moving a single pebble, along one of the edges of the graph to an empty vertex. A legal move is a transformation on the set of legal configurations. Let $BC(\cdot)$ be a legal configuration and $BC'(\cdot)$ be the configuration that results from $BC(\cdot)$ after a legal move. Then, there exist two adjacent vertices, $u, v \in V$ (i.e., $(u, v) \in E$), such that $BC'(u) = BC(v)$, $BC'(v) = BC(u) = 0$, and $BC'(w) = BC(w)$ for all $w \in V \setminus \{u, v\}$. In this move the pebble $BC(v)$ is moved from vertex v to vertex u.

O. Goldreich et al.: Studies in Complexity and Cryptography, LNCS 6650, pp. 1–5, 2011.

A sequence of Moves: A sequence of t moves is a sequence of legal board configurations, denoted $BC_0(\cdot), BC_1(\cdot), BC_2(\cdot), \ldots, BC_t(\cdot)$, such that for $i = 1, \ldots, t$ it holds that $BC_i(\cdot)$ is the result of applying a legal move to $BC_{i-1}(\cdot)$. The configuration $BC_0(\cdot)$ is called the beginning configuration of the above sequence, and the configuration $BC_t(\cdot)$ is called the finishing configuration of the above sequence.

Solutions: A pair of legal board configurations is said to have a solution if there exists a sequence of moves beginning at the first and finishing at the second.

2 Prior Work

Kornhauser, Miller and Spirakis [3] showed that, for any nonseparable graph $G(V, E)$, *if a pair of legal board configurations has a solution, then it has a solution by $O(|V|^3)$ moves.* Furthermore, they showed that *such a solution* (by $O(|V|^3)$ moves) *can be found in $O(|V|^3)$ time.* A natural algorithmic question arises:

> *Given a pair of legal board configurations that does have a solution, Is it feasible to find the shortest solution?*

We answer this question negatively, proving that finding such a solution is NP-Hard.

3 The NP-Completeness Result

In order to discuss the problem of finding the shortest solution to a solvable pair of legal board configurations, we introduce the following decision problem, herafter referred to as the Shortest Move Sequence (SMS) Problem:

Input: A nonseparable, simple, undirected graph $G(V, E)$; a pair, $B(\cdot)$ and $F(\cdot)$, of legal board configuration; and an integer K.

Question: Is there a sequence of K (or less) legal moves beginning at $B(\cdot)$ and finishing at $F(\cdot)$?

We prove the following result.

Theorem: *The Shortest Move Sequence (SMS) problem is NP-Complete.*

Proof: First note that SMS is in NP (since, w.l.o.g., $K = O(|V|^3)$). We prove that SMS is complete by reducing 3-Exact-Cover (3XC) to it. Recall that the 3XC is defined as follows:

Input: A set $U = \{e_i\}_{i=1}^{3n}$ and a collection $S = \{s_j\}_{j=1}^{m}$ of 3-element subsets (3-subsets) of U.

Question: Is there a subcollection, $S' \subseteq S$, such that every element in U occurs in exactly one member of S'?

If existing, such a collection, S', is called an *exact cover*. (Also, $|S| = n$.)

Recall that Karp has proved that 3XC is NP-complete (see [1]). Now, given an instance of 3XC, denoted $(U = \{e_i\}_{i=1}^{3n}, S = \{s_j\}_{j=1}^{m})$, we construct the following SMS instance:

- Let $V = U^0 \cup U^1 \cup S \cup \{t\}$, where $U^\sigma = \{e^\sigma : e \in U\}$ for $\sigma \in \{0,1\}$.
  The vertices e^0 and e^1 will be associated with the element $e \in U$. The vertices in S will be associated with the corresponding 3-subsets. The vertex t will be called the *temporary vertex*.
- Let $E = E_{3XC} \cup \{(t,s) : s \in S\} \cup \{(e^0, e^1) : e \in U\}$, where

$$E_{3XC} = \{(e^\sigma, s) : \sigma \in \{0,1\} \wedge e \in U \wedge e \in s\}.$$

  The edges in E_{3XC} encode the description of the 3XC instance. Note that $(e^\sigma, s) \in E_{3XC}$ iff the element $e \in U$ appears in the 3-subset $s \in S$.
- Let $B(e_i^\sigma) = 2i - 1 + \sigma$, for $1 \leq i \leq 3n$ and $\sigma \in \{0,1\}$, and $B(s_j) = 6n + j$, for $1 \leq j \leq m$. Let $B(t) = 0$.
  In the begin configurations t is empty while the pebbles are placed in a "canonical" order. In particular, the pebbles $2i - 1$ and $2i$, which are associated with the element e_i (for $1 \leq i \leq 3n$), are placed in vertices e_i^0 and e_i^1, respectively. The pebble $6n + j$, which is associated with the 3-subset s_j (for $1 \leq j \leq m$), is placed in vertex s_j.
- Let $F(e_i^\sigma) = 2i - \sigma$, for $1 \leq i \leq 3n$ and $\sigma \in \{0,1\}$, and $F(s_j) = 6n + j$, for $1 \leq j \leq m$. Let $F(t) = 0$.
  In the finish configurations t is still empty and the pebbles in the vertices that are associated with the 3-subsets remain invariant w.r.t the begin configuration. The pebbles associated with each element of U are switched w.r.t the begin configuration.
- Finally, let $K = 11n$.

Having presented our reduction it remains to show that it is indeed valid.

Assume that the 3XC instance has an exact cover, denoted $S' = \{s_{i_j}\}_{j=1}^{n}$. Let $f : \{1, 2, \ldots, n\} \times \{1, 2, 3\} \to \{1, 2, \ldots, 3n\}$ such that $e_{f(j,k)}$ is the k-th element in the 3-subset s_{i_j} (where the order on the elements in each 3-subset is induced by an ordering of U). Note that $s_{i_j} = \{e_{f(j,1)}, e_{f(j,2)}, e_{f(j,3)}\}$ and $U = \{e_{f(j,k)} : 1 \leq j \leq n \wedge 1 \leq k \leq 3\}$. Then, following is a solution to the corresponding SMS instance:

for $j = 1$ to n *do* *begin*
    move pebble $6n + i_j$ from s_{i_j} to t;
    for $k = 1$ to 3 *do* *begin*
        move pebble $2f(j,k) - 1$ from $e_{f(j,k)}^0$ to s_{i_j};
        move pebble $2f(j,k)$ from $e_{f(j,k)}^1$ to $e_{f(j,k)}^0$;
        move pebble $2f(j,k) - 1$ from s_{i_j} to $e_{f(j,k)}^1$;
        [Comment: At this stage, for every $k \in \{1, 2, 3\}$,
        the pebbles $2f(j,k) - 1$ and $2f(j,k)$ are switched.]
    end
    move pebble $6n + i_j$ from t to s_{i_j};

[Comment: At this stage all pebbles associated to elements in s_{i_j} are switched and all the pebbles associated with 3-subsets are back in place.]
end

One can easily verify that the foregoing procedure transforms the begin configuration into the finish configuration in $(1 + 3 \cdot 3 + 1) \cdot n = K$ moves

Assume, on the other hand, that the SMS instance has a solution in no more than $K = 11n$ moves. Let us denote this solution (i.e., sequence of moves) by Q. Recall that in each move a single pebble is moved (to an empty vertex). The following facts concerning Q can be easily verified:

Fact 1: Switching pebble $2i - 1$ with pebble $2i$ $(1 \leq i \leq 3n)$ requires at least two moves of one of these pebbles and at least one move of the other pebble. Furthermore, this switching requires that at least one of these pebbles passes through a vertex associated with a 3-subset that contains the element e_i.

The main part follows from the fact that each move must be to a previously empty vertex, and the furthermore part follows by the graph's structure.

Fact 2: If some pebble passes through a 3-subset vertex s_j $(1 \leq j \leq m)$ during Q, then the pebble $6n + j$ must have been moved during Q.

Let M denote the set of pebbles that are associated with 3-subsets that moved during Q. Using Facts 1 and 2, we get.

Fact 3: The number of moves in Q is at least $3 \cdot 3n + 2 \cdot |M|$.

Recall the number of moves (in Q) is at most $K = 11n$. Thus:

Fact 4: $|M| \leq n$.

Fact 5: The collection $C = \{s_j : 6n + j \in M\}$ constitutes a cover of the set U. That is, for every element $e \in U$, there exists a 3-subset $s \in C$ such that $e \in s$.

Note that $2i - 1$ has been switched with $2i$, for each $1 \leq i \leq 3n$, and by Facts 1 and 2 this implies that for some j such that $e_i \in s_j$ it holds that $6n + j \in M$.

Combining Facts 4 and 5, we conclude that C is an exact cover of the 3XC instance. This completes the proof of the theorem. ∎

Acknowledgements. I am grateful to Shimon Even, Dan Kornhauser, Silvio Micali and Gary Miller for very helpful discussions.

References

1. Garey, M.R., Johnson, D.S.: Computers and Intractability: A Guide to the Theory of NP-Completeness, p. 221. Freeman, San Francisco (1979)
2. Johnson, D.S.: The NP-Completeness Column: An Ongoing Guide. J. of Algorithms 4, 397–411 (1983)
3. Kornhauser, D.M., Miller, G., Spirakis, P.: Coordinating Pebble Motion on Graphs, the Diameter of Permutation Groups, and Applications. In: Proc. of the 25th FOCS, pp. 241–250 (1984)
4. Wilson, R.W.: Graphs, Puzzles, Homotopy, and Alternating Groups. J. of Comb. Th. (B) 16, 86–96 (1974)

Proving Computational Ability

Mihir Bellare and Oded Goldreich

Abstract. We investigate extending the notion of a proof of knowledge
to a proof of the ability to perform some computational task. We provide
some definitions and protocols for this purpose.

Keywords: Proofs of Knowledge, Zero-Knowledge, Cryptographic
Protocols.

This work was completed in August 1992, and earlier versions of it were posted
on the authors' webpages. The current revision is intentionally minimal.

1 Introduction

We extend the idea of proving "knowledge" of a string to encompass a notion
of proving the "ability to perform some task." Specifically, we wish to formalize
what it means to "prove the ability to compute a function f on some instance
distribution D."

Motivation. The aforementioned notion might have many uses, and two of them
are described here. Suppose Alice possess a trapdoor, $t(x)$, to a (publically
known) trapdoor permutation f_x and wishes to identify herself to Bob, by demonstrating ability to invert f_x. The proof of ability should be zero-knowledge so to
prevent Bob from latter impersonating Alice. Admittingly, in this case Alice
can establish her identity by directly proving, in a zero-knowledge manner, her
knowledge of the trapdoor $t(x)$ (which corresponds to the index x of f_x). Still
it may be cheaper to prove ability to invert f_x (e.g., by using a trivial protocol in which the prover inverts f_x on instances chosen by the verifier). This is
particularly valid in case Alice posseses special purpose hardware, in which the
trapdoor is hard-wired, making it very easy for her to invert the function on
inputs of her choice. A second application is for a party to prove possesion of
vast computing power by conducting very difficult tasks (e.g., inverting one-way
functions).

Related Work. This is an extension of our previous work on proofs of knowledge
[1] in which we try to generalize those ideas to the setting of proving computational ability. Proofs of knowledge are first mentioned in [5] and have been seeing
definitional refinements [3,6,2] culminating in the notions of [1,4]. We assume the
reader is somewhat familiar with the notion.

Proofs of computational ability were first discussed by Yung [7]. We adhere to
the same basic and natural idea (namely, that computational ability of a prover is

O. Goldreich et al.: Studies in Complexity and Cryptography, LNCS 6650, pp. 6–12, 2011.
© Springer-Verlag Berlin Heidelberg 2011

certified if some extractor can use the prover as a black box to solve the problem itself) but our approach is more general. For example whereas an assumption on the problem hardness is made in [7] it is not made here; we consider notions of distribution-free and distribution-dependent ability; following [1] we define an analogue of "knowledge error"; and following [1] we avoid some weaknesses inherited from earlier definitions of proofs of knowledge.

2   Definitions

For greater generality, we will consider relations rather than functions. By a family of relations we mean a sequence $\{R_x\}_{x \in \{0,1\}^*}$, where $R_x \subseteq \{0,1\}^{|x|} \times \{0,1\}^*$ for each x. For simplicity we restrict our attention to polynomially bounded families; that is, we assume there is a polynomial p such that $(z,y) \in R_x$ implies $|z| = |x|$ and $|y| \leq p(|x|)$. Following the notation used in [1], we denote $R_x(z) \overset{\text{def}}{=} \{ y : (z,y) \in R_x \}$ and $L_{R_x} \overset{\text{def}}{=} \{ z : \exists y \text{ such that } (z,y) \in R_x \}$. Prover and verifier will interact on common input x, with the goal of the interaction being for the prover to "convince" the verifier that he has the "ability to solve R_x."

We need to address the meaning of both of the phrases in quotes above. We will first define what it means for a machine to "solve a relation" (or a family of relations), and only next will we define what is a "proof of ability" to do so.

The standard meaning of efficiently solving a relation, $S \subseteq \{0,1\}^* \times \{0,1\}^*$, is the existence of an efficient algorithm that, on input z, outputs $y \in S(z)$, called a solution to z, if such exists. This is a notion of worst case. Instead, we adopt a notion of average case by which we consider a probability distribution on the inputs and require that the algorithm is efficient on the average (with respect to the input distribution). An even more liberal notion is derived by allowing the solver to ask for alternative inputs, which are generated according to the same distribution (and independently of previous inputs), until it can present a solution to any of the inputs.

Notation: Let $S \subseteq \{0,1\}^* \times \{0,1\}^*$. Then, $\mathrm{dom}(S) \overset{\text{def}}{=} \{ z \in \{0,1\}^* : S(z) \neq \emptyset \}$ is the domain of S.

Definition 2.1 (solving relations): *Let $S \subseteq \{0,1\}^* \times \{0,1\}^*$ be a relation, and D be a distribution on $\mathrm{dom}(S)$. Suppose $t \in \mathbb{N}$ and let $M(\cdot)$ be a machine.*

- *We say that* machine $M(\cdot)$ solves S under D in expected t steps *if, on input $(z_1, z_2, ..., z_t)$, with each z_i drawn independently according to D, machine M halts within expected t steps and outputs a pair (z_i, y) so that $y \in S(z_i)$. (The expectation here is over the random choices of M as well as the t-product of the distribution D.)*

- *We say that* machine $M(\cdot)$ strongly solves S under D in expected t steps *if, on input z, drawn according to D, machine M halts within expected t steps with output $y \in S(z)$. (The expectation here is over the random choices of M as well as the distribution D.)*

Conventions: If a machine has several inputs, we may fix some of them to obtain a machine on the remaining inputs. Likewise, for an oracle machine, we may fix the oracle and consider the resulting machine. Specifically, suppose that the oracle machine $M(\cdot, \cdot, \cdot)$ has three inputs, then $M^A(x, y, \cdot)$ denotes the machine with one input whose output on input z is $M^A(x, y, z)$.

Let $\mathcal{R} = \{R_x\}_{x \in \{0,1\}^*}$ be a family of relations. We say that $\mathcal{D} = \{D_x\}_{x \in \{0,1\}^*}$ is an input distribution for $\mathcal{R}$ if for every x, it holds that D_x is a distribution on $\mathrm{dom}(R_x)$. We are now ready to define proofs of ability to solve (repectively, ability to strongly solve) a family of relations under a family of distributions.

Definition 2.2 (proof of ability): *Let $\mathcal{R} = \{R_x\}_{x \in \{0,1\}^*}$ be a family of relations, and $\mathcal{D} \stackrel{\mathrm{def}}{=} \{D_x\}_{x \in \{0,1\}^*}$ be an input distribution for $\mathcal{R}$. Let $\kappa\colon \{0,1\}^* \to [0,1]$. We say that an interactive function, V, is a* verifier of the ability to solve *(resp., strongly solve),* $\mathcal{R}$ under $\mathcal{D}$ with error κ *if the following two conditions hold.*

- non-triviality: *There exists an interactive function P^* so that for all x, all possible interactions of V with P^* on common input x are accepting; that is, $\Pr[\mathrm{tr}_{P^*, V^{D_x}}(x) \in \mathrm{ACC}_V(x)] = 1$, where $\mathrm{tr}_{A,B}(x)$ denotes B's view of the interaction with P on common input x, and $\mathrm{ACC}_B(x)$ denotes the views that convince B* (i.e., make it accept).
- validity: *There exists a constant $c > 0$ and a probabilistic oracle machine $K(\cdot, \cdot, \cdot)$ such that for every interactive function P, every $x \in \{0,1\}^*$ and every $\gamma \in \mathrm{ACC}_V(x)$, machine $K^{P_x}(x, \gamma, \cdot)$ satisfies the following condition: if $p(x) \stackrel{\mathrm{def}}{=} \Pr[\mathrm{tr}_{P, V^{D_x}}(x) \in \mathrm{ACC}_V(x)] > \kappa(x)$ then machine $K^{P_x}(x, \gamma, \cdot)$ solves (resp., strongly solves) R_x under D_x in an expected number of steps bounded by*

$$\frac{|x|^c}{p(x) - \kappa(x)}$$

The oracle machine K is called an ability extractor *(resp.,* strong ability extractor*) under $\mathcal{D}$.*

Hence an *ability extractor* is given a sequence of instances, each independently selected according to D_x, and is supposed to output a solution to one of these instances within the specified (expected) time bound. A *strong ability extractor* is given a single instance, selected according to D_x, and is supposed to output a solution to this instances within the specified (expected) time bound. (In both cases, solutions are with respect to R_x.)

Relation to Proofs of Knowledge. We note that proofs of knowledge (as per [1, Def. 3.1]) are a special case of proofs of ability. To justify this claim, given a binary relation R we define the family of relations $\mathcal{R} = \{R_x\}$ so that $R_x = \{(x,y) : (x,y) \in R\}$. Clearly, $\mathrm{dom}(R_x)$ is the singleton $\{x\}$ if $R(x) \neq \emptyset$ and $\emptyset$ otherwise. Let D_x be the distribution on $\mathrm{dom}(R_x)$ which, in the former case, assigns the entire probability mass to x (and is undefined in the latter case). Clearly $\mathcal{D} = \{D_x\}$ is an input distribution for $\mathcal{R}$. It is easy to see that if V is a verifier of the ability to solve $\mathcal{R}$ under $\mathcal{D}$ (with error κ) then V is also a knowledge verifer for R (with knowledge error κ).

On the Dependence on the Distribution D_x. Definition 2.2 refers to a specific input distribution. Clearly, *both* the ability-verifier and the ability-extractor may depend on this distribution, and this dependency seems inevitable. However, the dependency on the input distribution can be "uniform" in the sense that both verifier and extractor can be fixed machines with access to a random source that generates the input distribution. We call such a proof of ability distribution-free.

The foregoing notion is defined as follows. Let $\mathcal{D}$ be a family of distributions for some $\mathcal{R}$, and let M be an (interactive and/or oracle) probabilistic machine. A $\mathcal{D}$-source augmentation of machine M is a machine that, on input x, in addition to the standard behaviour of M can obtain elements draw independently from distribution D_x (at the cost of reading them).

Definition 2.3 (distribution-free proof of ability): *Let $\mathcal{R} = \{R_x\}_{x \in \{0,1\}^*}$ be a family of relations, and let $\kappa \colon \{0,1\}^* \to [0,1]$.*

- *We say that an interactive machine, V, is a* distribution-free *verifier of the ability to solve $\mathcal{R}$* with error κ *if for every input distribution, denoted $\mathcal{D}$, for $\mathcal{R}$, the $\mathcal{D}$-source augmentation of machine V constitutes a verifier of the ability to solve $\mathcal{R}$ under $\mathcal{D}$ with error κ.*
- *We say that a distribution-free verifier of the ability to solve $\mathcal{R}$ (with error κ) has a* distribution-free *ability extractor if there exists an oracle machine, K, such that the $\mathcal{D}$-source augmentation of machine K constitutes a ability extractor under $\mathcal{D}$.*

A definition of a *distribution-free strong ability extractor* is derived analogously.

3　Examples

To demonstrate the above definitions we consider two natural examples. Both examples refer to a familty of one-way permutations, $\{f_x\}_{x \in \{0,1\}^*}$. The string x is called the index of the permutation $f_x \colon \{0,1\}^{|x|} \to \{0,1\}^{|x|}$, and there exists an efficient algorithm that, on input index x and argument y, returns the value $f_x(y)$. We shall consider proofs of ability to invert $\{f_x\}$; intuitively, such ability requires either super-polynomial computational resources or knowledge of some trapdoor information (in case the collection has such trapdoors).

Example 1: Consider a verifier that, on common input x, sends the prover a single uniformly selected string $v \in \{0,1\}^{|x|}$, and accepts if and only if the prover answers with the inverse of v under f_x (i.e., with y satisfying $f_x(y) = v$). We show (below) that the foregoing verifier is an ability-verifier for inverting f_x under the uniform distribution.

Example 2: Consider a verifier that, on common input $x \in \{0,1\}^n$ ($n \in \mathbb{N}$), sends the prover $2n$ uniformly and indepedently selected strings, $v_1, ..., v_{2n} \in \{0,1\}^n$, and accepts if and only if the prover answers with the inverse of each of these v_i's under f_x (i.e., with $y_1, ..., y_{2n}$ satisfying $f_x(y_i) = v_i$, for every i). We show (below) that the foregoing verifier is a *strong* ability-verifier for inverting f_x on at least one out of $2|x|$ of uniformly selected instances.

Proposition 3.1. *The program described in Example 1 is an ability-verifier* (with error zero) *for solving* $\mathcal{R} = \{R_x\}$ *under* $\mathcal{D} = \{D_x\}$, *where*

- $R_x = \{(v, y) : v = f_x(y)\};$
- D_x *is uniform over the set of all strings of length* $|x|$.

Furthermore, if the verifier in Example 1, selects v *according to an arbitrary distribution* D_x, *then the system described constitutes a distribution-free proof of ability.*

Proof sketch: We present here only the case of uniform distribution, and focus on the validity condition. Consider an arbitrary, fixed prover, and let p_x denote the probability that the verifier is convinced by this prover on common input x. Here the probability space is over all choices of both the verifier and prover. Assume, without loss of generality, that $p_x > 2^{-|x|}$, otherwise the extractor satisfies the requirement merely by exhaustive search. Also, we may assume that the ability-extractor "knows" p_x since it may estimate p_x in expected time $\text{poly}(|x|)/p_x$ by repeated experiments. Let $q_x(v)$ denote the probability that the verifier is convinced conditioned on the event that it chose and sent v to the prover. Here the probability distribution is merely over the prover's random coins (in case it is probabilistic). Let $V_x(i)$ be the set of v's for which $q_x(v)$ is greater than 2^{-i} and smaller/equal to 2^{-i+1}. Clearly, there exists an $i \leq |x|$ such that

$$\frac{|V_x(i)|}{2^{|x|}} > \frac{p_x \cdot 2^i}{n}. \tag{1}$$

We are now ready to present the ability-extractor. Formally speaking, the extractor gets as input an index, x, and a sequence of independently and uniformly selected $|x|$-bit long strings, and its task is to invert f_x on one of them. However, to simplify the exposition, we prefer to think of these strings as being chosen by the extractor. Hence, on input x, the extractor executes $m \overset{\text{def}}{=} \lceil \log_2(1/p_x) \rceil$ copies of the following procedure, each with a different value of $i \in \{1, ..., m\}$. The ith copy consists of uniformly and independently selecting $M \overset{\text{def}}{=} \text{poly}(n)/(p_x \cdot 2^i)$ values, $v_1, ..., v_M \in \{0, 1\}^n$, and executing the following sub-procedure on each of them. The sub-procedure with value v_j invokes the prover's program (as oracle), on input x and message v_j, for $\text{poly}(n) \cdot 2^i$ times, each time checking whether the prover's answer is the inverse of v_j under f_x. Once a positive answer is obtained, the extractor halts with the corresponding value-inverse pair.

The extractor's expected running-time is bounded above by

$$\sum_{i=1}^{m} \frac{\text{poly}(n)}{p_x \cdot 2^i} \cdot \left(\text{poly}(n) \cdot 2^i\right) = \frac{\text{poly}(n)}{p_x}.$$

To evaluate the performace of the above extractor, consider the ith copy, where i satisfies Equation (1). With overwhelmingly high probability (i.e., greater than $1 - 2^{-n}$), one of the v_j's chosen in this copy satisfies $q_x(v_j) \geq 2^{-i}$. In this case, with overwhelmingly high probability, the extractor inverts f_x on this v_j.

The exponentially small error probabilities can be eliminated by running an exhaustive search algorithm (for inverting f_x) in parallel to the entire algorithm described above. The proposition follows.   ∎

Proposition 3.2. *The program described in Example 2 is a strong ability-verifier (with error zero) for solving* $\mathcal{R} = \{R_x\}$ *under* $\mathcal{D} = \{D_x\}$, *where*

- $R_x = \{(v_1, ..., v_{2|x|}, y) : \exists i \text{ s.t. } v_i = f_x(y)\}$;
- D_x *is uniform over the set of strings of length* $2|x|^2$.

Proof sketch: As in the proof of Proposition 3.1, we consider an arbitrary fixed prover and let p_x denote the probability that the verifier is convinced on common input x. As before, we may assume that $p_x > 2^{-|x|}$ and that the ability-extractor has a good estimate of p_x. Let $n \overset{\text{def}}{=} |x|$, and consider an $2n$-dimentional table in which the dimensions correspond to the $2n$ values chosen by the verifier. The $(v_1, ..., v_{2n})$-entry in the table equals the probability that the prover convinces the verifier (i.e., successfuly inverts f_x on v_1 through v_{2n}) conditioned on the event that the verifier sent message $(v_1, ..., v_{2n})$ to the prover. The probability here is merely on the prover's random choices. As in the proof of Proposition 3.1, we consider a partition of these probabilities to clusters of similar magnitude. It follows that there exists an $i < 2n$ such that at least a $p_{x,i} \overset{\text{def}}{=} p_x \cdot 2^i/2n$ fraction of the entries have value greater than 2^{-i}. We call these entries admissible. It follows that there exists a dimention k such that at least a $\sqrt[2n]{p_{x,i}/2} > \frac{1}{2}$ fraction of the rows in the kth dimention contain at least $p_{x,i}/2n$ admissible entries. We call such a (i, k) pair good.

We are now ready to present the *strong* ability-extractor. The extractor gets as input an index, x, and a uniformly chosen $2|x|^2$-long string $\bar{v} = (v_1, ..., v_{2n})$, where $v_j \in \{0,1\}^n$ and $n = |x|$. The extractor is suppose to find a solution to $\bar{v}$, and this amounts to inverting f_x on one of the v_j's. To this end the extractor executes $8n^3$ copies of the following procedure, each with a different triples (i, k, j), where $1 \leq i, k, j \leq 2n$. The $(i, k, j)^{\text{th}}$ copy of the procedure tries to invert f_x on v_j, using the parameters i and k. Specifically, the $(i, k, j)^{\text{th}}$ copy consists of repeatedly invoking the sub-procedure $A_{i,k}$ on input v_j, for at most $\lfloor \text{poly}(n)/p_{x,i} \rfloor$ times (where $p_{x,i} = p_x \cdot 2^i/2n$). On input v, the sub-procedure $A_{i,k}$ proceeds as follows.

1. Selects uniformly $2n$ strings of length n each. These strings are denoted $u_1, ..., u_{2n}$;
2. Invokes the (oracle to the) prover $\text{poly}(n) \cdot 2^i$ times, each time with input x and verifier's message $(u_1, ..., u_{k-1}, v, u_{k+1}, ..., u_{2n})$. The message consist of the sequence selected at Step 1, except that u_k is replaced by v.
3. If in one of these invocations, the prover answers with a $2n$-tuple $(y_1, ..., y_{2n})$ such that $f_x(y_k) = v$, then the extractor halts with output (v, y_k).

Clearly, the expected running-time of the foregoing extractor is at most $\sum_{i=1}^{2n} \text{poly}(|x|) 2^i/p_{x,i} = \text{poly}(|x|)/p_x$. To evaluate the performance of this extractor, consider a good pair (i, k). By definition of a good pair, it follows that at least one half of the rows in the kth direction contain at least $\rho_{x,i} \overset{\text{def}}{=} p_x \cdot 2^i/(2n)^2$

entries on which the prover convinces the verifier with probability at least 2^{-i}. Let us denote the set of n-bit strings corresponding to these rows by $S_{x,k}$. It follows that for every $v \in S_{x,k}$, the sub-procedure $A_{i,k}$ inverts f_x on v with probability at least $\rho_{x,i} - 2^{-n}$. Hence, when invoking $A_{i,x}$ on $v \in S_{x,k}$ for $\text{poly}(n)/\rho_{x,i}$ times, with overwhelming probability (i.e., probability greater than $1 - 2^{-n}$), we invert f_x on v. The final observarion is that, since $|S_{x,k}| \geq \frac{1}{2} \cdot 2^n$, the probability that none of $2n$ indepedently and uniformly selected n-bit strings hits $S_{x,k}$ is exponentially vanishing (i.e., smaller than 2^{-n}). As in the proof of Proposition 3.1, this exponentially small error can be elliminated. It follows that the extractor strongly solve R_x under D_x.   ∎

Acknowledgements. Work done while the first author was at the IBM T.J. Watson Research Center (New York), and the second author was at the Techion (Israel).

References

1. Bellare, M., Goldreich, O.: On defining proofs of knowledge. In: Brickell, E.F. (ed.) CRYPTO 1992. LNCS, vol. 740, pp. 390–420. Springer, Heidelberg (1993)
2. Brassard, G., Crépeau, C., Laplante, S., Léger, C.: Computationally Convincing Proofs of Knowledge. In: Jantzen, M., Choffrut, C. (eds.) STACS 1991. LNCS, vol. 480, Springer, Heidelberg (1991)
3. Feige, U., Fiat, A., Shamir, A.: Zero-Knowledge Proofs of Identity. Journal of Cryptology 1, 77–94 (1988)
4. Feige, U., Shamir, A.: Witness Indistinguishability and Witness Hiding Protocols. In: Proceedings of the Twenty Second Annual Symposium on the Theory of Computing, pp. 416–426. ACM, New York (1990)
5. Goldwasser, S., Micali, S., Rackoff, C.: The Knowledge Complexity of Interactive Proof Systems. SIAM J. on Computing 18(1), 186–208 (1989) (Preliminary Version in the 7th STOC, 1985)
6. Tompa, M., Woll, H.: Random Self-Reducibility and Zero-Knowledge Interactive Proofs of Possession of Information. University of California (San Diego) Computer Science and Engineering Dept. Technical Report Number CS92-244 (June 1992) (Preliminary Version in the 27th FOCS, pp. 472–482, 1987)
7. Yung, M.: Zero-knowledge proofs of computational power. In: Quisquater, J.-J., Vandewalle, J. (eds.) EUROCRYPT 1989. LNCS, vol. 434, pp. 196–207. Springer, Heidelberg (1990)

On Constructing 1-1 One-Way Functions

Oded Goldreich, Leonid A. Levin, and Noam Nisan

Abstract. We show how to construct length-preserving 1-1 one-way functions based on popular intractability assumptions (e.g., RSA, DLP). Such 1-1 functions should not be confused with (infinite) families of (finite) one-way permutations. What we want and obtain is a single (infinite) 1-1 one-way function.

Keywords: One-Way Functions, RSA, Discrete Logarithm Problem.

This work was conducted in the summer of 1994. An early version of it appeared as TR95-029 of *ECCC*. Section 4 has been revised and improved by relying on subsequent advances regarding primality testing [1]. Specifically, we replace the randomized primality tester of [3] (which builds upon [13,17]) by the deterministic primality tester of Agrawal, Kayal, and Saxena [1]. Various footnotes indicate these (as well as other significant) deviations from the aforementioned early version.

1 Introduction

Given any one-way permutation (i.e., a length preserving 1-1 one-way function), one can easily construct an efficient pseudorandom generator. The construction follows the scheme given by Blum and Micali [4], using the fact that every one-way function has a hard-core bit [8]. Specifically, assume that f is such a function and let b be a hard core-bit for it (e.g., starting with a function f', we may define $f(x,r) \stackrel{\text{def}}{=} (f'(x),r)$ and $b(x,r)$ as the inner-product mod 2 of the strings x and r when viewed as binary vectors of length $|x| = |r|$). Then, on input a seed s, the pseudorandom generator outputs the sequence $b(s), b(f(s)), b(f(f(s))), b(f^3(s)), \ldots$

Pseudorandom generators can be constructed also based on arbitrary one-way functions [12]; yet, the known construction is very complex and inefficient.[1] In fact, it is of no practical value. The construction in [9], which uses arbitrary *regular* one-way functions is more attractive in these respects, yet it is far less attractive than the simple construction outlined above. A similar situation occurs with respect to the construction of digital signature schemes (cf., [14] vs [19]). In general, 1-1 one-way functions currently offer simpler and more practical constructions (of more complex primitives) than offered by general one-way functions.

---

[1] The same applies also to subsequent improvements, currently culminating in [11].

O. Goldreich et al.: Studies in Complexity and Cryptography, LNCS 6650, pp. 13–25, 2011.
© Springer-Verlag Berlin Heidelberg 2011

These facts were our initial motivation for trying to construct length-preserving 1-1 one-way functions. Such functions should not be confused with what is commonly referred to (especially in the "Crypto Community") as "one-way permutations", which are actually infinite sets of finite functions – see definitions below. What we want is a single infinite function that is both length-preserving and 1-1 (and needless to say one-way). We show how to construct such 1-1 one-way functions based on popular intractability assumptions such as the intractability of DLP and inverting RSA.

Indeed, some (but not all) of the constructions that use length-preserving 1-1 one-way functions can be modified such that families of one-way permutations can be used instead. Still the question of whether the former (i.e., length-preserving 1-1 one-way functions) exists is of both theoretical and practical importance.

2 One-Way Functions and Families

We start by recalling the standard definitions.

Definition 2.1 (one-way functions): *Let* $f : \{0,1\}^* \to \{0,1\}^*$ *be a* length preserving *function that is polynomial-time computable.*

- (strongly one-way): *f is called* (strongly) **one-way** *if for any probabilistic polynomial-time algorithm A, any positive polynomial p and all sufficiently large n, it holds that*

$$\mathrm{Prob}[A(f(x)) \in f^{-1}f(x)] < \frac{1}{p(n)}$$

where the probability is taken uniformly over $x \in \{0,1\}^n$, *and the internal coin tosses of algorithm A.*

- (weakly one-way): *f is called* **weakly one-way** *if there exists a positive polynomial p such that for any probabilistic polynomial-time algorithm A and all sufficiently large n, it holds that*

$$\mathrm{Prob}[A(f(x)) \notin f^{-1}f(x)] > \frac{1}{p(n)}$$

where the probability is as above.

Recall that $f : \{0,1\}^* \to \{0,1\}^*$ is 1-1 if $f(x) \neq f(y)$ for all $x \neq y$. In the case that $f(x) \neq f(y)$ for all but a negligible fraction of the pairs (x,y) we say that f is **almost 1-1**. Namely, an almost 1-1 function f satisfies, for every positive polynomial p and all sufficiently large n, it holds that

$$\mathrm{Prob}[f(x) = f(y)] < \frac{1}{p(n)}$$

where the probability is taken uniformly and independently over all $x, y \in \{0,1\}^n$.

Definition 2.2 (family of one-way permutations – simplified version): *An infinite set of finite permutations, $\mathbf{F} = \{f_i : D_i \stackrel{1\text{-}1}{\rightarrow} D_i\}_{i \in I}$, is called a* family of one-way permutations *if the following conditions hold*

- (efficient evaluation): *There exists a polynomial-time algorithm that on input an index (of a permutation) $i \in I$ and a domain element $x \in D_i$ returns $f_i(x)$.*
- (efficient index selection): *There exists a probabilistic algorithm S that on input n, runs for $\mathrm{poly}(n)$ time and returns a uniformly distributed index of length n (i.e., an i uniformly distributed in $I \cap \{0,1\}^n$).*
- (efficient domain sampling): *There exists a probabilistic polynomial-time algorithm D that on input an index $i \in I$, returns a uniformly distributed element of D_i.*
- (one-wayness): *For any probabilistic polynomial-time algorithm A, any positive polynomial p and all sufficiently large n, it holds that*

$$\mathrm{Prob}[A(i, f_i(x)) = x] < \frac{1}{p(n)}$$

where the probability is taken uniformly over $i \in I \cap \{0,1\}^n$, $x \in D_i$, and the internal coin tosses of algorithm A.

In the non-simplified version, both the aforementioned probabilistic algorithms (i.e., S and D) are allowed to produce output with only noticeable probability (i.e., probability at least $1/\mathrm{poly}(n)$). Furthermore, given these algorithms have produced an output, the output is allowed to be wrong (i.e., out of the target set or non-uniformly distributed) with negligible probability (e.g., with probability at most 2^{-n}). Our transformations will take advantage of the first relaxation, but not of the second.[2]

Analogously to Definition 2.1, families of permutations can be defined to be *weakly* one-way, rather than (strongly) one-way.

3   Transforming One-Way Families into Functions

Clearly, any family of one-way permutations can be converted into a single one-way function; namely, $f(r, s) \stackrel{\mathrm{def}}{=} f_i(x)$, where $i = S(n, r)$ is the index selected using coin-tosses r and $x = D(i, s) \in D_i$ is the element selected on input i and coin-tosses s. (Padding can be applied, if necessary, to make f length preserving.) However, this procedure does not necessarily yield a 1-1 function; furthermore, for most natural examples such as RSA, DLP, etc., the resulting function will be many-to-one.

---

[2] In the earlier version of this work, we also took advantage of the second relaxation. This was done in order to account for the probability of error that was present in the probabilistic primality tests that we used. The need to accommodate error is currently eliminated by using the deterministic primality test of [1].

An alternative construction, which does yield a 1-1 one-way function, is possible under some additional conditions, as demonstrated below. In fact, the conditions are defined to make this natural construction work and the thrust of this paper is in demonstrating that these conditions can be met under reasonable and popular assumptions (see next section).

3.1   The Conditions

Let **F** be a family of one-way permutations and that let $q(n)$ denote the number of coins flipped by the index-selection algorithm S on input n. We consider the following conditions that **F** may satisfy.

Definition 3.1 (additional conditions)

– Augmented one-wayness:[3] *For any probabilistic polynomial-time algorithm A, any positive polynomial p and all sufficiently large n, it holds that*

$$\mathrm{Prob}[A(r, f_{S(n,r)}(x)) = x] < \frac{1}{p(n)}$$

where the probability is taken uniformly over $r \in \{0,1\}^{q(n)}$, $x \in D_{S(n,r)}$, and the internal coin tosses of algorithm A.
(Namely, the permutations are hard to invert even when the inverting algorithm is given the random coins used to generate the index of the permutation.)

– Canonical domain sampling:[4] *The domain-sampling algorithm may consist of uniformly selecting a string of specific (easy to determine) length and testing whether the string resides in the domain. In other words, we require*
  • (recognizable domain): *There exists a polynomial-time algorithm that on input an index $i \in I$ and a string x decides if $x \in D_i$.*
  • (noticeable domain): *There exists a polynomial-time computable function $l : \mathbb{N} \to \mathbb{N}$ and a positive polynomial $p(\cdot)$ so that $D_i \subseteq \{0,1\}^{l(n)}$ and $|D_i| > \frac{1}{p(n)} \cdot 2^{l(n)}$*

3.2   The Construction

Given a family of one-way permutations that satisfies the additional conditions, we explicitly construct a 1-1 one-way function as follows.

---

[3] Note that this condition is different from the notion of *enhanced* one-wayness as defined in [6, Apdx. C.1]. Specifically, here the inverting algorithm gets the coins that were used by the *index selection* algorithm, whereas in [6, Apdx. C.1] the inverting algorithm gets the coins that are used by the *domain sampling* algorithm.

[4] Interestingly, this condition was rediscovered in [10] as an alternative to the enhanced one-wayness condition of [6, Apdx. C.1]. Actually, Haitner [10] only requires noticeable domains (and refers to collections of permutations that satisfy this condition by the term *collections having dense domains*).

Construction 3.1 (simple version): *Let* $\mathbf{F}$ *be a family of permutations with an index-selection algorithm* S *that uses* $q(\cdot)$ *coins and having domains* D_i*'s that are subsets of* $\{0,1\}^{l(|i|)}$, *for some function* $l(\cdot)$. *We construct the function* f *as follows*

$$f(r,s) \stackrel{\text{def}}{=} \begin{cases} (r, f_i(s)) & \text{if } s \in D_i, \text{ where } i \stackrel{\text{def}}{=} S(n,r) \\ (r,s) & \text{otherwise} \end{cases}$$

where $r \in \{0,1\}^{q(n)}$ *and* $s \in \{0,1\}^{l(n)}$.

Proposition 3.1 (analysis of Construction 3.1): *The function* f *is 1-1 and length preserving. If* $\mathbf{F}$ *is a family of one-way permutations satisfying the additional conditions of Definition 3.1, then* f *is weakly one-way. The latter holds even if* $\mathbf{F}$ *is only weakly one-way* (as long as it satisfies the additional conditions).

Proof: By definition f is length-preserving. Let G_n be the set of pairs $(r,s) \in \{0,1\}^{q(n)} \times \{0,1\}^{l(n)}$ such that $s \in D_{S(n,r)}$ holds and let B_n be the set of the other pairs (i.e., $B_n = (\{0,1\}^{q(n)} \times \{0,1\}^{l(n)}) \setminus G_n$). The key observation is that if $(r,s) \in G_n$, then for $i = S(n,r)$ it holds that $s \in D_i$. Furthermore, in that case, $f_i(s) \in D_i$ and $f(r,s) = (r, f_i(s)) \in G_n$ follows. On the other hand, if $(r,s) \in B_n$, then $f(r,s) = (r,s) \in B_n$. Thus, f maps G_n (resp., B_n) to itself and furthermore the mapping induced on G_n (rep., B_n) is 1-1. It follows that f is 1-1.

The function f is polynomial-time computable by virtue of the first two efficiency conditions of $\mathbf{F}$ and the additional 'recognizable domain' condition. By the additional 'noticeable domain' condition we know that G_n forms a noticeable fraction of $G_n \cup B_n$ and by the 'augmented one-wayness' condition we infer that f is hard to invert on G_n. Thus, we conclude that f is weakly one-way. In fact, the latter conclusion remain valid even if the family of permutations $\mathbf{F}$ is only weakly one-way. $\qquad\square$

Remark: The function f (constructed above) may be only weakly one-way, since it equals the identity transformation for a part of its domain (and this part may have a noticeable measure). To get a (strongly) one-way function, one may apply the transformation in [7] (cf. [5, Sec. 2.6]) to the function f. (In fact, degenerate versions of the transformation in [7] suffice for this purpose; see Section 5.)

Handling the non-simplified version of Definition 2.2. The above construction is stated with respect to the simplified definition of a family of one-way permutations. Recall that in the non-simplified version, the index-selecting algorithm, S, is only required to have an output with noticeable probability (i.e., the probability is at least $1/p(n)$, where p is some fixed positive polynomial). Furthermore, S is allowed to err (i.e., have output not in I) with a negligible probability. For the general case, we redefine the function f as follows.

Construction 3.2 (general version): *Let* $\mathbf{F} = \{f_i : D_i \stackrel{1\text{-}1}{\rightarrow} D_i\}_{i \in I}$ *be a family of permutations with an index-selecting algorithm, S, that produces output with* *noticeable probability and errs with negligible probability. We construct the function f as follows*

$$f(r,s) \stackrel{\text{def}}{=} \begin{cases} (r, f_i(s)) & \text{if } i \stackrel{\text{def}}{=} S(n,r) \neq \perp \text{ and } s \in D_i \\ (r,s) & \text{otherwise} \end{cases}$$

where the convention is that if on input n and coin tosses $r \in \{0,1\}^{q(n)}$ the algorithm S halts with no output, then $S(n,r) \stackrel{\text{def}}{=} \perp \notin \{0,1\}^*$.

Proposition 3.2 (analysis of Construction 3.2): *The function f is length preserving and almost 1-1. Furthermore, f is 1-1 if S never errs. If $\mathbf{F}$ is a family of one-way permutations satisfying the additional conditions of Definition 3.1, then f is weakly one-way. The latter holds even if $\mathbf{F}$ is only weakly one-way (and satisfies the additional conditions).*

Proof: In case algorithm S never errs, the proof is similar to the proof of Proposition 3.1 (i.e., G_n is redefined as the set of all pairs (r,s) such that $i \stackrel{\text{def}}{=} S(n,r) \neq \perp$ and $s \in D_i$). Otherwise, we observe that the collision probability of f is bounded above by the probability that S errs (and outputs a string not in I). Since this happens with negligible probability, we are done.   □

4   Applying the Transformation

Using the transformation specified in the previous section, we show how to construct a 1-1 one-way function based on one of several popular intractability assumptions. To this end, we use these intractability assumptions in order to construct families of one-way permutations satisfying the additional conditions of Definition 3.1. Before presenting these constructions, we wish to stress an important aspect regarding them; namely, their (quantified) level of "security" (see next).

Security

The security of a one-way function f is a function, $s : \mathbb{N} \to \mathbb{N}$, specifying the amount of "work" required to invert f on inputs of given length. The work of an algorithm is defined as the product of the running-time (of the inverting algorithm) and the inverse of its success probability; namely, $w_A(n) \stackrel{\text{def}}{=} t_A(n) \cdot \frac{1}{p_A(n)}$, where $t_A(n)$ is the running time of A on f-images of length n and $p_A(n) \stackrel{\text{def}}{=}$ $\text{Prob}_{x \in \{0,1\}^n}[A(f(x)) \in f^{-1}f(x)]$ is its success probability.

Typical cryptographic constructions, and in particular our constructions, transform one object (in our case, a family of one-way permutations) of security $s(\cdot)$ into another object (in our case, a single 1-1 one-way function) of related security $s'(\cdot)$. The relation between s and s' is of key importance. A weak relation, which is usually easier to obtain, is that $s'(\text{poly}(n)) > s(n)/\text{poly}(n)$. Although this relation translates any super-polynomial security s into a super-polynomial security

s', it is of limited practical value. In order to use the resulting object of security s' one may needs to use very big instances. For example, if $s'(n^5) = s(n)$ and the original object is "secure in reality" for instance size 100 (bits), then the resulting object (of security s') will be "secure in reality" only for instances of size 10^{10} (and is thus unlikely to be of practical value). Thus, stronger relation between the security s of the original object and the security s' of the resulting object are of more value. In particular, it is desirable to have $s'(O(n)) > s(n)/\text{poly}(n)$, in which case we say that the transformation preserves the security.

Getting back to the constructions of the previous section, we note that the security of the resulting one-way 1-1 function f, on f-images of length $q(n) + l(n)$, is closely related to the security of the family of one-way permutations on f_i-images of length $l(n)$. (Recall, n denotes the length of the index of the permutation, $l(n)$ the length of the description of elements in the domain of the permutation, and $q(n)$ the randomness complexity of the index-selecting algorithm.) Thus, $s'(q(n)+l(n)) > s(l(n))/\text{poly}(n)$, where s denotes the security of the family $\mathbf{F}$ and s' the security of the function f. Therefore, *the smaller the polynomial $q(\cdot)$ is, the better security one gets*. It is particularly desirable to keep $q(n)$ linear in $l(n)$. All the constructions presented below achieve this goal. Consequently, *the one-way functions constructed below preserve the security of the intractability assumption on which they are based.* We remark that the (weak to strong one-way) transformation of [7] (mentioned in the Remark in Section 3) preserves security too.

Preliminaries: Selecting Prime Numbers

Prime numbers play a key role in all our constructions, and so efficient algorithms for selecting such numbers are of key importance to us. We will use two algorithms, the first being being the celebrated deterministic polynomial-time primality tester of Agrawal, Kayal, and Saxena [1],

Theorem 4.1 (primes are in $\mathcal{P}$): *There exists a deterministic polynomial-time primality tester; that is, an algorithm that decides whether a given integer is a prime number.*

The second algorithm is Bach's probabilistic polynomial-time algorithm that on input 1^n uniformly generates an n-bit long composite number along with its factorization [2]. A straightforward implementation of Bach's algorithm requires a super-linear number of coin tosses (i.e., a number of coin tosses that is super-linear in the length of the composite being generated). Here we claim an approximate version that uses a linear number of coin tosses. We say that a distribution X on n-bit long strings is almost uniformly distributed over $S \subseteq \{0,1\}^n$ if the variation distance between X and the uniform distribution over S is negligible (as a function of n).

Theorem 4.2 (randomness efficient generation of integers with known prime factorization): *There exists a probabilistic polynomial-time algorithm that, on input 1^n, uses $O(n)$ coin tosses to select a random number N almost uniformly in the interval $[2^{n-1}, 2^n - 1]$, and outputs the prime factorization of N.*

Proof: While it is possible to present a direct implementation of an approximate version of Bach's algorithm that uses only a linear number of coin tosses, the details are quite tedious. Hence, we prefer to invoke a general result of Nisan and Zuckerman [15] that asserts that *any probabilistic polynomial-time algorithm that uses linear space has an approximated version that uses a linear number of coin tosses*, where in our context approximation means that the output distribution of the new algorithm (on any fixed input) is statistically close to the output distribution of the original algorithm (i.e., the variation distance is negligible).[5] It is easy to see that Bach's algorithm utilizes linear space, and the theorem follows.   □

4.1   A Construction Based on RSA

The standard presentation of RSA [18] yields a family of permutations, which is believed to be one-way, but is certainly *not* one-way in the augmented sense of Definition 3.1. Here we refer to a family in which the indices are pairs (N, e), where N is the product of two primes of equal length and e is relatively prime to $\phi(N)$. The index is generated by randomly selecting these two primes, multiplying them and next selecting a proper e. Thus, giving these random choices away compromises the security of RSA, since given the prime factors it is easy to invert the function.

We consider, instead, the following family of weak one-way permutations. The indices in this family are pairs of integers (N, P) such that P is a prime and $|P| = |N|$. For each such pair we define a permutation over $\mathbb{Z}_N^*$, the multiplicative group modulo N; specifically, $f_{N,P}(x) \stackrel{\text{def}}{=} x^P \bmod N$. Note that we do not insist that N is a product of two primes of the same length. Yet, a noticeable fraction of the possible N's will have this form. Thus, if the standard RSA family is strongly one-way (for random exponent) then it is also (strongly) one-way for a prime exponent, and consequently the foregoing (non-standard) family of functions will be weakly one-way (due to the noticeable fraction of composites of the standard form). Since P is relatively-prime to $\phi(N)$, the functions in this family are in fact permutations over $\mathbb{Z}_N^*$. (Note that the index-selecting algorithm does not know $\phi(N)$, and so relative-primality of P and $\phi(N)$ is imposed by requiring that P is prime.)

We now show that the foregoing family satisfies the non-simplified requirements (from a family of one-way permutations) as well as the additional conditions in Definition 3.1. Among the efficiency conditions of Definition 2.2 only the

---

[5] The original result is stated for algorithms that output a single bit, but it extends trivially to algorithms to algorithms that output a linear number of bits, which is the case in our application (i.e., the aforementioned "context").

one referring to the index selection is problematic, yet it does hold when only requiring that output is produced merely with noticeable probability; specifically, we select two n-bit integers at random and check whether the second is prime, producing an output only if the answer is affirmative. Furthermore, $\mathbb{Z}_N^*$ is easily recognizable and has noticeable density with respect to $\{0,1\}^{|N|}$. This family is one-way in the augmented sense (under the "RSA assumption"), since the modulus is generated via an identity transformation from the coins of the index-selecting algorithm (and thus these coins add no knowledge to the inverter). It follows that we can apply Proposition 3.2 and derive a length-preserving 1-1 one-way function.

Definition 4.1 (standard RSA Assumption): *We say that* inverting RSA is intractable with security $s(\cdot)$ *if any algorithm for the* inverting task *uses work greater than $s(\cdot)$. The inverting task consists of finding x such that $y = x^e \bmod N$, when given N, e and y, where N is uniformly selected among all composites that are the product of two $(n/2)$-bit long primes, e is uniformly selected among the elements of the multiplicative group modulo $\phi(N)$, and y is uniformly selected among the elements of the multiplicative group modulo N.*

To justify our claim that the security (of the RSA Assumption) is preserved, we note that pairs (N, P) as required can be selected using $O(\|(N, P)\|)$ random bits.[6] Thus, we get

Corollary 4.1 (a length-preserving 1-1 one-way function based on RSA): *Suppose that inverting RSA is intractable with security $s(n)$. Then, there exists a length-preserving 1-1 one-way function with security $s'(O(n)) \stackrel{\text{def}}{=} s(n)/\text{poly}(n)$.*

4.2   A Construction Based on a Restricted DLP

Here we rely on the assumption that the Discrete Logarithm Problem (DLP) in the multiplicative group modulo P is hard also for the special case of primes of the form $P = 2Q + 1$, where Q is a prime. We also use the assumption that such primes form a noticeable fraction of the integers of the same length. Based on these assumptions, the following family of permutations is one-way. The indices in the family are pairs (P, g), where P is a prime of the above form and g is a primitive element modulo P. The index is selected by first selecting a prime of the above form and next using the known factorization of $\phi(P) = 2Q$ to test candidates for primitivity (see details below). For each index, (P, g), we define a permutation over $\mathbb{Z}_P^*$, the multiplicative group modulo P; specifically, $f_{P,g}(x) \stackrel{\text{def}}{=} g^x \bmod P$. Noting that $\mathbb{Z}_P^*$ is both 'noticeable' and easy to recognize,

---

[6] Currently, the random bits are merely used to select (N, P) uniformly among all pairs of n-bit long integers. Indeed, checking primality is done by using the deterministic algorithm guaranteed in Theorem 4.1 (whereas in the earlier versions Bach's randomness-efficient algorithm [3] was used for that purpose, which only resulted in an almost 1-1 function).

we can apply Proposition 3.2, provided that the index-selection process satisfies the augmented one-way condition.

To address the last concern as well as justify our claim that the resulting 1-1 one-way function preserves the security of the family, we need to specify the way in which the pairs (P, g) are selected. On input n we uniformly select an $(n - 1)$-bit integer, Q, and test Q and $P = 2Q + 1$ for primality. In case we are successful, we uniformly select $g \in \mathbb{Z}_P^*$ and test if it is primitive (mod P) by computing $g^{P-1} \bmod P$, $g^Q \bmod P$ and $g^2 \bmod P$. (If the first expression evaluates to 1 whereas the other two do not, then g is a primitive element modulo P.)[7] Thus, we use $|(P, g)|$ random bits to generate pairs (P, g), and these coins are identical to the pairs themselves. Combining this with the foregoing assumption regarding the density of primes of the desired form and the fact that in this case approximately half the elements of $\mathbb{Z}_P^*$ are primitive, we get

Corollary 4.2 (a length-preserving 1-1 one-way function based on restricted DLP): *Suppose that the restricted DLP is intractable with security $s(n)$ (as in Definition 4.2), and that the set of n-bit primes, P, for which $\phi(P)/2$ is prime, constitute a $1/\mathrm{poly}(n)$ fraction of the n-bit long integers. Then, there exists a length-preserving 1-1 one-way function with security $s'(O(n)) \stackrel{\text{def}}{=} s(n)/\mathrm{poly}(n)$.*

Definition 4.2 (restricted DLP Assumption): *We say that the restricted DLP is intractable with security $s(\cdot)$ if any algorithm for the following inverting task uses work greater than $s(\cdot)$. The inverting task consists of finding x such that $y = g^x \bmod P$, when given P, g and y, where P is uniformly selected among all n-bit primes for which $\phi(P)/2$ is prime, g is uniformly selected among the primitive elements modulo P, and y is uniformly selected among the elements of the multiplicative group modulo P.*

4.3   A Construction Based on the General DLP

Here we rely on a alternative assumption concerning the DLP. Specifically, we assume that the Discrete Logarithm Problem (DLP) in the multiplicative group modulo a prime P is hard also when given the factorization of $\phi(P)$. (Note that for primes of the special form $P = 2Q + 1$ the factorization of $\phi(P) = 2 \cdot Q$ is always known.) Furthermore, we shall assume that this DLP problem remains hard when given any $O(|P|)$ bits of information regarding P; that is, we assume that there are no trapdoors (of linear length) for the DLP in the multiplicative group modulo a prime. Making this assumption, we can waive the assumption made in the previous subsection concerning the density of primes of special form $P = 2Q + 1$, where Q is a prime.

Based on the foregoing intractability assumption, the following family of permutations is one-way. The indices in the family are pairs (P, g), where P

---

[7] Again, checking primality is done by using the deterministic algorithm guaranteed in Theorem 4.1. Indeed, this allows to obtain a 1-1 function (rather than an almost 1-1 function, as in earlier versions).

is a prime and g is a primitive element modulo P. The index is chosen by first generating a random prime P with known factorization of $\phi(P)$ (see details below), and next using this factorization to test candidates for primitivity. For each index, (P, g), we define a permutation over $\mathbb{Z}_P^*$ as before (i.e., $f_{P,g}(x) \overset{\text{def}}{=} g^x \bmod P$). Again, we shall apply Proposition 3.2 to the current family.

We have postponed the discussion of how to randomly generate primes P with known factorization of $\phi(P)$. Here is where we use Theorem 4.2, which asserts the existence of an adequate algorithm and furthermore one that uses a linear number of coin tosses. This yields an index-selection algorithm that selects pairs (P, g) using $O(|(P, g)|)$ random bits, which is instrumental to our claim that the resulting 1-1 one-way function preserves the security of the family. The fact that the coins used by this index-selection algorithm provide additional information on P is "covered" by the assumption formulated in Definition 4.3. Thus, we get:

Corollary 4.3 (a length-preserving 1-1 one-way function based on general DLP): *Suppose that DLP is intractable with security $s(n)$, even when the factorization of the order of the group is given* (as in Definition 4.3). *Then, there exists a length-preserving 1-1 one-way function with security $s'(O(n)) \overset{\text{def}}{=} s(n)/\mathrm{poly}(n)$.*

Definition 4.3 (DLP Assumption): *We say that the DLP is intractable with security $s(\cdot)$ if, for every randomized mapping Π such that $|\Pi(z)| = O(|z|)$, any algorithm for the following inverting task uses work greater than $s(\cdot)$. The inverting task consists of finding x such that $y = g^x \bmod P$, when given $\Pi(P)$ and a pair (g, y), where P is uniformly selected among all n-bit primes, g is uniformly selected among the primitive elements modulo P, and y is uniformly selected among the elements of the multiplicative group modulo P.*

The randomized mapping Π captures possible trapdoor information that may assist in inverting $f_{P,g}$, and the assumption asserts that the inverting task remains hard also in the presence of such information.[8] In particular, the randomized mapping Π may yield the coins used by the factored-number generating algorithm of Theorem 4.2. Thus, inverting $f_{P,g}$ is hard also in the augmented sense of Definition 3.1.

5   Conclusions and Open Problems

We have presented a method for constructing (strongly) one-way permutations. The method consists of three steps.

Step (1): Using well-known intractability assumptions to construct families of one-way permutations satisfying the additional properties specified in Definition 3.1.

---

[8] Indeed, the fact that Π is applied only to P and that $|\Pi(P)| = O(|P|)$ makes the assumption potentially weaker. On the other hand, the fact that $\Pi(P)$ may contain P allows us to omit P from the list of inputs to the inverting task.

Step (2): Using such a family to construct a weak one-way function.
Step (3): Transforming the resulting function into a strongly one-way function.

We consider the identification of the conditions in Definition 3.1 and the construction of families of one-way permutations satisfying these conditions to be the most important contributions of the current paper. Thus, most of the paper is dedicated to the implementation of Step (1), whereas Step (2) is obtained by Construction 3.2, and Step (3) is obtained by referring to [7].

Regarding Step (3), we remark that applying the general ("weak to strong") transformation of [7] seems an over-kill, since in our case the weakly one-way function f has a special structure (e.g., it is hard to invert almost on all points on which it is not the identity transformation). However, in our attempts to avoid using [7], we were not able to avoid using random walks on expander graphs (for the repeated attempts to generate a valid index and/or a sample in the corresponding domain). Since expander graphs are the only non-elementary component of [7], we see no point in presenting these alternatives here. Certainly, it will be better to avoid the use of expander graphs and perform Step (3) in a more efficient manner.

Another obvious open problem is to construct length-preserving 1-1 one-way functions based on the conjectured intractability of factoring.[9] To achieve this goal using our method one will need to construct a family of one-way permutations satisfying the additional properties specified in Definition 3.1. (The standard construction of a family of one-way permutations based on factoring [16] does not satisfy the augmented one-wayness condition.)

Acknowledgments. We would like to thank Eric Bach and Hugo Krawczyk for helpful discussions and comments.

References

1. Agrawal, M., Kayal, N., Saxena, N.: Primes is in P. Annals of Mathematics 160(2), 781–793 (2004)
2. Bach, E.: Analytic Methods in the Analysis and Design of Number-Theoretic Algorithms (ACM Distinguished Dissertation 1984). MIT Press, Cambridge (1985)
3. Bach, E.: Realistic Analysis of some Randomized Algorithms. In: 19th STOC, pp. 453–461 (1987)
4. Blum, M., Micali, S.: How to Generate Cryptographically Strong Sequences of Pseudo-Random Bits. SIAM J. on Computing 13, 850–864 (1984)
5. Goldreich, O.: Foundation of Cryptography: Basic Tools. Cambridge University Press, Cambridge (2001)
6. Goldreich, O.: Foundation of Cryptography: Basic Applications. Cambridge University Press, Cambridge (2004)

---

[9] We comment that, using Theorem 4.1, one can obtain a 1-1 function based on factoring, alas this function is not length preserving. Specifically, consider the function that maps a pair of integers of the form (x, y) to their multiple if $x < y$ and both numbers are prime, and maps it to (x, y) otherwise. Using an adequate encoding that distinguishes the two cases, this mapping is 1-1, but not length preserving.

7. Goldreich, O., Impagliazzo, R., Levin, L., Venkatesan, R., Zuckerman, D.: Security Preserving Amplification of Hardness. In: 31st FOCS, pp. 318–326 (1990)
8. Goldreich, O., Levin, L.: A Hard-Core Predicate for any One-way Function. In: 21st STOC, pp. 25–32 (1989)
9. Goldreich, O., Krawczyk, H., Luby, M.: On the Existence of Pseudorandom Generators. SIAM J. on Computing 22, 1163–1175 (1993)
10. Haitner, I.: Implementing Oblivious Transfer Using Collection of Dense Trapdoor Permutations. In: Naor, M. (ed.) TCC 2004. LNCS, vol. 2951, pp. 394–409. Springer, Heidelberg (2004)
11. Haitner, I., Reingold, O., Vadhan, S.: Efficiency Improvements in Constructing Pseudorandom Generator from any One-way Function. In: 42nd STOC, pp. 437–446 (2010)
12. Håstad, J., Impagliazzo, R., Levin, L.A., Luby, M.: A Pseudorandom Generator from any One-way Function. SICOMP 28(4), 1364–1396 (1990); Combines papers of Impagliazzo, Levin, and Luby (21st STOC, 1989) and J. Håstad (22nd STOC, 1990)
13. Miller, G.L.: Riemann's Hypothesis and tests for primality. JCSS 13, 300–317 (1976)
14. Naor, M., Yung, M.: Universal Hash Functions and their Cryptographic Applications. In: 21st STOC, pp. 33–43 (1989)
15. Nisan, N., Zuckerman, D.: Randomness is Linear in Space. JCSS 52(1), 43–52 (1996); Preliminary version in 25th STOC (1993)
16. Rabin, M.O.: Digitalized Signatures and Public Key Functions as Intractable as Factoring. MIT/LCS/TR-212 (1979)
17. Rabin, M.O.: Probabilistic algorithm for testing primality. Jour. of Number Theory 12, 128–138 (1980)
18. Rivest, R., Shamir, A., Adleman, L.: A Method for Obtaining Digital Signatures and Public Key Cryptosystems. CACM 21, 120–126 (1978)
19. Rompel, J.: One-way Functions are Necessary and Sufficient for Secure Signatures. In: 22nd STOC, pp. 387–394 (1990)
20. Solovay, R., Strassen, V.: A fast Monte-Carlo test for primality. SIAM Jour. on Computing 6, 84–85 (1977)

On the Circuit Complexity of Perfect Hashing

Oded Goldreich and Avi Wigderson

Abstract. We consider the size of circuits that *perfectly hash* an arbitrary subset $S \subset \{0,1\}^n$ of cardinality 2^k into $\{0,1\}^m$. We observe that, in general, the size of such circuits is exponential in $2k - m$, and provide a matching upper bound.

Keywords: Perfect Hashing, Circuit Complexity.

An early version of this work appeared as TR96-041 of *ECCC*. We later found out that, in contrast to our previous impression, the lower bound has been known. In fact, our lower bound argument is analogous to the one presented in [6, pp. 128-129]. The current revision is quite minimal.

Summary

We consider the problem of perfectly hashing an arbitrary subset $S \subset \{0,1\}^n$ of cardinality 2^k into $\{0,1\}^m$, where $k \leq m$. That is, given an arbitrary subset $S \subset \{0,1\}^n$ of cardinality 2^k, we seek a function $h : \{0,1\}^n \to \{0,1\}^m$ so that $h(x) \neq h(y)$ for every two distinct $x \neq y$ in S. Clearly, such a function always exists, the question is what is its complexity; that is, what is the size of the smallest circuit computing h. Two obvious upper bounds follow.

1. For every $S \subset \{0,1\}^n$, there is a circuit of size $|S| \cdot n$ that perfectly hashes S into $\{0,1\}^{\lceil \log_2 |S| \rceil}$.

   (The circuit is merely a look-up table for S.)
2. For every $S \subset \{0,1\}^n$, there is a circuit of size $\text{poly}(n)$ that perfectly hashes S into $\{0,1\}^{2\lceil \log_2 |S| \rceil}$.

   (The circuit implements a suitable function from a family of Universal$_2$ Hashing [2]. Such a family always contains perfect hashing functions for S [4].)

We show that these upper bounds are the best possible. That is:

Theorem 1 (lower bound): *For every n, k and $m \leq n - 1$, there exists a subset $S \subset \{0,1\}^n$ of cardinality 2^k such that perfectly hashing S into $\{0,1\}^m$ requires a circuit of size $\Omega(2^{2k-m}/n)$.*

Interestingly, this lower bound is tight for all values of $m \in [k, 2k]$ (and not merely for $m \in \{k, 2k\}$). That is:

O. Goldreich et al.: Studies in Complexity and Cryptography, LNCS 6650, pp. 26–29, 2011.
© Springer-Verlag Berlin Heidelberg 2011

Theorem 2 (matching upper bound):[1] *For every n, m, k where $k \leq m \leq 2k$, and every subset $S \subset \{0,1\}^n$ of cardinality 2^k, there exists a circuit of size $2^{2k-m} \cdot \text{poly}(n)$ that perfectly hashes S into $\{0,1\}^m$.*

1 Proof of Theorem 1

The proof follows by a simple counting argument, combining an upper bound on the number of circuits of given size with a lower bound on the size of a family of functions that can perfectly hash all subsets of size 2^k. Improved lower bounds for the latter appears in [3,5,7]. For sake of completeness, we prove a weaker bound, which is sufficient for our purposes, and present the argument in probabilistic terms.

Suppose, in contrary to Theorem 1, that for every subset $S \subset \{0,1\}^n$ of cardinality $K \overset{\text{def}}{=} 2^k$ there exists a circuit of size $o(2^{2k-m}/(2k-m))$ that perfectly hashes S into $\{0,1\}^k$. We will show that each circuit can serve as a perfect hashing for too few K-subsets, and hence there are too few circuits to perfectly hash all possible K-subsets. The main observation follows:

Lemma 1.1 (the fraction of sets that are perfectly hashed by any function): *For any $m \leq n-1$, let $C : \{0,1\}^n \to \{0,1\}^m$ be an arbitrary circuit, and let $S \subset \{0,1\}^n$ be a uniformly selected subset of cardinality $K = 2^k$. Then, the probability that C perfectly hashes S into $\{0,1\}^m$ is at most $2^{-\Omega(2^{2k-m})}$.*

Proof: Let $N \overset{\text{def}}{=} 2^n$ and $M \overset{\text{def}}{=} 2^m$. Clearly, we may assume that $k \leq m$ (as otherwise the probability is zero). Let $c_1, ..., c_M$ denote the sizes of the preimages of the various m-bit strings under C (i.e., $c_i = |C^{-1}(s_i)|$, where s_i denotes the ith (m-bit long) string by some order). Then, the probability we are interested in is

$$\frac{\sum_{I \subseteq [M]: |I|=K} \prod_{i \in I} \binom{c_i}{1}}{\binom{N}{K}} \leq \frac{\binom{M}{K} \cdot (N/M)^K}{\binom{N}{K}}$$

$$= \prod_{i=0}^{K-1} \frac{1 - (i/M)}{1 - (i/N)}$$

$$= \exp\left\{ -\sum_{i=1}^{K-1} \ln\left(1 + \frac{(i/M) - (i/N)}{1 - (i/M)}\right) \right\}$$

$$< \exp\left\{ -\sum_{i=1}^{K-1} ((i/M) - (i/N)) \right\}$$

$$= \exp\left\{ -\frac{K \cdot (K-1)}{2} \cdot \left(\frac{1}{M} - \frac{1}{N}\right) \right\}$$

which for $M \leq N/2$ yields $2^{-\Omega(K^2/M)}$. The lemma follows. ∎

---

[1] We stress that the circuits guaranteed here cannot, in general, be simply described; that is, this result is inherently nonuniform.

Deriving Theorem 1. Adding up the contribution of all possible circuits, while applying Lemma 1.1 to each of them, we conclude that if too few circuits are considered then not all K-subsets can be perfectly hashed. Specifically, there are $s^{O(s)}$ possible circuits of size s, and so we need $s^{O(s)} \cdot 2^{-\Omega(2^{2k-m})} \geq 1$. Theorem 1 follows.

2   Proof of Theorem 2

We consider two cases. In the case that $m \leq k + \log_2 n$, the theorem follows by constructing an obvious circuit that maps each string in S to its rank (in S) represented as an m-bit long string. This circuit has size $|S| \cdot n \leq 2^{2k-m} \cdot n^2$ (since $k \leq 2k - m + \log_2 n$), and the theorem follows.

The less obvious case is when $m \geq k + \log_2 n$. Here we use a family of n-wise independent functions mapping $\{0,1\}^n$ onto $\{0,1\}^\ell$, where $\ell \overset{\text{def}}{=} m - \log_2 n$. Function in such a family can be evaluated by poly(n)-size circuits (cf. [1]). We consider the collisions caused by a uniformly chosen function from this family applied to S. Specifically,

Lemma 2.1 (hashing by n-wise independence functions): *Let H be a family of functions $\{h : \{0,1\}^n \to \{0,1\}^\ell\}$ such that $\text{Prob}_{h \in H}[\wedge_{i=1}^n h(\alpha_i) = \beta_i] = 2^{-n\ell}$, for every n distinct $\alpha_1, ..., \alpha_n \in \{0,1\}^n$ and for every $\beta_1, ..., \beta_n \in \{0,1\}^\ell$. Then, for every $S \subset \{0,1\}^n$ of cardinality $2^k \leq 2^\ell$, there exists $h \in H$ such that*

1. *No value has more than n preimages under h; that is, $|h^{-1}(\beta) \cap S| \leq n$, for every $\beta \in \{0,1\}^\ell$.*
2. *At most $2^{2k-\ell}$ values have more than one preimage under h; that is, $|\{\beta \in \{0,1\}^\ell : |h^{-1}(\beta) \cap S| > 1\}| \leq 2^{2k-\ell}$.*

Proof: Fixing an arbitrary 2^k-subset, S, and uniformly selecting $h \in H$, we consider the probability that the two items (above) hold. Firstly, we consider the probability that h maps n elements of S to the same image. Using the n-wise independence of the family H, the probability of this event is bounded by

$$\binom{2^k}{n} \cdot 2^{-\ell n} < \frac{2^{kn}}{n!} \cdot 2^{-kn} < \frac{1}{2}$$

where the first inequality uses $\ell = m - \log_2 n \geq k$. Thus, the probability that Item (1) does not hold is less than $1/2$. Next, we consider the probability that Item (2) does not hold. We start by using the pairwise independence of H to note that the collision probability is $2^{-\ell}$ (i.e., $\text{Prob}_{h \in H}[h(\alpha_1) = h(\alpha_2)] = 2^{-\ell}$, for any $\alpha_1 \neq \alpha_2 \in \{0,1\}^n$). It follows that the expected number of h-images that have more than a single preimage in S is bounded above by the expected number of collisions; that is, by $\binom{2^k}{2} \cdot 2^{-\ell} < \frac{1}{2} \cdot 2^{2k-\ell}$. Applying Markov's Inequality, we conclude that the probability that Item (2) does not hold is less than $1/2$. The lemma follows.   ∎

Deriving Theorem 2. Fixing an arbitrary 2^k-subset, $S \subset \{0,1\}^n$, and using Lemma 2.1, we present a circuit that perfectly hashes S into $\{0,1\}^m$ (where $m \geq k + \log_2 n$). Our construction uses the double hashing paradigm (see, e.g., [4]). Let $h \colon \{0,1\}^n \to \{0,1\}^{m-\log_2 n}$ be as guaranteed by the lemma (w.r.t the set S). We define a perfect hashing function $f \colon \{0,1\}^n \to \{0,1\}^m$ for S by letting

$$f(\alpha) \overset{\text{def}}{=} h(\alpha) \circ \text{rank}_{S \cap h^{-1}(h(\alpha))}(\alpha)$$

where $\text{rank}_R(\alpha)$ is an $\log_2 n$-bit long string representing the rank of α among the elements of R. A circuit computing the function f is constructed as follows. For each β having more than a unique h-preimage in S, we maintain a table ranking these preimages in S. By Item (1) of Lemma 2.1 such a table need only contain n entries, whereas by Item (2) we only need $2^{2k-\ell}$ such tables. (If a string, α, does not appear in any of the tables, then $f(\alpha) = h(\alpha) \circ 0^{\log_2 n}$.) The size of the circuit is $\text{poly}(n) + 2^{2k-\ell} \cdot n^2 = \text{poly}(n) + 2^{2k-m} \cdot n^3$, and so Theorem 2 follows.

References

1. Alon, N., Babai, L., Itai, A.: A fast and Simple Randomized Algorithm for the Maximal Independent Set Problem. J. of Algorithms 7, 567–583 (1986)
2. Carter, L., Wegman, M.: Universal Classes of Hash Functions. J. Computer and System Sciences 18, 143–154 (1979)
3. Fredman, M., Komlós, J.: On the Size of Separating Systems and Perfect Hash Functions. SIAM J. Algebraic and Discrete Methods 5, 61–68 (1984)
4. Fredman, M., Komlós, J., Szemerédi, E.: Storing a Sparse Table with O(1) Worst Case Access Time. Journal of the ACM 31, 538–544 (1984)
5. Korner, J., Marton, K.: New Bounds for Perfect Hashing via Information Theory. Europ. J. Combinatorics 9, 523–530 (1988)
6. Mehlhorn, K.: Data Structures and Algorithms. EATCS Monographs on Theoretical Computer Science, vol. 1 (1984)
7. Nilli, A.: Perfect Hashing and Probability. Combinatorics, Probability and Computing 3, 407–409 (1994)

Collision-Free Hashing from Lattice Problems

Oded Goldreich, Shafi Goldwasser, and Shai Halevi

Abstract. In 1995, Ajtai described a construction of one-way functions whose security is equivalent to the difficulty of some well known approximation problems in lattices. We show that essentially the same construction can also be used to obtain collision-free hashing. This paper contains a self-contained proof sketch of Ajtai's result.

Keywords: Integer Lattices, One-EWay Functions, Worst-Case to Average-Case Reductions, Collision-Resistant Hashing.

An early version of this work appeared as TR96-042 of *ECCC*. The current revision is intentionally minimal.

1 Introduction

In 1995, Ajtai described a problem that is *hard on the average* if some well-known lattice problems are *hard to approximate in the worst case*, and demonstrated how this problem can be used to construct one-way functions [1]. We show that Ajtai's method can also be used to construct families of collision-free hash functions. Furthermore, a slight modification of this construction yields families of functions which are both universal and collision-free.

1.1 The Construction

The construction is very simple. For security parameter n, we pick a random $n \times m$ matrix M with entries from $\mathbb{Z}_q$, where m and q are chosen so that $n \log q < m < \frac{q}{2n^4}$ and $q = O(n^c)$ for some constant $c > 0$ (e.g., $m = n^2, q = n^7$). See Section 3 for a discussion of the choice of parameters. The hash function $h_M : \{0,1\}^m \to \mathbb{Z}_q^n$ is then defined, for $s = s_1 s_2 \cdots s_m \in \{0,1\}^m$, as

$$h_M(s) = Ms \bmod q = \sum_i s_i M_i \bmod q, \tag{1}$$

where M_i is the ith column of M.

Notice that h_M's input is m-bit long, whereas its output is $n \log q$ bits long. Since we chose the parameters such that $m > n \log q$, there are collisions in h_M. As we will argue below, however, it is infeasible to find any of these collisions unless some well known lattice problems have good approximation in the worst case. It follows that, although it is easy to find solutions for the equations $Ms \equiv 0$ (mod q), it seems hard to find binary solutions (i.e., a vector $s \in \{0,1\}^m$ in the solution space).

O. Goldreich et al.: Studies in Complexity and Cryptography, LNCS 6650, pp. 30–39, 2011.
© Springer-Verlag Berlin Heidelberg 2011

Remark. Using our notation, the candidate one-way function introduced by Ajtai is $f(M, s) \stackrel{\text{def}}{=} (M, h_M(s))$. We note that this function is *regular* (cf., [3]); that is, the number of preimage of any image is about the same. (Furthermore, for most M's the number of pre-images under h_M of almost all images is about the same.) To the best of our knowledge, it is easier (and more efficient) to construct a pseudo-random generator based on a regular one-way function than based on an arbitrary one-way function (cf., [3] and [4]).

1.2   A Modification

A family of hash functions is called *universal* if a function uniformly selected in the family maps every two images uniformly on its range in a pairwise indepedent manner [2]. To obtain a family of functions that is both universal and collision-free, we slightly modify the foregoing construction. First we set q to be a *prime* of the desired size. Then, in addition to picking a random matrix $M \in \mathbb{Z}_q^{n \times m}$, we also pick a random vector $r \in \mathbb{Z}_q^n$. The function $h_{M,r} : \{0,1\}^m \to \mathbb{Z}_q^n$ is then defined, for $s = s_1 \cdots s_m \in \{0,1\}^m$, as

$$h_M(s) = Ms + r \mod q = r + \sum_i s_i M_i \mod q. \qquad (2)$$

The modified construction resembles the standard construction of universal hash functions [2], with calculations done over $\mathbb{Z}_q$ instead of over $\mathbb{Z}_2$.

2   Formal Setting

In this section we give a brief description of some well known lattice problems, outline Ajtai's reduction, and our version of it.

2.1   Lattices

Definition 1. *Given a set of n linearly independent vectors in $\mathbb{R}^n$, denoted $V = \langle v_1, \cdots, v_n \rangle$, we define the* lattice spanned *by V as the set of all possible linear combinations of the v_i's with integral coefficients; that is,*

$$L(V) \stackrel{\text{def}}{=} \left\{ \sum_i a_i v_i \; : \; a_i \in \mathbb{Z} \text{ for all } i \right\} \qquad (3)$$

We call V the basis *of the lattice $L(V)$. We say that a set of vectors, $L \subset \mathbb{R}^n$, is a* lattice *if there is a basis V such that $L = L(V)$.*

It is convenient to view a lattice L in $\mathbb{R}^n$ as a "tiling" of the space $\mathbb{R}^n$ using small parallelepipeds, with the lattice points being the vertices of these parallelepipeds. The parallelepipeds themselves are spanned by some basis of L. We call the parallelepiped that are spanned by the "shortest basis of L" (the one whose vectors have the shortest Euclidean norm) the basic cells of the lattice L. See Figure 1 for an illustration of these terms in a simple lattice in $\mathbb{R}^2$.

A lattice in $\boldsymbol{R}^2$

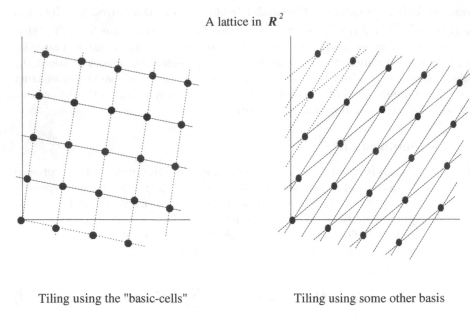

Tiling using the "basic-cells"   Tiling using some other basis

Fig. 1. Tiling of a simple lattice in $\mathbb{R}^2$ with two different bases

Computational Problems Regarding Lattices. Finding "short vectors" (i.e., vectors with small Euclidean norm) in lattices is considered a hard problem. There are no known efficient algorithms to find, given an arbitrary basis of a lattice, either the shortest non-zero vector in the lattice, or another basis for the same lattice whose longest vector is as short as possible. No efficient algorithms are known for approximation versions of these problems as well. The approximation versions being considered here are the following:

(W1) Given an arbitrary basis B of a lattice L in $\mathbb{R}^n$, approximate (up to a polynomial factor in n) the length of the shortest vector in L.

(W2) Given an arbitrary basis B of a lattice L in $\mathbb{R}^n$, find another basis of L whose length is at most polynomially (in n) larger than that of the smallest basis of L (where the length of a basis is the length of its longest vector).

We choose 'W' for the foregoing notation to indicate that we will be interested in the *worst-case complexity* of these problems. The best known algorithms for these problems are the L^3 algorithm and Schnorr algorithm. The L^3 algorithm, due to Lenstra, Lenstra and Lovász [5] approximates these problems to within a ratio of $2^{n/2}$ in the worst case, and Schnorr's algorithm [6] improves this to $(1 + \varepsilon)^n$ for any fixed $\varepsilon > 0$. Another problem, which can be shown to be equivalent to the above approximation problems (cf. [1]), is the following:

(W3) Given an arbitrary basis B of a lattice L, find a set of n linearly independent lattice vectors, whose length is at most polynomially (in n) larger than the length of the smallest set of n linearly independent lattice vectors. (Again, the length of a set of vectors is the length of its longest vector.)

A few remarks about (**W3**) are in order:

1. Note that not every linearly independent set of n lattice points is a basis for that lattice. For example, if $V = \{v_1, v_2\}$ span some lattice in $\mathbb{R}^2$, then the set $\{2v_1, v_2\}$ is a linearly independent set of 2 vectors that does not span $L(V)$, since we cannot represent v_1 as an integral linear combination of $2v_1$ and v_2.

2. In the sequel we reduce the security of our construction to the difficulty of solving Problem (W3). It will be convenient to use the following notation: For a given polynomial $Q(\cdot)$, denote by $(\text{W3})_Q$ the problem of approximating the smallest independent set in an n-dimensional lattice up to a factor of $Q(n)$.

2.2 Ajtai's Reduction

In his paper Ajtai described the following problem:

Problem (A1): For parameters $n, m, q \in \mathbb{N}$ such that $n \log q < m \le \frac{q}{2n^4}$ and $q = O(n^c)$, for some constant $c > 0$.

Input: A matrix $M \in \mathbb{Z}_q^{n \times m}$.
Output: A vector $x \in \mathbb{Z}_q^m \setminus \{0^m\}$ such that $Mx \equiv 0 \pmod{q}$ and $\|x\| < n$ (where $\|x\|$ denotes the Euclidean norm of x).

Here, we used 'A' (in the notation) to indicate that we will be interested in the *average-case complexity* of this problem. Ajtai proved the following theorem, reducing the worst-case complexity of (W3) to the average-case complexity of (A1).

Ajtai's Theorem [1]: *Suppose that it is possible to solve a uniformly selected instance of Problem (A1) in expected $T(n, m, q)$-time, where the expectation is taken over the choice of the instance as well as the coin-tosses of the solving algorithm. Then, it is possible to solve Problem (W3) in expected $\text{poly}(|I|) \cdot T(n, \text{poly}(n), \text{poly}(n))$ time on every n-dimensional instance I, where the expectation is taken over the coin-tosses of the solving algorithm.*

Remark. Ajtai [1] has noted that the theorem remain valid also when Problem (A1) is relaxed so that the desired output is allowed to have Euclidean norm of up to $\text{poly}(n)$ (i.e., one requires $\|x\| \le \text{poly}(n)$ rather than $\|x\| < n$).

2.3 Our Version

We observe that one can use essentially the same proof to show that the following problem is also hard on the average.

Problem (A2): For parameters $n, m, q \in \mathbb{N}$ as in (A1).

Input: A matrix $M \in \mathbb{Z}_q^{n \times m}$.
Output: A vector $x \in \{-1, 0, 1\}^m \setminus \{0^m\}$ such that $Mx \equiv 0 \pmod{q}$.

Theorem 1: *Suppose that it is possible to solve a* uniformly selected *instance of Problem (A2) in expected $T(n, m, q)$-time, where the expectation is taken over the choice of the instance as well as the coin-tosses of the solving algorithm. Then, it is possible to solve Problem (W3) in expected $\mathrm{poly}(|I|) \cdot T(n, \mathrm{poly}(n), \mathrm{poly}(n))$ time on every n-dimensional instance I, where the expectation is taken over the coin-tosses of the solving algorithm.*

Proof: By the foregoing Remark, Ajtai's Theorem holds also when modifying Problem (A1) such that the output is (only) required to have Euclidean norm of up to m. Once so modified, Problem (A1) becomes more relaxed than Problem (A2) and so the current theorem follows.   □

For the sake of self-containment we sketch the main ideas of the proof of Ajtai's Theorem (equivalently, of Theorem 1) in Section 4. The reader is referred to [1] for further details.

3   Constructing Collision-Free Hash Functions

The security of our proposed collision-free hash functions follows directly from Theorem 1. Below, we spell out the argument and discuss the parameters.

3.1   The Functions and Their Security

Recall our construction of a family of collision-free hash functions:

Picking a hash-function
    To pick a hash-function with security-parameters n, m, q (where $n \log q < m \leq \frac{q}{2n^4}$ and $q = O(n^c)$), we pick a random matrix $M \in \mathbb{Z}_q^{n \times m}$.
Evaluating the hash function
    Given a matrix $M \in \mathbb{Z}_q^{n \times m}$ and a string $s \in \{0, 1\}^m$, compute

$$h_M(s) = Ms \bmod q = \sum_i s_i M_i \bmod q.$$

The collision-free property is easy to establish assuming that Problem (A2) is hard on the average. That is:

Theorem 2: *Suppose that given a uniformly chosen matrix, $M \in \mathbb{Z}_q^{n \times m}$, it is possible to find in (expected) $T(n, m, q)$-time two vectors $x \neq y \in \{0, 1\}^m$ such that $Mx \equiv My \pmod{q}$. Then, it is possible to solve a uniformly selected instance of Problem (A2) in (expected) $T(n, m, q)$-time.*

Proof: If we can find two binary strings $s_1 \neq s_2 \in \{0, 1\}^m$ such that $Ms_1 \equiv Ms_2 \pmod{q}$, then we have $M(s_1 - s_2) \equiv 0 \pmod{q}$. Since $s_1, s_2 \in \{0, 1\}^m$, we have $x \stackrel{\text{def}}{=} (s_1 - s_2) \in \{-1, 0, 1\}^m$, which constitutes a solution to Problem (A2) for the instance M.   □

3.2 The Parameters

The proof of Theorem 1 imposes restrictions on the relationship between the parameters n, m and q. First of all, we should think of n as the security parameter of the system, since we derive the difficulty of solving Problem (A2) by assuming the difficulty of approximating some problems over n-dimensional lattices.

The condition $m > n \log q$ is necessary for two reasons. The first is simply because we want the output of the hash function to be shorter than its input. The second is that when $m < n \log q$, a random instance of problem (A2) typically does not have a solution at all, and the reduction procedure in the proof of Theorem 1 falls apart.

The conditions $q = O(n^c)$ and $m < q/2n^4$ also come from the proof of Theorem 1. Their implications for the security of the system are as follows:

- The larger q is, the stronger the assumption that needs to be made regarding the complexity of problem (W3). Namely, the security proof shows that (A2) with parameters n, m, q is hard to solve on the average, if the problem (W3)$_{(qn^6)}$ is hard in the worst case, where (W3)$_{(qn^6)}$ is the problem of approximating the shortest independent set of a lattice up to a factor of qn^6. Thus, for example, if we worry (for a given n) that an approximation ratio of n^{15} is feasible, then we better choose $q < n^9$. Also, since we know that approximation within exponential factor is possible, we must always choose q to be sub-exponential in n.
- By the above, the ratio $R \overset{\text{def}}{=} \frac{q/n^4}{m}$ must be strictly bigger than 1 (above, for simplicy, we stated $R > 2$). The larger R is, the better the reduction becomes: In the reduction from (W3) to (A2) we need to solve several random (A2) problems to obtain a solution to one (W3) problem. The number of instances of (A2) problem which need to be solved depends on R. Specifically, this number behaves roughly like $n^2/\log R$. This means that when $q/n^4 = 2m$ we need to solve about n^2 instances of (A2) per any instance of (W3), which yields a ratio of $O(n^2)$ between the time it takes to break the hashing scheme and the time it takes to solve a worst-case (W3) problem. On the other hand, when R approaches 1 the number of iterations (in the reduction) grows rapidly (and tends to infinity).

Notice also that the inequalities $n \log q < m < \frac{q}{n^4}$ implies a lower bound on q, namely $\frac{q}{\log q} > n^5$, which means that $q = \Omega(n^5 \log n)$.

4 Self-contained Sketch of the Proof of Theorem 1

At the heart of the proof is the following procedure for solving (W3): It takes as inputs a basis $B = \langle b_1, \cdots, b_n \rangle$ for a lattice and a set of n linearly independent lattice vectors $V = \langle v_1, \cdots, v_n \rangle$, with $|v_1| \leq |v_2| \leq \cdots \leq |v_n|$. The procedure produces another lattice vector w, such that $|w| \leq |v_n|/2$ and w is linearly independent of $v_1, \cdots, v_{n-1}$. We can then replace the vector v_n with w and repeat this process until we get a "very short independent set". When invoking this procedure, we denote by S the length of the vector v_n (which is the longest vector in V).

In the sequel we describe this procedure and show that as long as S is more than n^c times the size of the basic lattice-cell (for some constant $c > 0$), the procedure succeeds with high probability. Therefore we can repeat the process until the procedure fails, and then conclude that (with high probability) the length of the longest vector in V is not more that n^c times the size of the basic lattice-cell. For the rest of this section we will assume that S is larger than n^c times the size of the basic lattice-cell.

The procedure consists of five steps: We first construct an "almost cubic" parallelepiped of lattice vectors, which we call a pseudo-cube. Next, we divide this pseudo-cube into q^n small parallelepipeds (not necessarily of lattice vectors), which we call sub-pseudo-cubes. We then pick some random lattice points in the pseudo-cube (see Step 3) and consider the location of each point with respect to the partition of the pseudo-cube into sub-pseudo-cubes (see Step 4). Each such location is represented as a vector in $\mathbb{Z}_q^n$ and the collection of these vectors forms an instance of Problem (A2). A solution to this instance yields a lattice point that is pretty close to a "corner" of the pseudo-cube. Thus, our final step consists of using the solution to this (A2) instance to compute the "short vector" w. Below we describe each of these steps in more details.

1. Constructing a "pseudo-cube". The procedure first constructs a parallelepiped of lattice vectors that is "almost a cube". This can be done by taking a sufficiently large cube (say, a cube with side length of $n^3 S$), expressing each of the cubes' basis vectors as a linear combination of the v_i's, and then rounding the coefficients in this combination to the nearest integers. Denote the vectors thus obtained by $f_1, \cdots, f_n$ and the parallelepiped that is spanned by them by C. The f_i's are all lattice vectors, and their distance from the basis vectors of the "real cube" is very small compared to the size of the cube.[1] Hence the parallelepiped C is very "cube-like". We call this parallelepiped a pseudo-cube.

2. Dividing the pseudo-cube into "sub-pseudo-cubes". We then divide C into q^n equal sub-pseudo-cubes, each of which can be represented by a vector in $\mathbb{Z}_q^n$ as follows:

$$\text{for every } T = \begin{pmatrix} t_1 \\ \vdots \\ t_n \end{pmatrix} \in \mathbb{Z}_q^n, \quad \text{define } C_T \stackrel{\text{def}}{=} \left\{ \sum_i \alpha_i f_i \; : \; \frac{t_i}{q} \leq \alpha_i < \frac{t_i + 1}{q} \right\}.$$

For each sub-pseudo-cube C_T, we call the vector $o_T = \sum_i \frac{t_i}{q} f_i$ the origin of C_T (i.e., o_T is the vector in C_T that is closest to the origin). We note that any vector in $v \in C_T$ can be written as $v = o_T + \delta$ where δ is the location of v inside the sub-pseudo-cube C_T. See Figure 2 for an illustration of that construction (with $n = 2, q = 3$).

---

[1] The f_i's can be as far as $Sn/2$ away from the basis vectors of the real cube, but this is still much smaller than the size of the cube itself.

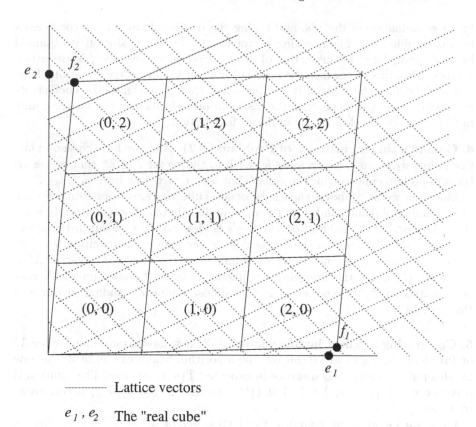

— — — — Lattice vectors

e_1, e_2 The "real cube"

f_1, f_2 The pseudo-cube

Fig. 2. The basic construction in the proof of Theorem 1 (for $q = 3$)

The parameter q was chosen such that each C_T is "much smaller" than S. That is, the side-length of each sub-pseudo-cube C_T is $Sn^3/q \leq S/2nm$. On the other hand, with this choice, each C_T is still much larger than the basic lattice cell (since S is much bigger than the size of the basic cell). This, together with the fact that the C_T's are close to being cubes, implies that each C_T contains approximately the same number of lattice points.

3. Choosing random lattice points in C. We then choose m random lattice points $u_1, \cdots u_m \in C$. To do that, we use the basis $B = \{b_1, \cdots, b_n\}$ of the lattice. To choose each point, we take a linear combination of the basis vectors b_i with large enough integer coefficients (say, in the range $[0, 2^{n^c} \cdot \max(S, |B|)]$ for some constant c). This gives us some lattice point p.

We then "reduce p mod C". By this we mean that we look at a tiling of the space $\mathbb{R}^n$ with the pseudo-cube C, and we compute the location vector of p in its surrounding pseudo-cube. Formally, this is done by representing p as a linear combination of the f_i's, and taking the fractional part of the coefficients in this combination. The resulting vector is a lattice point, since it is obtained by subtracting integer combination of the f_i's from p, whereas the f_i's are lattice vectors. Also, this vector must lie inside C, since it is a linear combination of the f_i's with coefficients in $[0, 1)$. It can be shown that if we choose the coefficients from a large enough range, then the distribution induced over the lattice points in C is statistically close to the uniform distribution.

4. Constructing an instance of Problem (A2). After we have chosen m lattice points $u_1, \cdots, u_m$, we compute for each u_i the vector $T_i \in \mathbb{Z}_q^n$ that represent the sub-pseudo-cube in which u_i falls. That is, for each i we have $u_i \in C_{T_i}$.

Since, as we said above, each sub-pseudo-cube contains approximately the same number of points, and since the u_i's are distributed almost uniformly in C, then the distribution induced on the C_{T_i}'s is close to the uniform distribution, and so the distribution over the T_i's is close to the uniform distribution over $\mathbb{Z}_q^n$.

We now consider the matrix whose columns are the vectors T_i, that is, $M = (T_1|T_2|\cdots|T_m)$. By the foregoing argument, it is an "almost uniform" random matrix in $\mathbb{Z}_q^{n \times m}$, and so, it is an "almost uniform" random instance of Problem (A2).

5. Computing a "short lattice vector". We now have a random instance M of Problem (A2), and so we can use the algorithm whose existence we assume in Theorem 1 to solve this instance in expected $T(n, m, q)$ time. The solution is a vector $x = \{x_1, \cdots, x_m\} \in \{-1, 0, 1\}^m$ such that $Mx = \sum_i x_i T_i$ is congruent to 0 mod q.

Once we found x, we compute the lattice vector $w' = \sum_{i=1}^m x_i u_i$. Let us examine the vector w': Recall that we can represent each u_i as the sum of $o_i \overset{\text{def}}{=} o_{T_i}$ (the origin vector of C_{T_i}) and δ_i (the location of u_i inside C_{T_i}). Thus,

$$w' = \sum_{i=1}^m x_i u_i = \sum_{i=1}^m x_i o_i + \sum_{i=1}^m x_i \delta_i.$$

A key observation is that since $\sum_i x_i T_i \equiv \bar{0} \pmod{q}$, "reducing the vector $(\sum_i x_i o_i)$ mod C" yields the all-zeros vector; that is, $(\sum_i x_i o_i) \bmod C = \bar{0}$. To see why this is the case, recall that each $o_i = o_{T_i}$ has the form $\sum_j \frac{t_i(j)}{q} f_j$, where $t_i(j) \in \{0, ..., q-1\}$ is the jth component of T_i. Now, the hypothesis $\sum_i x_i t_i(j) \equiv 0 \pmod{q}$ for $j = 1, .., n$, yields that

$$\sum_i x_i o_{T_i} = \sum_i x_i \sum_j \frac{t_i(j)}{q} f_j = \sum_j \frac{\sum_i x_i t_i(j)}{q} f_j = \sum_j c_j f_j$$

where all c_j's are integers. Since "reducing the vector $\sum_i x_i o_{T_i}$ mod C" means subtracting from it an integer linear combination of f_j's, the resulting vector is $\bar{0}$. Thus, "reducing w' mod C" we get $\sum_{i=1}^m x_i \delta_i$; that is,

$$w' \bmod C = \sum_{i=1}^m x_i \delta_i.$$

Since each δ_i is just the location of some point inside the sub-pseudo-cube C_{T_i}, the size of each δ_i is at most $n \cdot S/2mn = S/2m$. Moreover as $x_i \in \{-1, 0, 1\}$ for all i we get

$$\left\| \sum_i x_i \delta_i \right\| \leq \sum_i |x_i| \cdot \|\delta_i\| \leq m \cdot \frac{S}{2m} = \frac{S}{2}.$$

This means that the lattice vector w' mod C is close up to $\frac{S}{2}$ to one of the "corners" of C. Thus, all we need to do is to find the difference vector between the lattice vector w' mod C and that corner (which is also a lattice vector). Doing that is very similar to reducing w' mod C: We express w' as a linear combination of the f_i's, but instead of taking the fractional part of the coefficients, we take the difference between these coefficients and the closest integers. This gives us the "promised vector" w, a lattice vector whose length is at most $S/2$.

The only thing left to verify is that with high probability, w can replace the largest vector in V (i.e., it is linearly independent of the other vectors in V). To see that, notice that the vector x does not depend on the exact choice of the u_i's, but only on the choice of their sub-pseudo-cubes C_{T_i}'s. Thus, we can think of the process of choosing the u_i's as first choosing the C_{T_i}'s, next computing the x_i's and only then choosing the δ_i's.

Assume (w.l.o.g.) that we have $x_1 \neq 0$. Let us now fix all the δ_i's except δ_1 and then pick δ_1 so as to get a random lattice point in C_{T_1}. Thus, the probability that w falls in some fixed subspace of $\mathbb{R}^n$ (such as the one spanned by the $n-1$ smallest vectors in V), equals the probability that a random point in C_{T_1} falls in such subspace. Since C_{T_1} is a pseudo-cube that is much larger than the basic cell of L, this probability is very small.

Acknowledgments. We thank Dan Boneh and Jin Yi Cai for drawing our attention to an error in a previous version of this note.

References

1. Ajtai, M.: Generating Hard Instances of Lattice Problems. In: 28th ACM Symposium on Theory of Computing, Philadelphia, pp. 99–108 (1996)
2. Carter, L., Wegman, M.: Universal Classes of Hash Functions. J. Computer and System Sciences 18, 143–154 (1979)
3. Goldreich, O., Krawczyk, H., Luby, M.: On the existence of pseudorandom generators. SIAM J. on Computing 22(6), 1163–1175 (1993)
4. Håstad, J., Impagliazzo, R., Levin, L.A., Luby, M.: A Pseudorandom Generator from any One-way Function. SIAM J. on Computing 28(4), 1364–1396 (1990); Combines papers of Impagliazzo et al. (21st STOC, 1989) and Håstad (22nd STOC, 1990)
5. Lenstra, A.K., Lenstra, H.W., Lovász, L.: Factoring Polynomials with Rational Coefficients. Mathematische Annalen 261, 515–534 (1982)
6. Schnorr, C.P.: A more efficient algorithm for a lattice basis reduction. Journal of Algorithms 9, 47–62 (1988)

Another Proof That $\mathcal{BPP} \subseteq \mathcal{PH}$ (and More)

Oded Goldreich and David Zuckerman

Abstract. We provide another proof of the Sipser–Lautemann Theorem by which $\mathcal{BPP} \subseteq \mathcal{MA}$ ($\subseteq \mathcal{PH}$). The current proof is based on strong results regarding the amplification of $\mathcal{BPP}$, due to Zuckerman (1996). Given these results, the current proof is even simpler than previous ones. Furthermore, extending the proof leads to two results regarding $\mathcal{MA}$: $\mathcal{MA} \subseteq \mathcal{ZPP}^{\mathcal{NP}}$ (which seems to be new), and that two-sided error $\mathcal{MA}$ equals $\mathcal{MA}$. Finally, we survey the known facts regarding the fragment of the polynomial-time hierarchy that contains $\mathcal{MA}$.

Keywords: BPP, The Polynomial-Time Hierarchy, Interactive Proof Systems (AM and MA), Randomness–Efficient Error Reduction (Amplification).

An early version of this work appeared as TR97-045 of *ECCC*. The current revision is quite minimal.

1 Introduction

Non-trivial results, showing containment of fundamental complexity classes in one another, are quite rare. One of the first such results is Sipser's Theorem [14] by which $\mathcal{BPP}$ is contained in the Polynomial-Time Hierarchy. A simpler proof, placing $\mathcal{BPP}$ even lower in this hierarchy, was presented by Lautemann [11]. Although not stated in these (subsequently introduced) terms, Lautemann's proof actually establishes the following:

Theorem 1 (The Sipser–Lautemann Theorem): $\mathcal{BPP} \subseteq \mathcal{MA}$.

See definitions in next section.

The contents of this note, In this note, we present an alternative proof of the Sipser–Lautemann Theorem. Our proof relies on powerful results regarding randomness–efficient error reduction (a.k.a amplification) for $\mathcal{BPP}$. Given these powerful results, our proof is almost a triviality.

Using similiar arguments, we show that $\mathcal{MA} \subseteq \mathcal{ZPP}^{\mathcal{NP}}$ (re-establishing a theorem of Zachos and Heller [16] by which $\mathcal{BPP} \subseteq \mathcal{ZPP}^{\mathcal{NP}}$). It follows that $\mathcal{NP}^{\mathcal{BPP}} \subseteq \mathcal{ZPP}^{\mathcal{NP}}$. To the best of our knowledge, these results were not known before.

In summary, the purpose of this note is three-fold: Firstly to demonstrate the power of the currently known results regarding randomness–efficient error reduction. We believe that these results have not been fully assimilated into

O. Goldreich et al.: Studies in Complexity and Cryptography, LNCS 6650, pp. 40–53, 2011.

complexity theory and are yet to be exploited by it. Secondly we wish to focus attention on the fragment of the polynomial-time hierarchy that contains $\mathcal{MA}$. It seems that this fragment gives rise to some challenges which may be within our current reach. Finally, we take the oppertunity to prove the aforementioned new result.

Organization. The core of this work (i.e., the alternative proof of the Sipser–Lautemann Theorem) is presented in Sections 2 and 3.1. This alternative proof is further discussed in Section 3.2, and applied in the context of two-sided MA in Section 3.3. The same proof strategy is then applied to show that $\mathcal{MA}$ is contained in $\mathcal{ZPP}^{\mathcal{NP}}$ (see Section 4). Finally, we conclude with a brief survey of the complexity classes around $\mathcal{MA}$ (see Section 5).

2 Background

(For further background, see Section 5.)

2.1 BPP and Randomness-Efficient Error Reduction

Definition 1 (the class BPP): *For any set S, we denote by χ_S the characteristic function of the set; that is, $\chi_S(x) = 1$ if $x \in S$ and $\chi_S(x) = 0$ otherwise. A set S is in $\mathcal{BPP}$ if there exists a probabilistic polynomial-time machine M such that for every $x \in \{0, 1\}^*$*

$$\mathrm{Prob}[M(x) \neq \chi_S(x)] \leq \frac{1}{3}$$

where the probability is taken uniformly over the internal coin tosses of M.

The error probability in the foregoing procedure can be reduced by repetitions (a process hereafter referred to as *amplification*). The obvious way of doing so transforms a machine (as above) that, on input x, uses $p(|x|)$ coins into a machine having error probability at most $2^{-t(|x|)}$ that uses $O(t(|x|) \cdot p(|x|))$ coins (for any polynomial t). More efficient amplification procedures, utilizing Expander Random Walks, yield the same error bound while using only $p(|x|) + (4 + o(1)) \cdot t(|x|)$ coins (see survey [6]). In particular, *for any constant $c > 4$, using a sufficiently large polynomial t, we get a procedure that uses $c \cdot t(|x|)$ coins and has error probability at most $2^{-t(|x|)}$.* An alternative construction due to Zuckerman [17] provides, *for any constant $c > 1$ and sufficiently large polynomial t, a procedure that uses $c \cdot t(|x|)$ coins and has error probability at most $2^{-t(|x|)}$.* What is remarkable in the last procedure is that the number of coins used is essentially the logarithm of the error bound. Put in other words, the number of "bad" coin sequences can be made any (constant) root of the total number of coin sequences. In particular,

Theorem 2 (Zuckerman's randomness-efficient amplification of BPP [17]): *For any set S in $\mathcal{BPP}$, there exists a polynomial-time recognizable binary relation R and a polynomial p such that*

$$|\{r \in \{0, 1\}^{p(|x|)} : R(x, r) \neq \chi_S(x)\}| < 2^{p(|x|)/3}.$$

2.2   The Complexity Class MA

Definition 2 (the class MA): *A set S is in $\mathcal{MA}$ if there exists a polynomial-time recognizable 3-ary relation V and polynomials p, q such that*

- *If $x \in S$, then there exists $w \in \{0,1\}^{q(|x|)}$ such that for every $r \in \{0,1\}^{p(|x|)}$ it holds that $V(x, w, r) = 1$.*
- *If $x \notin S$, then for every $w \in \{0,1\}^{q(|x|)}$ it holds that*

$$\text{Prob}_r[V(x, w, r) = 1] \leq \frac{1}{2}$$

where the probability is taken uniformly over all $r \in \{0,1\}^{p(|x|)}$.

The class $\mathcal{MA}$, introduced by Babai [1], consists of sets having a Merlin–Arthur proof system: The prover (Merlin) sends a certificate (denoted w above) to the verifier (Arthur) who assesses it probabilistically (by tossing coins r and applying the predicate V). Merlin–Arthur proof systems are a degenerate type of interactive proof systems (introduced by Goldwasser, Micali and Rackoff [8] and Babai [1]). Actually, in a Merlin–Arthur proof system there is no real interaction. Instead, it is instructive to view $\mathcal{MA}$ as *the* randomized version of $\mathcal{NP}$: Here the "certificates" (for membership) can be verified via a randomized procedure and errors may occur (yet with bounded probability).

3   A Proof of the Sipser–Lautemann Theorem

3.1   The Proof Itself

Using Zuckerman's efficient amplification of BPP, we present the following MA proof system. Specifically, we will refer to the relation R and the polynomial p guaranteed in Theorem 2.

The protocol. On input x, both parties compute $m = p(|x|)$, and proceed as follows.

1. Merlin tries to select $r' \in \{0,1\}^{m/2}$ such that $R(x, r'r'') = 1$ for all $r'' \in \{0,1\}^{m/2}$. Merlin sends r' to Arthur.
2. Upon receiving r', Arthur selects $r'' \in \{0,1\}^{m/2}$ uniformly and accepts if and only if $R(x, r'r'') = 1$.

Analysis of the foregoing protocol. If $x \in S$, then there are at most $2^{m/3}$ possible r's for which $R(x, r) = 0$. Thus there are at most $2^{m/3}$ prefixes $r' \in \{0,1\}^{m/2}$ for which some r'' exists so that $R(x, r'r'') = 0$. Merlin may just select any of the other $2^{m/2} - 2^{m/3}$ prefixes and make Arthur always accept. On the other hand, if $x \notin S$, then there are at most $2^{m/3}$ possible r's for which $R(x, r) = 1$. Thus, for each $r' \in \{0,1\}^{m/2}$, it holds that

$$\text{Prob}_{r'' \in \{0,1\}^{m/2}}[R(x, r'r'') = 1] \leq \frac{2^{m/3}}{2^{m/2}} \ll \frac{1}{2}.$$

3.2 Discussion

Let us review our proof strategy. Starting with Theorem 2, we partitioned the space of all (2^m) possible coin-tosses outcomes into $(2^{m/2})$ subsets of equal size. We then used the following two facts:

1. The number of bad outcomes is smaller than the number of subsets (and so there exists a subset with no bad outcomes). This was used to analyze the case $x \in S$.
2. The number of bad outcomes is much smaller than the size of each subset (and so each subset contains a majority of good outcomes). This was used to analyze the case $x \notin S$.

Thus, what we have used is the fact that number of bad outcomes is much smaller than the square root of the total number of outcomes. We stress that the fact that any BPP-machine can be transformed into a machine for which the foregoing holds (i.e., Theorem 2) is highly non-trivial. We believe that this fact (or known generalizations of it) may find further applications in complexity theory.

Comparison to Lautemann's proof. Recall that Lautemann's proof has the prover send the verifier $t = m/\log_2 m$ strings, $s_1, ..., s_t$, and the verifier tosses coins $r \in \{0,1\}^m$ and accepts iff $R(x, r \oplus s_i) = 1$ holds for some i. The existence of an appropriate sequence of strings is proven by an elementary probabilistic argument. Actually, s_1 may be any fixed string (e.g., 0^m) and so needs not be sent (by the prover). We observe that IF we start with R as guaranteed by Theorem 2, then $t = 2$ suffices. This gets us very close to the proof above. In fact, the probabilistic argument of Lautemann reduces to the trivial counting argument above. Thus, using Theorem 2 allows also a simplification of Lautemann's argument, although the proof presented earlier is believed to be simpler: Technically speaking, we have the prover send only $m/2$ bits (rather than m required in the simplified Lautemann's argument), the verifier tosses only $m/2$ coins (again, rather than m), and the predicate R is evaluated only once (rather than twice).

3.3 Two-Sided Error Equals One-Sided Error for MA

Both Lautemann's proof and our proof can be extended to show that a two-sided error version of $\mathcal{MA}$ equals the one-sided error defined above. (This provides an alternative proof to the one presented in [15].) We mention that interactive proof systems with zero error collapse to $\mathcal{NP}$, whereas for all (higher than MA) levels of the interactive proof hierarchy, the two-sided error version equals the one-sided one [5].

Definition 3 (two-sided version of MA): *A set S is in $\mathcal{MA}_2$ if there exists a polynomial-time recognizable 3-ary relation V and polynomials p, q such that*

*   *If $x \in S$, then there exists $w \in \{0,1\}^{q(|x|)}$ such that*

$$\mathrm{Prob}_r[V(x, w, r) = 1] \geq \frac{2}{3}.$$

– *If $x \notin S$, then for every $w \in \{0,1\}^{q(|x|)}$ it holds that*

$$\text{Prob}_r[V(x,w,r) = 0] \geq \frac{2}{3}.$$

In both cases, the probability is taken uniformly over all $r \in \{0,1\}^{p(|x|)}$.

Theorem 3 [15, Thm 2(i)]: $\mathcal{MA} = \mathcal{MA}_2$.

Proof: Clearly, $\mathcal{MA} \subseteq \mathcal{MA}_2$, and so we focus on showing that $\mathcal{MA}_2 \subseteq \mathcal{MA}$. Let S be an arbitrary set in $\mathcal{MA}_2$. For every $x \in S$, we consider w as guaranteed by the first condition of Definition 3, whereas for $x \notin S$ we consider any $w \in \{0,1\}^{q(|x|)}$. Both Lautemann's proof and our proof extend to promise problems in BPP, and in particular to the following BPP promise problem, $\Pi = (\Pi_{\text{YES}}, \Pi_{\text{NO}})$, where

$$\Pi_{\text{YES}} \stackrel{\text{def}}{=} \left\{ (x,w) : \text{Prob}_r[V(x,w,r) = 1] \geq \frac{2}{3} \right\}$$

$$\Pi_{\text{NO}} \stackrel{\text{def}}{=} \{(x,w) : x \notin S\}$$

$$\subseteq \left\{ (x,w) : \text{Prob}_r[V(x,w,r) = 0] \geq \frac{2}{3} \right\}$$

In particular, the amplification technique of Zuckerman applies also to this case and so we obtain a predicate V' and a polynomail q' such that

$$\forall (x,w) \in \Pi_{\text{YES}} \quad |\{r \in \{0,1\}^{q'(|x|)} : V'(x,w,r) = 0\}| < 2^{q'(|x|)/3} \qquad (1)$$

$$\forall (x,w) \in \Pi_{\text{NO}} \quad |\{r \in \{0,1\}^{q'(|x|)} : V'(x,w,r) = 1\}| < 2^{q'(|x|)/3} \qquad (2)$$

Thus, we augment the MA-protocol of Section 3.1 as follows. On input x, with $m = q'(|x|)$, Merlin sends (w, r'), where $|r'| = m/2$, and Arthur uniformly selects $r'' \in \{0,1\}^{m/2}$ and accepts if and only if $V'(x, w, r'r'') = 1$. As before, in case $x \in S$, by sending an adequate (w, r'), Merlin can make Arthur accept for every choice of r''; whereas, in case $x \notin S$, for any choice of (w, r'), Arthur accepts with negligible probability. It follows that $S \in \mathcal{MA}$. ∎

4   MA Is Contained in ZPP with an NP-Oracle

The machines in the following definition may halt with a non-Boolean output (which may be interpreted as abstaining from a decision regarding membership).

Definition 4 (the class ZPP): *A set S is in $\mathcal{ZPP}$ if there exists a probabilistic polynomial-time machine M such that for every $x \in \{0,1\}^*$*

$$\text{Prob}[M(x) = \chi_S(x)] \geq \frac{1}{2}$$

$$\text{Prob}[M(x) = 1 - \chi_S(x)] = 0$$

where the probability is taken uniformly over the internal coin tosses of M.

Thus, the ZPP machine either gives the correct answer or gives no answer at all (i.e., a non-Boolean output is interpreted as no output). Clearly $\mathcal{ZPP} = \mathcal{RP} \cap \text{co}\mathcal{RP}$ (actualy, $\mathcal{ZPP}$ is sometimes defined this way).

4.1 BPP Is Contained in ZPP with an NP-Oracle

We start by providing an alternative proof to a result of Zachos and Heller.

Theorem 4 [16, P. 132, Cor. 3]: $\mathcal{BPP} \subseteq \mathcal{ZPP}^{\mathcal{NP}}$.

Proof: Using the same amplification and notations as in Section 3.1, we construct a probabilistic polynomial-time oracle machine, M, that on input x operates as follows (where $m = p(|x|)$):

1. Selects $\sigma \in \{0, 1\}$ uniformly (as guess for $\chi_S(x)$);
2. Selects $r' \in \{0, 1\}^{m/2}$ uniformly;
3. Queries the oracle on whether (x, σ, r') is in the following $\text{co}\mathcal{NP}$ set

$$\{(y, \tau, u) : \forall v \in \{0, 1\}^{|s|}, R(y, uv) = \tau\}. \tag{3}$$

4. If the oracle answers YES, then the machine outputs σ. Otherwise it halts with no output.

Recall that by the foregoing amplification, for any x, the following holds:

– For each r', it holds that

$$|\{r'' \in \{0, 1\}^{m/2} : R(x, r'r'') \neq \chi_S(x)\}| < 2^{m/2},$$

and so the oracle never answers YES on query $(x, 1 - \chi_S(x), r')$. Thus, the machine never outputs the wrong answer.
– On the other hand, it holds that

$$\text{Prob}_{r'}[\forall r'' \in \{0, 1\}^{m/2}, R(x, r'r'') = \chi_S(x)] > \frac{1}{2}$$

and so with probability at least $1/4$, over the choices of σ and r', the oracle answers YES (and the machine produces a (correct) 0-1 output).

Using straightforward amplification, the theorem follows. ∎

4.2 Extension to MA

Combining ideas from the last two proofs, we obtain.

Theorem 5 (seemingly new): $\mathcal{MA} \subseteq \mathcal{ZPP}^{\mathcal{NP}}$.

Observing that $\mathcal{NP}^{\mathcal{BPP}} \subseteq \mathcal{MA}_2$ (see Fact 6), and using Theorems 3 and 5, we conclude that $\mathcal{NP}^{\mathcal{BPP}} \subseteq \mathcal{ZPP}^{\mathcal{NP}}$.

Fact 6 (folklore): $\mathcal{NP}^{\mathcal{BPP}} \subseteq \mathcal{MA}_2$.

Proof: Let $S \in \mathcal{NP}^{\mathcal{BPP}}$. Then, for every $x \in S$, we instruct Merlin to send a transcript of an accepting computation of the non-deterministic polynomial-time oracle-machine, and instruct Arthur to verify the validity of transcript as well as the correctness of the the oracle answers (by running a probabilistic decision procedure of negligible two-sided error). ∎

Proof of Theorem 5: Let $S \in \mathcal{MA}$ and consider the same promise problem $\Pi = (\Pi_{\text{YES}}, \Pi_{\text{NO}})$ as in the proof of Theorem 3. Furthermore, consider the set $\Pi'_{\text{YES}} \subseteq \Pi_{\text{YES}}$ that consists of all pairs (x, w) such that for all $r \in \{0, 1\}^{p(|x|)}$ it holds that $V(x, w, r) = 1$, and recall that for every $x \in S$ there exists $w \in \{0, 1\}^{q(|x|)}$ such that $(x, w) \in \Pi'_{\text{YES}}$.

We construct a probabilistic polynomial-time oracle machine, M, that on input x and access to an NP-oracle, first attempts to find w such that $(x, w) \in \Pi_{\text{YES}}$, and next verifies that $(x, w) \in \Pi_{\text{YES}}$ indeed holds. Following is a detailed description of the operation of M (as well as key observations towards its analysis). On input x, where $n = q(|x|)$ and $k = p(|x|)$, machine M proceeds as follows.

Step 1: **Attempting to find a good** w. The machine uniformly selects $r_1, ..., r_{2n} \in \{0, 1\}^k$, and queries the NP-oracle on whether there exists a $w \in \{0, 1\}^n$ such that $\wedge_{i=1}^{2n} V(x, w, r_i) = 1$. If the answer is NO, then M halts with output 0, otherwise M iteratively recovers the bits of such a string w (by $|w|$ additional queries) and proceeds to the next step. Specifically, all queries have the form $(x, w', r_1, ..., r_{2n})$, and each such query is answered by a YES if and only if there exists a $w'' \in \{0, 1\}^{n-|w'|}$ such that $\wedge_{i=1}^{2n} V(x, w'w'', r_i) = 1$.
Note that if $x \in S$, then a string w such that $\wedge_{i=1}^{2n} V(x, w, r_i) = 1$ exists (e.g., consider w such that $(x, w) \in \Pi'_{\text{YES}}$), and so Step 1 must be completed while finding such a string w. On the other hand, for each $(x, w) \notin \Pi_{\text{YES}}$, the probability that $\wedge_{i=1}^{2n} V(x, w, r_i) = 1$ holds, where $r_1, ..., r_{2n}$ are selected uniformly in $\{0, 1\}^k$, is at most $(2/3)^{2n}$, and it follows that

$$\text{Prob}_{r_1,...,r_{2n}}[\exists w \text{ s.t. } (x, w) \notin \Pi_{\text{YES}} \text{ and } \wedge_{i=1}^{2n} V(x, w, r_i) = 1]] \leq 2^n \cdot (2/3)^{2n},$$

which is exponentially vanishing (in n).

Step 2: **Verifying that** w **is good** (i.e., $(x, w) \in \Pi_{\text{YES}}$). The machine treats (x, w) as an input to the promise problem Π and proceeds as in the proof of Theorem 4. Specifically, by using the same amplification as in the proof of Theorem 3, we obtain a verification procedure V' that satisfies Eq. (1)-(2). Letting $m = q'(|x|)$, machine M selects an $m/2$-bit long random prefix r', and queries the NP-oracle on whether all $m/2$-bit long suffixes make the predicate V' evaluate to 1 (i.e., whether for every r'' it holds that $V'(x, w, r'r'') = 1$). If the oracle answers YES, then M halts with output 1; otherwise, M halts with no output. (We stress that we never output 0 in this step.)

If $x \in S$, then Step 1 never halts but rather always yields a string w (for Step 2). Furthermore, with overwhelmingly high probability, the string w satisfies $(x, w) \in \Pi_{\text{YES}}$. Thus, with overwhelmingly high probability, Step 2 accepts.

On the other hand, if $x \notin S$, then with overwhrlmingly high probability Step 1 halts (with output 0). Furthermore, if the procedure continued to Step 2 with some string w, then $(x, w) \in \Pi_{\text{NO}}$ (since $x \notin S$). In this case, the oracle will always answer NO, and M will halt with no output. Thus, for any x, the machine never errs, and with overwhelmingly high probability it produces the correct output.   ∎

5 The Bigger Picture: Complexity Classes around MA

(For a wider perspective on interactive proofs, see [7, Sec. 9.1].)

5.1 Definitions

All the (binary and 3-ary) relations that are mentioned in the following definitions are only satisfied by arguments of polynomially related length (i.e., all tuples in a relation have arguments that are of length that is polynomial in the length of the first argument). Likewise, all quantifiers range over arguments of such lengths.

Definition 5 (traditional classes – classes of the 1970's):

- A set S is in $\Sigma_2^P = \mathcal{NP}^{\mathcal{NP}}$ (resp., $\Pi_2^P = \text{co}\mathcal{NP}^{\mathcal{NP}}$) if there exists a polynomial-time recognizable 3-ary relation R such that

$$S = \{x : \exists y \forall z \; R(x, y, z) = 1\}$$
$$(resp., \quad S = \{x : \forall y \exists z \; R(x, y, z) = 1\}).$$

- A set S is in $\Delta_2^P = \mathcal{P}^{\mathcal{NP}}$ if there exists a deterministic polynomial-time oracle machine M and a set $S' \in \mathcal{NP}$ such that $x \in S$ iff $M^{S'}(x) = 1$ ($\forall x$).
- A set S is in $\mathcal{RP}$ if there exists a probabilistic polynomial-time machine M such that

$$x \in S \implies \text{Prob}[M(x) = 1] \geq \frac{1}{2}$$
$$x \notin S \implies \text{Prob}[M(x) = 1] = 0$$

For any class $\mathcal{C}$, we define $\text{co}\mathcal{C} \overset{\text{def}}{=} \{\{0,1\}^* \setminus S : S \in \mathcal{C}\}$.

Definition 6 ($\mathcal{AM}$ [1] – a class of the 1980's): A set S is in $\mathcal{AM}$ if there exists a polynomial-time recognizable 3-ary relation V and polynomials p, q such that

- If $x \in S$, then for every $r \in \{0,1\}^{p(|x|)}$ there exists $w \in \{0,1\}^{q(|x|)}$ such that $V(x, r, w) = 1$.
- If $x \notin S$, then it holds that

$$\text{Prob}_r[\exists w \text{ s.t. } V(x, r, w) = 1] \leq \frac{1}{2}$$

where the probability is taken uniformly over all $r \in \{0,1\}^{p(|x|)}$.

In other words, the class $\mathcal{AM}$, introduced by Babai [1], consists of sets having an Arthur–Merlin proof systems: The verifier (Arthur) challenges the prover (Merlin) with a random query, denoted r, and given the prover's answer (denoted w) makes a decision using the predicate V. Thus, in contrast to Merlin–Arthur systems (where Arthur just (probabilistically) evaluates the validity of a "written proof"), in Arthur–Merlin systems we have a real interaction between the prover and the verifier. The class $\mathcal{AM}$ coincides with the class of sets having constant-round interactive proof systems [1,9]. Thus, it is the lowest level of the hierarchy of "real" interactive proofs [1,8] (i.e., interactive proofs that, unlike $\mathcal{NP}$ and $\mathcal{MA}$, are really interactive).

Definition 7 ($\mathcal{S}_2^P$ [4,13] – a class of the 1990's): S is in $\mathcal{S}_2^P$ if there exists a polynomial-time recognizable 3-ary relation R such that for every $x \in \{0,1\}^*$

$$\exists y \forall z \quad R(x,y,z) = \chi_S(x) \tag{4}$$
$$\exists z \forall y \quad R(x,y,z) = \chi_S(x) \tag{5}$$

The class $\mathcal{S}_2^P$ was introduced independently by Canetti [4] and Russell and Sundaram [13] with the motivation of providing a low "symmetric alternation class" that contains $\mathcal{BPP}$. Indeed, Canetti [4] has extended Lautemann's proof to show that $\mathcal{BPP} \subseteq \mathcal{S}_2^P$, whereas Russell and Sundaram [13] showed that $\mathcal{MA} \subseteq \mathcal{S}_2^P$ (and thus $\mathcal{BPP} \subseteq \mathcal{S}_2^P$).

5.2   Known Inclusions

We recall some known inclusions between the aforementioned classes. For sake of self-containment, we present proofs as well. Recall that, $\mathcal{BPP} \subseteq \mathcal{MA}$, by Theorem 1. We start with some simple *syntactical facts*:

1. $\mathcal{P} \subseteq \mathcal{RP} \subseteq \mathcal{NP} \subseteq \mathcal{MA}$.
2. $\mathcal{RP} \subseteq \mathcal{BPP}$.
3. $\mathcal{RP} \subseteq \text{co}\mathcal{MA}$ (equiv., $\text{co}\mathcal{RP} \subseteq \mathcal{MA}$).[1]
4. $\mathcal{NP} \cup \text{co}\mathcal{NP} \subseteq \mathcal{P}^{\mathcal{NP}}$.
5. $\mathcal{AM} \subseteq \Pi_2^P$.
6. $\mathcal{S}_2^P \subseteq \Sigma_2^P \cap \Pi_2^P$.
   (Actually, the transparent syntactical facts are the inclusion $\mathcal{S}_2^P \subseteq \Sigma_2^P$ and the closure of $\mathcal{S}_2^P$ under complement.)
7. $\mathcal{ZPP}^{\mathcal{NP}} \subseteq \Sigma_2^P \cap \Pi_2^P$.
   (Here the transparent facts are $\mathcal{ZPP}^{\mathcal{NP}} \subseteq \mathcal{RP}^{\mathcal{NP}} \subseteq \mathcal{NP}^{\mathcal{NP}} = \Sigma_2^P$.)

We now turn to three non-trivial results.

Proposition 7 [1]: $\mathcal{MA} \subseteq \mathcal{AM}$.

Proof: We use a naive amplification to reduce the error probability in the Merlin–Arthur game so to obtain error that is substantially smaller than the reciprocal of the number of possible Merlin messages. Specifically, we obtain a polynomial-time recognizable 3-ary relation V and polynomials p, q such that

---

[1] This syntactical fact can also be derived from $\mathcal{RP} \subseteq \mathcal{BPP}$, by using $\mathcal{BPP} \subseteq \mathcal{MA}$.

1. If $x \in S$, then there exists $w_0 \in \{0,1\}^{q(|x|)}$ such that for every $r \in \{0,1\}^{p(|x|)}$ it holds that $V(x, w_0, r) = 1$.
2. If $x \notin S$, then for every $w \in \{0,1\}^{q(|x|)}$ it holds that

$$\mathrm{Prob}_r[V(x, w, r) = 1] < \frac{1}{2} \cdot 2^{-q(|x|)}.$$

Thus,

$$\mathrm{Prob}_r[\exists w \in \{0,1\}^{q(|x|)} : V(x, w, r) = 1] \leq \sum_{w \in \{0,1\}^{q(|x|)}} \mathrm{Prob}_r[V(x, w, r) = 1]$$

$$< \frac{1}{2}.$$

We construct an Arthur–Merlin proof system (defined by a new predicate V') by merely reversing the order of moves in the foregoing proof system, and using essentially the same decision predicate as above: That is, we let $V'(x, r, w) \overset{\text{def}}{=} V(x, w, r)$. This potentially makes the task of Merlin easier, and so we need only worry about the case $x \notin S$ (which we handle easily using the above bound). Specifically, for the case $x \in S$, we may use the string w_0 (guaranteed in Item 1) as Merlin's response to any challenge r (and so $V'(x, r, w_0) = V(x, w_0, r) = 1$ for all r's). For the case $x \notin S$ we use the bound in Item 2 and so $\mathrm{Prob}_r[\exists w \in \{0,1\}^{q(|x|)} : V'(x, r, w) = 1] < 0.5$. The proposition follows. ∎

Proposition 8 [13]: $\mathcal{MA} \subseteq \mathcal{S}_2^P$.

Proof: We use the same amplification as in the previous proof. Here we write the case of $x \notin S$ as

$$\forall w \in \{0,1\}^{q(|x|)} \quad |\{r \in \{0,1\}^{p(|x|)} : V(x, w, r) = 1\}| < 2^{p(|x|) - q(|x|)} - 1$$

We define a relation R (for the class $\mathcal{S}_2^P$) such that $R(x, y, z) = 1$ if $|y| = |z| = q(|x|)$ and at least one of the following two conditions holds:

1. $y = w0^{p(|x|) - q(|x|)}$ and $V(x, w, z) = 1$.
2. $z = w0^{p(|x|) - q(|x|)}$ and $V(x, w, y) = 1$.

Clearly, this predicate is symmetric with respect to y and z; that is, condition (1) holds iff condition (2) holds. Thus, we only show, for any x, the existence of a string y such that, for all z's, $R(x, y, z) = \chi_S(x)$. Let us shorthand $m = p(|x|)$ and $n = q(|x|)$. For $x \in S$ there exists $w \in \{0,1\}^n$ such that for all $r \in \{0,1\}^m$ it holds that $V(x, w, r) = 1$. Thus, there exists $y = w0^{m-n} \in \{0,1\}^m$ such that for all $z \in \{0,1\}^m$ it holds that $R(x, y, z) = 1$. We now turn to the case where $x \notin S$: In this case,

$$|\{r : \exists w \text{ s.t. } V(x, w, r) = 1\}| \leq \sum_{w \in \{0,1\}^n} |\{r : V(x, w, r) = 1\}|$$

$$< 2^n \cdot (2^{n-m} - 1)$$

$$= 2^m - 2^n.$$

Thus, there exists $r \in \{0,1\}^m \setminus \{0,1\}^n 0^{m-n}$ such that for every $w \in \{0,1\}^n$ it holds that $V(x, w, r) = 0$. Given such an r, we prove that for all z's $R(x, r, z) = 0$. This holds since $R(x, r, z) = 1$ requires either r ending with 0^{m-n} (which does not hold by our choice) or $z = w0^{n-m}$ with $V(x, w, r) = 1$ (which again cannot hold). ∎

Proposition 9 [13]: $\mathcal{P}^{\mathcal{NP}} \subseteq \mathcal{S}_2^P$.

Proof: Let S be an arbitrary set in $\mathcal{P}^{\mathcal{NP}}$, and let M be a (deterministic) polynomial-time oracle machine recognizing S when given access to the NP-complete set S'. We say that a string τ is a **valid transcript** of $M(x)$ if there exists *some* oracle such that τ describes the computation of M on input x and access to this oracle. Note that the oracle's answers in a valid transcript of $M(x)$ do *not* necessarily agree with the set S'. A valid transcript is said to be **supported** by a sequence of pairs $\bar{s}$ if for each oracle query q in the transcript τ that was answered by 1 there is a pair (q, w) in $\bar{s}$, where w is an NP-witness for membership of q in S'. A valid transcript is said to be **consistent** with a sequence of pairs $\bar{s}$ if for each oracle query q in the transcript τ that was answered by 0 there is no pair (q, w) in $\bar{s}$, where w is an NP-witness for membership of q in S'. We consider a fixed parsing of strings into pairs $(\tau, \bar{s})$, where $\bar{s}$ is a sequence of pairs.

We are now ready to define a relation R (for the class $\mathcal{S}_2^P$): For $y = (\tau, \bar{s})$ and $z = (\tau', \bar{s}')$, we let $R(x, y, z) \stackrel{\text{def}}{=} \sigma$ if at least one of the following two conditions holds:

1. τ is a valid transcript of $M(x)$ with output σ, supported by $\bar{s}$ and consistent with $\bar{s}'$.
2. τ' is a valid transcript of $M(x)$ with output σ, supported by $\bar{s}'$ and consistent with $\bar{s}$.

In case none of the conditions hold, $R(x, y, z)$ may be defined arbitrarily. Intuitively, the quantification $\exists y \forall z$ guarantees that the transcript contained in y records correct oracle answers (since positive answers must be supported by NP-witnesses, whereas negative answers must be unrefutable by NP-witnesses to the opposite). Formally, we have to prove that R is well-defined, and that the actual execution transcript is both supportable and unrefutable (i.e., consistent with all valid sequences).

We first show that R is well-defined (i.e., it can not be the case that τ and τ' are both valid, supported and consistent but with different outputs). Here we use the fact that M is deterministic and so given the same oracle answers it must yield the same output. Also, if two valid transcripts differ on some oracle answer, then it cannot be that both transcripts are supported and consistent with respect to the same two sequences of pairs.[2] Finally, observe that for every

---

[2] Consider the first conflicting answer and suppose, without loss of generality, that in transcript τ the answer is 1. Since τ is supported by a sequence of pairs $\bar{s}$, it cannot be the case that τ' (in which the answer to the same query is 0) is consistent with $\bar{s}$.

x, there exists a pair $(\tau, \bar{s})$ with output $\chi_S(x)$ such that τ is a valid transcript of $M(x)$, supported by $\bar{s}$ and consistent with any possible sequence of pairs. ∎

5.3   Conjectured Separations

Below we list some well-known conjectures.

Conjecture 1 (the leading conjecture of TOC): $\mathcal{P} \neq \mathcal{NP}$.

Conjecture 2 (most widely believed): $\mathcal{NP} \not\subseteq \mathcal{BPP}$.

Conjecture 3 (most widely believed): $\mathcal{NP} \neq \mathrm{co}\mathcal{NP}$.

Conjecture 4 (widely believed): *The Polynomial-Time Hierarchy does not collapse.*

Conjecture 4 implies the following (see [3]):

Conjecture 5 (widely believed): $\mathrm{co}\mathcal{NP} \not\subseteq \mathcal{AM}$.

We believe that Conjecture 5 is interesting on its own; indeed, it is a natural extension of Conjecture 3.

5.4   Conjectured Inclusions

What we know combined with what is widely believed is depicted in Figure 1. We note that some of the inclusions that were not conjectured to be separations are believed to be equalities or "close to it". In particular, it is widely believed that $\mathcal{BPP}$ is very close to $\mathcal{P}$. This belief is supported, among other things, by the conjecture that (uniform) exponential-time cannot be computed by subexponential-size (non-uniform) circuits [2,10]. We note that the latter conjecture holds provided there exist strong one-way functions (i.e., polynomial-time computable functions that cannot be inverted on typical images by subexponential-sized circuits).

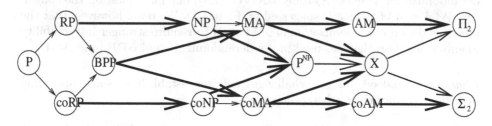

Fig. 1. Arrows indicate containment between classes, with $\mathcal{C}_1 \to \mathcal{C}_2$ indicating that $\mathcal{C}_1 \subseteq \mathcal{C}_2$. Bolder (and bigger) arrows indicate conjectured gaps between the classes. The symbol X is a placeholder for either $\mathcal{S}_2^P$ or $\mathcal{ZPP}^{\mathcal{NP}}$ (and we do not know how these two classes are related).

The Derandomization of BPP versus the Derandomization of MA. We note that results about derandomization of $\mathcal{BPP}$ are likely to imply results on the derandomization of $\mathcal{MA}$. This holds provided that the former results extend also to the generalization of $\mathcal{BPP}$ to promise problems. We note that all known derandomization results have this feature. In the next proposition co$\mathcal{RP}$ denotes the class of *promise problems* of the form $\Pi = (\Pi_{\text{YES}}, \Pi_{\text{NO}})$, where there exists a probabilistic polynomial time machine M such that

$$x \in \Pi_{\text{YES}} \Longrightarrow \text{Prob}[M(x) = 1] = 1$$
$$x \in \Pi_{\text{NO}} \Longrightarrow \text{Prob}[M(x) = 1] \leq \frac{1}{2}$$

Proposition 10 (folklore): *Suppose that* co$\mathcal{RP} \subseteq \text{DTIME}(t(n))$, *for a time constructible function* $t : \mathbb{N} \to \mathbb{N}$. *Then,* $\mathcal{MA} \subseteq \cup_{i \in \mathbb{N}} \text{NTIME}(t(n^i))$.

Proof: Each set $L \in \mathcal{MA}$ gives rise to a promise problem $\Pi = (\Pi_{\text{YES}}, \Pi_{\text{NO}})$, where

$$\Pi_{\text{YES}} \stackrel{\text{def}}{=} \{(x, w) : \forall r \in \{0,1\}^{p(|x|)} \ \ V(x, w, r) = 1\}$$
$$\Pi_{\text{NO}} \stackrel{\text{def}}{=} \{(x, w) : x \notin L\}$$

with V and p as in Definition 2. Note that, for every $x \in L$ there exists $w \in \{0,1\}^{q(|x|)}$ such that $(x, w) \in \Pi_{\text{YES}}$, whereas for every $x \notin L$ and every $w \in \{0,1\}^{q(|x|)}$ it holds that $(x, w) \in \Pi_{\text{NO}}$. Also, for every $(x, w) \in \Pi_{\text{NO}}$ it holds that

$$\text{Prob}_{r \in \{0,1\}^{p(|x|)}}[V(x, w, r) = 1] \leq \frac{1}{2}.$$

We conclude that $\Pi \in$ co$\mathcal{RP}$. Now, using the hypothesis, we have $\Pi \in \text{DTIME}(t(n + q(n)))$, and so $L \in \text{NTIME}(t(n + q(n)))$. The proposition follows. ∎

On the Derandomization of MA (a comment added in revision). In light of recent derandomization results regarding $\mathcal{AM}$ (cf. [12]), one may question the conjecture $\mathcal{MA} \neq \mathcal{AM}$ (which is suggested by Figure 1). We note, however, that the aforementioned derandomization of $\mathcal{AM}$ seem to require stronger intractability assumptions than the ones used in the derandomization of $\mathcal{BPP}$ (and $\mathcal{MA}$).

Challenges. Indeed, all our challenges call for establishing some appealing inclusions (rather than separations).

1. Try to put $\mathcal{BPP}$ in $\mathcal{P}^{\mathcal{NP}}$. (Recall that $\mathcal{BPP}$ in in $\mathcal{ZPP}^{\mathcal{NP}}$.)
2. Try to put $\mathcal{MA}$ in $\mathcal{P}^{\mathcal{NP}}$. (This certainly implies (1).)
3. Try to put $\mathcal{RP}$ in co$\mathcal{NP}$. (Recall that $\mathcal{RP}$ is in co$\mathcal{MA}$.)
4. Try to put $\mathcal{AM}$ in $\Sigma_2^P \cap \Pi_2^P$.

References

1. Babai, L.: Trading Group Theory for Randomness. In: 17th STOC, pp. 421–429 (1985)
2. Babai, L., Fortnow, L., Nisan, N., Wigderson, A.: BPP has Subexponential Time Simulations unless EXPTIME has Publishable Proofs. Complexity Theory 3, 307–318 (1993)
3. Boppana, R., Håstad, J., Zachos, S.: Does Co-NP Have Short Interactive Proofs? IPL 25, 127–132 (1987)
4. Canetti, R.: On BPP and the Polynomial-time Hierarchy. IPL 57, 237–241 (1996)
5. Fürer, M., Goldreich, O., Mansour, Y., Sipser, M., Zachos, S.: On Completeness and Soundness in Interactive Proof Systems. In: Micali, S. (ed.) Advances in Computing Research (Randomness and Computation), vol. 5, pp. 429–442 (1989)
6. Goldreich, O.: A Sample of Samplers – A Computational Perspective on Sampling. This volume. See also ECCC, TR97-020, TR97-020 (May 1997)
7. Goldreich, O.: Computational Complexity: A Conceptual Perspective. Cambridge University Press, Cambridge (2008)
8. Goldwasser, S., Micali, S., Rackoff, C.: The knowledge Complexity of Interactive Proofs. SIAM J. on Computing 18(1), 186–208 (1989)
9. Goldwasser, S., Sipser, M.: Private Coins versus Public Coins in Interactive Proof Systems. In: Micali, S. (ed.) Advances in Computing Research (Randomness and Computation), vol. 5, pp. 73–90 (1989)
10. Impagliazzo, R., Wigderson, A.: P=BPP if E requires exponential circuits: Derandomizing the XOR Lemma. In: 29th STOC, pp. 220–229 (1997)
11. Lautemann, C.: BPP and the Polynomial Hierarchy. IPL 17, 215–217 (1983)
12. Miltersen, P.B., Vinodchandran, N.V.: Derandomizing Arthur-Merlin Games using Hitting Sets. Computational Complexity 14(3), 256–279 (2005); Preliminary version in 40th FOCS (1999)
13. Russell, A., Sundaram, R.: Symmetric Alternation Captures BPP. Journal of Computational Complexity (1995) (to appear); Preliminary version in Technical Report MIT-LCS-TM-54
14. Sipser, M.: A Complexity Theoretic Approach to Randomness. In: 15th STOC, pp. 330–335 (1983)
15. Zachos, S., Fürer, M.: Probabilistic Quantifiers vs. Distrustful Adversaries. In: Nori, K.V. (ed.) FSTTCS 1987. LNCS, vol. 287, pp. 443–455. Springer, Heidelberg (1987)
16. Zachos, S., Heller, H.: A decisive characterization of BPP. Information and Control 69(1-3), 125–135 (1986)
17. Zuckerman, D.: Simulating BPP Using a General Weak Random Source. Algorithmica 16, 367–391 (1996)

Strong Proofs of Knowledge

Oded Goldreich

Abstract. The concept of proofs-of-knowledge, introduced in the seminal paper of Goldwasser, Micali and Rackoff, plays a central role in various cryptographic applications. An adequate formulation, which enables modular applications of proofs of knowledge inside other protocols, was presented by Bellare and Goldreich. However, this formulation depends in an essential way on the notion of expected (rather than worst-case) running-time. Here we present a seemingly more restricted notion that maintains the main feature of the prior definition while referring only to machines that run in strict probabilistic polynomial-time (rather than to expected polynomial-time).

Keywords: Proof of Knowledge, Zero-Knowledge.

This work was completed in May 1998, and was integrated in the author's work *Foundation of Cryptography* as [7, Sec. 4.7.6]. The current text is based on a private memo from May 1998, whereas the postscript section (Sec. 4) is recent (and confirms speculations raised in the original memo).

1 Introduction

The reader is referred to [3] for a discussion of the intuitive notion of a proof-of-knowledge (cf., [11]), and the previous attempts to define it [4,13], cumlinating in the definition presented in [3]. We also assume that the reader is familiar with the definition given in [3].

The definition given in [3] relies in a fundamental way on the notion of *expected* running-time. Throughout the years we remained bothered by this feature, and while working on [6] we decided to look for an alternative. Specifically, we present a more stringent definition in which the knowledge extractor is required to run in *strict* polynomial-time (rather than in *expected* polynomial-time). We call proof systems for which this more stringent definition holds, *strong proofs of knowledge* (in contrast to ordinary *proofs of knowledge* as defined in [3]).

There are two reasons to prefer strong proofs of knowledge over ordinary ones. Firstly, we feel more comfortable with the notion of strict polynomial-time than with the notion of expected polynomial-time. For example, it is intuitively unclear why a machine which runs for time 2^n on an 2^{-n} fraction of its coin-tosses (and in linear time otherwise) should be considered fundamentally different than a machine which runs for time 2^{n^2} on the same fraction. Secondly, it seems much more convinient to work (i.e., to compose) strict polynomial-time computations rather than expected polynomial-time ones. (For further discussion of this issue, the interested reader is directed to [9].)

O. Goldreich et al.: Studies in Complexity and Cryptography, LNCS 6650, pp. 54–58, 2011.

Unfortunately, there seems to be a loss in going from ordinary proofs of knowledge to strong ones: Not all proofs of knowledge are known to be *strong proofs of knowledge*. Furthermore, we conjecture that there are proofs of knowledge that are not *strong proofs of knowledge* (see Section 4). Still, zero-knowledge strong-proofs-of-knowledge do exist for all NP-relations, provided that one-way functions exist.

2   The Definition

We assume that the reader is familiar with the definition of a proof of knowledge (as presented in [3]) as well as with the underlying motivation.

Definition 1 (System of strong proofs of knowledge): *Let R be a binary relation. We say that an efficient strategy V is a* **strong knowledge verifier** *for the relation R if the following two conditions hold.*

- Non-triviality: *There exists an interactive machine P such that for every $(x, y) \in R$ all possible interactions of V with P on common-input x and auxiliary-input y are accepting.*
- Strong Validity: *There exists a negligible function $\mu : \mathbb{N} \mapsto [0, 1]$ and a probabilistic* (strict) *polynomial-time oracle machine K such that for every strategy P and every $x, y, r \in \{0, 1\}^*$, machine K satisfies the following condition:*

  Let $P_{x,y,r}$ be a prover strategy, in which the common input x, auxiliary input y and random-coin sequence r have been fixed, and denote by $p(x)$ the probability that the interactive machine V accepts, on input x, when interacting with the prover specified by $P_{x,y,r}$. Now, if $p(x) > \mu(|x|)$ then, on input x and access to oracle $P_{x,y,r}$, with probability at least $1 - \mu(|x|)$, machine K outputs a solution s for x. That is:

  $$\text{If } p(x) > \mu(|x|), \text{ then } \Pr[(x, K^{P_{x,y,r}}(x)) \in R] > 1 - \mu(|x|). \quad (1)$$

  The oracle machine K is called a **strong knowledge extractor**.

An interactive pair (P, V) so that V is a strong knowledge verifier for a relation R and P is a machine satisfying the non-triviality condition (with respect to V and R) is called a **system for strong proofs of knowledge** *for the relation R.*

Thus, it is required that whenever $p(x) > \mu(|x|)$ (i.e., whenever the prover convinces the verifier with non-negligible probability), the extractor fails with negligible probability. Our choice to bound the failure probability of the extractor by the specific negligible function μ (which serves mainly as bound on $p(x)$) is rather arbitrary. What is important is to have this failure probability be a negligible function of $|x|$. Actually, in case membership in the relation R can be determined in polynomial-time, one may reduce the failure probability from $1 - \frac{1}{\text{poly}(n)}$ to $2^{-\text{poly}(n)}$, while maintaining the polynomial running-time of the extractor. Finally, we note that the extractor presented in the next section has failure probability 0.

3 On the Existence of Strong Proofs of Knowledge

Some zero-knowledge proof (of knowledge) systems for NP are in fact strong proofs of knowledge. In particular, consider n sequential repetitions of the following basic proof system for the *Hamiltonian Cycle* (HC) problem (which is NP-complete). We consider directed graphs (and the existence of directed Hamiltonian cycles), and employ a commitment scheme $\{C_n\}$ as above.

Construction 2 (Basic proof system for HC):

- Common Input: *a directed graph* $G = (V, E)$ *with* $n \stackrel{\text{def}}{=} |V|$.
- Auxiliary Input to Prover: *a directed Hamiltonian Cycle,* $C \subset E$, *in* G.
- Prover's first step (P1): *The prover selects a random permutation,* π, *of the vertices of* G, *and commits to the entries of the adjacency matrix of the resulting permuted graph. That is, it sends an* n-*by-*n *matrix of commitments such that the* $(\pi(i), \pi(j))^{\text{th}}$ *entry is* $C_n(1)$ *if* $(i, j) \in E$, *and* $C_n(0)$ *otherwise.*
- Verifier's first step (V1): *The verifier uniformly selects* $\sigma \in \{0, 1\}$ *and sends it to the prover.*
- Prover's second step (P2): *If* $\sigma = 0$, *then the prover sends* π *to the verifier along with the revealing (i.e., preimages) of all* n^2 *commitments. Otherwise, the prover reveals to the verifier only the commitments to the* n *entries that correspond to* C; *that is, it reveals the* $(\pi(i), \pi(j))^{\text{th}}$ *entry if and only if* $(i, j) \in C$. *(By revealing a commitment* c, *we mean supply a preimage of* c *under* C_n; *i.e., a pair* (σ, r) *so that* $c = C_n(\sigma, r)$.)
- Verifier's second step (V2): *If* $\sigma = 0$, *then the verifier checks that the revealed graph is indeed isomorphic, via* π, *to* G. *Otherwise, the verifier just checks that all revealed values are 1 and that the corresponding entries form a simple* n-*cycle. (Of course in both cases, the verifier checks that the revealed values do fit the commitments.) The verifier accepts if and only if the corresponding condition holds.*

The reader may easily verify that sequentially repeating the basic protocol for n times yields a zero-knowledge proof system for HC, with soundness error 2^{-n}. We argue that the resulting system is also a strong proof of knowledge of the Hamiltonian cycle. Intuitively, the key observation is that each application of the basic proof system results in one of two possible situations depending on the verifier's choice, σ. In case the prover answers correctly in both cases, we can retrieve an Hamiltonian cycle in the input graph. On the other hand, in case the prover fails in both cases, the verifier will reject regardless of what the prover does from this point on. This observation suggests the following construction of a strong knowledge extractor (where we refer to repeating the basic proof systems n times and set $\mu(n) = 2^{-n}$).

Strong Knowledge Extractor for Hamiltonian Cycle: On input G and access to the prover-strategy oracle P^*, we proceed in n iterations, starting with $i = 1$. Initially, T (the transcript so far), is empty.

1. Obtain the matrix of commitments, M, from the prover strategy (i.e., $M \leftarrow P^*(T)$).
2. Obtain the prover's answer to both possible verifier moves; that is, for every $\sigma \in \{0, 1\}$, obtain the corresponding answer $A_\sigma \leftarrow P^*(T, \sigma)$. Each of these answers may be correct (i.e., passing the corresponding verifier check) or not.
3. If both answers are correct, then we recover a Hamiltonian cycle. In this case the extractor outputs the cycle and halts.
4. In case a single answer, say the one for value σ, is correct and $i < n$, we let $T \leftarrow (T, \sigma)$, and proceed to the next iteration (i.e., $i \leftarrow i + 1$). Otherwise, we halt with no output.

Note that we reach iteration i only if and only if in each of the prior $i-1$ iterations a single verifier choice is answered correctly (and we have appended this choice in T). Hence, if the extractor halts with no output in iteration $i < n$, then the verifier (in the real interaction) accepts with probability zero (since in iteration i both verifier choices yield incorrect answers). Similarly, if the extractor halts with no output in iteration n, then the verifier (in the real interaction) accepts with probability at most 2^{-n} (since at most one choine is answered correctly). Thus, whenever $p(G) > 2^{-n}$, the extractor succeeds in recovering a Hamiltonian cycle (with probability 1).

4   Postscript

This section was added in the current revision and provides some support for conjectures made explicitly or implicitly in the original text.

Regarding our conjecture that there exist proofs-of-knowledge that are not strong proofs-of-knowledge, partial evidence is provided by subsequent work of Barak, Lindell, and Vadhan [1,2]. Both work refer to constant-round zero-knowledge protocols (for sets outside $\mathcal{BPP}$), and the seperation relies on the existence of such protocols (under standard computational assumptions) that are (ordinary) proofs of knowledge for NP-relations.

1. Barak and Lindell [1] show that such protocols cannot have a strict probabilistic polynomial-time black-box extractor, which implies that they cannot be proven to be strong proofs-of-knowledge in a black-box manner. (Still, recall that non-black-box extractors may exist.)
2. Barak, Lindell, and Vadhan [2] show that if (exponentially) strong one-way permutations exist, then such prtotocols cannot have a strict probabilistic polynomial-time extractor, which implies that they cannot be strong proofs-of-knowledge.

The existence of constant-round zero-knowledge protocols that are (ordinary) proofs of knowledge for NP-relations can be based on standard intractability assumptions: See Feige and Shamir [5] for the case of argument systems and Lindell [12] for the case of proof systems.

References

1. Barak, B., Lindell, Y.: Strict Polynomial-Time in Simulation and Extraction. SIAM J. on Comput. 33(4), 783–818 (2004)
2. Barak, B., Lindell, Y., Vadhan, S.: Lower Bounds for Non-Black-Box Zero-Knowledge. J. of Comp. and Sys. Sci. 72(2), 321–391 (2006)
3. Bellare, M., Goldreich, O.: On defining proofs of knowledge. In: Brickell, E.F. (ed.) CRYPTO 1992. LNCS, vol. 740, pp. 390–420. Springer, Heidelberg (1993)
4. Feige, U., Fiat, A., Shamir, A.: Zero-Knowledge Proofs of Identity. J. of Crpto. 1, 77–94 (1988)
5. Feige, U., Shamir, A.: Zero Knowledge Proofs of Knowledge in Two Rounds. In: Brassard, G. (ed.) CRYPTO 1989. LNCS, vol. 435, pp. 526–544. Springer, Heidelberg (1990)
6. Goldreich, O.: Secure Multi-Party Computation. Unpublished manuscript (1998), Superseded by (8, Chap. 7), http://www.wisdom.weizmann.ac.il/?oded/foc.html
7. Goldreich, O.: Foundation of Cryptography: Basic Tools. Cambridge University Press, Cambridge (2001)
8. Goldreich, O.: Foundation of Cryptography: Basic Applications. Cambridge University Press, Cambridge (2004)
9. Goldreich, O.: On Expected Probabilistic Polynomial-Time Adversaries – A suggestion for restricted definitions and their benefits. J. of Crypto. 23(1), 1–36 (2010)
10. Goldreich, O., Micali, S., Wigderson, A.: Proofs that Yield Nothing but their Validity or All Languages in NP Have Zero-Knowledge Proof Systems. J. of the ACM 38(1), 691–729 (1991); Preliminary Version in 27th FOCS (1986)
11. Goldwasser, S., Micali, S., Rackoff, C.: The Knowledge Complexity of Interactive Proof Systems. SIAM J. on Comput. 18, 186–208 (1989); Preliminary Version in 27th FOCS (1986)
12. Lindell, Y.: Constant-Round Zero-Knowledge Proofs of Knowledge. ECCC, TR11-003 (January 2011)
13. Tompa, M., Woll, H.: Random Self-Reducibility and Zero-Knowledge Interactive Proofs of Possession of Information. University of California (San Diego), Computer Science and Engineering Department, Technical Report Number CS92-244 (June 1992); Preliminary version in 28th FOCS, pp. 472–482 (1987)

Simplified Derandomization of BPP Using a Hitting Set Generator

Oded Goldreich, Salil Vadhan, and Avi Wigderson

Abstract. A hitting-set generator is a deterministic algorithm that generates a set of strings such that this set intersects every dense set that is recognizable by a small circuit. A polynomial time hitting-set generator readily implies $\mathcal{RP} = \mathcal{P}$, but it is not apparent what this implies for $\mathcal{BPP}$. Nevertheless, Andreev et al. (ICALP'96, and JACM 1998) showed that a polynomial-time hitting-set generator implies the seemingly stronger conclusion $\mathcal{BPP} = \mathcal{P}$. We simplify and improve their (and later) constructions.

Keywords: Derandomization, RP, BPP, one-sided error versus two-sided error.

This work is considered the final version of [7]. An early version of this work appeared as TR00-004 of *ECCC*. The current revision is quite minimal.

1 Introduction

The relation between randomized computations with one-sided error and randomized computations with two-sided error is one of the most interesting questions in the area. Specifically, we refer to the relation betwen $\mathcal{RP}$ and $\mathcal{BPP}$. In particular, *does $\mathcal{RP} = \mathcal{P}$ imply $\mathcal{BPP} = \mathcal{P}$?*

1.1 An Affirmative Partial Answer

The breakthrough paper of Andreev et al. [2] (and its sequel [3]) gave a natural setting in which the answer to the foregoing question is YES. The setting is a specific natural way to prove $\mathcal{RP} = \mathcal{P}$, namely via "hitting-set generators" (see exact definition below). Informally, such a generator outputs a set of strings that hits every large efficiently-recognizable set (e.g., the witness set of a positive input of an $\mathcal{RP}$ set). Having such a generator that runs in polynomial time enables a trivial deterministic simulation of an $\mathcal{RP}$ algorithm by using each of the generator's outputs as the random pad of the given algorithm.

The main result of [2] was that such a generator (which immediately yields deromization of 1-sided error algorithms) actually suffices for derandomizing 2-sided error algorithms. In particular, the existence of polynomial-time hitting set generators implies $\mathcal{BPP} = \mathcal{P}$.

O. Goldreich et al.: Studies in Complexity and Cryptography, LNCS 6650, pp. 59–67, 2011.

Definition 1 (hitting set generator):[1] *An algorithm,* G, *is called a* hitting set generator for circuits *if, on input* $n, s \in \mathbb{N}$ (given in unary), *it generates as output a set of n-bit strings,* $G(n, s)$, *such that every circuit of size* s (on n input bits) *that accepts at least half its inputs, accepts at least one element from the set* $G(n, s)$.

Since s is the essential complexity parameter $(n \leq s)$, we let $t_G(s)$ denote the running time of the generator G on input (n, s), and $N_G(s)$ denote the size of its output set. Clearly $N_G(s) \leq t_G(s)$. The result of Andreev *et al.* [2] is the following:

Theorem 2 (derandomization via hitting sets [2]): *If there exists a hitting-set generator* G *running in time* t_G, *then* $\mathcal{BPP} \subseteq \mathcal{DTIME}(\text{poly}(t_G(\text{poly}(n))))$.

Indeed, the most important special case (i.e., $t_G(s) = \text{poly}(s)$) is the following:

Corollary 3 (Theorem 2, specialized [2]): *If* G *runs in polynomial time, then* $\mathcal{BPP} = \mathcal{P}$.

1.2   In Quest of Simplifications

Our main result is a simple proof of Theorem 2. Explaining what "simple" means is not so simple. We start by explaining how the given generator (assumed in the hypothesis of Theorem 2) is used in [2] (and in subsequent works [3,4]) to derandomize $\mathcal{BPP}$. Indeed, later proofs (of [3] and then [4]) were much simpler than [2], but while proving Corollary 3, they fell short of proving Theorem 2.[2]

Warning: The following discussion is carried out on an intuitive (and somewhat vague) level. Readers who do not like such discussions may skip the rest of the introduction, and proceed directly to the formal proof (presented in Sections 2 and 3).

The Two Different Uses of the Hitting Set Generator in [2]. The proof in [2] uses the generator in two ways. The first use is literally as a producer of a hitting set for all sufficiently dense and efficiently recognized sets. The second use (which is more subtle) is as a hard function. Indeed, observe that the existence of such

---

[1] In other settings, (pseudorandom) generators are defined as outputting a single string. In terms of Definition 1 this means that, on input an index $i \in \{1, ..., |G(n, s)|\}$ (viewed as a seed), the generator outputs the ith string in $G(n, s)$. However, in the current context, the current convention (which in the standard terms means considering the set of all possible outputs of the generator) seems simpler to work with.

[2] We note, however, that both [3] and [4] use their techniques to study other aspects of the relationship between one-sided and two-sided error (i.e., aspects not addressed by Theorem 2). In particular, Buhrman and Fortnow [4] resolve the promise-problem analogue of the question "Does $\mathcal{RP} = \mathcal{P}$ imply $\mathcal{BPP} = \mathcal{P}$?" in the positive. See Section 1.3.

a generator G immediately implies the existence of a function on $O(\log t_G(s))$ bits that is computable in time $t_G(s)$ but cannot be computed by circuits of size s (or else a contradiction is reached by considering a circuit that that accepts a vast majority of the strings that are not generated by G).[3] These two uses of G are combined in a rather involved way for the derandomization of $\mathcal{BPP}$.

It is interesting to note that for the case of $t_G(s) = \text{poly}(s)$, the aforementioned hard function can be plugged into the pseudorandom generator of [8], to yield $\mathcal{BPP} = \mathcal{P}$ as in Corollary 3. (Note, however, that [8] was not available to the authors of [2] at the time (the two papers are independent).) Moreover, [8] is far from "simple", it does use the computational consequence (which we are trying to avoid), and anyhow it is not strong enough to yield Theorem 2.

The Two-Level Use of the Hitting Set Generator in [3]. A considerably simpler proof was given in [3]. There, the generator is used only in its "original capacity" as a hitting set generator, without explicitly using any computational consequence of its existence. In some sense, this proof is more clearly a "black-box" use of the output set of the generator. However, something was lost. The running time of the derandomization is replaced by $\text{poly}(t_G(t_G(\text{poly}(n))))$.

On the one hand, this is not too bad. For the interesting case of $t_G(s) = \text{poly}(s)$ (which implies $\mathcal{RP} = \mathcal{P}$), they still get the consequence $\mathcal{BPP} = \mathcal{P}$, as in Corollary 3 (since iterating a polynomial function twice results in a polynomial). On the other hand, if the function t_G grows moderately such that $t_G(t_G(n)) = 2^n$, then we have as assumption a highly nontrivial derandomization of $\mathcal{RP}$, but the consequence is a completely trivial derandomization of $\mathcal{BPP}$.

In our opinion, the best way to understand the origin of the iterated application of the function t_G in the aforementioned result is explained in the recent paper of [4], which further simplifies the proof of [3]. They remind the reader that the proofs [9,10] putting $\mathcal{BPP}$ in $\Sigma^2 \cap \Pi^2$ actually gives much more. In fact, viewed appropriately, it almost begs (with hindsight) the use of hitting sets! The key is, that in both the $\forall\exists$ and $\exists\forall$ expressions for the $\mathcal{BPP}$ set, the "witnesses" for the existential quantifier are abundant. Put differently, $\mathcal{BPP} \subseteq \mathcal{RP}^{\text{pr}\mathcal{RP}}$, where $\text{pr}\mathcal{RP}$ is the promise-problem version of $\mathcal{RP}$. But if you have a hitting set, you can use it first to derandomize the "oracle" part of the right-hand side. This leaves us with an $\mathcal{RTIME}(t_G(\text{poly}(n))$ machine, which can again be derandomized (using hitting sets for $t_G(\text{poly}(n))$ size circuits).

In short, the "two quantifier" representation of $\mathcal{BPP}$ leads to a two-level recursive application of the generator. It seems hopeless to reduce the number of quantifiers to one in the $\mathcal{BPP} \subseteq \Sigma^2 \cap \Pi^2$ result. So another route has to be taken in order to prove Theorem 2 in a similar "direct" (or "black-box") manner, but without incurring the penalty arising from this two-level recursion.

Our Proof. We eliminate the two-level recursion, obtaining a single application of the hitting set generator, by "increasing the dimension to two" in the following

---

[3] For example, consider a circuit C that accepts a $(2\log_2 t_G(s))$-bit string, z, if and only if z is not a prefix of any string in $G(n, s)$.

sense. Inspired by Lautemann's proof [9] of $\mathcal{BPP} \subseteq \Sigma^2 \cap \Pi^2$, we consider, for each input to a given $\mathcal{BPP}$ algorithm that uses $\ell(n)$ random coins, a $2^{\ell(n)}$-by-$2^{\ell(n)}$ Boolean matrix such that the $(a,b)^{\text{th}}$ entry represents the decision of the algorithm when using the random pad $a \oplus b$.[4] In this matrix, the fraction of incorrect answers in each row (resp., column) is small. The hitting set is used to select a small subset of the rows and a small subset of the columns, and the entries of this submatrix determine the result. Specifically, we will look for "enough" (yet few) rows that are monochromatic, and decide accordingly. The correctness and efficiency of the test are spelled out in Lemma 6, which is essentially captured by the following simple Ramsey-type result.

Proposition 4 (log-size dominating sets): *For every n-vertex graph, either the graph or its complement has a dominating set of size $\lceil \log_2(n+1) \rceil$. Furthermore, such a set can be found in polynomial time.*

(Proposition 4 is seemingly new and may be of independent interest.)[5]
    We end by observing that (like the previous results) our result holds in the context of promise problems. Hence, the existence of hitting set generators provides an efficient way for approximately counting the fraction of inputs accepted by a given circuit within additive polynomial fraction.

1.3   Perspective

As described above, Buhrman and Fortnow [4] prove that $\mathcal{BPP} \subseteq \text{pr}\mathcal{RP}^{\text{pr}\mathcal{RP}}$, and actually $\text{pr}\mathcal{BPP} = \text{pr}\mathcal{RP}^{\text{pr}\mathcal{RP}}$. It follows immediately that $\text{pr}\mathcal{RP} = \text{pr}\mathcal{P}$ implies $\text{pr}\mathcal{BPP} = \text{pr}\mathcal{P}$, resolving the main question of this area for promise classes! This result suggests two natural extensions that remain open. The first is to obtain an analogue of their result for the standard classes of decision problems, $\mathcal{RP}$ and $\mathcal{BPP}$. (In [4], it is shown that such an extension cannot relativize.) The second possible extension is to "scale" the result upwards. In fact, from the hypothesis $\text{pr}\mathcal{RP} \subseteq \mathcal{DTIME}(t(n))$, they obtain the conclusion $\text{pr}\mathcal{BPP} \subseteq \mathcal{DTIME}(\text{poly}(t(t(\text{poly}(n)))))$. Theorem 2, as proved in [2] and in this paper, replaces the composition $t(t(\cdot))$ with a single $t(\cdot)$ for the (very) special case when the derandomization of $\text{pr}\mathcal{RP}$ is via a hitting-set generator.

2   The Derandomization Procedure

Given $L \in \mathcal{BPP}$, consider a probabilistic polynomial-time algorithm A for L. Let $\ell = \ell(n)$ be a fixed polynomial denoting the number of coin tosses made by A on

---

[4] In a preliminary version of this work [7], we considered a different matrix such that its $(a,b)^{\text{th}}$ entry represents the decision of the algorithm when using the random pad $a \circ b$. For that matrix to have the desired properties, it was necessary to first perform drastic error reduction (using extractors) on the $\mathcal{BPP}$ algorithm, where this strategy was inspired by [6]. The main simplification here is in avoiding this strategy.

[5] Proposition 4 corresponds to the special case of Lemma 6 that refers to symmetric matrices. Note that this special case actually suffices for our application.

inputs of length n; similarly, define $s = s(n)$ such that the computation of A on inputs of length n can be implemented by circuits of size $s(n)$. We assume that A has error probability at most $1/2\ell(n)$; this can be achieved by straightforward amplification of any $\mathcal{BPP}$ algorithm for L.

Let $A(x, r)$ denote the output of algorithm A on input $x \in \{0,1\}^n$ and random-tape contents $r \in \{0,1\}^{\ell(n)}$. Our derandomization procedure, described next, utilizes a hitting-set generator G as defined earlier (i.e., in Definition 1).

Derandomization procedure: On input $x \in \{0,1\}^n$, let A, ℓ, and s be as above.

1. Invoking the hitting-set generator G, obtain $H \leftarrow G(\ell, \ell \cdot s)$. That is, H is a hitting set for circuits of size $\ell \cdot s$ and input length ℓ. Denote the elements of H by $e_1, ..., e_N$, where $N \stackrel{\text{def}}{=} N_G(s)$ and each e_i is in $\{0,1\}^\ell$.
2. Construct an N-by-N matrix, $M = (v_{i,j})$, such that $v_{i,j} = A(x, e_i \oplus e_j)$. That is, run A with all possible random-pads composed by XORing each of the possible pairs of strings in H. (We merely use the fact that $a \oplus b$ is easy to compute and that for any a the mapping $b \mapsto a \oplus b$ is 1-1, and similarly for any b and $a \mapsto a \oplus b$.)
3. Using the procedure guaranteed by Lemma 6 (of Section 3 (below)), determine whether for every ℓ columns there exists a row on which all these columns have 1-value. If this procedure accepts, then accept, else reject. That is, accept if and only if

$$\forall c_1, ..., c_\ell \in [N] \; \exists r \in [N] \text{ s.t. } \wedge_{i=1}^\ell (v_{c_i, r} = 1). \tag{1}$$

We first show that if $x \in L$, then Eq. (1) holds; and, analogously, if $x \notin L$, then

$$\forall r_1, ..., r_\ell \in [N] \; \exists c \in [N] \text{ s.t. } \wedge_{i=1}^\ell (v_{r_i, c} = 0). \tag{2}$$

Note that the foregoing description, by itself, does not establish the correctness of the procedure. Neither did we specify how to efficiently implement Step 3, To that end we use a general technical lemma (indeed Lemma 6) that implies that *it cannot be the case that both Eq. (1) and Eq. (2) hold.* Furthermore, this lemma asserts that one can efficiently determine which of the two conditions does not hold. These aspects are deferred to the next section. But first we prove the foregoing implications.

Proposition 5 (on the matrix constructed in Step 2): *If $x \in L$ (resp., $x \notin L$), then Eq. (1) (resp., Eq. (2)) holds.*

Proof: We shall prove a slightly more general statement. Let χ_L be the characteristic function of L (i.e., $\chi_L(x) = 1$ if $x \in L$ and $\chi_L(x) = 0$ otherwise). Then, we next prove that, for every $x \in \{0,1\}^n$, for every ℓ rows (resp., columns) there exists a column (resp., row) on which the value of the matrix is $\chi_L(x)$.

Fixing the input $x \in \{0,1\}^n$ to algorithm A, we consider the circuit C_x which takes an ℓ-bit input r and outputs $A(x, r)$ (i.e., evaluates A on input x and coins r). By our hypothesis regarding the error probability of A, it holds that

$$\Pr_{r \in \{0,1\}^\ell}[C_x(r) \neq \chi_L(x)] \leq \frac{1}{2\ell}.$$

It follows that for every $y_1, ..., y_\ell \in \{0,1\}^\ell$, it holds that

$$\Pr_{z \in \{0,1\}^\ell}[(\forall i) \; C_x(y_i \oplus z) = \chi_L(x)] \geq \frac{1}{2}. \tag{3}$$

Let $\overline{y} = (y_1, ..., y_\ell)$, and consider the circuit $C_{x,\overline{y}}(z) \overset{\text{def}}{=} \wedge_{i=1}^\ell (C_x(y_i \oplus z) = \chi_L(x))$. Then, by the Eq. (3), it holds that $\Pr_z[C_{x,\overline{y}}(z) = \chi_L(x)] \geq 1/2$. On the other hand, the size of $C_{x,\overline{y}}$ is merely ℓ times the size of C_x, which was at most s. Thus, by definition of the hitting-set generator G, the set $H = \mathrm{G}(\ell, \ell \cdot s)$ must contain a string z such that $C_{x,\overline{y}}(z) = \chi_L(x)$.

The foregoing holds for any $\overline{y} = (y_1, ..., y_\ell)$. Thus, for every $y_1, ..., y_\ell \in H \subseteq \{0,1\}^\ell$, there exists $z \in H$ such that $A(x, y_i \oplus z) = C_x(y_i \oplus z) = \chi_L(x)$ holds for every $i \in [\ell]$. Thus, we have proved that, for every ℓ rows in M, there exists a column (in M) on which the value of the matrix is $\chi_L(x)$.

A similar argument applies to sets of ℓ columns in M. Specifically, for every $z_1, ..., z_\ell \in \{0,1\}^\ell$, it holds that

$$\Pr_{y \in \{0,1\}^\ell}[(\forall i) \; C_x(y \oplus z_i) = \chi_L(x)] \geq \frac{1}{2}. \tag{4}$$

Again, we conclude that, for every $z_1, ..., z_\ell \in H$, there exists $y \in H$ such that $C_x(y \oplus z_i) = \chi_L(x)$ holds for every $i \in [\ell]$. Thus, for every ℓ columns in M there exists a row (in M) on which the value of the matrix is $\chi_L(x)$. The proposition follows. □

Digest. The foregoing procedure is a simplified version of the procedure given in the preliminary version of this work [7]. Specifically, inspired by [6], the argument in [7] relies on the explicit constructions of extractors for drastic error reduction of the $\mathcal{BPP}$ algorithm. Here, we only use a mild (and trivial) error reduction. This difference stems from the fact that the matrix considered in [7] is different (i.e., the $(a, b)^{\text{th}}$ entry in the matrix considered in [7] represents the decision of the algorithm when using the random pad $a \circ b$ (rather than $a \oplus b$)). In contrast, Steps 1 and 3 of the foregoing derandomization procedure are identical to the steps in [7], and so is Lemma 6. Thus, our argument relies on two essential ingredients: The first ingredient, adopted from [3], is the use of auxiliary circuits (depending on C_x but not identical to it), in order to argue that a hitting-set must have certain strong properties with respect to C_x. The second ingredient is the constructive combinatorial result given by Lemma 6. (A third ingredient, which consists of using extractors as in [7], is eliminated here.)

3   Correctness and Efficiency of the Derandomization

Proposition 5 shows that for every x either Eq. (1) or Eq. (2) holds. But, as stated before, it is not even clear that Eq. (1) and Eq. (2) cannot hold simultaneously. This is asserted next.

Lemma 6 (a generic technical lemma): *For every $k \geq \log_2(n+1)$, every n-by-n Boolean matrix either has k rows whose bit-wise disjunction is the all 1's row, or k columns whose bit-wise conjunction is the all 0's column. Moreover, there is a (deterministic) polynomial-time algorithm that given such a matrix find such a set.*

We prove the lemma momentarily. But first let use show that Eq. (1) and Eq. (2) cannot hold simultaneously. We first note that in our case $n = N = N_G(s)$ and $k = \ell$, and furthermore $n < 2^\ell$ (since there is no point in having $G(\ell, \cdot)$ contain more than $2^{\ell-1} + 1 < 2^\ell$ strings of length $\ell > 1$). The claim then follows by applying the following corollary to Lemma 6.

Corollary 7 (corollary to Lemma 6): *For every n-by-n Boolean matrix and every $k \geq \log_2(n+1)$, it is impossible that both the following conditions hold:*

1. *For every k rows, there exists a column such that all the k rows have a 0-entry in this column.*
2. *For every k columns, there exists a row such that all the k columns have a 1-entry in this row.*

Furthermore, assuming one of the foregoing conditions holds, deciding which one holds can be done in (deterministic) polynomial-time.

Proof (of Corollary 7): Suppose Item (1) holds. Then, the bit-wise disjunction of every k rows contains a 0-entry, and so it cannot be the all 1's row. Likewise, if Item (2) holds then the bit-wise conjunction of every k columns contains a 1-entry, and so it cannot be the all 0's column. Thus, the case in which both items holds stands in contradiction to Lemma 6. Furthermore, finding a set as in the lemma indicates which of the two items does not hold. □

Proof of Lemma 6: Let $B = (b_{r,c})$ be an arbitray n-by-n Boolean matrix, and consider the following iterative procedure, initiated with $C_0 = [n]$ and $R = \emptyset$. For $i = 1, 2, \ldots$, take a row r not in R that has at least $|C_{i-1}|/2$ 1's in C_{i-1} (i.e., $r \in [n] \setminus R$ such that $|\{c \in C_{i-1} : b_{r,c} = 1\}| \geq |C_{i-1}|/2$). Add r to R, and let C_i be the part of C_{i-1} that had 0's in row r (i.e., $C_i \stackrel{\text{def}}{=} \{c \in C_{i-1} : b_{r,c} = 0\}$). We get stuck if for any i, no row in the current set $[n] \setminus R$ has at least $|C_{i-1}|/2$ 1's in C_{i-1}. Otherwise, we terminate when $C_i = \emptyset$

If we never get stuck, then we generated at most $\log_2(n+1) \leq k$ rows (since $|C_i| \leq |C_{i-1}|/2$, which implies that $|C_{\lceil \log_2(n+1) \rceil}| < 1$). Furthermore, the bit-wise disjunction of these rows is the all 1's row (i.e., for the final R and every $c \in [n]$, it holds that $\vee_{r \in R} b_{r,c} = 1$), since the ith row in R has 1-entries in every column in $C_{i-1} \setminus C_i$, and the last C_i is empty.

On the other hand, if we got stuck at iteration i, then we let $S = C_i$ and note that every row of B has at least $|S|/2$ 0's in the columns S. (This includes the rows in the current R, which each have 0's in all the columns in $S \subset C_{i-1} \subset \cdots \subset C_0$.) In this case, an argument analogous to Adlemam's proof [1] that

$\mathcal{RP} \subseteq \mathcal{P}/\text{poly}$ implies that there exist a set of k columns C that contains a 0-entry in every row (i.e., for every $r \in [n]$, it holds that $\wedge_{c \in C} b_{r,c} = 0$).[6]

Turning to the algorithmics, note that the foregoing procedure for constructing R, S and C is implementable in polynomial-time. Thus, in case the "row" procedure was completed successfully, we may output the set of rows R, and otherwise the set C of columns.   □

Proof of Theorem 2: Proposition 5 shows that for every x either Eq. (1) or Eq. (2) holds, and furthermore that the former (resp., latter) holds whenever $x \in L$ (resp., $x \notin L$). As mentioned above, by applying Corollary 7, it follows that only one of these equations may hold. Using the decision procedure guaranteed by Corollary 7, we implement Step 3 in our derandomized procedure, and Theorem 2 follows.   □

A Finer Analysis. For a $\mathcal{BPP}$ algorithm that uses ℓ coin tosses and can be implemented by circuits of size s, our derandomization only invokes the hitting-set generator with parameters $(\ell, s \cdot \ell)$, and otherwise it runs in polynomial time. However, if the algorithm only has constant error probability, we must first reduce the error to $1/2\ell$, which increases these parameters somewhat. Using standard error reduction (running the algorithm $O(\log \ell)$ times independently and ruling by majority), we obtain the following more quantitative version of our result:

Theorem 8 (Theorem 2, refined): *Suppose there is a hitting set generator* G *such that* $G(\ell, s)$ *is computable in time* $t(\ell, s)$. *Let* L *be a problem with a constant-error* $\mathcal{BPP}$ *algorithm that, on inputs of length n, uses $\ell = \ell(n)$ coin tosses and can be implemented by circuits of size $s = s(n)$. Then,*

$$L \in \mathcal{DTIME}(\text{poly}(t(\ell', s \cdot \ell'))),$$

where $\ell' = O(\ell \log \ell)$.

We comment that, by using random walks on expanders for error reduction, one can replace $t(\ell', s \cdot \ell')$ by $t(\ell'', s \cdot \ell')$, where $\ell'' = \ell + O(\log \ell) \ll \ell'$.

Acknowledgments. The second author thanks Adam Klivans for explaining [7] to him.

---

[6] Let us spell out the argument in the current setting. We initiate an iterative process of picking columns from S such that at each iteration we pick a column that covers the largest number of 0's in the remaining rows. That is, we initialize $R_0 = [n]$ and $C = \emptyset$, and for $j = 1, 2, ...$, take a column c not in C that has a maximal number of 0's in R_{i-1}, add c to C, and let R_j be the part of R_{j-1} that has 1's in column c (i.e., $R_j \overset{\text{def}}{=} \{r \in R_{i-1} : b_{r,c} = 1\}$). The point is that, by our hypothesis, for the current C, the submatrix $R_{j-1} \times (S \setminus C)$ contains at least $|R_{j-1}| \cdot |S|/2$ 0's, and therefore there exists a column $c \in S \setminus C$ such that $|\{r \in R_{i-1} : b_{r,c} = 0\}| \geq |R_{j-1}|/2$.

References

1. Adleman, L.: Two theorems on random polynomial-time. In: 19th FOCS, pp. 75–83 (1978)
2. Andreev, A.E., Clementi, A.E.F., Rolim, J.D.P.: A new general derandomization method. Journal of the Association for Computing Machinery (J. of ACM) 45(1), 179–213 (1998); Hitting Sets Derandomize BPP. In: XXIII International Colloquium on Algorithms, Logic and Programming, ICALP 1996 (1996)
3. Andreev, A.E., Clementi, A.E.F., Rolim, J.D.P., Trevisan, L.: Weak Random Sources, Hitting Sets, and BPP Simulations. To appear in SICOMP (1997); Preliminary version in 38th FOCS, pp. 264–272 (1997)
4. Buhrman, H., Fortnow, L.: One-sided versus two-sided randomness. In: Proceedings of the 16th Symposium on Theoretical Aspects of Computer Science. LNCS, Springer, Berlin (1999)
5. Even, S., Selman, A.L., Yacobi, Y.: The Complexity of Promise Problems with Applications to Public-Key Cryptography. Inform. and Control 61, 159–173 (1984)
6. Goldreich, O., Zuckerman, D.: Another proof that BPP ⊆ PH (and more). In: Goldreich, O., et al.: Studies in Complexity and Cryptography. LNCS, vol. 6650, pp. 40–53. Springer, Heidelberg (1997)
7. Goldreich, O., Wigderson, A.: Improved derandomization of BPP using a hitting set generator. In: Hochbaum, D.S., Jansen, K., Rolim, J.D.P., Sinclair, A. (eds.) RANDOM 1999 and APPROX 1999. LNCS, vol. 1671, pp. 131–137. Springer, Heidelberg (1999)
8. Impagliazzo, R., Wigderson, A.: P=BPP unless E has Subexponential Circuits: Derandomizing the XOR Lemma. In: 29th STOC, pp. 220–229 (1997)
9. Lautemann, C.: BPP and the Polynomial Hierarchy. Information Processing Letters 17, 215–217 (1983)
10. Sipser, M.: A complexity-theoretic approach to randomness. In: 15th STOC, pp. 330–335 (1983)
11. Zuckerman, D.: Simulating BPP Using a General Weak Random Source. Algorithmica 16, 367–391 (1996)

On Testing Expansion in Bounded-Degree Graphs

Oded Goldreich and Dana Ron

Abstract. We consider testing graph expansion in the bounded-degree graph model. Specifically, we refer to algorithms for testing whether the graph has a second eigenvalue bounded above by a given threshold or is far from any graph with such (or related) property.

We present a natural algorithm aimed towards achieving the foregoing task. The algorithm is given a (normalized) eigenvalue bound $\lambda < 1$, oracle access to a bounded-degree N-vertex graph, and two additional parameters $\epsilon, \alpha > 0$. The algorithm runs in time $N^{0.5+\alpha}/\text{poly}(\epsilon)$, and accepts any graph having (normalized) second eigenvalue at most λ. We believe that the algorithm rejects any graph that is ϵ-far from having second eigenvalue at most $\lambda^{\alpha/O(1)}$, and prove the validity of this belief under an appealing combinatorial conjecture.

Keywords: Property Testing, Graph Expansion.

This work appeared as TR00-020 of *ECCC*. It is based on a research project pursued in the years 1998–99, which was stuck at the gap outlined in Section 4.2. The current revision is intentionally minimal, because the original publication has triggered several subsequent works, which directly address the topic of this work (see [8] and the references therein) or are indirectly inspired by it (see, e.g., [4]). For further discussion of subsequent work, see Section 5.

1 Introduction

This memo reports partial results regarding the task of testing whether a given bounded-degree graph is an expander. That is, we refer to the "bounded-degree model" of testing graph properties as formulated in [5]. In this model, the (randomized) algorithm is given integers d and N, a distance parameter ϵ (as well as some problem-specific parameters), and oracle access to a N-vertex graph G with degree bound d; that is, query $(v, i) \in [N] \times [d]$ is answered by the ith neighbor of v in G (or by a special symbol in case v has less than i neighbors). For a predetermined property $\mathcal{P}$, the algorithm is required to accept (with probability at least 2/3) any graph having property $\mathcal{P}$, and reject (with probability at least 2/3) any graph that is ϵ-far from having property $\mathcal{P}$, where distance between graphs is defined as the fraction of edges (over dN) on which the graphs differ.

Loosely speaking, the specific property considered here is being an expander. More precisely, for a given bound $\lambda < 1$, we consider the property, denoted $\mathcal{E}_\lambda$, of having a normalized by d adjacency matrix with second eigenvalue at most

O. Goldreich et al.: Studies in Complexity and Cryptography, LNCS 6650, pp. 68–75, 2011.
© Springer-Verlag Berlin Heidelberg 2011

λ. Actually, we further relax the property testing formulation (as in [9]): Using an additional parameter $\lambda' \geq \lambda$, we only require that

- the algorithm must accept (with probability at least 2/3) any graph having property $\mathcal{E}_\lambda$ (i.e., having second eigenvalue at most λ); and
- the algorithm must reject (with probability at least 2/3) any graph that is ϵ-far from having property $\mathcal{E}_{\lambda'}$ (i.e., from any graph that has second eigenvalue at most λ').

Setting $\lambda' = \lambda$ we regain the more strict formulation of testing whether a graph has second eigenvalue at most λ.

We mention that the $\Omega(\sqrt{N})$ lower bound on "testing expansion" (presented in [5]) continues to hold for the relaxed formulation, provided that $\lambda' < 1$. This is the case because the lower bound is established by showing that any $o(\sqrt{N})$-query algorithm fails to distinguish between a very good expander and an unconnected graph with several huge connected components.[1]

In view of the foregoing, we shall be content with any sub-linear time algorithm for testing expansion. Below, we present a parameterized family of algorithms. For any $\alpha > 0$, the algorithm has running-time $n^{0.5+\alpha}/\mathrm{poly}(\epsilon)$ and is supposed to satisfy the foregoing requirement with $\lambda' = \lambda^{\alpha/7}$. Unfortunately, we only prove that this is indeed the case provided that a certain combinatorial conjecture (presented in Section 4.2) holds.

2   Conventions and Notation

We consider N-vertex graphs of degree bound d, which should be thought of as fixed. We consider the stochastic matrix representing a canonical random walk on this graph, where canonical is anything reasonable (e.g., go to each neighbor with probability $1/2d$). The eigenvalues below refer to this matrix.

Recall that λ denotes the claimed second eigenvalue (i.e., we need to accept graphs having second eigenvalue at most λ), and ϵ denotes the distance parameter (i.e., we need to reject graphs that are ϵ-far from having second eigenvalue at most λ', where $\lambda' > \lambda$ is related to λ).

The algorithm presented next is parameterized by a small constant $\alpha > 0$ that determines both its complexity (i.e., $O(N^{0.5+\alpha}/\mathrm{poly}(\epsilon))$) and its performance (i.e., $\lambda' = \lambda^{\alpha/O(1)}$). To be of interest, the algorithm must use $\alpha < 0.5$.

3   The Algorithm

We set $L = \frac{1.5 \ln N}{\ln(1/\lambda)}$. This guarantees that a graph with second eigenvalue at most λ mixes well in L steps (i.e., the deviation in max-norm of the end probability from the uniform distribution is at most $N^{-1.5}$). The following algorithm evaluates the distance of the end probability (of an L-step random walk starting at a fixed vertex) from the uniform probability distribution. It is based on the fact that the uniform distribution over a set has the smallest possible collision probability, among all distributions over this set.

---

[1] In the latter case, the graph has (normalized) second eigenvalue equal 1.

Repeat the following steps $t \stackrel{\text{def}}{=} \Theta(1/\epsilon)$ times:

1. Select uniformly a start vertex, denoted s.
2. Perform $m \stackrel{\text{def}}{=} \Theta(N^{0.5+\alpha}/\epsilon)$ random walks of length L, starting from vertex s.
3. Count the number of pairwise collisions between the endpoints of the fore-going m walks.
4. If the count is greater than $\frac{1+0.5 \cdot N^{-\alpha/2}}{N} \cdot \binom{m}{2}$, then `reject`.

If all repetitions were completed without rejection, then `accept`.

Comment: Random walks were used before in the context of testing graph properties (in the bounded-degree model). Specifically, $\widetilde{O}(\sqrt{N}/\text{poly}(\epsilon))$ such walks were used by the bipartitness tester of [6]. Needless to say, random walks seem much more natural here.

4   Analysis

Fixing any start vertex s, we denote by $p_{s,v}$ the probability that a random walk of length L starting at s ends in v. The collision probability of L-walks starting at s is given by

$$\sum_v p_{s,v}^2 \geq \frac{1}{N}. \tag{1}$$

By our choice of L, if the graph has eigenvalue at most λ, then (for any starting vertex s) the collision probability of L-walks starting at s is very close to $1/N$ (i.e., is smaller than $(1/N) + (1/N^2)$).

4.1   Approximation of the Collision Probabilities

The first issue to address is the approximation to Eq. (1) provided by Steps (2)–(3) of the algorithm.

Lemma 1 *With probability at least $1-(1/3t)$, the (normalized) empirical count[2] computed in Steps (2)–(3) is within a factor of $1 \pm \frac{1}{4} \cdot N^{-\alpha/2}$ of the value of Eq. (1).*

Thus, with probability at least $2/3$, all approximations provided by the algorithms are within a factor of $1 \pm \frac{1}{4} \cdot N^{-\alpha/2}$ of the correct value.

Proof: For every $i < j$, define a 0-1 random variable $\zeta_{i,j}$ such that $\zeta_{i,j} = 1$ if the endpoint of the ith path is equal to the endpoint of the jth path. Clearly, $\mu \stackrel{\text{def}}{=} E[\zeta_{i,j}] = \sum_v p_{s,v}^2$, for every $i < j$. Using Chebyshev's inequality we bound the probability that the count provided by Steps (2)–(3) deviates from its (correct) expected value. Let $P \stackrel{\text{def}}{=} \{(i,j) : 1 \leq i < j \leq m\}$ and $\delta = \frac{1}{4} \cdot N^{-\alpha/2}$. Then:

---

[2] That is, the number of pairwise collisions divided by $\binom{m}{2}$.

$$\Pr\left[\left|\sum_{(i,j)\in P}\zeta_{i,j}-|P|\cdot\mu\right|>\delta\cdot|P|\cdot\mu\right]\le\frac{\mathrm{Var}[\sum_{(i,j)\in P}\zeta_{i,j}]}{(\delta\cdot|P|\cdot\mu)^2}.\qquad(2)$$

Denote $\overline{\zeta}_{i,j}\stackrel{\text{def}}{=}\zeta_{i,j}-\mu$. The rest of the proof needs to deal with the fact that the random variables associated with P are *not* pairwise independent. Specifically, for four *distinct* i,j,i',j', indeed $\zeta_{i,j}$ and $\zeta_{i',j'}$ are independent, and so $\mathrm{E}[\overline{\zeta}_{i,j}\overline{\zeta}_{i',j'}]=\mathrm{E}[\overline{\zeta}_{i,j}]\cdot\mathrm{E}[\overline{\zeta}_{i',j'}]=0$; but for $i<j\ne k$ the random variables $\zeta_{i,j}$ and $\zeta_{i,k}$ are *not* independent (since they both depend on the same i^{th} walk). Still, it holds that

$$\mathrm{Var}\left[\sum_{(i,j)\in P}\zeta_{i,j}\right]=\mathrm{E}\left[\left(\sum_{(i,j)\in P}\overline{\zeta}_{i,j}\right)^2\right]$$

$$=\sum_{(i,j)\in P}\mathrm{E}\left[\overline{\zeta}_{i,j}^2\right]+5\cdot\sum_{1\le i<j<k\le m}\mathrm{E}\left[\overline{\zeta}_{i,j}\overline{\zeta}_{i,k}\right]$$

$$<|P|\cdot\mu+m^3\cdot\sum_v p_{s,v}^3.$$

since $\zeta_{i,j}\zeta_{i,k}=1$ if and only if all three random walks end at the same vertex. Using $(\sum_v p_{s,v}^3)^{1/3}\le(\sum_v p_{s,v}^2)^{1/2}$, and $m^2<3\cdot|P|$, we obtain

$$\mathrm{Var}\left[\sum_{(i,j)\in P}\zeta_{i,j}\right]\le|P|\cdot\mu+(3|P|)^{3/2}\cdot\mu^{3/2}<6\cdot(|P|\cdot\mu)^{3/2}.\qquad(3)$$

Combining Eq. (2) and (3), we obtain

$$\Pr\left[\left|\sum_{(i,j)\in P}\zeta_{i,j}-|P|\cdot\mu\right|>\delta\cdot|P|\cdot\mu\right]<\frac{6}{\delta^2\cdot(|P|\cdot\mu)^{1/2}}.$$

Using $\mu\ge 1/N$ and $|P|>\frac{m^2}{4}=\Theta(\frac{N^{1+2\alpha}}{\epsilon^2})$, the denominator is at least $\delta^2\cdot\Theta(\frac{N^\alpha}{\epsilon})$. Recalling that $\delta=\frac{1}{4}\cdot N^{-\alpha/2}$ and $t=O(1/\epsilon)$, the lemma follows. ∎

As an immediate corollary we get:

Corollary 2. *If the graph has second eigenvalue at most λ, then the foregoing algorithm accepts it with probability at least $2/3$.*

Another immediate corollary is the following:

Corollary 3. *Suppose that for at least a $\epsilon/O(1)$ fraction of the vertices s in G the collision probability of L-walks starting at s is greater than $\frac{1+0.8N^{-\alpha/2}}{N}$. Then, the algorithm rejects with probability at least $2/3$.*

Thus, if a graph passes the test (with probability greater than $1/3$), then it must have less than $(\epsilon/O(1)) \cdot N$ *exceptional vertices*; that is, vertices s for which the collision probability of L-walks starting at s is greater than $\frac{1+0.8N^{-\alpha/2}}{N}$.

Comment: Note that by changing parameters in the algorithm (i.e., $t = \Theta(N^\alpha/\epsilon)$ and $m = \Theta(N^{0.5+2\alpha}/\epsilon)$), we can make the fraction of exceptional vertices smaller than $\epsilon N^{-\alpha}$. This may help in closing the gap (described in Section 4.2), and only increases the complexity from $N^{0.5+\alpha}/\text{poly}(\epsilon)$ to $N^{0.5+3\alpha}/\text{poly}(\epsilon)$.

4.2   The Gap

We believe that the following conjecture (or something similar to it) is true.

Conjecture: Let G be an N-vertex graph of degree-bound d. Suppose that *for all but at most $\epsilon/O(1)$ fraction* of the vertices s in G the collision probability of L-walks starting at s is at most $\frac{1+0.8N^{-\alpha/2}}{N}$. Then, G is ϵ-close to a N-vertex graph (of degree-bound d) in which the collision probability of L-walks starting at any vertex *is at most* $\frac{1+N^{-\alpha/2}}{N}$.

The conjecture is very appealing: Suppose that you add ϵdN edges connecting at random the exceptional vertices to the rest of the graph. Ignoring for a moment the issue of preserving the degree bounds, this seems to work – but we cannot prove it. Indeed, one can show that the previously exceptional vertices now enjoy rapid mixing, but it is not clear that the added edges will not cause harm to the mixing properties of the previously non-exceptional vertices.

4.3   Finishing It Off

Once the gap is closed, we have the following situation: If the algorithm rejects with probability smaller than $2/3$, then the input graph is ϵ-close to a graph in which the collision probability of L-walks starting at *any vertex* is at most $\frac{1+N^{-\alpha/2}}{N}$. But the excess of the collision probability beyond $1/N$ is nothing but the square of the distance, in norm 2, of the probability vector $(p_{s,v})_{v \in [N]}$ from the uniform probability vector (i.e., $(\sum_v p_{s,v}^2) - (1/N) = \sum_v (p_{s,v} - (1/N))^2$). Thus, for every s the distance, in norm 2, of the probability vector $(p_{s,v})_{v \in [N]}$ from the uniform probability vector is at most $\sqrt{\frac{N^{-\alpha/2}}{N}} = N^{-(0.5+\beta)}$, where $\beta = \alpha/4$.

The plan now is to "reverse" the standard *eigenvalue to rapid-mixing connection*. That is, infer from the rapid-mixing feature that the graph has a small second eigenvalue. Such a lemma has appeared in [7]:

Lemma 4 (Lemma 4.6 in [7]): *Consider a regular connected graph on N vertices, let A be its normalized adjacency matrix and λ_2 denote the absolute value of the second eigenvalue of A. Let ℓ be an integer and Δ_ℓ denote an upper bound on the maximum, taken over all possible start vertices s, of the difference in* *norm 2 between the distribution induced by an ℓ-step random walk starting at s and the uniform distribution. Then $\lambda_2 \leq (N \cdot \Delta_\ell)^{1/\ell}$.*

Note that by the foregoing, we have $\Delta_L < N^{-(0.5+\beta)}$. This does not give anything useful when applying the lemma directly. Instead, we apply the lemma after bounding Δ_ℓ for $\ell = O(L)$. (This strategy may be an oversight, but that's how we argue it now.)

Claim 5. *Let Δ_ℓ be define as in Lemma 4. Then $\Delta_{k\ell} \leq (\sqrt{N} \cdot \Delta_\ell)^k$, for every integer k.*

Proof: Let $B = A^\ell$ be the stochastic matrix representing an ℓ-step random walk, and let $\vec{e}_1, ..., \vec{e}_N$ denote probability vectors in which all the mass is on one vertex. Let $\vec{\nu}$ denote the uniform probability vector. Then Δ_ℓ (resp., $\Delta_{k\ell}$) equals the maximum of $\|B\vec{e}_i - \vec{\nu}\|$ (resp., $\|B^k\vec{e}_i - \vec{\nu}\|$) taken over all the $\vec{e}_i$'s.

Considering the basis of $\vec{e}_i$'s, let $\vec{z}$ be an arbitrary zero-sum vector (such as $\vec{e}_i - \vec{\nu}$). That is, $\vec{z}$ is written in the basis of $\vec{e}_i$'s as $\vec{z} = \sum_i z_i \vec{e}_i$, and $\sum_i z_i = 0$. We obtain

$$\|B\vec{z}\| = \left\| B\left(\sum_i z_i \vec{e}_i \right) - \sum_i z_i B\vec{\nu} \right\|$$

$$= \left\| \sum_i z_i B(\vec{e}_i - \vec{\nu}) \right\|$$

$$\leq \sum_i \|z_i B(\vec{e}_i - \vec{\nu})\|$$

$$= \sum_i |z_i| \cdot \|B(\vec{e}_i - \vec{\nu})\|$$

$$\leq \left(\sum_i |z_i| \right) \cdot \Delta_\ell.$$

Since $\sum_i |z_i| \leq \sqrt{N} \cdot \sqrt{\sum_i z_i^2} = \sqrt{N} \cdot \|\vec{z}\|$, we get

$$\|B\vec{z}\| \leq \sqrt{N} \cdot \Delta_\ell \cdot \|\vec{z}\|.$$

Using $B\vec{\nu} = \vec{\nu}$, we get for every i

$$\|B^k\vec{e}_i - \vec{\nu}\| = \|B(B^{k-1}\vec{e}_i - \vec{\nu})\|$$
$$\leq \Delta_\ell \cdot \sqrt{N} \cdot \|B^{k-1}\vec{e}_i - \vec{\nu}\|$$
$$< \left(\Delta_\ell \cdot \sqrt{N} \right)^k$$

and the claim follows.   ∎

Combining Lemma 4 and Claim 5, we obtain the following

Corollary 6. *Suppose that for every s the distance, in norm 2, of the probability vector* $(p_{s,v})_{v \in [N]}$ *from the uniform probability vector is at most* $N^{-(0.5+\beta)}$. *Then, for every constant* $\gamma < 2\beta/3$, *the second eigenvalue of the graph is at most* λ^γ.

So once the gap is filled, we are done (using $\beta = \alpha/4$ and $\gamma \approx 2\beta/3$).

Proof: Let λ' be the second eigenvalue of the graph. Then, for every k we have

$$\lambda' \le (N \cdot \Delta_{kL})^{1/kL} \qquad\qquad \text{[Lemma 4]}$$

$$\le \left(N \cdot \left(\sqrt{N} \cdot \Delta_L \right)^k \right)^{1/kL} \qquad \text{[Claim 5]}$$

$$\le \left(N \cdot \left(N^{-\beta} \right)^k \right)^{1/kL} \qquad \text{[hypothesis]}$$

$$= \exp\left(\tfrac{(1-k\beta)\cdot \ln N}{kL} \right).$$

Substituting for $L = \frac{1.5 \ln N}{\ln(1/\lambda)}$, we get

$$\frac{(1 - k\beta) \cdot \ln N}{kL} = \frac{(1 - k\beta) \cdot \ln N}{k \cdot ((1.5 \ln N)/\ln(1/\lambda))}$$

$$= -\left(\frac{2\beta}{3} - \frac{2}{3k} \right) \cdot \ln(1/\lambda)$$

$$< -\gamma \cdot \ln(1/\lambda),$$

for sufficiently large k (since $\gamma < 2\beta/3$). We get $\lambda' < \lambda^\gamma$, and the corollary follows. ∎

Comment: We have $\lambda' \le \lambda^\gamma$ for any $\gamma < 2\beta/3 = \alpha/6$ (e.g., $\gamma = \alpha/7$ will do). One may be able to increase the exponent (i.e., γ) somewhat, but a linear dependency (of the exponent γ) on α seems unavoidable (under the current approach).

5   Subsequent Work

Subsequent works, culminating in [8], have addressed the problem of testing expansion of graphs. These subsequent works refer to a combinatorial definition of graph expansion, rather than to the algebraic definition of eigenvalues. Although both definitions are related (see [1,2] or [3, Sec. 9.2]), the translation is not tight. Still, for some values of $\lambda < \lambda' < 1$, these works resolve the open problem raised in our work.

In addition, the current work has inspired work on testing distributions, as initiated in [4]. Specifically, these works use the observation that the empirical collision count of $O(\sqrt{N})$ samples taken from a distribution over $[N]$ provides an approximation to the distance of this distribution from the uniform distribution.

References

1. Alon, N.: Eigenvalues and expanders. Combinatorica 6, 83–96 (1986)
2. Alon, N., Milman, V.D.: λ_1, Isoperimetric Inequalities for Graphs and Superconcentrators. J. Combinatorial Theory, Ser. B 38, 73–88 (1985)
3. Alon, N., Spencer, J.H.: The Probabilistic Method, 2nd edn. John Wiley & Sons, Inc., Chichester (2000)
4. Batu, T., Fortnow, L., Rubinfeld, R., Smith, W.D., White, P.: Testing that Distributions are Close. In: 41st FOCS, pp. 259–269 (2000)
5. Goldreich, O., Ron, D.: Property Testing in Bounded Degree Graphs. In: Proc. of the 29th ACM Symp. on Theory of Computing, pp. 406–415 (1997)
6. Goldreich, O., Ron, D.: A sublinear Bipartite Tester for Bounded Degree Graphs. Combinatorica 19(2), 1–39 (1999)
7. Goldreich, O., Sudan, M.: Computational Indistinguishability: A Sample Hierarchy. JCSS 59, 253–269 (1999)
8. Kale, S., Seshadhri, C.: Testing expansion in bounded degree graphs. In: 35th ICALP, pp. 527–538 (2008); Preliminary version appeared as TR07-076, ECCC (2007)
9. Parnas, M., Ron, D.: Testing the diameter of graphs. In: Hochbaum, D.S., Jansen, K., Rolim, J.D.P., Sinclair, A. (eds.) RANDOM 1999 and APPROX 1999. LNCS, vol. 1671, pp. 85–96. Springer, Heidelberg (1999)

Candidate One-Way Functions Based on Expander Graphs

Oded Goldreich

Abstract. We suggest a candidate one-way function using combinatorial constructs such as expander graphs. These graphs are used to determine a sequence of *small* overlapping subsets of input bits, to which a hard-wired random predicate is applied. Thus, the function is extremely easy to evaluate: All that is needed is to take multiple projections of the input bits, and to use these as entries to a look-up table. It is feasible for the adversary to scan the look-up table, but we believe it would be infeasible to find an input that fits a given sequence of values obtained for these *overlapping* projections.

The conjectured difficulty of inverting the suggested function does not seem to follow from any well-known assumption. Instead, we propose the study of the complexity of inverting this function as an interesting open problem, with the hope that further research will provide evidence to our belief that the inversion task is intractable.

Keywords: One-Way Functions, Expander Graphs.

This work appeared as TR00-090 of *ECCC*. The current revision is intentionally minimal, because the original publication has triggered several subsequent works (although less than we have hoped). For further discussion of these subsequent works and some afterthoughts, see Section 6.

1   Introduction

In contrary to the present attempts to suggest a practical private-key encryption scheme to replace the DES, we believe that attempts should focus on suggesting practical one-way functions and pseudorandom functions. Being a simpler object, one-way functions should be easier to construct, and such constructions may later yield directly or indirectly a variety of other applications (including private-key encryption schemes).

The current attempts to suggest a practical private-key encryption scheme in place of the DES seem quite ad-hoc: Not only that they cannot be reduced to any well-known problem, but (typically) they do not relate to a computational problem of natural appeal. Thus, the study of these suggestions is of limited appeal (especially from a conceptual point of view).

In this manuscript, we propose a general scheme for constructing one-way functions. We do not believe that the complexity of inverting the resulting function follows from some well-known intractability assumptions. We believe that

O. Goldreich et al.: Studies in Complexity and Cryptography, LNCS 6650, pp. 76–87, 2011.
© Springer-Verlag Berlin Heidelberg 2011

the complexity of inverting this function is a new interesting open problem, and hope that other researcher will be able to obtain better understanding of this problem.

In addition to the abstract presentation, we propose several concrete instantiations of our proposal. It seems to us that a reasonable level of "security" (i.e., hardness to invert) may be achieved at very modest input lengths. Specifically, on input length at the order of a couple of hundreds of bits, inverting the function may require complexity (e.g., time) beyond 2^{100}.

Style and Organization. This write-up is intended to two different types of readers, since we believe that it is relevant to two different research communities (i.e., computational complexity and applied cryptography). Consequently, we provide an asymptotic presentation as well as suggestions for concrete parameters. The basic suggestion is presented in Sections 2 and 3. Concrete instantiations of this suggestion are proposed in Section 4. Concluding comments appear in Section 5.

2   The Basic Suggestion

We construct a (uniform) collection of functions $\{f_n : \{0,1\}^n \rightarrow \{0,1\}^n\}_{n \in \mathbb{N}}$. Our construction utilizes a collection of $\ell(n)$-subsets, $S_1, ..., S_n \subset [n] \stackrel{\text{def}}{=} \{1, ..., n\}$, and a predicate $P : \{0,1\}^{\ell(n)} \rightarrow \{0,1\}$. Jumping ahead, we hint that:

1. The function ℓ is relatively small: Theoretically speaking, $\ell = O(\log n)$ or even $\ell = O(1)$. In practice ℓ should be in the range $\{7, ..., 16\}$, whereas n should range between a couple of hundreds and a couple of thousands.
2. We prefer to have $P : \{0,1\}^\ell \rightarrow \{0,1\}$ be a random predicate. That is, it will be randomly selected, fixed, and "hard-wired" into the function. For sure, P should *not* be linear, nor depend on few of its bit locations.
3. The collection $S_1, ..., S_n$ should be *expanding*; specifically, for *some* k, every k subsets should cover at least $k + \Omega(n)$ elements of $\{1, ..., n\}$. The complexity of the inversion problem (for f_n constructed based on such a collection) seems to be exponential in the "net expansion" of the collection (i.e., the cardinality of the union minus the number of subsets).

For $x = x_1 \cdots x_n \in \{0,1\}^n$ and $S \subset [n]$, where $S = \{i_1, i_2, ..., i_t\}$ and $i_j < i_{j+1}$, we denote by x_S the projection of x on S; that is, $x_S = x_{i_1} x_{i_2} \cdots x_{i_t}$. Fixing P and $S_1, ..., S_n$ as above, we define

$$f_n(x) \stackrel{\text{def}}{=} P(x_{S_1}) P(x_{S_2}) \cdots P(x_{S_n}). \tag{1}$$

Note that we think of ℓ as being relatively small (i.e., $\ell = O(\log n)$), and aim at having f_n be univertible within time $2^{n/O(1)}$. Thus, the hardness of inverting f_n cannot be due to the hardness of inverting P. Instead, the hardness of inverting f_n is supposed to come from the combinatorial properties of the collection of sets $C = \{S_1, ..., S_n\}$ (as well as from the combinatorial properties of predicate P).

2.1   The Preferred Implementation

Our preference is to have P be a fixed randomly chosen predicate, which is hard-wired into the algorithm for evaluating f_n. Actually, one better avoid some choices; see next section. (In case $\ell = \Theta(\log n)$ bad choices are rare enough.) In practice, we think of ℓ in the range $\{7, ..., 16\}$, and so hard-wiring a (random) predicate defined on $\{0,1\}^\ell$ is quite feasible. The ℓ-subsets will be determined by combinatorial constructions called expander graphs. At this point the reader may think of them too as being hard-wired into the algorithm. On input $x \in \{0,1\}^n$, the algorithm for computing f_n proceeds as follows:

1. For $i = 1, .., n$, projects x on S_i, forming the ℓ-bit long string $x^{(i)}$.
2. For $i = 1, .., n$, by accessing a look-up table for P, determines the bit $y_i = P(x^{(i)})$.

The output is the n-bit long string $y_1 y_2 \cdots y_n$.

(Note that the n operations, in each of the foregoing two steps, can be performed in parallel.)

2.2   An Alternative Implementation

An alternative to having P "hard-wired" to the algorithm (as above) is to have it appear as part of the input (and output). That is, letting $\langle P \rangle$ denote the 2^ℓ-bit string that fully specifies P, we have

$$f'_n(\langle P \rangle, x) \stackrel{\text{def}}{=} (\langle P \rangle, P(x_{S_1})P(x_{S_2}) \cdots P(x_{S_n})) \tag{2}$$

Thus, P is essentially random, since the inversion problem is considered with respect to a uniformly chosen input. This implementation is more appealing from a theoretical point of view, and in such a case one better let $\ell = \log_2 n$ (rather than $\ell = O(1)$).[1]

2.3   Determining Suitable Collections

As hinted above, the collection of ℓ-subsets, $C = \{S_1, ..., S_n\}$, is to be determined by a suitable combinatorial construction known as expander graphs. The reason for this choice will become more clear from the analysis of one obvious attack (presented in Section 3.2). The specific correspondence (between expanders and subsets) depends on whether one uses the bipartite or non-bipartite formulation of expander graphs:

Bipartite formulation: In   this   case   one   considers   a   bipartite   graph $B = ((U, V), E)$, where (U, V) is a partition of the vertex set, with $|U| = |V|$,

---

[1] Our main concern at the time was to make the new assumption as strong as possble. From that perspective, we preferred $\ell = O(\log n)$ over $\ell = O(1)$.

and $E \subset U \times V$ is typically sparse. The expanding property of the graph provides, for every $U' \subset U$ (of size at most $|U|/2$), a lower bound on $|\Gamma(U')| - |U'|$ (in terms of $|U'|$), where $\Gamma(U') = \{v : \exists u \in U'$ s.t. $(u,v) \in E\}$.

Our collection of subsets will be defined as $C = \{S_u\}_{u \in U}$, where $S_u = \{v : (u,v) \in E\}$.

Non-bipartite formulation: In this case one considers a graph $G = (V, E)$, so that for every $V' \subset V$ (of size at most $|V|/2$), a suitable lower bound on $|\Gamma(V') \setminus V'|$ holds, where $\Gamma(V') = \{v : \exists v' \in V'$ s.t. $(v', v) \in E\}$.

Our collection of subsets is defined as $C = \{S_v\}_{v \in V}$, where $S_v = \{w : (v, w) \in E\} \cup \{v\}$.

In both cases, the lower bound provided by the expansion property on the size of the neighbor set is linear in the size of the specific vertex set; e.g., for the non-bipartite formulation it holds that $|\Gamma(V') \setminus V'| \geq c \cdot |V'|$ for some constant $c > 0$ and every admissible V'.

3 Avoiding Obvious Weaknesses

Considering a few obvious attacks, we rule out some obviously bad choices of the predicate P and the collection C.

3.1 The Choice of the Predicate

We start by discussing two bad choices (for the predicate P), which should be avoided.

Linear Predicates. It is certainly bad to use a linear predicate P (i.e., $P(\sigma_1 \cdots \sigma_\ell) = p_0 + \sum_{i=1}^{\ell} p_i \sigma_i$, for some $p_0, p_1, ..., p_\ell \in \{0, 1\}$). Under a linear P, the question of inverting f_n, regardless of what collection of subsets C is used, boils down to solving a linear system (of n equations in n variables), which is easy. Having a predicate P that is close to a linear predicate is dangerous too.

Horn Predicates. Likewise, one should avoid having any predicate that will make the system of equations (or conditions) solvable in polynomial-time. The only other type of easily solvable equations are these arising from Horn formulae (e.g., an OR of all variables).

Degenerate Predicates. The rest of our analysis refers to the collection of sets that determine the inputs to which the predicate P is applied. For this analysis to be meaningful, the predicate should actually depend on all bits in its input (i.e., be non-degenerated).

Good Predicates. We believe that most predicates are good for our purpose. In particular, we suggest to use a uniformly chosen predicate.

3.2   The Choice of the Collection

Since the inverting algorithm can afford to consider all preimages of the predicate P, it is crucial that the inversion of f_n cannot be performed by interactively inverting P. To demonstrate this point, consider the case $\ell = 1$ and the collection $\{S_1, ..., S_n\}$ such that $S_i = \{i\}$. In this case the S_i's are disjoint and we can recover the preimage by inverting P on each of the bits of the image, separately from the others. For a less trivial example, consider the case where the collection C consists of $n/2\ell$ sub-collections, each having 2ℓ subsets of some distinct set of 2ℓ elements. In this case, inversion can be performed in time $O(n \cdot 2^{2\ell})$ by considering each of these disjoint sets (of 2ℓ elements) separately. Recall that we wish the complexity of inversion to be exponential in n (and not in ℓ, which may be a constant).

In general, a natural inverting algorithm that should be defeated is the following: On input $y = f_n(x)$, the algorithm proceeds in n steps, maintaining a list of *partially specified* preimages of y under f_n. Initially, the list consists of the unique fully-undetermined string $*^n$. In the first step, depending on the first bit of $y = y_1 \cdots y_n$, we form the list L_1 of strings over $\{*, 0, 1\}$ such that, for every $z \in L_1$, it holds that $P(z_{S_1}) = y_1$ and $z_{[n]\backslash S_1} = *^{n-\ell}$, where $[m] \overset{\text{def}}{=} \{1, ..., m\}$ (and $z_{S_1} \in \{0,1\}^\ell$). In the $i + 1^{\text{st}}$ step, we extend L_i to L_{i+1} in the natural manner:

- Let $U' = \cup_{j=1}^{i} S_j$ and $U = \cup_{j=1}^{i+1} S_j$.
- For every $z' \in L_i$, we consider all $2^{|U\backslash U'|}$ strings $z \in \{*, 0, 1\}^n$ satisfying
  1. $z_{U'} = z'_{U'}$,
  2. $z_{U\backslash U'} \in \{0, 1\}^{|U\backslash U'|}$, and
  3. $z_{[n]\backslash U} = *^{n-|U|}$.
  The string z is added to L_{i+1} if and only if $P(z_{S_{i+1}}) = y_{i+1}$.

Thus (when invoked on input y), for every i, it holds that

$$L_i = \left\{ z \in \{*, 0, 1\}^n : \begin{array}{l} z_k = * \text{ if and only if } k \in [n] \backslash \cup_{j=1}^{i} S_j \\ \text{and} \\ P(z_{S_j}) = y_j \text{ for } j = 1, ..., i \end{array} \right\}.$$

The average running-time of this algorithm is determined by the expected size of the list at step i, for the worst possible i. Letting $U = \cup_{j=1}^{i} S_j$, consider the set

$$A_{\sigma_1 \cdots \sigma_i} \overset{\text{def}}{=} \left\{ z \in \{*, 0, 1\}^n : \begin{array}{l} z_k = * \text{ if and only if } k \in [n] \backslash U \\ \text{and} \\ P(z_{S_j}) = \sigma_j \text{ for } j = 1, ..., i \end{array} \right\},$$

and let X be uniformly distributed over $\{0, 1\}^n$. Then, the expected size of L_i equals

$$\sum_{\alpha \in \{0,1\}^i} \mathbf{Pr}[f(X)_{[i]} = \alpha] \cdot |A_\alpha| = \sum_{\alpha \in \{0,1\}^i} \mathbf{Pr}[\exists z \in A_\alpha \text{ s.t. } X_U = z_U] \cdot |A_\alpha|$$

$$= \sum_{\alpha \in \{0,1\}^i} \frac{|A_\alpha|}{2^{|U|}} \cdot |A_\alpha| = 2^{-|U|} \cdot \sum_{\alpha \in \{0,1\}^i} |A_\alpha|^2$$

$$\geq 2^{-|U|} \cdot \frac{\left(2^{|U|}\right)^2}{2^i} = 2^{|U|-i}$$

where the inequality is due to the fact that the minimum value of $\sum_i z_i^2$, taken over M $(= 2^i)$ non-negative z_i's summing to N $(= 2^{|U|})$, is obtained when the z_i's are equal, and the value itself is $M \cdot (N/M)^2 = N^2/M$.

Note that the algorithm needs not proceed by the foregoing order of sets (i.e., $S_1, S_2, S_3, ...$). In general, for every 1-1 permutation π over $[n]$, we may proceed by considering in the ith step the set $S_{\pi(i)}$. Still, the complexity of this (generalized) algorithm is at least exponential in

$$\min_\pi \left\{ \max_i \left\{ |\cup_{j=1}^i S_{\pi(j)}| - i \right\} \right\} \tag{3}$$

We should thus use a collection such that Eq. (3) is big (i.e., bounded below by $\Omega(n)$).

Bad collections. Obviously, it is a bad idea to have $S_j = \{j+1, ..., j+\ell\}$, since in this case we have $| \cup_{j=1}^i S_j | - i \leq \ell - 1$ for every i. It also follows that we cannot use $\ell \leq 2$, since in this case one can always find an order π such that Eq. (3) is bounded above by $\ell - 1$.

Good collections. An obvious lower bound on Eq. (3) is obtained by the expansion property of the collection $C = \{S_j\}$, where the expansion of C is defined as

$$\max_k \min_{I : |I|=k} \left\{ |\cup_{j \in I} S_j| - k| \right\}. \tag{4}$$

A natural suggestion is to determine the collection C according to the neighborhood sets of an *expander graph*. Loosely speaking, known constructions of expander graphs allow to let ℓ be a small constant (in the range $\{7, ..., 16\}$), while guaranteeing that Eq. (4) is a constant fraction of n.

4   Concrete Parameters for Practical Use

If we go for random predicates, then we should keep ℓ relatively small (say, $\ell \leq 16$), since our implementation of the function must contain a 2^ℓ-size table look-up for P. (Indeed, $\ell = 8$ poses no difficulty, and $\ell = 16$ requires a table of 64K bits which seems reasonable.) For concrete security we will be satisfied with time complexities such as 2^{80} or so. Our aim is to have n as small as possible (e.g., a couple of hundreds).

The issue addressed below is which expander to use. It is somewhat "disappointing" that for some specific parameters we aim for, we cannot use the "best" known explicit constructions.

Below we use the bipartite formulation of expanders. By *expansion* we mean a lower bound established on the quantity in Eq. (4). Recall that the time complexity is conjectured to be exponential in this bound.

Random Construction. This yields the best results, but the "cost" is that with small probability we may select a bad construction. (The fact that we need to hard-wire the construction into the function description is of little practical concern, since we are merely talking of hard-wiring $n \cdot \ell \cdot \log_2 n$ bits, which for the biggest n and ℓ considered below merely means hard-wiring 20K bits.) Alternatively, one may incorporate the specification of the construction in the input of the one-way function, at the cost of augmenting the input by $n \cdot \ell \cdot \log_2 n$ (where the original input is n-bit long). Specific values that may be used are tabulated in Figure 1.[2]

degree (i.e., ℓ)	#vertices (i.e., n)	expansion	error prob.
10	256	77	2^{-81}
12	256	90	2^{-101}
14	256	103	2^{-104}
16	256	105	2^{-152}
8	384	93	2^{-83}
10	384	116	2^{-121}
12	384	139	2^{-141}
8	512	130	2^{-101}
10	512	159	2^{-151}
12	512	180	2^{-202}

Fig. 1. The parameters of a random construction

The last column (i.e., *error prob.*) states the probability that a random construction (with given n and ℓ) does not achieve the stated expansion. Actually, we only provide upper bounds on these probabilities.

Alon's Geometric Expanders [4]. These constructions do not allow $\ell = O(\log n)$, but rather ℓ that is polynomially related to n. Still for our small numbers we get meaningful results, when using $\ell = q + 1$ and $n = q^2 + q + 1$, where q is a prime power. Specific values that may be used are tabulated in Figure 2.[3] Note that these are all the suitable values for Alon's construction (with $\ell \leq 16$); in particular, ℓ uniquely determines n and $\ell - 1$ must be a prime power.

---

[2] The expansion was computed in a straightforward manner. The key issue is to provide, for any fixed k and h, a good upper bound on the probability that a specific set of k vertices has less than h neighbors.

[3] The expansion is computed from the eigenvalues, as in [5]. Actually, we use the stronger bound provided by [4, Thm. 2.3] rather than the simpler (and better known) bound. Specifically, the lower bounds in [4, Thm. 2.3] are on the size of the neighborhood of x-subsets, and so we should subtract x from them, and maximize over all possible x's. (We use the stronger lower bound of $n - \frac{(n-x)(\ell n+1)}{\ell n+1+(n-\ell-2)x}$, rather than the simpler bound of $n - \frac{n^{3/2}}{x}$, both provided in [4, Thm. 2.3].)

degree (i.e., ℓ)	#vertices (i.e., n)	expansion	comment
10	91	49	expansion too low
12	133	76	quite good
14	183	109	very good

Fig. 2. The parameters of Geometric expanders

The Ramanujan Expanders of Lubotzky, Phillips, and Sarnak [10].
Until one plays with the parameters governing this construction, one may not
realize how annoying these may be with respect to an actual use: The difficulty
is that there are severe restrictions regarding the degree and the number of
vertices,[4] making $n \approx 2000$ the smallest suitable choice. Admissible values are
tabulated in Figure 3.[5]

Parameters			Results		
p	q	bipartite?	ℓ	n	expansion (+ comment)
13	5	NO	15	120	20 (unacceptable)
5	13	NO	7	2184	160 (better than needed)
13	17	YES	14	2448	392 (better than needed)

Fig. 3. The parameters of Ramanujan expanders

Note that $p = 5$ and $p = 13$ are the only admissible choices for $\ell \leq 16$. Larger
values of q may be used, but this will only yield larger value of n.

Using the Simple Expander of Gaber–Galil [8]. Another nasty surprise
is that the easy-to-handle expander of Gaber–Galil performs very poorly on
our range of parameters. This expander has degree 7 (i.e., $\ell = 7$), and can be
constructed for any $n = m^2$, where m is an integer. But its expansion is $(c/2) \cdot n$,
where $c = 1 - \sqrt{3/4} \approx 0.1339746$, and so to achieve expansion above 80 we need
to use $n = 1225$. See Figure 4.

A Second Thought. In some applications having n on the magnitude of a couple of thousands may be acceptable. In such a case, the explicit constructions
of Lubotzky, Phillips, and Sarnak [10] and of Gaber and Galil [8] become relevant. In view of the lower degree and greater flexibility, we would prefer the
construction of Gaber–Galil.

---

[4] Specifically, $\ell = p + 1$ and $n = (q^3 - q)/2$, where p and q are different primes, both
congruent to 1 mod 4, and p is a square mod q. For the non-bipartite case, p is a
non-square mod q, and $n = q^3 - q$. Recall that for non-bipartite graphs ℓ equals the
degree plus 1 (rather than the degree).

[5] The expansion is computed from the eigenvalues, as in [5].

degree (i.e., ℓ)	#vertices (i.e., n)	expansion	comment
7	400	27	expansion way too low
7	1225	83	good
7	1600	108	very good
7	2500	168	beyond our requirements

Fig. 4. The parameters of Gaber–Galil expanders

5   Concluding Remarks

This was the last section of the original write-up.

5.1   Variations

One variation is to use either a specific predicate or predicates selected at random from a small domain, rather than using a truly random predicate (as in the foregoing presentation). The advantage of these suggestions is that the description of the predicate is shorter, and so one may use larger values of ℓ. Two specific suggestions follow:

1. Use the predicate that partitions its input into two equal length strings and takes their inner product modulo 2. That is, $P(z_1, ..., z_{2t}) = \sum_{i=1}^{t} z_i z_{t+i}$ mod 2.

   In this case, the predicate is described without reference to ℓ, and so any value of ℓ can be used (in practice). This suggestion is due to Adi Shamir.
2. Use a random low-degree ℓ-variant polynomial as a predicate. Specifically, we think of a random ℓ-variant polynomial of degree $d \in \{2, 3\}$ over the finite field of two elements, and such a polynomial can be described by $\binom{\ell}{d}$ bits. In practice, even for $d = 3$, we may use $\ell = 32$ (since the description length in this case is less than 6K bits).

On the other extreme, for sake of simplifying the analysis, one may use different predicates in each application (rather than using the same predicate in all applications).

5.2   Directions for Investigation

1. *The combinatorial properties of the function f_n.* Here we refer to issues such as under what conditions is f_n 1-to-1 or merely "looses little information"; that is, how is $f_n(X_n)$ distributed, when X_n is uniformly selected in $\{0, 1\}^n$. One can show that if the collection $(S_1, ..., S_n)$ is sufficiently expending (as defined above), then $f_n(X_n)$ has min-entropy $\Omega(n)$; i.e., $\Pr[f_n(X_n) = \alpha] < 2^{-\Omega(n)}$, for every $\alpha \in \{0, 1\}^n$. We seek min-entropy bounds of the form $n - O(\log n)$.

2. *What happens when f_n is iterated?* Assuming that f_n "looses little informa-tion", iterating it may make the inverting task even harder, as well as serves as a starting point for the next item.[6]

3. Modifying the construction to obtained a "keyed"-function with the hope that the result is a pseudorandom function (cf. [9]). The idea is to let the key specify the (random) predicate P. We stress that this modification is applied to the iterated function, not to the basic one.[7] We suggest using $\Theta(\log n)$ iteration; in practice 3–5 iterations should suffice.

Our construction is similar to a construction that was developed by Alekhnovich et. al. [3] in the context of proof complexity. Their results may be applicable to show that certain search method that are related to resolution will require exponential-time to invert our function [Avi Wigderson, private communication, 2000].[8]

5.3 Inspiration

Our construction was inspired by the construction of Nisan and Wigderson [11]; however, we deviate from the latter work in two important aspects:

1. Nisan and Wigderson reduce the security of their construction to the hard-ness of the predicate in use. In our construction, the predicate is not complex at all (and our hope that the function is hard to invert can not arise from the complexity of the predicate). That is, we hope that the function is harder to invert than the predicate is to compute.[9]

2. The set system used by Nisan and Wigderson has different combinatorial properties than the systems used by us. Specifically, Nisan and Wigderson ask for small intersections of each pair of sets, whereas we seek expansion properties (of a type that cannot be satisfied by pairs of sets).

Our construction is also reminiscent of a sub-structure of of the DES; specifically, we refer to the mapping from 32-bit long strings to 32-bit long strings induced by the eight S-boxes. However, the connection within input bits and output bits is far more complex in our case. Specifically, in the DES, each of the 8 (4-bit) output strings is a function (computed by an S-box) of 6 (out of the 32) input

---

[6] An additional motivation for iterating f_n is to increase the dependence of each output bit on the input bits. A dependency of each output bit on all output bits is considered by some researchers to be a requirement from a one-way function; we beg to differ.

[7] We note that applying this idea to the original function will definitely fail. In that case, by using 2^ℓ queries (and inspecting only one bit of the answers) we can easily retrieve the key P.

[8] This conjecture was proved in [7]. We mention that in the original posting of this work we expressed the opinion that *this direction requires further investigation*.

[9] We comment that it is not clear whether the Nisan and Wigderson construction can be broken within time comparable to that of computing the predicate; their paper only shows that it cannot be broken substantially faster.

bits. The corresponding 8 subsets have a very simple structure; the ith subset holds bit locations $\{4(i-1)+j : j = 0, ..., 5\}$, where $i = 1, ..., 8$ and 32 is identified with 0. Indeed, inverting the mapping induced on 32-bit strings is very easy.[10] In contrast, the complex relation between the input bits corresponding to certain output bits, in our case, defeat such a simple inversion attack. We stress that this complex (or rather expanding) property of the sets of input bits is the heart of our suggestion.

6   Subsequent Work and Afterthoughts

As evident from the Introduction (as well as from Section 4), our primary motivation for this proposal was to address a pratical concern. We hoped that the apparently low cost of a hardware implementation should make this proposal very appealing to practice. We further hoped that the simplicity of the proposal may encourage theoretically inclined researchers to study it, and that such research may generate more confidence in this proposal. From this perspective and for reasons outlined next, it felt preferrable to promote a setting of $\ell = O(\log n)$.

At the time, we were unclear as to what may be the best choice of a predicate P, and our feeling was that most predicates (or a random predicate) will do. Believing that it is best that P lacks any structure (and thus using a random P), the truthtable of P had to be hard-wired in the function (or appear as part of the input and output to the function). For that reason, $\ell = O(\log n)$ was an absolute limit. As clear from the text, we did not rule out the possibility of setting $\ell = O(1)$, but letting ℓ be as large as possible felt safer (cf. Footnote 1).

In contrast to our initial motivations, the complexity theory community becames more interested in the possibility of setting $\ell = O(1)$, as this yields a function in $\mathcal{NC}_0$. Interest in this aspect of the current work was fueled by the celebrated work of Applebaum, Ishai, and Kushilevitz [2], since one of their main results implies that $\mathcal{NC}_0$ contains one-way functions if and only if $\mathcal{NC}_1$ contains such functions. In fact, it is fair to say that the current work was practically rediscovered after [2].

Subsequent studies have shakened our confidence that a random predicate P is the best possible choice for our proposal. In particular, generalizing our proposal to functions with $m = O(n)$ output bits (rather than n output bits), Bogdanov and Qiao showed [6] that a necessary requirement for security is using a balanaced predicate P (i.e., P such that $|\{z \in \{0,1\}^{\ell} : P(z) = 1\}| = 2^{\ell-1}$). The use of balanaced predicates is also advocated in [1,7].

Acknowledgments. We are grateful to Noga Alon, Adi Shamir, Luca Trevisan, and Avi Wigderson for useful discussions.

---

[10] In an asymptotic generalization of the scheme, inversion takes time linear in the number of bits.

References

1. Applebaum, B., Barak, B., Wigderson, A.: Public-key cryptography from different assumptions. In: 42nd STOC, pp. 171–180 (2010)
2. Applebaum, B., Ishai, Y., Kushilevitz, E.: Cryptography in NC0. SICOMP 36(4), 845–888 (2006)
3. Alekhnovich, M., Ben-Sasson, E., Razborov, A., Wigderson, A.: Pseudorandom Generators in Propositional Proof Complexity. In: 41st FOCS, pp. 43–53 (2000)
4. Alon, N.: Eigenvalues, Geometric Expanders, Sorting in Rounds, and Ramsey Theory. Combinatorica 6, 207–219 (1986)
5. Alon, N., Milman, V.D.: λ_1, Isoperimetric Inequalities for Graphs and Superconcentrators. J. Combinatorial Theory, Ser. B 38, 73–88 (1985)
6. Bogdanov, A., Qiao, Y.: On the Security of Goldreich's One-Way Function. In: Dinur, I., Jansen, K., Naor, J., Rolim, J. (eds.) APPROX 2009. LNCS, vol. 5687, pp. 392–405. Springer, Heidelberg (2009)
7. Cook, J., Etesami, O., Miller, R., Trevisan, L.: Goldreich's One-Way Function Candidate and Myopic Backtracking Algorithms. In: Reingold, O. (ed.) TCC 2009. LNCS, vol. 5444, pp. 521–538. Springer, Heidelberg (2009)
8. Gaber, O., Galil, Z.: Explicit Constructions of Linear Size Superconcentrators. JCSS 22, 407–420 (1981)
9. Goldreich, O., Goldwasser, S., Micali, S.: How to Construct Random Functions. JACM 33(4), 792–807 (1986)
10. Lubotzky, A., Phillips, R., Sarnak, P.: Ramanujan Graphs. Combinatorica 8, 261–277 (1988)
11. Nisan, N., Wigderson, A.: Hardness vs Randomness. JCSS 49(2), 149–167 (1994)

Using the FGLSS-Reduction to Prove Inapproximability Results for Minimum Vertex Cover in Hypergraphs

Oded Goldreich

Abstract. Using known results regarding PCP, we present simple proofs of the inapproximability of vertex cover for hypergraphs. Specifically, we show that

1. Approximating the size of the minimum vertex cover in $O(1)$-regular hypergraphs to within a factor of 1.99999 is NP-hard.
2. Approximating the size of the minimum vertex cover in 4-regular hypergraphs to within a factor of 1.49999 is NP-hard.

Both results are inferior to known results (by Trevisan (2001) and Holmerin (2001)), but they are derived using much simpler proofs. Furthermore, these proofs demonstrate the applicability of the FGLSS-reduction in the context of reductions among combinatorial optimization problems.

Keywords: Complexity of approximation, combinatorial optimization problems, Vertex Cover, PCP, regular hypergraphs.

An early version of this work appeared as TR01-102 of *ECCC*. A discussion of subsequent works is deferred to Section 5.

1 Introduction

This note was inspired by a work of Dinur and Safra [5], which was new at the time this work was completed. Specifically, what we take from their work is the realization that the so-called FGLSS-reduction is actually a general paradigm that can be applied in various ways and achieve various purposes.

The FGLSS-reduction, introduced by Feige, Goldwasser, Lovász, Safra and Szegedy [7], is typically understood as a reduction from sets having certain PCP systems to approximation versions of Max-Clique (or Max Independent Set). The reduction maps inputs (either in or out of the set) to graphs that represent the pairwise consistencies among possible views of the corresponding PCP verifier. It is instructive to think of these possible verifier views as of possible partial solutions to the problem of finding an oracle that makes the verifier accept.

Dinur and Safra apply the same underlying reasoning to derive graphs that represent pairwise consistencies between partial solutions to other combinatorial problems [5]. In fact, they use two different instantiations of this reasoning. Specifically, in one case they start with the vertex-cover problem and consider

O. Goldreich et al.: Studies in Complexity and Cryptography, LNCS 6650, pp. 88–97, 2011.

the restrictions of possible vertex-covers to all possible $O(1)$-subsets of the vertex set. The partial solutions in this case are the vertex-covers of the subgraphs induced by all possible $O(1)$-subsets, and pairwise consistency is defined in the natural way. Thus, we claim that in a sense, the work of Dinur and Safra [5] suggests that the FGLSS-reduction is actually a general paradigm that can be instantiated in various ways. Furthermore, the goal of applying this paradigm may vary too. In particular, the original instantiatation of the FGLSS-reduction by Feige *et. al.* [7] was aimed at linking the class PCP to the complexity of approximating combinatorial optimization problems. In contrast, in the work of Dinur and Safra [5] one instantiation is aimed at deriving instances of very low "degree" (i.e., co-degree at most 2), and the other instantiation is aimed at moving the "gap location" (cf. [16] and further discussion below).

We fear that the complexity of the work of Dinur and Safra [5] may cause researchers to miss the foregoing observation (regarding the wide applicability of the FGLSS-reduction). This would be unfortunate, because we believe in the potential of that observation. In fact, this note grew out of our fascination with the foregoing observation and our attempt to find a simple illustration of it.

Our Concrete Results: Combining known results regarding PCP with the FGLSS-reduction, we present simple proofs of inapproximability results regarding the *minimum vertex cover problem for hypergraphs*. Specifically, we show that:

1. For every constant $\epsilon > 0$, approximating the size of the minimum vertex cover in $O(1)$-regular hypergraphs to within a $(2 - \epsilon)$-factor is NP-hard (see Section 3). In fact, the hypergraphs we use are $O((1/\epsilon)^{o(1)})$-regular.
   This result is inferior to Holmerin's result [12], by which approximating vertex cover in 4-regular hypergraphs to within a $(2 - \epsilon)$-factor is NP-hard. We also mention Trevisan's result [17] by which, for every constant k, approximating vertex cover in k-regular hypergraphs to within a $\Omega(k^{1/19})$-factor is NP-hard. Clearly, in terms of achieving a bigger inapproximation factor, Trevisan's result is superior, but in terms of achieving an inapproximation result for k-regular graphs when k is small (e.g., $k < 2^{19}$) it seems that our result is better.
2. For every constant $\epsilon > 0$, approximating the size of the minimum vertex cover in 4-regular hypergraphs to within a $(1.5 - \epsilon)$-factor is NP-hard (see Section 4).
   Again, this result is inferior to Holmerin's result [12].

We mention that our work was done independently of Holmerin's work [12], but after the publication of Trevisan's work [17].

2 Preliminaries

This section contains a review of the notion of a vertex cover in a hypergraph and the notion of free-bit complexity. We also recall the FGLSS-reduction and discuss its relation to the vertex cover problem in graphs.

Vertex Covers in Hypergraphs. A k-regular hypergraph is a pair (V, E) such that E is a collection of k-subsets (called hyper-edges) of V; that is, for every $e \in E$ it holds that $e \subseteq V$ and $|e| = k$. For a k-regular hypergraph $H = (V, E)$ and $C \subseteq V$, we say that C is a vertex cover of H if for every $e \in E$ it holds that $e \cap C \neq \emptyset$.

Free-bit Complexity and the Class $\mathcal{FPCP}$. We assume that the reader is familiar with the basic PCP-terminology (cf. [1,2,3] and [8, Sec. 2.4]). (For sake of simplicity we consider non-adaptive verifiers.) We say that the free-bit complexity of a PCP system is bounded by $f : \mathbb{N} \to \mathbb{R}$ if on every input x and any possible random-pad ω used by the verifier, there are at most $2^{f(|x|)}$ possible sequence of answers that the verifier may accept (on input x and random-pad ω). Clearly, the free-bit complexity of a PCP system is bounded by the number of queries it makes, but the former may be much lower. Free-bit complexity is a key parameter in the FGLSS-reduction. For functions $c, s : \mathbb{N} \to [0, 1]$, $r : \mathbb{N} \to \mathbb{N}$ and $f : \mathbb{N} \to \mathbb{R}$, we denote by $\mathcal{FPCP}_{c,s}[r, f]$ the class of sets having PCP systems of completeness bound c, soundness bound s, randomness complexity r and free-bit complexity f. In particular, for every input x in the set, there exist an oracle that makes the verifier accept with probability at least $c(|x|)$, whereas for every input x not in the set and every oracle the verifier accepts with probability at most $s(|x|)$.

The FGLSS-Graph. For $S \in \mathcal{FPCP}_{c,s}[r, f]$, the FGLSS-reduction maps x to a graph G_x having $2^{r(|x|)}$ layers, each having at most $2^{f(|x|)}$ vertices. The vertices represent possible views of the verifier, where the $N \stackrel{\text{def}}{=} 2^{r(|x|)}$ layers correspond to all possible choices of the random-tape and the vertices in each layer correspond to the up-to-$2^{f(|x|)}$ possible sequences of answers that the verifier may accept. The edges represent *inconsistencies* among these views. In particular, each layer consists of a clique (because only one sequence of answers is possible for a fixed random-tape and a fixed oracle). If the random-tapes $\omega_1, \omega_2 \in \{0, 1\}^{r(|x|)}$ both lead the verifier to make the same query q (and both answers are acceptable), then the corresponding layers will have edges between vertices encoding views in which different answers are given to query q. In the case that $x \in S$ the graph G_x will have an independent set of size $c(|x|) \cdot N$, whereas in the case that $x \notin S$ the maximum independent set in G_x has size at most $s(|x|) \cdot N$. Thus, the inapproximability factor for the maximum independent set problem shown by such a reduction is $c(|x|)/s(|x|)$, and the fact the maximum independent set is always at most a $2^{-f(|x|)}$ fraction of the size of G_x does not effect the gap. However, inapproximability factor for the minimum vertex cover shown by such a reduction is

$$\frac{2^{f(|x|)} \cdot N - s(|x|) \cdot N}{2^{f(|x|)} \cdot N - c(|x|) \cdot N} = \frac{2^{f(|x|)} - s(|x|)}{2^{f(|x|)} - c(|x|)} < \frac{2^{f(|x|)}}{2^{f(|x|)} - 1}. \tag{1}$$

This is the reason that, while the FGLSS-reduction allows to establish quite optimal inapproximability factors for the maximum independent set problem, it failed so far to establish optimal inapproximability factors for the minimum vertex cover problem (although, it was used by Hastad [10] in deriving the 7/6 hardness factor by using Eq. (1) with $f = 2$, $c \approx 1$ and $s = 1/2$). In a sense, the gap between the size of the maximum independent set of G_x for $x \in S$ versus for $x \notin S$ is at the "right" location for establishing inapproximability factors for the maximum independent set problem, but is at the "wrong" location for establishing inapproximability factors for the minimum vertex cover problem. Thus, what we do below is "move the gap location": Specifically, in Section 3, we take a maximum independent set gap of $c2^{-f}$ versus $s2^{-f}$ (which means a minimum vertex cover gap of $1 - c2^{-f}$ versus $1 - s2^{-f}$), and transform it into a minimum vertex cover gap of $(2 - c) \cdot 2^{-f}$ versus $(2 - s) \cdot 2^{-f}$.

3   A $2 - \epsilon$ Hardness Factor for $O(1)$-Regular Hypergraphs

We start with the usual FGLSS-graph, denoted G, derived from the FGLSS-reduction as applied to input x of a $\mathcal{FPCP}_{1-\epsilon,s}[\log, f]$ scheme (for a set in $\mathcal{NP}$). For simplicity, think of f as being a constant such that 2^f is an integer. Without loss of generality, each layer of G has $\ell = 2^f$ vertices.

We now apply the "FGLSS paradigm" by considering vertex-covers of G, and their projection on each layer. Such projections (or "partial assignments") have either ℓ or $\ell - 1$ vertices. We focus on the good vertex covers, having exactly $\ell - 1$ vertices in (almost) each layer. Thus, for each $(\ell - 1)$-subset of each layer, we introduce a vertex in the hypergraph, to be denoted H. We also introduce hyper-edges so to reflect the inconsistencies of the various partial (i.e. layer-projected) vertex covers of G. This construction, presented next, will provide a correspondance between vertex covers of G and vertex covers of H.

The Construction of the Hypergraph H. For each layer $L = (v_1, ..., v_\ell)$ in G, we introduced a corresponding layer in H containing ℓ vertices such that each H-vertex corresponds to an $(\ell - 1)$-subset of L; that is, we introduce ℓ vertices that correspond to $L \setminus \{v_1\}, ..., L \setminus \{v_\ell\}$. For each pair of layers $L' = (v'_1, ..., v'_\ell)$ and $L'' = (v''_1, ..., v''_\ell)$, if (v'_i, v''_j) is an edge in G, then we introduce the $2 \cdot (\ell - 1)$-hyperedge containing the H-vertices that correspond to the subsets $\{L' \setminus \{v'_k\} : k \neq i\}$ and $\{L'' \setminus \{v''_k\} : k \neq j\}$; that is, the hyper-edge consists of all the H-vertices of these two layers except for the two H-vertices that correspond to the subsets $L' \setminus \{v'_i\}$ and $L'' \setminus \{v''_j\}$. In addition, for each layer in H, we introduce an ℓ-size hyper-edge containing all ℓ vertices of that layer.

To get rid of the non-regularity of this construction, we augment each layer with a sets of $\ell - 2$ auxiliary vertices, and replace the abovementioned ℓ-size hyper-edge by a hyper-edge containing all vertices of that layer (i.e., the original ℓ vertices as well as the $\ell - 2$ auxiliary vertices). We refer to these hyper-edges as intra-layer ones. This completes the construction of H.

Motivation to the Analysis. Consider a generic vertex cover, C, of G, and let S denote the set of all vertices of H that correspond to the $(\ell - 1)$-subsets of

C. Note that C contains ℓ vertices of some layer of G if and only if S contains all vertices of the corresponding layer in H, and in this case all edges (resp., hyper-edges) adjacent to this layer are covered. Thus, we focus on layers of G that contain $\ell - 1$ vertices of C, and note that (for each such layer) S contains a single vertex of H that resides in the corresponding layer. Let $L' = (v'_1, ..., v'_\ell)$ and $L'' = (v''_1, ..., v''_\ell)$ be two such layers of G, and let v'_i and v''_j denote the two vertices that are missing from C (which implies that (v'_i, v''_j) is not an edge in G). Then, $L' \setminus \{v'_i\}$ and $L'' \setminus \{v''_j\}$ are in S, and they cover all the hyper-edges that connect L' and L'', because $\{L' \setminus \{v'_k\} : k \neq i\} \cup \{L'' \setminus \{v''_k\} : k \neq j\}$ is not a hyper-edge in H.

The Actual Analysis. Fixing any input x, we consider the corresponding FGLSS-graph $G = G_x$, and the hypergraph $H = H_x$ derived from G by following the above construction. Let N denote the number of layers in G (and H).

Claim 3.1. *If x is a yes-instance, then the hypergraph H_x has a vertex-cover of size at most $(1 + \epsilon) \cdot N$.*

Proof: Since x is a yes-instance, the graph $G = G_x$ has an independent set (IS) of size at least $(1 - \epsilon) \cdot N$. Consider this IS or actually the corresponding vertex-cover (i.e., VC) of G. Call a layer in G good if it has $\ell - 1$ vertices in this VC, and note that at least $(1 - \epsilon) \cdot N$ layers are good. We create a vertex-cover for $H = H_x$ as follows. For each good layer, place in C the corresponding H-vertex; that is, the H-vertex corresponding to the $(\ell - 1)$-subset (of this layer in G) that is in the VC of G. For the rest of the layers (i.e., the non-good layers), place in C any two H-vertices of each (non-good) layer.

In total we placed in C at most $(1 - \epsilon)N + 2\epsilon N = (1 + \epsilon)N$ vertices. We show that C is a vertex cover of H by considering all possible hyper-edges, bearing in mind the correspondence between layers of G and layers of H.

- Each intra-layer hyper-edge of H (which consists of all vertices of that layer) is definitely covered, because we placed in C at least one H-vertex from each layer.
- Each hyper-edge connecting H-vertices from two good layers is covered. This is shown by considering the edge, denoted (u, v), of G that is "responsible" for the introduction of each hyper-edge (in H).[1] Since we started with a vertex cover of G, either u or v must be in that cover. Suppose, without loss of generality, that u is in the VC of G. Then, we must have placed in C one of the H-vertices that corresponds to a $(\ell - 1)$-subset that contains u. But, then, this H-vertex covers the said hyper-edge (because, by construction, the latter hyper-edge contains all $(\ell - 1)$-subsets that contain u).
- Each hyper-edge that contains H-vertices from at least one non-good layer is covered, because we placed in C two H-vertices from each non-good layer,

---

[1] A hyper-edge that correspons to layers L' and L'' has the form $\{L' \setminus \{w\} : w \neq u\} \cup \{L'' \setminus \{w\} : w \neq v\}$, where $u \in L'$ and $v \in L''$. Furthermore, (u, v) must be an edge in G.

whereas each hyper-edge containing H-vertices of some layer contains all but at most one vertex of that layer.

The claim follows.   ∎

Claim 3.2. *If x is a no-instance, then every vertex-cover of the hypergraph H_x has size at least $(2 - s(|x|)) \cdot N$.*

Proof: Consider any vertex cover C of H. Note that due to the intra-layer hyper-edges, C must contain at least one vertex in each layer. Furthermore, without loss of generality, C contains only original vertices (rather than the $\ell-2$ auxiliary vertices added to each layer). Denote by C' the set of layers that have a *single* vertex in C. Then, $|C| \geq |C'| + 2(N - |C'|) = 2N - |C'|$. The claim follows by proving that $|C'| \leq sN$, where $s \stackrel{\text{def}}{=} s(|x|)$.

Suppose, towards the contradiction, that $|C'| > sN$. We consider the set of G-vertices, denoted I, that correspond to the (single) H-vertices in these layers; that is, for layer L (in C') such that C contains the H-vertex (which corresponds to) $L \setminus \{v\}$, place $v \in G$ in I. We show that I is an independent set in G (and so derive a contradiction to $G = G_x$ not having an independent set of size greater than sN, because x is a no-instance). Specifically, for every $u, v \in I$, we show that (u, v) cannot be an edge in G. Suppose (u, v) is an edge in G, then the corresponding hyper-edge in H cannot be covered by C; that is, the hyper-edge $\{L \setminus \{w\} : w \neq u\} \cup \{L' \setminus \{w\} : w \neq v\}$ (which must be introduced due to the edge (u, v)) cannot be covered by the H-vertices that correspond to the $(\ell-1)$-subsets $L \setminus \{u\}$ and $L' \setminus \{v\}$. The claim follows.   ∎

Conclusion: Starting from a $\mathcal{FPCP}_{1-\epsilon,s}[\log, f]$ system for $\mathcal{NP}$, we have shown that the minimum vertex-cover in $(2^{f+1} - 2)$-regular hypergraphs is NP-hard to approximate to a $(2-s)/(1+\epsilon)$-factor. Now, if we start with any $\mathcal{FPCP}_{1,s}[\log, f]$ for $\mathcal{NP}$, with $s \approx 0$, then we get a hardness result for a factor of $2 - s \approx 2$. Any $\mathcal{NP} \subseteq \mathcal{PCP}[\log, O(1)]$ result (starting from [1]) will do for this purpose, because a straightforward error-reduction will yield $\mathcal{NP} \subseteq \mathcal{FPCP}_{1,s}[\log, O(1)]$, for any $s > 0$. The (amortized) free-bit complexity only effects the growth of the hyper-edge size as a function of the deviation of the hardness-factor from 2. Specifically, if we start with an "amortized free-bit complexity zero" result (i.e., $\mathcal{NP} \subseteq \mathcal{FPCP}_{1,s}[\log, o(\log_2(1/s))]$ for every $s > 0$), then we get a factor of $2 - s$ hardness for $(1/s)^{o(1)}$-regular hypergraphs. That is, starting with Hastad's first such result [9] (or from the simplest one currently known [11]), and applying the foregoing reasoning, we obtain our first little result:

Theorem 3.3. *For every constant $\epsilon > 0$, approximating the size of the minimum vertex cover in $(1/\epsilon)^{o(1)}$-regular hypergraphs to within a $(2 - \epsilon)$-factor is NP-hard.*

Alternatively, if we start with Hastad's "maxLIN3 result" [10] (i.e., the result $\mathcal{NP} \subseteq \mathcal{FPCP}_{1-\epsilon,0.5}[\log, 2]$ for every $\epsilon > 0$), then we get a hardness factor of $(2 - 0.5)/(1 + \epsilon) \approx 1.5$ for 6-regular hypergraphs. Below we show that the same hardness factor holds also for 4-regular hypergraphs (by starting with the same "maxLIN3 result" [10] but capitalizing on an additional feature of it).

4   A 1.5 − ε Hardness Factor for 4-Regular Hypergraphs

We start with the FGLSS-graph derived from applying the FGLSS-reduction to Hastad's "maxLIN3 system" [10]; that is, the $\mathcal{FPCP}_{1-\epsilon,0.5}[\log, 2]$ system for $\mathcal{NP}$ ($\forall \epsilon > 0$). The key observation is that, *in this system*, for any two queries, all four answer pairs are possible (as accepting configurations).[2] This observation is relied upon when establishing (below) simple structural properties of the derived FGLSS-graph.

As before, there will be a correspondence between the vertex set of G and the vertex set of H. Here it is actually simpler to just identify the two sets. So it just remains to specify the hyper-edges of H. Again, we place (intra-layer) hyper-edges between all (i.e., four) vertices of each layer. As for the construction of inter-layer hyper-edges, we consider three cases regarding each pair of layers:

1. *The trivial case*: If there are no edges between these two layers in G, then there would be no hyper-edges between these layers in H. This case corresponds to the case that these two layers correspond to two random-tapes that induce two query sets with empty intersection.

2. *The interesting case* is when these two layers correspond to two random-tapes that induce two query sets having a *single* query, denoted q, in common. Relying on the property of the starting PCP system, it follows that both answers are possible to this query and that each possible answer is represented by two vertices in each corresponding layer. Accordingly, we denote the vertices of the first layer by $u_1^0, u_2^0, u_1^1, u_2^1$, where u_i^b is the ith configuration in this layer in which query q is answered by the bit b. Similarly, denote the vertices of the second layer by $v_1^0, v_2^0, v_1^1, v_2^1$. (We stress that this notation is used only for determining the hyper-edges between the current pair of layers, and when considering a different pair of layers a different notation may be applicable.) In this case we introduce the two hyper-edges $\{u_1^0, u_2^0, v_1^1, v_2^1\}$ and $\{u_1^1, u_2^1, v_1^0, v_2^0\}$.

   Intuition: Note that the edges in G [*sic*] between these two layers are two $K_{2,2}$'s (i.e., for each $b \in \{0, 1\}$, between the two u_i^b's on one side and the two v_i^{1-b}'s on the other side). These two $K_{2,2}$'s enforce that if some u_i^b is in some IS, then v_j^{1-b} is not in the IS. For a H-VC having a single vertex in each layer, the (two) hyper-edges will have the same effect.

3. *The annoying case* is when these two layers (correspond to two random-tapes that induce two query sets that) have *two* or more queries in common. In this case, we label the vertices in these two layers according to these two answers; that is, we denote the four vertices of the first layer by $u_{0,0}, u_{0,1}, u_{1,0}, u_{1,1}$, where $u_{a,b}$ is the unique configuration in this layer in which these two queries are answered by a and b, respectively. Similarly, denote the vertices of the second layer by $v_{0,0}, v_{0,1}, v_{1,0}, v_{1,1}$. (Again, this notation is used only for determining the hyper-edges between the current pair of layers.) In this case,

---

[2] Recall that the number of queries is typically higher than the free-bit complexity. Indeed, the aforementioned system makes *three queries* and has free-bit complexity *two*.

we introduce four hyper-edges between these two layers, each has one vertex of the first layer and the three "non-matching" vertices of the second layer; that is, the hyper-edges are $\{u_{a,b}, v_{a,1-b}, v_{1-a,b}, v_{1-a,1-b}\}$, for $a, b \in \{0, 1\}$. Intuition: The pair $(u_{a,b}, v_{a',b'})$ is an edge in G if and only if either $a \neq a'$ or $b \neq b'$. Similarly, the pair $(u_{a,b}, v_{a',b'})$ participates in an hyper-edge of H if and only if either $a \neq a'$ or $b \neq b'$.

This completes the construction. Note that $H = H_x$ is a 4-regular hypergraph.

Claim 4.1. *If x is a yes-instance, then the hypergraph H_x has a vertex-cover of size at most $(1 + 3\epsilon) \cdot N$, where N denotes the number of layers.*

Proof: Since x is a yes-instance, the graph $G = G_x$ has an independent set (IS) of size $(1 - \epsilon)N$. Consider such an IS, denoted I. Call a layer in G *good* if it has a vertex in I, and note that at least $(1 - \epsilon)N$ layers are good. Augment I by the set of all vertices residing in non-good layers. In total we took at most $(1 - \epsilon)N + 4\epsilon N = (1 + 3\epsilon)N$ vertices. We show that these vertices cover all hyper-edges of H.

- The intra-layer hyper-edges are definitely covered (since we took at least one vertex from each layer).
- Each hyper-edge connecting vertices from two good layers is covered.
  This is shown by considering each of the two non-trivial cases (in the construction). In the interesting case, I (having a single vertex in each good layer) must have a single vertex in *each* $K_{2,2}$. But then this vertex covers the corresponding hyper-edge. In the annoying case, I (having a single vertex in each good layer) must contain vertices with matching labels in these two layers. But then these two vertices cover all four hyper-edges, because each hyper-edge contains a (single) vertex of each label.
- Hyper-edges containing H-vertices from non-good layers are covered trivially (because we took all vertices of each non-good layer).

The claim follows. ∎

Claim 4.2. *If x is a no-instance, then every vertex-cover of the hypergraph H_x has size at least $1.5 \cdot N$.*

Proof: Consider a cover C of H. Note that (due to the intra-layer hyper-edges) C must contain at least one vertex in each layer. Denote by C' the set of layers that have a single vertex in C. Then, $|C| \geq |C'| + 2(N - |C'|)$. The claim follows by proving that $|C'| \leq 0.5N$. Suppose, towards the contradiction, that $|C'| > 0.5N$. Consider the set of vertices, denoted I, that correspond to these layers (i.e., for a layer in C' consider the layer's vertex that is in C). We show that I is an independent set in G (and so we derive contradiction).

Suppose (towards the contradiction) that $u, v \in I$ and (u, v) is an edge in G. In the interesting case, this (i.e., (u, v) being an edge in G) means that u and

v are in the same hyper-edge in H, and being the only vertices in C that are in these layers, no vertex covers the other (vertex-disjoint) hyper-edge between these layers. In the annoying case, this (i.e., (u,v) being an edge in G) means that u and v do not have the same label and one of the four hyper-edges in H cannot be covered by them; specifically, without loss of generality, suppose that u is in the first layer, then neither $v = v_{a,b}$ nor $u \neq u_{a,b}$ covers the hyper-edge $\{u_{a,b}, v_{a,1-b}, v_{1-a,b}, v_{1-a,1-b}\}$. ∎

Conclusion: Starting from the abovementioned $\mathcal{NP} \subseteq \mathcal{FPCP}_{1-\epsilon,0.5}[\log, 2]$ result of Hastad [10], we have shown that the minimum vertex-cover in 4-regular hypergraphs is NP-hard to approximate to a factor of $1.5/(1 + 3\epsilon)$. Let us state this as our second little result:

Theorem 4.3. *For every constant $\epsilon > 0$, approximating the size of the minimum vertex cover in 4-regular hypergraphs to within a $(1.5-\epsilon)$-factor is NP-hard.*

5   Subsequent Work

As hinted in the introduction, our motivation in this memo was to draw attention to the wide applicability of the FGLSS-reduction, and the specific results obtained were merely a good excuse to do so. Recall that all our results are inferior to Holmerin's independently achieved result [12], by which approximating vertex cover in 4-regular hypergraphs to within a $(2 - \epsilon)$-factor is NP-hard. Thus, the fact that also the latter result was subsequently improved is not relevant to the main motivation of the current work. Nevertheless, we briefly review some of the related results that appear after the current work was completed, differentiating between what was known already in 2001 and what is known in 2010.

Original Postscript (2001). Following this work, Holmerin has applied related FGLSS-type reductions to different PCP systems and obtained improved inapproximability results for vertex cover in hypergraphs [14]. Specifically, for every constant $\epsilon > 0$, he showed that:

1. Approximating the size of the minimum vertex cover in k-regular hypergraphs to within a factor of $\Omega(k^{1-\epsilon})$ is NP-hard.
2. Approximating the size of the minimum vertex cover in 3-regular hypergraphs to within a factor of $1.5 - \epsilon$ is NP-hard.

Additional Postscript (2010). The results reported in the original postscript were further improved by subsequent works, culminating in the following two results:

1. For every constant $\epsilon > 0$, approximating the size of the minimum vertex cover in k-regular hypergraphs to within a factor of $k - 1 - \epsilon$ is NP-hard [4].
2. Assuming the Unique Game Conjecture (UGC), for every constant $\epsilon > 0$ and every integer $k \geq 2$, it is hard to approximate the size of the minimum vertex cover in k-regular hypergraphs to within a factor of $k - \epsilon$ [15].[3]

---

[3] Indeed, the case $k = 2$ drew most attention.

Acknowledgments. We are grateful to Johan Hastad for referring us to the works of Trevisan [17] and Holmerin [12].

References

1. Arora, S., Lund, C., Motwani, R., Sudan, M., Szegedy, M.: Proof Verification and Intractability of Approximation Problems. JACM 45, 501–555 (1998); Preliminary Version in 33rd FOCS (1992)
2. Arora, S., Safra, S.: Probabilistic Checkable Proofs: A New Characterization of NP. JACM 45, 70–122 (1998); Preliminary Version in 33rd FOCS (1992)
3. Bellare, M., Goldreich, O., Sudan, M.: Free Bits, PCPs and Non-Approximability – Towards Tight Results. SICOMP 27(3), 804–915 (1998)
4. Dinur, I., Guruswami, V., Khot, S., Regev, O.: A New Multilayered PCP and the Hardness of Hypergraph Vertex Cover. SICOMP 34(5), 1129–1146 (2005)
5. Dinur, I., Safra, S.: The Importance of Being Biased. manuscript, See also (2001)
6. Dinur, I., Safra, S.: The importance of being biased. In: 34th STOC, pp. 33–42 (2002)
7. Feige, U., Goldwasser, S., Lovász, L., Safra, S., Szegedy, M.: Approximating Clique is almost NP-complete. JACM 43, 268–292 (1996); Preliminary Version in 32nd FOCS (1991)
8. Goldreich, O.: Modern Cryptography, Probabilistic Proofs and Pseudorandomness. Algorithms and Combinatorics series, vol. 17. Springer, Heidelberg (1999)
9. Hastad, J.: Clique is hard to approximate within $n^{1-\epsilon}$. Acta Mathematica 182, 105–142 (1999); Preliminary versions in 28th STOC (1996) and 37th FOCS (1996)
10. Hastad, J.: Some optimal in-approximability results. In: 29th STOC, pp. 1–10 (1997)
11. Hastad, J., Khot, S.: Query efficient PCPs with Perfect Completeness. In: 42nd FOCS, pp. 610–619 (2001)
12. Holmerin, J.: Vertex Cover on 4-regular Hypergraphs is Hard to Approximate within $2 - \epsilon$ TR01-094, ECCC (2001), See also [13]
13. Holmerin, J.: Vertex Cover on 4-regular Hypergraphs is Hard to Approximate within $2-\epsilon$. In: 34th STOC, pp. 544–552 (2002)
14. Holmerin, J.: Improved Inapproximability Results for Vertex Cover on k-Uniform Hypergraphs. In: Widmayer, P., Triguero, F., Morales, R., Hennessy, M., Eidenbenz, S., Conejo, R. (eds.) ICALP 2002. LNCS, vol. 2380, pp. 1005–1016. Springer, Heidelberg (2002)
15. Khot, S., Regev, O.: Vertex Cover Might be Hard to Approximate to within $2-\epsilon$. JCSS 74(3), 335–349 (2008); Preliminary Version in 18th Conf. on Comput. Complex (2003)
16. Petrank, E.: The Hardness of Approximations: Gap Location. Computational Complexity 4, 133–157 (1994)
17. Trevisan, L.: Non-approximability Results for Optimization Problems on Bounded-Degree Instances. In: 33rd STOC, pp. 453–461 (2001)

The GGM Construction Does NOT Yield Correlation Intractable Function Ensembles

Oded Goldreich

Abstract. We consider the function ensembles emerging from the construction of Goldreich, Goldwasser and Micali (GGM), when applied to an *arbitrary* pseudoramdon generator. We show that, in general, such functions fail to yield correlation intractable ensembles. Specifically, it may happen that, given a description of such a function, one can easily find an input that is mapped to zero under this function.

Keywords: Cryptography, Correlation Intractability.

An early version of this work appeared as TR96-042 of *ECCC*. The current revision is quite minimal.

1 Introduction

The general context of this work is the so-called Random Oracle Methodolody, or rather its critical review, undertaken by Canetti, Goldreich and Halevi [CGH98], Loosely speaking, this methodology suggests to design cryptographic schemes in a two-step process. In the first step, an ideal scheme is designed in an ideal model in which all parties (including the adversary) have access to a random orcale. In the second step, the ideal scheme is realized by replacing the random oracle by a fully-specified function (selected at random in some function emsemble (see Definition 1)), while providing all parties with a description of the function.

Canetti, Goldreich, and Halevi [CGH98] showed that, in general, this methodology may lead to the design of insecure schemes. That is, in general, it may be that the ideal scheme is secure in the ideal model (in which all parties have access to a random orcale), but replacing the random oracle by any function ensemble yields an insecure scheme. Their analysis is based on the notion of correlation intractability, which seems a very minimal requirement from such a replacement. Loosely speaking, a function f is correlation intractable with respect to a sparse binary relation R if it is infeasible (given a description of f) to find x such that $(x, f(x)) \in R$. The point is that the sparseness condition implies that when given access to a random oracle $\mathcal{O}$ it is infeasible to find x such that $(x, \mathcal{O}(x)) \in R$, and so we should require the same from the function f. Before proceeding, let use clarify two of the aforementioned notions.

1.1 Function Ensembles and Correlation Intractability

A function ensemble is a collection of finite functions, where each function has a finite description (viewed as its index in the ensemble). The functions map strings

O. Goldreich et al.: Studies in Complexity and Cryptography, LNCS 6650, pp. 98–108, 2011.

of certain length to strings of another length, where these lengths are determined as a function of the index length. For simplicity, we consider a (natural) special case in which the input and output lengths are equal.

Definition 1 (function ensembles): *Let $\ell : \mathbb{N} \to \mathbb{N}$. A function ensemble with length ℓ is a set of functions $F = \{f_s\}_{s \in \{0,1\}^*}$ such that each function f_s maps $\ell(|s|)$-bit long strings to $\ell(|s|)$-bit long strings. That is:*

$$F \stackrel{\text{def}}{=} \{f_s : \{0,1\}^{\ell(|s|)} \to \{0,1\}^{\ell(|s|)}\}_{s \in \{0,1\}^*}. \tag{1}$$

An imprortant requirement, which we avoid here, is that the function ensemble be *efficiently computable* (i.e., that there exists an efficient algorithm A such that for every $s \in \{0,1\}^*$ and every $x \in \{0,1\}^{\ell(|s|)}$ it holds that $A(s,x) = f_s(x)$).

Turning to the notion of correlation intractabity, we again consider a (natural) specail case (of a more general definition from [CGH98]). Loosely speaking, a function ensemble F is *correlation intractable with respect to a binary relation R* if every feasible adversary, given a uniformly distributed $s \in \{0,1\}^k$, fails to find an $x \in \{0,1\}^{\ell(|s|)}$ such that $(x, f_s(x)) \in R$, except with negligible probability.

Definition 2 (correlation intractabity): *Let F be as in Definition 1.*

- *Let $R \subseteq \cup_k \{0,1\}^{\ell(k)} \times \{0,1\}^{\ell(k)}$. We say that F is* correlation intractable with respect to R *if for every probabilistic polynomial-time algorithm A it holds that*

$$\Pr_{s \in \{0,1\}^k}[(A(s), f_s(A(s)) \in R] = \mu(k),$$

where the probability is taken uiformly over $s \in \{0,1\}^k$ and the internal coin tosses of A, and μ is some negligible function (i.e., for every positive polynomial p, and all sufficiently large k, it holds that $\mu(k) < 1/p(k)$).
- *Let R be as in Part 1. We say that R is* sparse *if*

$$\max_{x \in \{0,1\}^{\ell(k)}} \left\{ \left| \{y \in \{0,1\}^{\ell(k)} : (x,y) \in R\} \right| \right\} = \mu(k) \cdot 2^{\ell(k)},$$

where μ is some negligible function.
- *We say that F is* correlation intractable *if it is correlation intractable with respect to every sparse relation.*

Note that Part 2 implies that a random oracle is correlation intractable with respect to R (in the sense that for every probabilistic polynomial-time oracle machine M it holds that $\Pr[(M^{\mathcal{O}}(1^k), \mathcal{O}(M^{\mathcal{O}}(1^k)) \in R] = \mu(k)$, where $\mathcal{O} : \{0,1\}^{\ell(k)} \to \{0,1\}^{\ell(k)}$ denotes a random function).

Canetti, Goldreich, and Halevi [CGH98] showed that that no function ensembles (with length $\ell(k) \geq k$) are correlation intractable. In particular, they showed that the function ensemble $F = \{f_s\}$ is not correlation intractable with respect to the "diagonalization" relation $D = \{(x, f_{x'}(x)) : x \in \{0,1\}^*\}$, where x' is a prefix (of adequate length) of x (i.e., $|x| = \ell(|x'|) \geq |x'|$).

1.2 Our Results

In view of the foregoing, we focus on function ensembles with length $\ell : \mathbb{N} \to \mathbb{N}$ such that $\ell(k) \le k$ (and recall that for $\ell(k) < k/2$ no negative results are known). Furthermore, we will focus on the special case of "constant" relations; that is, relations of the form $R = \{(x, y) : x \in \{0, 1\}^* \wedge y \in S \cap \{0, 1\}^{|x|}\}$, for some (sparse) set $S \subset \{0, 1\}^*$. We investigate natural candidates for function ensembles that may be correlation intractable in such a restricted sense. Note that in this case, correlation intractability means the infeasiblity of finding an input x such that $f_s(x) \in S$, where s is given to us as input.

The Failure of Generic Pseudorandom Functions. One natural candidate for restricted notions of correlation intractability is provided by pseudorandom function ensembles (as defined in [GGM84]). However, these ensembles may fail (w.r.t correlation intractability), because they guarantee nothing with respect to adversaries that are given the function's description (i.e., s). Indeed, in general, pseudorandom function ensembles may not be correlation intractable w.r.t some *very simple* relations (e.g., $R_0 = \{(x, 0^{|x|}) : x \in \{0, 1\}^*\}$): The reason is that any pseudorandom function ensemble $\{f_s\}$ can be modified into a pseudorandom function ensemble $\{f'_{r,s}\}$ such that $f'_{r,s}(x) = 0^{|x|}$ if $x = r$ and $f'_{r,s}(x) = f_s(x)$ otherwise. Thus, given the description (r, s) of a function, we can easily find an input x (i.e., $x = r$) such that $(x, f'_{r,s}(x)) \in R_0$.

The Failure of the GGM Construction. Our main interest here is in a specific (natural) construction of pseudorandom functions (based on pseudorandom generators). That is, while one may argue that the aforementioned failure of generic pseudorandom functions is due to a contrived example, we show that a natural construction of pseudorandom functions fails (i.e., it is not correlation intractable w.r.t some simple relations such as the aforementioned R_0). Specifically, we refer to the construction of pseudorandom functions due to Goldreich, Goldwasser, and Micali [GGM84]. Recall that in their construction, hereafter referred to as the GGM construction, a function $f_s : \{0, 1\}^{\ell(|s|)} \to \{0, 1\}^{|s|}$ is define based on a (length doubling) pseudorandom generator G such that

$$f_s(x) \stackrel{\text{def}}{=} G_{x_\ell}(G_{x_{\ell-1}}(\cdots G_{x_1}(s) \cdots)), \tag{2}$$

where $G(z) = G_0(z)G_1(z)$, $\ell \stackrel{\text{def}}{=} \ell(|s|)$, and $x = x_1 \cdots x_\ell \in \{0, 1\}^\ell$. A length preserving version of f_s is obtained by considering only the $\ell(|s|)$-bit long prefix of $f_s(x)$. (Recall that we assume here that $\ell(k) \le k$.) Our main result is:

Theorem 3 (main result): *If there exists pseudorandom generators, then there exists a pseudorandom generator G such that the function ensemble resulting from applying Eq. (2) to G is not correlation intractable with respect to the relation $R_0 = \{(x, 0^{|x|}) : x \in \{0, 1\}^*\}$.*

That is, although the resulting function ensemble is pseudorandom (cf. [GGM84]), given the description s of a function in the ensemble, one can find in polynomial-time an input x such that $f_s(x) = 0^{|x|}$. The result can be easily extended to hitting other relations. The rest of the paper is devoted to establishing Theorem 3.

2 The Overall Plan and an Abstraction

The first observation is that 0^ℓ is likely to have a preimage under f_s, and the central idea is that, *for a carefully constructed G*, this preimage is easy to find when given s. Intuitively, G is constructed such that (1) either $G_0(s)$ or $G_1(s)$ is likely to have a longer all-zero prefix than s, and (2) it is always the case that either $G_0(s)$ or $G_1(s)$ has an all-zero prefix that is at least as long as the one in s.

Notation. (At this point, the reader may think of n as equal k.)[1] For $t = 0, ..., n - 1$, let $S_t \stackrel{\text{def}}{=} \{0^t 1\gamma : \gamma \in \{0,1\}^{n-(t+1)}\}$ be the set of n-bit long strings having a (maximal) all-zero prefix of length t. Let P_t be the set of strings $\alpha\beta \in \{0,1\}^{2n}$ such that $\alpha, \beta \in \cup_{i=0}^{t} S_i$ and either $\alpha \in S_t$ or $\beta \in S_t$. That is:

$$P_t \stackrel{\text{def}}{=} \{\alpha\beta : \alpha, \beta \in (\cup_{i=0}^{t} S_i) \wedge (\alpha \in S_t \vee \beta \in S_t)\} \tag{3}$$

$$= \{\alpha\beta : (\alpha, \beta \in S_t) \vee (\alpha \in S_t \wedge \beta \in \cup_{i=0}^{t-1} S_i) \vee (\alpha \in \cup_{i=0}^{t-1} S_i \wedge \beta \in S_t)\}. \tag{4}$$

Our aim is to construct a pseudorandom generator G such that for every $t \leq \ell$ and $\alpha \in S_t$ it holds that $G(\alpha) \in \cup_{i \geq t} P_i$, and for a constant fraction of $\alpha \in S_t$ it holds that $G(\alpha) \in \cup_{i \geq t+1} P_i$. Intuitively, given $s_\lambda \stackrel{\text{def}}{=} s$ we may find an $x = x_1 \cdots x_\ell$ such that $f_s(x)$ has an all-zero prefix of length $\Omega(\ell)$, by iteratively inspecting both parts of $G(s_{x_1 \cdots x_i})$ for the current $s_{x_1 \cdots x_i}$ and setting x_{i+1} such that $s_{x_1 \cdots x_i x_{i+1}} \stackrel{\text{def}}{=} G_{x_{i+1}}(s_{x_1 \cdots x_i})$ is the part with a longer all-zero prefix.

The Desired Random Mapping. In order to implement and analyze the foregoing idea, we first introduce a random process $\Pi : \{0,1\}^n \rightarrow \{0,1\}^{2n}$ with the intention of satisfying the following three properties:

1. $\Pi(U_n) \equiv U_{2n}$, where U_m denotes the uniform distribution on $\{0,1\}^m$.
2. For every $t \leq \ell$ and $\alpha \in S_t$, it holds that $\Pi(\alpha) \in \cup_{i \geq t} P_i$.
3. For every $t \leq \ell$ and $\alpha \in S_t$, it holds that $\Pr[\Pi(\alpha) \in \cup_{i \geq t+1} P_i] > c$, where $c > 0$ is a universal constant.

One natural way to define Π is to proceed in iterations, starting with $t = 0$. In each iteration, we map seeds in S_t to outcomes in P_t until P_t gets enough probability mass, and map the residual probability mass to $\cup_{i \geq t+1} P_i$ (first to P_{t+1}, next to P_{t+2}, etc). In order to satisfy the foregoing Conditions 1 and 2, it must hold that, for every t, the fraction of n-bit seeds residing in $\cup_{i=0}^{t} S_i$ is at least as big as the fraction of $2n$-bit long outcomes in $\cup_{i=0}^{t} P_i$. In fact, to satisfy Condition 3 the former must be sufficiently bigger than the latter. (Actually, we shall see that Condition 3 follows from the other two conditions.)

We now turn to the analysis of the desired process Π. Let $s_t \stackrel{\text{def}}{=} \Pr[U_n \in S_t] = 2^{-(t+1)}$, and $p_t \stackrel{\text{def}}{=} \Pr[U_{2n} \in P_t]$. By Eq. (3)-(4), it holds that $p_t = s_t^2 + 2s_t \sum_{i=0}^{t-1} s_i$. The following technical claim will play a key role in our analysis.

---

[1] At a later point, it will become clear why we chose to use n rather than k here.

Claim 4 (central technical claim): *For every $t \geq 0$:*

1. $\sum_{i=0}^{t} p_i = \left(\sum_{i=0}^{t} s_i \right)^2$.
2. $\sum_{i=0}^{t} s_i = \frac{1}{1-2^{-(t+1)}} \cdot \sum_{i=0}^{t} p_i > \left(1 + 2^{-(t+1)} \right) \cdot \sum_{i=0}^{t} p_i$.
3. $\Delta_t \stackrel{\text{def}}{=} \sum_{i=0}^{t} s_i - \sum_{i=0}^{t} p_i > \frac{1}{2} \cdot p_{t+1}$. *Furthermore,* $\Delta_t > (1 - 2^{-t}) \cdot p_{t+1}$.

Part 3 is not used in the actual analysis, and so its proof is moved to the Appendix.

Proof: We first establish Part 1:

$$\sum_{i=0}^{t} p_i = \sum_{i=0}^{t} \left(s_i^2 + 2s_i \sum_{j=0}^{i-1} s_j \right)$$

$$= \sum_{i,j \in \{0,\dots,t\}} s_i s_j$$

$$= \left(\sum_{i=0}^{t} s_i \right)^2.$$

Combining Part 1 and $\sum_{i=0}^{t} s_i = \sum_{i=0}^{t} 2^{-(i+1)} = 1 - 2^{-(t+1)}$, we get $\sum_{i=0}^{t} s_i = \left(1 - 2^{-(t+1)} \right)^{-1} \cdot \sum_{i=0}^{t} p_i$. Part 2 follows (using $(1-\epsilon)^{-1} > 1 + \epsilon$ for $\epsilon > 0$). ■

Using Claim 4, it follows that by the time we get to deal with seeds in S_t ($t \geq 1$), we have already spend a probability mass of $\sum_{i=0}^{t-1} s_i - \sum_{i=0}^{t-1} p_i > \frac{1}{2} p_t$ towards covering P_t. Thus, some seeds in S_{t-1} are mapped to P_t (or to $\cup_{i>t} P_i$). The following claim implies that seeds in S_{t-1} are actually mapped to either P_{t-1} or P_t (but never to $\cup_{i>t} P_i$).

Claim 5 (another technical claim): $\sum_{i=0}^{t} s_i = \sum_{i=0}^{t+1} p_i - 2^{-(2t+4)} < \sum_{i=0}^{t+1} p_i$

Proof: Using Part 1 of Claim 4 (and $s_j = 2^{-(j+1)}$), we get:

$$\sum_{i=0}^{t+1} p_i = \left(\sum_{i=0}^{t+1} s_i \right)^2$$

$$= \left(1 - 2^{-(t+2)} \right)^2$$

$$= 1 - 2^{-(t+1)} + 2^{-(2t+4)}$$

$$= 2^{-(2t+4)} + \sum_{i=0}^{t} s_i$$

and the current claim follows. ■

The Implementation of Π. Given Claims 4 and 5, we explicitly define the process Π. On input $\alpha \in S_0$, with probability $p_0/s_0 = 1/2$, we output a uniformly selected element of P_0, otherwise we output a uniformly selected element of P_1. For $t \geq 1$, on input $\alpha \in S_t$, we first compute $\Delta_{t-1} = \sum_{i=0}^{t-1} s_i - \sum_{i=0}^{t-1} p_i$. (Note that by Claims 4 and 5 it holds that $0 < \Delta_{t-1} < p_t$, and $p_t - \Delta_{t-1} = s_t - \Delta_t < s_t$ follows.) With probability $(p_t - \Delta_{t-1})/s_t$, we output a uniformly selected element of P_t, otherwise we output a uniformly selected element of P_{t+1}. Indeed, $0 < (p_t - \Delta_{t-1})/s_t < 1$. Thus, Π is well-defined.

Note that Π can be implemented in probabilistic polynomial-time. Combining Claims 4 and 5, we get:

Proposition 6 (Π satisfies the desired properties):

1. $\Pi(U_n) \equiv U_{2n}$, where U_m denotes the uniform distribution on $\{0, 1\}^m$.
2. For every $t \leq \ell$ and $\alpha \in S_t$, it holds that $\Pi(\alpha) \in P_t \cup P_{t+1}$.
3. For every $t \leq \ell$ and $\alpha \in S_t$, it holds that $\Pr[\Pi(\alpha) \in P_{t+1}] \geq 1/2$.

Part 3 (which follows from Part 3 of Claim 4) is not used in the actual analysis and is only given for intuition.

Proof: Part 2 is immediate by the construction. It is also clear that $\Pi(U_n)$ is uniform over each of the P_t's. Thus, to prove Part 1 it suffices to show that, for every t, it holds that $\Pr[\Pi(U_n) \in P_t] = p_t$. In proving this, we use Part 2 (i.e., $\Pi(\alpha) \in P_t \cup P_{t+1}$ for every $\alpha \in S_t$). We first consider the case of $t = 0$, and get

$$\Pr[\Pi(U_n) \in P_0] = \Pr[U_n \in S_0] \cdot \Pr[\Pi(U_n) \in P_0 | U_n \in S_0]$$
$$= s_0 \cdot \frac{p_0}{s_0} = p_0.$$

For $t \geq 1$ (using $\Delta_{-1} \overset{\text{def}}{=} 0$ in case $t = 1$), we have

$$\Pr[\Pi(U_n) \in P_t] = \Pr[U_n \in S_t] \cdot \Pr[\Pi(U_n) \in P_t | U_n \in S_t]$$
$$+ \Pr[U_n \in S_{t-1}] \cdot \Pr[\Pi(U_n) \in P_t | U_n \in S_{t-1}]$$
$$= s_t \cdot \frac{p_t - \Delta_{t-1}}{s_t} + s_{t-1} \cdot \left(1 - \frac{p_{t-1} - \Delta_{t-2}}{s_{t-1}}\right)$$
$$= p_t - \Delta_{t-1} + s_{t-1} - p_{t-1} + \Delta_{t-2}$$
$$= p_t,$$

since $\Delta_{t-1} = \Delta_{t-2} + s_{t-1} - p_{t-1}$.

Part 3 follows by noting that for every $\alpha \in S_t$ (with $t \geq 1$),

$$\Pr[\Pi(\alpha) \in P_{t+1}] = 1 - \frac{p_t - \Delta_{t-1}}{s_t}$$
$$= \frac{\sum_{i=0}^{t} s_t - \sum_{i=0}^{t} p_i}{s_t}$$
$$> \frac{(1 - 2^{-t}) \cdot s_t}{s_t} \geq \frac{1}{2}$$

where the strict inequality is due to $\Delta_t > (1 - 2^{-t}) \cdot 2^{-(t+1)} = (1 - 2^{-t}) \cdot s_t$ (which is established in the first paragraph of the Appendix). For $\alpha \in S_0$, it holds that $\Pr[\Pi(\alpha) \in P_1] = 1 - (p_0/s_0) = 1/2$. ∎

The Randomly-labeled Tree: We consider a depth ℓ binary tree with nodes labeled by n-bit long strings. The root is labeled with a uniformly selected string, and if a node is labeled with α then its children are labeled with the corresponding parts of $\Pi(\alpha)$. (The root is said to be in level 0 and the 2^ℓ leaves are in level ℓ.)

Using induction on $i = 0, 1..., \ell$ (and relying on Part 1 of Proposition 6), it follows that the nodes at level i are assigned independently and uniformly distributed labels. Specifically, suppose that the claim holds for level i, then using Part 1 of Proposition 6 the claim holds for level $i + 1$. On the other hand, by Part 2 of Proposition 6, the labels along each path from the root to a leaf belong to S_j's such that the sequence of j's increases by at most one unit at each step.

Now, on the one hand, with probability $s_0 + s_1 = 3/4$, the (level 0) root has a label in $S_0 \cup S_1$. On the other hand, with probability $1 - (1 - s_\ell)^{2^\ell} = 1 - (1 - 2^{-(\ell+1)})^{2^\ell} > 0.39$, there exists a (level ℓ) leaf with label in S_ℓ. We conclude that, with probability at least $0.39 - 0.25 = 0.14$, the root has label in $S_0 \cup S_1$ and there exist a leaf with a label in S_ℓ. Furthermore, due to the mild-increasing property of the label sequence along each path, the ith intermediate node on the path from the root to this leaf must have a label in $S_i \cup S_{i+1}$.[2] On the other hand, the expected number of level i nodes with label in $S_i \cup S_{i+1}$ is $2^i \cdot (2^{-(i+1)} + 2^{-(i+2)}) = 3/4$. Thus, except with exponentially vanishing probability, level i contains less than n nodes with label in $S_i \cup S_{i+1}$. To summarize, with probability at least 0.13, the following good event holds:

1. The root has label in $S_0 \cup S_1$.
2. There exist a leaf with a label in S_ℓ. Furthermore, the ith intermediate node on the path from the root to this leaf has a label in $S_i \cup S_{i+1}$.
3. For every $i \leq \ell$, level i has at most n nodes that have a label in $S_i \cup S_{i+1}$.

The following search procedure is "geared towards" the foregoing good event.

The (Ideal) Search Procedure: Starting at the root, proceed in a DFS-like manner according to the following rule: *If the currently reached node is at level i and has a level not in $S_i \cup S_{i+1}$, then backtrack immediately, else develop it according to the standard DFS-rule.* If we ever reach a leaf having a label in S_ℓ, then the search is considered successful.

Assuming that the good event holds, the search is successful. Furthermore, in this case the search has visited at most $2n$ nodes at each level (i.e., the children of parents that were DFS-developed), and so the complexity is bounded

---

[2] Recall that a node with label in S_j has children with labels in $\cup_{k=0}^{j+1} S_k$. Since the root has label in $S_0 \cup S_1$, each node at level i has a label in $\cup_{k=0}^{i+1} S_k$. Furthermore, since the specific leaf on the said path has a label in S_ℓ, the ith intermediate node on the said path cannot have a label in $\cup_{k=0}^{i-1} S_k$.

by $O(\ell \cdot n)$. In fact, the complexity analysis depends only on the third condition (in the definition of a good event), and thus holds except for with exponentially vanishing probability.

3   The Actual Construction

Recall that we have given a probabilistic polynomial-time implementation of Π. We now consider a deterministic polynomial-time algorithm Π' satisfying $\Pi'(\alpha, U_m) \equiv \Pi(\alpha)$, where $m = \text{poly}(|\alpha|)$. Next, using suitable pseudorandom generators G' and G'' (i.e., $G' : \{0,1\}^n \to \{0,1\}^m$ and $G'' : \{0,1\}^n \to \{0,1\}^{4n}$), we replace $\Pi' : \{0,1\}^{n+m} \to \{0,1\}^{2n}$ by $\Pi'' : \{0,1\}^{n+2n} \to \{0,1\}^{2\cdot(n+2n)}$ such that

$$\Pi''(\alpha, r'r'') = ((\alpha_1, r_1), (\alpha_2, r_2)) \tag{5}$$
$$\text{where } (\alpha_1, \alpha_2) = \Pi'(\alpha, G'(r')) \text{ and } (r_1, r_2) = G''(r'') \tag{6}$$

That is, $|r_1| = |r_2| = |r'r''|$ and $|r'| = |r''| = |\alpha|$.

Theorem 7 (Theorem 3, specialized): *Let $\ell : \mathbb{N} \to \mathbb{N}$ such that $\ell(k) \le k$ and let $G \stackrel{\text{def}}{=} \Pi''$. Then:*

1. *G is a pseudorandom generator.*
2. *Let $f'_s : \{0,1\}^{\ell(|s|)} \to \{0,1\}^{|s|}$ be defined by applying Eq. (2) to G, and let $f_s : \{0,1\}^{\ell(|s|)} \to \{0,1\}^{\ell(|s|)}$ be defined by letting $f_s(x)$ equal the $\ell(|s|)$-bit long prefix of $f'_s(x)$. Then, the function ensemble $\{f_s\}_{s\in\{0,1\}^*}$ is not correlation intractable with respect to the relation $R_0 = \{(x, 0^{|x|}) : x \in \{0,1\}^*\}$. That is, there exists a probabilistic polynomial-time algorithm that given a uniformly distributed $s \in \{0,1\}^n$, finds with probability at least $1/10$ a string $x \in \{0,1\}^{\ell(|s|)}$ such that $f_s(x) = 0^{\ell(|s|)}$.*

Theorem 3 follows.

Proof: In order to prove Part 1 we first observe that $\Pi'(U_n, U_m) \equiv U_{2n}$. Letting U_n, U'_n, U''_n denote independent random variables each uniformly distributed in $\{0,1\}^n$, we recall that $\Pi''(U_n, U'_n U''_n) = ((Z_1, R_1), (Z_n, R_n))$, where $(Z_1, Z_2) \stackrel{\text{def}}{=} \Pi'(U_n, G'(U'_n))$ and $(R_1, R_2) \stackrel{\text{def}}{=} G''(U''_n)$. Thus, $\Pi''(U_n, U'_n U''_n)$ is computationally indistinguishable from $((Z'_1, R'_1), (Z'_n, R'_n))$, where $(Z'_1, Z'_2) \stackrel{\text{def}}{=} \Pi'(U_n, U_m)$ and (R'_1, R'_2) is uniformly distributed over $\{0,1\}^{2n} \times \{0,1\}^{2n}$. It follows that $G(U_{3n}) \equiv \Pi''(U_n, U'_n U''_n)$ is computationally indistinguishable from $((U'_n, U'_{2n}), (U''_n, U''_{2n}))$. Since G is computable in polynomial-time, and $|G(U_{3n})| = 6n$, Part 1 follows.

In order to prove Part 2, we consider an algorithm that on input $s \in \{0,1\}^{3n}$ invokes the ideal search procedure described at the end of Section 2, while providing it with labels of an imaginary depth $\ell = \ell(n)$ binary tree as follows. The label of the root is the n-bit long prefix of s, and the $2n$-bit long suffix is called the **secret** of the root. If an internal node has label $\alpha \in \{0,1\}^n$ and secret $s's'' \in \{0,1\}^{2n}$, then its children will have labels corresponding to the two n-bit long parts of $\Pi'(\alpha, G'(s'))$ and secrets corresponding to the two $2n$-bit long parts of $G''(s'')$. We stress that the search procedure is only given the labels of nodes (at its request), but it is not given the nodes' secrets. Note that the way in which we label the nodes corresponds to the way the function ensemble $\{f_s\}$ is defined (using $G = \Pi''$).

Recall that the search procedure succeeds with probability at least 0.13 on the randomly-label tree, called the **ideal setting**, where the children of a node labeled by α are assigned labels that corresponding to the two n-bit long parts of $\Pi'(\alpha, U_m)$. Our aim is to show that approximately the same must occur in the foregoing **real setting**, where the tree is labeled according to Π'' (or, equivalently, according to $\Pi'(\cdot, G'(\cdot))$ and $G''(\cdot)$). To prove this claim, consider a **hybrid setting** in which *all nodes are associated uniformly distributed secrets* (rather than secrets derived by applying G'' to the second part of their parent's secret), and the children of a node labeled by α are assigned labels that corresponding to the two n-bit long parts of $\Pi'(\alpha, G'(s'))$, where s' is the first part of the parent's secret (and the second part is never used). We observe that:

1. The success probability of the search in the ideal setting is approximately the same as its success in the hybrid setting.

   Otherwise, we derive a contradiction to the hypothesis that G' is a pseudo-random generator. Specifically, we will show how to distinguish $n \cdot \ell$ samples of the distribution $G'(U_n)$ from $n \cdot \ell$ samples of the distribution U_m. Given a sequence of samples, we run the search procedure while feeding it with labels generated on-the-fly as follows.

   – The root is assigned a uniformly distributed label, and labels that were assigned to nodes are used whenever the node is visited.

   – When reaching a node (e.g., the root) for the first time, we assign labels to its children by using the next unused sample. Specifically, if the new node has label $\alpha \in \{0,1\}^n$ and the next sample in the input sequence is $s' \in \{0,1\}^m$ then we assign its children (as labels) the corresponding parts of $\Pi'(\alpha, s') \in \{0,1\}^{2n}$.

   Note that when the input sequence is taken from U_m, the foregoing process describes the ideal setting, whereas when the input sequence is taken from $G'(U_n)$ we get the hybrid setting.

2. The success probability of the search in the real setting is approximately the same as its success in the hybrid setting.

   Otherwise, we derive a contradiction to the hypothesis that G'' is a pseudo-random generator by considering ℓ additional hybrid settings. For $i = 1, ..., \ell$, the ith hybrid (or i-hybrid) consists of running the foregoing search while feeding it with labels generated on-the-fly as follows. The label of a node al level $j < i$ is generated as in the hybrid setting; that is, these nodes are assigned uniformly distributed secrets (and the children of such a node labeled by α are assigned labels that corresponding to the two n-bit long parts of $\Pi'(\alpha, G'(s'))$, where s' is the first part of the parent's secret). On the other hand, the label of a node al level $j \geq i$ is generated as in the real setting; that is, these nodes are assigned secrets that are derived from the second part of their parent's secret (and are assigned labels exactly as in case $j < i$). That is, if a node at level $j - 1$ has secret $s's''$, then its children are always labeled according to $\Pi'(\alpha, G'(s'))$, whereas the secrets that they are assigned are either uniformly distributed or derived from $G''(s'')$ depending on whether $j < i$ or $j \geq i$. Note that the ℓ-hybrid corresponds to the hybrid setting, whereas the 1-hybrid corresponds to the real setting. Thus, it suffices to show that for every $i \in \{1, ..., \ell - 1\}$, the i-hybrid and $(i + 1)$-hybrid are computationally indistinguishable. This is shown by using a potential distinguisher to violate the pseudorandomness of G''.

Given a distinguisher of the i-hybrid and $(i + 1)$-hybrid, we will show how to distinguish $n \cdot \ell$ samples of the distribution $G''(U_n)$ from $n \cdot \ell$ samples of the distribution U_{4n}. Specifically, given a sequence of samples, we run the search procedure while feeding it with secrets and labels generated on-the-fly as follows. When required to provide a label to a newly visited node we always provide the label according to $\Pi'(\alpha, G'(s'))$, where s' is the first part of the parent's secret (and α is the parent's label). The important issue is the generation of secrets:

- Nodes at level $j \leq i$ are assigned uniformly distributed secrets.
- Nodes at level $j \geq i + 2$ are assigned secrets according to $G''(s'')$ where s'' is the second part of their parent's secret.
- Nodes at level $i + 1$ are assigned secrets (on the fly) that equal the corresponding part of the next unused sample in the input sequence; that is, when a node at level i is first visited, its two children are assigned secrets according to the two parts of the next unused sample.

Note that when the input sequence is taken from U_{4n}, the foregoing process describes the $(i + 1)$-hybrid, whereas when the input sequence is taken from $G''(U_n)$ we get the i-hybrid (although the secrets at level $i + 1$ do not fit the second part of the secrets at level i but rather a re-randomization of the latter).

Combining the two foregoing observations, we conclude that in the real setting the search procedure is successful with probability at least 0.1. Using the correspondence of the real setting to an attack on the function ensemble $\{f_s\}$, Part 2 (and so the entire theorem) follows.   ∎

Acknowledgments. The question was originally posed by Silvio Micali (in the early 1990's if I recall correctly), and re-posed by Boaz Barak in Summer 2001. I am grateful to both of them.

References

[CGH98]   Canetti, R., Goldreich, O., Halevi, S.: The Random Oracle Methodology. In: 30th STOC, pp. 209–218 (1998) (revisited)

[GGM84]   Goldreich, O., Goldwasser, S., Micali, S.: How to Construct Random Functions. JACM 33(4), 792–807 (1986)

Appendix: Proof of Part 3 of Claim 4

Using Part 2, we have

$$\sum_{i=0}^{t} s_i - \sum_{i=0}^{t} p_i > 2^{-(t+1)} \cdot \sum_{i=0}^{t} p_i$$

$$= 2^{-(t+1)} \cdot \left(\sum_{i=0}^{t} s_i \right)^2$$

$$= 2^{-(t+1)} \cdot \left(1 - 2^{-(t+1)} \right)^2$$

$$> 2^{-(t+1)} \cdot \left(1 - 2^{-t} \right).$$

On the other hand,

$$p_{t+1} = s_{t+1}^2 + 2 s_{t+1} \sum_{i=0}^{t} s_i$$

$$= s_{t+1} \cdot \left(s_{t+1} + 2 \sum_{i=0}^{t} s_i \right)$$

$$= 2^{-(t+2)} \cdot \left(2^{-(t+2)} + 2 \cdot \left(1 - 2^{-(t+1)} \right) \right)$$

$$= 2^{-(t+1)} \cdot \left(1 - 2^{-(t+1)} + 2^{-(t+3)} \right)$$

$$= 2^{-(t+1)} \cdot \left(1 - \frac{3}{8} \cdot 2^{-t} \right).$$

Combining $\Delta_t = \sum_{i=0}^{t} s_i - \sum_{i=0}^{t} p_i > 2^{-(t+1)} \cdot (1 - 2^{-t})$ with $p_{t+1} = 2^{-(t+1)} \cdot \left(1 - \frac{3}{8} \cdot 2^{-t} \right)$, we get

$$\Delta_t > \frac{1 - 2^{-t}}{1 - \frac{3}{8} \cdot 2^{-t}} \cdot p_{t+1}$$

$$= \left(1 - \frac{\frac{5}{8} \cdot 2^{-t}}{1 - \frac{3}{8} \cdot 2^{-t}} \right) \cdot p_{t+1}$$

$$> \left(1 - \frac{\frac{5}{8} \cdot 2^{-t}}{1 - \frac{3}{8}} \right) \cdot p_{t+1}$$

$$= \left(1 - 2^{-t} \right) \cdot p_{t+1}.$$

Thus, $\Delta_t > \frac{1}{2} p_{t+1}$, provided $t \geq 1$. For $t = 0$, we note that $\Delta_0 = s_0 - p_0 = \frac{1}{2} - \frac{1}{4} = \frac{1}{4}$ whereas $p_1 = \frac{5}{16}$ and so $\Delta_0 = \frac{4}{5} \cdot p_1$. Part 3 follows. ∎

From Logarithmic Advice to Single-Bit Advice

Oded Goldreich, Madhu Sudan, and Luca Trevisan

Abstract. Building on Barak's work (*Random'02*), Fortnow and San-thanam (*FOCS'04*) proved a time hierarchy for probabilistic machines with one bit of advice. Their argument is based on an implicit translation technique, which allow to translate separation results for short (say logarithmic) advice (as shown by Barak) into separations for a single-bit advice. In this note, we make this technique explicit, by introducing an adequate translation lemma.

Keywords: Machines that take advice, separations among complexity classes.

An early version of this work appeared as TR04-093 of *ECCC*. The current revision is quite minimal.

1 Introduction and High Level Description

Trying to address the open problem of providing a probabilistic time hierarchy, Barak [1] presented a time hierarchy for slightly non-uniform probabilistic machines. Specifically, he showed that, in presence of double-logarithmic advice, there exists a hierarchy of probabilistic polynomial-time. Subsequently, Fortnow and Santhanam [2] showed that a similar hierarchy holds in the presence of a single-bit advice. Their argument is based on an implicit translation technique, which allow to translate separation results for short (say logarithmic) advice into separations for a single-bit advice. In this note, we make this technique explicit, by introducing an adequate translation lemma and showing that applying it to Barak's result [1] yields the aforementioned result of [2].

Interestingly (as in [2]), we rely on the fact that Barak [1] actually shows a time separation that holds even when the more time-restricted machine is given a somewhat longer advice. In contrast, arguably, the more natural statement of such results refers to machines that use the same advice length.[1]

The basic idea underlying the proof in [2] is that short advice can be incorporated in the (length of the) instance of a padded set, while using a single bit of advice to indicate whether or not the resulting instance length encodes a valid advice. For this to work, the length of the resulting instance should indicate a

---

[1] That is, in order to show, say, that $\text{BPtime}(n^3)/1$ is not contained in $\text{BPtime}(n^2)/1$, we use the fact that Barak showed that $\text{BPtime}(n^6)/\log n$ is not contained in $\text{BPtime}(n^4)/2\log n$ (rather than that $\text{BPtime}(n^6)/\log n$ is not contained in $\text{BPtime}(n^4)/\log n$).

O. Goldreich et al.: Studies in Complexity and Cryptography, LNCS 6650, pp. 109–113, 2011.

unique length of the original instance as well as a value of a corresponding advice (for this instance length).

Suppose we wish to treat a set S that is decidable (within some time bound) using *eight* bits of advice. Viewing the possible values of the advice as integers in $\{0, 1, ..., 255\}$, we define a (padded) set S' as follows: the string $x0^{255|x|+i}$ is in S' if and only if $x \in S$ and i is an adequate advice for instances of length $|x|$. Note that S' can be decided using a single bit of advice that indicates whether the instance length encodes a valid advice for S. Specifically, the advice bit for length m instances (of S') is 1 if and only if $m \bmod 256$ is a valid advice for instances of length $\lfloor m/256 \rfloor$ (of S). Thus, on input $y = x0^{255|x|+i}$, where $i \in \{0, ..., 255\}$, we accept if and only if the advice bit is 1 and the original machine accepts x when given advice i.

Note that we should also show that if S is undecidable using less time (and, say, *nine* bits of advice), then S' is correspondingly hard (even using a single bit of advice). This is shown by using a machine for deciding S' as a subroutine for deciding S, while using part of the advice (given for deciding S) for determining an adequate instance for S'. In other words, we present a non-uniform reduction of S to S', where the non-uniformity is accounted for by the longer advice allowed in deciding S.

2   Preliminaries

We consider advice-taking probabilistic machines, denoting by $M(a, x)$ the output distribution of machine M on input x and advice a. We denote by BPtime$(T)/A$ the class of sets decidable by advice-taking probabilistic machines of time complexity T and advice complexity A. That is, $S \in$ BPtime$(T)/A$ if there exists a probabilistic machine M and a sequence of strings $(a_n)_{n \in \mathbb{N}}$ such that the following conditions hold:

1. For every $n \in \mathbb{N}$, it holds that $|a_n| = A(n)$.
2. For every $x \in \{0, 1\}^*$, on input x and advice $a_{|x|}$, machine M makes at most $T(|x|)$ steps.
3. For every $x \in \{0, 1\}^*$, it holds that $\Pr[M(a_{|x|}, x) = \chi_S(x)] \geq 2/3$, where $\chi_S(x) = 1$ if $x \in S$ and $\chi_S(x) = 0$ otherwise.

We assume that the machine model supports some trivial computations with little overhead. Specifically, we refer to computing the square root of the length of the input in linear time. Our results hold with minor modifications in case the machine model is less flexible (e.g., if computing the square root of the length of the input requires quadratic time).

To simplify the presentation, we will associate binary strings with the integers that they represents. That is, the ℓ-bit long binary string $\sigma_{\ell-1} \cdots \sigma_0$ will be associated with the integer $\sum_{j=0}^{\ell-1} \sigma_j \cdot 2^j$. Thus, when writing $0^{\sigma_{\ell-1} \cdots \sigma_0}$, we mean a binary string consisting of $\sum_{j=0}^{\ell-1} \sigma_j \cdot 2^j$ zeros.

3   Detailed Technical Presentation

We state our translation lemma for probabilistic machines, and note that an analogous lemma holds for deterministic (and non-deterministic) machines.

Lemma 1 (Translation Lemma): *Suppose that S is a set that is decided by some advice-taking probabilistic machine M in time $T_M(n)$ using $A_M(n) \leq \lfloor \log_2 n \rfloor$ bits of advice, where n denotes the length of the instance of S. Suppose further that S is not decided by any $a(n)$-advice probabilistic machine in time $t(n)$, where $a(n) \geq A_M(n)$. Then, there exists a set $S' = S'_M$ that is decided in probabilistic time T' using a single bit of advice, where $T'(m) = T_M(\lfloor \sqrt{m} \rfloor) + m$, but is not decidable by any $(a(\lfloor \sqrt{m} \rfloor) - A_M(\lfloor \sqrt{m} \rfloor))$-advice probabilistic machine in time $t(\lfloor \sqrt{m} \rfloor) - m$, where m denotes the length of the instance of S'.*

Needless to say, the lemma can be generalized to handle $A_M(n) = O(\log n)$, in which case $\lfloor \sqrt{m} \rfloor$ should be replaced by $m^{1/O(1)}$.

3.1   Using the Translation Lemma

Before proving the Translation Lemma, let us spell-out its main implication.

Corollary 2 (reducing non-uniformity in BPtime separations): *Let $T, A, t, a :$ $\mathbb{N} \rightarrow \mathbb{N}$ such that $a(n) \geq A(n)$. If BPtime$(T)/A$ contains sets not in BPtime$(t)/a$, then BPtime$(T')/1$ contains sets not in BPtime$(t')/a'$, where $T'(m) \stackrel{\text{def}}{=} T(\lfloor \sqrt{m} \rfloor) + m$, $t'(m) \stackrel{\text{def}}{=} t(\lfloor \sqrt{m} \rfloor) - m$ and $a'(m) \stackrel{\text{def}}{=} a(\lfloor \sqrt{m} \rfloor) - A(\lfloor \sqrt{m} \rfloor)$.*

For example, we can apply Corollary 2 to Barak's result [1] that asserts the existence of a set S in, say, $($BPtime$(n^6)/\log \log n) \setminus ($BPtime$(n^4)/\log n)$. Doing so, we conclude that there exists a set in $($BPtime$(m^3)/1) \setminus ($BPtime$(m^2)/(0.5 \log m - \log \log m))$, which in particular implies BPtime$(m^2)/1 \subset$ BPtime$(m^3)/1$. Thus, we can translate Barak's separations, which refer to probabilistic machines with logarithmic advice, into separations that refer to probabilistic machines with a single bit of advice, as established by Fortnow and Santhanam [2]. (This consequence is not surprising, because the Translation Lemma makes explicit the ideas in [2].)

Note that in order to obtain an interesting consequence out of Corollary 2, we need $a(n) \geq A(n) + 1$. In contrast, using $a(n) = A(n)$ implies that BPtime$(T')/1$ contains sets not in BPtime(t'), which holds regardless of the hypothesis and for any choice of $T' > 0$ and t' (even for $t' \gg T'$).

3.2   Proving the Translation Lemma

Recall that M decides S in time T_M, using advice of length A_M, where $A_M(n) \leq \lfloor \log_2 n \rfloor$. Fixing a sequence of advice strings $(a_n)_{a \in \mathbb{N}}$ for machine M, we define S' depending on this sequence. Specifically,

$$S' \stackrel{\text{def}}{=} \{x0^{(|x|-1)|x|+a_{|x|}} : x \in S\}. \tag{1}$$

That is, $y = x0^{(|x|-1)|x|+i} \in S'$ if and only if it holds that $x \in S$ and $a_{|x|} = i$. Observe that $|x0^{(|x|-1)|x|+i}| = |x|^2 + i$ and that, for every $m \in \{n^2 + 0, ..., n^2 + 2^{A_M(n)} - 1\}$ (which in turn is contained in $\{n^2, ..., (n+1)^2 - 1\}$), it holds that $\lfloor \sqrt{m} \rfloor = n$. In what follows, n (resp., m) will always denote the length of instances to S (resp., S').

We first show that S' is decidable by a probabilistic machine M' taking one bit of advice and running in time $T_M(\lfloor \sqrt{m} \rfloor) + m$. Machine M' checks whether its input $y \in \{0,1\}^m$ has the form $x0^{(n-1)n+i}$, where $|x| = n = \lfloor \sqrt{m} \rfloor$ and $i < n$, and otherwise rejects y up-front. Given the advice bit σ_m, machine M' always rejects if $\sigma_m = 0$ and invokes M on input x and advice i (viewed as an $A_M(n)$-bit long string) otherwise. Thus, M' accepts $y = x0^{(|x|+1)|x|+i}$ using advice σ_m if and only if $\sigma_m = 1$ and M accepts x using advice i. The advice (bit) σ_m regarding m-bit inputs is determined in correspondence to the aforementioned parsing: the advice bit is 1 if and only if $m = \lfloor \sqrt{m} \rfloor^2 + a_{\lfloor \sqrt{m} \rfloor}$. Indeed, this setting of the advice σ_m guarantees that M' accepts $y = x0^{(|x|-1)|x|+i}$ if and only if $x \in S$ and $i = a_{|x|}$. Thus, using adequate advice, M' decides S'. Indeed, as required, the running time of M' is $m + T_M(\lfloor \sqrt{m} \rfloor)$, where m steps are used to parse y (into x and i) and $T_M(|x|)$ steps are used to emulate $M(i, x)$.

We next show that S' is not decidable by any probabilistic machine that runs in time $t(\lfloor \sqrt{m} \rfloor) - m$ and takes a $(a(\lfloor \sqrt{m} \rfloor) - A_M(\lfloor \sqrt{m} \rfloor))$-bit long advice. Actually, for any monotonically non-decreasing functions t' and a', we will show that if S' is decidable by some probabilistic machine that runs in time $t'(m)$ and takes $a'(m)$ bits of advice, then S is decidable by a probabilistic machine that runs in time $t''(n) = t'(n^2 + n) + n^2$ and takes $a''(n) = A_M(n) + a'(n^2 + n)$ bits of advice.[2] Suppose that M' is a machine deciding S' as in the hypothesis, and let $\mathrm{adv}_{M'}(m)$ be the advice it uses for m-bit inputs. Then consider the following machine M'' (designed to decide S) whose advice on inputs of length n is the pair $a''_n = (a_n, \mathrm{adv}_{M'}(n^2 + a_n))$. On input x and advice (i, j), machine M'' invokes M' on input $x0^{(|x|-1)|x|+i}$ with advice j. Thus, M'' accepts x when given the (adequate) advice $a''_{|x|}$ if and only if M' accepts $x0^{(|x|-1)|x|+a_{|x|}}$ when given the advice $\mathrm{adv}_{M'}(|x|^2 + a_{|x|})$. It follows that M'' decides S, and does so within the stated complexities. ∎

Digest: We defined S' based not only on S but rather based on an adequate advice sequence $(a_n)_{n \in \mathbb{N}}$ that vouches that $S \in \mathrm{BPtime}(T)/A$ (via a machine M). Once S' is defined, the proof proceeds in two steps:

1. Relying on the hypothesis that M decides S in time T using advice of length A, we establish that $S' \in \mathrm{BPtime}(T')/1$, where $T'(m) = T(\lfloor \sqrt{m} \rfloor) + m$.

   The advice-bit for S' is used in order to facilitate the partition of the instances of S' into two sets: a set of instances $x0^{(|x|-1)|x|+i}$ that satisfy

---

[2] Indeed, suppose that $t'(m) = t(\lfloor \sqrt{m} \rfloor) - m$ and $a'(m) = a(\lfloor \sqrt{m} \rfloor) - A_M(\lfloor \sqrt{m} \rfloor)$, then $t''(n) = t'(n^2 + n) + n^2 = (t(\lfloor \sqrt{n^2 + n} \rfloor) - (n^2 + n)) + n^2 < t(n)$ and $a''(n) = A_M(n) + a'(n^2 + n) = A_M(n) + (a(n) - A_M(n)) = a(n)$, in contradiction to the lemma's hypothesis.

$i = a_{|x|}$, and a set of instances that do not satisfy this condition. Machine M is invoked only for instances of the first type, and instances of the second type are rejected up-front.

2. Assuming that $S' \in \mathrm{BPtime}(t')/a'$, we establish that $S \in \mathrm{BPtime}(t)/a$, where $t(n) = t'(n^2 + n) + n^2$ and $a(n) = A(n) + a'(n^2 + n)$.

This is done by "reducing" the problem of "deciding S with $a(n)$ bits of advice" to the problem of "deciding S' with $a'(m)$ bits of advice", while the reduction itself uses $A(n) = a(n) - a'(m)$ bits of advice.

4   Subsequent Work

We mention a subsequent related work by van Melkebeek and Pervyshev [3], which provides a direct proof of a more general result. We still feel that there is interest in the approach taken in the current work (i.e., the translation lemma and its proof).

References

1. Barak, B.: A Probabilistic-Time Hierarchy Theorem for Slightly Non-uniform Algorithms. In: Rolim, J.D.P., Vadhan, S.P. (eds.) RANDOM 2002. LNCS, vol. 2483, pp. 194–208. Springer, Heidelberg (2002)
2. Fortnow, L., Santhanam, R.: Hierarchy theorems for probabilistic polynomial time. In: 45th FOCS, pp. 316–324 (2004)
3. van Melkebeek, D., Pervyshev, K.: A Generic Time Hierarchy for Semantic Models with One Bit of Advice. Computational Complexity 16, 139–179 (2007)

On Probabilistic versus Deterministic Provers in the Definition of Proofs of Knowledge

Mihir Bellare and Oded Goldreich

Abstract. This article points out a gap between two natural formulations of the concept of a proof of knowledge, and shows that in all natural cases (e.g., NP-statements) this gap can be bridged. The aforementioned formulations differ by whether they refer to (all possible) *probabilistic* or *deterministic* prover strategies. Unlike in the rest of cryptography, in the current context, the obvious transformation of probabilistic strategies to deterministic strategies does not seem to suffice *per se*. The source of trouble is "bad interaction" between the expectation operator and other operators, which appear in the definition of a proof of knowledge (reviewed here).

Keywords: Proof of Knowledge, Probabilistic Proof Systems, Probabilism versus Determinism, Expected Running Time.

An early version of this work appeared as TR06-136 of *ECCC*. The current revision is quite minimal.

1   Introduction

The concept of a "proof of knowledge" was informally introduced by Goldwasser, Micali and Rackoff [4], and plays an important role in the design of cryptographic schemes and protocols (see, e.g., [2,3]). This article refers to the common formulation of the aforementioned concept, which was given in [1].

Loosely speaking, the definition of a proof of knowledge requires the existence of a "knowledge extractor" that, when given access to any strategy, outputs the relevant information within (expected) time that inversely proportional to the probability that the given strategy convinces the knowledge verifier. Schematically, the definition of a proof of knowledge *requires something with respect to any strategy*.

The issue addressed in this article is the following. Usually, in definitions of the aforementioned type, it does not matter whether one quantifies over all probabilistic strategies or over all deterministic strategies. The reason is that, usually, satisfying the more restricted definition (which refers only to all deterministic strategies) immediately implies satisfying the general definition (which refers to all probabilistic strategies). Unfortunately, this does not seem to be the case in the current setting (of the definition of proofs of knowledge).

O. Goldreich et al.: Studies in Complexity and Cryptography, LNCS 6650, pp. 114–123, 2011.

1.1   The Source of Trouble

In this subsection we provide a high-level description of the technical problem addressed in this work. We re-iterate this explanation, using more precise style after presenting the relevant definitions (in Section 2).

To clarify the source of trouble, let us first consider one of the many settings in which the problem does not arise; specifically, we consider the setting of zero-knowledge. In this case, the ability to simulate (in a black-box manner) any deterministic verifier strategy, implies the ability to simulate any probabilistic verifier strategy. The same holds also when we restrict attention to strategies that can be implemented by polynomial-size circuits. The reason is that given any probabilistic strategy, we may consider all residual deterministic strategies (obtained by all possible fixing of the strategy's coins), and obtain the desired simulation (for the probabilistic strategy) by combining all the corresponding simulations (i.e., of the residual deterministic strategies).

This simple argument (*per se*) fails when applied in the current context (of proofs of knowledge). Indeed, we can consider all residual deterministic prover strategies that emerge from a given probabilistic prover strategy, and we can combine the corresponding extraction procedures, but the combined procedure does not necessarily run in time that is inversely proportional to the probability that this prover convinces the verifier. For example, suppose that on input x, with probability $\frac{1}{2}$ (over the choice of the prover's coins), the residual prover convinces the verifier with probability $2^{-|x|}$ (where the probability here is over the verifier's moves), and otherwise the residual prover convinces the verifier with probability 1. Then, in the first case extraction may run in (expected) time related to $2^{|x|}$, whereas in the second case it runs for polynomial-time. It follows that the extraction for the original probabilistic prover strategy runs in (expected) time that is related to $\frac{1}{2} \cdot 2^{|x|}$. But this probabilistic prover strategy convinces the verifier with probability exceeding $\frac{1}{2}$. (Thus, this extractor does not run in time that is inversely proportional to the success probability of the probabilistic prover strategy.)

1.2   On the Importance of Relating the Two Definitions

Needless to say, when faced with two natural definitions we wish to know whether they are equivalent. Furthermore, we note that the two different definitions have appeared in the literature: For example, the definition in [1] refers to any probabilistic prover strategy, while the definition in [2, Sec. 4.7] only refers to (arbitrary) deterministic strategies (see further discussion in Section 2). Thus, equating the two definitions (which appear in two central texts on this subject) becomes even more important (as it aims at eliminating a source of confusion in the current literature).

In addition to the foregoing generic and abstract motivation, there is also a concrete motivation to our study. It is typically easier to deal with deterministic strategies than with probabilistic ones, and thus relating the two definitions yields a useful methodology (i.e., demonstrating the "proof of knowledge" property with respect to deterministic strategies and deriving it for free with respect to probabilistic strategies). For example, we note that in [1, Apdx E] the "proof of knowledge" property (of the Graph Isomorphism protocol) is only demonstrated with respect to deterministic strategies, and this demonstration does not seem to extend to probabilistic strategies.[1]

Let us stress that in many applications the relevant prover strategies are in fact probabilistic. This is the case whenever proof-of-knowledge are the end goal (or close to it as in identification schemes), because in these cases the prover strategy represents an arbitrary adversarial behavior.[2]

1.3   Our Result

We show that the aforementioned gap (between the two natural formulations of the concept of a proof of knowledge), can be bridged in all natural cases (e.g., for NP-statements). The basic idea is that, instead of using (in the extraction) a single residual deterministic prover (derived by fixing random coins to the original probabilistic strategy), we employ numerous such residual deterministic strategies. Specifically, we invoke in parallel many executions of the knowledge-extractor (for deterministic strategies), and provide each of these invocations oracle access to a different residual deterministic strategy. These parallel executions are emulated in a specific manner (as detailed in Section 3) in order to ensure the desired extraction property.

2   Formal Setting

Let us start by recalling the definitional schema that underlies the two definitions that we study. Generalizing the treatment in [1] and [2, Sec. 4.7.1], we shall refer to an arbitrary class of potential (prover) strategies, denoted S. Indeed, the treatment of [1] is obtained by letting S be the class of all (probabilistic) strategies, whereas the treatment of [2, Sec. 4.7.1] is obtained by letting S be the class of all *deterministic* strategies.

2.1   Preliminaries

We first recall the basic setting, which consists of strategies (for parties in protocols) and a formulation of potential knowledge.

---

[1] It seems that the authors of [1] overlooked this point. They either did not notice that the argument is restricted to deterministic strategies or assumed that the demonstration can be easily extended to probabilistic strategies. We mention that the argument presented in [1, Apdx E] applies to any three-move Arthur-Merlin protocol for NP that has the following strong soundness property: Given any two accepting transcripts (for the same input) that start with the same Merlin message but differ on Arthur's message, one can efficiently find a corresponding NP-witness.

[2] In contrast, in other applications, where proofs-of-knowledge are used as a tool (and the corresponding knowledge-extractor is used by some simulator), it suffices to consider deterministic prover strategies (because these are derived from residual deterministic strategies that are derived in the course of the security analysis).

Strategies. Loosely speaking, deterministic strategies are functions that specify the next message to be sent by a party, based on its private input (which is hardwired in them) and as a function of the messages it has received so far. General (probabilistic) strategies are similar, except that the next message may also depend on a random input that is presented to these strategies. Formally, a (probabilistic) strategy σ is a function from $\{0,1\}^* \times \{0,1\}^*$ to $\{0,1\}^*$ such that $\sigma(\omega, \overline{\gamma})$ denotes the message to be sent by the corresponding party given that its random input equals ω, and the sequence of messages received so far equals $\overline{\gamma}$. Note that the strategy depends also on private inputs of the corresponding party, to which the outside world has no direct access. (These private inputs are hardwired in σ and do not appear explicitly in our notation.)

For a probabilistic strategy σ, we often consider residual deterministic strategies of the form $\sigma_\omega = \sigma(\omega)$ obtained by fixing the value of the random input to ω (i.e., $\sigma_\omega(\overline{\gamma}) = \sigma(\omega, \overline{\gamma})$).

The Two Perceptions of Strategies. Strategies will be used both as oracles and as specifying the actions of interactive machines. Specifically, we mean the following:

- When we discuss the interaction between parties on a common input, we incorporate this common input in each of the two strategies. The interaction of a strategy σ with a strategy σ' is the sequence of messages exchanged between the residual deterministic strategies σ_ω and $\sigma'_{\omega'}$, where ω and ω' are uniformly distributed. This sequence equals $\alpha_1, \beta_1, \alpha_2, \beta_2, \ldots$ such that $\alpha_{i+1} = \sigma(\omega, (\beta_1, \ldots, \beta_i))$ and $\beta_i = \sigma'(\omega', (\alpha_1, \ldots, \alpha_i))$.
- When using σ as an oracle, the oracle machine may issue arbitrary queries, which need not be consistent with the way that σ interact with any interactive machine. In particular, these queries may relate to different values of random input ω, all chosen at the discretion of the oracle machine.

The second item represents a relaxation of the common interpretation of the definition of *using a probabilistic strategy as an oracle oracle*, and thus a short discussion is in place. The common interpretation of this notion is that the user (i.e., the oracle machines) is given oracle access to a (single) residual deterministic strategy (i.e., σ_ω) that is obtained from σ by fixing a uniformly distributed ω. In fact, all prior constructions of knowledge extractors used this interpretation. We believe, however, that the more liberal interpration suggested above (i.e., by which the user is given oracle access to σ itself) is consistent with the simulation paradigm and is adequate in all reasonable applications. Actually, the knowledge extractor constructed in this work refers to an intermediate interpretation (of using a probabilistic strategy σ as an oracle). By this interpretation the oracle machine may is given access to several residual deterministic strategies (i.e., several σ_ω's) that are derived from the same probabilistic strategy by the selection of independently and uniformly distributed values of the random input ω.

The Relevant Knoweledge. We capture the relevant knowledge by a binary relation $R \subseteq \{0,1\}^* \times \{0,1\}^*$ such that, on common input x, the "claimed knowledge" refers to knowledge of a string in $R(x) \overset{\text{def}}{=} \{y : (x, y) \in R\}$. The archetypical case is of NP-relations; that is, relations R that are polynomially bounded (i.e., $(x, y) \in R$ implies $|y| \leq \text{poly}(|x|)$) and are polynomial time recognizable (i.e., there exists a polynomial-time algorithm A such that $A(x, y) = 1$ if and only if $(x, y) \in R$). We denote by S_R the set of strings for which a "claim of knowledge" is not bluntly wrong; that is, $S_R \overset{\text{def}}{=} \{x : R(x) \neq \emptyset\}$.

2.2   The Actual Definitions

Our focus will be on the validity condition of the following definition, but for sake of completeness we state also the non-triviality condition.

Definition 1 (schema for defining proofs of knowledge): *Let R be a binary relation, and $\kappa : \{0, 1\}^* \to [0, 1]$. We say that an interactive machine V is a* knowledge verifier *for the relation R with respect to a class of strategies S (and* knowledge error κ) *if the following two conditions hold.*

Non-triviality: *For every $x \in S_R$, there exists a strategy $\sigma \in S$ such that the verifier V always accepts when interacting with σ on common input x.*

Validity (with error κ): *There exists a probabilistic oracle machine K and a polynomial q such that, for every strategy $\sigma \in S$ and every x, machine K satisfies the following condition:*

> *If when interacting with σ, on common input x, the verifier V accepts with probability $p_x > \kappa(x)$, then on input x when given oracle access to σ machine K outputs a string in $R(x)$ within an expected number of steps upper-bounded by*

$$\frac{q(|x|)}{p_x - \kappa(x)}. \tag{1}$$

> *Note that the value of p_x depends on V, the strategy σ, and the common input x. The probability space to which p_x refers is that of all possible coin tosses of the strategies V and σ. Likewise, the probability space underlying Eq. (1) consists of all possible coin tosses of the machine K and the strategy σ.*

The oracle machine K is called a (universal) knowledge extractor, *and κ is called the* knowledge error *function.*

In particular, it follows that $x \notin S_R$ implies $p_x \leq \kappa(x)$. We stress that, on input x and when given oracle access to a strategy σ that convinces V to accept x with probability exceeding $\kappa(x)$, the knowledge extractor always outputs a string in $R(x)$; that is, in this case, $\Pr[K^\sigma(x) \notin R(x)] = 0$. However, when the said probability does not exceed $\kappa(x)$, all bets are off. Nevertheless, if R is an NP-relation then we may assume, without loss of generality, that for every x and every σ it holds that $\Pr[K^\sigma(x) \notin (R(x) \cup \{\bot\})] = 0$, where $\bot$ indicates halting without output. We now turn to the definitions studied in this article.

Definition 2 (the two definitions):

Following Definition 3.1 in [1]: *We say that V is a* knowledge verifier *for the relation R with knowledge error κ if Definition 1 holds with S being the set of all possible* (probabilistic) *strategies.*

Following Definition 4.7.2 in [2]: *We say that V is a* restricted knowledge verifier *for the relation R with knowledge error κ if Definition 1 holds with S being the set of all possible* deterministic *strategies.*

The two definitions differ only in the scope of strategies considered: [1, Def. 3.1] refers to all possible (probabilistic) strategies, whereas [2, Def. 4.7.2] refers only to all possible *deterministic* strategies.[3] Nevertheless, we show that in all natural cases (e.g., NP-relations) the restricted definition implies the general one.

2.3   Our Result

Before stating this result formally, let us point out why it is not as obvious as analogous results regarding related definitions.[4] Suppose that V is a *restricted knowledge-verifier* (with knowledge error $\kappa = 0$) and let K be the corresponding knowledge extractor. Given a probabilistic strategy σ, the straightforward attempt to extract knowledge from σ consists of invoking K while providing it with oracle access to the residual deterministic strategy σ_ω, where ω is uniformly distributed. The problem is that the probability that σ_ω convinces V, denoted $p(\omega)$, may deviate arbitrarily from the probability that σ convinces V, denoted p. That is, the random variable $p(\omega)$ may behave arbitrarily subject (only) to the condition $p = \mathrm{E}_\omega[p(\omega)]$ (and, of course, $p(\omega) \in [0,1]$). This, in turn, implies that the expected running-time of K^{σ_ω} (taken also over the random choice of ω) is not necessarily inversely proportional to p. For example, it may be that $\Pr[p(\omega) = 2^{-n}] = 1/2$ and $\Pr_\omega[p(\omega) = 1] = 1/2$, and in this case the expected running-time of K^{σ_ω} may be 2^n while $\mathrm{E}_\omega[p(\omega)] > 1/2$. Indeed, in general, it does not necessarily hold that $\mathrm{E}_\omega[1/p(\omega)] \leq \mathrm{poly}(n) \cdot \mathrm{E}_\omega[p(\omega)]$. Nevertheless, we prove the following.

Theorem 3 (main result): *Let V be a* restricted *knowledge verifier for R with knowledge error κ, where the length of the binary expansion of $\kappa(x)$ is polynomial in $|x|$. Suppose that the corresponding knowledge extractor, K, never outputs a wrong answer; that is, for every x and σ, it holds that $\Pr[K^\sigma(x) \notin R(x) \cup \{\perp\}] =$*

---

[3] Unfortunately, these facts are not perfectly clear in the original texts: The formulation of [1, Def. 3.1] refers to all possible "interactive functions", yet the latter are defined in [1, Def. 2.1] as arbitrary *probabilistic* strategies. The formulation of [2, Def. 4.7.2] refers to all residual deterministic strategies that can be obtained by fixing the random input of some probabilistic strategy, but in retrospect the latter condition is a red herring (and does not help in extending this definition to the general case of [1, Def. 3.1]).

[4] Recall that simulation-security with respect to arbitrary (polynomial-size) deterministic adversaries typically implies simulation-security with respect to arbitrary probabilistic (polynomial-time) adversaries.

0, *where* ⊥ *indicates halting without output. Then,* V *is a knowledge verifier for* R *with knowledge error* κ.

Theorem 3 asserts that, under the additional assumptions regarding κ and K, the restricted definition (i.e., [2, Def. 4.7.2]) implies the general definition (i.e., [1, Def. 3.1]). As illustrated by the forgoing discussion, the corresponding knowledge extractor (for [1, Def. 3.1]) is not K (or the minor modification of K discussed above). We note that the two additional assumptions (regarding κ and K) can be easily met in case that R is an NP-relation. Details follows.

Recall that if R is an NP-relation, then we can check the output of K, and thus (on input x) we can always avoid outputting a string that is not in $R(x)$. This eliminates the additional assumption regarding K. As for the additional condition regarding κ, it can always be enforced by possiblly increasing κ a little; that is, by resetting $\kappa(x)$ to $\lceil 2^{q(|x|)} \cdot \kappa(x) \rceil / 2^{q(|x|)}$, where q is an arbitrary polynomial. Furthermore, in the case that R is an NP-relation, we may reset $\kappa(x)$ to $\kappa'(x) \stackrel{\text{def}}{=} \lfloor 2^{q(|x|)} \cdot \kappa(x) \rfloor / 2^{q(|x|)}$, for a sufficiently large polynomial q (by taking advantage of the fact that, for any $x \in S_R$, a string in $R(x)$ can be found in time $\exp(q(|x|))$).[5]

3   Proof of Theorem 3

Recall that the source of trouble is that for a uniformly distributed value of the random input, the success probability of the corresponding residual deterministic strategy (w.r.t convincing V) may be very different from the success probability of the original probabilistic strategy. This may lead to overwhelmingly long runs of the knowledge extractor (i.e., runs that contribute to the total expected running-time more than we can allow). The basic idea is to truncate such overwhelmingly long runs, and rely on the existence (in sufficient probability measure) of runs that are not overwhelmingly long.

Let us illustrate this idea by referring to the foregoing example, where $\Pr[p(\omega) = 2^{-n}] = 1/2$ and $\Pr[p(\omega) = 1] = 1/2$ (and $\kappa = 0$).[6] In this case, $p = E_\omega[p(\omega)] > 1/2$, and so our extraction procedure should run in expected polynomial-time. Thus, we invoke K providing it with oracle access to σ_ω, where ω is uniformly distributed among all possible random inputs, and truncate the

---

[5] This fact allows for handing the case that the probability that σ convinces V to accept x (i.e., p_x) is very close to $\kappa(x)$ in the sense that $p_x - \kappa'(x)$ is significantly larger than $p_x - \kappa(x)$. We first note that in this case $p_x < \kappa(x) + 2^{-q(|x|)}$ (as otherwise $p_x - \kappa(x) \geq 2^{-q(|x|)}$ and $p_x - \kappa'(x) < p_x - \kappa(x) + 2^{-q(|x|)} \leq 2 \cdot (p_x - \kappa(x))$). Thus, in this case (where $(p_x - \kappa(x))^{-1} < 2^{q(|x|)}$), we can afford running the standard exhaustive search algorithm (which runs in time $2^{q(|x|)}$) in parallel to the given knowledge extractor. On the other hand, if $p_x - \kappa'(x) = O(p_x - \kappa(x))$, then $(p_x - \kappa(x))^{-1} = O((p_x - \kappa'(x))^{-1})$. Thus, given an knowledge extractor of error κ, we obtain a knowledge extractor of error κ'.

[6] Throughout the text, n denotes the length of the common input x, which we often omit from the notation.

execution after a polynomial number of steps has elapsed. If an output was obtained in this execution attempt, then we output it, otherwise we repeat the experiment again. Note that, with probability $1/2$, the residual strategy σ_ω satisfies $p(\omega) = 1$, in which case K^{σ_ω} is expected to halt in polynomial-time with the desired output. Otherwise (i.e., $p(\omega) = 2^{-n}$), the (truncated) execution of K^{σ_ω} may be useless, but it will not cause much harm (since it is suspended after a polynomial number of steps).

In the foregoing example we relied on a good *a priori* knowledge of the distribution of $p(\omega)$, which may not be available in general. Thus, in general, we shall employ a somewhat more sophisticated argument. Following is a rough sketch of the general argument, where we still assume for simplicity that $\kappa = 0$. One key observation is that there exists an integer i such that $\Pr_\omega[p(\omega) \approx 2^{-i}]$ is linearly related to $2^i \cdot p$ (where $p = E_\omega[p(\omega)]$). We do not know this i and so we run, in parallel, numerous processes one per each of the relevant values of i. In the ith process (i.e., the one related to the value i), we repeatedly attempt extraction with deterministic residual provers (derived by random fixings of ω), but truncate each attempt after $\text{poly}(n) \cdot 2^i$ steps. Thus, for the correct value of i, the ith relevant process will succeed in extraction within the allowed expected number of steps (i.e., it is expected to make $\text{poly}(n)/(2^i \cdot p)$ attempts, each running for $\text{poly}(n) \cdot 2^i$ steps, and thus the total expected running time is $\text{poly}(n)/p$).

We now turn to a rigorous description of the actual knowledge extractor for probabilistic strategies. We fix an arbitrary $x \in S_R$, but omit it from most subsequent notations. Fixing an arbitrary randomized strategy σ, we consider an arbitrary choice of the strategy's coins, ω, and denote the residual strategy by σ_ω. In the rest, we will refer to selecting such ω's and providing oracle access to the corresponding σ_ω, but we need not select these ω's ourselves; it suffices to have the ability of providing oracle access to numerous random and independent "incarnations" of σ that correspond to such choices of ω's.

Let $p(\omega)$ denote the probability that verifier accepts when interacting with σ_ω, on common input x. By the hypothesis, if $p(\omega) > \kappa(x)$, then the knowledge extractor K, given oracle to σ_ω, outputs a string in $R(x)$ in expected time $q(|x|)/(p(\omega) - \kappa(x))$, where q is a fixed (universal) polynomial. As before, we let $p = E_\omega[p(\omega)]$, and assume, without loss of generality, that $p > \kappa(x)$ (because otherwise noting is required). In addition, let $\kappa = \kappa(x)$ and let $\ell = \text{poly}(|x|)$ denote an upper-bound on *the length of the random input used by V on common input x*. It follows that for every choice of ω (which determines a residual strategy σ_ω) it holds that $2^\ell \cdot p(\omega)$ is an integer (because the relevant probability space is uniformly distributed over 2^ℓ possibilities). Recalling that κ has a binary expansion of length $\text{poly}(|x|)$, we assume, without loss of generality, that $2^\ell \cdot \kappa$ is also an integer. It follows that if $p(\omega) \le \kappa + 2^{-\ell-1}$, then $p(\omega) \le \kappa$.

We consider a partition of $(\kappa + 2^{-\ell-1}, \kappa + 1]$ into $\ell + 1$ intervals such that the ith interval is $I_i = (\kappa + 2^{-i}, \kappa + 2^{-i+1}]$. We claim that there exists $i \in [\ell + 1]$ such that

$$\Pr_\omega[p(\omega) \in I_i] \ge \frac{2^i \cdot (p - \kappa)}{4(\ell + 1)} \tag{2}$$

This claim follows, because otherwise we derive a contradiction as follows (where in the first inequality we use the fact that $p(\omega) \le \kappa + 2^{-\ell-1}$ implies $p(\omega) \le \kappa$):

$$E_\omega[p(\omega)] \le \Pr_\omega[p(\omega) \le \kappa + 2^{-\ell-1}] \cdot \kappa + \sum_{i=1}^{\ell+1} \Pr_\omega[p(\omega) \in I_i] \cdot (\kappa + 2^{-i+1})$$

$$= \kappa + \sum_{i=1}^{\ell+1} \Pr_\omega[p(\omega) \in I_i] \cdot 2^{-i+1}$$

$$< \kappa + \sum_{i=1}^{\ell+1} \frac{2^i \cdot (p - \kappa)}{4(\ell+1)} \cdot 2^{-i+1}$$

$$= \kappa + \frac{p - \kappa}{2}$$

where the second inequality uses the contradiction hypothesis (by which Eq. (2) is violated for every $i \in [\ell + 1]$). Recalling that $p = E_\omega[p(\omega)]$, we obtain $p < \kappa + (p - \kappa)/2$, which contradicts the hypothesis $p > \kappa$.

The new extraction procedure consists of running $\ell + 1$ processes in parallel. The ith process successively invokes *time-bounded* executions of the knowledge extractor K, providing each such invocation with oracle access to a random and independent incarnation of σ (i.e., residual strategies σ_ω for uniformly and independently ditributed values of ω). The time-bound used in the ith process is $2 \cdot q(|x|) \cdot 2^i$, where the q is the polynomial guaranteed for K. Thus, if $p(\omega) \ge \kappa + 2^i$ then, with probability at least $1/2$, it holds that $K^{\sigma_\omega}(x)$ halts in $2 \cdot q(|x|) \cdot 2^i$ steps (because the expected number of steps is $q(|x|) \cdot 2^i$). Once any of these $\ell + 1$ processes outputs some string y, the entire parallel-process terminates and y is used as output.

Recall that by the theorem's hypothesis, whenever K outputs a string y it is the case that $y \in R(x)$. Thus, we confine ourselves to analyzing the expected running-time of the foregoing extraction process. Considering an arbitrary value i that satisfies Eq. (2), we observe that the ith process succeed after making an expected number of $2 \cdot \left(\frac{2^i \cdot (p-\kappa)}{4(\ell+1)} \right)^{-1}$ trials. Thus, the overall time spent by the new extractor has expectation

$$(\ell + 1) \cdot \frac{2 \cdot 4(\ell+1)}{2^i \cdot (p - \kappa)} \cdot (2 \cdot q(|x|) \cdot 2^i) = \frac{O(\ell^2 \cdot q(|x|))}{p - \kappa} = \frac{\text{poly}(|x|)}{p - \kappa}$$

and the theorem follows.

4   Concluding Remarks

We have established the equivalence of [1, Def. 3.1] and [2, Def. 4.7.2] while relying on the following three (reasonable) conventions (or assumptions):

1. We assumed that the pharse "given oracle access to a probabilistic strategy σ" means ability to query several (rather than one) residual deterministic strategies of the form σ_ω, where the ω's are uniformly and independently distributed.
2. We assumed that the knowledge-extractor never outputs a wrong string (i.e., a string not in $R(x)$), regardless of which input x and which strategy σ it is given access to.
3. We assumed that the knowledge error function κ is nice in the sense that, for every x, the binary expansion of $\kappa(x)$ has length polynomial in $|x|$.

We believe that these assumptions do not impair the applicability of our result. Still we wonder whether (some of) these assumptions can be eliminated.

References

1. Bellare, M., Goldreich, O.: On Defining Proofs of Knowledge. In: Brickell, E.F. (ed.) CRYPTO 1992. LNCS, vol. 740, pp. 390–420. Springer, Heidelberg (1993)
2. Goldreich, O.: Foundation of Cryptography – Basic Tools. Cambridge University Press, Cambridge (2001)
3. Goldreich, O.: Foundation of Cryptography – Basic Applications. Cambridge University Press, Cambridge (2004)
4. Goldwasser, S., Micali, S., Rackoff, C.: The Knowledge Complexity of Interactive Proof Systems. SIAM Journal on Computing 18, 186–208 (1989); Preliminary Version in 17th STOC (1985)

On the Average-Case Complexity of Property Testing

Oded Goldreich

Abstract. Motivated by a study of Zimand (*22nd CCC*, 2007), we consider the average-case complexity of property testing (focusing, for clarity, on testing properties of Boolean strings). We make two observations:

1. In the context of average-case analysis with respect to the uniform distribution (on all strings of a fixed length), property testing is trivial. Specifically, either the YES-instances (i.e., instances having the property) or the NO-instances (i.e., instances that are far from having the property) are exponentially rare, and thus the tester may just reject (resp., accept) obliviously of the input.

2. Turning to average-case derandomization with respect to distributions that assigns noticeable probability mass to both YES-instances and NO-instances, we identify a natural class of distributions and testers for which average-case derandomization results can be obtained directly (i.e., without using randomness extractors). Furthermore, the resulting deterministic algorithm may preserve the non-adaptivity of the original tester. (In contrast, Zimand's argument utilizes a strong type of randomness extractors and introduces adaptivity into the testing process.)

Keywords: Property Testing, Average-Case Complexity.

An early version of this work appeared as TR07-057 of *ECCC*. The current revision is quite minimal.

1 Introduction

The starting point of this article is Zimand's study of possible derandomizations of randomized sublinear-time algorithms [Z]. Zimand showed that randomized sublinear-time algorithms can be derandomized yielding deterministic algorithms of polynomially-related complexity that err on a negligible fraction of the instances. Specifically, he showed that, for some fixed $\alpha > 0$, any randomized algorithm of time-complexity T such that $T(n) < n^{\alpha}$ can be emulated by a $\mathrm{poly}(T)$-time deterministic algorithm that errs on at most an $\exp(-\Omega(T \log T))$ fraction of the instances. Needless to say, Zimand's work (as well as the current article) refers to a "direct access" model of computation in which each bit of the input can be read at unit cost. Zimand noted the relevance of his work to property testing, but our view is that this aspect of his work should be evaluated with great care. Articulating this view is the main motivation of the current article.

O. Goldreich et al.: Studies in Complexity and Cryptography, LNCS 6650, pp. 124–135, 2011.

1.1   Average-Case with Respect to the Uniform Distribution

In discussing the theoretical significance of his work, Zimand says "it shows that the properties that can be checked in sublinear time depend, except for a few inputs, on just a few bits of the input and the locations of these bits can be found very fast." [1] We fear that such a phrasing does not put adequate emphasis on the exception clause (i.e., "except for a few inputs"). Furthermore, in our opinion, *the crux of property testing is dealing with non-typical* (i.e., *exceptional*) *inputs*, whereas dealing with random inputs is typically uninteresting.

We first note that average-case analysis with respect to the uniform distribution is not adequate in the context of testing properties of strings, which in turn cover almost all types of property testing problems (e.g., testing graph properties in the adjacency matrix model). The reason is that property testing problems are special type of promise problems [2] in which one should distinguish instances having the property from instances that are far from any string having the property. However, as shown in Section 2, *for every property of n-bit strings either the first set* (i.e., instances having the property) *or the second set* (i.e., instances far from having the property) *has exponentially vanishing density.* In the first (resp., second) case, *a trivial tester* that rejects (resp., accepts) every input (without reading a single bit) *is correct on all but a exponentially vanishing fraction of all inputs*, where the exceptional cases consists of all the YES-instances (resp., all the "far-away" instances).

Indeed, the average-case complexity of promise problems is meaningful only with respect to *distributions that assign noticeable probability mass to both* YES-*instances and* NO-*instances* (because otherwise a trivial algorithm as above will do). However, the uniform distribution cannot satisfy the latter condition in the case of promise problems that correspond to property testing (of Boolean strings).

1.2   A Direct Average-Case Derandomization for Many Natural Cases

We thus turn to average-case derandomization with respect to distributions that assigns noticeable probability mass to both YES-instances and NO-instances (i.e., "far-away" instances). While Zimand's approach *may* be applicable to this context too [3], we identify a natural class of distributions and testers for which

---

[1] See last paragraph of [Z, Sec. 1.0].

[2] Recall that promise problems [ESY] are represented as pairs of non-intersecting sets $A, B \subseteq \{0, 1\}^*$ and solving such problems requires distinguishing inputs in A from inputs in B, while an arbitrary answer is allowed for inputs that are neither in A nor in B. For such a promise problem we say that a string in $A \cup B$ satisfy the promise (while strings outside $A \cup B$ violate the promise).

[3] As shown in Section 2, such distributions must have min-entropy at most $n - \Omega(n)$, while [Z] does not provide results for this range of paramters. Still it is possible that the basic approach of [Z] coupled with a suitable randomness extractor (possibly tailored for this application) may be applicable to such distributions.

average-case derandomization results can be obtained directly. Furthermore, we believe that our analysis provides a more illuminating account of what is actually going on.

We recall that, in continuation to [GW], Zimand [Z] emulates the computation of the original randomized tester by applying a (special type of) randomness extractor to the input, and replacing the coin tosses of the original tester with corresponding outputs of the extractor. Consequently, even if the original tester is non-adaptive (as is the case with many natural property testers), the resulting deterministic algorithm is adaptive (because the emulation step depends on the bits read in the randomness-extraction step). In contrast, we show that, in many natural cases, *an average-case derandomization can be obtained by arbitrarily fixing the coins of the original tester.*

To illustrate the point, let us consider the problem of testing whether a given Boolean string has a majority of 1-values (or is far from any such string). In this case, we may obtain a deterministic algorithm by inspecting the value of the *first* few bits in the string, where this algorithm decides correctly on almost all n-bit strings that have a number of 1-values that is bounded away from $n/2$; that is, ruling by the majority of the inspected bits, we decide correctly on almost all elements in the set of n-bit strings having Hamming weight outside the interval $[0.49n, 0.51n]$. Furthermore, any fixed set of sufficiently many bit positions can be used for this purpose. For a general treatment, see Section 3.

We illustrate the general treatment by considering the special case of testing graph properties in the adjacency matrix model (as in, e.g., [GGR]). In this setting (but also in other natural settings), the natural property testers use their randomness solely for determining the bit positions to be examined in the input. Furthermore, at the cost of squaring the query complexity, we may assume that any graph property can be tested by using randomness in such a restricted manner [GT]. In Section 3, we show that *a deterministic tester that inspects the subgraph induced by any fixed set of vertices* (of adequate size) *errs rarely with respect to any distribution on labeled graphs that is invariant under isomorphism.*

1.3   Additional Comments

We note that, in many cases, it is easier to construct property testers that work only on typical objects drawn from natural distributions rather than to construct standard testers that work on all objects. This fact is mildly reflected by the results shown in Section 3, where we convert standard (randomized) testers into *deterministic* "average-case testers"; that is, here getting rid of randomization is considered a simplification, but the query complexity of the resulting tester is not smaller than the query complexity of the original tester.[4] However, in many natural examples (see one below), we can also reduce the query complexity. Details follow.

Let us first emphasize the fact that, when considering worst-case complexity, *randomness is essential for testing natural properties* (see, e.g., [GS], and note

---

[4] Actually, the the query complexity of the resulting tester is somewhat larger than the query complexity of the original tester.

that this is an unconditional result). Indeed, this result stand in contrast to the aforementioned average-case testing results, and provides a formal sense in which "average-case testing" is easier than standard (worst-case) testing. However, we claim that things go beyond this sense: Detecting random objects that are far from a property is typically easier than detecting arbitrary objects that are far from this property.

Consider, for example, the notoriously hard problem of testing triangle-freeness in the adjacency matrix model. As shown by Alon [A], testing triangle-freeness requires a number of queries that is super-polynomial in the reciprocal of the proximity parameter, denoted ϵ. In contrast, for a random graph of edge density ϵ and any three vertices, with probability ϵ^3, the subgraph induced by these three vertices is a triangle.

Reservations Regarding Our Own Opinions. The direct average-case derandomizations presented in Section 3 refer to distributions that are invariant under natural reshuffling of the presentation of the studied objects (e.g., in the case of labeled graph we considered distribution that are invariant under isomorphism). Although such distributions arise naturally in many cases, distributions that lack this feature are natural in other cases. For example, consider a distribution over real-valued vectors (or matrices) that is obtained by the following two-step process: First a vector (resp., a matrix) is selected according to an arbitrary distribution, and then each of its entries is pertubed at random and independently of anything else. The resulting distribution may not satisfy any of the invariances considered in Section 3, but it does have high min-entropy. Recalling that various natural properties of vectors (resp., matrices) can be tested in probabilistic sublinear time (cf., e.g., [EKKRV, FK]), we note that Zimand's approach [Z] *may*[5] be applicable in this case (and if so yield average-case derandomization of natural appeal).

2   Average-Case with Respect to the Uniform Distribution

We start by recalling the setting of property testing (cf., e.g., [G, R]), when specialized to bit strings (of fixed length). We comment that other finite objects can be naturally represented by such generic strings, and thus corresponding properties can be naturally cast in this framework. The most notable example is property testing of graphs in the adjacency matrix model (as introduced in [GGR]).

For a generic length parameter n, we consider the set of all strings over $\{0,1\}^n$, and an arbitrary property $P_n \subseteq \{0,1\}^n$. Property testing with respect to a distance parameter $\epsilon > 0$ corresponds to distinguishing inputs in P_n from inputs in $\Gamma_\epsilon(P_n)$, where

$$\Gamma_\epsilon(P_n) \stackrel{\text{def}}{=} \{x \in \{0,1\}^n : \forall z \in P_n \ \Delta(x,z) > \epsilon \cdot n\} \tag{1}$$

---

[5] See Footnote 3.

and $\Delta(x_1 \cdots x_n, z_1 \cdots z_n) = |\{i : x_i \neq z_i\}|$ denotes the number of bits on which $x = x_1 \cdots x_n$ and $z = z_1 \cdots z_n$ disagree.[6] That is, property testing with respect to ϵ corresponds to deciding the promise problem $(\mathrm{P}_n, \Gamma_\epsilon(\mathrm{P}_n))$. However, as we shall see, with respect to the uniform distribution on $\{0,1\}^n$, this promise problem is trivial on the average. That is:

Theorem 2.1 ([AS, Thm. 7.5.3], reformulated): *For every constant $\epsilon > 0$ there exists a constant $c > 0$ such that for every n if $|\mathrm{P}_n| \geq 2^{-cn} \cdot 2^n$, then $|\Gamma_\epsilon(\mathrm{P}_n)| \leq 2^{-cn} \cdot 2^n$. More generally, if $|\mathrm{P}_n| \geq \rho \cdot 2^n$ and $\epsilon \geq \sqrt{\frac{8\ln(1/\rho)}{n}}$, then $|\Gamma_\epsilon(\mathrm{P}_n)| \leq \rho \cdot 2^n$.*

Indeed, Theorem 2.1 can be reformulated by referring to a uniformly distributed $x \in \{0,1\}^n$. This reformulation (of the special case of constant $\epsilon > 0$) asserts that (for some constant $c > 0$) either $\Pr_x[x \in \mathrm{P}_n] < 2^{-cn}$ or $\Pr_x[x \in \Gamma_\epsilon(\mathrm{P}_n)] \leq 2^{-cn}$. In the first case, a tester that always reject is correct on all but at most a 2^{-cn} fraction of the n-bit inputs, whereas in the second case the same holds for a tester that always accepts. Thus, *property testing is trivial on the average with respect to any distribution that has min-entropy* $m \overset{\text{def}}{=} n - o(n)$ (i.e., a distribution X_n such that of every x it holds that $\Pr[X_n = x] \leq 2^{-m}$).[7]

Proof: The theorem is merely a reformulation of a well-known result regarding the volume of balls around sets. Specifically, let $\mathcal{B}_d(S)$ denote the set of n-bit long strings that are at distance at most d from some string in S (i.e., $\mathcal{B}_d(S) \overset{\text{def}}{=} \{x \in \{0,1\}^n : \exists y \in S \text{ s.t. } \Delta(x,y) \leq d\}$). Then, Theorem 7.5.3 in [AS] (see proof in the Appendix) asserts that *if $|S| \geq e^{-\lambda^2/2} \cdot 2^n$, then $|\mathcal{B}_{2\lambda\sqrt{n}}(S)| \geq (1 - e^{-\lambda^2/2}) \cdot 2^n$*. Using $S = \mathrm{P}_n$ and $\lambda = \sqrt{2\ln(1/\rho)}$, where $\rho = |\mathrm{P}_n|/2^n$, we get $|\mathcal{B}_{\sqrt{8n\ln(1/\rho)}}(\mathrm{P}_n)| \geq (1 - \rho) \cdot 2^n$. Noting that $\Gamma_\epsilon(\mathrm{P}_n) = \{0,1\}^n \setminus \mathcal{B}_{\epsilon n}(\mathrm{P}_n)$, the general claim follows. The special case follows by noting that $\rho = 2^{-cn}$ implies $\sqrt{(8\ln(1/\rho))/n} = \sqrt{8c/\log e}$ (and so using $c = \epsilon^2/8$ will do). ∎

Generalization. We note that Theorem 2.1 generalizes to properties of sequences over any alphabet Σ. That is, for any property $\mathrm{P}_n \subseteq \Sigma^n$, it holds that *if $|\mathrm{P}_n| \geq \rho \cdot |\Sigma|^n$ and $\epsilon \geq \sqrt{\frac{8\ln(1/\rho)}{n}}$, then $|\Gamma_\epsilon(\mathrm{P}_n)| \leq \rho \cdot |\Sigma|^n$*, where $\Gamma_\epsilon(\mathrm{P}_n)$ denotes the set of n-long sequences over Σ that are ϵ-far from every sequence in P_n. (See further details in the Appendix.)

3   A Direct Average-Case Derandomization for Many Natural Cases

In this section we show that, in many interesting settings of property testing, average-case derandomization results can be obtained more directly than by

---

[6] An alternative exposition may refer to Boolean functions of the form $f : [n] \to \{0,1\}$. In this case $\Delta(f,g) = |\{i : f(i) \neq g(i)\}|$.

[7] In fact, we may allow min-entropy $m = n - (cn/2)$, where c is the constant in Theorem 2.1. For such a distribution X_n (of min-entropy $n - (cn/2)$), it holds that either $\Pr[X_n \in \mathrm{P}_n] \leq 2^{-cn/2}$ or $\Pr[X_n \in \Gamma_\epsilon(\mathrm{P}_n)] \leq 2^{-cn/2}$.

following the approach suggested by Zimand.[8] We start by considering the concrete setting of testing graph properties in the adjacency matrix model (as in [GGR]), and later generalize the treatment to other settings. Indeed, the setting of testing graph properties in the adjacency matrix model provides the most appealing application of the general approach to be described later.

3.1   On Testing Graph Properties in the Adjacency Matrix Model

Recall that in this model (for testing graph properties), n-vertex graphs are represented by Boolean strings of length n^2. For technical reasons, we prefer to represent such graphs as Boolean functions defined over the set of the $\binom{n}{2}$ (unordered) vertex-pairs, which is actually more natural (as well as non-redundant). Note that the set of all permutations over $[n]$ induces a transitive group of permutations over these pairs, where the permutation $\pi : [n] \to [n]$ induces a permutation that maps pairs of the form $\{i, j\}$ to $\{\pi(i), \pi(j)\}$. Indeed, any graph property is invariant under this group, which is hereafter referred to as the group of vertex-relabeling; that is, $G = ([n], E)$ has the property if and only if $\pi(G) = ([n], \{\{\pi(i), \pi(j)\} : \{i, j\} \in E\})$ has this property.

Theorem 3.1 *Let G_n be a graph property, referring to n-vertex graphs, and let X_n be any arbitrary distribution of n-vertex graphs that is invariant under the group of vertex-relabeling (i.e., for every permutation $\pi : [n] \to [n]$ it holds that X_n and $\pi(X_n)$ are identically distributed). Suppose that the promise problem $(\mathsf{G}_n, \Gamma_\epsilon(\mathsf{G}_n))$ can be decided correctly (in the worst case) by a probabilistic tester of query complexity $q(n, \epsilon)$ and error probability at most $1/3$. Then, for every $k < n/O(q(n, \epsilon)^2)$, there exists a deterministic algorithm of query complexity $O(k \cdot q(n, \epsilon)^2)$ that inspects only vertex pairs that correspond to the vertices $1, ..., O(k \cdot q(n, \epsilon))$ and is correct on a random input X_n with probability at least 2^{-k}.*

As will be clear from the proof, we may use any $O(k \cdot q(n, \epsilon))$ fixed vertices rather than the vertex set $\{1, ..., O(k \cdot q(n, \epsilon))\}$.

Proof: By [GT, Thm. 2], we may convert the original tester into a canonical tester that selects uniformly a set of $n' \stackrel{\text{def}}{=} O(q(n, \epsilon))$ vertices, denoted R, and accepts if and only if the subgraph induced by R has some predetermined (graph) property $\mathsf{G}'_{n'}$. By invoking the resulting (canonical) tester $t \stackrel{\text{def}}{=} O(k)$ times, we reduce its (worst-case) error probability to 2^{-k}. We claim that the resulting tester, denoted A, can be derandomized (for average-case performance) by merely using any *fixed* set of $t \cdot n'$ vertices rather than a *random* set of $t \cdot n'$ vertices as selected by A. We denote the resulting deterministic algorithm by D.

To prove the foregoing claim, we consider an arbitrary input graph G that satisfies the promise (i.e., either $G \in \mathsf{G}_n$ or G is ϵ-far from G_n). By the foregoing discussion we know that the probability that A errs on input G is at most 2^{-k}. Let π denote a uniformly distributed permutation of $[n]$, and consider the

---

[8] Here we ignore the question of the applicability of Zimand's approach to distributions of min entropy $n - \Omega(n)$; cf. Footnote 3.

graph $\pi(G)$ obtained from G by relabeling its vertices according to π. Note that $\pi(G) \in G_n$ if and only if $G \in G_n$ (and, likewise, $\pi(G)$ is ϵ-far from G_n iff G is ϵ-far from G_n). On the other hand, the distribution of the view of A on input G is identical to distribution of the view of D on input $\pi(G)$, because a random π maps any fixed set of vertices to a uniformly distributed set of vertices. We stress that the first probability space is defined over the coin tosses of A, whereas the second probability space is defined over the random relabeling π. We conclude that the probability that D errs on input $\pi(G)$ is at most 2^{-k}.

By the hypothesis that X_n is invariant under the group of vertex-relabeling, it follows that X_n can be described by a process in which one first selects a random graph G (possibly $G \leftarrow X_n$), and then outputs $\pi(G)$, where π is a uniformly distributed permutation of $[n]$. Note that if G violates the promise, then so does $\pi(G)$, whereas if G satisfies the promise, then the probability that D errs on input $\pi(G)$ is at most 2^{-k}. It follows that D errs on input X_n with probability at most 2^{-k}. ∎

3.2   Generalization

Theorem 3.1 can be extended in various ways. We first note that most natural testers (not only in the setting of testing graph properties in the adjacency matrix model) are "kind of canonical" in the sense that they select some random set of "pivots" and consider small sets of bit-locations as determined by these pivots. That is, randomization is only used in these testers for the selection of the pivots, which induce queries that are each uniformly distributed. Thus, the strategy of the proof of Theorem 3.1 can be applied, resulting in a deterministic algorithm that uses a fixed set of pivots and errs with probability at most 2^{-k} on any input distribution that is invariant under permutations that correspond to mapping among sets of pivots. To formalize the above discussion, we need some definitions.

We turn back to properties of n-bit strings, which we actually view as functions from $[n]$ to $\{0, 1\}$. More generally, we shall consider properties of functions from $[n]$ to an arbitrary alphabet Σ. For any set (or rather group) Π of permutations over $[n]$, we say that the property P_n (of such functions) is Π-invariant if for every $f : [n] \to \Sigma$ and every $\pi \in \Pi$ it holds that $f \in P_n$ if and only $(f \circ \pi) \in P_n$, where $(f \circ \pi)(i) = f(\pi(i))$ (for every $i \in [n]$). In the following definition, "normality" amounts to non-adaptivity augmented by the requirement that the final decision is deterministic and only depends on the oracle answers, whereas "Π-normality" corresponds to the mapping between the aforementioned pivots.

Definition 3.2 (normal testers): *Let Π be a permutation group over $[n]$ and P_n be a Π-invariant property. We say that a tester for P_n is normal if there exists a query-generating algorithm Q and a verdict predicate V such that on internal coins $\omega \in \{0, 1\}^r$ and oracle access to any $f : [n] \to \Sigma$ the tester accepts if and only if $V(f(i_1), ..., f(i_q)) = 1$, where $(i_1, ..., i_q) = Q(\omega)$. That is, the tester queries the function at locations $i_1, ..., i_q$, which are determined by $Q(\omega)$ and accepts if and only if the predicate V evaluates to 1 on the q-tuple of answers. We say that the tester is Π-normal if the following two conditions hold.*

1. *For every $\omega, \omega' \in \{0, 1\}^r$ there exists $\pi \in \Pi$ such that $Q(\omega') = \pi(Q(\omega))$, where $\pi(i_1, ..., i_q) = (\pi(i_1), ..., \pi(i_q))$.*
2. *For every $\omega \in \{0, 1\}^r$ and $\pi \in \Pi$ there exists $\omega' \in \{0, 1\}^r$ such that $Q(\omega') = \pi(Q(\omega))$.*

Note that, by definition, a normal tester is non-adaptive. The justification for referring to the two additional conditions by the term Π-normalily is provided by the following Fact 3.3. But let us first mention that, indeed, the canonical graph property testers (as defined in [GT] and used in the proof of Theorem 3.1) are normal. Furthermore, they are $\Pi^{(\mathrm{vr})}$-normal for the group $\Pi^{(\mathrm{vr})}$ of all vertex-relabeling. Other examples of normal testers are discussed at the end of this section.

Fact 3.3 *Let Π and P_n be as in Definition 3.2, and suppose that V and Q are as rerquired of a normal tester for P_n. Then, this tester is Π-normall if and only if for every $\omega \in \{0, 1\}^r$ and uniformly distributed $\pi \in \Pi$ it holds that $\pi(Q(\omega))$ is uniformly distributed in $S \stackrel{\text{def}}{=} \{Q(\omega') : \omega' \in \{0, 1\}^r\}$ (i.e., for every $\omega, \omega' \in \{0, 1\}^r$ it holds that $\mathrm{Pr}_{\pi \in \Pi}[\pi(Q(\omega)) = Q(\omega')] = 1/|S|$).*

Proof: Clearly, the latter ("distributional") condition implies the two condition in Definition 3.2. To see that the other direction, we show that Π-normality implies that, for any fixed $\omega, \omega', \omega'' \in \{0, 1\}^r$, it holds that $p_{\omega, \omega'} = p_{\omega, \omega''}$, where $p_{a,b} \stackrel{\text{def}}{=} \mathrm{Pr}_{\pi \in \Pi}[\pi(Q(a)) = Q(b)]$. The latter claim can be proved by fixing any permutation π_0 that satisfies $Q(\omega'') = \pi_0(Q(\omega'))$, and observing that a random permutation in Π can be written as $\pi_0 \circ \pi'$, where $\pi \in \Pi$ is uniformly distributed. Hence, $p_{\omega, \omega''} = \mathrm{Pr}_{\pi' \in \Pi}[(\pi_0 \circ \pi')(Q(\omega)) = Q(\omega'')]$, which equals $\mathrm{Pr}_{\pi' \in \Pi}[\pi'(Q(\omega)) = Q(\omega')]$. ∎

Theorem 3.4 (Theorem 3.1, generalized): *Let Π be a permutation group over $[n]$ and P_n be a Π-invariant property. Let X_n be a distribution over functions from $[n]$ to Σ such that for every such function f and every $\pi \in \Pi$ it holds $\mathrm{Pr}[X_n = f] = \mathrm{Pr}[X_n = f \circ \pi]$. Suppose that the promise problem $(\mathrm{P}_n, \Gamma_\epsilon(\mathrm{P}_n))$ can be decided correctly (in the worst case) by a Π-normal tester of query complexity $q(n, \epsilon)$ and error probability at most $1/3$. Then, for every $k < n/O(q(n, \epsilon))$, there exists a (non-adaptive) deterministic algorithm that inspects the function value at $O(k \cdot q(n, \epsilon))$ fixed and predetermined positions and is correct on a random X_n with probability at least 2^{-k}.*

A distribution X_n as in the hypothesis of Theorem 3.4 is called Π-invariant.

Proof: The deterministic algorithm, denoted D, is obtained by fixing the coins to the query-generating algorithm Q. For example, we may query the input function f at locations $(i_1, ..., i_q) = Q(0^r)$, and accept if and only if $V(f(i_1), ..., V(i_q)) = 1$. (Recall that V represents a fixed predicate.) As in the proof of Theorem 3.1, we actually apply this construction after reducing the error probability of the original tester to 2^{-k}.

To analyze the success probability of D on input X_n, we fix any function f and consider the function distribution $f \circ \pi$, where $\pi \in \Pi$ is uniformly distributed. As in the proof of Theorem 3.1, the distribution of the view of the original tester on input f is identical to distribution of the view of the deterministic algorithm D on the randomized input $f \circ \pi$. (Here we use Fact 3.3.) We conclude that if $f \in \mathsf{P}_n \cup \Gamma_\epsilon(\mathsf{P}_n)$, then the probability that D errs on the input distribution $f \circ \pi$ is at most 2^{-k}. Again, using the hypothesis that X_n is Π-invariant, we conclude that the probability that D errs on input X_n is at most 2^{-k}.   ∎

Corollaries. Indeed, Theorem 3.1 follows as a special case of Theorem 3.4 by invoking [GT, Thm. 2] (and referring to the group of vertex-relabeling permutations). Next, we illustrate the applicability of Theorem 3.4 to testing low-degree polynomials (see, e.g., [RS]) and to testing monotone functions (see, e.g., [GGLRS]).

- In the case of low-degree tests (see, e.g., [RS]), for some finite field F, we are given a function $f : F^m \to F$ and wish to test whether it is a low-degree polynomial. The standard test selects uniformly at random a line in F^m, queries some points that reside on fixed locations on this line and accepts if and only if an adequate interpolation condition holds. This tester is clearly normal. Furthermore, this tester is Π-normal, where Π is the group of all full-rank affine transformations of F^m (because such transformations define a transitive operation on the set of all pairs of different points).[9] Thus, Theorem 3.4 can be applied to any distribution of functions that is Π-invariant.

- In the case of testing monotonicity (see, e.g., [GGLRS]), for some ordered set S, we are given a function $f : S^m \to \mathbb{R}$ and wish to test whether it is monotone (i.e., whether $f(\alpha) \le f(\beta)$ for every $\alpha = (\alpha_1, ..., \alpha_m)$ and $\beta = (\beta_1, ..., \beta_m)$ such that $\alpha_i \le \beta_i$ for every $i \in [m]$). In the case that $S = \{0, 1\}$, the standard test selects uniformly at random two points in S^m that differ in a single coordinate, queries f on these two points, and accepts if and only if an adequate inequality holds. This tester is clearly normal. Furthermore, monotonicity is Π-invariant for the group Π that consists of all permutations $\pi : S^m \to S^m$ such that $\pi(\alpha_1, ..., \alpha_m) = (\alpha_{\pi'(1)}, ..., \alpha_{\pi'(m)})$ for some permutation $\pi' : [m] \to [m]$. Unfortunately, the foregoing tester is not Π-invariant, because the permutations in Π preserve the Hamming weight of strings in $\{0, 1\}^m$.

  In order to apply Theorem 3.4, we decouple the foregoing tester into m tests such that the i-th test selects uniformly an m-bit string α of Hamming weight i and queries f on this string and on a random string obtained from α by setting one of its 1-entries to zero. Each of these testers is Π-invariant, and so we may apply an adequate extension of Theorem 3.4 that refers to testing properties by a conjunction of several tests.

---

[9] Note that our notion of normality is closely related (but not identical) to the notion of linear invariances studied in [KS].

We comment that similar ideas can be applied even to non-adaptive testers, which seems essential to settings such as testing properties of bounded-degree graphs in the incidence list model (of [GR1]). For example, note that the testers presented in [GR1, GR2] only employ comparison-based computations; that is, they can described in terms of operations such as select a random vertex, select a random neighbor of a given vertex, and test equality of two given vertices.[10] Thus, the operation of these algorithms is maintained when we relabel the vertices. Consequently, they can be derandomized analogously to the proof of Theorem 3.1, resulting in an algorithm that uses a fixed set of vertices and a fixed set of neighbor indices.[11]

Acknowledgments. I am grateful to Omer Reingold and Ronen Shaltiel for extremely useful and insightful discussions. I am also grateful to Marius Zimand for correcting my initial impression by which [Z] can handle any source of linear min-entropy.

References

[A]   Alon, N.: Testing subgraphs of large graphs. Random Structures and Algorithms 21, 359–370 (2002)
[AS]   Alon, N., Spencer, J.H.: The Probabilistic Method, 2nd edn. John Wiley & Sons, Inc., Chichester (2000)
[EKKRV]   Ergun, F., Kannan, S., Kumar, S.R., Rubinfeld, R., Viswanathan, M.: Spot-Checkers. In: 30th STOC, pp. 259–268 (1998)
[ESY]   Even, S., Selman, A.L., Yacobi, Y.: The Complexity of Promise Problems with Applications to Public-Key Cryptography. Inform. and Control 61, 159–173 (1984)
[FK]   Frieze, A., Kanan, R.: Quick approximation to matrices and applications. Combinatorica 19(2), 175–220 (1999)
[G]   Goldreich, O.: A Brief Introduction to Property Testing.
[GGLRS]   Goldreich, O., Goldwasser, S., Lehman, E., Ron, D., Samorodnitsky, A.: Testing Monotonicity. Combinatorica 20(3), 301–337 (2000)
[GGR]   Goldreich, O., Goldwasser, S., Ron, D.: Property testing and its connection to learning and approximation. Journal of the ACM, 653–750 (July 1998)
[GK]   Goldreich, O., Kaufman, T.: Proximity Oblivious Testing and the Role of Invariances. ECCC, TR10-058
[GR1]   Goldreich, O., Ron, D.: Property testing in bounded degree graphs. Algorithmica, 302–343 (2002)
[GR2]   Goldreich, O., Ron, D.: A sublinear bipartite tester for bounded degree graphs. Combinatorica 19(3), 335–373 (1999)

---

[10] Many of these algorithms also use the operation of retrieving all neighbors of a given vertex, which can be emulated by successively selecting a random neighbor for sufficiently many times. We also note that in [GR1, GR2] the incidence-lists are sorted, but this is immaterial to the algorithms. For simplicity, here we refer to unsorted incidence-lists.

[11] Alternatively, the bounded-degree graph model can be handled by the formalism introduced in the subsequent work of [GK].

[GS]   Goldreich, O., Sheffet, O.: On the randomness complexity of property test-
        ing. Computational Complexity 19(1), 99–133 (2010); Extended abstract
        in Proc. of RANDOM 2007 (2007)
[GT]   Goldreich, O., Trevisan, L.: Three theorems regarding testing graph prop-
        erties. Random Structures and Algorithms 23(1), 23–57 (2003)
[GW]   Goldreich, O., Wigderson, A.: Derandomization that is rarely wrong from
        short advice that is typically good. In: Rolim, J.D.P., Vadhan, S.P. (eds.)
        RANDOM 2002. LNCS, vol. 2483, pp. 209–223. Springer, Heidelberg (2002)
[KS]   Kaufman, T., Sudan, M.: Algebraic Property Testing: The Role of Invari-
        ances. In: 40th STOC, pp. 403–412 (2008)
[R]   Ron, D.: Algorithmic and Analysis Techniques in Property Testing. Foun-
        dations and Trends in TCS 5(2), 73–205 (2010)
[RS]   Rubinfeld, R., Sudan, M.: Robust characterization of polynomials with ap-
        plications to program testing. SIAM Journal on Computing 25(2), 252–271
        (1996)
[Z]   Zimand, M.: On derandomizing probabilistic sublinear-time algorithms. In:
        The Proc. of the 22nd IEEE Conference on Computational Complexity, pp.
        1–9 (2007)

Appendix: Generalization of Theorem 2.1

We first detail the generalization of Theorem 2.1 to properties of sequences over
any alphabet Σ. This requires generalizing the definition of Γ_ϵ as follows (for
any $P_n \subseteq \Sigma^n$):

$$\Gamma_\epsilon(P_n) \overset{\text{def}}{=} \{x \in \Sigma^n : \forall z \in P_n \ \Delta(x, z) > \epsilon \cdot n\} \tag{2}$$

where $\Delta(x_1 \cdots x_n, z_1 \cdots z_n) = |\{i : x_i \neq z_i\}|$ denotes the number of position in
the sequence on which $x = x_1 \cdots x_n$ and $z = z_1 \cdots z_n$ disagree.

Theorem 2.1, generalized. *For any property $P_n \subseteq \Sigma^n$, it holds that if $|P_n| \geq$
$\rho \cdot |\Sigma|^n$ and $\epsilon \geq \sqrt{\frac{8\ln(1/\rho)}{n}}$, then $|\Gamma_\epsilon(P_n)| \leq \rho \cdot |\Sigma|^n$.*

Proof: The proof of Theorem 2.1 generalizes easily, because the proof of Theorem 7.5.3 in [AS] applies (without any change) also to the general case. For sake
of self-containment, we reproduce the proof of [AS, Thm. 7.5.3]. Indeed, the
original text refers to $\Sigma = \{0, 1\}$ but it actually holds for any finite Σ (provided
that Δ and Γ_ϵ are defined as above).

Fixing any $P_n \subseteq \Sigma^n$, define $\Delta_{P_n}(x) = \min_{z \in P_n}\{\Delta(x, z)\}$, and consider a uniformly distributed $\omega \in \Sigma^n$. Then, the theorem's statement can be reformulated
as asserting that *if* $\Pr_\omega[\Delta_{P_n}(\omega) = 0] \geq \rho$, *then* $\Pr_\omega[\Delta_{P_n}(\omega) > \sqrt{8n\ln(1/\rho)}] \leq \rho$. In
order to prove this claim, we introduce a martingale (cf. [AS, Chap. 7]), $\zeta_0, ..., \zeta_n$,
such that

$$\zeta_i = \zeta_i(\omega) = |\Sigma|^{-(n-i)} \cdot \sum_{r_{i+1}, ..., r_n \in \Sigma} \Delta_{P_n}(\omega_1 \cdots \omega_i r_{i+1} \cdots r_n) \tag{3}$$

where $\omega = \omega_1 \cdots \omega_n$. (Indeed, $\zeta_n(\omega) = \Delta_{\mathsf{P}_n}(\omega)$ and $\zeta_0 = \mathsf{E}_\omega[\zeta_n]$.) Note that actually ζ_i only depends on $\omega_1 \cdots \omega_i$. Indeed, the martingale condition holds (i.e., for every fixed $\omega_1 \cdots \omega_i$, it holds that $\mathsf{E}_{\omega_{i+1}}[\zeta_{i+1}|\zeta_i] = \zeta_i$) and $|\zeta_{i+1} - \zeta_i| \leq 1$ (because $|\Delta_{\mathsf{P}_n}(x) - \Delta_{\mathsf{P}_n}(x')| \leq \Delta(x, x')$). By the Martingale Tail Inequality (cf. [AS, Thm. 7.2.1]) we have

$$\Pr_\omega[\zeta_n < \zeta_0 - \lambda\sqrt{n}] < e^{-\lambda^2/2} \qquad (4)$$

$$\Pr_\omega[\zeta_n > \zeta_0 + \lambda\sqrt{n}] < e^{-\lambda^2/2} \qquad (5)$$

Setting $\lambda = \sqrt{2\log(1/\rho)}$ (so that $\rho = e^{-\lambda^2/2}$) and contrasting Eq. (4) with $\Pr[\zeta_n = 0] \geq \rho$, we conclude that $\zeta_0 \leq \lambda\sqrt{n}$. Thus, Eq. (5) implies $\Pr[\zeta_n > 2\lambda\sqrt{n}] < \rho$, and the theorem follows. ∎

A Candidate Counterexample to the Easy Cylinders Conjecture

Oded Goldreich

Abstract. We present a candidate counterexample to the easy cylinders conjecture, which was recently suggested by Manindra Agrawal and Osamu Watanabe (see *ECCC*, TR09-019). Loosely speaking, the conjecture asserts that any 1-1 function in $\mathcal{P}/\text{poly}$ can be decomposed into "cylinders" of sub-exponential size that can each be inverted by some polynomial-size circuit. Although all popular one-way functions have such easy (to invert) cylinders, we suggest a possible counterexample. Our suggestion builds on the candidate one-way function based on expander graphs (see *ECCC*, TR00-090), and essentially consists of iterating this function polynomially many times.

Keywords: One-Way Functions, Trapdoor Permutations, P/poly.

A version of this work appeared as TR09-028 of *ECCC*.

1 The Easy Cylinders Conjecture

Manindra Agrawal and Osamu Watanabe [2, Sec. 4] have recently suggested the following interesting conjecture. The conjecture refers to the notion of an **easy cylinder**, defined next, and asserts that *every 1-1 and length-increasing function in $\mathcal{P}/\text{poly}$ has easy cylinders*.

Definition 1 (easy cylinders, simplified[1]): *A length function $\ell : \mathbb{N} \to \mathbb{N}$ is* admissible *if the mapping $n \mapsto \ell(n)$ can be computed in* $\text{poly}(n)$-*time and there exists a constant $\epsilon > 0$ such that $\ell(n) \in [n^\epsilon, n - n^\epsilon]$. A function f has* easy cylinders *if for some admissible length function ℓ there exists mappings $\sigma_1, \sigma_2 : \{0,1\}^* \to \{0,1\}^*$ such that the following conditions hold:*

1. *For every x, it holds that $|\sigma_1(x)| = \ell(|x|)$ and $|\sigma_2(x)| = |x| - \ell(|x|)$.*
2. *The function $\sigma(x) = (\sigma_1(x), \sigma_2(x))$ is 1-1, polynomial-time computable and polynomial-time invertible. The* cylinders *defined by σ_1 consists of the collection of sets $\{\sigma_1^{-1}(x')|_n : x' \in \{0,1\}^{\ell(n)}\}_{n \in \mathbb{N}}$, where $\sigma_1^{-1}(x')|_n \overset{\text{def}}{=} \{x \in \{0,1\}^n : \sigma_1(x) = x'\}$.*
   Each such set (i.e., $\sigma_1^{-1}(x')|_n$) is called a cylinder.

---

[1] Our formulation is a special case of the formulation in [2], but we believe that our candidate counterexample also holds for the definition in [2].

O. Goldreich et al.: Studies in Complexity and Cryptography, LNCS 6650, pp. 136–140, 2011.
© Springer-Verlag Berlin Heidelberg 2011

3. For every $n \in \mathbb{N}$ and $x' \in \{0,1\}^{\ell(n)}$, there exists a $\mathrm{poly}(n)$-size circuit $C = C_{x'}$ such that for every $x \in \sigma_1^{-1}(x')|_n$ it holds that $C(f(x)) = \sigma_2(x)$.
   Thus, the circuit C (effectively)[2] inverts f on the cylinder $\sigma_1^{-1}(x')|_n$.

That is, when restricted to any such cylinder, the function f is easy to invert.

Needless to say, the existence of easy cylinders is interesting only in the case that f is not polynomial-time invertible. Agrawal and Watanabe noted that all popular candidates one-way functions have easy cylinders. Generalizing their observations (and going somewhat beyond them), we first present four classes of functions that are conjectured to be one-way and still have easy cylinders. Next (in Section 3), we present our candidate counterexample.

2   Four Classes of Functions That Have Easy Cylinders

The first class generalizes the multiplication function (i.e., $(x', x'') \mapsto x' \cdot x''$). This class consists of (polynomial-time computable) functions f having the form $f(x) = g(\sigma_1(x), \sigma_2(x))$ such that the σ_i's satisfy the first two conditions in Definition 1 and the mapping $(x', x'') \mapsto (x', g(x', x''))$ is easy to invert (by an efficient algorithm, denoted I). That is, whereas the mapping $(x', x'') \mapsto g(x', x'')$ may be hard to invert, augmenting the output with x' (i.e., considering $(x', x'') \mapsto (x', g(x', x'')))$ makes the mapping easy to invert. Clearly, the cylinders defined by σ_1 are easy (since we can let $C_{\sigma_1(x)}(f(x))$ output the second element in the pair $I(\sigma_1(x), f(x))$).

The second class consists of functions that are derived from collections of finite one-way functions having a dense index set and dense domains.[3] For example, consider the DLP-based collection that consists of the functions $\{f_{p,g} : \mathbb{Z}_p \to \mathbb{Z}_p\}_{(p,g)}$, where p is prime, g is a generator of the multiplicative group modulo p, and $f_{p,g}(z) = g^z \bmod p$. For simplicity, we consider collections of the form $\{f_\alpha : \{0,1\}^{|\alpha|} \to \{0,1\}^{|\alpha|}\}_{\alpha \in I}$, where the index set I is dense (i.e., $|I \cap \{0,1\}^n| > 2^n/\mathrm{poly}(n)$). The one-wayness condition means that, for a typical $\alpha \in I$, the function f_α is hard to invert, and so the "natural" cylinders defined by $\sigma_1(\alpha, z) = \alpha$ are not easy. Nevertheless, the function $F(\alpha, z) = (\alpha, f_\alpha(z))$, which is (weakly) one-way, has easy ("unnatural") cylinders that are defined by $\sigma_1(\alpha, z) = z$; specifically, it is trivial to extract $\sigma_2(\alpha, z) = \alpha$ from $F(\alpha, z) = (\alpha, f_\alpha(z))$. (Indeed, in these easy cylinders, the "hard to invert part of F" is fixed.)

The third class consists of functions that are derived from collections of trapdoor one-way permutations. Unlike in the previous class, in the current case a non-trivial index-sampling algorithm, denoted I, must exist. This algorithm

---

[2] For any $x \in \sigma_1^{-1}(x')|_n$, an f-preimage of $y = f(x)$ is obtained by computing $\sigma^{-1}(x', C(y))$.

[3] Indeed, we consider a restricted case of [4, Def. 2.4.3]. On the other hand, note that any collection of finite one-way functions with dense domains can be converted into a collection of finite one-way functions over the set of all strings of a fixed length. Thus, we may freely use the latter.

samples the index set along with corresponding trapdoors; that is, the coins used to sample an index-trapdoor pair cannot be used as the index (because the trapdoor must be hard to recover from the index). Let $I_1(r)$ denote the index sampled on coins r, and let $I_2(r)$ denote the corresponding trapdoor (and suppose that the domains are dense as before, which indeed restricts [4, Def. 2.4.4]). Then, the function $F(r, z) = (I_1(r), f_{I_1(r)}(z))$ is (weakly) one-way, but it has easy cylinders that are defined by $\sigma_1(r, z) = r$; specifically, we use the circuit $C_r(F(r, z)) = f_{I_1(r)}^{-1}(z)$, which in turn uses the trapdoor $I_2(r)$ that corresponds to the index $I_1(r)$. (Note that the cylinders defined by $\sigma_1(r, z) = z$ are not easy in this case, since I_1 is hard to invert!)

The last class consists of all functions that are computable in $\mathcal{NC}_0$; that is, functions in which each output bit depends on a constant number of input bits. Recall that this class is widely conjectured to contain one-way functions (cf., the celebrated work of Applebaum, Ishai, and Kushilevitz [1]). For every such function $f : \{0,1\}^n \rightarrow \{0,1\}^n$, if we let σ_1 be the projection of the n-bit input on $n - n^{1/3}$ random coordinates, then, with high probability, we obtain easy cylinders.[4] The reason is that, with high probability, no output bit of the function is influenced by more than one of the $n^{1/3}$ remaining coordinates (and so the residual function $f(x)$ obtained after fixing the value of $\sigma_1(x)$ is essentially a projection).

3   Our Candidate Counterexample to the Conjecture

We note that the last class of functions (i.e., $\mathcal{NC}_0$) contains the candidate one-way function suggested by us [3]. However, *we believe that iterating this function for a polynomial* (or even linear) *number of times yields a function that has no easy cylinders*. For sake of self-containment, we recall the proposal of [3], hereafter referred to as the basic function.

The Basic Function. We consider a collection of finite functions $\{f_n : \{0,1\}^n \rightarrow \{0,1\}^n\}_{n \in \mathbb{N}}$ such that f_n is based a collection of $d(n)$-subsets, $S_1, ..., S_n \subset [n] \stackrel{\text{def}}{=} \{1, ..., n\}$, and a predicate $P : \{0,1\}^{d(n)} \rightarrow \{0,1\}$ (as follows).

1. The function d is relatively small; that is, $d = O(\log n)$ or even $d = O(1)$, but $d > 2$.
2. The predicate $P : \{0,1\}^d \rightarrow \{0,1\}$ should be thought of as being a random predicate. That is, it will be randomly selected, fixed, and "hard-wired" into the function. For sure, P should *not* be linear, nor depend on few of its bit locations.

---

[4] In fact, the argument remain intact as long as $\ell(n) = n - o(n^{1/2})$ (rather than $\ell(n) = n - n^{1/3}$). Actually, using $n - o(n^{2/3})$ random coordinates would work too, since then (w.h.p.) no output bit of the function is influenced by more than two of the $o(n^{2/3})$ remaining coordinates (and so a 2SAT solver can invert the residual function on each of the individual cylinders).

3. The collection $S_1, ..., S_n$ should be expanding: specifically, for *some* k, the union of every k subsets should cover at least $k + \Omega(n)$ elements of $[n]$ (i.e., for every $I \subset [n]$ of size k it holds that $|\bigcup_{i \in I} S_i| \geq k + \Omega(n)$). Specifically, it is suggested to have S_i be the set of neighbors of the ith vertex in a d-regular expander graph.

For $x = x_1 \cdots x_n \in \{0, 1\}^n$ and $S \subset [n]$, where $S = \{i_1, i_2, ..., i_t\}$ and $i_j < i_{j+1}$, we denote by x_S the projection of x on S; that is, $x_S = x_{i_1} x_{i_2} \cdots x_{i_t}$. Fixing P and $S_1, ..., S_n$ as above, we define the function

$$f_n(x) \overset{\text{def}}{=} P(x_{S_1}) P(x_{S_2}) \cdots P(x_{S_n}). \tag{1}$$

Note that we think of d as being relatively small (i.e., $d = O(\log n)$), and hope that the complexity of inverting f_n is related to $2^{n/O(1)}$. Indeed, the hardness of inverting f_n cannot be due to the hardness of inverting P, but is rather supposed to arise from the combinatorial properties of the collection of sets $\{S_1, ..., S_n\}$ (as well as from the combinatorial properties of predicate P). In general, the conjecture is that the complexity of the inversion problem (for f_n constructed based on such a collection) is exponential in the "net expansion" of the collection (i.e., the cardinality of the union minus the number of subsets).

We note that a non-uniform complexity version of this basic function (or rather the sequence of f_n's) may use possibly different predicates (i.e., different P_i's) for the different n applications of P in Eq. 1.

The Iterated Function – the Vanilla Version. The candidate counterexample, F, is defined by $F(x) = f_{|x|}^{p(|x|)}(x)$, where p is some fixed polynomial (e.g., $p(n) = n$) and $f_n^{i+1}(x) = f_n(f_n^i(x))$ (and $f_n^1(x) = f_n(x)$). We conjecture that this function has no easy cylinders.

The Iterated Function, Revisited. One possible objection to the foregoing function F as a counterexample to the easy cylinder conjecture is that F is unlikely to be 1-1. Although we believe that the essence of the easy cylinder conjecture is unrelated to the 1-1 property, we point out that this property may be obtained by suitable modifications. One possible modification that may yield a 1-1 function is obtained by prepending the application of F with an adequate expanding function (e.g., a function that stretches n-bit long strings to $m(n)$-bit long strings, where m is a polynomial or even a linear function). Specifically, for a function $m : \mathbb{N} \to \mathbb{N}$ such that $m(n) \in [2n, \text{poly}(n)]$, we define $g_n : \{0, 1\}^n \to \{0, 1\}^{m(n)}$ analogously to Eq. 1 (i.e., here we use an expanding collection of $m(n)$ subsets), and let $F'(x) = F(g_{|x|}(x))$; that is, for every $x \in \{0, 1\}^n$, we have $F'(x) = f_{m(n)}^{p(m(n))}(g_n(x))$.

4   Conclusion

Starting with the aforementioned non-uniform complexity version of the basic function f_n, and applying different incarnations of this function in the different iterations, we actually obtain a rather generic counterexample. Alternatively, we may directly consider functions $F_n : \{0,1\}^n \to \{0,1\}^{m(n)}$ such that the function F_n has a poly(n)-sized circuit. Note that such a circuit may be viewed as a composition of polynomially many circuits in $\mathcal{NC}_0$, which in turn may be viewed as basic functions. Furthermore, a random poly(n)-sized circuit is likely to be decomposed to $\mathcal{NC}_0$ circuits that correspond to basic functions in which the collection of sets (of input bits that influence individual output bits) are expanding. Needless to say, we believe that generic polynomial-size circuits have no easy cylinders.

It seems that the existence of easy cylinders in all popular candidate one-way functions is due to the structured nature of these candidates. Such a structure will not exist in the generic case, and so *we conjecture that the Easy Cylinders Conjecture is false.*

References

1. Applebaum, B., Ishai, Y., Kushilevitz, E.: Cryptography in NC0. SICOMP 36(4), 845–888 (2006)
2. Agrawal, M., Watanabe, O.: One-Way Functions and the Isomorphism Conjecture. ECCC, TR09-019 (2009)
3. Goldreich, O.: Candidate One-Way Functions Based on Expander Graphs. In: Goldreich, O., et al.: Studies in Complexity and Cryptography. LNCS, vol. 6650, pp. 76–87. Springer, Heidelberg (2011)
4. Goldreich, O.: Foundation of Cryptography: Basic Tools. Cambridge University Press, Cambridge (2001)

From Absolute Distinguishability to Positive Distinguishability

Zvika Brakerski and Oded Goldreich

Abstract. We study methods of converting algorithms that distinguish pairs of distributions with a gap that has an *absolute value* that is noticeable into corresponding algorithms in which the gap is always *positive* (and noticeable). Our focus is on designing algorithms that, in addition to the tested string, obtain a fixed number of samples from each distribution. Needless to say, such algorithms can not provide a very reliable guess for the sign of the original distinguishability gap, still we show that even guesses that are noticeably better than random are useful in this setting.

Keywords: Computational Indistinguishability, Statistical Indistinguishability.

A version of this work appeared as TR09-031 of *ECCC*.

1 The Problem and Its Solutions

This work addresses a generic technical problem that arises in the context of trying to establish the computational indistinguishability of certain pairs of probability ensembles. The problem refers to the fact that computational (and also statistical) indistinguishability is defined in terms of the absolute difference between probabilities, whereas it is typically easier to manipulate the difference itself. Thus, we seek a method of converting a non-negligible absolute difference into a non-negligible difference; that is, we wish the difference itself (rather than its absolute value) to be positive.

1.1 A Motivational Example

Many security definitions are formulated by referring to two pairs of *probability ensembles that are indexed by strings*, and requiring that these pairs of probability ensembles are computationally indistinguishable (see, e.g., the definitions of computational zero-knowledge [2, Sec. 4.3.1.2] and secure two-party computation [3, Sec. 7.2]). Such a probability ensemble $\{Z_\alpha\}_{\alpha \in S}$ consists of (an infinite number of) "random variables" Z_α's, which are each distributed over some finite set (related to its index, α). Two such ensembles, $\{X_\alpha\}_{\alpha \in S}$ and $\{Y_\alpha\}_{\alpha \in S}$, are said to be computationally indistinguishable if for every probabilistic polynomial-time algorithm D it holds that

$$g_D(\alpha) \stackrel{\text{def}}{=} |\Pr[D(\alpha, X_\alpha) = 1] - \Pr[D(\alpha, Y_\alpha) = 1]| \tag{1}$$

O. Goldreich et al.: Studies in Complexity and Cryptography, LNCS 6650, pp. 141–155, 2011.
© Springer-Verlag Berlin Heidelberg 2011

is negligible as a function of $|\alpha|$ (i.e., for every positive polynomial p and all sufficiently long α, the value of $g_D(\alpha)$ is upper bounded by $1/p(|\alpha|)$).

The aforementioned formulation mandates that the value of $g_D(\alpha)$ is small for every $\alpha \in S$. A weaker requirement, which suffices in practice, is that it is infeasible to find $\alpha \in S$ for which the value of $g_D(\alpha)$ is not small. This requirement may be formulated as mandating that for every probabilistic polynomial-time algorithm F, representing a potential finder that given 1^n outputs an n-bit long string $\alpha \in S$, the expected value of $g_D(\alpha)$ (when defined as in Eq. (1)) is negligible (as a function of n); that is, $E[g_D(F(1^n))]$ is negligible in n. This condition means that

$$\sum_\alpha \Pr[F(1^n) = \alpha] \cdot |\Pr[D(\alpha, X_\alpha) = 1] - \Pr[D(\alpha, Y_\alpha) = 1]| \qquad (2)$$

is negligible as a function of n.

When trying to establish a condition as in Eq. (2) it is often easier to establish a corresponding condition in which the absolute value operator is dropped. Indeed, suppose that for every F and D as above it holds that

$$\sum_\alpha \Pr[F(1^n) = \alpha] \cdot (\Pr[D(\alpha, X_\alpha) = 1] - \Pr[D(\alpha, Y_\alpha) = 1]) \qquad (3)$$

is negligible (as a function of n). Can we infer that Eq. (2) holds too?

In the case that both ensembles are polynomial-time sampleable, a positive answer is implicit in many works. Essentially, given a probabilistic polynomial-time algorithm D such that Eq. (2) is not negligible, one derives a probabilistic polynomial-time algorithm D' such that Eq. (3) is not negligible by estimating the difference between $\Pr[D(\alpha, X_\alpha) = 1]$ and $\Pr[D(\alpha, Y_\alpha) = 1]$ and flipping D's output if the estimated difference is negative. Thus, the construction of D' depends also on g_D (which determines the adequate level of approximation). In particular, the time complexity of D' is (polynomially) related to g_D. *Our goal is to get rid of this dependency; in particular, we wish to avoid the aforementioned approximation.*

1.2   A Generic Problem and One Solution

The generic problem we face is converting an algorithm D that distinguishes X_α and Y_α (i.e., $|\Pr[D(\alpha, X_\alpha) = 1] - \Pr[D(\alpha, Y_\alpha) = 1]|$ is noticeable) into an algorithm D' that on input (α, X_α) outputs 1 with probability that is noticeably higher than $\Pr[D(\alpha, Y_\alpha) = 1]$. We stress that we wish this transformation to hold for every α, whereas it may be that for some α's the difference $\Pr[D(\alpha, X_\alpha) = 1] - \Pr[D(\alpha, Y_\alpha) = 1]$ is positive while for other α's the difference is negative. Clearly, D' must know something about X_α and Y_α in order for this to be possible, and indeed we provide D' with samples taken from X_α and Y_α (or, actually, with algorithms for sampling these distributions).

Thus, the problem we face is actually the following one. We are given a probabilistic polynomial-time algorithm D and sampling algorithms for two ensembles,

$\{X_\alpha\}_{\alpha\in S}$ and $\{Y_\alpha\}_{\alpha\in S}$ (i.e., probabilistic polynomial-time algorithms X and Y such that on any input α it holds that $X_\alpha \equiv X(\alpha)$ and $Y_\alpha \equiv Y(\alpha)$). Our task is to construct a probabilistic polynomial-time algorithm D' such that for some function $\rho : (0,1] \to (0,1]$ it holds that

$$\Pr[D'(\alpha, X_\alpha)\,{=}\,1] - \Pr[D'(\alpha, Y_\alpha)\,{=}\,1] \;\geq\; \rho\,(|\Pr[D(\alpha, X_\alpha)\,{=}\,1] - \Pr[D(\alpha, Y_\alpha)\,{=}\,1]|)\,.$$
$$(4)$$

We stress that the r.h.s of Eq. (4) refers to the *absolute* difference between two probabilities, whereas the l.h.s refers to a corresponding difference that is not taken in absolute value and yet is required to be positive (whenever the former difference is positive).

We seek a universal transformation of D into D', whereas this transformation may use a predetermined number of auxiliary samples of the two distributions. That is, the resulting algorithm D' is given as input a single sample that is drawn from one of two distributions (i.e., either from X_α or from Y_α), but in addition it can obtain (a predetermined number of) samples from each of the two distributions. Like D, algorithm D' should distinguish the two cases (which correspond to the source of its input). We stress that we wish the complexity of D' (and specifically the number of auxiliary samples it obtains) to be independent of $g_D(\alpha)$. We note that such a transformation (of D into D') may be useful also in other settings. One such generic example is provided by *settings in which the notion of negligible probability being considered is significantly smaller than the reciprocal of the complexity of the distinguishers* (e.g., consider polynomial-time distinguishers coupled with (sub-)exponentially small distinguishing gaps).

A Simple Transformation. One solution to the foregoing problem is to let D' estimate the sign of $\Pr[D(\alpha, X_\alpha)\,{=}\,1] - \Pr[D(\alpha, Y_\alpha)\,{=}\,1]$ by using a single sample of X_α and a single sample of Y_α. (Although this estimate is quite poor, it can be shown to suffice.) Specifically, on input (α and) z (where z is taken from either X_α or Y_α), algorithm D' proceeds as follows:

1. Ignoring its ("main") input (i.e., z), algorithm D' obtains a single sample x of X_α and a single sample y of Y_α, and computes $\sigma \leftarrow D(\alpha, x)$ and $\tau \leftarrow D(\alpha, y)$;
2. If $\sigma > \tau$, then D' invokes D on its input (i.e., z), and outputs $D(\alpha, z)$. If $\sigma < \tau$, then D' outputs $1 - D(\alpha, z)$. Otherwise (i.e., $\sigma = \tau$), algorithm D' outputs the outcome of a fair coin toss.

(Indeed, we have assumed here, without loss of generality, that D always outputs a Boolean value.)[1] Intuitively, $\sigma - \tau$ provides a probabilistic guess of the sign of $\Pr[D(\alpha, X_\alpha)\,{=}\,1] - \Pr[D(\alpha, Y_\alpha)\,{=}\,1]$, and (as we show next) using this guess in the obvious manner yields the desired result.

Proposition 1.1 (analysis of the simple transformation): *Let D and D' be as above. Then,*

$$\Pr[D'(\alpha, X_\alpha)\,{=}\,1] - \Pr[D'(\alpha, Y_\alpha)\,{=}\,1] \;=\; (|\Pr[D(\alpha, X_\alpha)\,{=}\,1] - \Pr[D(\alpha, Y_\alpha)\,{=}\,1]|)^2\,.$$

---

[1] In general, the distinguishing gap of D is defined in terms of the probability that D outputs 1, and so any non-1 output may be considered as a 0.

Proof: For the analysis of the performance of D', we consider an algorithm D'', which may output any number in $[0, 1]$, such that

$$D''(\alpha, z) \stackrel{\text{def}}{=} \frac{1}{2} \cdot \left(1 + \text{sign}(D(\alpha, X_\alpha) - D(\alpha, Y_\alpha)) \cdot (-1)^{D(\alpha, z)+1}\right), \quad (5)$$

where $\text{sign}(r) = 1$ if $r > 0$ (resp., $\text{sign}(r) = -1$ if $r < 0$), and $\text{sign}(0) = 0$. Recall that in Step 2 of $D'(\alpha, z)$, the output is set to $D(\alpha, z)$ if $\sigma > \tau$, to $1 - D(\alpha, z)$ if $\sigma < \tau$, and is random if $\sigma = \tau$. Using $D(\alpha, z) \in \{0, 1\}$ and assuming $\sigma \neq \tau$, the output of $D'(\alpha, z)$ can be written as $(1 + \text{sign}(\sigma - \tau) \cdot (-1)^{D(\alpha, z)+1})/2$. Thus, $D'(\alpha, z)$ outputs 1 with probability $\text{E}[D''(\alpha, z)]$, and it suffices to analyze the l.h.s of the following equality

$$\text{E}[D''(\alpha, X_\alpha)] - \text{E}[D''(\alpha, Y_\alpha)] = \Pr[D'(\alpha, X_\alpha) = 1] - \Pr[D'(\alpha, Y_\alpha) = 1]. \quad (6)$$

Wishing to substitute Eq. (5) in Eq. (6), we denote by X'_α and Y'_α independent copies of X_α and Y_α, and analyze Eq. (6) as follows.

$$
\begin{aligned}
g_{D''}(\alpha) &\stackrel{\text{def}}{=} \text{E}[D''(\alpha, X_\alpha)] - \text{E}[D''(\alpha, Y_\alpha)] \\
&= \frac{1}{2} \cdot \text{E}\left[1 + \text{sign}(D(\alpha, X'_\alpha) - D(\alpha, Y'_\alpha)) \cdot (-1)^{D(\alpha, X_\alpha)+1}\right] \\
&\quad - \frac{1}{2} \cdot \text{E}\left[1 + \text{sign}(D(\alpha, X'_\alpha) - D(\alpha, Y'_\alpha)) \cdot (-1)^{D(\alpha, Y_\alpha)+1}\right] \\
&= \frac{1}{2} \cdot \text{E}\left[\text{sign}(D(\alpha, X'_\alpha) - D(\alpha, Y'_\alpha))\right] \cdot \text{E}\left[(-1)^{D(\alpha, X_\alpha)+1} - (-1)^{D(\alpha, Y_\alpha)+1}\right]
\end{aligned}
$$

where the last equality uses the statistical independence of (X'_α, Y'_α) and (X_α, Y_α). Denoting $p = \Pr[D(\alpha, X_\alpha) = 1]$ and $q = \Pr[D(\alpha, Y_\alpha) = 1]$, we use $\text{E}[(-1)^{D(\alpha, X_\alpha)+1}] = p - (1 - p) = 2p - 1$ and $\text{E}[(-1)^{D(\alpha, Y_\alpha)+1}] = 2q - 1$, and get

$$
\begin{aligned}
g_{D''}(\alpha) &= (p - q) \cdot \text{E}\left[\text{sign}(D(\alpha, X_\alpha) - D(\alpha, Y_\alpha))\right] \\
&= (p - q) \cdot (\Pr[D(\alpha, X_\alpha) > D(\alpha, Y_\alpha)] - \Pr[D(\alpha, X_\alpha) < D(\alpha, Y_\alpha)]) \\
&= (p - q) \cdot (p \cdot (1 - q) - (1 - p) \cdot q),
\end{aligned}
$$

which equals $(p - q)^2$. $\blacksquare$

1.3 Other Transformations

Two natural questions arise:

1. Is the foregoing construction of D' optimal (with respect to all constructions that use a single auxiliary sample from each of the two distributions)?
2. Can we do better if we obtain k auxiliary samples from each of the two distributions (rather than one auxiliary sample from each of the two distributions)? How good can such a construction be?

Before answering these questions we note that no construction (which is given a single *test* sample from one of the two distribution) can outperform the variation distance between the tested distributions, (i.e., $|p-q|$, where $p = \Pr[D(\alpha, X_\alpha) = 1]$ and $q = \Pr[D(\alpha, Y_\alpha) = 1]$). This holds also when we have full information of the two tested distributions. Turning back to the foregoing questions, we answer them as follows.

Theorem 1.2 (Main Result): *For every $k \geq 1$, the best construction that uses k auxiliary samples from each of the two distributions is the one that rules analogously to Eq. (5), when applying the sign function to the difference between the average values of D on the k samples of each of the two distributions. That is, on input an index α, a main input z, and $2k$ auxiliary samples, denoted $x_1, ..., x_k, y_1, ..., y_k$, where $x_1, ..., x_k$ are samples of X_α and $y_1, ..., y_k$ are samples of Y_α, the optimal algorithm D' outputs 1 with probability $(1 + \delta \cdot (-1)^{D(\alpha, z) + 1})/2$, where*

$$\delta \stackrel{\text{def}}{=} \text{sign} \left(\sum_{i=1}^k D(\alpha, x_i) - \sum_{i=1}^k D(\alpha, y_i) \right) \in \{-1, 0, 1\}.$$

In other words, algorithm D outputs

- $D(\alpha, z)$ *if* $\sum_{i=1}^k D(\alpha, x_i) > \sum_{i=1}^k D(\alpha, y_i)$,
- $1 - D(\alpha, z)$ *if* $\sum_{i=1}^k D(\alpha, x_i) < \sum_{i=1}^k D(\alpha, y_i)$, *and*
- *the outcome of a fair coin toss otherwise.*

This algorithm yields a gap that equals the minimum of $\Omega(\sqrt{k}) \cdot (p - q)^2$ and $(1 - \epsilon_{p,q}(k)) \cdot |p - q|$, where $\epsilon_{p,q}(k) = \exp(-\Omega((p - q)^2 \cdot k))$.

Note that for $k = o(1/(p-q)^2)$ the said gap is $\Omega(\sqrt{k}) \cdot (p - q)^2$, whereas for $k = \omega(1/(p-q)^2)$ we approach the ultimate value of $|p - q|$. We stress that *the foregoing result holds both in the computational setting and in the information theoretic setting.*

2 The General Treatment

Let X and Y be 0-1 random variables (representing $D(\alpha, X_\alpha)$ and $D(\alpha, Y_\alpha)$, respectively), and let X_i's (resp., Y_i's) be independent copies of X (resp., Y) representing additional samples available to us. We seek a randomized process $\Pi : \{0, 1\}^{2k+1} \to \{0, 1\}$ such that

$$\mathrm{E}[\Pi(X_1, ..., X_k, Y_1, ..., Y_k, X)] - \mathrm{E}[\Pi(X_1, ..., X_k, Y_1, ..., Y_k, Y)] \tag{7}$$

is maximized (as a function of $\delta = |\mathrm{E}[X] - \mathrm{E}[Y]|$, when maximizing over all possible 0-1 random variables X and Y that are at statistical distance δ). Indeed, the probability that $\Pi(a_1, ..., a_k, b_1, ..., b_k, c) = 1$ is determined by the function $f : \{0, 1\}^{2k+1} \to [0, 1]$ such that

$$f(a_1, ..., a_k, b_1, ..., b_k, c) \stackrel{\text{def}}{=} \Pr[\Pi(a_1, ..., a_k, b_1, ..., b_k, c) = 1].$$

Thus, it suffices to seek a function $f : \{0, 1\}^{2k+1} \to [0, 1]$ that maximizes

$$\mathrm{E}[f(X_1, ..., X_k, Y_1, ..., Y_k, X)] - \mathrm{E}[f(X_1, ..., X_k, Y_1, ..., Y_k, Y)] \tag{8}$$

(as a function of $\delta = |\mathrm{E}[X] - \mathrm{E}[Y]|$). Let us formally define a more general optimization problem.

The General Question (and Its Accompanied Notation). For a function $f : \{0,1\}^{2k+1} \to [0,1]$ and a pair $(p,q) \in [0,1]$, we denote by $\mathcal{V}_{(p,q)}(f)$ the value of Eq. (8), where X and Y are 0-1 random variables that satisfy $p = E[X]$ and $q = E[Y]$. Now, for any (possibly infinite) set (or class) of pairs in $[0,1]$, denoted $\mathcal{C}$, and any function $f : \{0,1\}^{2k+1} \to [0,1]$, we denote $\mathcal{V}_{\mathcal{C}}(f) \overset{\text{def}}{=} \min_{(p,q)\in\mathcal{C}}\{\mathcal{V}_{(p,q)}(f)\}$. We seek a function f for which $\mathcal{V}_{\mathcal{C}}(f)$ is maximal.

Summary of Our Results (and Their Organization). First, we will show that, without loss of generality, the function $f(x_1, ..., x_k, y_1, ..., y_k, z)$ may only depend on $s \overset{\text{def}}{=} \sum_{i\in[k]} x_i$, $t \overset{\text{def}}{=} \sum_{i\in[k]} y_i$ and z, and furthermore that it can take a specific *canonical form* (see Section 2.1). Next, in Section 2.2, we will show that, in all natural cases (i.e., for "symmertic" classes), *the canonical form can be further simplified to depend only on* $\text{sign}(s-t)$ *and* z. Actually, *this will yield a single optimal function.* Lastly, in Section 2.3, we will analyze the performance of this function.

2.1   Canonical Functions

We will first show that it suffices to consider functions f of the form

$$f(a_1, ..., a_k, b_1, ..., b_k, c) = \frac{1 + g\left(\sum_{i\in[k]} a_i\,,\sum_{i\in[k]} b_i\right)\cdot(-1)^c}{2} \tag{9}$$

where $g : \mathbb{N}^2 \to [-1, +1]$. We call such an f canonical. Note that the normalization (i.e., shifting by 1 and dividing by 2) is used to map $[-1, +1]$ to $[0, 1]$. (Note that an additive shift of f leaves the value of Eq. (8) intact, whereas multiplying f by any factor has the same effect on the value of Eq. (8).)

Definition 2.1 (dominating strategies) *We say that f' dominates f (w.r.t $\mathcal{C}$) if for every $(p,q) \in \mathcal{C}$ it holds that $\mathcal{V}_{(p,q)}(f') \geq \mathcal{V}_{(p,q)}(f)$.*

Proposition 2.2 (strong optimality): *For every $\mathcal{C}$ and every $f : \{0,1\}^{2k+1} \to [0,1]$ there exists a canonical function that dominates f (w.r.t $\mathcal{C}$).*

Proof: Given any function f, we consider the function f' such that for every $a, b \in \{0, 1, ..., k\}$ and $c \in \{0, 1\}$, the value $f'(a,b,c)$ equals the average of $f(a_1, ..., a_k, b_1, ..., b_k, c)$ taken over all $(a_1, ..., a_k), (b_1, ..., b_k) \in \{0,1\}^k$ that satisfy $\sum_{i\in[k]} a_i = a$ and $\sum_{i\in[k]} b_i = b$. Then, for every (p,q), we have $\mathcal{V}_{(p,q)}(f') = \mathcal{V}_{(p,q)}(f)$, because each permuation of any fixed sequence $(v_1, ..., v_k)$ is as likely to be the outcome of k independently and identically distributed samples. Next, note that the value of f' at any $(a,b) \in \{0, 1, ..., k\}^2$ and $c \in \{0, 1\}$ (i.e., the value $f'(a,b,c)$) can be written as

$$\frac{1 + (-1)^c}{2}\cdot f'(a,b,0) + \frac{1 - (-1)^c}{2}\cdot f'(a,b,1)$$

$$= \frac{1}{2}\cdot (f'(a,b,0) + f'(a,b,1)) + \frac{(-1)^c}{2}\cdot (f'(a,b,0) - f'(a,b,1))$$

$$= g_0(a,b) + g_1(a,b)\cdot(-1)^c$$

where $g_0(a,b) = (f'(a,b,0)+f'(a,b,1))/2$ and $g_1(a,b) = (f'(a,b,0)-f'(a,b,1))/2$. Note that $g_1(a,b) \in [-0.5,+0.5]$ and that replacing $g_0(a,b)$ by 0.5 does not change the value of $\mathcal{V}_{(p,q)}(f')$. Thus, setting $f''(a,b,c) = (1+2g_1(a,b)\cdot(-1)^c)/2$, we obtain a canonical function f'' that dominates f (because $\mathcal{V}_{(p,q)}(f'') = \mathcal{V}_{(p,q)}(f') = \mathcal{V}_{(p,q)}(f)$). ∎

Conclusion and Notation. At this point we can limit our search for good functions (i.e., functions that maximize Eq. (8)) to canonical functions. Thus, for every function $g : \mathbb{N}^2 \times \{0,1\} \to [-1,+1]$ and every $k \in \mathbb{N}$, we define $f_g^{(k)}$ as in Eq. (9), and consider the value $\mathcal{V}_{(p,q)}(f_g^{(k)})$. To estimate $\mathcal{V}_{(p,q)}(f_g^{(k)})$, we let X and Y be 0-1 random variables with $\mathrm{E}[X] = p$ and $\mathrm{E}[Y] = q$ and get

$$\mathcal{V}_{(p,q)}(f_g^{(k)}) = \frac{1}{2} \cdot \mathrm{E}\left[g\left(\sum_{i\in[k]} X_i, \sum_{i\in[k]} Y_i \right) \cdot (-1)^X \right] \tag{10}$$

$$-\frac{1}{2} \cdot \mathrm{E}\left[g\left(\sum_{i\in[k]} X_i, \sum_{i\in[k]} Y_i \right) \cdot (-1)^Y \right]. \tag{11}$$

Using the independence of X,Y and the X_i's and Y_i's, we rewrite Eq. (10)&(11) as

$$\mathcal{V}_{(p,q)}(f_g^{(k)}) = \frac{1}{2} \cdot \mathrm{E}\left[g\left(\sum_{i\in[k]} X_i, \sum_{i\in[k]} Y_i \right) \right] \cdot \mathrm{E}\left[(-1)^X - (-1)^Y \right]. \tag{12}$$

Recalling that $\mathrm{E}[(-1)^X] = (1-p) - p = 1-2p$ and $\mathrm{E}[(-1)^Y] = 1-2q$, we get $\mathrm{E}[(-1)^X - (-1)^Y] = 2(q-p)$ and so

$$\mathcal{V}_{(p,q)}(f_g^{(k)}) = (q-p) \cdot \mathrm{E}[g(X',Y')], \tag{13}$$

where $X' = \sum_{i\in[k]} X_i$ and $Y' = \sum_{i\in[k]} Y_i$. Denoting $B(p,i,k) = \binom{k}{i} \cdot p^i \cdot (1-p)^{k-i}$, we get

$$\mathcal{V}_{(p,q)}(f_g^{(k)}) = (q-p) \cdot \sum_{i,j\in\{0,1,\dots,k\}} B(p,i,k) \cdot B(q,j,k) \cdot g(i,j). \tag{14}$$

2.2 Symmetric Classes

We focus on symmetric classes of pairs, where $\mathcal{C}$ is symmetric if for every $(p,q) \in \mathcal{C}$ it also holds that $(q,p) \in \mathcal{C}$. In contrast, if $\mathcal{C}$ contains only pairs (p,q) such that $p > q$, then we may set $k = 0$ and use the identity function (because $\mathrm{E}[X] - \mathrm{E}[Y] = p - q = \mathtt{StatDiff}(X,Y)$). We show that, for symmetric classes, the "sign of the difference" function (i.e., $\mathrm{sd}(a,b) = \mathtt{sign}(b-a) \in \{-1,0,+1\}$) is optimal as a function g.

Proposition 2.3 (optimality): *For every symmetric* $\mathcal{C}$ *and every* $k \in \mathbb{N}$ *and* $g :$ $\mathbb{N}^2 \to [-1, +1]$, *it holds that* $\mathcal{V}_{\mathcal{C}}(f_{\text{sd}}^{(k)}) \geq \mathcal{V}_{\mathcal{C}}(f_g^{(k)})$, *where* $\text{sd}(a, b) = \text{sign}(b - a)$.

Recall that $\text{sign}(d) = -1$ if $d < 0$ (resp., $\text{sign}(d) = 1$ if $d > 0$), and $\text{sign}(0) = 0$.

Proof: Let $(p, q) \in \mathcal{C}$ be such that $\mathcal{V}_{(p,q)}(f_{\text{sd}}^{(k)}) = \mathcal{V}_{\mathcal{C}}(f_{\text{sd}}^{(k)})$. Then, by definition of $\mathcal{V}_{\mathcal{C}}(f_g^{(k)})$ and the fact that $(q, p) \in \mathcal{C}$ (which follows by the symmetry of $\mathcal{C}$), it holds that

$$\mathcal{V}_{\mathcal{C}}(f_g^{(k)}) \leq \frac{\mathcal{V}_{(p,q)}(f_g^{(k)}) + \mathcal{V}_{(q,p)}(f_g^{(k)})}{2}.$$

On the other hand, by the choice of $(p, q) \in \mathcal{C}$, it holds that $\mathcal{V}_{\mathcal{C}}(f_{\text{sd}}^{(k)}) \geq \mathcal{V}_{(p,q)}$ $(f_{\text{sd}}^{(k)})$. Furthermore, $\mathcal{V}_{(p,q)}(f_{\text{sd}}^{(k)}) = \mathcal{V}_{(q,p)}(f_{\text{sd}}^{(k)})$, because by Eq. (13) we have

$$\begin{aligned}
\mathcal{V}_{(p,q)}(f_{\text{sd}}^{(k)}) &= (q - p) \cdot \text{E}[\text{sd}(X', Y')] \\
&= (q - p) \cdot \text{E}[\text{sign}(Y' - X')] \\
&= (p - q) \cdot \text{E}[\text{sd}(Y', X')] \\
&= \mathcal{V}_{(q,p)}(f_{\text{sd}}^{(k)}).
\end{aligned}$$

Thus, it suffices to show that

$$\mathcal{V}_{(p,q)}(f_{\text{sd}}^{(k)}) + \mathcal{V}_{(q,p)}(f_{\text{sd}}^{(k)}) \geq \mathcal{V}_{(p,q)}(f_g^{(k)}) + \mathcal{V}_{(q,p)}(f_g^{(k)}). \tag{15}$$

For every $a, b \in \{0, 1, ..., k\}$, we shall show that replacing $g(a, b)$ by $\text{sign}(b - a)$ may only increase the value of $\mathcal{V}_{(p,q)}(f_g^{(k)}) + \mathcal{V}_{(q,p)}(f_g^{(k)})$. Let us start by recalling Eq. (14), which yields

$$\begin{aligned}
&\mathcal{V}_{(p,q)}(f_g^{(k)}) + \mathcal{V}_{(q,p)}(f_g^{(k)}) \\
&= (q - p) \cdot \sum_{i,j \in \{0,1,...,k\}} B(p, i, k) B(q, j, k) \cdot g(i, j) \\
&\quad + (p - q) \cdot \sum_{i,j \in \{0,1,...,k\}} B(q, i, k) B(p, j, k) \cdot g(i, j) \\
&= (q - p) \cdot \sum_{i,j \in \{0,1,...,k\}} [B(p, i, k) B(q, j, k) - B(q, i, k) B(p, j, k)] \cdot g(i, j).
\end{aligned}$$

Clearly, for $i = j$ we have $B(p, i, k) B(q, j, k) = B(q, i, k) B(p, j, k)$. For $i < j$ (resp., $j < i$), it holds that $B(p, i, k) B(q, j, k) > B(q, i, k) B(p, j, k)$ if and only if $p < q$ (resp., $q < p$). The latter claim seems self-evident, yet we provide a detailed proof next (for the case $p, q \in (0, 1)$).

$$\begin{aligned}
B(p, i, k) B(q, j, k) &= \binom{k}{i} \cdot p^i \cdot (1 - p)^{k-i} \cdot \binom{k}{j} \cdot q^j \cdot (1 - q)^{k-j} \\
&= \binom{k}{i} \cdot (1 - p)^k \cdot \binom{k}{j} \cdot (1 - q)^k \cdot (p/(1 - p))^i \cdot (q/(1 - q))^j
\end{aligned}$$

Thus, $\frac{B(p,i,k)B(q,j,k)}{B(q,i,k)B(p,j,k)}$ equals

$$\frac{(p/(1-p))^i \cdot (q/(1-q))^j}{(q/(1-q))^i \cdot (p/(1-p))^j} = \frac{(q/(1-q))^{j-i}}{(p/(1-p))^{j-i}}$$

Note that we have $p < q$ iff $(p/(1-p)) < (q/(1-q))$, and so $p < q$ iff $(p/(1-p))^{j-i} < (q/(1-q))^{j-i}$. It follows that $p < q$ iff $B(p,i,k)B(q,j,k) > B(q,i,k)B(p,j,k)$.

Recall that for $i < j$, it holds that $B(p,i,k)B(q,j,k) - B(q,i,k)B(p,j,k) > 0$ if and only if $q > p$. Thus, in this case, we maximize

$$(q-p) \cdot [B(p,i,k)B(q,j,k) - B(q,i,k)B(p,j,k)] \cdot g(i,j) \qquad (16)$$

by setting $g(i,j) = 1$ (because the first two factors have the same sign). Similarly, for $j > i$, it holds that $B(p,i,k)B(q,j,k) - B(q,i,k)B(p,j,k) > 0$ if and only if $q < p$, and so the maximization requires $g(i,j) = -1$. Indeed, for $i = j$, any setting of $g(i,j)$ will do. Thus, an optimal setting of $g(i,j)$ is $\mathtt{sign}(j-i)$, which equals $\mathtt{sd}(i,j)$. The claim follows. ∎

2.3   The Performance of the Function $f_{\mathtt{sd}}^{(k)}$

We now turn to evaluating the performance of the optimal function; that is, we evaluate $\mathcal{V}_{(p,q)}(f_{\mathtt{sd}}^{(k)})$. Recall that

$$\mathcal{V}_{(p,q)}(f_{\mathtt{sd}}^{(k)}) = (q-p) \cdot \sum_{i,j \in \{0,1,\ldots,k\}} B(p,i,k)B(q,j,k) \cdot \mathtt{sd}(i,j)$$

$$= (p-q) \cdot \sum_{i,j \in \{0,1,\ldots,k\}} B(p,i,k)B(q,j,k) \cdot \mathtt{sign}(i-j)$$

which yields $\mathcal{V}_{(p,q)}(f_{\mathtt{sd}}^{(k)}) = (p-q) \cdot v_{p,q}^{(k)}$, where

$$v_{p,q}^{(k)} \stackrel{\text{def}}{=} \mathrm{E}\left[\mathtt{sign}\left(\sum_{i \in [k]} X_i - \sum_{i \in [k]} Y_i\right)\right] \qquad (17)$$

such that the X_i's (resp., Y_i's) are 0-1 i.i.d with expectation p (resp., q). Letting $T_i = X_i - Y_i$, we rewrite Eq. (17) as $\mathrm{E}[\mathtt{sign}(\sum_{i \in [k]} T_i)]$, which equals

$$\Pr\left[\sum_{i \in [k]} T_i > 0\right] - \Pr\left[\sum_{i \in [k]} T_i < 0\right]. \qquad (18)$$

Note that $\mathrm{E}[T_i] = p-q$ and $\mathrm{Var}[T_i] = p(1-p)+q(1-q)$. Thus, it is apparent that $\mathcal{V}_{(p,q)}(f_{\mathtt{sd}}^{(k)})$ grows with k, unless either $\{p,q\} = \{0,1\}$ or $p = q$ (in which case $\mathcal{V}_{(p,q)}(f_{\mathtt{sd}}^{(k)}) = |p-q|$ for every $k \geq 1$), and that $\lim_{k \to \infty} \mathcal{V}_{(p,q)}(f_{\mathtt{sd}}^{(k)}) = |p-q|$.

All that remains is to determines the behavior of $\mathcal{V}_{(p,q)}(f_{\mathrm{sd}}^{(k)})$ as a function of k, which calls for analyzing Eq. (18). It should come at little surprise that all we can offer is functional relations (e.g., relating $\mathcal{V}_{(p,q)}(f_{\mathrm{sd}}^{(k+1)})$ to $\mathcal{V}_{(p,q)}(f_{\mathrm{sd}}^{(k)})$), approximations, and close expressions for small values of k. We start with the latter.

The cases of $k = 1$ and $k = 2$. For small k, we can write explicit expressions for Eq. (18); for example, for $k = 1$ Eq. (18) yields $\Pr[T_1 > 0] - \Pr[T_1 < 0] = p(1-q) - q(1-p) = p - q$, and so $\mathcal{V}_{(p,q)}(f_{\mathrm{sd}}^{(1)}) = (p-q)^2$. For $k = 2$, we have

$$
\begin{aligned}
\Pr[T_1 + T_2 > 0] - \Pr[T_1 + T_2 < 0] &= \Pr[T_1 + T_2 = 2] + 2\Pr[T_1 = 1 \wedge T_2 = 0] \\
&\quad - (\Pr[T_1 + T_2 = -2] + 2\Pr[T_1 = -1 \wedge T_2 = 0]) \\
&= p^2(1-q)^2 + 2p(1-q)(pq + (1-p)(1-q)) \\
&\quad - (q^2(1-p)^2 + 2q(1-p)(pq + (1-p)(1-q))) \\
&= (1 + (1-p)(1-q) + pq) \cdot (p-q)
\end{aligned}
$$

and so $\mathcal{V}_{(p,q)}(f_{\mathrm{sd}}^{(2)}) = (1 + (1-p)(1-q) + pq) \cdot (p-q)^2$ (see an alternative proof following the statement of Proposition 2.4). Thus, the improvement of the case of $k = 2$ over the case of $k = 1$ is a factor of $(1 + (1-p)(1-q) + pq)$, which is greater than 1 unless $\{p, q\} = \{0, 1\}$ (where a single sample is as good as k samples, for any $k > 1$).

The general case of $k > 1$. We now turn to a general analysis of Eq. (18) (and $\mathcal{V}_{(p,q)}(f_{\mathrm{sd}}^{(k)})$). Specifically, we consider the increase in the value of Eq. (18) when going from k to $k + 1$; that is, we define

$$
\Delta_{(p,q)}(k) \overset{\text{def}}{=} \mathrm{E}\left[\mathtt{sign}\left(\sum_{i \in [k+1]} T_i\right)\right] - \mathrm{E}\left[\mathtt{sign}\left(\sum_{i \in [k]} T_i\right)\right] \tag{19}
$$

and note that $\mathcal{V}_{(p,q)}(f_{\mathrm{sd}}^{(k+1)}) = \mathcal{V}_{(p,q)}(f_{\mathrm{sd}}^{(k)}) + (p-q) \cdot \Delta_{(p,q)}(k)$.

Proposition 2.4 (the growth of $\mathcal{V}_{(p,q)}(f_{\mathrm{sd}}^{(k)})$ as a function of k): *For every $k \geq 1$, it holds that* $\Delta_{(p,q)}(k) = (p-q) \cdot \Pr[S_k = 0]$, *where* $S_k \overset{\text{def}}{=} \sum_{i \in [k]} T_i$.

It follows that $\mathcal{V}_{(p,q)}(f_{\mathrm{sd}}^{(k+1)}) = \mathcal{V}_{(p,q)}(f_{\mathrm{sd}}^{(k)}) + (p-q)^2 \cdot \Pr[S_k = 0]$, and so $\mathcal{V}_{(p,q)}(f_{\mathrm{sd}}^{(k+1)}) \geq \mathcal{V}_{(p,q)}(f_{\mathrm{sd}}^{(k)})$, where equality holds if and only if $\{p, q\} = \{0, 1\}$ (when ignoring the case of $p = q$). Proposition 2.4 can also be used to re-establish $\mathcal{V}_{(p,q)}(f_{\mathrm{sd}}^{(2)}) = (1 + pq + (1-p)(1-q)) \cdot (p-q)^2$, since $\mathcal{V}_{(p,q)}(f_{\mathrm{sd}}^{(1)}) = (p-q)^2$ and $\Pr[S_1 = 0] = pq + (1-p)(1-q)$.

Proof: Starting with Eq. (19), we have

$$
\Delta_{(p,q)}(k) = \mathrm{E}[\mathtt{sign}(S_k + T_{k+1})] - \mathrm{E}[\mathtt{sign}(S_k)]
$$

$$= \sum_{s\in\{-1,0,1\}} \Pr[S_k = s] \cdot \mathrm{E}[\mathrm{sign}(s + T_{k+1}) - \mathrm{sign}(s)]$$

$$= \Pr[S_k = 0] \cdot (\Pr[T_{k+1} = 1] - \Pr[T_{k+1} = -1])$$

$$+ \Pr[S_k = -1] \cdot \Pr[T_{k+1} = 1] - \Pr[S_k = 1] \cdot \Pr[T_{k+1} = -1].$$

By symmetry (e.g., consider the case of $k = 1$), it is rather self-evident that $\Pr[S_k = -1] \cdot \Pr[T_{k+1} = 1] = \Pr[S_k = 1] \cdot \Pr[T_{k+1} = -1]$, yet we provide a detailed proof next.

$$\Pr[S_k = -1] \cdot \Pr[T_{k+1} = 1] = p(1-q) \cdot \sum_{j=1}^{k} B(p, j-1, k) B(q, j, k)$$

$$= p(1-q) \cdot \sum_{j=1}^{k} \binom{k}{j-1} p^{j-1}(1-p)^{k-j+1} \binom{k}{j} q^{j}(1-q)^{k-j}$$

$$= \sum_{j=1}^{k} \binom{k}{j-1} p^{j}(1-p)^{k+1-j} \binom{k}{j} q^{j}(1-q)^{k-j+1}$$

$$= (1-p)q \sum_{j=1}^{k} \binom{k}{j-1} p^{j}(1-p)^{k-j} \binom{k}{j} q^{j-1}(1-q)^{k-j+1}$$

$$= (1-p)q \cdot \sum_{j=1}^{k} B(p, j, k) B(q, j-1, k)$$

$$= \Pr[S_k = 1] \cdot \Pr[T_{k+1} = -1].$$

Hence, $\Delta_{(p,q)}(k) = \Pr[S_k = 0] \cdot (\Pr[T_{k+1} = 1] - \Pr[T_{k+1} = -1])$, and the claim follows (since $\Pr[T_{k+1} = 1] - \Pr[T_{k+1} = -1] = p - q$). $\blacksquare$

Another expression for $\mathcal{V}_{(p,q)}(f_{\mathrm{sd}}^{(k)})$. Proposition 2.4 yields another expression for $\mathcal{V}_{(p,q)}(f_{\mathrm{sd}}^{(k)})$:

$$\mathcal{V}_{(p,q)}(f_{\mathrm{sd}}^{(k)}) = \mathcal{V}_{(p,q)}(f_{\mathrm{sd}}^{(1)}) + (p-q) \cdot \sum_{\ell=1}^{k-1} \Delta_{(p,q)}(\ell) \tag{20}$$

$$= (p-q)^2 + (p-q)^2 \cdot \sum_{\ell=1}^{k-1} \Pr[S_\ell = 0] \tag{21}$$

Note that for $\{p, q\} = \{0, 1\}$ this expression (i.e., Eq. (21)) equals 1 (for any $k \geq 1$), whereas for $p = q$ it equals 0. In all other cases (i.e., $0 < (p-q)^2 < 1$) Eq. (21) grows with k. Using $\Pr[S_\ell = 0] = \sum_{j=0}^{\ell} B(p, j, \ell) B(q, j, \ell)$, we get

$$\mathcal{V}_{(p,q)}(f_{\mathrm{sd}}^{(k)}) = (p-q)^2 + (p-q)^2 \cdot \sum_{\ell=1}^{k-1} \sum_{j=0}^{\ell} \binom{\ell}{j}^2 (pq)^j ((1-p)(1-q))^{\ell-j} \tag{22}$$

In the special case of $p = 0$, Eq. (22) yields

$$\mathcal{V}_{(0,q)}(f_{\text{sd}}^{(k)}) = q^2 + q^2 \cdot \sum_{\ell=1}^{k-1} (1-q)^\ell$$

$$= q^2 + q \cdot ((1-q) - (1-q)^k)$$

which converges to $q = |p - q|$ when $k \to \infty$. Similarly, $\mathcal{V}_{(1,q)}(f_{\text{sd}}^{(k)})$ converges to $1 - q = |p - q|$ (where $p = 1$). Note that in these cases convergence occurs with $k \gg |p - q|^{-1}$. As we shall see next, in the other cases (i.e., $p, q \in (0, 1)$), convergence occurs with $k \gg |p - q|^{-2}$.

Approximating $\mathcal{V}_{(p,q)}(f_{\text{sd}}^{(k)})$ *When* $p, q \in (0, 1)$. The hidden constants in the approximation given next depend on the distance of p and q from the boundaries of $(0, 1)$; that is, the constants in the Θ-notation depends on $\min(p, q, 1-p, 1-q)$.

Proposition 2.5 (the approximate value of $\mathcal{V}_{(p,q)}(f_{\text{sd}}^{(k)})$): *For any fixed* $p, q \in (0, 1)$ *and every* $k > 2$, *it holds that* $\mathcal{V}_{(p,q)}(f_{\text{sd}}^{(k)}) = v \cdot |p - q|$, *where* $v = \Theta(\sqrt{k}) \cdot |p - q|$ *if* $k \leq 5(p - q)^{-2}$ *and* $v \geq 1 - \exp(-(p-q)^2 k/3)$ *otherwise.*

Proof: We shall approximate $\mathcal{V}_{(p,q)}(f_{\text{sd}}^{(k)})$ by using Eq. (17) (rather than Eq. (22)). Recall that by Eq. (17) we have

$$\mathcal{V}_{(p,q)}(f_{\text{sd}}^{(k)}) = (p - q) \cdot \text{E}[\text{sign}(S_k)] \tag{23}$$

where $S_k = \sum_{i=1}^k T_i$ (and $T_i = X_i - Y_i$). We assume, without loss of generality, that $p > q$ and lower bound the value of $\text{E}[\text{sign}(S_k)]$, using $\text{E}[T_i] = p - q$. We distinguish three cases according to the relation between k and $p - q$:

Case 1: $k \geq 5(p - q)^{-2}$. In this case we use the (standard additive) Chernoff Bound, and derive

$$\text{E}[\text{sign}(S_k)] = \Pr[S_k > 0] - \Pr[S_k < 0]$$
$$> 1 - 2 \cdot \Pr[S_k \leq 0]$$
$$> 1 - 2 \cdot \exp\left(-\frac{(p-q)^2 \cdot k}{2}\right).$$

This establishes the relevant part of the claim (i.e., $\mathcal{V}_{(p,q)}(f_{\text{sd}}^{(k)}) = v \cdot |p - q|$, where $v = 1 - 2\exp(-(p-q)^2 k/2) > 1 - \exp(-(p-q)^2 k/3)$). The following complemantary two cases are distinguished according to a constant $c \geq 5$ that depends only on $\gamma_{p,q} \overset{\text{def}}{=} \sqrt{p(1 - p) + q(1 - q)}$.

Case 2: $k \in [c \cdot (p - q)^{-1}, 5(p - q)^{-2}]$. In this case we use the Berry–Esseen estimate of the Central Limit Theorem (cf., e.g., [1, Sec. XVI.5]). Specifically, we approximate $\text{E}[\text{sign}(S_k)]$ by $\text{E}[\text{sign}(\widetilde{S}_k)]$, where $\widetilde{S}_k$ is the normal distribution approximation of S_k; that is,

$$\widetilde{S}_k \overset{\text{def}}{=} k \cdot (p - q) + \sqrt{k} \cdot \gamma_{p,q} \cdot \text{N}(0, 1),$$

where $N(0,1)$ denotes the normal distribution (with mean 0 and variance 1), and $\sqrt{k}\cdot\gamma_{p,q}$ replaces $\sqrt{\mathrm{Var}[S_k]}=\sqrt{k}\cdot\sqrt{p(1-p)+q(1-q)}$. More formally, we use the fact that for every r it holds that that

$$|\Pr[S_k>r]-\Pr[\widetilde{S}_k>r]| < \epsilon \stackrel{\mathrm{def}}{=} \frac{3\rho}{\gamma_{p,q}{}^3\sqrt{k}}$$

where $\rho = \mathrm{E}[|T_1-(p-q)|^3] < 2\cdot\gamma_{p,q}{}^2$. It follows that

$$
\begin{aligned}
\mathrm{E}[\mathtt{sign}(S_k)] &= \Pr[S_k>0]-\Pr[S_k<0]\\
&= \Pr[\widetilde{S}_k>0]-\Pr[\widetilde{S}_k<0]\pm 2\epsilon\\
&= 2\Pr[\widetilde{S}_k>0]-1\pm 2\epsilon.
\end{aligned}
\tag{24}
$$

Now, we analyze $\Pr[\widetilde{S}_k>0]$ via

$$\Pr[(p-q)k+\sqrt{k}\gamma_{p,q}\cdot N(0,1) > 0] = \Pr\left[N(0,1) > -\frac{p-q}{\gamma_{p,q}}\cdot\sqrt{k}\right]$$

Setting $r\stackrel{\mathrm{def}}{=}(p-q)\sqrt{k}\leq 1$, it follows that $\Pr[N(0,1) > -r/\gamma_{p,q}]=0.5+\Theta(r)$. So Eq. (24) yields $\Theta(\sqrt{k}\cdot(p-q))-\Theta(k^{-1/2})$, which is lower bounded by $\Theta(\sqrt{k}\cdot(p-q))$, when using $k\geq c\cdot(p-q)^{-1}$ (where c is large enough w.r.t the above hidden constants). It follows $\mathcal{V}_{(p,q)}(f_{\mathtt{sd}}^{(k)})=\Theta(\sqrt{k})\cdot(p-q)^2$, which establishes the other part of the claim for the current case.

Case 3: $k\leq c\cdot(p-q)^{-1}$. It suffices to establish that $\mathcal{V}_{(p,q)}(f_{\mathtt{sd}}^{(k)})=\Theta(\sqrt{k})\cdot(p-q)^2$, for $k\leq(p-q)^{-1}$. This is done by writing T_i as $T_i'+(1-T_i')\cdot T_i''$, where $T_i'\in\{0,1\}$ and $T_i''\in\{-1,0,1\}$ are independent random variables satisfying $\Pr[T_i'=1]=p-q$ and $\Pr[T_i''=1]=\Pr[T_i''=-1]=\frac{q-pq}{1-(p-q)}$. Letting $S_k'=\sum_{i\in[k]}T_i'$ and $S_k''=\sum_{i\in[k]}T_i''$, we have

$$\mathrm{E}[\mathtt{sign}(S_k)] = \sum_{j=0}^{k}\Pr[S_k'=j]\cdot\mathrm{E}[\mathtt{sign}(S_{k-j}''+j)]$$

$$= \sum_{j=0}^{k}\Pr[S_k'=j]\cdot\left(\mathrm{E}[\mathtt{sign}(S_{k-j}'')]+2\cdot\Pr[0\leq S_{k-j}''<j]\right) \tag{25}$$

where S_{k-j}'' represents the sum of the $k-j$ variables T_i'' that correspond to the indices i that satisfy $T_i'=0$ (i.e., S_{k-j}'' represents $\sum_{i\in I}T_i''$, where $I=\{i:T_i'=0\}$). Since $\mathrm{E}[\mathtt{sign}(S_{k-j}'')]=0$ (becuase $\mathrm{E}[T_i'']=0$), Eq. (25) simplifies to

$$2\cdot\sum_{j=1}^{k}\Pr[S_k'=j]\cdot\Pr[0\leq S_{k-j}''<j]. \tag{26}$$

The lower bound in the claim (i.e., $v=\Omega(\sqrt{k}\cdot(p-q))$) follows once we prove that $\Pr[S_k'=1]\cdot\Pr[S_{k-1}''=0]=\Omega(\sqrt{k}\cdot(p-q))$. We start by noting that

$$\Pr[S'_k = 1] \cdot \Pr[S''_{k-1} = 0] = k \cdot (p-q)(1 - (p-q))^{k-1} \cdot \Pr[S''_{k-1} = 0] \quad (27)$$
$$> \frac{(p-q)k}{3} \cdot \Pr[S''_{k-1} = 0]$$

In order to estimate $\Pr[S''_{k-1} = 0]$, we write S''_{k-1} as the difference of $\sum_{i \in [k-1]} X''_i$ and $\sum_{i \in [k-1]} Y''_i$, where the X''_i's and Y'''_i's are iid 0-1 random variables (i.e., $p'' = \Pr[X''_i = 1]$ satisfies $p''(1 - p'') = \frac{(1-p)q}{1-(p-q)}$). We get

$$\Pr[S''_{k-1} = 0] \geq \sum_{j=(k-1)p'' \pm \sqrt{k-1}} \Pr\left[\sum_{i \in [k-1]} X''_i = j\right] \cdot \Pr\left[\sum_{i \in [k-1]} Y''_i = j\right]$$

$$= \sum_{j=(k-1)p'' \pm \sqrt{k-1}} \Pr\left[\sum_{i \in [k-1]} X''_i = j\right]^2$$

$$> \frac{\Pr\left[\sum_{i \in [k-1]} X''_i = (k-1)p'' \pm \sqrt{k-1}\right]^2}{2\sqrt{k-1} + 1}$$

$$> \frac{\Pr\left[\sqrt{(k-1)\gamma_{p'',p''}} \cdot N(0,1) = \pm\sqrt{k-1}\right]^2 - o(1)}{2\sqrt{k-1} + 1}$$

where the last inequality uses the Berry–Esseen estimate of the Central Limit Theorem. Observing that $\Pr[N(0,1) = \pm 1/\gamma_{p'',p''}] = \Omega(1)$, it follows that $\Pr[S''_{k-1} = 0] = \Omega(1/\sqrt{k-1})$, and so Eq. (27) is $\Omega((p-q)k/\sqrt{k-1})$ (and the same holds w.r.t Eq. (26)). To upper bound Eq. (26), we note that it can be upper bounded by

$$2 \cdot \sum_{j=1}^{k} \Pr[S'_k = j] \cdot j \cdot \Pr[S''_{k-j} = 0] < 2 \cdot \sum_{j=1}^{k} \binom{k}{j} \cdot (p-q)^j \cdot j \cdot \Pr[S''_{k-j} = 0]$$

$$= O((p-q)k \cdot \Pr[S''_{k-1} = 0])$$

and the claim follows because $\Pr[S''_{k-1} = 0] = O(1/\sqrt{k})$. This establishes $\mathcal{V}_{(p,q)}(f_{\text{sd}}^{(k)}) = \Theta(\sqrt{k}) \cdot (p-q)^2$ also in the current case.

The proposition follows.   ∎

3   Conclusion

The obvious way of using statistical information (e.g., a binary guess that is positively correlated with the correct value) is to amplify the confidence level of the information and use it as if it were certainly correct. The current work studies an alternative method of using statistical information and shows that in some settings using unreliable information directly works quite well. This was demonstrated already in Section 1.2, whereas the rest of this work studies the question of how to make the best use of multiple independent copies of such statistical information.

Acknowledgments. We are grateful to Ofer Zeitouni and Dana Ron for useful discussions.

References

1. Feller, W.: An Introduction to Probability Theory and Its Applications, 2nd edn., vol. II. John Wiley & Sons, Chichester (1972)
2. Goldreich, O.: Foundation of Cryptography – Basic Tools. Cambridge University Press, Cambridge (2001)
3. Goldreich, O.: Foundation of Cryptography – Basic Applications. Cambridge University Press, Cambridge (2004)

Testing Graph Blow-Up

Lidor Avigad and Oded Goldreich

Abstract. Referring to the query complexity of testing graph properties in the adjacency matrix model, we advance the study of the class of properties that can be tested non-adaptively within complexity that is inversely proportional to the proximity parameter. Arguably, this is the lowest meaningful complexity class in this model, and we show that it contains a very natural class of graph properties. Specifically, for every fixed graph H, we consider the set of all graphs that are obtained by a (possibly unbalanced) blow-up of H. We show a non-adaptive tester of query complexity $\widetilde{O}(1/\epsilon)$ that distinguishes graphs that are a blow-up of H from graphs that are ϵ-far from any such blow-up.

Keywords: Property Testing, Adaptivity vs Non-adaptivity, One-sided vs Two-sided Error, Graph Properties, Graph Blow-Up.

This work is based on the M.Sc. thesis of the first author [A], which was completed under the supervision of the second author.

1 Introduction

The general context of this work is that of testing graph properties in the adjacency matrix representation (as initiated in [GGR]). In this model graphs are viewed as (symmetric) Boolean functions over a domain consisting of all possible vertex-pairs (i.e., an N-vertex graph $G = ([N], E)$ is represented by the function $g : [N] \times [N] \to \{0,1\}$ such that $\{u,v\} \in E$ if and only if $g(u,v) = 1$). Consequently, an N-vertex graph represented by the function $g : [N] \times [N] \to \{0,1\}$ is said to be ϵ-far from some predetermined graph property if more than $\epsilon \cdot N^2$ entries of g must be modified in order to yield a representation of a graph that has this property. We refer to ϵ as the proximity parameter, and the complexity of testing is stated in terms of ϵ and the number of vertices in the graph (i.e., N).

Interestingly, many natural graph properties can be tested within query complexity that depends only on the proximity parameter; see [GGR], which presents testers with query complexity poly$(1/\epsilon)$, and [AFNS], which characterizes the class of properties that are testable within query complexity that depends only on the proximity parameter (where this dependence may be an arbitrary function of ϵ). A well-known open problem in this area is to characterize the class of graph properties that can be tested within query complexity poly$(1/\epsilon)$. We mention that such a characterization has been obtained in the special case of induced subgraph freeness properties [AS], but the general case seems quite difficult.

O. Goldreich et al.: Studies in Complexity and Cryptography, LNCS 6650, pp. 156–172, 2011.
© Springer-Verlag Berlin Heidelberg 2011

In light of this state of affairs, it was suggested in [GR08] to try to characterize lower query complexity classes, and in particular the class of graph properties that can be tested non-adaptively within query complexity $\widetilde{O}(1/\epsilon)$. As a first step towards this goal, it was shown in [GR08, Sec. 6] that, for every constant c, the set of graphs that each consists of at most c isolated cliques is such a property.

In this work we significantly extend the latter result by showing that the class of graph properties that can be tested non-adaptively within query complexity $\widetilde{O}(1/\epsilon)$ contains all graph blow-up properties. For any fixed graph $H = ([h], F)$, we say that a graph $G = ([N], E)$ is a blow-up of H if the vertices of G can be clustered in up to h clusters such that the edges between these clusters reflect the edge relation of H. That is, vertices in the ith and j^{th} cluster are connected in G if and only if $(i, j) \in F$. Note that, unlike in the case of balanced blow-up (cf. [GKNR]), the clusters are not required to have equal size.[1] Also note that the "collection of c cliques" property studied in [GR08, Sec. 6] can be cast as the property of being a blow-up of a c-vertex clique (by considering the complement graph).

Theorem 1.1 (main result): *For every fixed H, the property of being a blow-up of H is testable by $\widetilde{O}(1/\epsilon)$ non-adaptive queries. Furthermore, the tester has one-sided error* (i.e., it always accepts graphs that are blow-ups of H) *and runs in* $\text{poly}(1/\epsilon)$-*time.*

We mention that the aforementioned property cannot be tested by $o(1/\epsilon)$ queries, even when adaptivity and two-sided error are allowed (see [GR08, Prop. 6.1]). We also mention that, by [GR08, Prop. 6.2], a tester of $\widetilde{O}(1/\epsilon)$ query complexity cannot be canonical (i.e., it cannot rule by inspecting an induced subgraph).

Additional results. We also consider the complexity of testing "balanced blow-up" properties, showing that the two-sided error query complexity is quadratic in $1/\epsilon$ for both adaptive and non-adaptive testers; see Proposition 2.4. Finally, we present proximity oblivious testers (cf. [GR09]) for any (general) blow-up property; see Theorem 5.2.

Techniques. Theorem 1.1 is proved by presenting a suitable tester and analyzing it. Recall that this tester cannot be canonical; indeed, this tester selects at random a sample of $\widetilde{O}(1/\epsilon)$ vertices, but it inspects (or queries) only $\widetilde{O}(1/\epsilon)$ of the vertex pairs in this sample. Consequently, the tester (and the analysis) has to deal with partial knowledge of the subgraph induced by the sample. A pivotal notion regarding such partial views is of "inconsistency" between vertices (w.r.t a given partial view), which means that these vertices have different neighbor sets and thus cannot be placed in the same cluster (of a blow-up of H (or any other graph)). Specifically, the tester considers all sets of up to $h + 1$ pairwise inconsistent vertices, and accepts if and only if each such set (along with the

---

[1] We note that testing balanced blow-up properties requires $\Omega(1/\epsilon^2)$ queries. For details, see Section 2.2.

known incidence relations) can be embedded in H. As usual, the technically challenging part is analyzing the behavior of the tester on arbitrary graphs that are far from being blow-ups of H. Our analysis proceeds in iterations, where in each iteration some progress is made, but this progress is not necessarily reflected by a growing number of incidence constraints but rather in the decreasing density of the violations reflected in the incidence constraints. This progress is captured in Lemma 4.4 (which refers to notions introduced in Section 4.1). Here we merely stress that the number of iterations is polylogarithmic in ϵ^{-1} rather than being $O(h^2)$. (The degree of the polylogarithmic function depends on h.)

Organization. The core of this paper is presented in Sections 3 and 4, which contain a description of the tester and its analysis, respectively. (Indeed, this part establishes Theorem 1.1.) Section 2 provides preliminaries, which may be skipped by the experts, as well as a side discussion (and result) regarding "balanced blow-up" properties. Section 5 provides another secondary discussion; this one regarding proximity oblivious testers.

2   Preliminaries

In this section we review the definition of property testing, when specialized to graph properties in the adjacency matrix model. We also define the blow-up properties (and discuss the case of balanced blow-up).

2.1   Basic Notions

For an integer n, we let $[n] \overset{\text{def}}{=} \{1, ..., n\}$. A generic N-vertex graph is denoted by $G = ([N], E)$, where $E \subseteq \{\{u, v\} : u, v \in [N]\}$ is a set of (unordered) pairs of vertices.[2] Any set of (such) graphs that is closed under isomorphism is called a **graph property**. By oracle access to such a graph $G = ([N], E)$ we mean oracle access to the Boolean function that answers the query $\{u, v\}$ (or rather $(u, v) \in [N] \times [N]$) with the bit 1 if and only if $\{u, v\} \in E$. At times, we look at E as a subset of $[N] \times [N]$; that is, we often identify E with $\{(u, v) : \{u, v\} \in E\}$.

Definition 2.1 (property testing for graphs in the adjacency matrix model): *A* **tester** *for a graph property Π is a probabilistic oracle machine that, on input parameters N and ϵ and access to an N-vertex graph $G = ([N], E)$, outputs a binary verdict that satisfies the following two conditions.*

1. *If $G \in \Pi$, then the tester accepts with probability at least $2/3$.*
2. *If G is ϵ-far from Π, then the tester accepts with probability at most $1/3$, where G is ϵ-**far** from Π if for every N-vertex graph $G' = ([N], E') \in \Pi$ it holds that the symmetric difference between E and E' has cardinality that is greater than ϵN^2.*

---

[2] Thus, we consider *simple* graphs, with no self-loops nor parallel edges.

If the tester accepts every graph in Π with probability 1, then we say that it has one-sided error. *A tester is called* non-adaptive *if it determines all its queries based solely on its internal coin tosses* (and the parameters N and ϵ); *otherwise it is called* adaptive.

The query complexity of a tester is the number of queries it makes to any N-vertex graph oracle, as a function of the parameters N and ϵ. We say that a tester is efficient if it runs in time that is polynomial in its query complexity, where basic operations on elements of $[N]$ are counted at unit cost. We note that all testers presented in this paper are efficient, whereas the lower-bounds hold also for non-efficient testers.

We shall focus on properties that can be tested within query complexity that only depends on the proximity parameter, ϵ. Thus, the query-complexity upper-bounds that we state hold for any values of ϵ and N, but will be meaningful only for $\epsilon > 1/N^2$ or so. In contrast, the lower-bounds (e.g., of $\Omega(1/\epsilon)$) cannot possibly hold for $\epsilon < 1/N^2$, but they will indeed hold for any $\epsilon > N^{-\Omega(1)}$. Alternatively, one may consider the query-complexity as a function of ϵ, where for each fixed value of $\epsilon > 0$ the value of N tends to infinity.

2.2   The Blow-Up Properties

Following the discussion in the introduction, we first define the blow-up properties that are the subject of our study.

Definition 2.2 (graph blow-up): *We say that the graph $G = ([N], E)$ is a* blow-up *of the graph $H = ([h], F)$ if there is an h-way partition $(V_1, ..., V_h)$ of the vertices of G such that for every $i, j \in [h]$ and $(u, v) \in V_i \times V_j$ it holds that $(u, v) \in E$ if and only if $(i, j) \in F$. We stress that the V_i's are not required to be of equal size and that some of them may be empty. We denote by $\mathcal{BU}(H)$ (resp., $\mathcal{BU}_N(H)$) the set of all graphs (resp., N-vertex graphs) that are blow-ups of H.*

In contrast to Definition 2.2, let us briefly consider the more rigid (and popular) definition of a *balanced* blow-up.

Definition 2.3 (balanced blow-up): *We say that the graph $G = ([N], E)$ is a* balanced blow-up *of the graph $H = ([h], F)$ if there is an h-way partition $(V_1, ..., V_h)$ of the vertices of G such that the following two conditions hold:*

1. *For every $i, j \in [h]$ and $(u, v) \in V_i \times V_j$ it holds that $(u, v) \in E$ if and only if $(i, j) \in F$.*
2. *For every $i \in [h]$ it holds that $|V_i| \in \{\lfloor N/h \rfloor, \lceil N/h \rceil\}$.*

We denote by $\mathcal{BBU}(H)$ (resp., $\mathcal{BBU}_N(H)$) the set of all graphs (resp., N-vertex graphs) that are balanced blow-ups of H.

It is easy to see that, except for trivial cases (i.e., when H consists of isolated vertices), balanced blow-up cannot be tested with one-sided error and complexity that does not depend on the size of the graph. The two-sided error testing complexity of this property is $\Theta(1/\epsilon^2)$, as shown next.

Proposition 2.4 (on the complexity of testing balanced blow-up): *For every* $H = ([h], F)$ *such that* $F \neq \emptyset$, *testing the property* $\mathcal{BBU}(H)$ *requires* $\Omega(1/\epsilon^2)$ *queries even if adaptive testers of two sided error are allowed. On the other hand, for any* $H = ([h], F)$, *there exists a non-adaptive tester of query complexity* $O(1/\epsilon^2)$ *(and two-sided error) for the property* $\mathcal{BBU}(H)$.

Proof: The lower bound follows directly from the known lower bounds on estimating the average (cf. [CEG]). Specifically, distinguishing Boolean functions defined over $[N]$ and having an average value of 0.5 from Boolean functions having an average of $0.5 - \epsilon$ can be reduced to distinguishing N-vertex graphs that consist of two isolated cliques of the same size from graphs that consist of two isolated cliques of sizes $(0.5 - \epsilon) \cdot N$ and $(0.5 + \epsilon) \cdot N$, respectively. (Given oracle access to a function $f : [N] \to \{0, 1\}$ consider the graph $G = ([N], \{(u, v) : f(u) = f(v)\})$.)

In describing the tester, we first assume that $H = ([h], F)$ is not a blow-up of any smaller graph H'. Also, anticipating the extension to the general case, we generalize the balanced blow-up property into a **proportional blow-up** property. Here, for a fixed graph $H = ([h], F)$ and sequence of densities $\bar{\rho} = (\rho_1, .., \rho_h)$, the graph G is a $\bar{\rho}$-blow-up of H if Definition 2.3 holds with Condition 2 replaced by $|V_i| \in \{\lfloor \rho_i N \rfloor, \lceil \rho_i N \rceil\}$. The non-adaptive tester for $\bar{\rho}$-blow-up of H, where H is not a blow-up of any smaller graph, proceeds as follows (on input a graph G):

1. Select uniformly a sample of $\widetilde{O}(1/\min_i\{\rho_i\})$ vertices, denoted B, which will be used as a basis for clustering in Step 2. Select uniformly a sample of $O(|B|/\epsilon^2)$ vertices, denoted S. Finally, select uniformly a sample of $O(h^2/\epsilon)$ vertex pairs in $S \times S$, denoted T.
2. Query all pairs $(u, v) \in (B \times S) \cup T$, and cluster the vertices in S according to their neighbors in B. That is, for every $v \in [N]$, let $\mathrm{sg}_B(v) \stackrel{\text{def}}{=} \{u \in B : (u, v) \in E\}$, and, for every set $B' \subseteq B$, let $S_{B'} \stackrel{\text{def}}{=} \{v \in S : \mathrm{sg}_B(v) = B'\}$.
3. If the number of non-empty sets $S_{B'}$ exceeds h, then reject. Otherwise, consider all possible 1-1 mappings from $C \stackrel{\text{def}}{=} \{B' : S_{B'} \neq \emptyset\}$ to $[h]$, and for each such mapping ϕ determine whether or not the following two conditions hold.
   (a) For every $B' \in C$ it holds that $|S_{B'}| = (1 \pm \epsilon/2) \cdot \rho_{\phi(B')} \cdot |S|$.
   (b) For every $(u, v) \in T$ it holds that $(u, v) \in E$ if and only if $(\phi(\mathrm{sg}_B(u)), \phi(\mathrm{sg}_B(v))) \in F$.
   The test accepts if and only if there exists a mapping ϕ that satisfies both the above conditions.

The number of queries performed by the tester is $O(|B|^2/\epsilon^2) = O(1/\epsilon^2)$. We first consider what happens if G is a $\bar{\rho}$-blow-up of H. In this case, with high probability, (1) the sample B contains at least one representative from each cluster of G, and (2) for each $i \in [h]$ the sample S contains $(1 \pm \epsilon/2) \cdot \rho_i \cdot |S|$ representatives of the ith cluster. In this case, the tester accepts. We now turn to the case that $G = ([N], E)$ is ϵ-far from being a $\bar{\rho}$-blow-up of H. In this case, for any choice of B, we can consider the clustering of the entire graph according to sg_B, and denote the h largest clusters by $V_1, ..., V_h$ (where some of these V_i's may be empty). Letting $V \stackrel{\text{def}}{=} \bigcup_{i \in [h]} V_i$, we note that if $|V| < (1 - \epsilon/2) \cdot N$, then with high probability we reject at the onset of Step 3 due to seeing more than h clusters in the sample.[3] Otherwise, we consider all possible mappings of the vertices of the h largest clusters to $[h]$. For each such mapping $\psi : V \to [h]$ such that $\phi(u) = \phi(v)$ iff $u, v \in V_i$ for some i, either there exists an $i \in [h]$ such that $|V_i| \not\in (1 \pm \epsilon/4) \cdot \rho_i N$ or there exist at least $\epsilon N^2/4$ violating pairs (i.e., vertex pairs $(u, v) \in V \times V$ that have an edge relation in G that does not fit the edge relation of $(\psi(u), \psi(v))$ in H). In the first case, with high probability, the sample S will contain a deviating fraction of vertices from V_i, whereas in the second case, with high probability, the sample T will hit some of these violations.[4] In either cases, with high probability, the tester will reject. This completes the treatment of the case (of $\overline{\rho}$-blow-up) of a graph $H = ([h], F)$ that is not a blow-up of any smaller graph.

Finally, suppose that $H([h], F)$ is a blow-up of some smaller graph H', and suppose that H' is minimal (i.e., it is not a blow-up of any smaller graph). Then, testing the property $\mathcal{BBU}(H)$ reduces to testing a proportional blow-up property regarding H', where the proportions are determined according to the blow-up of H' into H (and the densities are multiples of $1/h$). ∎

3   The $\mathcal{BU}(H)$-Tester and Its Basic Features

Recall that a tester of the type we seek (i.e., a non-adaptive tester of $\widetilde{O}(1/\epsilon)$ query complexity) cannot operate by inspecting an induced subgraph, because by [GR08, Prop. 6.2] such a subgraph will have to be induced by $\Omega(1/\epsilon)$ vertices, which would yield query complexity $\Omega(1/\epsilon^2)$. Thus, like in [GR08, Sec. 6.2], our non-adaptive tester operates by using a less straightforward querying procedure. Specifically, it does select a sample of $\widetilde{O}(1/\epsilon)$ vertices, but does *not* query all vertex pairs.

Algorithm 3.1 (testing $\mathcal{BU}(H)$, for a fixed graph $H = ([h], F)$): *On input parameters, N and ϵ, and access to an oracle $g : [N] \times [N] \to \{0, 1\}$, representing a graph $G = ([N], E)$, the algorithm sets $\ell = \log_2(1/\epsilon) + O(\log \log(1/\epsilon))$ and proceeds as follows.*

1. *For every $i \in [\ell]$, it selects uniformly a sample of $\text{poly}(\ell) \cdot 2^i$ vertices, denoted T_i.*
   Denote $T = \bigcup_{i \in [\ell]} T_i$.
2. *For every $i, j \in [\ell]$ such that $i + j \leq \ell$, the algorithm queries all pairs in $T_i \times T_j$.*

---

[3] If $|V_h| \geq (\epsilon/2h) \cdot N$, then with high probability S will contain a vertex from each V_i as well as a vertex that does not belong to V. On the other hand, if $|V_h| \leq (\epsilon/2h) \cdot N$, then with high probability S will contain $h + 1$ vertices from different clusters in $[N] \setminus V$.

[4] Note that a $1/h^2$ fraction of these foregoing violations can be attributed to one of $2 \cdot \binom{h}{2}$ events that correspond to the existence or non-existence of edges between some pair of clusters.

3. *The algorithm accepts if and only if the answers obtained in Step 2 are consistent with some blow-up of H. That is, let $K : T \times T \to \{0, 1, *\}$ be a partial description of the subgraph of G induced by T such that $K(u, v) = g(u, v)$ if query (u, v) was made in Step 2, and otherwise $K(u, v) = *$. Then, the acceptance condition seeks a mapping $\phi : T \to [h]$ such that if $K(u, v) = 1$ then $(\phi(u), \phi(v)) \in F$ and if $K(u, v) = 0$ then $(\phi(u), \phi(v)) \notin F$.*

Indeed, at this point we ignore the computational complexity of implementing Step 3. We shall return to this issue at the end of the current section. But, first, let us note that the query complexity of Algorithm 3.1 is

$$\sum_{i,j : i+j \le \ell} \text{poly}(\ell) \cdot 2^{i+j} = \text{poly}(\ell) \cdot 2^\ell = \widetilde{O}(1/\epsilon). \tag{1}$$

It is also clear that Algorithm 3.1 is non-adaptive and that it accepts every $G \in \mathcal{BU}(H)$ with probability 1 (i.e., it has one-sided error). The bulk of this work (see Section 4) is devoted to showing that if G is ϵ-far from $\mathcal{BU}(H)$, then Algorithm 3.1 rejects it with probability at least $2/3$.

Relaxing the acceptance condition of Algorithm 3.1. A straightforward implementation of Step 3 amounts to considering all $h^{|T|}$ mappings of T to $[h]$, and checking for each such mapping ϕ whether the clustering induced by ϕ fits the graph H. Relaxing the acceptance condition (used in Step 3 of Algorithm 3.1) yields a more time-efficient algorithm. Actually, the relaxed acceptance condition (defined next) seems easier to analyze than the original one. The notion of *pairwise inconsistent rows* (of K) is pivotal to this relaxed acceptance condition. (Indeed, it will be instructive to think of K as a matrix, and to view rectangular restrictions of K as sub-matrices.)

Definition 3.2 (pairwise inconsistent rows): *Let $K' : R \times C \to \{0, 1, *\}$ be a sub-matrix of $K : T \times T \to \{0, 1, *\}$; that is, $R, C \subseteq T$ and $K'(r, c) = K(r, c)$ for every $(r, c) \in R \times C$. Then, the rows $r_1, r_2 \in R$ are said to be* inconsistent *(wrt K') if there exists a column $c \in C$ such that $K'(r_1, c)$ and $K'(r_2, c)$ are different Boolean values (i.e., $K'(r_1, c), K'(r_2, c) \in \{0, 1\}$ and $K'(r_1, c) \neq K'(r_2, c)$). A set of rows of K' is called* pairwise inconsistent *(wrt K') if each pairs of rows is inconsistent (wrt K').*

Another pivotal notion, which was alluded to before, is the notion of being consistent with some blow-up of H, which we now term H-*mappability*.

Definition 3.3 (H-mappable sub-matrices): *Let $K' : R \times C \to \{0, 1, *\}$ be a sub-matrix of $K : T \times T \to \{0, 1, *\}$. We say that K' is H-mappable if there exists a mapping $\phi : R \to [h]$ such that if $K'(u, v) = 1$ then $(\phi(u), \phi(v)) \in F$ and if $K'(u, v) = 0$ then $(\phi(u), \phi(v)) \notin F$. We call such a ϕ an H-mapping of K' (or R) to $[h]$.*

Note that if K is H-mappable, then every two *inconsistent* rows of K must be mapped (by ϕ as in Definition 3.3) to different vertices of H. In particular, if a sub-matrix $K' : R \times C \to \{0, 1, *\}$ of K has pairwise inconsistent rows, then any H-mapping of K to $[h]$ must be injective. Hence, if K contains more than h pairwise inconsistent rows, then K is not H-mappable.

Definition 3.4 (the relaxed acceptance condition (of Algorithm 3.1)): *The relaxed algorithm accept if and only if each set of pairwise inconsistent rows in K is H-mappable. That is, for every set R of pairwise inconsistent rows in K, we check whether the sub-matrix $K' : R \times T \to \{0, 1, *\}$ is H-mappable, where the pairwise inconsistency condition mandates that this mapping of R to $[h]$ is 1-1. In particular, if K has more than h pairwise inconsistent rows, then the relaxed acceptance condition fails.*

Note that the relaxed acceptance condition can be checked by considering all s-subsets of T, for all $s \le h + 1$. For each such subset that consists of pairwise inconsistent rows, we consider all possible 1-1 mappings of this subset to $[h]$, and check consistency with respect to H. This can be performed in time $\binom{|T|}{h+1} \cdot (h!) < |T|^{h+1} = \text{poly}(1/\epsilon)$, where the polynomial depends on h.

Clearly, if $G \in \mathcal{BU}(H)$, then for every $T \subseteq [N]$ it holds that the corresponding matrix K satisfies Definition 3.4. Thus, the relaxed algorithm always accepts graphs in $\mathcal{BU}(H)$. Section 4 is devoted to showing that if G is ϵ-far from $\mathcal{BU}(H)$, then the relaxed algorithm rejects with high probability.

4   The Acceptance Condition and Graphs That Are Far from $\mathcal{BU}(H)$

In light of the above, Theorem 1.1 follows from the fact that the relaxed version of Algorithm 3.1 (which uses the condition in Definition 3.4) rejects with very high probability any graph G that is ϵ-far from $\mathcal{BU}(H)$. This fact is established next.

Lemma 4.1 (main lemma): *Suppose that $G = ([N], E)$ is ϵ-far from $\mathcal{BU}_N(H)$, and let $T = \bigcup_{i \in [\ell]} T_i$ be selected at random as in Step 1 of Algorithm 3.1. Then, with probability at least $2/3$, there exists a set $R \subset T$ of pairwise inconsistent rows in the corresponding matrix $K : T \times T \to \{0, 1, *\}$ that is not H-mappable.*

Before embarking on the actual proof of Lemma 4.1, we provide a very rough outline.

Outline of the proof of Lemma 4.1. Our very rough plan of action is to partition the selection of T (and each of its parts, i.e., $T_0, T_1, ..., T_\ell$) into $p(\ell) \stackrel{\text{def}}{=} 2\ell^h$ many phases such that in the jth phase we select at random samples $T_0^j, T_1^j, ..., T_\ell^j$ such that $|T_i^j| = \text{poly}(\ell) \cdot 2^i$. Thus, we let each T_i equal $\bigcup_{j=1}^{p(\ell)} T_i^j$, but we shall consider the queries as if they are made in phases such that in the jth phase we only consider queries between $T^j \stackrel{\text{def}}{=} \bigcup_{i \in [\ell]} T_i^j$ and $T^{[j]} \stackrel{\text{def}}{=} \bigcup_{k \le j} T^k$. Letting $K^j : T^{[j]} \times T^{[j]} \to \{0, 1, *\}$ denote the partial information obtained on G in the first j phases, we consider a certain set R^j of pairwise inconsistent rows of K^j. If this set R^j is not H-mappable, then we are done. Otherwise, we show that, with high probability over the choice of the sample T^{j+1}, we obtain a new set R^{j+1} of pairwise inconsistent rows such that R^{j+1} has a higher *index* than R^j, where the indices refer to an order over sequences of length at most h over $[\ell]$. Since the number of such sequences is $\sum_{k \in [h]} \ell^k < p(\ell)$, with high probability, this process must reach a set R^j that is not H-mappable, and so we are done.

Needless to say, the crucial issue is the progress achieved in each phase; that is, the fact that at each phase j the index of the new set R^{j+1} is higher than the index of the old set R^j. Intuitively, this progress is achieved because the current (H-mappable) set R^j induces a clustering of all vertices of G that extends this H-mapping, whereas this clustering must contain many vertex pairs that violate the edge relation of H. The sample taken in the current phase (i.e., T^{j+1}) is likely to hit these violations, and this gives rise to a set R^{j+1} with higher index.

4.1   Basic Notions and Notations

In addition to the foregoing notations, T_i^j, T^j and $T^{[j]}$, we shall use the following notations.

- A pair (R, C) is called a j-basic pair if $C \subseteq T^{[j]}$ and $R \subseteq C$. Indeed, j-basic pairs correspond to restrictions of the sample available at phase j (i.e., $T^{[j]}$).
- The j-index of a vertex $v \in T^{[j]}$, denoted $\mathtt{idx}^j(v)$, is the smallest index i such that $v \in T_i^{[j]}$, where $T_i^{[j]} \stackrel{\text{def}}{=} \bigcup_{k \le j} T_i^k$. (Note that $\mathtt{idx}(\cdot)$ depends on T, but this dependence is not shown in the notation.)
  A key observation is that for every $u, v \in T$, it holds that $K(u, v) = g(u, v)$ if and only if $\mathtt{idx}^{p(\ell)}(u) + \mathtt{idx}^{p(\ell)}(v) \le \ell$. Otherwise, $K(u, v) = *$ (indicating that (u, v) was not queried by Algorithm 3.1).
  We comment that, with extremely high probability, for each j and $v \in T^{[j]}$, there exists a unique $i \in [\ell]$ and $k \in [j]$ such that $v \in T_i^k$. Thus, for any $v \in T^{[j]}$, we may assume that $\mathtt{idx}^{j+1}(v) = \mathtt{idx}^j(v)$.
- The indices of individual vertices in $T^{[j]}$ are the basis for defining the index of sets in $T^{[j]}$. Specifically, the j-index of a set $S \subseteq T^{[j]}$, denoted $\mathtt{idx}^j(S)$, is the multi-set consisting of all values $\mathtt{idx}^j(v)$ for $v \in S$. It will be instructive to consider an ordered version of this multi-set; that is, we redefine $\mathtt{idx}^j(S)$ as $(i_1, ..., i_{|S|})$ such that (1) for every $k < |S|$ it holds that $i_k \ge i_{k+1}$, and (2) for every $i \in [\ell]$ it holds that $|\{k \in [|S|] : i_k = i\}| = |\{v \in S : \mathtt{idx}^j(v) = i\}|$.
- We consider a natural lexicographic order over sequences, denoted $\succ$, such that for two (monotonically non-increasing) sequences of integers, $a = (a_1, ..., a_m)$ and $b = (b_1, ..., b_n)$, it holds that $a \succ b$ if
  - either there exists $i \le \min(n, m)$ such that $(a_1, ..., a_{i-1}) = (b_1, ..., b_{i-1})$ and $a_i > b_i$.
  - or $m > n$ and $(a_1, ..., a_n) = (b_1, ..., b_n)$.
  Note that $\succ$ is a total order on the set of monotonically non-increasing (finite) sequences of integers.

As hinted in the overview, a key notion in our analysis is the notion of a clustering of the vertices of G that is induced by an H-mapping of some small subset of vertices. Actually, the clustering is induced by a partial knowledge sub-matrix $K' : R \times C \to \{0, 1, *\}$ as follows.

Definition 4.2 (the clustering induced by K'): *Let $K' : R \times C \to \{0, 1, *\}$ be a sub-matrix of $K : T \times T \to \{0, 1, *\}$ such that K' has pairwise inconsistent rows. Then, for every $r \in R$, we denote by $V_r(K')$ the set of vertices $v \in [N]$ that are consistent with row r in K'. That is,*

$$V_r(K') \overset{\text{def}}{=} \{v \in [N] : (\forall c \in C) \, g(v, c) \cong K'(r, c)\} \tag{2}$$

where, for $\sigma, \tau \in \{0, 1, *\}$, we write $\sigma \cong \tau$ if either $\sigma = \tau$ or $\sigma = *$ or $\tau = *$. The vertices that are inconsistent with all rows, are placed in the leftover set $L(K') \overset{\text{def}}{=} [N] \setminus \bigcup_{r \in R} V_r(K')$.

Indeed, rows $r_1, r_2 \in R$ are inconsistent wrt K' (as per Definition 3.2) if there exists a column $c \in C$ such that $K'(r_1, c) \not\cong K'(r_2, c)$ (which means that $K'(r_1, c)$ and $K'(r_2, c)$ are both in $\{0, 1\}$ but are different). Thus, the hypothesis that the rows of K' are pairwise inconsistent implies that the sets in Eq. (2) are disjoint. Hence, the clustering in Definition 4.2 is indeed a partition of the vertex set of G (since $v \in L(K')$ if for every $r \in R$ there exists $c \in C$ such that $g(v, c) \not\cong K'(r, c)$). This motivates our focus on sub-matrices having pairwise inconsistent rows. The following definition adds a requirement (regarding such sub-matrices) that refers to the relation between the index of row r and the density of the corresponding set $V_r(K')$.

Definition 4.3 (nice pairs): *Let (R, C) be a j-basic pair and $K' : R \times C \to \{0, 1, *\}$ be the corresponding sub-matrix of K. We say that (R, C) is a j-nice pair if the following two conditions hold.*

1. *R is pairwise inconsistent with respect to K'.*

2. *For every $r \in R$ it holds that $\text{ind}^j(r) \le \rho(V_r(K')) + 1$, where $\rho(S) \overset{\text{def}}{=} \lceil \log(N/|S|) \rceil$.*

As a sanity check, suppose that $r \in R$ was selected in phase j (i.e., $r \in T^j$). Then, it is very likely that r (or some other member of $V_r(K')$) is selected in $T^j_{\rho(V_r(K'))-1}$, because $T^j_{\rho(V_r(K'))-1}$ is a random set of cardinality $\text{poly}(\ell) \cdot 2^{\rho(V_r(K'))-1} = \text{poly}(\ell) \cdot N/|V_r(K')|$.

For each phase j, we shall show the existence of a j-nice pair. Furthermore, we shall show that the corresponding set of rows has a higher index than all sets of rows associated with previous phases. The furthermore claim is the crux of the analysis, and is captured by the Progress Lemma presented in Section 4.2. But let us first establish the mere existence of j-nice pairs. Indeed, for every $j \ge 1$, we may pick an arbitrary $r \in T_1^1$, and consider the j-nice pair $(\{r\}, \{r\})$, while noting that $\text{idx}^1(r) = 1$ and $\rho(V_r(K')) \ge 0$ (where $K' : \{r\} \times \{r\} \to \{0, 1, *\}$).

4.2   The Progress Lemma

Recall that $G = ([N], E)$ is ϵ-far from $\mathcal{BU}(H)$, where $H = ([h], F)$. Furthermore, we consider the partial view $K : T \times T \to \{0, 1, *\}$ obtained by Algorithm 3.1, where $T = \bigcup_{i \in [\ell], j \in [p(\ell)]} T_i^j$ is the random sample selected. Throughout the rest of this section, we say that an event has negligible probability if it occurs with probability that vanishes faster than any polynomial in ϵ. Since we shall consider only $\mathrm{poly}(\ell)$ many events, we can safely ignore these negligible probabilities.[5] We say that an event occurs with overwhelmingly high probability if the probability that it does *not* occur is negligible.

Lemma 4.4 (Progress Lemma): *Let (R, C) be a j-nice pair and $K' : R \times C \to \{0, 1, *\}$ be the corresponding sub-matrix of K. If K' is H-mappable then, with overwhelmingly high probability over the choice of T^{j+1}, there exists a $(j+1)$-nice pair (R', C') such that $\mathrm{ind}^{j+1}(R') \succ \mathrm{ind}^j(R)$.*

Recalling that a (trivial) 1-nice pair always exists and that the number of possible indices is smaller than $p(\ell)$, we conclude that, with overwhelmingly high probability (over the choice of T), there exists a $j < p(\ell)$ and a j-nice pair that is not H-mappable. Lemma 4.1 follows. Thus, all that remains is proving Lemma 4.4, which we undertake next.

Proof: We consider the partition induced by K', as per Definition 4.2, and consider two cases regarding the size of $L \overset{\mathrm{def}}{=} L(K')$:

Case 1: $\rho(L) \le \ell$. In this case (i.e., $|L| \ge 2^{-\ell} \cdot N$), with overwhelmingly high probability, the sample T^{j+1} contains a vertex $u \in L(K')$. Using this u, we shall obtain a $(j + 1)$-nice pair with a set of rows that has a higher index than R. Intuitively, since $(g(u, c))_{c \in C}$ is inconsistent with all rows of K', we may add u as a row to K' while possibly omitting rows of K' that are consistent with $(K(u, c))_{c \in C}$ (see below), obtaining a sub-matrix that has a larger index (than the index of K'). The detailed analysis of this case is presented in Claim 4.4.2.

Case 2: $\rho(L) > \ell$. In this case (i.e., $|L| < 2^{-\ell} \cdot N < \epsilon N/2$), the partition induced by $(V_r(K'))_{r \in R}$ contains many pairs that violate the edge relation of H, since the number of pairs adjacent at L is smaller than $\epsilon N^2/2$. We shall show that, with overwhelmingly high probability, the sample T^{j+1} contains a vertex w such that augmenting K' with the column corresponding to w yields a sub-matrix K'' such that $\rho(L(K'')) < \ell$. Intuitively, pairs of vertices in $V(K') \overset{\mathrm{def}}{=} \bigcup_{r \in R} V_r(K')$ that violate the edge relation of H, yield vertices w that effectively shrink $V(K')$ in the sense that adding w as a column to K' moves many vertices from $V(K')$ to $L(K'')$. In particular, we shall show that $|L(K'')| = \Omega(\epsilon N/\ell)$, which means that $\rho(L(K'')) < \log_2(O(\ell)/\epsilon) < \ell$.

---

[5] In fact, it would have sufficed to define as negligible any probability that vanishes faster than any polynomial in $1/\ell$ (i.e., faster than any polylogarithmic function of ϵ).

At this point we may proceed as in Case 1. (Formally, in this case, the $j+1$st phase is partitioned into two sub-phases, where in each sub-phase we use half of each of the samples T_i^{j+1}.) The detailed analysis of this case is presented in Claim 4.4.3.

Our analysis of the two cases combines straightforward probabilistic arguments with manipulations of sub-matrices. The latter manipulations include adding rows and columns and truncating the sub-matrix so as to leave only rows that have an index that is lower-bounded by some value. It is thus instructive to discuss these three operations first.

Adding an arbitrary column from T^{j+1}. Suppose that (R, C) is j-nice with a corresponding sub-matrix K'. Then, adding any column $v \in T^{j+1}$ to K' results in a sub-matrix K'' such that the corresponding pair $(R, C \cup \{v\})$ is $(j+1)$-nice. Clearly, adding a column may only add inconsistencies, and so the pairwise inconsistency condition of K' is preserved. For any $r \in R$, the densities of $V_r(\cdot)$ may only drop when moving from K' to K'', and so $\mathrm{ind}^j(r) \le \rho(V_r(K')) + 1$ implies $\mathrm{ind}^{j+1}(r) \le \rho(V_r(K'')) + 1$.

Adding a row that belongs to $L(K') \cap T_{\rho(L(K'))}^{j+1}$. It is tempting to think that if (R, C) is j-nice, then adding any row $v \in T_{\rho(L(K'))}^{j+1} \cap L(K') \cap C$ to K' results in a sub-matrix K'' such that the corresponding pair $(R \cup \{v\}, C)$ is $(j+1)$-nice. It is true that $\mathrm{ind}^{j+1}(r) \le \rho(V_r(K''))+1$ holds for each row r, including the added row v (because $\mathrm{ind}^{j+1}(v) = \rho(L(K'))$ and $\rho(V_v(K'')) \ge \rho(L(K'))$, since $V_v(K'') \subseteq L(K')$). However, although for every $r \in R$ there exists $c \in C$ such that $g(v, c) \not\cong K'(r, c)$ (since $v \notin V_r(K')$), it not necessarily the case that the row v in K is inconsistent with all rows in K' (i.e., it may be the case that, for some $r \in R$ and each $c \in C$, it holds that $K(v, c) \cong K'(r, c)$, since $K(v, c) \in \{g(v, c), *\}$ and $* \cong K'(r, c)$). Coping with this problem, which arises from the fact that K may have $*$-values, leads us to introduce the following truncation operator.

Truncating at an added row. Suppose that (R, C) is j-nice with a corresponding sub-matrix K', and let $v \in L(K') \cap T_{\rho(L(K'))}^{j+1}$. Then, consider first adding v as a new row and column to K', and then leaving in the resulting sub-matrix only the rows that have a $(j+1)$-index that is at least as large as the one of v (i.e., row r remains if and only if $\mathrm{ind}^{j+1}(r) \ge \mathrm{ind}^{j+1}(v)$). We claim that these rows are pairwise inconsistent, and thus the resulting sub-matrix is $(j+1)$-nice.

It suffices to prove that the new row v (of K) is inconsistent with any row that was left from K'; that is, fixing any $r \in R$ such that $\mathrm{ind}^{j+1}(r) \ge \mathrm{ind}^{j+1}(v)$, we claim that there exists $c \in C$ such that $K(v, c) \not\cong K'(r, c)$. Since $v \in L(K')$, we know that there exists $c \in C$ such that $g(v, c) \not\cong K'(r, c)$, which implies that $K'(r, c) \in \{0, 1\}$, which in turn implies $\mathrm{ind}^j(r) + \mathrm{ind}^j(c) \le \ell$ (by definition of K). Now, using $\mathrm{ind}^{j+1}(v) \le \mathrm{ind}^{j+1}(r) \le \mathrm{ind}^j(r)$, we get $\mathrm{ind}^{j+1}(v) + \mathrm{ind}^j(c) \le \ell$, which implies that $K(v, c) = g(v, c)$. Recalling that $g(v, c) \not\cong K'(r, c)$, we obtain $K(v, c) \not\cong K'(r, c)$, and the claim follows.

Note that the truncation of $K' : R \times C \to \{0, 1, *\}$ at the added row v always contains the new row v, and that it may result in $|R| + 1$ rows (i.e., no "real truncation"). Another key feature of the truncation-at-an-added-row operation is that it yields a set of rows with an index larger than the index of R.

Claim 4.4.1 (the effect of truncation): *Suppose that (R, C) is j-nice with a corresponding sub-matrix K', and let $v \in L(K') \cap T^{j+1}_{\rho(L(K'))}$. Then, truncating the sub-matrix that corresponds to $(R \cup \{v\}, C \cup \{v\})$ at row v yields a $(j+1)$-nice pair with a row set having an index larger than $\mathrm{ind}^j(R)$.*

Proof: The first part of this claim was already established above. Denoting the resulting set of rows by R', we need to prove that $\mathrm{ind}^{j+1}(R') \succ \mathrm{ind}^j(R)$. If $R' = R \cup \{u\}$ then the claim is trivial, and so we consider the case that $\mathrm{ind}^{j+1}(R') = (i_1, ..., i_t)$, where $t \leq |R|$ and $i_t = \mathrm{ind}^{j+1}(v)$. This means that a non-trivial truncating took place, and that all omitted rows had index smaller than i_t, which implies that $(i_1, ..., i_t) \succ \mathrm{ind}^{j+1}(R)$ (because $\mathrm{ind}^{j+1}(R) = (i_1, ..., i_{t-1}, d_t, ..., d_{|R|})$ with $d_t < i_t$). $\square$

Claim 4.4.2 (case 1): *Suppose that (R, C) is j-nice and that $\rho(L) \leq \ell$, where $L = L(K')$. Then, with overwhelmingly high probability (over the choice of $T^{j+1}_{\rho(L(K'))}$), the sample $T^{j+1}_{\rho(L(K'))}$ contains a vertex $u \in L(K')$ such that adding u to K' (both as a row and a column) and truncating the resulting sub-matrix at row u yields a $(j + 1)$-nice pair (R', C') such that $\mathrm{ind}^{j+1}(R') \succ \mathrm{ind}^j(R)$.*

Proof: With overwhelmingly high probability, the sample $T^{j+1}_{\rho(L(K'))}$ contains a vertex $u \in L(K')$, while using any such vertex yields the desired result (due to Claim 4.4.1). $\square$

Claim 4.4.3 (case 2): *Suppose that (R, C) is j-nice and that the corresponding sub-matrix K' is H-mappable. Further suppose that $\rho(L) > \ell$, where $L = L(K')$. Then, with overwhelmingly high probability (over the choice of T^{j+1}), the sample T^{j+1} contains a vertex w such that adding the column w to K' yields a $(j + 1)$-nice pair $(R, C \cup \{w\})$ such that the corresponding sub-matrix K'' satisfies $\rho(L(K'')) \leq \ell$*

Proof: We combine the hypothesis that G is ϵ-far from $\mathcal{BU}(H)$ with the hypothesis that K' is H-mappable, and denote the corresponding H-mapping by $\phi : R \to [h]$. Extending this mapping to $V(K') \stackrel{\text{def}}{=} \bigcup_{r \in R} V_r(K')$ such that $\phi(v) = \phi(r)$ for every $v \in V_r(K')$, and using the hypothesis that $|L(K')| < 2^{-\ell}N < \epsilon N/2$, we conclude that there are at least $\epsilon N^2/2$ vertex pairs that violate the edge relation of H (i.e., pairs $(u, v) \in V(K') \times V(K')$ such that $(u, v) \in E$ iff $(\phi(u), \phi(v)) \notin F$). Actually, we should consider all $h!$ possible injections (from R to $[h]$), and apply the argument to each of them, but this only increases the error probability by a factor of $h!$. These violations can be of one of the following two types.

1. Edges $(u, v) \in E$ such that $(\phi(u), \phi(v)) \notin F$. If the number of such pairs exceeds $\epsilon N^2/4$, then we select a pair $(r, s) \in R \times R$ such that there exist at least $\epsilon N^2/4h^2$ pairs $(u, v) \in E$ for which $(\phi(u), \phi(v)) = (\phi(r), \phi(s)) \notin F$.

2. Non-edges $(u, v) \notin E$ such that $(\phi(u), \phi(v)) \in F$. If the number of such pairs exceeds $\epsilon N^2/4$, then we select a pair $(r, s) \in R \times R$ such that there exist at least $\epsilon N^2/4h^2$ pairs $(u, v) \notin E$ for which $(\phi(u), \phi(v)) = (\phi(r), \phi(s)) \in F$.

Fixing (r, s) as above we have at least $\epsilon N^2/4h^2$ violating pairs in $V_r(K') \times V_s(K')$. Next, we select an integer $m \in [\ell]$ such that there exists a set $W \subseteq V_r(K')$ of cardinality $2^{-m} \cdot N$ and every $w \in W$ participates in at least $\epsilon 2^m N/4h^2\ell > 2^{-(\ell-m-3)} \cdot N$ violating pairs (with vertices of $V_s(K')$). Clearly, $\rho(W) = m$, and so with overwhelmingly high probability T_m^{j+1} contains a vertex $w \in W$. Adding any such w as a column to K', we obtain a sub-matrix K'' and claim that $\rho(L(K'')) \leq \ell - m \leq \ell$. Specifically, we shall show that *every $u \in V_s(K')$ such that (u, w) is a violating pair must be in $L(K'')$*, and infer that $\rho(L(K'')) \leq \ell - m$ by recalling that the number of such violating pairs in which w participates exceeds $2^{-(\ell-m-3)} \cdot N$.

Thus, letting U^w denote the set of all $u \in V_s(K')$ such that (u, w) is a violating pair, we prove that $U^w \subseteq L(K'')$. Let u be an arbitrary vertex in $U^w \subseteq V_s(K')$ (and recall that $w \in W \cap T_m^{j+1} \subseteq V_r(K')$). Our argument proceeds as follows.

1. We first note that $\mathsf{ind}^j(r) \leq \rho(V_r(K'))+1$ (by the nicety condition), whereas $\rho(V_r(K')) \leq \rho(W) = m$.
   Similarly, $\mathsf{ind}^j(s) \leq \rho(V_s(K')) + 1$, whereas $\rho(V_s(K')) \leq \rho(U^w) \leq \ell - m - 3$ (since $V_s(K') \supseteq U^w$ and $|U^w| > 2^{-(\ell-m-3)} \cdot N$).
2. Combining the two foregoing facts, we conclude that $\mathsf{ind}^j(r) + \mathsf{ind}^j(s) \leq \ell$, which implies that $K'(r, s) = g(r, s)$.
3. Since $w \in V_r(K')$, it must be that $g(w, s) \cong K'(r, s)$, which implies $g(w, s) = g(r, s)$ (when combined with $K'(r, s) = g(r, s)$). Since ϕ is an H-mapping it must be that $g(s, w) = g(s, r)$ fits the edge relation of $(\phi(s), \phi(w)) = (\phi(s), \phi(r))$ with respect to H.
4. On the other hand, since (u, w) is a violating pair, the value $g(u, w)$ does not fit the edge relation of $(\phi(u), \phi(w)) = (\phi(s), \phi(r))$ with respect to H.
5. Combining Items 3 and 4, we infer that $g(u, w) \neq g(s, w)$, which implies $g(u, w) \ncong K''(s, w)$ (because $K''(s, w) = g(s, w)$ by virtue of $\mathsf{ind}^{j+1}(s) + \mathsf{ind}^{j+1}(w) \leq (\ell - m - 2) + m < \ell$, where $w \in T_m^{j+1}$ by the hypothesis). Thus, u is not in $V_s(K'')$, although it is in $V_s(K')$.
6. We observe that, for every $r \in R \setminus \{s\}$, vertex $u \in V_s(K')$ is not in $V_r(K'') \subseteq V_r(K')$, since the rows of K' are pairwise inconsistent.
7. Combining Items 5 and 6, we conclude that $u \notin \bigcup_{r \in R} V_r(K'')$, and hence $u \in L(K'')$.

Thus, we have established that $U^w \subseteq L(K'')$. As noted above, it follows that $\rho(L(K'')) \leq \rho(U^w) \leq \ell - m - 3 < \ell$. This completes the proof of Claim 4.4.3. $\square$

Completing the proof of Lemma 4.4. In accordance with the motivating discussion, we now complete the proof of the lemma by using the two latter claims. Specifically, if Case 1 holds (i.e., $\rho(L(K')) \leq \ell$), then we invoke Claim 4.4.2 and are done. Otherwise, Case 2 holds (i.e., $\rho(L(K')) > \ell$), and we take the following two steps. Recall that, as stated in the beginning of the proof, in this case (i.e., Case 2) we partition the sample T^{j+1} into two parts, and use a different part in each step. In the first step we apply Claim 4.4.3 to the first part, and get into Case 1; that is, we obtain a new K' such that $\rho(L(K')) \leq \ell$. (Note that the new K' has the same row set as the original one, and the latter set maintains its index.) Next, in the second step, we apply Claim 4.4.2 to the resulting K' and the second part of the sample, and are done.   ■

5 Proximity Oblivious Testing of Blow-Up

In this section we derive, for every fixed graph H, a constant-query proximity oblivious tester of $\mathcal{BU}(H)$. That is, we refer to the following definition of [GR09], when specialized to the dense graph model.

Definition 5.1 (proximity oblivious testing for graphs in the adjacency matrix model): *A* proximity oblivious tester *for a graph property Π is a probabilistic oracle machine that, on input parameter N and access to an N-vertex graph $G = ([N], E)$, outputs a binary verdict that satisfies the following two conditions.*

1. *If $G \in \Pi$, then the tester accepts with probability 1.*
2. *There exists a monotone function $\rho : (0, 1] \to (0, 1]$ such that, for every graph $G = ([N], E) \notin \Pi$, it holds that the tester rejects G with probability at least $\rho(\delta_\Pi(G))$, where $\delta_\Pi(G)$ denotes the (relative) distance of G from the set of N-vertex graphs that are in Π.*

The function ρ is called the detection probability *of the tester.*

Combining Lemma 4.1 and the ideas underlying [GR09, Thm. 6.3], we obtain.

Theorem 5.2 *For every fixed graph $H = ([h], F)$, there exists a $O(h^2)$-query proximity oblivious tester of $\mathcal{BU}(H)$. Furthermore, the tester has detection probability $\rho(\epsilon) = \epsilon^{O(h)}$.*

This extends the result of [GR09, Prob. 4.11], which corresponds to the special case in which H is a h-vertex clique. We also mention that, for constant-query proximity oblivious testers of $\mathcal{BU}(H)$, detection probability of the form $\rho(\epsilon) = \epsilon^{\Omega(h)}$ is essential (cf. [GR09, Prob. 4.3]).

Proof: While a direct application of [GR09, Thm. 6.3] would yield a detection bound of $\rho(\epsilon) = \epsilon^{O(h^2)}$, we obtain a quantitative improvement by using a version of [GR09, Thm. 6.3] that is specialized to the dense graph model. This version refers to *any graph property Π having a standard tester T* (of error probability $1/3$) *that satisfies the following three conditions*:

1. *T is non-adaptive;*
2. *for a monotonically non-decreasing $\nu : (0, 1] \to \mathbb{N}$, on proximity parameter ϵ, the queries of T refer to at most $\nu(\epsilon)$ vertices; and*
3. *for some fixed $s \in \mathbb{N}$, the tester T rejects if and only if it sees a partial view of some s-vertex subgraph that cannot occur in any graph in Π. (Such a partial view is called a witness for non-membership.)*

In such a case, Π has an $\binom{s}{2}$-query proximity-oblivious tester with detection probability at least $\rho(\epsilon) = \Omega(\epsilon/\nu(\epsilon/2)^s)$. We mention that a direct application of [GR09, Thm. 6.3] would have yielded a detection bound of $\rho(\epsilon) = \Omega(\epsilon/q(\epsilon/2)^{\binom{s}{2}})$, where $q < \nu^2$ denotes the query complexity of the original tester.

The foregoing claim is easily proved by following the ideas that underly the proof of [GR09, Thm. 6.3]. Specifically, the proximity oblivious tester select $i \in \{1, ..., \lceil \log_2 N \rceil\}$ with probability 2^{-i}, invokes the query-generator procedure of T on input ((alleged) proximity parameter) 2^{-i}, selects uniformly s vertices among those that appear in the generated queries, makes (only) the corresponding $\binom{s}{2}$ queries, and accept if and only if the induced subgraph is not a witness for non-membership. Clearly, the resulting tester rejects any graph that is 2^{-i}-far from Π with probability at least $2^{-i} \cdot \frac{2}{3} \cdot \left(\frac{\mu(2^{-i})}{s}\right)^{-1}$.

It remains to show that, when applied to $\Pi = \mathcal{BU}(H)$, the (non-adaptive) tester in Algorithm 3.1 (when using the relaxed condition of Definition 3.4) rejects based on a witness for non-membership that contains $O(h)$ vertices. Essentially, this holds since the condition in Definition 3.4 refers to a set of at most $h + 1$ pairwise inconsistent rows that are not H-mappable, whereas (as shown next) only $n - 1$ columns are required in order to establish that n rows are pairwise inconsistent. Thus, it suffices to augment the set of rows R by at most $|R| - 1$ additional vertices, and derive a witness for non-membership that contains at most $2h + 1$ vertices.

Lastly, we prove that $n - 1$ columns suffice for establishing the fact that n rows are pairwise inconsistent. Starting with a row r of the largest index, we pick an arbitrary column that witnesses the inconsistence of row r with some other row r'. This column c partitions the set of rows to two non-trivial sets: the set of rows having the same value as r on column c, and the set of rows having the opposite value on this column. (Note that all rows have a binary value on column c, since we started with a row r of largest index.) The process continues, separately, with each of these two sets, and the key observation is that each split requires only one (possibly new) column. ∎

6 Conclusions

We have shown a non-adaptive tester of query complexity $\widetilde{O}(1/\epsilon)$ for $\mathcal{BU}(H)$. The degree of the polynomial in the polylogarithmic factor that is hidden in the $\widetilde{O}()$ notation is $h + O(1)$, where h is the number of vertices in H. We wonder whether the query complexity can be reduced to $p(h \log(1/\epsilon)) \cdot \epsilon^{-1}$, where p is a fixed polynomial. We mention that such a dependence on h was obtained in [GR08, Sec. 6.2] for the special case in which H is an h-clique. Furthermore, we wonder whether non-adaptive testing of $\mathcal{BU}(H)$ is possible in query complexity $\text{poly}(h) \cdot \epsilon^{-1}$. We mention that such a result is only known for $h = 2$ (cf. [GR08, Sec. 6.1]), whereas an *adaptive* tester of query complexity $O(h^2/\epsilon)$ is known (cf. [A, Sec. 4]).

Acknowledgments. We are grateful to Dana Ron and to the anonymous reviewers of *RANDOM'11* for comments regarding previous versions of this work.

References

[AFKS]   Alon, N., Fischer, E., Krivelevich, M., Szegedy, M.: Efficient Testing of Large Graphs. Combinatorica 20, 451–476 (2000)

[AFNS]   Alon, N., Fischer, E., Newman, I., Shapira, A.: A Combinatorial Characterization of the Testable Graph Properties: It's All About Regularity. In: 38th STOC, pp. 251–260 (2006)

[AS]   Alon, N., Shapira, A.: A Characterization of Easily Testable Induced Subgraphs. Combinatorics Probability and Computing 15, 791–805 (2006)

[A]   Avigad, L.: On the Lowest Level of Query Complexity in Testing Graph Properties. Master thesis, Weizmann Institute of Science (December 2009)

[CEG]   Canetti, R., Even, G., Goldreich, O.: Lower Bounds for Sampling Algorithms for Estimating the Average. In: IPL, vol. 53, pp. 17–25 (1995)

[GGR]   Goldreich, O., Goldwasser, S., Ron, D.: Property testing and its connection to learning and approximation. Journal of the ACM, 653–750 (July 1998)

[GKNR]   Goldreich, O., Krivelevich, M., Newman, I., Rozenberg, E.: Hierarchy theorems for property testing. In: Dinur, I., Jansen, K., Naor, J., Rolim, J. (eds.) APPROX 2009. LNCS, vol. 5687, pp. 504–519. Springer, Heidelberg (2009)

[GR08]   Goldreich, O., Ron, D.: Algorithmic Aspects of Property Testing in the Dense Graphs Model. ECCC, TR08-039 (2008)

[GR09]   Goldreich, O., Ron, D.: On Proximity Oblivious Testing. ECCC, TR08-041 (2008); Extended Abstract in the Proceedings of the 41st STOC (2009)

[GT]   Goldreich, O., Trevisan, L.: Three theorems regarding testing graph properties. Random Structures and Algorithms 23(1), 23–57 (2003)

[RS]   Rubinfeld, R., Sudan, M.: Robust characterization of polynomials with applications to program testing. SIAM Journal on Computing 25(2), 252–271 (1996)

Proximity Oblivious Testing and the Role of Invariances

Oded Goldreich and Tali Kaufman

Abstract. We present a general notion of properties that are characterized by local conditions that are invariant under a sufficiently rich class
of symmetries. Our framework generalizes two popular models of testing
graph properties as well as the algebraic invariances studied by Kaufman and Sudan (STOC'08). Our focus is on the case that the property
is characterized by a constant number of local conditions and a rich set
of invariances.

We show that, in the aforementioned models of testing graph properties, characterization by such invariant local conditions is closely related to proximity oblivious testing (as defined by Goldreich and Ron,
STOC'09). In contrast to this relation, we show that, in general, characterization by invariant local conditions is neither necessary nor sufficient
for proximity oblivious testing. Furthermore, we show that easy testability is *not* guaranteed even when the property is characterized by local
conditions that are invariant under a 1-transitive group of permutations.

Keywords: Property Testing, Graph Properties, Locally Testable Codes,
Sparse Linear Codes, The Long-Code

A version of this work appeared as TR10-058 of *ECCC*.

1 Introduction

In the last couple of decades, the area of property testing has attracted much
attention (see, e.g., a couple of recent surveys [15,16]). Loosely speaking, property
testing typically refers to sub-linear time probabilistic algorithms for deciding
whether a given object has a predetermined property or is far from any object
having this property. Such algorithms, called testers, obtain local views of the
object by making adequate queries; that is, the object is seen as a function and
the testers get oracle access to this function (and thus may be expected to work
in time that is sub-linear in the size of the object).

While a host of fascinating results and techniques has emerged, the desire for a
comprehensive understanding of what makes some properties easy to test (while
others are hard to test) is far from being satisfied.[1] Two general approaches
that seem to have a potential of addressing the question (of "what makes testing
possible") were suggested recently.

---

[1] This assertion is not meant to undermine significant successes of several characterization projects, most notably the result of [1].

O. Goldreich et al.: Studies in Complexity and Cryptography, LNCS 6650, pp. 173–190, 2011.
© Springer-Verlag Berlin Heidelberg 2011

1. Restricting attention to the class of *proximity oblivious testers*, which are constant-query testers that reject any object with probability proportional (but not necessarily linearly proportional) to its distance from the predetermined property. Indeed, the characterization of proximity oblivious testers, in two central models of graph properties, obtained in [9], addresses the foregoing question: *graph properties have proximity oblivious testers if and only if they can be characterized in terms of adequate local conditions.*[2]
2. But even before [9], an approach based on *adequately invariant local conditions* was put forward in [13]. It was shown that in the context of testing algebraic properties, *a sufficient condition for testability* (which in fact yields proximity oblivious testers) *is that the property can be characterized in terms of local conditions that are invariant in an adequate sense.* (Here and throughout this paper, a local condition means a condition that refers to the value of the function at a constant number of points.)

Thus, these two approaches have a very similar flavor, but they are very different at the actual details. On the one hand, the definition of *proximity oblivious testers* does not refer to any structure of the underlying domain of functions, and the local conditions in the two graph models do not refer explicitly to any invariance. However, invariance under relabeling of the graph's vertices is implicit in the entire study of graph properties (since the latter are defined in terms of such invariance). On the other hand, the *linear invariances* considered in [13] presume that the functions' domain can be associated with some vector space and that the properties are invariant under linear transformations of this vector space.

Thus, the first task that we undertake is providing a definition of a general notion of "characterization by invariant local conditions", where at the very minimum this general definition should unify the notions underlying [9,13]. Such a definition is presented in Section 2. Loosely speaking, a property P is characterized by invariant local conditions if P is charaterized by a set C of local conditions (i.e., $f \in P$ iff f satisfies all conditions in C) and C is generated by a *constant* number of local conditions coupled with a set of actions that preserves P (i.e., the invariances).

Given such a definition, a natural conjecture that arises, hereafter referred to as the invariance conjecture, is that a property has a constant-query proximity-oblivious tester if and only if it can be characterized by invariant local conditions. This conjecture is rigorously formulated within our definitional framework (see Section 2.2) and the current work is devoted to its study. The main results of our study may be stated informally as follows:

1. The invariance conjecture holds in the context of testing graph properties in the dense graph model (see Theorem 3.1).
2. The invariance conjecture holds in the context of testing graph properties in the bounded-degree graph model if and only if all local properties are non-propagating (see Theorem 4.1 and [9, Open Problem 5.8]).

---

[2] We warn that the picture is actually not that clean, because in the case of the bounded-degree model the notion of adequacy includes some technical condition, termed non-propagation.

3. In general, the invariance conjecture fails in both directions.
   (a) Characterization by invariant local conditions is not necessary for proximity oblivious testing. This is demonstrated both by linear properties (see Theorem 5.1) and by the dictatorship property (see Theorem 5.2).
   (b) Characterization by invariant local conditions is not sufficient for proximity oblivious testing (see Theorem 5.3). This is demonstrated by the property called Eulerian orientation (which refers to the orientation of the edges of a cyclic grid, cf. [6]).

Thus, there are natural settings in which the invariance conjecture holds, but there are also natural settings in which it fails (in each of the possible directions).

The technical angle. Items 1 and 2 are established by relying on corresponding results of [9], while our contribution is in observing that the local conditions stated in [9] (in terms of subgraph freeness) coincide with local conditions that are invariant under graph isomorphisms. Actually, to rule out characterizations by other possible invariances (i.e., invariances other than graph isomorphism), we also use the canonization technique of [10, Thm. 2]. In the two examples of Item 3a we rely on the fact that these properties were shown to have (proximity oblivious) testers in [12] and [3], respectively. Thus, in both these cases, our contribution is showing that these properties *cannot* be characterized by invariant local conditions. In Item 3b we rely on a lower bound established in [6] (regarding testing Eulerian orientations of cyclic grids), and our contribution is in observing that this property *can* be characterized by invariant local conditions.

We mention that the property used towards establishing Item 3b is invariant under a 1-transitive[3] permutation group. Thus, even such an invariance feature does not guarantee easy testability (i.e., a standard tester of query complexity that only depends on the proximity parameter). Furthermore, this holds even when all local conditions are generated by a single local condition (closed under the said invariance).

Terminology. Throughout the text, when we say proximity oblivious testing we actually mean proximity oblivious testing *in a constant number of queries*. The definition of proximity oblivious testing appears in the appendix.

Organization. In Section 2 we provide a definitional framework that captures the foregoing discussion. In particular, this framework includes a general definition of the notion of characterizations by invariant local conditions and a formal statement of the invariance conjecture. In Section 3 we show that the invariance conjecture holds in the context of testing graph properties in the dense graph model, and in Section 4 we present an analogous conditional (or partial) result for the bounded-degree graph model. The failure of the invariance conjecture is demonstrated in Section 5, and possible conclusions are discussed in Section 6.

---

[3] A permutation group G over D is called 1-transitive if for every $e, e' \in D$ there exists a $\pi \in G$ such that $\pi(e) = e'$.

2   General Framework

For simplicity, we consider properties of finite functions defined over a finite domain D and having a finite range R, whereas an asymptotic treatment requires considering properties that are infinite sequences of such properties (i.e., a sequence of the type $(P_n)_{n \in \mathbb{N}}$ where P_n is a set of functions from D_n to R_n).[4] Still, we shall just write P, D, R, and (in order for our asymptotic statements to make sense) one should think of P_n, D_n, R_n. In particular, when we say that some quantity is a "constant", we actually think of D as growing (along with P and possibly R), while the said quantity remains fixed. Thus, in the rest of our presentation, D and R should be considered as generic sets having a variable size, although they will be often omitted from definitions and notations.

The simplified form of the invariant condition. We start by outlining a simplified version of the condition that we seek, regarding a property P (of functions $D \to R$):

1. P is closed under the action of some permutation group G, which is defined over D, and
2. P has a characterization via a constant number of "generic" constraints of constant size such that a function f is in P iff all actual constraints obtained by having G act on the generic constraints are satisfied.

In other words, P can be characterized by a set of constraints that are generated by some permutation group G acting on a constant number of constant-size constraints.

We stress that the foregoing permutation group G is chosen arbitrarily, and may depend on P (and not only on a natural class of properties to which P belongs). Thus, if P is a graph property, then G need not be the group that preserves *all* graph properties (i.e., the vertex-relabeling group), but rather may be any group that extends the vertex-relabeling group. For example, if P is the property of having more edges than non-edges, then G may be the symmetric group of all (unordered) vertex pairs, which in particular contains the vertex-relabeling group as a subgroup.

2.1   Characterization by Generated Constraints

We now generalize and clarify the above discussion. First we need to define what we mean by a constraint. A constraint will be a pair consisting of domain elements and a Boolean predicate applied to the corresponding values, and it is satisfied by a function f if applying the predicate to the f-values at the specified locations yields the Boolean value 1 (representing **true**).

Definition 2.1 (constraints): *A* constraint *is a pair* $((e_1, ..., e_c), \phi)$ *such that* $e_1, ..., e_c$ *are distinct elements in* D, *and* $\phi : R^c \to \{0, 1\}$ *is an arbitrary predicate. We say that the foregoing is a constraint of arity* c *(or a* c-constraint*). A function* $f : D \to R$ *is said to satisfy the foregoing constraint if* $\phi(f(e_1), ..., f(e_c)) = 1$.

---

[4] The reader may think of $n = |D_n|$, but it is helpful not to insist that $D_n = [n]$. On the other hand, the set R_n may be independent of n (cf., e.g., the case of Boolean functions).

Note that at this point the predicate ϕ may depend on the sequence of elements $(e_1, ..., e_c)$. Such a dependence will not exist in the case that a large set of constraints is generated based on few constraints (as in Definition 2.3).

The next notion is of characterization by a set of constraints. A property P of functions is characterized by a set of constraints if f is in P if and only f satisfies all the constraints in the set.

Definition 2.2 (characterization by constraints): *Let C be a set of constraints and P be a property. We say that P is characterized by C if for every $f : D \rightarrow R$ it holds that $f \in$ P if and only if f satisfies each constraint in C.*

Next, we consider the set of constraints generated by the combination of (1) a fixed set of constraints, (2) a group of permutations over D, and (3) a group of permutations over R. For starters, the reader is advised to think of the second group as of the trivial group containing only the identity permutation. In general, we shall consider a subset of the set of all pairs consisting of a permutation as in (2) and a permutation as in (3).

Definition 2.3 (generated constraints): *Let C be a finite set of c-constraints, and M be a set of pairs consisting of a permutation over D and a permutation over R (i.e., for any $(\pi, \mu) \in M$ it holds that π is a permutation of D and μ is a permutation of R). The set of constraints generated by C and M, denoted* $\text{CONS}(C, M)$, *is defined by*

$$\text{CONS}(C, M) \stackrel{\text{def}}{=} \{((\pi(e_1), ..., \pi(e_c)), \phi \circ \mu^{-1}) : ((e_1, ..., e_c), \phi) \in C , (\pi, \mu) \in M\}$$
(1)

where $\phi \circ \mu^{-1}(v_1, ..., v_c)$ denotes $\phi(\mu^{-1}(v_1), ..., \mu^{-1}(v_c))$.

Note that saying that f satisfies $((\pi(e_1), ..., \pi(e_c)), \phi \circ \mu^{-1})$ means that

$$(\phi \circ \mu^{-1})(f(\pi(e_1)), ..., f(\pi(e_c))) = \phi(\mu^{-1}(f(\pi(e_1))), ..., \mu^{-1}(f(\pi(e_c)))) = 1,$$

which means that $\mu^{-1} \circ f \circ \pi$ satisfies the constraint $((e_1, ..., e_c), \phi)$. Regarding the use of $\mu^{-1} \circ f \circ \pi$ rather than $\mu \circ f \circ \pi$, see the discussion following Definition 2.5.

Notation: As in Definition 2.3, it will be convenient to generalize functions to sequences over their domain. That is, for any function F defined over some set S, and for any $e_1, ..., e_t \in S$, we denote the sequence $(F(e_1), ..., F(e_t))$ by $F(e_1, ..., e_t)$. Throughout the text, id will be used to denote the identity permutation, where the domain is understood from the context.

2.2 The Invariance Condition

Returning to the condition outlined initially, let us now formulate it as follows. We consider a group of pairs (π, μ) such that π is a permutation over D and μ is a permutation over R with a group operation that corresponds to component-wise composition of permutations (i.e., $(\pi_1, \mu_1) \odot (\pi_2, \mu_2) = (\pi_1 \circ \pi_2, \mu_1 \circ \mu_2)$, where $\odot$ denotes the group operation). We call such a group a group of permutation pairs, and note that it need not be a direct product of a group of permutation over D and a group of permutations over R.

Definition 2.4 (the invariance condition): *A property* P *satisfies the invariance condition if there exists a constant, denoted c, a finite set of c-constraints, denoted C, and a group, denoted M, of permutation pairs over D × R such that* P *is characterized by* CONS(C, M). *In this case, we also say that* P *satisfies the invariance condition w.r.t M.*

The invariance condition and covering the domain. *We confine our discussion to the case that the domain contains only elements that are influential w.r.t the property* P; that is, for every $e \in D$, there exists $f_1 \in$ P and $f_0 \notin$ P such that $f_1(x) = f_0(x)$ for every $x \in D \setminus \{e\}$. Observe that if property P satisfies the invariance condition w.r.t M, then M induces a transitive permutation group on a constant fraction of D. This follows because the permutation group (over D) induced by M must map a constant number of elements (i.e., those appearing in the constraint set C) to all elements of D.

The main question. We are interested in the relation between satisfying the invariance condition and having a proximity oblivious tester (of constant-query complexity). One natural conjecture, hereafter referred to as the invariance conjecture, is that *a property satisfies the invariance condition if and only if it has a proximity oblivious tester.* Weaker forms of this conjecture refer to its validity within various models of property testing. This leads us to ask what "models of property testing" are.

2.3 Models of Property Testing

Natural models of property testing can be defined by specifying the domain and range of functions (i.e., D and R) as well as the closure features of the properties in the model.[5] We elaborate below (and mention that this view was elaborated independently by Sudan [18]).

For example, the model of testing graph properties in the adjacency matrix representation, introduced in [7], refers to $D = \binom{[N]}{2}$ and $R = \{0, 1\}$ as well as to the permutation group over D that is defined by all relabeling of $[N]$. Specifically, an N-vertex graph is represented by the Boolean function $g : \binom{[N]}{2} \rightarrow \{0, 1\}$ such that $g(\{u, v\}) = 1$ if and only if u and v are adjacent in the graph. Here an adequate closure feature gives rise to graph properties, where P is a graph property if, for every such function g and every permutation ψ over $[N]$, it holds that $g \in$ P iff $g_\psi \in$ P, where $g_\psi(\{u, v\}) \stackrel{\text{def}}{=} g(\{\psi(u), \psi(v)\})$.

In general, closure features are defined by groups of pairs of permutations, just as those in Definition 2.4.

Definition 2.5 (closure features): *Let M be as in Definition 2.4. We say that a property* P *is closed under M if, for every $(\pi, \mu) \in M$, it holds that $f \in$ P if and only if $\mu \circ f \circ \pi^{-1} \in$ P.*

---

[5] In addition, one may consider sub-models that are obtained by requiring the functions in such a model to satisfy some auxiliary properties.

Note that $\mu \circ f \circ \pi^{-1}$ (rather than $\mu \circ f \circ \pi$) is indeed the natural choice, since f maps D to R whereas the new function $f' = \mu \circ f \circ \pi^{-1}$ is meant to map $\pi(D)$ to $\mu(R)$; thus, when f' is applied to $e' = \pi(e)$ this results in first recovering e, next applying f, and finally applying μ.

Definition 2.6 (closure-based models of property testing): *The* model of M *consists of the class of all properties that are closed under M.*

For example, the model of testing graph properties in the adjacency matrix equals the *model of M*, where M is the set of all pairs $(\pi, \mathtt{id})$ such that π : $\binom{[N]}{2} \to \binom{[N]}{2}$ is induced by a permutation of $[N]$ (i.e., there exists a permutation ψ over $[N]$ such that $\pi(\{u,v\}) = \{\psi(u), \psi(v)\}$, for all $\{u,v\} \in D = \binom{[N]}{2}$). We comment that not all "popular models of property testing" can be reduced to Definition 2.6, but nevertheless Definition 2.6 is a good starting point; that is, various models can be naturally defined as subclasses of the class of all properties that are closed under some group M (where typically in such cases the subclass are characterized by a set of constraints that are generated as in Definition 2.3).[6]

We observe that closure under M is a necessary condition for satisfying the invariance condition with respect to M.

Proposition 2.7. *If* P *satisfies the invariance condition w.r.t M, then* P *is closed under M.*

Proof: For any $f \in$ P and $(\pi_0, \mu_0) \in M$, consider $f' \stackrel{\text{def}}{=} \mu_0 \circ f \circ \pi_0^{-1}$. We shall show that $f \in$ P if and only if $f' \in$ P. Suppose that P is characterized by $\mathtt{CONS}(C, M)$, and consider an arbitrary constraint in $\mathtt{CONS}(C, M)$. By definition (of being generated from (C, M)), this constraint has the form $(\pi(e_1), ..., \pi(e_c)), \phi \circ \mu^{-1})$, where $((e_1, ..., e_c), \phi) \in C$ and $(\pi, \mu) \in M$. Our aim is to show that f' satisfies this constraint if and only if f satisfies some related constraint in $\mathtt{CONS}(C, M)$, where the two constraints are related via (π_0, μ_0).

We start by looking at the value of $(\phi \circ \mu^{-1})(f'(\pi(e_1)), ..., f'(\pi(e_c)))$, which we shorthand as $(\phi \circ \mu^{-1})(f'(\pi(e_1, ..., e_c)))$. Plugging-in the definition of f', what we now look at is $(\phi \circ \mu^{-1})((\mu_0 \circ f \circ \pi_0^{-1})(\pi(e_1, ..., e_c)))$, which may be written as $\phi(\mu^{-1} \circ \mu_0 \circ f \circ \pi_0^{-1} \circ \pi(e_1, ..., e_c))$, which in turn equals $\phi((\mu^{-1} \circ \mu_0) \circ f \circ (\pi_0^{-1} \circ \pi)(e_1, ..., e_c))$. That is, we consider whether f satisfies the constraint $((\pi_0^{-1} \circ \pi)(e_1, ..., e_c), \phi \circ (\mu^{-1} \circ \mu_0))$, which can be written as $((\pi_0^{-1} \circ \pi)(e_1, ..., e_c), \phi \circ (\mu_0^{-1} \circ \mu)^{-1})$. But this constraint is in $\mathtt{CONS}(C, M)$, since it is generated from $((e_1, ..., e_c), \phi) \in C$ by using the pair $(\pi_0^{-1} \circ \pi, \mu_0^{-1} \circ \mu) \in M$. Thus, f' satisfies the constraint generated (from $((e_1, ..., e_c), \phi)$) by $(\pi_0^{-1} \circ \pi, \mu_0^{-1} \circ \mu)$ if and only if f satisfies the constraint generated (from it) by (π, μ). It follows that f' satisfies all constraints in $\mathtt{CONS}(C, M)$ if and only if f satisfies all constraints in $\mathtt{CONS}(C, M)$. ∎

---

[6] Indeed, an alternative formulation of the model of testing graph properties in the adjacency matrix representation is obtained by starting from $D = [N] \times [N]$ and M that equals all pairs $(\pi, \mathtt{id})$ such that $\pi(u, v) = (\psi(u), \psi(v))$, for some permutation ψ over $[N]$ and all $(u, v) \in D = [N] \times [N]$. In such a case, we consider the subclass of symmetric functions (i.e., functions g such that $g(u, v) = g(v, u)$ for all $(u, v) \in D$).

3   The Invariance Conjecture Holds in the Dense Graph Model

We prove that the invariance conjecture holds in the special case of graph properties in the adjacency matrix representation model (a.k.a the dense graph model). Recall that in this model, an N-vertex graph is represented by the (symmetric) Boolean function $g : [N] \times [N] \to \{0,1\}$ such that $g(u,v) = 1$ if and only if u and v are adjacent in the graph.

We rely on a recent result of [9], which states that (in this model) P has a proximity oblivious tester if and only if it is a subgraph-freeness property. We observe that being a subgraph-freeness property is equivalent to satisfying the invariance condition *with respect to the canonical set*, where the canonical set has the form $M = M' \times \{\text{id}\}$ such that M' is the group of permutations over vertex-pairs that is induced by vertex-relabeling.[7] (Indeed, the canonical set is the very set that defines the current model; see Section 2.3). So it is left to show that P *satisfies the invariance condition if and only if* P *satisfies the invariance condition with respect to the canonical set*. We thus get

Theorem 3.1. *Suppose that* P *is a set of Boolean functions over the set of unordered pairs over* $[N]$ *such that* P *is closed under relabeling of the base set (i.e.,* P *is a graph property that refers to the adjacency representation of graphs). Then,* P *has a proximity oblivious tester if and only if* P *satisfies the invariance condition. Furthermore, if* P *satisfies the invariance condition, then it satisfies this condition with respect to the canonical set.*

Proof: The key observation is that, in this model, *a property satisfies the invariance condition with respect to the canonical set if and only if it is a subgraph-freeness property*, where throughout this proof subgraph-freeness means not having certain induced graphs (which are specified in a forbidden set). The backward direction (i.e., from subgraph-freeness to the invariance condition) follows by observing that every subgraph-freeness property satisfies the invariance condition with respect to the canonical set, because it can be generated by the predicate that forbids certain unlabeled graphs (e.g., not having $F = ([n], E_F)$ as an induced subgraph is captured by the constraint $((\{1,2\}, .., \{1,n\}, ..., \{n-1,n\}), \phi)$ such that $\phi(a_{1,2}, ..., a_{n-1,n}) = 1$ if and only if F is not represented by $(a_{i,j})_{i,j}$). In proving the other direction (i.e., from the invariance condition to subgraph-freeness), observe that the "base" constraints may be viewed as a predicate on an unlabeled induced subgraph; that is, the constraint $((\{i_1, j_1\}, .., \{i_c, j_c\}), \phi)$ can be viewed as forbidding all induced subgraphs that are consistent with some $(a_{i_k, j_k})_{k \in [c]}$ such that $\phi(a_{i_1, j_1}, ..., a_{i_c, j_c}) = 0$.

Another important observation is that *if* P *satisfies the invariance condition then it does so with the canonical pair*. This observation is proven as follows. Let P be characterized by $\text{CONS}(C, M)$, where M is not necessarily the canonical set. Then, we view $\text{CONS}(C, M)$ (or rather the uniform distribution over it)

---

[7] Note that M' is a permutation group over $\binom{[N]}{2}$; it contains only permutations of the form π_ψ such that $\pi_\psi(\{u,v\}) = \{\psi(v), \psi(u)\}$, where ψ is an arbitrary permutation over $[N]$.

as a ((possibly "weak") non-adaptive) tester with one-sided error; that is, this tester always accepts any graph in P and its error probability (on no-instances) is strictly less than 1 (i.e., it accepts graphs that are not in P with probability is at most $1 - |\text{CONS}(C, M)|^{-1}$). Applying [10, Thm. 2], we obtain a tester with similar one-sided error that only inspects the graph induced by a random constant-size vertex-set. (Indeed, the transformation in [10, Thm. 2] preserves the detection probability no matter how small it is.) The latter tester gives rise to a characterization of P that can be generated by the decision predicate of this tester coupled with the group of vertex-relabeling; that is, P satisfies the invariance condition with the canonical set.

The current theorem now follows by combining the two foregoing observations with [9, Thm. 4.7]. Specifically, by [9, Thm. 4.7], P *has a proximity oblivious tester if and only if it is a subgraph freeness property*. By the first observation, P is a subgraph freeness property if and only if P satisfies the invariance condition with the canonical set, whereas (by the second observation) P satisfies the invariance condition if and only if P satisfies the invariance condition with respect to the canonical set. ∎

4 The Invariance Conjecture in the Bounded-Degree Graph Model

The next natural challenge is proving a result analogous to Theorem 3.1 for the bounded-degree graph model (introduced in [8]). Unfortunately, only a partial result is established here, because of a difficulty that arises in [9, Sec. 5] (regarding "non-propagation"), to be discussed below.

But first, we have to address a more basic difficulty that refers to fitting the bounded-degree graph model within our framework (i.e., Section 2.3). Recall that the standard presentation of the bounded-degree model represents an N-vertex graph of maximum degree d by a function $g : [N] \times [d] \to \{0, 1, ..., N\}$ such that $g(v, i) = u \in [N]$ if u is the ith neighbor of v and $g(v, i) = 0$ if v has less than i neighbors. This creates technical difficulties, which can be resolved in various ways.[8] The solution adopted here is to modify the representation of the bounded-degree graph model such that N-vertex graphs are represented by functions from $[N]$ to subsets of $[N]$. Specifically, such a graph is represented by a function $g : [N] \to 2^{[N]}$ such that $g(v)$ is the set of neighbors of vertex v. Furthermore, we are only interested in functions g that describe undirected graphs, which means that $g : [N] \to 2^{[N]}$ should satisfy $u \in g(v)$ iff $v \in g(u)$ (for every $u, v \in [N]$).

---

[8] The problem is that here it is important to follow the standard convention of allowing the neighbors of each vertex to appear in arbitrary order (as this will happen under relabeling of vertex names), but this must allow us to permute over $[d]$ without distinguishing vertices from the 0-symbol. One possibility is to give up the standard convention by which the vertices appear first and 0-symbols appear at the end of the list. We choose a different alternative.

Theorem 4.1. *Suppose that* P *is a set of functions from* $[N]$ *to* $\{S \subset [N] : |S| \leq d\}$ *that corresponds to an undirected graph property; in particular,* P *is closed under the following canonical set* M_0 *defined by* $(\pi, \mu) \in M_0$ *if and only if* π *is a permutation over* $[N]$ *and* μ *acts analogously on sets (i.e.,* $\mu(S) = \{\pi(v) : v \in S\}$*).* [9] *Then:*

1. *If* P *has a proximity oblivious tester, then it satisfies the invariance condition.*
2. *If* P *satisfies the invariance condition, then it satisfies it with respect to the canonical set, and it follows that* P *is a generalized subgraph freeness property (as defined in [9, Def. 5.1]).*

Recall that by [9, Sec. 5], if P is a generalized subgraph freeness property *that is non-propagating*, then P has a proximity oblivious tester. But it is unknown whether each generalized subgraph freeness property is non-propagating. (We note that this difficulty holds even for properties that satisfy the invariance condition with respect to the canonical set.)[10]

Proof: As in the dense graph model (i.e., Theorem 3.1), the key observation is that a property in this model satisfies the invariance condition with respect to the canonical set if and only if it is a generalized subgraph-freeness property (as defined in [9, Def. 5.1]). Thus, Part (1) follows immediately from [9, Thm. 5.5], and the point is proving Part (2).[11]

Suppose that P is characterized by $\mathtt{CONS}(C, M)$. Viewing the uniform distribution over $\mathtt{CONS}(C, M)$ as a (very weak) one-sided error non-adaptive tester, we apply a "canonicalization" procedure that is analogous to [10, Thm. 2], and obtain a (very weak) tester that inspects the neighborhoods of c randomly distributed vertices. This yields a characterization of P by $\mathtt{CONS}(\{((1, ..., c), \phi)\}, M_0)$, where ϕ is this tester's decision predicate. So we are done. ∎

5   The Invariance Conjecture Fails in Some Cases

We show that, in general, the invariance condition is neither necessary nor sufficient for the existence of proximity oblivious testers (POTs).

---

[9] Recall that we also assume that for every $g \in$ P it holds that $u \in g(v)$ iff $v \in g(u)$ (for every $u, v \in [N]$). We note that this extra property is easy to test.

[10] In fact, the negative example in [9, Prop. 5.4] can arise in our context. Specifically, consider the set of constraints generated by the constraint $((1, 2), \phi)$ such that $\phi(S_1, S_2) = 1$ iff both (1) $|\{i \in \{1, 2\} : S_i = \emptyset\}| \neq 1$ and (2) $|S_1| \in \{0\} \cup \{2i - 1 : i \in \mathbb{N}\}$. (Indeed, condition (1) mandates that if the graph contains an isolated vertex then it contains no edges, whereas condition (2) mandates that all non-isolated vertices have odd degree.)

[11] The point (i.e., Part (2)) is showing that if P satisfies the invariance condition, then it satisfies it with respect to the canonical set. We mention that the transformation from the possibly adaptive character of a proximity oblivious tester to the non-adaptive character of the invariance condition (equivalently, generalized subgraph-freeness) is performed in [9, Thm. 5.5].

5.1   The Invariance Condition Is Not Necessary for POT

We present two examples (i.e., properties) that demonstrate that satisfying the invariance condition is not necessary for having a proximity oblivious tester. Both examples are based on sparse linear codes that have (proximity oblivious) codeword tests (i.e., these codes are locally testable). In both cases, the key observation is that satisfying the invariance condition with respect to M (as in Definition 2.4) requires that M is "rich enough" since the domain permutations should map a fixed number of elements to all the domain elements. On the other hand, Proposition 2.7 requires that the property is closed under M, whereas this is shown to be impossible in both examples. In the first example, presented next, the property will be shown to be closed only under the trivial pair $(\mathrm{id}, \mathrm{id})$.

Theorem 5.1. *There exists a property, denoted* P, *of Boolean functions such that* P *has a proximity oblivious tester but does not satisfy the invariance condition. Furthermore,* P *is a linear property; that is, if* $f_1, f_2 \in P$ *then* $f_1 + f_2 \in P$, *where* $(f_1 + f_2)(x) = f_1(x) \oplus f_2(x)$ *for every* x.

Proof: We consider a random linear property of dimension $\ell = O(\log n)$. That is, for uniformly selected functions $g_1, ..., g_\ell : [n] \to \{0, 1\}$, we consider the property $P_n = \{\sum_{i \in I} g_i : I \subseteq [\ell]\}$. Actually, we repeat this selection for every value of n, obtaining the property $P = (P_n)_{n \in \mathbb{N}}$. It was shown in [12] that, with high probability over these random choices, the property P has a POT. We shall show that, with high probability over these random choices, the property P does not satisfy the invariance condition.

The key observation is that satisfying the invariance condition with respect to M (as in Definition 2.4) requires that M is non-trivial (i.e., contains a non-trivial pair), because otherwise P_n is characterized by a fixed (i.e., independent of n) number of constraints, which is highly improbable for random g_i's. On the other hand, Proposition 2.7 requires that P_n is closed under M, which is highly improbable when M is non-trivial. Specifically, we will show that with high probability (over the choice of P_n), for every non-trivial (π, μ), there exists $f \in P_n$ such that $\mu \circ f \circ \pi^{-1} \notin P$. We distinguish between two cases: (1) the case that π is not the identity permutation but μ is the identity permutation, and (2) the case that μ is not the identity permutation (which implies that $\mu(b) = 1 - b$ for every $b \in \{0, 1\}$).

Claim 5.1.1. *Let π be a permutation such that $m \stackrel{\text{def}}{=} |\{i \in [n] : \pi(i) \neq i\}| > 0$. Then, for a random P_n, the probability that $\{f \circ \pi : f \in P_n\} = P_n$ is less than* $2^{-m\ell/4}$.

Note that the number of permutations that satisfy the hypothesis is smaller than $\binom{n}{m} \cdot (m!) < 2^{m \log_2 n}$. Thus, the aggregated probability for the aforementioned Case (1) is a small constant (i.e., $\sum_{m > 0} 2^{-m \cdot ((\ell/4) - \log_2 n)}$ is smaller than, say, 0.01).

Proof: As a warm-up we upper bound the probability that $g \circ \pi = g$, where $g : [n] \to \{0, 1\}$ is uniformly distributed. For $g \circ \pi = g$ to hold, g must be constant on each cycle of π. Denoting the number of cycles by $c \le m/2$, it follows that $\Pr_g[g \circ \pi = g] = 2^{-m+c} \le 2^{-m/2}$. The argument extends to the case that we wish $g \circ \pi = g + f$ to hold for an arbitrary fixed f and a random g. Specifically, consider a cycle of π, denoted $i_1, ..., i_t$. Then, $\Pr_g[(\forall j \in [t-1])\, g(i_{j+1}) = g(i_j) + f(i_j)] = 2^{-(t-1)}$. It is even easier to prove that $\Pr_g[g \circ \pi = f] \le 2^{-m/2}$, since actually $\Pr_g[g \circ \pi = f] = 2^{-n}$. We now turn to upper-bound the probability that $\{f \circ \pi : f \in P_n\} = P_n$, by upper-bounding

$$\Pr_{g_1,...,g_\ell}[(\forall i \in [\ell])\, g_i \circ \pi \in P_n] = \Pr_{g_1,...,g_\ell}\left[\forall i \in [\ell]\, \exists I_i \subseteq [\ell] \text{ s.t. } g_i \circ \pi = \sum_{j \in I_i} g_j\right]$$

$$\le \sum_{I_1,...,I_\ell \subseteq [\ell]} \Pr_{g_1,...,g_\ell}\left[\forall i \in [\ell]\, g_i \circ \pi = \sum_{j \in I_i} g_j\right] \quad (2)$$

We break the sum in Eq. (2) into two parts, separating the single term that corresponds to $(I_1, ..., I_\ell) = (\{1\}, ..., \{\ell\})$ from all other terms. The contribution of the first term to Eq. (2) is upper-bounded by $(2^{-m/2})^\ell$, because $\Pr_{g_1,...,g_\ell}[\forall i \in [\ell]\, g_i \circ \pi = g_i]$ equals $\prod_{i=1}^{\ell} \Pr_{g_i}[g_i \circ \pi = g_i]$. For each other term corresponding to $(I_1, ..., I_\ell) \ne (\{1\}, ..., \{\ell\})$, we pick an arbitrary i such that $I_i \ne \{i\}$, and note that $\Pr_{g_1,...,g_\ell}[g_i \circ \pi = \sum_{j \in I_i} g_j]$ equals 2^{-n}, since g_i is uniformly distributed even when fixing the value of $\sum_{j \in I_i} g_j$. Furthermore, this assertion holds even if we only select g_i and $f_i = \sum_{j \in I_i} g_j$ at random (where in case $I_i = \emptyset$ we mean setting $f_i \equiv 0$). We now consider an iterative process starting with $i_1 = i$, such that at the first step we select uniformly g_{i_1} and $f_{i_1} = \sum_{j \in I_{i_1}} g_j$. Recall that we have $\Pr_{g_{i_1}, f_{i_1}}[g_{i_1} \circ \pi = f_{i_1}] = 2^{-n}$. For $k = 2, ..., \ell/2$, at the kth step we set i_k such that g_{i_k} is independent of $g_{i_1}, ..., g_{i_{k-1}}$ and $f_{i_1}, ..., f_{i_{k-1}}$ (where $f_i = \sum_{j \in I_i} g_j$), and uniformly select g_{i_k} and f_{i_k} (unless f_{i_k} was already determined, in which case it is left unchanged). Note that such an i_k exists as long as $k \le \ell/2$, but I_{i_k} need not be different than $\{i_k\}$. Then, the probability that $g_{i_k} \circ \pi = \sum_{j \in I_{i_k}} g_j$, conditioned on the values of $g_{i_1}, ..., g_{i_{k-1}}$ and $f_{i_1}, ..., f_{i_{k-1}}$, is at most $2^{-m/2}$, where the probability is taken merely over the choice of g_{i_k} (and possibly f_{i_k}). Thus, the contribution of this generic term to Eq. (2) is upper-bounded by $2^{-n} \cdot (2^{-m/2})^{(\ell/2)-1}$. Using the union bound, we upper-bound the contribution of all these $(2^\ell)^\ell - 1$ terms by

$$2^{\ell^2} \cdot 2^{-(n-(m/2))} \cdot (2^{-m/2})^{\ell/2}, \quad (3)$$

which is upper-bounded by $2^{-(m\ell/4)-1}$ (because $2^{\ell^2} \cdot 2^{-(n-(m/2))} < 1/2$). The claim follows (because $2^{-m\ell/2} < 2^{-(m\ell/4)-1}$). $\square$

Claim 5.1.2. *Let $\mu(b) = 1 - b$. Then, for a random P_n, the probability that there exists a permutation π such that $\{\mu \circ f \circ \pi^{-1} : f \in P_n\} = P_n$ is negligible as a function of n (i.e., it vanishes faster than any polynomial fraction (in n)).*

Proof: It suffices to show that, while the all-zero function is in P_n, with very high probability the constant-one function is not in P_n. This is the case because, with overwhelmingly high probability, for every non-empty $I \subseteq [\ell]$ it holds that $|\{j \in [n] : \sum_{i \in I} g_i(j) = 1\}|$ is in $(1 \pm o(1)) \cdot n/2$. $\square$

Combining Claims 5.1.1 and 5.1.2, we conclude that with high constant probability P is not closed under any non-trivial pair. Recalling the initial discussion, the theorem follows. $\blacksquare$

Testing the Long-Code (a.k.a dictatorship tests). We refer to the property $P = (P_n)$, where for $n = 2^\ell$, it holds that $f : \{0,1\}^\ell \to \{0,1\}$ is in P_n if and only if there exists $i \in [\ell]$ such that $f(\sigma_1 \cdots \sigma_\ell) = \sigma_i$. Such a function f is a dictatorship (determined by bit i) and can be viewed as the ith codeword in the long-code (i.e., the long-code encoding of i). Note that this property is closed under the pair (π, id), where π is a permutation π over $\{0,1\}^\ell$, if and only if there exists a permutation ϕ over $[\ell]$ such that $\pi(\sigma_1 \cdots \sigma_\ell) = \sigma_{\phi(1)} \cdots \sigma_{\phi(\ell)}$. (An analogous consideration applies to pairs (π, flip), where $\text{flip}(\sigma) = 1 - \sigma$ for every $\sigma \in \{0,1\}$.) We shall show that these are the only pairs under which the dictatorship property is closed, and it will follow that the dictatorship property violates the invariance condition.

Theorem 5.2. *The dictatorship property violates the invariance condition, although it has a proximity oblivious tester.*

Proof: The fact that the dictatorship property has a proximity oblivious tester is established in [3,14].[12] We shall show that this property violates the invariance condition because it is not closed under pairs (π, μ) unless π either preserves the (Hamming) weight of the strings or preserves this weight under flipping.

Indeed, the notion of (Hamming) weight is pivotal to this proof, where the weight of a string $\alpha \in \{0,1\}^\ell$, denoted $\text{wt}(\alpha)$, is defined as the number of bit positions that contain a one (i.e., $\text{wt}(\sigma_1 \cdots \sigma_\ell) \overset{\text{def}}{=} |\{i \in [\ell] : \sigma_i = 1\}|$). We first claim that if P_n is closed under (π, μ) then $\text{wt}(\pi(\alpha))$ equals either $\text{wt}(\alpha)$ or $\ell - \text{wt}(\alpha)$ for every $\alpha \in \{0,1\}^\ell$. (These two cases correspond to whether $\mu = \text{id}$ or $\mu = \text{flip}$ (i.e., $\mu(\sigma) = 1 - \sigma$).)

Suppose that π maps some ℓ-bit string α to a string β that has a different weight (i.e., $\text{wt}(\beta) \neq \text{wt}(\alpha)$). Then, $|\{f \in P_n : f(\alpha) = 1\}| = \text{wt}(\alpha)$, because for every $f \in P_n$ there exists a different $i \in [\ell]$ such that $f(\sigma_1 \cdots \sigma_\ell) = \sigma_i$. Similarly, $|\{f \circ \pi : f \in P_n \wedge (f \circ \pi)(\alpha) = 1\}| = \text{wt}(\beta)$, since $(f \circ \pi)(\alpha) = f(\beta)$. Using $\text{wt}(\alpha) \neq \text{wt}(\beta)$, we infer that $P_n \neq \{f \circ \pi : f \in P_n\}$, since each set contains a different number of functions that evaluate to 1 at the point α. This handles the case of $\mu = \text{id}$, and the case of $\mu = \text{flip}$ is handled similarly (i.e., if π maps some ℓ-bit string α to a string β such that $\text{wt}(\beta) \neq \ell - \text{wt}(\alpha)$, then $P_n \neq \{\mu \circ f \circ \pi : f \in P_n\}$).

Having established the above, we note that if P had satisfied the invariance condition then the corresponding M would have mapped a fixed number of elements to all domain elements. But this fixed number of domain elements (i.e., ℓ-bit long strings) have a fixed number of weights, whereas (by Proposition 2.7

---

[12] The longcode test of [3] only refers to the case that ℓ is a power of 2.

and the above) the set M may only contain pairs (π, μ) such that π preserves (or "complements") the weight of strings. This contradicts the requirement that all $\ell + 1$ different weights must be covered by the generated constraints, and the theorem follows.   ∎

5.2   The Invariance Condition Is Not Sufficient for POT

We next demonstrate that the invariance condition does not suffice for obtaining a proximity oblivious tester. Actually, the following example also shows that the invariance condition does not suffice for the standard definition of testing (with query complexity that only depends on the proximity parameter).

Theorem 5.3. *There exists a property, denoted* P, *of Boolean functions such that* P *satisfies the invariance condition but has no proximity oblivious tester. Furthermore, the invariant condition holds with respect to a single constraint that refers to four domain elements, and a group of domain permutations that is 1-transitive. Moreover,* P *cannot be tested (in the standard sense) within query complexity that only depends on the proximity parameter.*

Proof: We use a lower bound of [6] that refers to the query complexity of testing Eulerian orientations of fixed (and highly regular) bounded-degree graphs. Specifically, [6, Thm. 9.14] proves an $\Omega(\log \ell)$ query lower bound on the complexity of testing whether the orientation of an ℓ-by-ℓ cyclic grid is Eulerian. It follows that this property has no POT, while we shall see that it satisfies the invariance condition.

   We represent the orientation of the ℓ-by-ℓ cyclic grid by two functions $h, v :$ $\mathsf{Z}_\ell \times \mathsf{Z}_\ell \to \{0, 1\}$ such that $h(i, j)$ represents the orientation of the horizontal edge between the vertices (i, j) and $(i, j+1)$, whereas $v(i, j)$ represents the orientation of the vertical edge between the vertices (i, j) and $(i + 1, j)$, and the arithmetics is of Z_ℓ (i.e., modulo ℓ). Specifically, $h(i, j) = 1$ (resp., $v(i, j) = 1$) indicates an orientation from (i, j) to $(i, j + 1)$ (resp., $(i + 1, j)$). (Needless to say, we can pack both functions in a single function; for example, $f(1, i, j) = h(i, j)$ and $f(0, i, j) = v(i, j)$.)

   The key observation is that the Eulerian orientation property can be characterized by 4-constraints that are generated from a single constraint. Specifically, this property is characterized by the set of 4-constraints $\{h(i, j) + v(i, j) = h(i, j - 1) + v(i - 1, j) : i, j \in \mathsf{Z}_\ell\}$, where the constraint $h(i, j) + v(i, j) = h(i, j - 1) + v(i - 1, j)$ mandates that exactly two of the four edges of vertex (i, j) are oriented outwards. Finally, note that this set of constraints is generated by the single constraint $h(1, 1) + v(1, 1) = h(1, 0) + v(0, 1)$ and the set of mappings $\{(\pi_{r,s}, \mathtt{id})\}$, where $\pi_{r,s}(i, j) = (i + r, j + s)$. The main claim follows.

   The only part of the furthermore claim that requires elaboration is the claim that the group of domain permutations is 1-transitive. To show this we explicitly consider the packing of the aforementioned two functions in a single function $f : \{0, 1\} \times \mathsf{Z}_\ell \times \mathsf{Z}_\ell \to \{0, 1\}$ such that $f(1, i, j) = h(i, j)$ and $f(0, i, j) = v(i, j)$. We redefine the domain permutations $\pi_{r,s}$ such that $\pi_{r,s}(\sigma, i, j) = (\sigma, i + r, j + s)$

and introduce an auxiliary permutation π' such that $\pi'(\sigma, i, j) = (1 - \sigma, j, i)$. Observe that a generic constraint (now written as $f(1, i, j) + f(0, i, j) = f(1, i, j-1) + f(0, i-1, j))$ is preserved under the auxiliary permutation π'. The full claim now follows.   ∎

6   Conclusions

While the invariance conjecture holds in two natural models of testing graph properties, it was shown to fail in other settings. These failures, described in Section 5, are of three different types.

1. As shown in Theorem 5.1, proximity oblivious testers exist also for properties that are only closed under the identity mapping. That is, a strong notion of testability is achievable also in the absence of any invariants.
2. As shown in Theorem 5.2, the existence of proximity oblivious testers for properties that do not satisfy the invariance condition is not confined to unnatural properties and/or to properties that lack any invariance.
3. As shown in Theorem 5.3, the invariance condition does not imply the existence of a standard tester of query complexity that only depends on the proximity parameter. (Note that the non-existence of such testers implies the non-existence of proximity oblivious testers.) Furthermore, this holds even if the invariance condition holds with respect to a group of domain permutations that is 1-transitive and the set of local conditions is generated by a single condition (closed under this permutation group).

Our feeling is that the fact that the invariance condition is not necessary for proximity oblivious testing is less surprising than the fact that the former is insufficient for the latter. Giving up on the necessity part, we wonder whether a reasonable strengthening of the invariance condition may suffice for proximity oblivious testing.

A natural direction to consider is imposing additional restrictions on the group of domain permutations. As indicated by Theorem 5.3, requiring this group to be 1-transitive does not suffice, and so one is tempted to require this group to be 2-transitive[13] (as indeed suggested in [11] w.r.t standard testing).[14] Recalling that if P is closed under a 2-transitive group (over the domain) then P is self-correctable (and thus consists of functions that are pairwise far apart), one may also wonder about only requiring 1-transitivity but restricting attention to properties that consist of functions that are pairwise far apart. We mention that the property used in the proof of Theorem 5.3 contains functions that are close to one another.

---

[13] A permutation group G over D is called 2-transitive if for every $(e_1, e_2), (e_1', e_2') \in \binom{D}{2}$ there exists a $\pi \in G$ such that $\pi(e_1) = e_1'$ and $\pi(e_2) = e_2'$.

[14] Recall that here we refer to a set of local conditions that is generated by a *constant* number of local condition (closed under a 2-transitive permutation group). In contrast, Ben-Sasson *et al.* [4] have recently shown that a set of local conditions that is generated by a *non-constant* number of local condition (closed under a 2-transitive permutation group) can yield a non-testable property.

Actually, restricting attention to properties that are closed under a 1-transitive group of domain permutations, we may return to the question of necessity and ask whether the existence of proximity oblivious testers in this case implies the invariance condition. Note that our proofs of Theorems 5.1 and 5.2 rely on the fact that the corresponding group is not 1-transitive (e.g., in the first case the group action is trivial and in the second case it has a non-constant number of orbits).

An alternative perspective. We mention that Sudan's perspective on the role of invariance (cf. [18,19]) is different from the one studied in the current work. In particular, Sudan suggests to view the role invariance as a theme (or a technique, akin to others surveyed in [16,19]), which is indeed surveyed in [19, Sec. 5]. From this perspective, Sudan [19, Sec. 6] views our work as pointing out inherent limitations on the applicability of the "theme of invariances", and concludes that "despite the limitations, invariances have signifficant unifying power (even if they do not explain everything)."

Acknowledgments. We are grateful to Dana Ron for useful discussions. We also thank the anonymous reviewers of *RANDOM'11* for comments regarding a previous write-up of this work.

References

1. Alon, N., Fischer, E., Newman, I., Shapira, A.: A Combinatorial Characterization of the Testable Graph Properties: It's All About Regularity. In: 38th STOC, pp. 251–260 (2006)
2. Alon, N., Shapira, A.: A Characterization of Easily Testable Induced Subgraphs. Combinatorics Probability and Computing 15, 791–805 (2006)
3. Bellare, M., Goldreich, O., Sudan, M.: Free bits, PCPs and non-approximability – towards tight results. SIAM Journal on Computing 27(3), 804–915 (1998)
4. Ben-Sasson, E., Maatouk, G., Shpilka, A., Sudan, M.: Symmetric LDPC codes are not necessarily locally testable. ECCC, TR10-199 (2010)
5. Blum, M., Luby, M., Rubinfeld, R.: Self-Testing/Correcting with Applications to Numerical Problems. JCSS 47(3), 549–595 (1993)
6. Fischer, E., Lachish, O., Newman, I., Matsliah, A., Yahalom, O.: On the Query Complexity of Testing Orientations for Being Eulerian. In: Goel, A., Jansen, K., Rolim, J.D.P., Rubinfeld, R. (eds.) APPROX and RANDOM 2008. LNCS, vol. 5171, pp. 402–415. Springer, Heidelberg (2008), Full version available from http://www.cs.technion.ac.il/~oyahalom
7. Goldreich, O., Goldwasser, S., Ron, D.: Property testing and its connection to learning and approximation. Journal of the ACM, 653–750 (July 1998)
8. Goldreich, O., Ron, D.: Property Testing in Bounded Degree Graphs. Algorithmica 32(2), 302–343 (2002)
9. Goldreich, O., Ron, D.: On Proximity Oblivious Testing. ECCC, TR08-041 (2008); Also in the Proceedings of the 41st STOC (2009)
10. Goldreich, O., Trevisan, L.: Three theorems regarding testing graph properties. Random Structures and Algorithms 23(1), 23–57 (2003)

11. Grigorescu, E., Kaufman, T., Sudan, M.: 2-Transitivity is Insufficient for Local Testability. In: 23rd CCC, pp. 259–267 (2008)
12. Kaufman, T., Sudan, M.: Sparse Random Linear Codes are Locally Testable and Decodable. In: The Proceedings of the 48th FOCS, pp. 590–600 (2007)
13. Kaufman, T., Sudan, M.: Algebraic Property Testing: The Role of Invariances. In: 40th STOC, pp. 403–412 (2008)
14. Parnas, M., Ron, D., Samorodnitsky, A.: Testing basic boolean formulae. SIAM Journal on Discrete Math. 16(1), 20–46 (2002)
15. Ron, D.: Property Testing: A Learning Theory Perspective. Foundations and Trends in Machine Learning 1(3), 307–402 (2008)
16. Ron, D.: Algorithmic and Analysis Techniques in Property Testing. In: Foundations and Trends in TCS (to appear)
17. Rubinfeld, R., Sudan, M.: Robust characterization of polynomials with applications to program testing. SIAM Journal on Computing 25(2), 252–271 (1996)
18. Sudan, M.: Invariance in Property Testing. ECCC, TR10-051 (2010)
19. Sudan, M.: Testing Linear Properties: Some General Themes. ECCC, TR11-005 (2011)

Appendix: Property Testing and Proximity Oblivious Testers

We first recall the standard definition of property testing.

Definition A.1 (property tester): *Let $P = \bigcup_{n \in \mathbb{N}} P_n$, where P_n is a set of functions defined over the domain D_n. A* tester *for property P is a probabilistic oracle machine T that satisfies the following two conditions:*

1. *The tester accepts each $f \in P$ with probability at least 2/3; that is, for every $n \in \mathbb{N}$ and $f \in P_n$ (and every $\epsilon > 0$), it holds that $\Pr[T^f(n,\epsilon)=1] \geq 2/3$.*
2. *Given $\epsilon > 0$ and oracle access to any f that is ϵ-far from P, the tester rejects with probability at least 2/3; that is, for every $\epsilon > 0$, every $n \in \mathbb{N}$ and f over D_n, if $\delta_P(f) > \epsilon$, then $\Pr[T^f(n,\epsilon)=0] \geq 2/3$, where $\delta_P(f) \stackrel{\text{def}}{=} \min_{g \in P_n}\{\delta(f,g)\}$ and $\delta(f,g) \stackrel{\text{def}}{=} |\{e \in D_n : f(e) \neq g(e)\}|/|D_n|$.*

If the tester accepts every function in P with probability 1, then we say that it has one-sided error; *that is, T has one-sided error if for every $f \in P$ and every $\epsilon > 0$, it holds that $\Pr[T^f(n,\epsilon)=1]=1$. A tester is called* non-adaptive *if it determines all its queries based solely on its internal coin tosses* (and the parameters n and ϵ); *otherwise it is called* adaptive.

The query complexity of a tester is measured in terms of the size parameter, n, and the proximity parameter, ϵ. In this paper we focus on the case that the complexity only depends on ϵ (and is independent of n).

Turning to the definition of proximity-oblivious testers, we stress that they differ from standard testers in that they do not get a proximity parameter as input. Consequently, assuming these testers have sublinear complexity, they can only be expected to reject functions not in P with probability that is related to the distance of these functions from P. This is captured by the following definition.

Definition A.2 (proximity-oblivious tester): *Let* $P = \bigcup_{n \in \mathsf{N}} P_n$ *be as in Definition A.1. A* proximity-oblivious tester *for* P *is a probabilistic oracle machine* T *that satisfies the following two conditions:*

1. *The machine* T *accepts each function in* P *with probability 1; that is, for every* $n \in \mathsf{N}$ *and* $f \in P_n$, *it holds that* $\Pr[T^f(n) = 1] = 1$.
2. *For some* (monotone) *function* $\rho : (0,1] \rightarrow (0,1]$, *each function* $f \notin P$ *is rejected by* T *with probability at least* $\rho(\delta_P(f))$, *where* $\delta_P(f)$ *is as in Definition A.1.*

The function ρ *is called the* detection probability *of the tester* T.

In general, the query complexity of a proximity-oblivious tester may depend on the size parameter, n, but in this paper we focus on the case that this complexity is constant.

Note that a proximity-oblivious tester with detection probability ρ yields a standard (one-sided error) property tester of query complexity $O(1/\rho)$.

In a World of P=BPP

Oded Goldreich

Abstract. We show that proving results such as $\mathcal{BPP} = \mathcal{P}$ essentially necessitate the construction of suitable pseudorandom generators (i.e., generators that suffice for such derandomization results). In particular, the main incarnation of this equivalence refers to the standard notion of uniform derandomization and to the corresponding pseudorandom generators (i.e., the standard uniform notion of "canonical derandomizers"). This equivalence bypasses the question of which hardness assumptions are required for establishing such derandomization results, which has received considerable attention in the last decade or so (starting with Impagliazzo and Wigderson [*JCSS*, 2001]).

We also identify a natural class of search problems that can be solved by deterministic polynomial-time reductions to $\mathcal{BPP}$. This result is instrumental to the construction of the aforementioned pseudorandom generators (based on the assumption $\mathcal{BPP} = \mathcal{P}$), which is actually a reduction of the "construction problem" to $\mathcal{BPP}$.

Caveat: Throughout the text, we abuse standard notation by letting $\mathcal{BPP}, \mathcal{P}$ etc denote classes of promise problems. We are aware of the possibility that this choice may annoy some readers, but believe that promise problem actually provide the most adequate formulation of natural decisional problems.[1]

Keywords: BPP, derandomization, pseudorandom generators, promise problems, search problems, FPTAS, randomized constructions.

An earlier version of this work appeared as TR10-135 of *ECCC*.

1 Introduction

We consider the question of whether results such as $\mathcal{BPP} = \mathcal{P}$ *necessitate* the construction of suitable pseudorandom generators, and conclude that the answer is essentially positive. By suitable pseudorandom generators we mean generators that, in particular, imply that $\mathcal{BPP} = \mathcal{P}$. Thus, in a sense, the *pseudorandom generators approach to the BPP-vs-P Question is complete*; that is, if the question can be resolved in the affirmative, then this answer follows from the existence of suitable pseudorandom generators.

The foregoing equivalence bypasses the question of which hardness assumptions are required for establishing such derandomization results (i.e., $\mathcal{BPP} = \mathcal{P}$),

---

[1] Actually, the common restriction of general studies of feasibility to decision problems is merely a useful methodological simplification.

O. Goldreich et al.: Studies in Complexity and Cryptography, LNCS 6650, pp. 191–232, 2011.

which is a question that has received considerable attention in the last decade or so (see, e.g., [17,15,19]). Indeed, the current work would have been obsolete if it were the case that the known answers were tight in the sense that the hardness assumptions required for derandomization would suffice for the construction of the aforementioned pseudorandom generators. See further discussion in Section 1.5.

1.1 What Is Meant by Suitable Pseudorandom Generators?

The term pseudorandom generator is actually a general paradigm spanning vastly different notions that range from general-purpose pseudorandom generator (a la Blum, Micali, and Yao [2,27]) to special-purpose generators (e.g., pairwise-independence ones [3]). The common theme is that the generators are deterministic devices that *stretch* short random seeds into longer sequences that *look random* in some sense, and that their operation is *relatively efficient*. The specific incarnations of this general paradigm differ with respect to the specific formulation of the three aforementioned terms; that is, they differ with respect to the requirements regarding (1) the amount of stretching, (2) the sense in which the output "looks random" (i.e., the "pseudorandomness" property), and (3) the complexity of the generation (or rather the stretching) process.

Recall that general-purpose pseudorandom generators operate in (some fixed) polynomial-time while producing outputs that look random to *any* polynomial-time observers. Thus, the observer is more powerful (i.e., runs for more time) than the generator itself. One key observation of Nisan and Wigderson [20] is that using such general-purpose pseudorandom generators is an over-kill when the goal is to derandomize complexity classes such as $\mathcal{BPP}$. In the latter case (i.e., for derandomizing $\mathcal{BPP}$) it suffices to have a generator that runs in exponential time (i.e., time exponential in its seed's length), since our deterministic emulation of the resulting randomized algorithm is going to incur such a factor in its running-time anyhow.[2] This leads to the notion of a *canonical derandomizer*, which fools observers of fixed complexity, while taking more time to produce such fooling sequences.

Indeed, the aforementioned "suitable pseudorandom generators" are (various (standard) forms of) canonical derandomizers. Our starting point is the non-uniform notion of canonical derandomizers used by Nisan and Wigderson [20], but since we aim at "completeness results" (as formulated above), we seek uniform-complexity versions of it. Three such versions are considered in our work, and two are shown to be sufficient and necessary for *suitable derandomizations of $\mathcal{BPP}$*.

The last assertion raises the question of *what is meant by a suitable derandomization of $\mathcal{BPP}$*. The first observation is that any reasonable notion of a canonical derandomizer is also applicable to promise problems (as defined in [4]), and so

---

[2] Recall that the resulting (randomized) algorithm uses the generator for producing the randomness consumed by the original (randomized) algorithm, which it emulates, and that our deterministic emulation consists of invoking the resulting (randomized) algorithm on all possible random-pads.

our entire discussion refers to $\mathcal{BPP}$ as a class of promise problems (rather than a class of standard decision problems).[3]

The second observation is that standard uniform-complexity notions of canonical derandomizers would not allow to place $\mathcal{BPP}$ in $\mathcal{P}$, because rare instances that are hard to find may not lead to a violation of the pseudorandomness guarantee. The known fix, used by Impagliazzo and Wigderson in [17], is to consider "effective derandomization" in the sense that each problem $\Pi \in \mathcal{BPP}$ is approximated by some problem $\Pi' \in \mathcal{P}$ such that it is hard to find instances in the symmetric difference of Π and Π'. Our main result refers to this notion (see Sections 4.2–4.3): Loosely speaking, it asserts that canonical derandomizers (of exponential stretch) exist if and only if $\mathcal{BPP}$ is effectively in $\mathcal{P}$. We stress that this result refers to the standard notion of uniform derandomization and to the corresponding canonical derandomizers (as in [17] and subsequent works (e.g. [24])).

We also consider a seemingly novel notion of canonical derandomizers, which is akin to notions of auxiliary-input one-way functions and pseudorandom generators considered by Vadhan [26]. Here the generator is given a target string and the distribution that it produces need only be pseudorandom with respect to efficient (uniform) observers that are given this very string as an auxiliary input. We show that such canonical derandomizers (of exponential stretch) exist if and only if $\mathcal{BPP} = \mathcal{P}$; for details, see Section 4.4.

1.2 Techniques

Our starting point is the work of Goldreich and Wigderson [10], which studied pseudorandomness with respect to (uniform) deterministic observers. In particular, they show how to construct, for every polynomial p, a generator of exponential stretch that works in time polynomial in its output and fools all deterministic p-time tests of the next-bit type (a la [2]). They observe that an analogous construction with respect to general tests (i.e., deterministic p-time distinguishers) would yield some non-trivial derandomization results (e.g., any unary set in $\mathcal{BPP}$ would be placed in $\mathcal{P}$). Thus, they concluded that there is a fundamental gap between probabilistic and deterministic polynomial-time observers.[4]

Our key observation is that the gap between probabilistic observers and deterministic ones essentially disappears if $\mathcal{BPP} = \mathcal{P}$. Actually, the gap disappeared with respect to certain ways of constructing pseudorandom generators, and the

---

[3] Indeed, as stated upfront, we believe that, in general, promise problem actually provide the most adequate formulation of natural decisional problems (cf. [9, Sec. 2.4.1]). Furthermore, promise problems were considered in the study of derandomization when converse results were in focus (cf. [15]). An added benefit of the use of classes of promise problems is that $\mathcal{BPP} = \mathcal{P}$ implies $\mathcal{MA} = \mathcal{NP}$.

[4] In particular, they concluded that Yao's result (by which fooling next-bit tests implies pseudorandomness) may not hold in the (uniform) deterministic setting (or, actually, may be hard to establish in that context). Indeed, recall that the next-bit tests derived (in Yao's argument) from general tests (i.e., distinguishers) are probabilistic.

construction of [10] can be shown to fall into this category. We actually prefer a more direct approach, which is more transparent and amenable to variations. Specifically, we consider a straightforward probabilistic polynomial-time construction of a pseudorandom generator; that is, we observe that a random function (with exponential stretch) enjoys the desired pseudorandomness property, but of course the problem is that it cannot be constructed deterministically.

At this point, we define a search problem that consists of finding a suitable function (or rather its image), and observe that this problem is solvable in probabilistic polynomial-time. Using the fact that the suitability of candidate functions can be checked in probabilistic polynomial-time, we are able to *deterministically reduce* (in polynomial-time) this search problem to a (decisional) problem in $\mathcal{BPP}$. Finally, using the hypothesis (i.e., $\mathcal{BPP} = \mathcal{P}$), we obtain the desired (deterministic) construction.

1.3   Additional Results

The foregoing description alluded to the possibility that $\mathcal{BPP} = \mathcal{P}$ (which refers to promise problems of decisional nature) extends to search problems; that is, that $\mathcal{BPP} = \mathcal{P}$ implies that a certain class of probabilistic polynomial-time solvable search problems can be emulated deterministically. This fact, which is used in our construction of canonical derandomizers, is proven as part of our study of "BPP-search problems" (and their relation to decisional BPP problems), which seems of independent interest and importance. Other corollaries include the conditional (on $\mathcal{BPP} = \mathcal{P}$) transformation of any probabilistic FPTAS into a deterministic one, and ditto for any probabilistic polynomial-time method of contructing and verifying objects of a predetermined property. (For details see Section 3.)

Also begging are extensions of our study to general "stretch vs derandomization time" trade-off (akin to the general "hardness vs randomness" trade-off) and to the derandomization of classes such as $\mathcal{AM}$. The first extension goes through easily (see Section 5), whereas we were not able to pull off the second (see Section 6).

1.4   Reflection

Recalling that canonical derandomizers run for more time than the distinguishers that they are intended to fool, it is tempting to say that the existence of such derandomizers may follow by diagonalization-type arguments. Specifically, for every polynomial p, it should be possible to construct in (larger) polynomial time, a set of (poly(n) many) strings $S_n \subset \{0,1\}^n$ such that a string selected uniformly in S_n is $p(n)$-time indistinguishable from a totally random n-bit string.

The problem with the foregoing prophecy is that it is not clear how to carry out such a diagonalization. However, it was observed in a couple of related works (i.e., [17,10]) that a *random choice will do*. The problem, of course, is that we need our construction to be deterministic; that is, a deterministic construction should be able to achieve this "random looking" fooling effect. Furthermore, it is not a priori clear that the hypothesis $\mathcal{BPP} = \mathcal{P}$ may help us here, since $\mathcal{BPP} = \mathcal{P}$ refers to decisional problems.[5] Indeed, it seems that the interesting question of *determining the class of problems (e.g., search problems) that can be solved by deterministic polynomial-time reductions to $\mathcal{BPP}$* was not addressed before. Still, as stated above, we show that the aforemention "construction problem" belongs to this class, and thus the hypothesis $\mathcal{BPP} = \mathcal{P}$ allows us to derandomize the foregoing arguement.

In any case, the point is that $\mathcal{BPP} = \mathcal{P}$ enables the construction of the aforementioned type of (suitable) pseudorandom generators; that is, the very pseudorandom generators that imply $\mathcal{BPP} = \mathcal{P}$. Thus, our main result asserts that these pseudorandom generators exist if and only if $\mathcal{BPP} = \mathcal{P}$, which in our opinion is not a priori obvious. Furthermore, our proof uncovers a very tight connection between the construction of such pseudorandom generators and $\mathcal{BPP} = \mathcal{P}$. In particular, $\mathcal{BPP} = \mathcal{P}$ yields a very simple construction of such pseudorandom generators, which in turn can be seen as fulfillining the foregoing (diagonalization) prophecy.

1.5   Related Work

This work takes for granted the "hardness versus randomness" paradigm, pioneered by Blum and Micali [2], and its application to the derandomization of complexity classes such as $\mathcal{BPP}$, as pioneered by Yao [27] and revised by Nisan and Wigderson [20]. The latter work suggests that a suitable notion of a pseudorandom generator – indeed, the aforementioned notion of a canonical derandomizer – provides the "King's (high)way" to derandomization of $\mathcal{BPP}$. This view was further supported by subsequent work such as [16,17,25], and the current work seems to suggest that this King's way is essentially the only way.

As stated up-front, this work does not address the question of which hardness assumptions are required for establishing such derandomization results (i.e., $\mathcal{BPP} = \mathcal{P}$). Recall that this question has received considerable attention in the last decade or so, starting with the aforementioned work of Impagliazzo and Wigderson in [17], and culminating in the works of Impagliazzo, Kabanets, and Wigderson [15,19]. We refer the interested reader to [23, Sec. 1.1-1.3] for an excellent (and quite updated) overview of this line of work.

---

[5] For example, obviously, even if $\mathcal{BPP} = \mathcal{P}$, there exist no deterministic algorithms for uniformly selecting a random solution to a search problem (or just tossing a coin). Interestingly, while problems of uniform generation cannot be solved deterministically, the corresponding problems of approximating the number of solutions can be solved deterministically (sometimes in polynomial-time, especially when assuming $\mathcal{BPP} = \mathcal{P}$). This seems to contradict the celebrated equivalence between these two types of problems [18] (cf. [9, §6.4.2.1]), except that the relevant direction of this equivalence is established via probabilistic polynomial-time reductions (which are inherently non-derandomizable). Going beyond the strict boundaries of complexity, we note that $\mathcal{BPP} = \mathcal{P}$ would not eliminate the essential role of randomness in cryptography (e.g., in the context of zero-knowledge (cf. [8, Sec. 4.5.1]) and secure encryption (cf. [11])).

Note that the foregoing discussion refers to three possible events: The first event is the existence of a good derandomization (e.g., $\mathcal{BPP} = \mathcal{P}$), the second is the existence of certain pseudorandom generators (i.e., canonical derandomizers), and the third is the existence of certain lower bound (i.e., hardness results). The main thread of past work (e.g., [2,27,20,16,17]) goes from hardness assumptions to pseudorandom generators and further to good derandomization (e.g., $\mathcal{BPP} = \mathcal{P}$). Later work such as [17,15] partially reverse the the hardness to derandomization implication, whereas our work only refers and reverses the second leg of the main thread (i.e., showing that $\mathcal{BPP} = \mathcal{P}$ implies certain pseudorandom generators). We comment that the reversing of the first leg (i.e., showing that pseudorandom generators imply hardness) is folklore (see, e.g., [9, Exer. 8.24]). All these implications are depicted in Figure 1.

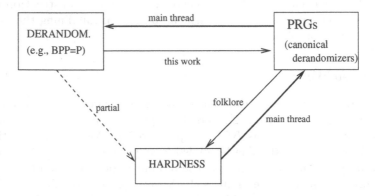

Fig. 1. The three related events. The solid arrows show implications that hold for the full range of parameters, whereas the dashed arrow shows a partial implication that does not suffice for the "high end" (i.e., for pseudorandom generators that suffice for $\mathcal{BPP} = \mathcal{P}$).

Actually, both the aforementioned works [17,15] imply results that are in the spirit of our main result, but these results refer to weak notions of derandomization, and their proofs are fundamentally different. The work of Impagliazzo and Wigderson [17] refers to the "effective infinitely often" containment of $\mathcal{BPP}$ in $\mathcal{SUBEXP}$, whereas the work of Impagliazzo, Kabanets, and Wigderson [15] refers to the (standard) containment of $\mathcal{BPP}$ in $\mathcal{NSUBEXP}/n^\epsilon$. In both cases, the derandomization hypotheses are shown to imply corresponding hardness results (i.e., functions in $\mathcal{EXP}$ that are not in $\mathcal{BPP}$ or functions in $\mathcal{NEXP}$ having no polynomial size circuits, resp.), which in turn yield "correspondingly canonical" derandomizers (i.e., canonical w.r.t effectively placing $\mathcal{BPP}$ in $\mathcal{SUBEXP}$ infinitely often or placing $\mathcal{BPP}$ in $\mathcal{NSUBEXP}$, resp.).[6] Thus, in both cases,

---

[6] Note that in case of [17] the generators are pseudorandom only infinitely often, whereas in the case of [15] the generators are computable in *non-deterministic* polynomial-time (with short advice). In both cases, the generators have polynomial stretch.

the construction of these generators (based on the relevant derandomization hypothesis) follows the "hardness versus randomness" paradigm (and, specifically, the Nisan–Wigderson framework [20]). In contrast, our constructions bypass the "hardness versus randomness" paradigm.

We also mention that the possibility of reversing the pseudorandomness-to-derandomization transformation was studied by Fortnow [5]. In terms of his work, our result indicates that in some sense Hypothesis III implies Hypothesis II.

There is a remote similarity between our search to decision reduction (see Section 3.2) and one part of the work of Aaronson *et. al.* [1]. Our reduction relies on the fact that additive error approximation of certain probabilities can be done in $\mathcal{BPP}$, and these approximations are use in our search process. Interestingly, our main application is for constructing an adequate pseudorandom set, which may be viewed as a diagonalization (w.r.t certain class of algorithms). The argument in [1, Sec. 3] relies on the fact that a multiplicative factor approximation of certain set sizes can be done in $\mathcal{AM}$, and uses these approximations to diagonalized over a certain class of circuits. (These two processes were discovered independently.)

Finally, we mention that the relation between derandomizing probabilistic search and decision classes was briefly mentioned by Reingold, Trevisan, and Vadhan in the context of $\mathcal{RL}$; see [22, Prop. 2.7].

1.6 Organization

The rather standard conventions used in this work are presented in Section 2. In Section 3 we take a close look at "BPP search problems" and their relation to $\mathcal{BPP}$. The relation between derandomizations of $\mathcal{BPP}$ and various forms of pseudorandom generators is studied in Section 4, and ramified in Section 5. A few open problems that arise naturally from this work are discussed in Section 6. The appendix presents two prior proofs of our main result, which may be of interest.

2   Preliminaries

We assume a sufficiently strong model of computation (e.g., a 2-tape Turing machine), which allows to do various simple operations very efficiently. Exact complexity classes such as $\mathrm{DTIME}(t)$ and $\mathrm{BPTIME}(t)$ refer to such a fixed model. We shall say that a problem Π is in $\mathrm{DTIME}(t)$ (resp., in $\mathrm{BPTIME}(t)$) if there exists a deterministic (resp., probabilistic) t-time algorithm that solves the problem *on all but finitely many inputs*.

We assume that all polynomials, time bounds, and stretch functions are monotonically increasing functions from $\mathbb{N}$ to $\mathbb{N}$, which means, in particular, that they are injective. Furthermore, we assume that all these functions are time-constructible (i.e., the mapping $n \mapsto f(n)$ can be computed in less than $f(n)$ steps).

Promise Problems. We rely heavily on the formulation of promise problems (introduced in [4]). We believe that, in general, the formulation of promise problems is far more suitable for any discussion of feasibility results. The original formulation of [4] refers to decision problems, but we shall also extend it to search problem. In the original setting, a promise problem, denoted $\langle P, Q \rangle$, consists of a **promise** (set), denoted P, and a **question** (set), denoted Q, such that the problem $\langle P, Q \rangle$ is defined as *given an instance $x \in P$, determine whether or not $x \in Q$.* That is, the solver is required to distinguish inputs in $P \cap Q$ from inputs in $P \setminus Q$, and nothing is required in case the input is outside P. Indeed, an equivalent formulation refers to two disjoint sets, denoted Π_{YES} and Π_{NO}, of YES- and NO-instances, respectively. We shall actually prefer to present promise problems in these terms; that is, as pairs $(\Pi_{\text{YES}}, \Pi_{\text{NO}})$ of disjoint sets. Indeed, standard decision problems appear as special cases in which $\Pi_{\text{YES}} \cup \Pi_{\text{NO}} = \{0, 1\}^*$. In the general case, inputs outside of $\Pi_{\text{YES}} \cup \Pi_{\text{NO}}$ are said to **violate the promise**.

Unless explicitly stated otherwise, all "decisional problems" discussed in this work are actually promise problems, and $\mathcal{P}, \mathcal{BPP}$ etc denote the corresponding classes of promise problems. For example, $(\Pi_{\text{YES}}, \Pi_{\text{NO}}) \in \mathcal{BPP}$ if *there exists a probabilistic polynomial-time algorithm A such that for every $x \in \Pi_{\text{YES}}$ it holds that $\Pr[A(x) = 1] \geq 2/3$, and for every $x \in \Pi_{\text{NO}}$ it holds that $\Pr[A(x) = 0] \geq 2/3$.*

Standard Notation. For a natural number n, we let $[n] \overset{\text{def}}{=} \{1, 2, ..., n\}$ and denote by U_n a random variable that is uniformly distributed over $\{0, 1\}^n$. When referring to the probability that a uniformly distributed n-bit long string hits a set S, we shall use notation such as $\Pr[U_n \in S]$ or $\Pr_{r \in \{0,1\}^n}[r \in S]$.

Negligible, Noticeable, and Overwhelmingly High Probabilities. A function $f : \mathbb{N} \to [0, 1]$ is called **negligible** if is decreases faster than the reciprocal of any positive polynomial (i.e., for every positive polynomial p and all sufficiently large n it holds that $f(n) < 1/p(n)$). A function $f : \mathbb{N} \to [0, 1]$ is called **noticeable** if it is lower bound by the reciprocal of some positive polynomial (i.e., for some positive polynomial p and all sufficiently large n it holds that $f(n) > 1/p(n)$). We say that the probability of an event is **overwhelmingly high** if the probability of the complement event is negligible (in the relevant parameter).

3   Search Problems

Typically, search problems are captured by binary relations that determine the set of valid instance-solution pairs. For a binary relation $R \subseteq \{0, 1\}^* \times \{0, 1\}^*$, we denote by $R(x) \overset{\text{def}}{=} \{y : (x, y) \in R\}$ the set of valid solutions for the instance x, and by $S_R \overset{\text{def}}{=} \{x : R(x) \neq \emptyset\}$ the set of instances having valid solutions. Solving a search problem R means that given any $x \in S_R$, we should find an element of $R(x)$ (whereas, possibly, we should indicate that no solution exists if $x \notin S_R$).

3.1 The Definition

The definition of "BPP search problems" is supposed to capture search problems that can be solved efficiently, when random steps are allowed. Intuitively, we do not expect randomization to make up for more than an exponential blow-up, and so the naive formulation that merely asserts that solutions can be found in probabilistic polynomial-time is not good enough. Consider, for example, the relation R such that $(x, y) \in R$ if $|y| = |x|$ and for every $i < |x|$ it holds that $M_i(x) \neq y$, where M_i is the ith deterministic machine (in some fixed enumeration of such machines). Then, the search problem R can be solved by a probabilistic polynomial-time algorithm (which, on input x, outputs a uniformly distributed $|x|$-bit long string), but cannot be solved by any deterministic algorithm (regardless of its running time).

What is missing in the naive formulation is any reference to the "complexity" of the solutions found by the solver, let alone to the complexity of the set of all valid solutions. The first idea that comes to mind is to just postulate the latter; that is, confine ourselves to the class of search problems for which valid instance-solution pairs can be efficiently recognized (i.e., R, as a set of pairs, is in $\mathcal{BPP}$).

Definition 3.1 (BPP search problems, first attempt): *A BPP-search problem is a binary relation R that satisfies the following two conditions.*

1. *Membership in R is decidable in probabilistic polynomial-time.*
2. *There exists a probabilistic polynomial-time algorithm A such that, for every $x \in S_R$, it holds that $\Pr[A(x) \in R(x)] \geq 2/3$.*

We may assume, without loss of generality, that, for every $x \notin S_R$, it holds that $\Pr[A(x) = \bot] \geq 2/3$. Note that Definition 3.1 is robust in the sense that it allows for error reduction, which may not be the case if Condition 1 were to be avoided. A special case in which Condition 1 holds is when R is an NP-witness relation; in that case, the algorithm in Condition 1 is actually deterministic.

In view of our general interest in promise problems, and of the greater flexibility they offer, it makes sense to extend the treatment to promise problems. The following generalization allows a promise set not only at the level of instances, but also at the level of instance-solution pairs. Specifically, we consider disjoint sets of valid and invalid instance-solution pairs, require this promise problem to be efficiently decidable, and of course require that valid solutions be found whenever they exist.

Definition 3.2 (BPP search problems, revisited): *Let R_{YES} and R_{NO} be two disjoint binary relations. We say that $(R_{\text{YES}}, R_{\text{NO}})$ is a BPP-search problem if the following two conditions hold.*

1. *The decisional problem represented by $(R_{\text{YES}}, R_{\text{NO}})$ is solvable in probabilistic polynomial-time; that is, there exists a probabilistic polynomial-time*

algorithm V such that for every $(x,y) \in R_{\text{YES}}$ it holds that $\Pr[V(x,y)=1] \geq 2/3$, and for every $(x,y) \in R_{\text{NO}}$ it holds that $\Pr[V(x,y)=1] \leq 1/3$.

2. *There exists a probabilistic polynomial-time algorithm A such that, for every $x \in S_{R_{\text{YES}}}$, it holds that $\Pr[A(x) \in R_{\text{YES}}(x)] \geq 2/3$, where $R_{\text{YES}}(x) = \{y : (x,y) \in R_{\text{YES}}\}$ and $S_{R_{\text{YES}}} = \{x : R_{\text{YES}}(x) \neq \emptyset\}$.*

We may assume, without loss of generality, that, for every x such that $(x,y) \in R_{\text{NO}}$ ($\forall y$), it holds that $\Pr[A(x) = \bot] \geq 2/3$. Note that the algorithms postulated in Definition 3.2 allow to find valid solutions as well as distinguish valid solutions from invalid ones (while guaranteeing nothing for solutions that are neither valid nor invalid).

The promise problem formulation (of Definition 3.2) captures many natural "BPP search" problems that are hard to fit into the more strict formulation of Definition 3.1. Typically, this can be done by narrowing the set of valid solutions (and possibly extending the set of invalid solutions) such that the resulting (decisional) promise problem becomes tractable. Consider for example, a search problem R (as in Definition 3.1) for which the following stronger version of Condition 2 holds.

(2') There exists a noticeable function $\mathtt{ntc} : \mathbb{N} \to [0,1]$ such that, for every $x \in S_R$ there exists $y \in R(x)$ such that $\Pr[A(x)=y] > \mathtt{ntc}(|x|)$, whereas for every $(x,y) \notin R$ it holds that $\Pr[A(x)=y] < \mathtt{ntc}(|x|)/2$.

Then, we can define $R'_{\text{YES}} = \{(x,y) : \Pr[A(x) = y] > \mathtt{ntc}(|x|)\}$ and $R'_{\text{NO}} = \{(x,y) : \Pr[A(x) = y] < \mathtt{ntc}(|x|)/2\}$, and conclude that $R' = (R'_{\text{YES}}, R'_{\text{NO}})$ is a BPP-search problem (by using A also for Condition 1), which captures the original problem just as well. Specifically, solving the search problem R is trivially reducible to solving the search problem R', whereas we can distinguish between valid solutions to R' (which are valid for R) and invalid solutions for R (which are also invalid for R'). This is a special case of the following observation.

Observation 3.3 (companions): *Let $\Pi = (R_{\text{YES}}, R_{\text{NO}})$ and $\Pi' = (R'_{\text{YES}}, R'_{\text{NO}})$ be two search problems such that $S_{R'_{\text{YES}}} = S_{R_{\text{YES}}}$ and $R'_{\text{NO}} \supseteq (\{0,1\}^* \times \{0,1\}^*) \setminus R_{\text{YES}}$, which implies $R'_{\text{NO}} \supseteq R_{\text{NO}}$ and $R'_{\text{YES}} \subseteq R_{\text{YES}}$.[7] Then, solving the search problem $(R_{\text{YES}}, R_{\text{NO}})$ is trivially reducible to solving the search problem $(R'_{\text{YES}}, R'_{\text{NO}})$, whereas deciding membership in $(R'_{\text{YES}}, R_{\text{NO}})$ is trivially reducible to deciding*

---

[7] The first conclusion (i.e., $R'_{\text{NO}} \supseteq R_{\text{NO}}$) follows by the fact that $R_{\text{NO}} \subseteq (\{0,1\}^* \times \{0,1\}^*) \setminus R_{\text{YES}}$, whereas $R'_{\text{YES}} \subseteq R_{\text{YES}}$ follows since $R'_{\text{NO}} \subseteq (\{0,1\}^* \times \{0,1\}^*) \setminus R'_{\text{YES}}$. Observation 3.3 itself relies only on these conclusions (i.e., $R'_{\text{NO}} \supseteq R_{\text{NO}}$ and $R'_{\text{YES}} \supseteq R_{\text{YES}}$) as well as on $S_{R'_{\text{YES}}} \supseteq S_{R_{\text{YES}}}$; the stronger condition (i.e., $R'_{\text{NO}} \supseteq (\{0,1\}^* \times \{0,1\}^*) \setminus R_{\text{YES}}$) is used in other applications of the notion of companion problems (see the discussion following Theorem 3.5).

membership in $(R'_{\text{YES}}, R'_{\text{NO}})$. *We call* Π' *a* companion *of* Π, *and note that in general this notion is not symmetric.*[8]

The point of these reductions is that they allow using algorithms associated with Π' for handling Π. Specifically, we can search solutions with respect to Π' and test validity of solutions with respect to Π', while being guaranteed that nothing was lost (since we still find valid solutions for any $x \in S_{R_{\text{YES}}}$, any solution in $R'_{\text{YES}}(x) \subseteq R_{\text{YES}}(x)$ is recognized by us as valid, and any candidate solution in $R_{\text{NO}}(x) \subseteq R'_{\text{NO}}(x)$ is rejected as invalid). Furthermore, by the companion condition, candidate solutions that are not valid with respect to Π are also rejected (since they are invalid w.r.t Π'); that is, if $(x, y) \notin R_{\text{YES}}$ (although it needs not be in R_{NO}), then $(x, y) \in R'_{\text{NO}}$ (since $R'_{\text{NO}} \supseteq (\{0,1\}^* \times \{0,1\}^*) \setminus R_{\text{YES}}$).

The methodology alluded to above is demonstrated next in casting any probabilistic fully polynomial-time approximation scheme (i.e., FPTAS, cf. [13]) as a search-BPP problem. A (probabilistic) FPTAS for a quantity $q : \{0,1\}^* \to \mathbb{R}^+$ is an algorithm that on input x and $\epsilon > 0$ runs for $\text{poly}(n/\epsilon)$ steps and, with probability at least $2/3$, outputs a value in the interval $[(1 \pm \epsilon) \cdot q(x)]$. A straightforward casting of this approximation problem as a search problem refers to the binary relation Q such that $Q \overset{\text{def}}{=} \{(\langle x, 1^m \rangle, v) \in \mathbb{R}^+ : |v - q(x)| \leq q(x)/m\}$. In general, however, this does not yield a BPP-search problem, since Q may not be probabilistic polynomial-time recognizable. Instead, we consider the BPP-search problem $(R_{\text{YES}}, R_{\text{NO}})$ such that $(\langle x, 1^m \rangle, v) \in R_{\text{YES}}$ if $|v - q(x)| \leq q(x)/3m$ and $(\langle x, 1^m \rangle, v) \in R_{\text{NO}}$ if $|v - q(x)| > q(x)/m$. Indeed, on input $\langle x, 1^m \rangle$ we find a solution in $R_{\text{YES}}(\langle x, 1^m \rangle)$ by invoking the FPTAS on input x and $\epsilon = 1/3m$, and deciding the validity of a pair $(\langle x, 1^m \rangle, v)$ w.r.t $(R_{\text{YES}}, R_{\text{NO}})$ is done by obtaining a good approximation of $q(x)$ (and deciding accordingly).[9] Indeed, $(R_{\text{YES}}, R_{\text{NO}})$ is a companion of $(Q, \overline{Q})$, where $\overline{Q} = (\{0,1\}^* \times \{0,1\}^*) \setminus Q$. Thus, we obtain.

Observation 3.4 (FPTAS as BPP-search problems): *Let* $q : \{0,1\}^* \to \mathbb{R}^+$ *and suppose that there exists a probabilistic FPTAS for approximating* q; *that is, suppose that there exists a probabilistic polynomial-time algorithm* A *such that* $\Pr[|A(x, 1^m) - q(x)| \leq q(x)/m] \geq 2/3$. *Then, this approximation task is trivially*

---

[8] Actually, if $(R_{\text{YES}}, R_{\text{NO}})$ and $(R'_{\text{YES}}, R'_{\text{NO}})$ are companions of one another, then they are identical (since $R'_{\text{NO}} = R_{\text{NO}}$ and $R'_{\text{YES}} = R_{\text{YES}}$ must hold). Furthermore, in this case the promise is trivial, since $R'_{\text{NO}} = R_{\text{NO}} \supseteq (\{0,1\}^* \times \{0,1\}^*) \setminus R'_{\text{YES}}$ whereas $R_{\text{NO}} \subseteq (\{0,1\}^* \times \{0,1\}^*) \setminus R'_{\text{YES}}$. Also note that each problem is its own companion, and that problems with trivial promise have no other companion (i.e., if $(R'_{\text{YES}}, R'_{\text{NO}})$ is a companion of $(R, \overline{R})$, where $\overline{R} = (\{0,1\}^* \times \{0,1\}^*) \setminus R$, then $(R'_{\text{YES}}, R'_{\text{NO}}) = (R, \overline{R})$).

[9] That is, we invoke the FPTAS on input x and $\epsilon = 1/3m$, obtain a value $q'(x)$, which with probability at least $2/3$ is in $(1 \pm \epsilon) \cdot q(x)$, and accept if and only if $|v - q'(x)| \leq 2q(x)/3m$. Indeed, if $v \in R_{\text{YES}}(\langle x, 1^m \rangle)$, then with probability at least $2/3$ it holds that $|v - q'(x)| \leq |v - q(x)| + |q(x) - q'(x)| \leq 2q(x)/3m$, whereas if $v \in R_{\text{NO}}(\langle x, 1^m \rangle)$, then with probability at least $2/3$ it holds that $|v - q'(x)| \geq |v - q(x)| - |q(x) - q'(x)| > 2q(x)/3m$.

reducible to some search-BPP problem (i.e., the foregoing one). *Furthermore, the probabilistic time-complexity of the latter search problem is linearly related to the probabilistic time-complexity of the original approximation problem. Moreover, this search-BPP problem is a companion of the original approximation problem.*

3.2   The Reduction

One may expect that any BPP-search problem be *deterministically* reducible to some BPP decision problem. Indeed, this holds for the restricted definition of BPP-search problems as in Definition 3.1, but for the revised formulation of Definition 3.2 we only present a weaker result. Specifically, for every BPP-search problem $(R_{\mathrm{YES}}, R_{\mathrm{NO}})$, there exists $R \supseteq R_{\mathrm{YES}}$ such that $R \cap R_{\mathrm{NO}} = \emptyset$ and solving the search problem of R is deterministically reducible to some BPP decision problem.[10]

Theorem 3.5 (reducing search to decision): *For every BPP-search problem $(R_{\mathrm{YES}}, R_{\mathrm{NO}})$, there exists a binary relation R such that $R_{\mathrm{YES}} \subseteq R \subseteq (\{0,1\}^* \times \{0,1\}^*) \setminus R_{\mathrm{NO}}$ and solving the search problem of R is deterministically reducible to some decisional problem in $\mathcal{BPP}$, denoted Π. Furthermore, the time-complexity of the reduction is linear in the probabilistic time-complexity of finding solutions for $(R_{\mathrm{YES}}, R_{\mathrm{NO}})$, whereas the probabilistic time-complexity of Π is the product of a quadratic polynomial and the probabilistic time-complexity of the decision procedure guaranteed for $(R_{\mathrm{YES}}, R_{\mathrm{NO}})$.*

Applying Theorem 3.5 to a BPP-search problem $(R_{\mathrm{YES}}, R_{\mathrm{NO}})$ that is a companion of some search problem $(\Psi_{\mathrm{YES}}, \Psi_{\mathrm{NO}})$, we obtain a deterministic reduction of solving the search problem $(\Psi_{\mathrm{YES}}, \Psi_{\mathrm{NO}})$ to some promise problem in $\mathcal{BPP}$, because $S_{\Psi_{\mathrm{YES}}} = S_{R_{\mathrm{YES}}} \subseteq S_R$ whereas $R \subseteq (\{0,1\}^* \times \{0,1\}^*) \setminus R_{\mathrm{NO}} \subseteq \Psi_{\mathrm{YES}}$. The argument in depicted in Figure 2.

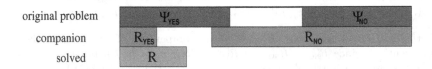

original problem	Ψ_{YES}		Ψ_{NO}
companion	R_{YES}	R_{NO}	
solved	R		

Fig. 2. The reduction applied to a companion of Ψ

Proof: Let A and V be the two probabilistic polynomial-time algorithms associated (by Definition 3.2) with the BPP-search problem $(R_{\mathrm{YES}}, R_{\mathrm{NO}})$, and let t_A and t_V denote their (probabilistic) time-complexities. Specifically, A is the

---

[10] Indeed, in the special case of Definition 3.1 (where $(R_{\mathrm{YES}}, R_{\mathrm{NO}})$ is a partition of the set of all pairs), it holds that $R = R_{\mathrm{YES}}$.

solution-finding algorithm guaranteed by Condition 2, and V is the decision procedure guaranteed by Condition 1. Denote by $A(x, r)$ the output of algorithm A on input x and internal coins $r \in \{0,1\}^{t_A(|x|)}$, and let $V((x, y), \omega)$ denote the decision of V on input (x, y) when using coins $\omega \in \{0,1\}^{t_V(|x|+|y|)}$. Now, define

$$R \stackrel{\text{def}}{=} \left\{ (x, y) : \Pr_{\omega \in \{0,1\}^{t_V(|x|+|y|)}}[V((x, y), \omega) = 1] \geq 0.4 \right\}, \tag{1}$$

and note that $R_{\text{YES}} \subseteq R$ and $R_{\text{NO}} \cap R = \emptyset$.

We now consider an auxiliary algorithm A'' such that $A''(x, r, \omega) \stackrel{\text{def}}{=} V((x, A(x, r)), \omega)$. Note that, for every x and r such that $(x, A(x, r)) \in R_{\text{YES}}$, it holds that $\Pr_\omega[A''(x, r, \omega) = 1] \geq 2/3$, and thus, for every $x \in S_{R_{\text{YES}}}$, it holds that $\Pr_{r,\omega}[A''(x, r, \omega) = 1] \geq 4/9$.

Given x, our strategy is to try to find r such that $A(x, r) \in R(x)$, by determining the bits of r one by one. We thus start with an empty prefix of r, denoted r', and in each iteration we try to extend r' by one bit. Assuming that $x \in S_{R_{\text{YES}}}$ (or rather that Eq. (2) holds for $r' = \lambda$), we try to *maintain the invariant*

$$\Pr_{r'' \in \{0,1\}^{m-|r'|}, \omega \in \{0,1\}^{\ell}}[A''(x, r'r'', \omega) = 1] \geq \frac{4}{9} - \frac{|r'|}{25m}, \tag{2}$$

where $m = t_A(|x|)$ and $\ell = t_V(|x| + m)$. Note that if this invariant holds for $r' \in \{0,1\}^m$, then necessarily $y \stackrel{\text{def}}{=} A(x, r') \in R(x)$ (since in this case Eq. (2) implies that $\Pr_\omega[V((x, y), \omega) = 1] \geq \frac{4}{9} - 0.04 > 0.4$).

Once a candidate solution $y = A(x, r')$ is found (using the corresponding $r' \in \{0,1\}^m$), we check whether y is a good solution, and output y if it is good and $\perp$ otherwise. Specifically, we test whether $(x, y) \in R$ or $(x, y) \in R_{\text{NO}}$ by making a single BPP-query (analogously to the next discussion, since for $(x, y) \in R_{\text{NO}}$ it holds that $\Pr_\omega[V((x, y), \omega) = 1] \leq 1/3$).

In view of the foregoing, we focus on the design of a single iteration. Our strategy is to rely on an oracle for the promise problem $\Pi_{A''}$ that consists of YES-instances $(x, 1^m, r')$ such that $\Pr_{r'',\omega}[A''(x, r'r'', \omega) = 1] \geq \frac{4}{9} - \frac{|r'|-1}{25m}$ and NO-instances $(x, 1^m, r')$ such that $\Pr_{r'',\omega}[A''(x, r'r'', \omega) = 1] < \frac{4}{9} - \frac{|r'|}{25m}$, where in both cases the probability is taken uniformly over $r'' \in \{0,1\}^{m-|r'|}$ (and $\omega \in \{0,1\}^{\ell}$). The oracle $\Pi_{A''}$ is clearly in $\mathcal{BPP}$ (e.g., consider a probabilistic polynomial-time algorithm that on input $(x, 1^m, r')$ estimates $\Pr_{r'',\omega}[A''(x, r'r'', \omega) = 1]$ up to an additive term of $1/50m$ with error probability at most $1/3$, by taking a sample of $O(m^2)$ random pairs (r'', ω)).

In each iteration, which starts with some prefix r' that satisfies Eq. (2), we make a single query to the oracle $\Pi_{A''}$; specifically, we query $\Pi_{A''}$ on $(x, 1^m, r'0)$. If the oracle answers positively, then we extend the current prefix r' with 0 (i.e., we set $r' \leftarrow r'0$), and otherwise we set $r' \leftarrow r'1$.

The point is that if $\Pr_{r'' \in \{0,1\}^{m-|r'|}, \omega}[A''(x, r'r'', \omega) = 1] \geq \frac{4}{9} - \frac{|r'|}{25m}$, then there exists $\sigma \in \{0,1\}$ such that $\Pr_{r''' \in \{0,1\}^{m-|r'|-1}, \omega}[A''(x, r'\sigma r''', \omega) = 1] \geq \frac{4}{9} - \frac{|r'|}{25m} = \frac{4}{9} - \frac{|r'\sigma|-1}{25m}$, which means that $(x, 1^m, r'\sigma)$ is a YES-instance. Thus, if Π answers negatively to the query $(x, 1^m, r'0)$, then $(x, 1^m, r'0)$ cannot be a

YES-instance, which implies that $(x, 1^m, r'1)$ is a YES-instance, and the invariance of Eq. (2) holds for the extended prefix $r'1$. On the other hand, if $\Pi = \Pi_{A''}$ answers positively to the query $(x, 1^m, r'0)$, then $(x, 1^m, r'0)$ cannot be a NO-instance, and the invariance of Eq. (2) holds for the extended prefix $r'0$. We conclude that each iteration of our reduction preserves the said invariance.

To verify the furthermore-part, we note that the reduction consists of $t_A(|x|)$ iterations, where in each iteration a query is made to Π and some very simple steps are taken. In particular, each query made is simply related to the previous one (i.e., can be obtained from it in constant time), and so the entire reduction has time complexity $O(t_A)$. The time complexity of Π on inputs of the form $y = (x, 1^m, r')$ is $O(m^2) \cdot O(t_V(|x| + m)) = O(|y|^2 \cdot t_V(|y|))$. The theorem follows.   ∎

Digest. The proof of Theorem 3.5 follows the strategy of reducing NP-search problems to $\mathcal{NP}$, except that more care is required in the process. This is reflected in the invariance stated in Eq. (2) as well as in the fact that we make an essential use of promise problems (in the oracle).

3.3   Applications

As stated in the introduction, Theorem 3.5 plays a central role in establishing our main result (i.e., the reversing of the "pseudorandomness to derandomization" implication). In this section, we explore a few additional applications of Theorem 3.5. In particular, we show that $\mathcal{BPP} = \mathcal{P}$ implies a host of derandomization results that refer to computational problems that are not of the decisional type. Indeed, we shall reduce these problems to BPP-search problems and apply Theorem 3.5.

Approximations. In light of the foregoing discussion (i.e., Observation 3.4), every approximation problem that has a probabilistic FPTAS can be deterministically reduced to $\mathcal{BPP}$. Thus:

Corollary 3.6 (implication for FPTAS): *If $\mathcal{BPP} = \mathcal{P}$, then every function that has a probabilistic fully polynomial-time approximation scheme* (FPTAS) *also has such a deterministic scheme. Furthermore, for every polynomial p, there exists a polynomial p' such that if the probabilistic scheme runs in time p, then the deterministic one runs in time p'.*

The furthermore part is proved by using the furthermore parts of Observation 3.3 and Theorem 3.5 as well as a completeness feature of $\text{BPTIME}(\cdot)$. Specifically, by combining the aforementioned reductions, we infer that the approximation problem (which refers to instances of the form $\langle x, 1^m \rangle$) is (deterministically) p_1-time reducible to a problem in $\text{BPTIME}(p_2)$, where $p_1(n) = O(p(n))$ and $p_2(n) = O(n^2 \cdot p(n))$. Next, we use the fact that $\text{BPTIME}(p_2)$ has a complete problem, where completeness holds under quadratic-time reductions (which prepend the input by the original problem's description and pad it with a quadratic number of zeros).[11] The point is that this complete problem only depends on p_2, which in turn is uniquely determined by p. The hypothesis (i.e., $\mathcal{BPP} = \mathcal{P}$) implies that this BPTIME(p_2)-complete problem is in DTIME(p_3) for some polynomial p_3, which is solely determined by p_2, and the claim follows for $p' = p_3 \circ p_1^2$. Indeed, we have also established *en passant* the following result, which is of independent interest.

Proposition 3.7 *If $\mathcal{BPP} = \mathcal{P}$, then, for every polynomial p, there exists a polynomial p' such that* BPTIME(p) $\subseteq$ DTIME(p').

Indeed, by the DTIME Hierarchy Theorem, it follows that, if $\mathcal{BPP} = \mathcal{P}$, then, *for every polynomial p, there exists a polynomial p'' such that* DTIME(p'') *contains problems that are not in* BPTIME(p).

Constructions of Varying Quality. While the foregoing discussion of approximation schemes is related to our previous proofs of the main result (see the Appendix), the following discussion is more related to the current proof (as presented in Section 4.2). We consider general construction problems, which are defined in terms of a quality function $q: \{0,1\}^* \to [0,1]$, when for a given n we need to construct an object $y \in \{0,1\}^n$ such that $q(y) = 1$. Specifically, we consider such construction problems that can be solved in probabilistic polynomial-time and have a FPTAS for evaluating the quality of candidate constructions. One interesting special case corresponds to rigid construction problems in which the function q is Boolean (i.e., candidate constructions have either value 0 or 1). In this special case (e.g., generating an n-bit long prime) the requirement that q has a FPTAS is replaced by requiring that the set $q^{-1}(1)$ is in $\mathcal{BPP}$.

Proposition 3.8 (derandomizing some constructions): *Consider a generalized construction defined via a quality function q that has a FPTAS, and let $R_q \overset{\text{def}}{=} \{((1^n, 1^m), y) : y \in \{0,1\}^n \land q(y) > 1 - (1/m)\}$. Suppose that there exists a probabilistic polynomial-time algorithm that solves the search problem of R_q. Then, if $\mathcal{BPP} = \mathcal{P}$, then there exists a deterministic polynomial-time algorithm that solves the search problem of R_q.*

For example, if $\mathcal{BPP} = \mathcal{P}$, then n-bit long primes can be found in deterministic poly(n)-time. On the other hand, the treatment can be generalized to constructions that need to satisfy some auxiliary specification, captured by an auxiliary input x (e.g., on input a prime $x = P$ find a quadratic non-residue mod P). In this formulation, $R_q \overset{\text{def}}{=} \{((x, 1^m), y) : q(x,y) > 1 - (1/m)\}$, where $q: \{0,1\}^* \times \{0,1\}^* \to [0,1]$ can also impose length restrictions on the desired construct.

Proof: Consider   the   BPP-search   problem $(\Pi_{\text{YES}}, \Pi_{\text{NO}})$,   where $\Pi_{\text{YES}} = \{((1^n, 1^m), y) : y \in \{0,1\}^n \land q(y) > 1 - (1/2m)\}$ and $\Pi_{\text{NO}} = \{((1^n, 1^m), y) :$

---

[11] The quadratic padding of x allows $p_2(|x|)$ steps of $M(x)$ to be emulated in time $\widetilde{O}(|M| \cdot p_2(|x|))$, which is upper-bounded by $p_2((|M| + |x|)^2)$, assuming that p_2 is (say) at least quadratic.

$y \in \{0,1\}^n \wedge q(y) \le 1-(1/m)\}$. Note that $(\Pi_{\mathrm{YES}}, \Pi_{\mathrm{NO}})$ is a companion of the search problem R_q, and apply Theorem 3.5. ∎

Corollary 3.9 (a few examples): *If* $\mathcal{BPP} = \mathcal{P}$, *then there exist deterministic polynomial-time algorithms for solving the following construction problems.*

1. *For any fixed* $c > 7/12$, *on input* N, *find a prime in the interval* $[N, N+N^c]$.
2. *On input a prime* P *and* 1^d, *find an irreducible polynomial of degree* d *over* GF(P).
   Recall that finding a quadratic non-residue modulo P *is a special case.*[12]
3. *For any fixed* $\epsilon > 0$ *and integer* $d > 2$, *on input* 1^n, *find a* d-*regular* n-*vertex graph with second eigenvalue having absolute value at most* $2\sqrt{d-1} + \epsilon$.

The foregoing items are based on the density of the corresponding objects in a natural (easily sampleable) set. Specifically, for Item 1 we rely on the density of prime numbers in this interval [14], for Item 2 we rely on the density of irreducible polynomials [7], and for Item 3 we rely on the density of "almost Ramanujan" graphs [6].[13] In all cases there exist deterministic polynomial-time algorithms for recognizing the desired objects.

4   Canonical Derandomizers

In Section 4.1 we present and motivate the rather standard notion of a canonical derandomizer, which is the notion to which most of this work refers to. Our main result, the reversing of the pseudorandomness-to-derandomization transformation is presented in Section 4.2. One tightening, which allows to derive an equivalence, is presented in Section 4.3, which again refers to a rather standard notion (i.e., of "effectively placing $\mathcal{BPP}$ in $\mathcal{P}$"). An alternative equivalence is derived in Section 4.4, which refers to a (seemingly new) notion of a *targeted canonical derandomizer*.

4.1   The Definition

We start by reviewing the most standard definition of canonical derandomizers (cf., e.g., [9, Sec. 8.3.1]). Recall that in order to "derandomize" a probabilistic polynomial-time algorithm A, we first obtain a functionally equivalent algorithm A_G that uses a pseudorandom generator G in order to reduce the randomness-complexity of A, and then take the majority vote on all possible executions of A_G (on the given input). That is, we scan all possible outcomes of the coin tosses of $A_G(x)$, which means that the deterministic algorithm will run in time that is exponential in the randomness complexity of A_G. Thus, it suffices to have a

---

[12] If the polynomial $X^2 + bX + c$ is irreducible, then so is $(X + (b/2))^2 + (c - (b/2)^2)$, and it follows that $-(c - (b/2)^2)$ is a quadratic non-residue.

[13] Recall that Ramanujan graphs are known to be constructable only for specific values of d and of n.

pseudorandom generator that can be evaluated in time that is exponential in its seed length (and polynomial in its output length).

In the standard setting, algorithm A_G has to maintain A's input-output behavior on all (but finitely many) inputs, and so the pseudorandomness property of G should hold with respect to distinguishers that receive non-uniform advice (which models a potentially exceptional input x on which $A(x)$ and $A_G(x)$ are sufficiently different). Without loss of generality, we may assume that A's running-time is linearly related to its randomness complexity, and so the relevant distinguishers may be confined to linear time. Similarly, for simplicity (and by possibly padding the input x), we may assume that both complexities are linear in the input length, $|x|$. (Actually, for simplicity we shall assume that both complexities just equal $|x|$, although some constant slackness seems essential.) Finally, since we are going to scan all possible random-pads of A_G and rule by majority (and since A's error probability is at most $1/3$), it suffices to require that for every x it holds that $|\Pr[A(x) = 1] - \Pr[A_G(x) = 1]| < 1/6$. This leads to the pseudorandomness requirement stated in the following definition.

Definition 4.1 (canonical derandomizers, standard version [9, Def, 8.14])[14]: *Let $\ell : \mathbb{N} \to \mathbb{N}$ be a function such that $\ell(n) > n$ for all n. A* canonical derandomizer *of stretch ℓ is a deterministic algorithm G that satisfies the following two conditions.*

(generation time): *On input a k-bit long seed, G makes at most $\text{poly}(2^k \cdot \ell(k))$ steps and outputs a string of length $\ell(k)$.*
(pseudorandomness): *For every (deterministic) linear-time algorithm D, all sufficiently large k and all $x \in \{0,1\}^{\ell(k)}$, it holds that*

$$| \Pr[D(x, G(U_k)) = 1] - \Pr[D(x, U_{\ell(k)}) = 1]| \; < \; \frac{1}{6}. \tag{3}$$

The algorithm D represents a potential distinguisher, which is given two $\ell(k)$-bit long strings as input, where the first string (i.e., x) represents a (non-uniform) auxiliary input and the second string is sampled either from $G(U_k)$ or from $U_{\ell(k)}$. When seeking to derandomize a linear-time algorithm A, the first string (i.e., x) represents a potential main input for A, whereas the second string represents a possible sequence of coin tosses of A (when invoked on a generic (primary) input x of length $\ell(k)$).

Towards a uniform-complexity variant. Seeking a uniform-complexity analogue of Definition 4.1, the first thing that comes to mind is the following definition.

Definition 4.2 (canonical derandomizers, a uniform version): *As Definition 4.1, except that the original pseudorandomness condition is replaced by*

(pseudorandomness, revised): *For every (deterministic) linear-time algorithm D, it is infeasible, given $1^{\ell(k)}$, to find a string $x \in \{0,1\}^{\ell(k)}$ such that Eq. (3)*

---

[14] To streamline our exposition, we preferred to avoid the standard additional step of replacing $D(x, \cdot)$ by an arbitrary (non-uniform) Boolean circuit of quadratic size.

does not hold. That is, for every probabilistic polynomial-time algorithm F such that $|F(1^{\ell(k)})| = \ell(k)$, there exists a negligible function negl *such that if $x \leftarrow F(1^{\ell(k)})$, then Eq. (3) holds with probability at least $1 - \mathtt{negl}(\ell(k))$.*

When seeking to derandomize a probabilistic (linear-time) algorithm A, the auxiliary algorithm F represents an attempt to find a string $x \in \{0,1\}^{\ell(k)}$ on which $A(x)$ behaves differently depending on whether it is fed with random bits (i.e., $U_{\ell(k)}$) or with pseudorandom ones produced by $G(U_k)$.

Note that if there exists a canonical derandomizer of exponential stretch (i.e., $\ell(k) = \exp(\Omega(k))$), then $\mathcal{BPP}$ is "effectively" in $\mathcal{P}$ in the sense that for every problem in $\mathcal{BPP}$ there exists a deterministic polynomial-time algorithm A such that it is infeasible to find inputs on which A errs. We hoped to prove that $\mathcal{BPP} = \mathcal{P}$ implies the existence of such derandomizers, but do not quite prove this. Instead, we prove a closely related assertion that refers to the following revised notion of a canonical derandomizer, which is implicit in [17]. In this definition, the finder F is incorporated in the distinguisher D, which in turn is an arbitrary probabilistic algorithm that is allowed some fixed polynomial-time (rather than being deterministic and linear-time).[15] (In light of the central role of this definition in the current work, we spell it out rather than use a modification on Definition 4.1 (as done in Definition 4.2).)

Definition 4.3 (canonical derandomizers, a revised uniform version): *Let ℓ, t : $\mathbb{N} \rightarrow \mathbb{N}$ be functions such that $\ell(n) > n$ for all n. A t-robust canonical derandomizer of stretch ℓ is a deterministic algorithm G that satisfies the following two conditions.*

(generation time (as in Definition 4.1)): *On input a k-bit long seed, G makes at most $\mathrm{poly}(2^k \cdot \ell(k))$ steps and outputs a string of length $\ell(k)$.*

(pseudorandomness, revised again): *For every probabilistic t-time algorithm D and all sufficiently large k, it holds that*

$$| \Pr[D(G(U_k)) = 1] - \Pr[D(U_{\ell(k)}) = 1] | < \frac{1}{t(\ell(k))} . \tag{4}$$

Note that, on input an $\ell(k)$-bit string, the algorithm D runs for at most $t(\ell(k))$ steps.

---

[15] Thus, Definition 4.2 and Definition 4.3 are incomparable (when the time bound t is a fixed polynomial). On the one hand, Definition 4.3 seems weaker because we effectively fix the polynomial time bound of F (which is incorporated in D). On the other hand, Definition 4.3 seems stronger because D itself is allowed to be probabilistic and run in time t (whereas in Definition 4.2 these privileges are only allowed to F, which may be viewed as a preprocessing step). Indeed, if $\mathcal{E}$ requires exponential size circuits, then there exist pseudorandom generators that satisfy one definition but not the other: On the one hand, this assumption yields the existence of a non-uniformly strong canonical pseudorandom generator (i.e., satisfying Definition 4.1) of exponential stretch [16] that is not p-robust (i.e., fails Definition 4.3), for some sufficiently large polynomial p. On the other hand, the assumption implies $\mathcal{BPP} = \mathcal{P}$, which leads to the opposite separation described at the end of Section 4.2.

The pseudorandomness condition implies that, for every linear-time D' and every probabilistic t-time algorithm F (such that $|F(1^n)| = n$ for every n), it holds that

$$| \Pr[D'(F(1^{\ell(k)}), G(U_k)) = 1] - \Pr[D'(F(1^{\ell(k)}), U_{\ell(k)}) = 1] | \; < \; \frac{1}{t(\ell(k))}. \quad (5)$$

Note that if, for every x, there exists a σ such that $\Pr[D'(x, U_{|x|}) = \sigma] \geq 1 - (1/3t(|x|))$ (as is the case when D' arises from an "amplified" BPP decision procedure), then the probability that $F(1^{\ell(k)})$ finds an instance $x \in \{0,1\}^{\ell(k)}$ on which $D'(x, G(U_k))$ leans in the opposite direction (i.e., $\Pr[D'(x, U_{|x|}) \neq \sigma] \geq 1/2$) is smaller than $3/t(\ell(k))$. A more general (albeit quantatively weaker) statement is proved next.

Proposition 4.4 (on the effect of canonical derandomizers): *For $t : \mathbb{N} \to \mathbb{N}$ such that $t(n) > (n \log n)^3$, let G be a t-robust canonical derandomizer of stretch ℓ. Let A be a probabilistic linear-time algorithm, and let A_G be as in the foregoing discussions (i.e., $A_G(x, s) = A(x, G(s))$). Then, for every probabilistic $(t/2)$-time algorithm F and all sufficiently large k, the probability that $F(1^{\ell(k)})$ hits the set $\nabla_{A,G}(k) \setminus B_A(k)$ is at most $40/t(\ell(k))^{1/3}$, where*

$$\nabla_{A,G}(k) \stackrel{\text{def}}{=} \left\{ x \in \{0,1\}^{\ell(k)} : \; | \Pr[A_G(x, U_k) = 1] - \Pr[A(x, U_{\ell(k)}) = 1] | \; > \; \frac{1}{3} \right\} (6)$$

$$B_A(k) \stackrel{\text{def}}{=} \left\{ x \in \{0,1\}^{\ell(k)} : \; \frac{1}{t(\ell(k))^{1/3}} < \Pr[A(x, U_{\ell(k)}) = 1] < 1 - \frac{1}{t(\ell(k))^{1/3}} \right\} (7)$$

That is, $B_A(\cdot)$ denotes the set of inputs x on which $A(x) = A(x, U_{|x|})$ is not "almost determined" and $\nabla_{A,G}(\cdot)$ denotes the set of inputs x on which there is a significant discrepancy between the distributions $A(x)$ and $A_G(x)$.

The forgoing discussion refers to the special case in which $B_A(k) = \emptyset$. In general, if A is a decision procedure of negligible error probability (for some promise problem),[16] then A_G is essentially as good as A, since it is hard to find an instance x that matters (i.e., one on which A's error probability is negligible) on which A_G errs (with probability greater than, say, 0.4). This leads to "effectively good" derandomization of $\mathcal{BPP}$. In particular, if G has exponential stretch, then $\mathcal{BPP}$ is "effectively" in $\mathcal{P}$ (see Theorem 4.9).

Proof: Suppose towards the contradiction that there exist algorithms A and F that violate the claim. For each $\sigma \in \{0,1\}$, we consider the following probabilistic t-time distinguisher, denoted D_σ. On input r (which is drawn from either $U_{\ell(k)}$ or $G(U_k)$), the distinguisher D_σ behaves as follows.

1. Obtains $x \leftarrow F(1^{|r|})$.

---

[16] That is, $B_A(\cdot)$ contains only instances that violate the promise.

2. Approximates $p(x) \stackrel{\text{def}}{=} \Pr[A(x, U_{|x|}) = \sigma]$, obtaining an estimate, denoted $\tilde{p}(x)$, such that $\Pr[|\tilde{p}(x) - p(x)| \leq t(|x|)^{-1/3}] = 1 - \text{negl}(|x|)$.

3. If $\tilde{p}(x) < 1 - 2t(|x|)^{-1/3}$, then D_σ halts with output 0.

4. Otherwise (i.e., $\tilde{p}(x) \geq 1 - 2t(|x|)^{-1/3}$), D_σ invokes A on (x, r), and outputs 1 if and only if $A(x, r) = \sigma$. (Indeed, the actual input r is only used in this step.)

We stress that D_σ only approximate the value of $p(x) = \Pr[A(x, U_{|x|}) = \sigma]$ (i.e., it does not approximate the value of $\Pr[A(x, G(U_{\ell-1(|x|)})) = \sigma]$, which would have required invoking G). Observe that D_σ runs for at most $t(|r|)$ steps, because the approximation of $p(x)$ amounts to $\tilde{O}(t(|r|)^{2/3})$ invocations of $A(x)$, whereas each invocation costs $O(|r|)$ time (including the generation of truly random coins for A).

Let $q_\sigma(k)$ denote the probability that, on an $\ell(k)$-bit long input, algorithm D_σ moves to the final (input dependent) step, and note that $q_\sigma(k)$ is independent of the specific input $r \in \{0, 1\}^{\ell(k)}$. Assuming that $|\tilde{p}(x) - p(x)| \leq t(|x|)^{-1/3}$ (for the string x selected at the first step), if the algorithm moves to the final step, then $p(x) > 1 - 3t(|x|)^{-1/3}$. (Similarly, if $p(x) > 1 - t(|x|)^{-1/3}$, then the algorithm moves to the final step.) Thus, the probability that $D_\sigma(U_{\ell(k)})$ outputs 1 is at least $(1 - \text{negl}(\ell(k))) \cdot q_\sigma(k) \cdot (1 - 3t(|x|)^{-1/3})$, which is greater than $q_\sigma(k) - 4t(|x|)^{-1/3}$. On the other hand, by the contradiction hypothesis, there exists a σ such that with probability at least $20t(\ell(k))^{-1/3}$, it holds that $F(1^{\ell(k)})$ hits the set $\nabla_{A,G}(k) \cap S_{\sigma,A}(k)$, where

$$S_{\sigma,A}(k) \stackrel{\text{def}}{=} \left\{ x \in \{0, 1\}^{\ell(k)} : \Pr[A(x, U_{\ell(k)}) = \sigma] \geq 1 - \frac{1}{t(\ell(k))^{1/3}} \right\} \quad (8)$$

In this case (i.e., when $x \in \nabla_{A,G}(k) \cap S_{\sigma,A}(k)$) it holds that $p(x) > 1 - t(|x|)^{-1/3}$ (since $x \in S_{\sigma,A}(k)$) and $\Pr[A(x, G(U_{\ell-1(|x|)})) = \sigma] < 2/3$ (since $x \in \nabla_{A,G}(k)$ and $p(x) > 1 - t(|x|)^{-1/3}$). It follows that the probability that $D_\sigma(G(U_k))$ outputs 1 is at most $(q_\sigma(k) - 20t(|x|)^{-1/3}) \cdot 1 + 20t(|x|)^{-1/3} \cdot 2/3 + \text{negl}(\ell(k))$, which is smaller than $q_\sigma(k) - 5t(|x|)^{-1/3}$. Thus, we derive a contradiction to the t-robustness of G, and the claim follows.   ∎

4.2   The Main Result

Our main result is that $\mathcal{BPP} = \mathcal{P}$ implies the existence of canonical derandomizers of exponential stretch (in the sense of Definition 4.3). We conclude that seeking canonical derandomizers of exponential stretch is "complete" with respect to placing $\mathcal{BPP}$ in $\mathcal{P}$. (The same holds w.r.t "effectively" placing $\mathcal{BPP}$ in $\mathcal{P}$, see Theorem 4.9.)

Theorem 4.5 (on the completeness of canonical derandomization): *If $\mathcal{BPP} = \mathcal{P}$, then, for every polynomial p, there exists a p-robust canonical derandomizer of exponential stretch.*

The proof of Theorem 4.5 is inspired by the study of pseudorandomness with respect to deterministic (uniform p-time) observers, which was carried out by Goldreich and Wigderson [10]. Specifically, for every polynomial p, they presented a polynomial-time construction of a sample space that fools any p-time *deterministic next-bit test*. They observed that an analogous construction with respect to general (deterministic p-time) tests (i.e., distinguishers) would yield some nontrivial derandomization results (e.g., any unary set in $\mathcal{BPP}$ would be placed in $\mathcal{P}$). Thus, they concluded that there is a fundamental gap between probabilistic and deterministic polynomial-time observers. Our key observation is that this gap may disappear if $\mathcal{BPP} = \mathcal{P}$. Specifically, the hypothesis $\mathcal{BPP} = \mathcal{P}$ allows us to derandomize a trivial "probabilistic polynomial-time construction" of a canonical derandomizer.

Proof: Our starting point is the fact that, for some exponential function ℓ, with very high probability, a random function $G : \{0,1\}^k \to \{0,1\}^{\ell(k)}$ satisfies the pseudorandomness requirement associated with $2p$-robust canonical derandomizers. Furthermore, given the explicit description of any function $G : \{0,1\}^k \to \{0,1\}^{\ell(k)}$, we can efficiently distinguish between the case that G is $2p$-robust and the case that G is not p-robust.[17] Thus, the construction of a suitable pseudorandom generator is essentially a BPP-search problem. Next, applying Theorem 3.5, we can deterministically reduce this construction problem to $\mathcal{BPP}$. Finally, using the hypothesis $\mathcal{BPP} = \mathcal{P}$, we obtain a deterministic construction. Details follow.

Let us fix an arbitrary polynomial p, and consider a suitable exponential function ℓ (to be determined later). Our aim is to construct a sequence of mappings $G : \{0,1\}^k \to \{0,1\}^{\ell(k)}$, for arbitrary $k \in \mathbb{N}$, that meets the requirements of a p-robust canonical derandomizer. It will be more convenient to construct a sequence of sets $S = \cup_{k \in \mathbb{N}} S_{\ell(k)}$ such that $S_n \subseteq \{0,1\}^n$, and let $G(i)$ be the ith string in $S_{\ell(k)}$, where $i \in [2^k] \equiv \{0,1\}^k$. (Thus, the stretch function $\ell : \mathbb{N} \to \mathbb{N}$ satisfies $\ell(\log_2 |S_n|) = n$, whereas we shall have $|S_n| = \text{poly}(n)$, which implies $\ell(O(\log n)) = n$ and $\ell(k) = \exp(\Omega(k))$.) The set S_n should be constructed in $\text{poly}(n)$-time (so that G is computable in $\text{poly}(2^k \cdot \ell(k))$-time), and the pseudorandomness requirement of G coincides with requiring that, for every probabilistic p-time algorithm D, and all sufficiently large n, it holds that[18]

$$\left| \Pr[D(U_n)=1] - \frac{1}{|S_n|} \cdot \sum_{s \in S_n} \Pr[D(s)=1] \right| < \frac{1}{p(n)}. \tag{9}$$

---

[17] Formally, the asymptotic terminology of p-robustness is not adequate for discussing finite functions mapping k-bit long strings to $\ell(k)$-bit strings. However, as detailed below, what we mean is distinguishing (in probabilistic polynomial-time) between the case that G is "$2p$-robust" with respect to a given list of p-time machines and the case that G is not "p-robust" with respect to this list.

[18] In [10, Thm. 2] the set S_n was only required to fool *deterministic* tests of the next-bit type.

Specifically, we consider an enumeration of (modified)[19] probabilistic p-time machines, and focus on fooling (for each n) the $p(n)$ first machines, where fooling a machine D means that Eq. (9) is satisfied (w.r.t this D). Note that, with overwhelmingly high probability, a random set S_n of size $K = \widetilde{O}(p(n)^2)$ satisfies Eq. (9) (w.r.t the $p(n)$ first machines). Thus, the following search problem, denoted CON$^{(p)}$, is solvable in probabilistic $\widetilde{O}(p(n)^2 \cdot n)$-time: *On input 1^n, find a K-subset S_n of $\{0,1\}^n$ such that Eq. (9) holds for each of the $p(n)$ first machines.*

Next, consider the following promise problem CC$^{(p)}$ (which is a companion of CON$^{(p)}$). The *valid instance-solution pairs of* CC$^{(p)}$ *are pairs* $(1^n, S_n)$ *such that for each of the first $p(n)$ machines Eq. (9) holds with $p(n)$ replaced by $2p(n)$*, and its *invalid instance-solution pairs are pairs* $(1^n, S_n)$ *such that for at least one of the first $p(n)$ machines Eq. (9) does not hold*. Note that CC$^{(p)}$ is a BPP-search problem (as per Definition 3.2), and that it is indeed a companion of CON$^{(p)}$ (as per Observation 3.3). Thus, by Theorem 3.5,[20] solving the search problem CON$^{(p)}$ is deterministically (polynomial-time) reducible to some promise problem in $\mathcal{BPP}$. Finally, using the hypothesis $\mathcal{BPP} = \mathcal{P}$, the theorem follows.   ■

Observation 4.6 (on the exact complexity of the construction): *Note that* (by Theorem 3.5) *the foregoing reduction of* CON$^{(p)}$ *to* $\mathcal{BPP}$ *runs in time $t(n) = \widetilde{O}(p(n)^2 \cdot n)$, whereas the reduction is to a problem in quartic-time, because the verification problem associated with* CC$^{(p)}$ *is in sub-quadratic probabilistic time.*[21] *Thus, assuming that probabilistic quartic-time is in* DTIME(p_4), *for some polynomial p_4* (see Proposition 3.7), *it follows that* CON$^{(p)}$ $\in$ DTIME(p'), *where $p'(n) = p_4(t(n))$.*

Observation 4.7 (including the seed in the output sequence): *The construction of the generator G (or the set S_n) can be modified such that for every $s \in \{0,1\}^k$ the k-bit long prefix of $G(s)$ equals s (i.e., the ith string in S_n starts with the $(\log_2 |S_n|)$-bit long binary expansion of i).*

---

[19] Recall that one cannot effectively enumerate all machines that run within some given time bound. Yet, one can enumerate all machines, and modify each machine in the enumeration such that the running-time of the modified machine respects the given time bound, while maintaining the functionality of the original machines in the case that the original machine respects the time bound. This is done by simply incorporating a time-out mechanism.

[20] See also the discussion just following the statement of Theorem 3.5, which asserts that if the search problem of a companion of Π is reducible to $\mathcal{BPP}$ then the same holds for Π.

[21] On input $(1^n, S)$ we need to compare the average performance of $p(n)$ machines on S versus their average performance on $\{0,1\}^n$, where each machine makes at most $p(n)$ steps. Recalling that $|S| = K = \widetilde{O}(p(n)^2)$, and that it suffices to get an approximation of the performance on $\{0,1\}^n$ up to an additive term of $1/2p(n)$, we conclude that the entire task can be performed in time $p(n) \cdot \widetilde{O}(p(n)^2) \cdot p(n) < (n + |S|n)^2$ (i.e., the number of machines times the number of experiments (which is $|S| + \widetilde{O}(p(n)^2)$) times the running time of one experiment).

Observation 4.7 implies that a (deterministic) polynomial-time distinguisher, which runs for more time than the foregoing generator, can distinguish the generator's output from a truely random sequence. Next, we show that, in certain cases, the distinguishing task is extremely easy (i.e., can be performed in sublinear time) if the distinguisher is provided with an auxiliary input that can be generated in polynomial-time independently of the tested string.

A Separation between Definition 4.2 and Definition 4.3: The p-robust canonical derandomizer constructed in the foregoing proof (or rather a small variant of it) does *not* satisfy the notion of a canonical derandomizer stated in Definition 4.2. Indeed, in this case, a (deterministic) polynomial-time finder F, which runs for more time than the foregoing generator, can find a string x that allows very fast distinguishing. Details follow.

The variant that we refer to is different from the one used in the proof of Theorem 4.5 only in the details of the underlying randomized construction. Instead of selecting a random set of $\widetilde{O}(p(n)^2)$ strings, we select $m = O(p(n)^3)$ strings in a pairwise independent manner. (This somewhat bigger set suffices to make the probabilistic argument used in the proof of Theorem 4.5 go through.) Furthermore, we consider a specific way of generating such an m-long sequence over $\{0,1\}^n$: For $b = \log_2 m$ and $t = n/b$, we generate an m-long sequence by selecting uniformly $(r_1, s_1), ..., (r_t, s_t) \in \{0,1\}^{2b}$, and letting the ith string in the m-long sequence be the concatenation of the t strings $r_1 + i \cdot s_1, ..., r_t + i \cdot s_t$ (where the arithmetics is of $\mathrm{GF}(2^b)$). (In the actual determintic construction of S_n (and G) a sutibale sequence $((r_1, s_1), ..., (r_t, s_t)) \in \{0,1\}^{2bt}$ is found and fixed, and the $G(i)$ equals the concatenation of the t strings $r_1 + i \cdot s_1, ..., r_t + i \cdot s_t$.) Referring to this specific construction, we propose the following attack:

- The finder F determines the set S_n (just as the generator does). In particular, F *determines the elements* r_1, s_1, r_2, s_2 *used in its construction*, finds $\alpha, \beta \in \mathrm{GF}(2^b)$ such that $\alpha s_1 + \beta s_2 = 0$ and $(\alpha, \beta) \neq (0,0)$, lets $\gamma = \alpha \cdot r_1 + \beta \cdot r_2$, and encodes (α, β, γ) in the $3b$-bit long prefix of x.
- On input x (viewed as starting with the $3b$-bit long prefix $(\alpha, \beta, \gamma) \in \mathrm{GF}(2^b)^3$) and a tested n-bit long string that is viewed as a sequence $(z_1, ..., z_t) \in \mathrm{GF}(2^b)^t$, the distinguisher D *output 1 if and only if* $\alpha \cdot z_1 + \beta \cdot z_2 = \gamma$.

Note that $D(x, G(U_k))$ is identically 1 (because $\alpha \cdot (r_1 + i \cdot s_1) + \beta \cdot (r_2 + i \cdot s_2)$ equals $\gamma = \alpha \cdot r_1 + \beta \cdot r_2$ for every $i \in [m]$), whereas $\Pr[D(x, U_{\ell(k)}) = 1] = 2^{-b}$ (because a fixed non-zero linear combination of two random elements of $\mathrm{GF}(2^b)$ is uniformly distributed in $\mathrm{GF}(2^b)$).

Non-Resilience to Multiple Samples. The foregoing example also demonstrates the non-resilience of Definition 4.3 to multiple samples. Specifically, consider a distinguisher D that obtains three samples, denoted $(z_1^{(1)}, ..., z_t^{(1)}), (z_1^{(2)}, ..., z_t^{(2)})$, and $(z_1^{(3)}, ..., z_t^{(3)})$ (each viewed as a t-long sequence over $\mathrm{GF}(2^b)$), and outputs 1 if and only if $(z_1^{(1)} - z_1^{(2)}) \cdot (z_2^{(2)} - z_2^{(3)}) = (z_2^{(1)} - z_2^{(2)}) \cdot (z_1^{(2)} - z_1^{(3)})$. Then, $D(G(i_1), G(i_2), G(i_3)) = 1$ for every $i_1, i_2, i_3 \in [2^k]$ (because $G(i_1)_j - G(i_2)_j =$

$(i_1 - i_2) \cdot s_j$ and $G(i_2)_j - G(i_3)_j = (i_2 - i_3) \cdot s_j$ for every $j \in [t]$, which implies that each of the two compared products equals $(i_1 - i_2)(i_2 - i_3) \cdot s_1 s_2$, whereas $D(U_{\ell(k)}^{(1)}, U_{\ell(k)}^{(2)}, U_{\ell(k)}^{(3)})$ equals 1 with probability 2^{-b} (because the two compared products are uniformly distributed in $GF(2^b)$ independently of one another).

4.3 A Tedious Tightening

Recall that we (kind of) showed that canonical derandomizers of exponential stretch imply that $\mathcal{BPP}$ is "effectively" contained in $\mathcal{P}$ (in the sense detailed in Definition 4.8), whereas $\mathcal{BPP} = \mathcal{P}$ implies the existence of the former. In this section we tighten this relationship by showing that the existence of canonical derandomizers of exponential stretch also follows from the hypothesis that $\mathcal{BPP}$ is "effectively" (rather than perfectly) contained in $\mathcal{P}$.

Definition 4.8 (effective containment): *Let $\mathcal{C}_1$ and $\mathcal{C}_2$ be two classes of promise problems, and let $t : \mathbb{N} \to \mathbb{N}$. We say that $\mathcal{C}_1$ is t-effectively contained in $\mathcal{C}_2$ if for every $\Pi \in \mathcal{C}_1$ there exists $\Pi' \in \mathcal{C}_2$ such that for every probabilistic t-time algorithm F and all sufficiently large n it holds that $\Pr[F(1^n) \in \nabla(\Pi, \Pi') \cap \{0,1\}^n] < 1/t(n)$, where $\nabla(\Pi, \Pi')$ denotes the symmetric difference between $\Pi = (\Pi_{\mathrm{YES}}, \Pi_{\mathrm{NO}})$ and $\Pi' = (\Pi'_{\mathrm{YES}}, \Pi'_{\mathrm{NO}})$ (i.e., $\nabla(\Pi, \Pi') \stackrel{\mathrm{def}}{=} \nabla(\Pi_{\mathrm{YES}}, \Pi'_{\mathrm{YES}}) \cup \nabla(\Pi_{\mathrm{NO}}, \Pi'_{\mathrm{NO}})$, where $\nabla(S, S') \stackrel{\mathrm{def}}{=} (S \setminus S') \cup (S' \setminus S))$.*

Theorem 4.9 *The following two conditions are equivalent.*

1. *For every polynomial p, it holds that $\mathcal{BPP}$ is p-effectively contained in $\mathcal{P}$.*
2. *For every polynomial p, there exists a p-robust canonical derandomizer of exponential stretch.*

Proof: We first prove that Condition 2 implies Condition 1. (Indeed, this assertion was made several times in the foregoing discussions, and here we merely detail its proof.)

Let $\Pi = (\Pi_{\mathrm{YES}}, \Pi_{\mathrm{NO}})$ be an arbitrary problem in $\mathcal{BPP}$, and consider the corresponding probabilistic linear-time algorithm A (of negligible error probability) derived for a padded version of Π, denoted $\Psi = (\Psi_{\mathrm{YES}}, \Psi_{\mathrm{NO}})$. Specifically, suppose that for some polynomial p_0, it holds that $\Psi_{\mathrm{YES}} = \{x0^{p_0(|x|)-|x|} : x \in \Pi_{\mathrm{YES}}\}$ and ditto for Ψ_{NO}. Now, for any polynomial p, consider the promise problem $\Psi' = (\Psi'_{\mathrm{YES}}, \Psi'_{\mathrm{NO}})$ such that

$$\Psi'_{\mathrm{YES}} \stackrel{\mathrm{def}}{=} \{x \in \Psi_{\mathrm{YES}} : \Pr[A_G(x) = 1] > 0.6\} \tag{10}$$

$$\Psi'_{\mathrm{NO}} \stackrel{\mathrm{def}}{=} \{x \in \Psi_{\mathrm{NO}} : \Pr[A_G(x) = 1] < 0.4\}, \tag{11}$$

where A_G is the algorithm obtained by combining A with a p-robust derandomizer G of exponential stretch ℓ (i.e., $A_G(x, s) = A(x, G(s))$, where $\ell(|s|) = |x|$). Then, Proposition 4.4 implies that for every probabilistic p-time algorithm F and all sufficiently large k, it holds that

$$\Pr[F(1^{\ell(k)}) \in \nabla(\Psi, \Psi') \cap \{0,1\}^{\ell(k)}] < \frac{40}{p(\ell(k))^{1/3}}, \tag{12}$$

because $\nabla(\Psi, \Psi') \cap \{0,1\}^{\ell(k)}$ is contained in $\nabla_{A,G}(k) \setminus B_A(k)$, where $\nabla_{A,G}(k)$ and $B_A(k)$ are as in Eq. (6) and Eq. (7), respectively. Now, since G has exponential stretch, it follows that the randomness complexity of A_G is logarithmic (in its input length). Thus, algorithm A_G runs in polynomial-time, and we can also fully derandomize it in polynomial-time (by invoking A_G on all possible random pads). Concluding that $\Psi' \in \mathcal{P}$, we further infer that the same holds with respect to the "unpadded version" of Ψ', denoted $\Pi' = (\Pi'_{\mathrm{YES}}, \Pi'_{\mathrm{NO}})$; that is, we refer to $\Pi'_{\mathrm{YES}} = \{x : x0^{p_0(|x|)-|x|} \in \Psi'_{\mathrm{YES}}\}$ and ditto for Π'_{NO}. Finally, since $\nabla(\Pi, \Pi') \cap \{0,1\}^n$ equals $\{x : x0^{p_0(|x|)-|x|} \in \nabla(\Psi, \Psi') \cap \{0,1\}^{p_0(n)}\}$, it follows that for every probabilistic $p \circ p_0$-time algorithm F and all sufficiently large n, it holds that $\Pr[F(1^n) \in \nabla(\Pi, \Pi') \cap \{0,1\}^n] < 40/p(p_0(n))^{1/3}$. Noting that the same applies to any $\Pi \in BPP$ (and any polynomial p), we conclude that $\mathcal{BPP}$ is $(p^{1/3}/40)$-effectively contained in $\mathcal{P}$, for every polynomial p. This completes the proof that Condition 2 implies Condition 1.

We now turn to proving the converse (i.e., that Condition 1 implies Condition 2). The idea is to go through the proof of Theorem 4.5, while noting that a failure of the resulting generator (which is supposed to be p-robust) yields contradiction to the p'-effective containment of $\mathcal{BPP}$ in $\mathcal{P}$, where p' is a polynomial that arises from the said proof. Specifically, note that the hypothesis $\mathcal{BPP} = \mathcal{P}$ is used in the proof of Theorem 4.5 to transform a probabilistic construction into a deterministic one. This transformation is actually a (deterministic) p^3-time[22] reduction (of the construction problem) to a fixed problem Π in $\mathrm{BPTIME}(p_\Pi) \subseteq \mathcal{BPP}$, where $p_\Pi(m) = m^4$. We also note that all queries made by the reduction have length $\Theta(2^k \cdot \ell(k))$ (see the proof of Theorem 3.5, and recall that $2^k = \widetilde{O}(p(\ell(k))^2))$. Thus, the reduction fails only if at least one of the queries made by it is answered incorrectly by the problem in $\mathcal{P}$ that is used to p'-effective place Π in $\mathcal{P}$. Let us suppose for a moment that the reduction never makes a query that violates the promise of Π. Then, randomly guessing the index of the (first wrongly answered) query ($i \in [p(\ell(k))^3]$), we may answer the prior $(i-1)$ queries by using the fixed BPP algorithm for Π, and hit an m-bit long instance in the symmetric difference with probability at least $1/p(n)^3$, where $n = \ell(k)$ and $m = \widetilde{\Theta}(p(n)^2 \cdot n)$. Thus, for a sufficiently large polynomial p', this contradicts the hypothesis that $\mathcal{BPP}$ is p'-effectively contained in $\mathcal{P}$. Specifically, on input 1^n, our probabilistic algorithm runs for time $p(n)^3 \cdot p_\Pi(p(n)^3) = p(n)^{15}$ and hits a bad m-bit long string (on which the derandomization fails) with probability at least $1/p(n)^3$, where $m = \widetilde{\Theta}(p(n)^2 \cdot n)$. Thus, setting $p'(m) = m^8$ suffices. (Formally, the claim follows by considering a modified algorithm that on input 1^m invokes the foregoing algorithm on input $1^{m^{1/8}}$.)

Recall, however, that the foregoing analysis relies on the unjustified assumption that the reduction never makes a query that violates the promise of Π. In general, when such a query is made, the answer of the deterministic algorithm

---

[22] See Observation 4.6, and use $\widetilde{O}(p(n)^2 \cdot n) \ll p(n)^3$, which holds for all practical purposes.

A (which p'-effectively places Π in $\mathcal{P}$) may be arbitrary and may not reflect the arbitrary distribution of the answer of the BPP algorithm, deboted B. By randomizing the reduction we may avoid this violation event (or rather bound the probability that it occurs), without effecting the behavior on queries that satisfy the promise. Before detailing how this is done, we stress that this modification will be performed only in the analysis, towards showing that failure of the original deterministic reduction when using algorithm A implies hitting a query on which A returns an incorrect answer. Turning back to the reduction to $\Pi = (\Pi_{\text{YES}}, \Pi_{\text{NO}})$ (which makes a number of queries that is smaller than the query length), we consider the problem $\Pi' = (\Pi'_{\text{YES}}, \Pi'_{\text{NO}})$ such that $(x, \alpha) \in \Pi'_{\text{YES}}$ (resp., $(x, \alpha) \in \Pi'_{\text{NO}}$) if and only if $\Pr[B(x) = 1] > \alpha + 1/20|x|$ (resp., $\Pr[B(x) = 1] < \alpha - 1/20|x|$). Clearly, $\Pi' \in \mathcal{BPP}$ (specifically, $\Pi' \in \text{BPTIME}(p_{\Pi'})$, where $p_{\Pi'}(m) = m^6$). In the reduction, we replace each query x by the query (x, α) such that α is selected at random uniformly in $[0.4, 0.6]$. Thus, for every x that satisfies the promise of Π and every $\alpha \in [0.4, 0.6]$, it holds that (x, α) satisfies the promise of Π'. On the other hand, with probability at least $1 - |x| \cdot (2 \cdot (20|x|)^{-1}/0.2) = 1/2$, all queries made by the reduction (to Π') satisfy the promise, since a query (x, α) violates the promise if and only if $|\Pr[B(x) = 1] - \alpha| \le 1/30|x|$. Now, let A' be a deterministic algorithm that p'-effectively places Π' in $\mathcal{P}$, and let $A(x) = A'(x, 0.5)$. (Indeed, we are using an algorithm derived from the algorithm for Π', rather than using the algorithm derived directly for Π.) Now, if A fails during the original deterministic reduction, then, with probability at least $2/3$, algorithm A' fails during the randomized reduction (i.e., answers some query incorrectly while all queries satisfy the promise). Hence, we derive a contradiction to the hypothesis that Π' is p'-effectively in $\mathcal{P}$ (via algorithm A').[23]   ∎

Comment. The second part of the foregoing proof actually establishes that *there exists a fixed polynomial p' such that if $\mathcal{BPP}$ is p'-effectively contained in $\mathcal{P}$, then, for every every polynomial p, there exists a p-robust canonical derandomizer of exponential stretch.* Thus, we obtain that $\mathcal{BPP}$ is p'-effectively contained in $\mathcal{P}$ if and only if for every polynomial p $\mathcal{BPP}$ is p-effectively contained in $\mathcal{P}$. We comment that this result can be proved directly by a padding argument.

4.4   A Different Tightening (Targeted Generators)

The use of uniform-complexity notions of canonical derandomizers does not seem to allow deriving perfect derandomization (of the type $\mathcal{BPP} = \mathcal{P}$). As we saw, the problem is that exceptional inputs (in the symmetric difference between the original problem and the one solved deterministically) need to be found in order to yield a violation of the pseudorandomness condition. An alternative approach may let the generator depend on the input for which we wish to derandomize the execution of the original probabilistic polynomial-time algorithm. This suggests the following notion of a targeted canonical derandomizer, where both the

---

[23] Note that here we use $p'(m) = m^{11}$, since the running-time of our probabilistic process is $p(n)^3 \cdot p_{\Pi'}(p(n)^3) = p(n)^{21}$, where $m = \widetilde{\Theta}(p(n)^2 \cdot n)$.

generator and the distinguisher are presented with the same auxiliary input (or "target").

Definition 4.10 (targeted canonical derandomizers): *Let $\ell : \mathbb{N} \to \mathbb{N}$ be a function such that $\ell(n) > n$ for all n. A targeted canonical derandomizer of stretch ℓ is a deterministic algorithm G that satisfies the following two conditions.*

(generation time): *On input a k-bit long seed and an $\ell(k)$-bit long auxiliary input, G makes at most $\mathrm{poly}(2^k \cdot \ell(k))$ steps and outputs a string of length $\ell(k)$.*

(pseudorandomness (targeted)): *For every (deterministic) linear-time algorithm D, all sufficiently large k and all $x \in \{0,1\}^{\ell(k)}$, it holds that*

$$| \Pr[D(x, G(U_k, x)) = 1] - \Pr[D(x, U_{\ell(k)}) = 1] | \; < \; \frac{1}{6}. \tag{13}$$

Definition 4.10 is a special case of related definitions that have appeared in [26, Sec. 2.4]. Specifically, Vadhan [26] studied auxiliary-input pseudorandom generators (of the general-purpose type [2,27]), while offering a general treatment in which pseudorandomness needs to hold for an arbitrary set of targets (i.e., $x \in I$ for some set $I \subseteq \{0,1\}^*$).[24] (On the other hand, Definition 4.1 is obatined from Definition 4.10 by mandating that G ignores s; i.e., $G(s, x) = G'(s)$.)

The notion of a targeted canonical derandomizer is not as odd as it looks at first glance. Indeed, the generator is far from being general-purpose (i.e., it is tailored to a specific x), but this merely takes to (almost) the limit the insight of Nisan and Wigderson regarding relaxations that are still useful towards derandomization [20]. Indeed, even if we were to fix the distinguisher D, constructing a generator that just fools $D(x, \cdot)$ is not straightforward, because we need to find a suitable "fooling set" deterministically (in polynomial-time).

Theorem 4.11 (another equivalence): *Targeted canonical derandomizers of exponential stretch exist if and only if $\mathcal{BPP} = \mathcal{P}$.*

Proof: Using any targeted canonical derandomizer of exponential stretch we obtain $\mathcal{BPP} = \mathcal{P}$, where the argument merely follows the one used in the context of non-uniformly strong canonical derandomizers (i.e., canonical derandomizers in the sense of Definition 4.1). Turning to the opposite direction, we observe that the construction undertaken in the proof of Theorem 4.5 can be carried out with respect to the given auxiliary-input. In particular, the fixed auxiliary-input is merely passed among the various algorithms, and the argument remains intact. (See further discussion in Observation 4.12.) The theorem follows. ∎

Observation 4.12 (super-exponential stretch): *In contrast to the situation with respect to the prior notions of canonical derandomizers* (of Definitions 4.1–

---

[24] His treatment vastly extends the original notion of auxiliary-input one-way functions put forward in [21].

4.3),[25] *targeted canonical derandomizer of super-exponential stretch may exist. Indeed, they exists if and only if targeted canonical derandomizer of exponential stretch exist. To see this note that the hypothesis $\mathcal{BPP} = \mathcal{P}$ allows to carry out the proof of Theorem 4.5 for any stretch function. Specifically, for any super-exponential function ℓ, when constructing the set $S_n \subset \{0,1\}^n$ it suffices to fool the first $g(n)$ (linear-time) machines, where g is any unbounded and non-decreasing function and fooling means keeping the distinguishability gap below $1/6$. Thus, $|S_n| = 2^{\ell^{-1}(n)}$ (which is $o(n)$) needs only satisfy $2 \cdot \exp(-2 \cdot (1/6)^2 \cdot |S_n|) \cdot g(n) < 1/3$, which calls for using a function g such that $g(n) \le 0.1 \cdot \exp(2 \cdot (1/6)^2 \cdot 2^{\ell^{-1}(n)})$. The claim follows.*

4.5   Relating the Various Generators

It is syntactically clear that any non-uniformly strong canonical derandomizer (as per Definition 4.1) satisfies both Definition 4.2 (the first uniform version of canonical derandomizers) and Definition 4.10 (the targeted version of canonical derandomizers). On the other hand, there are good reasons to believe that such a canonical derandomizer is not necessarily a p-robust canonical derandomizer (as per Definition 4.3, for some polynomial p).[26] However, using Theorems 4.9 and 4.11, we observe that the existence of a generator that satisfies either Definition 4.2 or Definition 4.10 implies, for every polynomial p, the existence of p-robust canonical derandomizer (as per Definition 4.3).

Corollary 4.13 *If there exists a targeted canonical derandomizer of exponential stretch, then for every polynomial p there exists a p-robust canonical derandomizer of exponential stretch. The same holds if the hypothesis refers to Definition 4.2.*

The various relations are depicted in Figure 3. A similar result can be proved for other (polynomially closed) families of stretch functions, by using the results of Section 5.

Proof: The existence of a targeted canonical derandomizer of exponential stretch implies that $\mathcal{BPP} = \mathcal{P}$ (see Theorem 4.11), which in turn implies the existence of a p-robust canonical derandomizer of exponential stretch (see Theorem 4.5 or Theorem 4.9). Starting with a generator that satisfies Definition 4.2, one can easily prove that, for every polynomial p', it holds that $\mathcal{BPP}$ is p'-effectively in $\mathcal{P}$, where the proof is actually more direct than the corresponding

---

[25] For Definitions 4.1 and 4.2 super-exponential stretch is impossible because we can encode in $x \in \{0,1\}^{\ell(k)}$ the list of all $(k+1)$-bit long strings that do not appear as a prefix of any string in $\{G(s) : s \in \{0,1\}^k\}$, which yields a linear-time distinguisher of gap at least $1/2$. In case of Definition 4.3, super-exponential stretch is impossible because of a distinguisher that output 1 if and only if the tested string starts with 0^{k+1}, and so has a distinguishing gap of at least $2^{-(k+1)}$. Indeed, in both cases we ruled out $\ell(k) \ge 2^{k+1}$.

[26] One such reason was noted in Footnote 15: If $\mathcal{E}$ requires exponential size circuits, then such a "separator" exists.

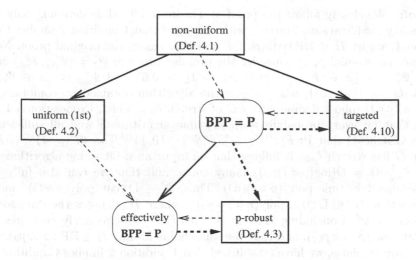

Fig. 3. Relations among various notions of canonical derandomizers (of exponential stretch). Solid arrows indicate syntactic implications (which hold for any generator), whereas dashed arrows indicate existential implications.

direction of Theorem 4.9. We are done by using the other direction of Theorem 4.9 (i.e., the construction of p-robust canonical derandomizer based on p'-effective containment of $\mathcal{BPP}$ in $\mathcal{P}$). ∎

5 Extension: The Full "Stretch vs Time" Trade-Off

In this section we extend the ideas of the previous section to the study to general "stretch vs derandomization time" trade-off (akin to the general "hardness vs randomness" trade-off). That is, here the standard hardness vs randomness trade-off takes the form of a trade-off between the stretch function of the canonical derandomizer and time complexity of the deterministic class containing $\mathcal{BPP}$. The robustness (resp., effectiveness) function will also be adapted accordingly.

Theorem 5.1 (Theorem 4.9, generalized): *For every function $t : \mathbb{N} \to \mathbb{N}$, the following two conditions are equivalent.*

1. *For every two polynomials p_0 and p, it holds that $\mathrm{BPTIME}(p_0)$ is $(p \circ t)$-effectively contained in $\mathrm{DTIME}(\mathrm{poly}(p \circ t \circ p_0))$.*
2. *For every polynomial p, there exists a $(p \circ t)$-robust canonical derandomizer of stretch $\ell_{pot} : \mathbb{N} \to \mathbb{N}$ such that $\ell_{pot}(k) \stackrel{\mathrm{def}}{=} (p \circ t)^{-1}(2^{\Omega(k)}) = t^{-1}(p^{-1}(2^{\Omega(k)}))$.*

Furthermore, the hidden constants in the Ω and poly *notation are independent of the functions t, p and p_0.*

Indeed, Theorem 4.9 follows as a special case (when setting $t(n) = n$), whereas for $t(n) \geq 2^n$ both conditions hold trivially. Note that for $t(n) = 2^{\epsilon n}$ (resp., $t(n) = 2^{n^\epsilon}$), we get $\ell_{pot}(k) = \Omega(k/\epsilon)$ (resp., $\ell_{pot}(k) = \Omega(k)^{1/\epsilon}$).

Proof: We closely follow the proof of Theorem 4.9, while detailing only the necessary modifications. Starting with the proof that Condition 2 implies Condition 1, we let $\Pi \in \mathrm{BPTIME}(p_0)$, Ψ and A be as in the original proof. Now, for any polynomial p, we consider the promise problem $\Psi' = (\Psi'_{\mathrm{YES}}, \Psi'_{\mathrm{NO}})$ such that $\Psi'_{\mathrm{YES}} = \{x \in \Psi_{\mathrm{YES}} : \Pr[A_G(x) = 1] > 0.6\}$ and $\Psi'_{\mathrm{NO}} = \{x \in \Psi_{\mathrm{NO}} : \Pr[A_G(x) = 1] < 0.4\}$, where A_G is the algorithm obtained by combining A with a $(p \circ t)$-robust derandomizer G of stretch ℓ_{pot}. Then, Proposition 4.4 implies that for every probabilistic $(p \circ t)$-time algorithm F and all sufficiently large k, it holds that $\Pr[F(1^{\ell(k)}) \in \nabla(\Psi, \Psi') \cap \{0,1\}^{\ell(k)}] < 40/(p \circ t)^{1/3}(\ell(k))$. Since G has stretch ℓ_{pot}, it follows that on input an n-bit string algorithm A_G uses $\ell_{pot}^{-1}(n) = O(\log(p \circ t)(n))$ many coins, and thus we can also fully derandomize it in time $\mathrm{poly}((p \circ t)(n))$. Thus, $\Psi' \in \mathrm{DTIME}(\mathrm{poly}(p \circ t))$, and it follows that $\Pi' \in \mathrm{DTIME}(\mathrm{poly}(p \circ t \circ p_0))$, where Π' denotes the "unpadded version" of Ψ'. Concluding that Π is $((p \circ t)^{1/3}/40)$-effectively contained in $\mathrm{DTIME}(\mathrm{poly}(p \circ t \circ p_0))$, and that the same holds for any $\Pi \in \mathrm{BPTIME}(p_0)$ and every polynomial p, we have established that Condition 2 implies Condition 1.

We now turn to proving the converse (i.e., that Condition 1 implies Condition 2). Again, we merely go through the proof of Theorem 4.5, except that here we construct a set S_n of size $\mathrm{poly}(p \circ t)(n)$. Specifically, the discrepancies we aim at are linearly related to $1/(p \circ t)(n)$, and we can afford spending time $\mathrm{poly}(p \circ t)(n)$ in the construction. We shall indeed make use of this allowance, since we can only rely on the (t'-effective) containment of $\mathrm{BPTIME}(p_0)$ in $\mathrm{DTIME}(\mathrm{poly}(p \circ t \circ p_0))$, where $t' = \mathrm{poly}(p \circ t) = \mathrm{poly}(p) \circ t$. The rest of the argument proceeds analogously to the proof of Theorem 4.9. We note that the aforementioned hypothesis regarding $\mathrm{BPTIME}(p_0)$ is only used when deterministically reducing (in time $\mathrm{poly}(p \circ t)$) the construction of S_n to a fixed problem Π in $\mathrm{BPTIME}(p_0)$, where $p_0(m) = m^4$ (as in the proof of Theorem 4.9). Thus, the reduction fails only if at least one of the queries made by it is answered incorrectly by the problem in $\mathcal{D} \overset{\text{def}}{=} \mathrm{DTIME}(\mathrm{poly}(p \circ t \circ p_0))$ that is used to t'-effective place Π in $\mathcal{D}$. Randomly guessing the the index of the (wrongly answered) query, we hit an m-bit long instance in the symmetric difference with probability at least $1/\mathrm{poly}(p(t(n)))$, where $m = \Omega(\ell(k))$, which contradicts the hypothesis that $\mathrm{BPTIME}(p_0)$ is t'-effectively contained in $\mathcal{D}$.[27]   ∎

6   Open Problems

We start by recalling the famous open problem regarding whether the a full derandomization of standard decision problems implies the same for promise problems. That is, assuming that any decision problem in $\mathcal{BPP}$ is in $\mathcal{P}$, does it follow that $\mathcal{BPP} = \mathcal{P}$?[28]

---

[27] Here, too, $m = \Theta(\widetilde{O}(p(t(n))^2 \cdot n)$ actually holds, and so it actually suffices to set $t'(m) = \mathrm{poly}(m)$.

[28] Formally, let $\mathcal{D}$ denote the set of all promise problems having a trivial promise; that is, a promise problem $(\Pi_{\mathrm{YES}}, \Pi_{\mathrm{NO}})$ is in $\mathcal{D}$ if $\Pi_{\mathrm{YES}} \cup \Pi_{\mathrm{NO}} = \{0,1\}^*$. Then, the question is whether $\mathcal{BPP} \cap \mathcal{D} = \mathcal{P} \cap \mathcal{D}$ implies $\mathcal{BPP} = \mathcal{P}$.

One problem that arises from the current work refers to the relationship between the two uniform definitions of canonical derandomizers (i.e., Definitions 4.2 and 4.3). Recall (see Section 4.5) that the existence of generators (of exponential stretch) that satisfy Definition 4.2 implies the existence of generators (of exponential stretch) that satisfy Definition 4.3, but the converse is not clear.

Another open problem refers to the deriving of analogous results regarding the derandomization of $\mathcal{AM}$ (or $\mathcal{AM} \cap \text{co}\mathcal{AM}$). Here the canonical derandomizer should be computable in *non-deterministic* poly$(2^k \cdot \ell(k))$-time, where computation by non-deterministic machines refers to the so called "single-value" model (see, e.g., [23] or [9, Def. 5.13]). The problem in reversing the "pseudorandomness to derandomization" connection refers to a tension between the distinguishers used to argue about the derandomization versus our need to handle them in the construction of the canonical derandomizer. We would welcome any result, even for a targeted version and even for derandomizing some subclass such as $\mathcal{AM} \cap \text{co}\mathcal{AM}$ or $\mathcal{SZK}$.

Finally, we return to the question raised in passing in Section 1.4. Specifically, we ask *which search problems can be solved by deterministic polynomial-time reductions to $\mathcal{BPP}$*. Denoting the class of such search problems by $\mathcal{C}$, we note that Theorem 3.5 implies that $\mathcal{C}$ contains all search problems that have a companion that is a BPP-search problem. The converse holds in the special case that the target of the reduction is a standard decision problem (and the reduced search problem has a trivial promise at the instance level (see below)). Let us consider the general case and see what happens. Suppose that the search problem $(R_{\text{YES}}, R_{\text{NO}})$ is reducible in deterministic polynomial-time to $\Pi \in \mathcal{BPP}$. Denoting the oracle machine effecting the reduction by M, we consider the search problem $(R'_{\text{YES}}, R'_{\text{NO}})$ such that $(x, y) \in R'_{\text{YES}}$ if $M^f(x) = y$ for some f that is consistent with Π and $(x, y) \in R'_{\text{NO}}$ otherwise.[29] The correctness of the reduction implies that $S_{R'_{\text{YES}}} \supseteq S_{R_{\text{YES}}}$ whereas $R'_{\text{NO}} \supseteq R_{\text{NO}}$, which means that if $S_{R_{\text{YES}}} \cup S_{R_{\text{NO}}} = \{0,1\}^*$, then $(R'_{\text{YES}}, R'_{\text{NO}})$ is a companion of $(R_{\text{YES}}, R_{\text{NO}})$. Now if Π is a standard decision problem, then f is unique; hence, $R'_{\text{YES}}(x)$ is a singleton if $x \in S_{R'_{\text{YES}}}$ and is empty otherwise (since $S_{R'_{\text{NO}}} = \{0,1\}^* \setminus S_{R'_{\text{YES}}}$). In this case membership of (x, y) in R'_{YES} can be easily tested by checking whether $M^\Pi(x) = y$. The same holds if the reduction is "smart" (i.e., avoids making queries that violate the promise, cf. [12]),[30] but in general it is not clear what happens.

Acknowledgments. We are grateful to Noga Alon, Or Meir, Madhu Sudan, and Avi Wigderson for useful discussions on related issues. We also thank Or Meir, Dieter van Melkebeek, Amnon Ta-Shma, and Salil Vadhan for their comments on early drafts of this work.

---

[29] Saying that f is consistent with $\Pi = (\Pi_{\text{YES}}, \Pi_{\text{NO}})$ means that $f(x) = 1$ for every $x \in \Pi_{\text{YES}}$, whereas $f(x) = 0$ for every $x \in \Pi_{\text{NO}}$. Indeed, the value of f on inputs that violate the promise (i.e., $x \notin \Pi_{\text{YES}} \cup \Pi_{\text{NO}}$) is arbitrary.

[30] We note, however, that the reduction used in the proof of Theorem 3.5 is not smart. Furthermore, we doudt that a smart reduction can be found.

References

1. Aydinlioglu, B., Gutfreund, D., Hitchcock, J.M., Kawachi, A.: Derandomizing Arthur-Merlin Games and Approximate Counting Implies Exponential-Size Lower Bounds. Computational Complexity (to appear)
2. Blum, M., Micali, S.: How to Generate Cryptographically Strong Sequences of Pseudo-Random Bits. In: SICOMP, vol. 13, pp. 850–864 (1984); Preliminary version in 23rd FOCS, pp. 80–91 (1982)
3. Chor, B., Goldreich, O.: On the Power of Two-Point Based Sampling. Jour. of Complexity 5, 96–106 (1989)
4. Even, S., Selman, A.L., Yacobi, Y.: The Complexity of Promise Problems with Applications to Public-Key Cryptography. Inform. and Control 61, 159–173 (1984)
5. Fortnow, L.: Comparing Notions of Full Derandomization. In: 16th CCC, pp. 28–34 (2001)
6. Friedman, J.: A Proof of Alon's Second Eigenvalue Conjecture. In: 35th STOC, pp. 720–724 (2003)
7. Gauss, C.F.: Untersuchungen Über Höhere Arithmetik, 2nd edn. Chelsea publishing company, New York (1981) (reprinted)
8. Goldreich, O.: Foundation of Cryptography: Basic Tools. Cambridge University Press, Cambridge (2001)
9. Goldreich, O.: Computational Complexity: A Conceptual Perspective. Cambridge University Press, Cambridge (2008)
10. Goldreich, O., Wigderson, A.: On Pseudorandomness with respect to Deterministic Observers. In: RANDOM 2000, Proceedings of the Satellite Workshops of the 27th ICALP. Carleton Scientific (Proc. in Inform. 8), pp. 77–84 (2000); See also ECCC, TR00-056
11. Goldwasser, S., Micali, S.: Probabilistic Encryption. JCSS 28(2), 270–299 (1984); Preliminary version in 14th STOC (1982)
12. Grollmann, J., Selman, A.L.: Complexity Measures for Public-Key Cryptosystems. In: SICOMP, vol. 17(2), pp. 309–335 (1988)
13. Hochbaum, D. (ed.): Approximation Algorithms for NP-Hard Problems. PWS (1996)
14. Huxley, M.N.: On the Difference Between Consecutive Primes. Invent. Math. 15, 164–170 (1972)
15. Impagliazzo, R., Kabanets, V., Wigderson, A.: In Search of an Easy Witness: Exponential Time vs Probabilistic Polynomial Time. JCSS 65(4), 672–694 (2002); Preliminary version in 16th CCC (2001)
16. Impagliazzo, R., Wigderson, A.: P=BPP if E requires exponential circuits: Derandomizing the XOR Lemma. In: 29th STOC, pp. 220–229 (1997)
17. Impagliazzo, R., Wigderson, A.: Randomness vs. Time: De-randomization under a uniform assumption. JCSS 63(4), 672–688 (2001); Preliminary version in 39th FOCS (1998)
18. Jerrum, M., Valiant, L., Vazirani, V.V.: Random Generation of Combinatorial Structures from a Uniform Distribution. In: TCS, vol. 43, pp. 169–188 (1986)
19. Kabanets, V., Impagliazzo, R.: Derandomizing Polynomial Identity Tests Means Proving Circuit Lower Bounds. Computational Complexity 13, 1–46 (2003); Preliminary version in 35th STOC (2003)
20. Nisan, N., Wigderson, A.: Hardness vs Randomness. JCSS 49(2), 149–167 (1994); Preliminary version in 29th FOCS (1988)

21. Ostrovsky, R., Wigderson, A.: One-Way Functions are Essential for Non-Trivial Zero-Knowledge. In: 2nd Israel Symp. on Theory of Computing and Systems, pp. 3–17. IEEE Comp. Soc. Press, Los Alamitos (1993)
22. Reingold, O., Trevisan, L., Vadhan, S.: Pseudorandom walks on regular digraphs and the RL vs. L problem. In: 38th STOC, pp. 457–466 (2006); See details in ECCC, TR05-022
23. Shaltiel, R., Umans, C.: Low-end Uniform Hardness vs Randomness Tradeoffs for AM. SICOMP 39(3), 1006–1037 (2009); Preliminary version in 39th STOC (2007)
24. Trevisan, L., Vadhan, S.: Pseudorandomness and Average-Case Complexity Via Uniform Reductions. Computational Complexity 16(4), 331–364 (2007); Preliminary version in 17th CCC (2002)
25. Umans, C.: Pseudo-random Generators for all Hardness. JCSS 67(2), 419–440 (2002); Preliminary version in 34th STOC (2002)
26. Vadhan, S.: An Unconditional Study of Computational Zero Knowledge. SICOMP 36(4), 1160–1214 (2006); Preliminary version in 45th FOCS (2004)
27. Yao, A.C.: Theory and Application of Trapdoor Functions. In: 23rd FOCS, pp. 80–91 (1982)

Appendices: Prior Proofs of the Main Result (Theorem 4.5)

The current proof of Theorem 4.5 is the second simplification we found: It is a third incarnation of the same underlying principles, but it hides the original inspiration to our ideas, which are rooted in [10]. Since we do have written records of these prior proofs, and since they may be of some interest, we decided to include them in the current appendix.

Our starting point was the work of Goldreich and Wigderson [10], which studied pseudorandomness with respect to (uniform) deterministic observers. In particular, they show how to construct, for every polynomial p, a generator of exponential stretch that works in time polynomial in its output and fools all deterministic p-time tests of the next-bit type (a la [2]). They observe that an analogous construction with respect to general (deterministic p-time) tests (or distinguishers) would yield some non-trivial derandomization results (e.g., any unary set in $\mathcal{BPP}$ would be placed in $\mathcal{P}$). Thus, they conclude that Yao's result[31] asserting that *fooling all efficient next-bit tests implies fooling all efficient distinguishers* relies on the fact that the class of test includes probabilistic p-time algorithms and not only deterministic ones.

Our key observation is that the gap between probabilistic next-bit tests and deterministic ones essentially disappears if $\mathcal{BPP} = \mathcal{P}$. Actually, the gap disappears if we generalize the notion of next-bit tests so to allow the (deterministic) tester to output a guess of the probability that the next bit equals 1 (rather than a guess for the actual value of the next bit), and consider the correlation between the corresponding random variables. Indeed, assuming that $\mathcal{BPP} = \mathcal{P}$, allows to

---

[31] Attributed to oral presentations of [27].

deterministically emulate a probabilistic p-time next bit test by a (generalized) deterministic p'-time next bit test, where p' is a polynomial that depends only on p. Plugging this into the construction of [10], which can be shown to fool also (generalized) deterministic p'-time next bit test, we obtain the desired generator (which produces ℓ-bit outputs in time $\mathrm{poly}(p'(\ell))$). A crucial point in the foregoing argument is that the next-bit test does not need to invoke the generator, which is not feasible because the generator runs for more time than the potential tests.

The foregoing argument led to the first proof, which is presented in Appendix A.2. Subsequently we found a more direct approach, which is presented in Appendix A.1. This approach is more transparent and amenable to variations than the first one (but less so in comparison to the proof presented in Section 4.2). Specifically, rather than working with (generalized) next-bit tests, we directly work with (probabilistic p-time) distinguishers, and *adapt the argument of [10] to apply in this context*. It turns out that in order for this to work, we only need to approximate the probability that a fixed probabilistic p-time distinguishers outputs 1 when presented with random $(\ell - i)$-bit long extensions of some fixed i-bit long strings, for $i = 1, ..., \ell$. Assuming that $\mathcal{BPP} = \mathcal{P}$, allows to deterministically approximate these probabilities (again, in p'-time, where $p' = \mathrm{poly}(p)$), and so we are done. Needless to say, the fact that such approximations suffices is specific to (our adaptation of) the construction of [10].

A.1   An Alternative Proof of Theorem 4.5 (via Derandomizing a FPTAS)

The alternative proof of Theorem 4.5 proceeds by generalizing the main idea that underlies the work of Goldreich and Wigderson [10], while using the hypothesis (i.e., $\mathcal{BPP} = \mathcal{P}$) to extend its scope to probabilistic (rather than deterministic) observers. Specifically, for every polynomial p, they presented a polynomial-time construction of a sample space that fools any p-time *deterministic next-bit test*. The construction is iterative, where in each iteration the next bit of each string in the sample space is determined such that the resulting space fools all relevant next-bit tests. Here we consider any (p-time) *probabilistic distinguisher*, and seek to determine the next bit so that the probability that this distinguisher output 1 (on a random extension of the current sample space) is approximately maintained. Towards this end, we need to approximate the probability that a fixed p-time *probabilistic* algorithm outputs 1 on a *random* extension of the current prefix. Our key observation is that, due to the hypothesis that $\mathcal{BPP} = \mathcal{P}$, this quantity can be approximated in deterministic polynomial-time. The use of this hypothesis is far from being surprising, since (as noted before) the conclusion of Theorem 4.5 implies that, in some "effective" sense, $\mathcal{BPP}$ does equal $\mathcal{P}$.

Proof: We follow the general outline of the proof of [10, Thm. 2], while commenting (mostly in footnotes) about the points of deviation. Let us fix an arbitrary polynomial p, and consider a suitable exponential function ℓ (to be determined later). Our aim is to construct a sequence of mappings $G: \{0,1\}^k \to$

$\{0,1\}^{\ell(k)}$, for arbitrary $k \in \mathbb{N}$, that meets the requirements of a p-robust canonical derandomizer. However, it will be more convenient to construct a sequence of sets $S = \cup_{k\in\mathbb{N}}S_{\ell(k)}$ such that $S_n \subseteq \{0,1\}^n$, and let $G(i)$ be the ith string in $S_{\ell(k)}$, where $i \in [2^k] \equiv \{0,1\}^k$. (Thus, the function $\ell:\mathbb{N}\to\mathbb{N}$ satisfies $\ell(\log_2 |S_n|) = n$, whereas we shall have $|S_n| = \text{poly}(n)$.) The set S_n should be constructed in $\text{poly}(n)$-time (so that G is computable in $\text{poly}(2^k \cdot \ell(k))$-time), and the pseudorandomness requirement of G coincides with requiring that, for every probabilistic p-time algorithm D, and all sufficiently large n, it holds that[32]

$$\left| \Pr[D(U_n)=1] - \frac{1}{|S_n|} \cdot \sum_{s\in S_n} \Pr[D(s)=1] \right| < \frac{1}{p(n)} \qquad (14)$$

Specifically, we consider an enumeration of (modified)[33] probabilistic machines running within time $p(n)$ on input of length n, and focus on fooling the $m = m(n) < p(n)$ first machines in the sense of Eq. (14). Let $\epsilon = 1/p(n)$, and M be a generic machines that we wish to fool.

We construct S_n in (roughly) n iterations, such that in iteration i we construct $S_{n,i} \subseteq \{0,1\}^i$. We start with $S_{n,k} = \{0,1\}^k$, where $k = 2\log_2(2nm/\epsilon)$, and let $K = 2^k$. In the $i + 1^{\text{st}}$ iteration, we consider the function $f_M : [K] \to [0,1]$ representing the probability that M outputs 1 on a random extension of each of the K strings in $S_{n,i}$; that is, $f_M(j) = \Pr[M(x^{(j)}U_{n-i})=1]$, where $x^{(j)}$ is the jth string in $S_{n,i} \subseteq \{0,1\}^i$. (The function f_M represents M's average output on all possible $(n - i)$-bit long extensions of all strings in $S_{n,i}$.)[34] Our aim is to find a vector $u \in \{0,1\}^K$ such that, for each machine M (among the first m machines), it holds that the average value of $\Pr[M(x^{(j)}u[j]U_{n-i-1})=1]$ is ϵ/n-close to the average value of $f_M(j)$; that is,

$$\left| \frac{1}{K} \sum_{j\in[K]} \Pr[M(x^{(j)}u[j]U_{n-i-1})=1] - \frac{1}{K} \cdot \sum_{j\in[K]} f_M(j) \right| < \frac{\epsilon}{n}. \qquad (15)$$

Once such a vector u is found, we extend $S_{n,i}$ into $S_{n,i+1}$ in the natural manner; that is,

$$S_{n,i+1} \stackrel{\text{def}}{=} \{x^{(j)}u[j] : \text{where } x^{(j)} \text{ is the } j^{\text{th}} \text{ string in } S_{n,i}\} \subset \{0,1\}^{i+1}. \qquad (16)$$

---

[32] In [10, Thm. 2] the set S_n was only required to fool *deterministic* tests of the next-bit type.

[33] Recall that one cannot effectively enumerate all machines that run within some given time bound. Yet, one can enumerate all machines, and modify each machine in the enumeration such that the running-time of the modified machine respects the given time bound, while maintaining the functionality of the original machines in the case that the original machine respects the time bound. This is done by simply incorporating a time-out mechanism.

[34] In contrast, in [10], the function f_M (which is denoted v_M there) represented M's attempt to guess the $i + 1^{\text{st}}$ bit of a string in S_n, based on the i-bit long prefix of that string. Furthermore, since in [10] the machine M is deterministic, the function f_M (there) can be constructed by invoking M on K different i-bit strings.

It follows that $S_n \stackrel{\text{def}}{=} S_{n,n}$ satisfies Eq. (14), because, for each of the aforementioned M's and for each $i \in [n-k, n-1]$, it holds that

$$\left| \frac{1}{|S_{n,i}|} \cdot \sum_{s \in S_{n,i}} \Pr[M(s)=1] - \frac{1}{|S_{n,i+1}|} \cdot \sum_{s \in S_{n,i+1}} \Pr[M(s)=1] \right| < \frac{\epsilon}{n} \qquad (17)$$

since the terms in the l.h.s are represented by the function f_M defined at the ith iteration, whereas the terms in the r.h.s correspond to the function f_M defined in the next iteration.

It remains to specify how a suitable vector $u \in \{0,1\}^K$ is found, in each iteration. This is done by using a pairwise independent sample space for strings of length K, while recalling that such spaces can be constructed in $\text{poly}(K)$-time (cf., e.g., [3]). Two issues arise:

1. Showing that such a sample space must always contain a suitable vector $u \in \{0,1\}^K$; that is, a vector $u \in \{0,1\}^K$ satisfies Eq. (15). This is quite an immediate consequence of the fact that, when defining $q_{M,j}(\sigma) \stackrel{\text{def}}{=}$ $\Pr[M(x^{(j)}\sigma U_{n-i-1})=1]$, we can write Eq. (15) as

$$\left| \frac{1}{K} \sum_{j \in [K]} q_{M,j}(u[j]) - \frac{1}{K} \cdot \sum_{j \in [K]} \sum_{\sigma \in \{0,1\}} \frac{q_{M,j}(\sigma)}{2} \right| < \frac{\epsilon}{n}. \qquad (18)$$

   Indeed, when u is selected from a pairwise-independent sample space of $\{0,1\}^K$, Eq. (18) holds with probability at least $1 - (1/((\epsilon/n)^2 K))$, and the claim follows whenever $(\epsilon/n)^2 K > m$.
   Actually, we shall use the fact that, with probability at least $1 - (1/((\epsilon/2n)^2 K))$, a modified version of Eq. (18) holds, where in the modification the upper bound ϵ/n is replaced by the (tighter) upper bound $\epsilon/2n$.

2. Showing that we can distinguish in deterministic polynomial-time whether a given vector $u \in \{0,1\}^K$ satisfies the aforementioned tighter form of Eq. (18) or violates Eq. (15).
   This issue hardly arose in [10], since there $f_M(j)$ referred to the output of a deterministic machine on fixed string (i.e., the jth string in $S_{n,i}$). But here $f_M(j)$ refers to the expected output of a probabilistic machine on a random extension of a fixed string. Nevertheless, $f_M(j)$ can be approximated to within an additive term of $\epsilon/4n$ by a simple probabilistic algorithm that merely invokes M sufficiently many (i.e., $O(n/\epsilon)^2$) times on such random extensions, and ditto for $q_{M,j}(\cdot)$. Now, using the hypothesis $\mathcal{BPP} = \mathcal{P}$ and applying Corollary 3.6,[35] we conclude that $q_{M,j}(\cdot)$ can be approximated well-enough (i.e., up to an additive term of $\epsilon/4n$) by a deterministic polynomial-time algorithm. Formally, the approximation problem is defined for inputs of

---

[35] Formally, there is a problem here since we do not have a FPTAS for the quantities $q_{M,j}(\cdot) \in [0,1]$, but we do have a FPTAS for the quantities $1 + q_{M,j}(\cdot) \in [1,2]$ and this suffices here.

the form $(M, 1^n, x)$, where M and n are as above and x is a string of length at most n (i.e., in our application, $x = x^{(j)}\sigma \in \{0,1\}^{i+1}$, where $x^{(j)}$ is the jth string in $S_{n,i}$).

Thus, in each iteration we can find a vector u as desired, an consequently we construct the desired set S_n in time that is polynomial in n. The theorem follows. ∎

Comment: The generator constructed in the foregoing proof does not satisfy the notion of a canonical derandomizer stated in Definition 4.2. Indeed, in this case, a (deterministic) polynomial-time finder F, which runs for more time than the foregoing generator, can find a string x that allows very fast distinguishing. Details follow.

Referring to the foregoing construction of a pseudorandom generator G, we show how to find in polynomial-time a string $x \in \{0,1\}^{\ell(k)}$ such that $(x, G(U_k))$ and $(x, U_{\ell(k)})$ are easy to tell apart. Actually, we refer to a specific implementation of the construction; that is, to a specific implementation of the pairwise independence sample space, where the aspect we rely on is that this sample space is a vector space of low dimension.

Recall that the set S_n, which is the support of $G(U_k)$ (for $n = \ell(k)$), is constructed by concatenating n vectors, where each vector is an 2^k-bit long sequence taken from a pairwise independence sample space. Specifically, consider the sample space generated by strings of length $k+1$ such that the jth coordinate of the vector generated by $r_1 \cdots r_{k+1} \in \{0,1\}^{k+1}$ equals $\sum_{h=1}^{k+1} j_h r_h$, where j_h is the hth bit in the $(k+1)$-bit long binary expansion of $j \in [2^k]$. Thus, each of the vectors used in the construction of S_n reside in the very same $(k+1)$-dimensional vector space.

Then, the finder F can just construct S_n (as done by G), and record the sequence $u_{k+1}, ..., u_n$ of vectors taken in each of the $n - k$ iterations (or rather their succinct representation). In fact, it suffices to record the indices of a subsequence, of length $d \leq k+2$, that sums-up to the all zero vector; that is, record $(i_1, ..., i_d)$ such that the sum of the vectors taken in iterations $i_1, ..., i_d$ equals the all-zero vector (i.e., $\sum_{h=1}^d u_{i_h} = 0$). Now, F just records such a sequence in a prefix of x, and the distinguisher D just outputs the XOR value of the bits that appear in these positions. Clearly, $D(x, z) = 0$ for every $z \in S_n$, whereas $\Pr[D(x, U_\ell) = 0] = 1/2$.

A.2   Another Alternative Proof of Theorem 4.5 (via next-bit tests)

The alternative proof of Theorem 4.5 follows by combining two results: The first is an unconditional result that asserts a generator that passes the next-bit test *with respect to deterministic observers*, and the second is a conditional *deterministic analogue* of the fact that next-bit unpredictability implies pseudorandomness. The first result is a technical generalization of [10, Thm. 2] (see Theorem A.2 below), whereas the second result uses the hypothesis that $\mathcal{BPP} = \mathcal{P}$. The use of this hypothesis should not come at a surprise, since as noted before the conclusion of Theorem 4.5 implies that in some "effective" sense $\mathcal{BPP}$ does equal $\mathcal{P}$. (Thus, as observed in [10], an unconditional implication of the foregoing type, would yield an unconditional derandomization of BPP.)

Next-bit unpredictability, generalized. Let us first present a natural generalization of the notion of next-bit unpredictability, which is pivotal to our argument. Recall that the standard notion refers to the probability of guessing the next bit in a sequence, when given its prefix. That is, in the standard formulation, the predictor outputs a bit, and the question is whether this bit matches the actual next bit. Now, consider a generalization in which the predictor output its estimate of the probability that the next bit is 1, and normalizing the "payoff" accordingly. That is, the generalized predictor outputs a probability $p \in [0, 1]$, and gets payoff $2p - 1$ if the outcome (of the next bit) is 1 and payoff $1 - 2p$ if the outcome is 0. (In general, the payoff is $(1 - 2p) \cdot (-1)^\sigma$, where σ denotes the outcome of the next bit.) Note that such a generalization allows the capture the intuitive case that the predictor wishes to pass on the guess, collect no potential gain but also suffer from no potential loss.

In the context of *probabilistic* algorithms, nothing is gained by this generalization, because a a probabilistic predictor may maintain the expected payoff by replacing the output $p \in [0, 1]$ with a biased coin toss (i.e., a random Boolean value that is biased with probability p towards 1).[36] But this generalization seems more powerful in the context of deterministic predictors, and we shall thus refer to it.

It will be more convenient to replace prediction probabilities with correlation, as we have already done implicitly above. Thus, the predictors will output a value in $v \in [-1, +1]$ (rather than in $[0, 1]$) and collect the payoff $v \cdot (-1)^\sigma$, where σ denotes the outcome of the next bit. Thus, v corresponds to $1 - 2p$ in the forgoing discussion, and $v \cdot (-1)^\sigma$ corresponds to the correlation of v with $(-1)^\sigma$.

Definition A.1 (next bit unpredictability, generalized): *For $\ell : \mathbb{N} \to \mathbb{N}$, let $G : \{0, 1\}^* \to \{0, 1\}^*$ be such that for every $s \in \{0, 1\}^*$ it holds that $|G(s)| = \ell(|s|)$.*

- *For $\epsilon : \mathbb{N} \to [0, 1]$, we say that $A : \mathbb{N} \times \{0, 1\}^* \to [-1, +1]$ has* correlation at most ϵ with the next bit of G *if for all sufficiently large k and all $i < \ell(k)$, it holds that*

$$\mathrm{E}[A(1^{\ell(k)}, x_1, ..., x_i) \cdot (-1)^{x_{i+1}}] \leq \epsilon(\ell(k)), \qquad (19)$$

  where $(x_1, ..., x_{\ell(k)}) \leftarrow G(s)$ for a uniformly selected $s \in \{0, 1\}^k$.
  We will often omit $1^{\ell(k)}$ from the list of inputs to A.
- *We say that G is* next-bit unpredictable by a class of algorithms $\mathcal{A}$ with respect to an advantage ϵ *if every algorithm in $\mathcal{A}$ has correlation at most ϵ with the next bit of G.*

---

[36] That is, the expected payoff with respect to a random variable $\zeta \in \{0, 1\}$ is maintained, because the original payoff is $\mathrm{E}[(1 - 2p) \cdot (-1)^\zeta]$, whereas the payoff of the resulting Boolean predictor is $p \cdot \mathrm{E}[(-1)^{\zeta+1}] + (1 - p) \cdot \mathrm{E}[(-1)^\zeta]$.

We say that G is next-bit *t*-unpredictable *if G is next-bit unpredictable with respect to advantage $1/t$ by the class of deterministic t-time algorithms that have output in $[-1, +1]$.*

By a straightforward extension of the ideas of [10], we obtain the following unconditional result.

Theorem A.2 ([10, Thm. 2], generalized): *For every polynomial p, there exist an exponential function ℓ and a deterministic algorithm G that satisfies the first condition of Definition 4.3 such that G is next-bit p-unpredictable.*

The original result was stated in terms of predicting probabilities, and corresponds to the special case in which the algorithm's output is always in $\{-1, +1\}$ (where, in this case, $E[A(x_1, ..., x_i) \cdot (-1)^{x_{i+1}}]$ equals $2 \cdot \Pr[A(x_1, ..., x_i) = (-1)^{x_{i+1}}] - 1$). The original proof extends in a straightforward manner; see Appendix A.3.

The following result presents a conditional transformation from next-bit unpredictability (by deterministic algorithms) to pseudorandomness (which holds with respect to probabilistic algorithms). This transformation relies on the fact that the hypothesis $\mathcal{BPP} = \mathcal{P}$ allows for derandomizing potential probabilistic next-bit predictors, which are obtained (in the usual way) from potential distinguishers.

Theorem A.3 (next-bit unpredictability implies pseudorandomness): *If $\mathcal{BPP} = \mathcal{P}$, then, for every polynomial p, there exists a polynomial p' such that the following holds: If G satisfies the first condition of Definition 4.3 and is next-bit p'-unpredictable, then G is a p-robust canonical derandomizer.*

Indeed, an unconditioned (but weaker) version of Theorem A.3 (i.e., one that does not assume $\mathcal{BPP} = \mathcal{P}$ (but considers only quadratic-time deterministic distinguishers)) was discussed in [10], and was observed to imply some derandomization (albeit weaker than the one stated in Proposition 4.4, since [10] could not allow probabilistic distinguishers in Definition 4.3). Goldreich and Wigderson saw this implication as evidence to the unlikeliness of proving such a version [10]. Our point, however, is that the assumption $\mathcal{BPP} = \mathcal{P}$ does allow to prove that next-bit unpredictability implies pseudorandomness (in an adequate sense (i.e., as in Definition 4.3)).

Proof: Suppose towards the contradiction that G is not pseudorandom in the sense of Definition 4.3. That is, there exists a probabilistic p-time distinguisher with a distinguishing gap of $\delta(k)$ that for infinitely many k is larger than $1/p(\ell(k))$. Applying the standard transformation from distinguishing to predicting (cf., e.g., [8, Sec. 3.3.5]), we obtain a probabilistic p-time predictor A that outputs a binary value (say in $\{-1, 1\}$) and has a correlation of at least $\epsilon(k) \stackrel{\text{def}}{=} \delta(k)/\ell(k)$ in guessing the next bit of G (wrt a prefix of some length).[37]

---

[37] As stressed in [10], the resulting predictor is probabilistic even if we start with a deterministic distinguisher.

More precisely, for infinitely many k, there exists $i < \ell(k)$, such that

$$\mathrm{E}[A(1^{\ell(k)}, x_1, ..., x_i) \cdot (-1)^{x_{i+1}}] \geq \epsilon(k), \tag{20}$$

where $(x_1, ..., x_{\ell(k)}) \leftarrow G(s)$ for a uniformly selected $s \in \{0,1\}^k$.

Next, we consider a probabilistic FPTAS for approximating the quantity q such that $q(1^{\ell(k)}, x) \stackrel{\mathrm{def}}{=} \mathrm{E}[A(1^{\ell(k)}, x)]$. Note that $q : \{0,1\}^i \to [-1, 1]$ has the same correlation that A has with the $i+1^{\mathrm{st}}$ bit of G, because for every x and every random variable $\zeta \in \{0,1\}$ it holds that $\mathrm{E}[q(x) \cdot (-1)^\zeta] = \mathrm{E}[A(x) \cdot (-1)^\zeta]$. Thus, our aim is to obtain a deterministic FPTAS for q, which we do by noting that the existence a probabilistic FPTAS is straightforward, and invoking Corollary 3.6. Details follow.

A probabilistic FPTAS for q is obtained by invoking A for $O(\ell/\epsilon^2)$ times and outputting the average value.[38] This yields an algorithm A' with output in $[-1, +1]$ such that for every x it holds that $\Pr[A'(x) = q(x) \pm \epsilon/4] > 2/3$. At this point we invoke Corollary 3.6 (or rather its furthermore part), and obtain a deterministic algorithm A'' that satisfies $A'(x) = q(x) \pm \epsilon/2$ for every x. Furthermore, A'' has polynomial running time, where the polynomial p' only depends on the polynomial p (since p determines the running time of A as well as ϵ). Note that for every x and every random variable $\zeta \in \{0,1\}$ it holds that $\mathrm{E}[A''(x) \cdot (-1)^\zeta] > \mathrm{E}[q(x) \cdot (-1)^\zeta] - \epsilon/2$, which implies that A'' has correlation at least $\epsilon(k)/2$ with the next-bit of $G(U_k)$ for infinitely many k. Since the entire argument can be applied to any p-time distinguisher, yielding a p'-time predictor of correlation greater than $1/p'$ (for the same p', which only depends on p), we derive a contradiction to the p'-unpredictability hypothesis. The theorem follows.   ∎

Comment: The generator constructed in the foregoing proof does not satisfy the notion of a canonical derandomizer stated in Definition 4.2. Indeed, in this case, a probabilistic polynomial-time finder F, which runs for more time than the foregoing generator, can find a string x that allows very fast distinguishing. The details are as at the end of Appendix A.2.

A.3   Proof of Theorem A.2

It will be more convenient to restate Theorem A.2 it terms of $\{-1, 1\}$. Thus, the generator outputs sequences over $\{-1, 1\}$ rather than sequences over $\{0, 1\}$. Also, it will be convenient to consider constructing the support of the generator, and assuming that it just outputs a strong that is uniformly distributed in its support. Since we are interested in generators of exponential stretch, these (support) sets have size that is polynomial in the length of the strings in them, since the seed length is logarithmic in these set sizes. Specifically, we refer to sets of the form

---

[38] Actually, this does not yield a FPTAS, but rather an approximation up to an additive term of $\epsilon/2$, We can present this as a FPTAS for the value of $q + 2 \in [1, 3]$, and so proceed via the FPTAS formalism.

$S = \cup_{n \in \mathbb{N}} S_n$, where $S_n \subset \{-1, 1\}^n$, and say that such a set is polynomial-time constructible if there exists a polynomial-time algorithm that on input 1^n outputs the list of all sequences in S_n.

Theorem A.4 (Theorem A.2, restated): *For every polynomial p, there exists a polynomial-time constructible set $S = \cup_{n \in \mathbb{N}} S_n$ such that, for every deterministic algorithm A of running-time p and output in $[-1, +1]$, and for all sufficiently large n and all $i < n$, it holds that $\mathrm{E}[A(x_1, ..., x_i) \cdot x_{i+1}] \leq 1/p(n)$, where $x = (x_1, ..., x_n)$ is uniformly selected in S_n.*

The following text was essentially reproduced from [10], while introducing the necessary adaptations (which are quite minor).

Proof: Consider an enumeration of (modified)[39] deterministic machines running within time $p(n)$ on input of length n, and suppose that we wish to fool the $m = m(n) < p(n)$ first machines in the sense that we wish to upper bound its correlation with the next bit better than ϵ, where $\epsilon = 1/p(n)$. Let M be one of the machines we wish to fool.

We construct S_n in (roughly) n iterations, such that in iteration i we construct $S_{n,i} \subseteq \{-1, 1\}^i$. We start with $S_{n,k} = \{-1, 1\}^k$, where $k = 2 \log_2(m/\epsilon)$, and let $K = 2^k$. In the $i + 1^{\mathrm{st}}$ iteration, we consider the vector $v_M \in [-1, +1]^K$ representing the output of M on each of the K possible i-bit long strings; that is, $v_M[j] = M(x^{(j)})$, where $x^{(j)}$ is the jth string in $S_{n,i} \subseteq \{-1, 1\}^i$. (This represents M's attempt to correlate its output with the $i + 1^{\mathrm{st}}$ bit of a string uniformly selected in S_n, based on the i-bit long prefix of that string.) Our aim is to find a vector $u \in \{-1, 1\}^K$ that has low correlation with all the v_M's. Once such a vector u is found, we extend $S_{n,i}$ into $S_{n,i+1}$ in the natural manner; that is,

$$S_{n,i+1} \overset{\mathrm{def}}{=} \{x^{(j)} u[j] : \text{where } x^{(j)} \text{ is the } j^{\mathrm{th}} \text{ string in } S_{n,i}\} \subset \{-1, 1\}^{i+1}. \quad (21)$$

It follows that $S_n \overset{\mathrm{def}}{=} S_{n,n}$ satisfies the small correlation requirement; that is, for each of the above M's and for each $i < n$, the correlation of $M(x_1 \cdots x_i)$ with x_{i+1}, when x is uniformly selected in S_n, is at most ϵ.

It remains to specify how a suitable vector $u \in \{-1, 1\}^K$ is found, in each iteration. This is done by using a pairwise independent sample space for strings of length K, while recalling that such spaces can be constructed in poly(K)-time (cf. [3]). Thus, it suffices to show that such a sample space must always contain a suitable vector $u \in \{-1, 1\}^K$. This is quite an immediate consequence of the following claim.

Claim: Let $v \in [-1, +1]^K$ be arbitrary, and u be a sequence of K uniformly-distributed pairwise-independent elements of $\{-1, 1\}$. Then, the probability that

---

[39] Recall that one cannot effectively enumerate all machines that run within some given time bound. Yet, one can enumerate all machines, and modify each machine in the enumeration such that the running-time of the modified machine respects the given time bound, while maintaining the functionality of the original machines in the case that the original machine respects the time bound. This is done by simply incorporating a time-out mechanism.

the correlation between u and v is greater than ϵ (i.e., $\sum_i u_i v_i > \epsilon K$) is strictly less than $\frac{1}{\epsilon^2 K}$.

Proof: For each $j \in [K]$, we define a random variable $\eta_j \in [-1, +1]$ such that $\eta_j \stackrel{\text{def}}{=} v[j] \cdot u[j]$. Since v is fixed and u is a sequence of K uniformly-distributed pairwise-independent bits, it follows that the η_j's are pairwise-independent and $\mathrm{E}(\eta_j) = 0$ for each $j \in [K]$. Using Chebyshev's Inequality, we have

$$\Pr\left[\left|\sum_{j \in [K]} \eta_j\right| \geq \epsilon \cdot K\right] \leq \frac{\mathrm{Var}(\sum_j \eta_j)}{(\epsilon K)^2}$$

$$= \frac{1}{\epsilon^2 K}$$

and the claim follows.   □

Since $\epsilon^2 K > m$, it follows that the aforementioned sample space contains a vector u that has correlation at most ϵ with each of the m vectors representing the m machines (i.e., the vectors v_M's). Thus, we construct the desired set S_n in time polynomial in n, and the theorem follows.   ■

Notes on Levin's Theory of Average-Case Complexity

Oded Goldreich

Abstract. In 1984, Leonid Levin initiated a theory of average-case complexity. We provide an exposition of the basic definitions suggested by Levin, and discuss some of the considerations underlying these definitions.

Keywords: Average-case complexity, reductions.

This survey is rooted in the author's (exposition and exploration) work [4], which was partially reproduded in [1]. An early version of this survey appeared as TR97-058 of *ECCC*. Some of the perspective and conclusions were revised in light of a relatively recent work of Livne [21], but an attempt was made to preserve the spirit of the original survey. The author's current perspective is better reflected in [7, Sec. 10.2] and [8], which advocate somewhat different definitional choices (e.g., focusing on *typical* rather than *average* performace of algorithms).

1 Introduction

The average complexity of a problem is, in many cases, a more significant measure than its worst case complexity. This has motivated the development of a rich area in algorithmic research – the probabilistic analysis of algorithms [14,16]. However, historically, this line of research focuses on the analysis of specific algorithms *with respect to specific, typically uniform, probability distributions*.

The general question of average case complexity was addressed for the first time by Levin [18]. Levin's work can be viewed as the basis for a theory of average NP-completeness, much the same way as Cook's [2] (and Levin's [17]) works are the basis for the theory of NP-completeness. Subsequent works [9,22,10,21] have provided additional complete problems. Other basic complexity problems, such as decision versus search, were studied in [1].

Levin's Average-Case Complexity Theory in a Nutshell. An average case complexity class consists of pairs, called distributional problems. Each such pair consists of a decision (resp., search) problem and a probability distribution on problem instances. We focus on the class $\text{DistNP} \overset{\text{def}}{=} \langle \text{NP}, \text{P-computable} \rangle$, defined by Levin [18], which is a distributional analogue of NP: It consists of NP decision problems coupled with distributions for which the accumulative measure is polynomail-time computable. That is, P-computable is the class of distributions

O. Goldreich et al.: Studies in Complexity and Cryptography, LNCS 6650, pp. 233–247, 2011.

for which there exists a polynomial time algorithm that on input x computes the total probability of all strings $y \leq x$. The easy distributional problems are those solvable in "average polynomial-time" (a notion which surprisingly require careful formulation). Reductions between distributional problems are defined in a way guaranteeing that if Π_1 is reducible to Π_2 and Π_2 is in average polynomial-time, then so is Π_1. Finally, it is shown that the class DistNP contains a complete problem.

Levin's Average-Case Theory, Revisited. Levin's laconic presentation [18] hides the fact that several non-trivial choices were in the development of the average-case complexity theory. We discuss some of this choices here. Firstly, we note that our motivation is providing a theory of efficient computation (as suggested above), rather than a theory of infeasible computations (e.g., as in Cryptography). Recall that a theory of useful-for-cryptography infeasible computations does exist (cf., [5,6]). A key difference between these two theories is that in Cryptography we seek problems for which one may generate instance-solution pairs such that solving the problem given only the instance is hard. In the theory of average-case complexity (considered below), we wish to draw the line between problems that are easy to solve and problems that are hard to solve (on the average), but we do not require an efficient procedure for generating hard (on the average) instances coupled with solutions.

Secondly, one has to admit that the class DistNP (i.e., specifically, the choice of distributions) is somewhat problematic. Indeed, P-computable distributions seem "simple" (albeit one may reconsider this view in light of [21]), but it is not clear if they exhaust all natural distributions. A much wider class, which certainly contains all natural distributions, is the class, denoted P-samplable, of all distributions having an efficient algorithm for generting instances (according to the distribution): Arguably, the instances of any problem that we may need to solve in real life are generated by some efficient process, and so the latter class of distributions (i.e., P-samplable) suffices as the scope of our theory [1]. But in this case it becomes even harder to argue that a distributional problem that refers to (a computational problem coupled wityh) an *arbitrary* P-samplable distribution is natural. Fortunately, it was showed in [13] that any distributional problem that is complete for DistNP=⟨NP, P-computable⟩, is also complete with respect to the class ⟨NP, P-samplable⟩. Thus, in retrospect, Levin's choice only makes the theory stronger: It requires to select complete distributional problems from the restricted class ⟨NP, P-computable⟩, whereas hardness holds with respect to the wider class ⟨NP, P-samplable⟩.

As hinted above, the definition of average polynomial-time is less straightforward than one may expect. The obvious attempt at formulation this notion leads to fundamental problems that, in our opinion, deem it inadequate. (For a detailed discussion of this point, the reader is referred to Appendix A.) We believe that once the failure of the obvious attempt is understood, Levin's definition (presented below) does look a natural one.

2   Definitions and Notations

In this section we present the basic definitions underlying the theory of average-case complexity. Most definitions originate from Levin's work [18], but the reader is advised to look for further explanations and motivating discussions elsewhere (e.g., [14,11,4]). An alternative formulation, which uses probability ensembles (rather than a single infinite distribution) as a pivot, is presented in [7, Sec. 10.2.1] and [8].

For sake of simplicity, we consider the standard lexicographic ordering of binary strings, but any other efficient enumeration of strings will do.[1] By writing $x < y$ we mean that the string x precedes y in lexicographic order, and $y - 1$ denotes the immediate predecessor of y. Also, we associate pairs, triples etc. of binary strings with single binary strings in some standard manner (i.e., a standard encoding).

Definition 1 (probability distribution functions): *A distribution function μ : $\{0,1\}^* \rightarrow [0,1]$ is a non-decreasing function from strings to the unit interval $[0,1]$ that converges to one (i.e., $\mu(0) \geq 0$, $\mu(x) \leq \mu(y)$ for each $x < y$, and $\lim_{x \to \infty} \mu(x) = 1$). The* density function *associated with the distribution function μ is denoted μ' and is defined by $\mu'(0) = \mu(0)$ and $\mu'(x) = \mu(x) - \mu(x-1)$ for every $x > 0$.*

Clearly, $\mu(x) = \sum_{y \leq x} \mu'(y)$. For notational convenience, we often describe distribution functions converging to some positive constant $c \neq 1$. In all the cases where we use this convention, it is easy to normalize the distribution such that it converges to one. An important example is the *uniform* distribution function μ_0 defined as $\mu_0'(x) = \frac{1}{|x|^2} \cdot 2^{-|x|}$. (A minor variation that does converge to 1 is obtained by letting $\mu_0'(x) = \frac{1}{|x| \cdot (|x|+1)} \cdot 2^{-|x|}$.)

Definition 2 (distributional problems): *A distributional decision problem (resp., distributional search problem) is a pair (D, μ) (resp., (S, μ)), where $D : \{0,1\}^* \rightarrow \{0,1\}$ (resp., $S \subseteq \{0,1\}^* \times \{0,1\}^*$) and $\mu : \{0,1\}^* \rightarrow [0,1]$ is a distribution function.*

In the sequel we consider mainly decision problems. Similar formulations for search problems can be easily derived.

2.1   Distributional-NP

Simple distributions are identified with the P-computable ones. The importance of restricting attention to simple distributions (rather than allowing arbitrary ones) is demonstrated in [1, Sec. 5.2].

Definition 3 (P-computable): *A distribution μ is in the class P-computable if there is a deterministic polynomial time Turing machine that on input x outputs the binary expansion of $\mu(x)$. (Indeed, the running time is polynomial in $|x|$.)*

---

[1] An efficient enumeration of strings is a 1-1 and onto mapping of strings to integers that can be computed and inverted in polynomial-time.

It follows that the binary expansion of $\mu(x)$ has length polynomial in $|x|$. A necessary condition for distributions to be of interest is their putting noticeable probability weight on long strings (i.e., for some polynomail, p, and sufficiently large n the probability weight assigned to n-bit strings should be at least $1/p(n)$). Consider to the contrary the density function $\mu'(x) \stackrel{\text{def}}{=} 2^{-3|x|}$. An algorithm of running time $t(x) = 2^{|x|}$ will be considered to have constant on the average running-time w.r.t this μ (since $\sum_x \mu'(x) \cdot t(|x|) = \sum_n 2^{-n} = 1$).

If the distribution function μ is in P-computable, then the corresponding density function, μ', is computable in time polynomial in $|x|$. The converse, however, is false, unless P = NP (see [11]). In spite of this remark we usually present the density function, and leave it to the reader to verify that the corresponding distribution function is in P-computable.

We now present the class of distributional problems that corresponds to (the traditional) class NP. Most of results in the literature refer to this class.

Definition 4 (DistNP): *A distributional problem (D, μ) belongs to the class Dis-tNP if D is an NP-predicate and μ is in P-computable. DistNP is also denoted $\langle NP,$ P-computable$\rangle$.*

A wider class of distributions, denoted P-samplable, gives rise to a wider class of distributional NP problems, which was discussed in the Introduction: A distribution μ is in the class P-samplable if there exists a polynomial p and a probabilistic algorithm A that outputs the string x with probability $\mu'(x)$ within $p(|x|)$ steps. That is, elements in a P-samplable distribution are generated in time polynomial in their length. We comment that any P-computable distribution is P-samplable, whereas the converse is false (provided one-way functions exist). For a detailed discussion see [1].

2.2   Average Polynomial-Time

The following definitions, regarding average polynomial-time, may seem obscure at first glance. Thus, it is important to point out that the naive formalizations of the corresponding notions suffer from serious problems such as not being closed under functional composition of algorithms, being model dependent, encoding dependent, etc. For a more detailed discussion, see Appendix A.

Definition 5 (polynomial on the average): *A function $f : \{0,1\}^* \to \mathbb{N}$ is polynomial on the average with respect to a distribution μ if there exists a constant $\epsilon > 0$ such that*

$$\sum_{x \in \{0,1\}^*} \mu'(x) \cdot \frac{f(x)^\epsilon}{|x|} < \infty.$$

The function $l(x) = f(x)^\epsilon$ is linear on the average w.r.t. μ.

Thus, *a function is polynomial on the average if it is bounded by a polynomial in a function that is linear on the average*. In fact, the basic definition is that of a function that is linear on the average (cf. [1]). The notion of polynomial on the average is the basis of the complexity class of distributional problems that are solvable in time that is polynomial on the average.

Definition 6 (Average-P): *A distributional problem (D, μ) is in the class Average-P if there exists an algorithm A solving D, so that the running time of A is polynomial on the average with respect to the distribution μ.*

We view the classes Average-P and DistNP as the average-case analogue of P and NP (respectively). We mention that if EXP $\neq$ NEXP (i.e., DTime($2^{O(n)}$) $\neq$ NTime($2^{O(n)}$)), then Average-P does not contain all of DistNP (see [1]).

2.3   Reducibility between Distributional Problems

We now present definitions of (average polynomial time) reductions of one distributional problem to another. Intuitively, such a reduction should be efficiently computable, yield a valid result and "preserve" the probability distribution. The purpose of the last requirement is to ensure that the reduction does not map very likely instances of the first problem to rare instances of the second problem. Otherwise, having a polynomial time on the average algorithm for the second distributional problem does not necessarily yield such an algorithm for the first distributional problem. Following is a definition of randomized Turing reductions. Definitions of deterministic and many-to-one reductions can be easily derived as special cases.

Definition 7 (randomized reductions): *We say that the probabilistic oracle Turing machine M randomly reduces the distributional problem (D_1, μ_1) to the distributional problem (D_2, μ_2) if the following three conditions hold.*

Efficiency: *Machine M is polynomial time on the average, where the average is taken over the distribution μ_1 and the internal coin tosses of M; that is, letting $t_M(x, r)$ denote the running time of M on input x and internal coin tosses r, we require that there exists $\epsilon > 0$ such that*

$$\sum_{x,r} \mu_1'(x)\mu_0'(r) \cdot \frac{t_M(x, r)^\epsilon}{|x|} < \infty,$$

where μ_0 is the uniform distribution.
Validity: *For every $x \in \{0, 1\}^*$,*

$$\Pr[M^{D_2}(x) = D_1(x)] \geq \frac{2}{3}$$

where $M^{D_2}(x)$ is the random variable (determined by M's internal coin tosses) that denotes the output of the oracle machine M on input x and access to oracle for D_2.
Domination: *There exists a constant $c > 0$ such that for every $y \in \{0, 1\}^*$ it holds that*

$$\mu_2'(y) \geq \frac{1}{|y|^c} \cdot \sum_{x \in \{0,1\}^*} \mathrm{Ask}_M(x, y) \cdot \mu_1'(x)$$

where $\mathtt{Ask}_M(x, y)$ *is the probability (taken over M's internal coin tosses) that machine M asks query y on input x.*

In the special case of *deterministic* Turing reductions the value of $M^{D_2}(x)$ is fully determined by x (rather than being a random variable) and $\mathtt{Ask}_M(x, y)$ is either 0 or 1 (rather than being any arbitrary rational in $[0, 1]$). In the case of a many-to-one deterministic reduction, for every x, there exists a unique y such that $\mathtt{Ask}_M(x, y) = 1$ holds.

Proposition: *If (D_1, μ_1) is deterministically (resp., randomly) reducible to (D_2, μ_2) and (D_2, μ_2) is solvable by a deterministic (resp., randomized) algorithm of running time that is polynomial on the average, then so is (D_1, μ_1).*

Proof: Given any reduction of (D_1, μ_1) to (D_2, μ_2) we consider the distribution μ_I of the queries of the reduction on random instances distributed according to μ_1; that is,

$$\mu_I(y) \stackrel{\text{def}}{=} \sum_{x \in \{0,1\}^*} \mathtt{Ask}_M(x, y) \cdot \mu_1'(x).$$

We next decouple the original reduction to a reduction of (D_1, μ_1) to (D_2, μ_I) (via the original transformation) and a reduction by the identity transformation of (D_2, μ_I) to (D_2, μ_2). (Note that each of these reductions satisfies Definition 7.) Thus, it suffices to establish the proposition for each of these two reductions.

1. Considering the reduction of (D_1, μ_1) to (D_2, μ_I), we note that when this reduction is invoked on inputs distributed according to μ_1, it makes queries that are distributed according to μ_I. Thus, if (D_2, μ_I) is solvable in polynomial-time on the avearge, then so is (D_1, μ_1).

2. Considering the reduction (by the identity transformation) of (D_2, μ_I) to (D_2, μ_2), it suffices to show that *if $t : \{0, 1\}^* \to \mathbb{N}$ is polynomial on the average w.r.t μ_2, then t is polynomial on the average w.r.t μ_I.*

   By the hypothesis regarding t and μ_2, for some for $\epsilon > 0$, it holds that $\sum_y \mu_2'(y) \frac{t(y)^\epsilon}{|y|} = O(1)$, whereas by the hypothesis that μ_2 dominates μ_I it holds that $\mu_I'(y) \leq |y|^c \mu_2'(y)$ (for all y). Let $G \stackrel{\text{def}}{=} \{y : t(y) \leq |y|^{2c/\epsilon}\}$, and split the sum $\sum_y \mu_I'(y) \frac{t(y)^{\epsilon/2c}}{|y|}$ according to whether $y \in G$ or not. The sum $\sum_{y \in G} \mu_I'(y) \frac{t(y)^{\epsilon/2c}}{|y|}$ is upper bounded by 1 (by using $t(y)^{\epsilon/2c} \leq |y|$ for $y \in G$); whereas

$$\sum_{y \notin G} \mu_I'(y) \frac{t(y)^{\epsilon/2c}}{|y|} \leq \sum_{y \notin G} |y|^c \mu_2'(y) \frac{t(y)^{\epsilon/2}}{|y|}$$

$$\leq \sum_{y \notin G} t(y)^{\epsilon/2} \mu_2'(y) \frac{t(y)^{\epsilon/2}}{|y|}$$

$$= \sum_{y \notin G} \mu_2'(y) \frac{t(y)^\epsilon}{|y|}$$

$$= O(1)$$

where the first inequality uses $\mu'_I \leq |y|^c \mu'_2(y)$ and the second inequality uses $|y|^c \leq t(y)^{\epsilon/2}$ (for $y \notin G$).

The proposition follows.   □

We also mention that reductions are transitive in the special case in which they are *honest*; that is, on input x they ask queries of length at least $|x|^\epsilon$, for some constant $\epsilon > 0$. All known reductions have this property. Finally, we spell out the resulting definition of DistNP-completeness.

Definition 8 (DistNP-completeness): *A distributional problem Π is DistNP-complete if $\Pi \in$ DistNP and every problem in DistNP is reducible to Π.*

We shall actually use the most restricted notion of a reduction; that is, unless stated otherwise, all the reductions we discuss are deterministic many-to-one reductions.

2.4 A Generic DistNP Complete Problem

The following distributional version of *Bounded Halting*, denoted $\Pi_{BH} = (BH, \mu_{BH})$, is known to be DistNP-complete (see Section 3).

Definition 9 (distributional Bounded Halting): *The distributional problem $\Pi_{BH} = (BH, \mu_{BH})$ consists of the following*

- Decision: $BH(M, x, 1^k) = 1$ *iff there exists a computation of the non-deterministic machine M on input x that halts within k steps.*
- Distribution: *The distribution μ_{BH} is defined in terms of its density function*

$$\mu'_{BH}(M, x, 1^k) \stackrel{\text{def}}{=} \frac{1}{|M|^2 \cdot 2^{|M|}} \cdot \frac{1}{|x|^2 \cdot 2^{|x|}} \cdot \frac{1}{k^2}$$

Note that μ'_{BH} is very different from the uniform distribution on binary strings (e.g., consider relatively large k), but this seems fair since part of its input is not binary. Nevertheless, as noted by Levin, one can obtain a variant of Π_{BH} that refers to the uniform distribution and is DistNP-complete with respect to randomized reduction. Specifically, we replace the unary time bound by a string of equal length, and assign each such string the same probability (see [7, §10.2.1.3] or [8, Sec. 2.3]).

3 DistNP-completeness of Π_{BH}

The proof, presented here, is due to Guretich [9]. (An alternative proof is implied by Levin's original paper [18].)

Perspective. In the traditional theory of $\mathcal{NP}$-completeness, the *mere* existence of complete problems is almost immediate. For example, it is extremely simple to show that the *Bounded Halting* problem is $\mathcal{NP}$-complete. Recall that Bounded Halting (BH) is defined over triples $(M, x, 1^k)$, where M is a non-deterministic machine, x is a binary string and k is an integer (given in unary). The decision problem is to determine whether there exists a computation of M on input x that halts within k steps. Clearly, Bounded Halting is in $\mathcal{NP}$ (here its crucial that k is given in unary). Let D be an arbitrary $\mathcal{NP}$ problem, and let M_D be the non-deterministic machine solving it in time $P_D(n)$ on inputs of length n, where P_D is a fixed polynomial. Then, the reduction of D to BH consists of the transformation $x \rightarrow (M_D, x, 1^{P_D(|x|)})$.

In the case of distributional-NP an analogous theorem is much harder to prove. The difficulty is that we have to reduce all DistNP problems (i.e., pairs consisting of decision problems and simple distributions) to one single distributional problem (i.e., Bounded Halting with a single simple distribution). If we apply reductions as above (and consider the induced probability distributions), then we will end up with many distributional versions of Bounded Halting. Furthermore the corresponding distribution functions will be very different and will not necessarily dominate one another. Instead, one should reduce a distributional problem, (D, μ), with an arbitrary P-computable distribution to a distributional problem with a fixed (P-computable) distribution (e.g. Π_{BH}). The difficulty in doing so is that the reduction should have the domination property.

Consider, for example, an attempt to reduce each problem in DistNP to Π_{BH} by using the standard transformation of D to BH (sketched above). This transformation fails when applied to distributional problems in which the distribution of (infinitely many) strings is much higher than the distribution assigned to them by the uniform distribution. In such cases, the standard reduction maps an instance x having probability mass $\mu'(x) \gg 2^{-|x|}$ to a triple $(M_D, x, 1^{P_D(|x|)})$ with much lighter probability mass (since $\mu'_{BH}(M_D, x, 1^{P_D(|x|)}) < 2^{-|x|}$). This violates the domination condition, and thus an alternative reduction is required.

The key to the alternative reduction is an (efficiently computable) encoding of strings taken from an *arbitrary* polynomial-time computable distribution by strings that have comparable probability mass under a *fixed* distribution (i.e., the uniform one). Specifically, this encoding will map x into a codeword of length that is at most the logarithm of $1/\mu'(x)$, which means that under the uniform distribution the codeword of x has probability weight approximately $\mu'(x)$. Accordingly, the reduction will map x to a triple $(M_{D,\mu}, x', 1^{|x|^{O(1)}})$, where $|x'| < \log_2(1/\mu'(x)) + O(1)$, and $M_{D,\mu}$ is a non-deterministic Turing machine that first retrieves x from x', and then applies the standard non-deterministic machine (i.e., M_D) of the problem D. Such a reduction will be shown to satisfy all three conditions (i.e. efficiency, validity, and domination). Thus, instead of forcing the structure of the original distribution μ on the target distribution μ_{BH}, the reduction will incorporate the structure of μ into the the reduced instance (i.e., in $M_{D,\mu}$). The following technical lemma is the basis of the reduction.

Coding Lemma: *Let μ be a polynomial-time computable distribution function. Then there exist a coding function C_μ satisfying the following three conditions.*

Compression: *For every x, it holds that*

$$|C_\mu(x)| \leq \min\left\{|x|, \log_2 \frac{1}{\mu'(x)}\right\} + 1.$$

Efficient Encoding: *The function C_μ is computable in polynomial-time.*
Unique Decoding: *The function C_μ is one-to-one (i.e., $C_\mu(x) = C_\mu(x')$ implies $x = x'$).*

Proof: The function C_μ is defined as follows. If $\mu'(x) \leq 2^{-|x|}$, then $C_\mu(x) = 0x$ (i.e., in this case x serves as its own encoding). Otherwise, if $\mu'(x) > 2^{-|x|}$, then $C_\mu(x) = 1z$, where z is the longest common prefix of the binary expansions of $\mu(x-1)$ and $\mu(x)$; that is, if $\mu(x-1)$ and $\mu(x)$ have binary expansions $0.\sigma_1\sigma_2\cdots$ and $0.\tau_1\tau_2\cdots$, respectively, then $z = \sigma_1\cdots\sigma_\ell$ such that $\sigma_1\cdots\sigma_\ell = \tau_1\cdots\tau_\ell$ whereas $\sigma_{\ell+1} = 0$ and $\tau_{\ell+1} = 1$ (e.g., if $\mu(1010) = 0.10000$ and $\mu(1011) = 0.10101111$, then $C_\mu(1011) = 1z$ with $z = 10$). Consequently, $0.z1$ is in the interval $(\mu(x-1), \mu(x)]$; that is, $\mu(x-1) < 0.z1 \leq \mu(x)$.

We now verify that C_μ satisfies the conditions of the lemma. We start with the compression condition. Clearly, if $\mu'(x) \leq 2^{-|x|}$, then $|C_\mu(x)| = 1 + |x| \leq 1 + \log_2(1/\mu'(x))$. On the other hand, suppose that $\mu'(x) > 2^{-|x|}$ and let $z = z_1 \cdots z_\ell$ be as above (i.e., the longest common prefix of the binary expansions of $\mu(x-1)$ and $\mu(x)$). Then,

$$\mu'(x) = \mu(x) - \mu(x-1) \leq \left(\sum_{i=1}^{\ell} 2^{-i} z_i + \sum_{i=\ell+1}^{\text{poly}(|x|)} 2^{-i}\right) - \sum_{i=1}^{\ell} 2^{-i} z_i < 2^{-|z|}$$

and $|z| \leq \log_2(1/\mu'(x))$ follows. Thus, $|C_\mu(x)| \leq 1 + \log_2(1/\mu'(x))$ in both cases, and if $\log_2(1/\mu'(x)) \geq |x|$ (i.e., $\mu'(x) \leq 2^{-|x|}$), then $|C_\mu(x)| = |x| + 1$. Clearly, C_μ can be computed in polynomial-time (by computing $\mu(x-1)$ and $\mu(x)$). Finally, note that C_μ is one-to-one by considering the two cases (i.e., $C_\mu(x) = 0x$ and $C_\mu(x) = 1z$), while using the fact that $\mu(x-1) < 0.z1 \leq \mu(x)$ (in the second case). $\square$

Towards the Reduction. For every distributional problem (D, μ) in DistNP, we introduce a non-deterministic machine $M_{D,\mu}$ that will be used in the following reduction of (D, μ) to $\Pi_{BH} = (BH, \mu_{BH})$ such that all instances (of D) are mapped to triples with first element $M_{D,\mu}$. The description of this machine, $M_{D,\mu}$, refers to the aforementioned coding function C_μ (as well as to the non-deterministic machine M_D associated with D). On input $y = C_\mu(x)$, machine $M_{D,\mu}$ computes $D(x)$, by first retrieving x from $C_\mu(x)$ (e.g., guess and verify), and next running the non-deterministic polynomial-time machine (i.e., M_D) that solves D.

The Reduction Itself. An instance x (of D) is mapped by the reduction to the triple $(M_{D,\mu}, C_\mu(x), 1^{P(|x|)})$, where $P(n) \stackrel{\text{def}}{=} P_D(n) + P_C(n) + n$ such that $P_D(n)$ is a polynomial bounding the running time of M_D on (acceptable) inputs of length n, and $P_C(n)$ is a polynomial bounding the running time of an algorithm for encoding inputs (of length n).

Proposition: *The foregoing mapping constitutes a reduction of (D, μ) to (BH, μ_{BH}).*

Proof: We verify the three requirements from a reduction.

- The transformation can be computed in polynomial-time.
  (Indeed, we rely on the fact that C_μ is polynomial-time computable.)
- By construction of $M_{D,\mu}$ it follows that $D(x) = 1$ if and only if there exists a computation of machine $M_{D,\mu}$ that on input $C_\mu(x)$ halts outputting 1 within $P(|x|)$ steps.
  (Recall that, on input $C_\mu(x)$, machine $M_{D,\mu}$ non-deterministically guesses x, verifies in $P_C(|x|)$ steps that x is encoded by $C_\mu(x)$, and non-deterministically "computes" $D(x)$.)
- To see that the distribution induced by the reduction is dominated by the distribution μ_{BH}, we first recall that the transformation $x \to C_\mu(x)$ is one-to-one. It suffices to consider instances of BH which have a preimage under the reduction (since instances with no preimage satisfy the condition trivially). All these instances are triples with first element $M_{D,\mu}$. By the definition of μ_{BH}

$$\mu'_{BH}(M_{D,\mu}, C_\mu(x), 1^{P(|x|)}) = c \cdot \frac{1}{P(|x|)^2} \cdot \frac{1}{|C_\mu(x)|^2 \cdot 2^{|C_\mu(x)|}} \qquad (1)$$

where $c = \frac{1}{|M_{D,\mu}|^2 \cdot 2^{|M_{D,\mu}|}}$ is a constant depending only on (D, μ).
By virtue of the Coding Lemma it holds that

$$\mu'(x) \leq 2 \cdot 2^{-|C_\mu(x)|}. \qquad (2)$$

Combing Eq. (1) and (2), we get

$$\mu'_{BH}(M_{D,\mu}, C_\mu(x), 1^{P(|x|)}) \geq c \cdot \frac{1}{P(|x|)^2} \cdot \frac{1}{|C_\mu(x)|^2} \cdot \frac{\mu'(x)}{2}$$
$$> \frac{c}{2 \cdot |M_{D,\mu}, C_\mu(x), 1^{P(|x|)}|^2} \cdot \mu'(x).$$

The proposition follows. $\qquad\qquad\qquad\qquad\qquad\qquad\qquad\qquad\qquad\qquad\qquad$ □

4 Conclusions

In general, a theory of average case complexity should provide

1. a specification of a broad class of *interesting* distributional problems;
2. a definition capturing the subclass of (distributional) problems that are *easy* on the average;
3. notions of reducibility that allow to infer the easiness of one (distributional) problem from the easiness of another;
4. and, of course, results...

It seems that the theory of average case complexity, initiated by Levin and further developed in [9,22,1,13,21], satisfies these expectations to some extent. Following is my evaluation regarding its "performance" with respect to each of the above.

1. The scope of the theory, originally restricted to P-computable distributions, has been significantly extended to cover all P-sampleable distributions (as suggested in [1]). The key result here is by Impagliazzo and Levin [13] whow proved that every language which is ⟨NP, P-computable⟩-complete is also ⟨NP, P-samplable⟩-complete. This important result makes the theory of average case very robust: It allows to reduce distributional problems from an utmost wide class to distributional problems with very restricted/simple type of distributions.

   Till Livne's result [21], my feeling was that the set of P-computable distributions may be too restricted in the sense that it may not allow to present DistNP-complete problems that refer to most natural NP decision problems. However, Livne showed that all natural NP decision problems do have distributional versions that are DistNP-complete, alas these versions turned out to be somewaht unnatural (i.e., the distribution does not seem simple and/or natural).[2]

2. The definition of average polynomial-time does seem strange at first glance, but it seems that it (or a similar alternative) does captures the intuitive meaning of "easy on the average".

   We mention that an alternative notion that refers to the *typical* running time rather than the *average* of (a quantity that is polynomially related to) the running time is considered in [7, Sec. 10.2.1] and [8]. Specifically, we may say that $f : \{0,1\}^* \rightarrow \mathbb{N}$ is typically polynomial with respect to a distribution μ if there exists a positive polynomial p such that, for every polynomial q, it holds that

$$\sum_{x \in \{0,1\}^*:f(x)>p(|x|)} q(|x|) \cdot \mu'(x) < \infty.$$

3. The notions of reducibility are both natural and adequate.

---

[2] For a discussion of the notion of a natural problem, the interested reader is referred to Appendix B.

4. Results did follow, but here indeed much more is expected. Currently, quite natural DistNP-complete problems are known for the following areas: Computability (e.g., Bounded-Halting) [9], Combinatorics (e.g., Tiling [18] and a generalization of graph coloring [22]), Formal Languages (cf., [9,4]), and Algebra (e.g., of matrix groups [10]). Furthermore, the aforementioned result of Livne [21] asserts that all natural NP problems have DistNP-complete versions. However, the challenge of finding a really natural distributional problem (e.g., subset sum with uniform distribution) that is complete in DistNP has not been met so far. It seems that what is still lacking are techniques for design of "distribution preserving" reductions.

In addition to their central role in the theory of average-case complexity, reductions that preserve uniform (or very simple) instance distribution are of general interest. Such reductions, unlike most known reductions used in the theory of NP-completeness, cannot map all instances to some "pathological" subcase.

Levin views the results in his paper [18] as an indication that all P-computable distributions are in fact related (or similar). Additional support to this statment is provided by his latter work [20].

Acknowledgements. I'm very grateful to Leonid Levin for many inspiring discussions. Special thanks also to Noam Livne for reconfiguring my views regarding P-computable distributions.

References

1. Ben-David, S., Chor, B., Goldreich, O., Luby, M.: On the Theory of Average Case Complexity. Journal of Computer and system Sciences 44(2), 193–219 (1992)
2. Cook, S.A.: The Complexity of Theorem Proving Procedures. In: Proc. 3rd ACM Symp. on Theory of Computing, pp. 151–158 (1971)
3. Garey, M.R., Johnson, D.S.: Computers and Intractability: A Guide to the Theory of NP-Completeness. W.H. Freeman and Company, New York (1979)
4. Goldreich, O.: Towards a Theory of Average Case Complexity (a survey). TR-531, Computer Science Department, Technion, Haifa, Israel (March 1988)
5. Goldreich, O.: Foundation of Cryptography: Basic Tools. Cambridge University Press, Cambridge (2001)
6. Goldreich, O.: Foundation of Cryptography: Basic Applications. Cambridge University Press, Cambridge (2004)
7. Goldreich, O.: Computational Complexity: A Conceptual Perspective. Cambridge University Press, Cambridge (2008)
8. Goldreich, O.: Average Case Complexity, Revisited. In: Goldreich, O., et al. (eds.) Studies in Complexity and Cryptography. LNCS, vol. 6650, pp. 424–452. Springer, Heidelberg (2011)
9. Gurevich, Y.: Complete and Incomplete Randomized NP Problems. In: Proc. of the 28th IEEE Symp. on Foundation of Computer Science, pp. 111–117 (1987)
10. Gurevich, Y.: Matrix Decomposition Problem is Complete for the Average Case. In: Proc. of the 31st IEEE Symp. on Foundation of Computer Science, pp. 802–811 (1990)

11. Gurevich, Y., McCauley, D.: Average Case Complete Problems (1987) (preprint)
12. Hartmanis, J., Stearns, R.E.: On the Computational Complexity of of Algorithms. Transactions of the AMS 117, 285–306 (1965)
13. Impagliazzo, R., Levin, L.A.: No Better Ways to Generate Hard NP Instances than Picking Uniformly at Random. In: Proc. of the 31st IEEE Symp. on Foundation of Computer Science, pp. 812–821 (1990)
14. Johnson, D.S.: The NP-Complete Column – an ongoing guide. Jour. of Algorithms 4, 284–299 (1984)
15. Karp, R.M.: Reducibility among Combinatorial Problems. In: Miller, R.E., Thatcher, J.W. (eds.) Complexity of Computer Computations, pp. 85–103. Plenum Press, New York (1972)
16. Karp, R.M.: Probabilistic Analysis of Algorithms. (1986) (manuscript)
17. Levin, L.A.: Universal Search Problems. Problemy Peredaci Informacii 9, 115–116 (1973); Translated in problems of Information Transmission 9, 265–266
18. Levin, L.A.: Average Case Complete Problems. SIAM Jour. of Computing 15, 285–286 (1986); Extended abstract appeared In: 16th STOC (1984)
19. Levin, L.A.: One-Way Function and Pseudorandom Generators. In: Proc. 17th ACM Symp. on Theory of Computing, pp. 363–365 (1985)
20. Levin, L.A.: Homogeneous Measures and Polynomial Time Invariants. In: Proc. 29th IEEE Symp. on Foundations of Computer Science, pp. 36–41 (1988)
21. Livne, N.: All Natural NPC Problems Have Average-Case Complete Versions. Computational Complexity (2006) (to appear); Preliminary version in ECCC, TR06-122 (2006)
22. Venkatesan, R., Levin, L.A.: Random Instances of a Graph Coloring Problem are Hard. In: Proc. 20th ACM Symp. on Theory of Computing, pp. 217–222 (1988)

Appendix A: Failure of a Naive Formulation

When asked to motivate his definition of average polynomial-time, Leonid Levin replies, non-deterministically, in one of the following three ways:

1. "This is *the* natural definition".
2. "This definition is *not important* for the results in my paper; only the definitions of reduction and completeness matter (and even they can be modified in many ways while preserving the results)".
3. "Any definition that *makes sense* is either equivalent or weaker".

For further elaboration on the first argument the reader is referred to Leonid Levin. The second argument is, off course, technically correct but unsatisfactory. We will need a definition of "easy on the average" when motivating the notion of a reduction and developing useful relaxations of it. The third argument is a thesis, which should be interpreted along Wittgenstein's suggestion to the teacher: "say nothing and restrict yourself to pointing out errors in the students' attempts to say something". We will follow this line of argument here by showing that the definition that seems natural to an average computer scientist suffers from serious problems and should be rejected.

Definition X (naive formulation of the notion of easy on the average): *A distributional problem (D, μ) is polynomial-time on the average if there exists an algorithm A solving D (i.e., on input x outputs D(x)) such that the running time of algorithm A, denoted t_A, satisfies*

$$\exists c > 0 \; \forall n \quad \sum_{x \in \{0,1\}^n} \mu'_n(x) \cdot t_A(x) < n^c \tag{3}$$

where $\mu'_n(x)$ is the conditional probability that x occurs given that an n-bit string occurs (i.e., $\mu'_n(x) = \mu'(x)/\sum_{y \in \{0,1\}^n} \mu'(y)$).

The Main Problem with Definition X. The problem that we consider most upsetting is that *Definition X is not robust under functional composition of algorithms.* Namely, if the distributional problem A can be solved in average polynomial-time given access to an oracle for B, and problem B can be solved in polynomial-time, then it does *not* follow that the distributional problem A can be solved in average polynomial-time.

For example, consider the uniform probability distribution (on inputs of each length) and an oracle Turing machine M that solves A when given access to oracle B. Suppose that M runs for $2^{\frac{n}{2}}$ steps on $2^{\frac{n}{2}}$ of the inputs of length n, and n^2 steps on all other inputs of length n. Furthermore, supposed that when M makes t steps, it asks a single query of length $\sqrt{t}$. Note that machine M is polynomial-time on the average. But now suppose that the algorithm for B has cubic running-time. The reader can verify that, although M itself (when given access to the oracle B) is polynomial-time on the average, combining M with the cubic running-time algorithm for B *does not* yield an algorithm that is polynomial-time on the average according to Definition X. It is easy to see that this problem does not arise when using the definition presented in Section 2.

The source of the above problem with Definition X is the fact that the definition of polynomial-on-the-average that underlies it is not closed under application of polynomials. Namely, if $t : \{0,1\}^* \to \mathbb{N}$ is polynomial on the average (according to Eq. (3)), with respect to some distribution, it does not follow that also $t^2(\cdot)$ is polynomial on the average (with respect to the same distribution).

The foregoing technical problem is also the source of the following problem, which Levin considers most upsetting: Definition X is *not* machine independent. This is the case because some of the simulations of one computational model on another square the running time (e.g., the simulation of two-tape Turing machines on a one-tape Turing machine, or the simulation of a RAM (Random Access Machine) on a Turing machine).

Having pointed out several weaknesses of Definition X, let us also doubt its "clear intuitive advantage" over the definition presented in Section 2. Definition X is derived from the formulation of worst case polynomial-time algorithms, which requires that $\exists c > 0 \; \forall n$ such that

$$\forall x \in \{0,1\}^n \quad t_A(x) < n^c. \tag{4}$$

Definition X was derived by applying the expectation operator to the Eq. (4), But why not make a very simple algebraic manipulation of Eq. (4) before applying the expectation operator? How about taking the c-th root of both sides and dividing by n; indeed, this yields that $\exists c > 0 \; \forall n$ it holds that

$$\forall x \in \{0,1\}^n \quad \frac{t_A(x)^{\frac{1}{c}}}{n} < 1. \tag{5}$$

But now, applying the expectation operator to the Eq. (5) leads to the definition presented in Section 2...

We conclude that both Definition X and the definition presented in Section 2 are obtained by applying the expectation operator to a worst-case inequality, where the two base inequalities (i.e., Eq. (4) and Eq. (5)) are easily related to one another. From this perspective it is hard to argue (a priori) that one application is more natural than another. However, a posteriori, it becomes evident that the definition presented in Section 2 demonstrates a better understanding of the effect of the expectation operator with respect to complexity measures.

Summary: Robustness under functional composition as well as machine independence seems to be essential for a coherent theory. These are among the primary reasons for the acceptability of P as capturing problems that can be solved efficiently. In going from worst case analysis to average case analysis we should not and would not like to lose these properties.

Appendix B: On the Notion of Natural Problems

Throughout this article, we made several references to the undefined notion of a natural computational problem. While most researchers have some intuition regarding this notion, we feel that an attempt to articulate this notion is in place.

We comment that one should not expect to see a formal definition of intuitive notions such as "simple" or "natural"; yet, this does not mean that we should not try to articulate our intuition about them.

The first idea that comes to mind is to say that a problem is natural if most researchers would say so. This empirically oriented definition seems workable, but it leaves us wondering as to what makes some problems natural whereas other problems are not natural; that is, why would most researchers agree on the foregoing classification of problems?

An appealing criterion was proposed by Livne [21]: *The extent to which a computational problem is natural, with respect to some result, is proportional to the amount of references to the said problem that are prior to the said result and occur in a different context.* Thus, for example, Satisfiability is a very natural problem with respect to the Cook-Levin Theorem [2,17], because this problem was defined and studied in numerous works and in different contexts (such as logic) prior to the Cook-Levin Theorem. To the contrary, the sequence of decision problems constructed in the proof of the Hierarchy Theorem of [12] is definitely unnatural, because these decision problems were first defined in this context (let alone that they are never mentioned outside the context of the Hierarchy Theorem).

Three XOR-Lemmas — An Exposition

Oded Goldreich

Abstract. We provide an exposition of three lemmas that relate general properties of distributions over bit strings to the exclusive-or (xor) of values of certain bit locations.

The first XOR-Lemma, commonly attributed to Umesh Vazirani (1986), relates the statistical distance of a distribution from the uniform distribution over bit strings to the maximum bias of the xor of certain bit positions. The second XOR-Lemma, due to Umesh and Vijay Vazirani (*19th STOC*, 1987), is a computational analogue of the first. It relates the pseudorandomness of a distribution to the difficulty of predicting the xor of bits in particular or random positions. The third Lemma, due to Goldreich and Levin (*21st STOC*, 1989), relates the difficulty of retrieving a string and the unpredictability of the xor of random bit positions. The most notable XOR Lemma – that is the so-called Yao XOR Lemma – is not discussed here.

We focus on the proofs of the aforementioned three lemma. Our exposition deviates from the original proofs, yielding proofs that are believed to be simpler, of wider applicability, and establishing somewhat stronger quantitative results. Credits for these improved proofs are due to several researchers.

Keywords: Vector spaces, Kroniker and Fourier bases, pseudorandomness, space-bounded computation, one-way functions, hard-core predicates and functions.

An earlier version of this survey, which was first drafted in July 1991, appeared as TR95-056 of *ECCC*. Section 1.6 was added in the current revision. Other than that, the current revision is quite minimal.

Preface

Unfortunately, the TCS community does not excell in its choice of names of various notions and phenomena. Consequently, we often find the same name used for several different issues. The name "XOR Lemma" is indeed a good example; at least four different technical statements are often referred to as XOR Lemmas. Indeed, the XOR operation features in each of these lemmas, but the actual context and contents of these lemmas vary.

The current article surveys three XOR lemmas, focusing on their proofs. As stated in the abstract, Yao's XOR-Lemma is not one of the XOR Lemmas surveyed here; the interested reader is referred to [11].

O. Goldreich et al.: Studies in Complexity and Cryptography, LNCS 6650, pp. 248–272, 2011.

1 The Information Theoretic XOR-Lemma

The Information Theoretic XOR-Lemma, commonly attributed to Umesh Vazirani, relates two measures of the "randomness" of distributions over n-bit long strings.

- The statistical difference from uniform; namely, the statistical difference (variation difference) between the "target" distribution and the uniform distribution over the set of all n-bit strings.
- The maximum bias of the xor of certain bit positions; namely, the bias of a 0-1 random variable obtained by taking the exclusive-or of certain bits in the "target" distribution.

It is well known that the statistical difference from uniform is bounded above by 2^n times the maximum bias of the xor's. Several researchers have noticed that the factor in the bound can be improved to $\sqrt{2^n}$. We provide a four line proof of this fact. We also explain the reason for the popularity of the worse bound.

As motivation to the XOR-Lemma, we point out that it has been used in numerous works (e.g., Vazirani [20], Naor and Naor [16]). In a typical application, one first derives an upper bound on the maxbias of the constructed distribution, and then the XOR-Lemma is applied to infer an upper bound on the statistical difference from the uniform distribution.

Credit: The proof presented here has appeared as an appendix in [2].

1.1 Formal Setting

Let π be a an arbitrary probability distribution over $\{0,1\}^n$ and let μ denote the uniform distribution over $\{0,1\}^n$ (i.e., $\mu(x) = 2^{-n}$ for every $x \in \{0,1\}^n$). Let $x = x_1 \cdots x_n$ and $N \stackrel{\text{def}}{=} 2^n$. The XOR-Lemma relates two "measures of closeness" of π and μ.

- The statistical difference ("variation difference") between π and μ; namely,

$$\text{stat}(\pi) \stackrel{\text{def}}{=} \frac{1}{2} \cdot \sum_x |\pi(x) - \mu(x)| \tag{1}$$

- The "maximum bias" of the exclusive-or of certain bit positions in strings chosen according to the distribution π; namely,

$$\text{maxbias}(\pi) \stackrel{\text{def}}{=} \max_{S \neq \emptyset} \{|\pi(\{x : \bigoplus_{i \in S} x_i = 0\}) - \pi(\{x : \bigoplus_{i \in S} x_i = 1\})|\} \tag{2}$$

The XOR-Lemma, commonly attributed to Umesh Vazirani [20][1], states that stat $(\pi) \leq N \cdot \text{maxbias}(\pi)$. The proof is based on viewing distributions as elements in an N-dimensional vector space and observing that the two measures considered

---

[1] The special case where the maxbias is zero appears in Chor et. al. [5].

by the lemma are merely two norms taken with respect to two different orthogonal bases (see Section 1.2). Hence, the XOR-Lemma follows from a (more general and quite straightforward) technical lemma that relates norms taken with respect to different orthonormal bases (see Section 1.3). It turns out this argument actually yields $\mathrm{stat}(\pi) \leq \sqrt{N} \cdot \mathrm{maxbias}(\pi)$, and it seems that the previously inferior bound of [20] was due to a less careful use of the same underlying ideas.

1.2   Preliminaries: The XOR-Lemma and Vector Spaces

Probability distributions over $\{0,1\}^n$ are functions from $\{0,1\}^n$ to the reals. Such functions form a N-dimensional vector space. We shall consider two alternative bases of this vector space.

The standard basis, denoted K, is the orthonormal basis defined by the Kroniker functions; that is, the Boolean functions $\{k_\alpha : \alpha \in \{0,1\}^n\}$, where $k_\alpha(x) = 1$ if $x = \alpha$. The statistical difference between two distributions equals (half) the norm L_1 of their difference taken in the above K basis.

On the other hand, the maxbias of a distribution equals the maximum Fourier coefficient of the distribution, which in turn corresponds to the max-norm (norm L_∞) of the distribution taken in a different basis. This basis is defined by the functions $\{b_S : S \subseteq \{1,2,...,n\}\}$, where $b_S(x) = (-1)^{\Sigma_{i \in S} x_i}$. Note that $b_S(x) = 1$ if the exclusive-or of the bits $\{x_i : i \in S\}$ is 0 and $b_S(x) = -1$ otherwise. The new basis is orthogonal but not orthonormal. We hence consider the normalized basis, denoted F, consisting of the functions $f_S = \frac{1}{\sqrt{N}} \cdot b_S$.

Notation: Let B be an orthonormal basis and r an integer. We denote by $\mathbf{N}_r^B(v)$ the norm L_r of v with respect to the basis B. Namely, $\mathbf{N}_r^B(v) = (\sum_{e \in B} \langle e, v \rangle^r)^{(1/r)}$, where $\langle e, v \rangle$ is the absolute value of the inner product of the vectors e and v. We denote by $\mathbf{N}_\infty^B(v)$ the limit of $\mathbf{N}_r^B(v)$ when $r \to \infty$ (i.e., $\mathbf{N}_\infty^B(v)$ is $\max_{e \in B}\{\langle e, v \rangle\}$).

Clearly, $\mathrm{stat}(\pi) = \frac{1}{2} \cdot \mathbf{N}_1^K(\pi - \mu)$ whereas $\mathrm{maxbias}(\pi) = \sqrt{N} \cdot \mathbf{N}_\infty^F(\pi - \mu)$. Following is a proof of the second equality. Let $\delta(x) = \pi(x) - \mu(x)$. Clearly, $\mathrm{maxbias}(\mu) = 0$ and hence $\mathrm{maxbias}(\pi) = \mathrm{maxbias}(\delta)$. Also $\sum_x \delta(x) = 0$. We get

$$\mathrm{maxbias}(\delta) = \max_{S \neq \emptyset} \{|\delta(\{x : b_S(x) = 1\}) - \delta(\{x : b_S(x) = -1\})|\}$$

$$= \max_{S \neq \emptyset} \left\{ \left| \sum_x b_S(x) \cdot \delta(x) \right| \right\}$$

$$= \sqrt{N} \cdot \max_S \left\{ \left| \sum_x f_S(x) \cdot \delta(x) \right| \right\}$$

$$= \sqrt{N} \cdot \mathbf{N}_\infty^F(\delta)$$

We now turn to the actual proof of the XOR Lemma.

1.3   Proof of the XOR-Lemma

The XOR-Lemma follows from the following technical lemma.

Technical Lemma: *For every two orthogonal bases A and B and every vector v, it holds that*

$$\mathbf{N}_1^A(v) \leq N \cdot \mathbf{N}_\infty^B(v). \tag{3}$$

This technical lemma has a three line proof:

For every orthogonal basis A,

$$\mathbf{N}_1^A(v) \leq \sqrt{N} \cdot \mathbf{N}_2^A(v). \tag{4}$$

For every pair of orthonormal bases A and B,

$$\mathbf{N}_2^A(v) = \mathbf{N}_2^B(v). \tag{5}$$

For every orthogonal basis B,

$$\mathbf{N}_2^B(v) \leq \sqrt{N} \cdot \mathbf{N}_\infty^B(v) \tag{6}$$

Indeed, the Technical Lemma (i.e., Eq. (3)) is obtained by combining Eq. (4)–(6). Next, using this Technical Lemma, we get:

XOR-Lemma (revised): $\mathrm{stat}(\pi) \leq \frac{1}{2} \cdot \sqrt{N} \cdot \mathrm{maxbias}(\pi)$.

Proof: By the above

$$\mathrm{stat}(\pi) = \frac{1}{2} \cdot \mathbf{N}_1^K(\pi - \mu) \leq \frac{1}{2} \cdot N \cdot \mathbf{N}_\infty^F(\pi - \mu) = \frac{1}{2} \cdot \sqrt{N} \cdot \mathrm{maxbias}(\pi).$$

∎

1.4   Discussion

The inferior bound, $\mathrm{stat}(\pi) \leq N \cdot \mathrm{maxbias}(\pi)$, has been derived by using one of the following two bounds instead of our Technical Lemma:

1. $\mathbf{N}_1^A(v) \leq \sqrt{N}\mathbf{N}_1^B(v) \leq \sqrt{N} \cdot N\mathbf{N}_\infty^B(v)$.
   The first inequality is proved similarly to the proof of our Technical Lemma (i.e., using $\mathbf{N}_2^B(v) \leq \mathbf{N}_1^B(v)$ instead of Eq. (6)). The second inequality is trivial. Each of the two inequalities is tight, but their combination is wasteful.
2. $\mathbf{N}_1^A(v) \leq N \cdot \mathbf{N}_\infty^A(v) \leq N \cdot \sqrt{N}\mathbf{N}_\infty^B(v)$.
   The second inequality is proved similarly to the proof of our Technical Lemma (i.e., using $\mathbf{N}_\infty^A(v) \leq \mathbf{N}_2^A(v)$ instead of Eq. (4)). The first inequality is trivial. Again, each of the inequalities is tight, but their combination is wasteful.

1.5   Variants

Using small variations on the foregoing argument, we obtain the following variants of the XOR-Lemma:

1. $\max_{x \in \{0,1\}^n}\{|\pi(x) - \mu(x)|\} \leq \mathrm{maxbias}(\pi)$.
2. $\mathrm{stat}(\pi) \leq \sqrt{\sum_{S \neq \emptyset} \mathrm{bias}_S(\pi)^2}$, where $\mathrm{bias}_S(\pi) = \sum_x b_S(x) \cdot \pi(x)$.

Proof: The first claim follows by using $\mathbf{N}_\infty^A(v) \leq \mathbf{N}_2^A(v)$ (instead of $\mathbf{N}_1^A(v) \leq \sqrt{N} \cdot \mathbf{N}_2^A(v)$), and obtaining $\mathbf{N}_\infty^K(\pi - \mu) \leq \sqrt{N} \cdot \mathbf{N}_\infty^F(\pi - \mu)$. The second claim follows by using $\mathbf{N}_1^A(v) \leq \sqrt{N} \cdot \mathbf{N}_2^B(v)$ and $\mathbf{N}_2^F(\pi - \mu) = \sqrt{\sum_{S \neq \emptyset} \mathrm{bias}_S(\pi)^2}$. In both parts we also use $\mathrm{bias}_\emptyset(\pi - \mu) = 0$.   ■

1.6   Generalization to GF(p), for Any Prime p

The entire treatment can be generalized to distributions over $\mathrm{GF}(p)^n$, for any prime p. In this case, we redefine $N \stackrel{\text{def}}{=} p^n$, and let $\mathrm{stat}(\pi)$ denote the statistical difference between π and the uniform distribution over $\mathrm{GF}(p)^n$ (cf. Eq. (1)). Letting ω denote the pth root of unity, we generalize Eq. (2) to

$$\mathrm{maxbias}(\pi) \stackrel{\text{def}}{=} \max_{\beta \in \mathrm{GF}(p)^n \setminus \{0\}^n} \left\{ \left| \sum_{e \in \mathrm{GF}(p)} \omega^e \, \pi\left(\left\{x : \sum_{i \in [n]} \beta_i x_i \equiv e \pmod{p}\right\}\right) \right| \right\}.$$

The Fourier basis is generalized analogously: The new basic consists of the functions $\{b_\beta : \beta \in \mathrm{GF}(p)^n\}$, where $b_\beta(x) = \omega^{\sum_{i \in [n]} \beta_i x_i}$. The normalized basis, denoted F, consists of the functions $f_\beta = N^{-1/2} \cdot b_\beta$.

Note that, in the case of $p = 2$, these definitions coincides with the definitions presented before. By following exactly the same manipulations as in the case of $p = 2$, we obtain the following generalization.

The XOR-Lemma, generalized to $\mathrm{GF}(p)$: *Let π be an arbitrary distribution over $\mathrm{GF}(p)^n$, and let μ denote the uniform distribution over $\mathrm{GF}(p)^n$. Then*

1. $\mathrm{stat}(\pi) \leq \frac{1}{2} \cdot \sqrt{N} \cdot \mathrm{maxbias}(\pi)$.
2. $\max_{x \in \{0,1\}^n}\{|\pi(x) - \mu(x)|\} \leq \mathrm{maxbias}(\pi)$.
3. $\mathrm{stat}(\pi) \leq \frac{1}{2} \cdot \sqrt{\sum_{beta \neq 0^n} \mathrm{bias}_\beta(\pi)^2}$, *where* $\mathrm{bias}_\beta(\pi) = \sum_x b_\beta(x) \cdot \pi(x)$.

2   The Computational XOR-Lemma

We provide an exposition of the computational XOR-Lemma. By computational XOR-Lemma we refer to the assertion that a distribution on "short" strings is pseudorandom if and only if the xor of any of its bits is unpredictable. This Lemma was first proved by Umesh and Vijay Vazirani. The proof we present here is taken from the paper of Goldreich and Levin. We demonstrate the applicability of the computational XOR-Lemma by using it to construct pseudorandom generators with linear expansion factor that are "secure" against small (yet linear) bounded space machines.

2.1   Introduction

This section is concerned the relation between *two types of computationally restricted tests of randomness*. To be more precise, we are concerned with the pseudorandomness of a random variable Y given some partial information represented by an related random variable X. For sake of simplicity we write $X = f(R)$ and $Y = g(R)$ where f and g are fixed functions and R is a random variable uniformly distributed on strings of some length. Throughout this section, *we assume that f and g are polynomial-time computable*.

Tests of the first type are algorithms that, on input a pair (x, y), output a single bit. We consider the probability that the test outputs 1 given that $x = f(r)$ and $y = g(r)$ where r is selected uniformly and compare it to the probability that the test outputs 1 given that $x = f(r)$ as before and y is selected (independently and) uniformly among the strings of length $|g(r)|$. We call the absolute value of the difference between these two probabilities the distinguishing gap of the test.

Tests of the second type are algorithms that, on input a string $f(r)$, output a single bit. The output is supposed to be the inner-product (mod 2) of the string $g(r)$ with some fixed string β (which is not all-zero). We consider the probability that the algorithm outputs the correct value given that r is selected uniformly. We call the absolute value of the difference between the success probability and the failure probability, the advantage of the algorithm. Note that the inner-product (mod 2) of $g(r)$ and β equals the exclusive-or of the bits in $g(r)$ that are located in positions corresponding to the 1 bits of β. Hence, tests of the second type try to predict the xor of bits in $g(r)$ that are in specified bit locations.

Vazirani and Vazirani [22] proved that if the tests are restricted to run in probabilistic polynomial-time and the length of $g(r)$ is logarithmic in the length of $f(r)$, then the two types of tests are equivalent in the following sense: There exists a test of the first type with a non-negligible distinguishing gap if and only if there exists a test of the second type with a non-negligible advantage[2]. A different proof has appeared in Goldreich and Levin [10]. The interesting direction is, of course, the assertion that if there exists a test of the first type with a non-negligible distinguishing gap, then there exists a test of the second type with a non-negligible advantage[3]. This assertion is hereafter referred to as the computational xor-lemma.

The purpose of this section is to present a clear proof of the computational xor-lemma and to point out its applicability to other resource bounded machines. Our presentation follows the proof presented in [10], where all obvious details are omitted. Hence, the only advantage of our presentation is in its redundancy (w.r.t [10]).

---

[2] A function $\mu : \mathbf{N} \to \mathbf{R}$ is non-negligible if there exists a polynomial p such that for all sufficiently large n we have $\mu(n) > 1/p(n)$.

[3] The opposite direction follows by noting that a test of a second type can be easily converted into a test of the first type: Just run the predicting algorithm and compare its outcome with the actual xor of the corresponding bits.

2.2   Proving the Computational XOR-Lemma

The proof proceeds via the counterpositive. That is, we show how to transform any test that distinguishes pairs $(f(r), g(r))$ from pairs $(f(r), y)$, where r and y are independently and uniformly distributed (among strings of adequate length), into a predictor of the xor of some bits of $g(r)$ from $f(r)$ such that the complexity of the predictor and its advantage are related to the complexity and distinshiong gap of the original tester. Actually, the construction yields a predictor that has a related advantage w.r.t a *random* subset of bits positions (rather than w.r.t *some* subset). The construction of the predictor and its analysis are captured by the following Technical Lemma.

In the following technical lemma, we present a particular algorithm, denoted G, that (given $f(r)$) tries to predict a specified xor of the bits of $g(r)$. The predictor G uses as subroutine a test, T, that (on input $f(r)$ and y) distinguishes a random y from $y = g(r)$. In particular, on input x and a subset S, the predictor selects y at random, runs the test T on inputs x and y, and output $\bigoplus_{i \in S} y_i$ if $T(x, y) = 1$ and the complement bit otherwise. The following lemma, lower bounds the advantage of the predictor G in terms of the distinguishing gap of the test T.

Technical Lemma (the core of the Computational XOR-Lemma): *Let f and g be arbitrary functions each mapping strings of the same length to strings of the same length. Let T be an algorithm* (of the first type). *Denote*

$$p \stackrel{\text{def}}{=} \Pr[T(f(r), g(r)) = 1] \tag{7}$$

and

$$q \stackrel{\text{def}}{=} \Pr[T(f(r), y) = 1], \tag{8}$$

where the probability is taken over all possible choices of $r \in \{0,1\}^m$ and $y \in \{0,1\}^{|g(r)|}$ with uniform probability distribution. Let G be an algorithm that, on input β and x, selects y uniformly in $\{0,1\}^{|\beta|}$, and outputs $T(x,y) \oplus 1 \oplus (y,\beta)_2$, where $(y,\beta)_2$ is the inner product modulo 2 of y and β. Then,

$$\Pr[G(\beta, f(r)) = (g(r), \beta)_2] = \frac{1}{2} + \frac{p-q}{2^{|\beta|} - 1}, \tag{9}$$

where the probability is taken over all possible choices of $r \in \{0,1\}^m$ and $\beta \in \{0,1\}^{|g(r)|} \setminus \{0\}^{|g(r)|}$ with uniform probability distribution.

A full proof of the Technical Lemma is presented in Section 2.3. Before turning to that proof, we show that this lemma implies the Computational XOR-Lemma. This demonstration is immediate by the following two comments.

1. Algorithm G has almost the same complexities as T, with the exception that G must toss few more coins (to select β). Hence, G is randomized even in case T is deterministic.

2. Clearly, there exists a non-zero string β for which $\Pr[G(\beta, f(r)) = (g(r), \beta)_2] \geq \frac{1}{2} + \frac{p-q}{2^{|\beta|}-1}$, where the probability is taken over all possible choices of $r \in \{0, 1\}^m$ with uniform probability distribution. A string β with approximately such a performance can be found by sampling a string β and evaluating the performance of algorithm G with β as its first input. This requires ability to compute the functions f and g on many randomly selected instances (and collect the statistics). One should verify that this added complexity can be afforded. On the other hand, one should note that finding an appropriate β (i.e. on which G has almost the average advantage) may not be required (see the first remark below).

The following Computational XOR-Lemma follows as an immediate corollary to the Technical Lemma.

Computational XOR-Lemma: *Let $\mathcal{C}$ be a class of randomized (or non-uniform) algorithms, such that $\mathcal{C}$ is closed under sequential application of algorithms and contains an algorithm for computing $|g(r)|$ from $f(r)$. Suppose that every algorithm in the class $\mathcal{C}$, given $f(r)$, can predict the xor of a (given) random subset of the bits of $g(r)$ with (average) success probability at most $\frac{1}{2} + \epsilon$. Then, for every algorithm, T, in the class $\mathcal{C}$ it holds that*

$$|\Pr[T(f(r), g(r)) = 1] - \Pr[T(f(r), y) = 1]| < 2^{|g(r)|} \cdot \epsilon$$

where r is selected uniformly in $\{0, 1\}^m$, the string y is selected uniformly and independently in $\{0, 1\}^{|g(r)|}$.

Remarks. As motivation to the Computational XOR-Lemma, we point out that it has been used in numerous works (e.g., Vazirani and Vazirani [22], Goldreich and Levin [10]). Another application of the Computational XOR-Lemma is presented in Section 2.4. In a typical application, the pseudorandomness of a short string is proved by showing that every xor of its bits is unpredictable (and using the Computational XOR-Lemma to argue that this suffices). Since it is typically the case that one can prove that the xor of a (given) *random* non-empty subset of the bits is unpredictable, the Computational XOR-Lemma can be used directly without finding an appropriate β (as suggested by a previous remark).

In case there are no computational restrictions on the tests, a stronger statement known as the XOR-Lemma can be proved: The statistical difference from uniform does not exceed $\sqrt{2^{|g(r)|}}$ times the maximum bias of a non-empty subset (see Sectioon 1).

2.3   Proof of the Technical Lemma

Our goal here is to evaluate the success probability of algorithm G. In the following analysis we denote $\Pr_x[P(x, y)]$ the probability that $P(x, y)$ holds when x is distributed according to a distribution to be understood from the context, and y is fixed. In the case that the predicate P depends on the test T, the probability will be taken also over the internal coin tosses of T. Hence, the coin tosses of T

are implicit in the notation. In contrast, the additional coin tosses of G, namely the string y, are explicit in the notation. Hence, we rewrite

$$p = \Pr_r[T(f(r), g(r)) = 1]$$
$$q = \Pr_{r,y}[T(f(r), y) = 1]$$

Recall that r is distributed uniformly on $\{0,1\}^m$, whereas y is distributed uniformly on $\{0,1\}^{|g(r)|}$. In the following analysis β is selected uniformly in $B \overset{\text{def}}{=} \{0,1\}^{|g(r)|} - 0^{|g(r)|}$. Our aim is to evaluate $\Pr_{r,\beta,y}[G(\beta, f(r)) = (g(r), \beta)_2]$.

We start by fixing any $r \in \{0,1\}^m$ and evaluating $\Pr_{\beta,y}[G(\beta, f(r)) = (g(r), \beta)_2]$. We define $\equiv_\beta$ (resp., $\not\equiv_\beta$) such that $y \equiv_\beta z$ hold iff $(y, \beta)_2 = (z, \beta)_2$ (resp., $y \not\equiv_\beta z$ iff $(y, \beta)_2 \neq (z, \beta)_2$). We let $n \overset{\text{def}}{=} |g(r)|$.

By the definition of G (i.e., $G(\beta, f(r)) = T(x, y) \oplus 1 \oplus (y, \beta)_2$, where $y \in \{0,1\}^{|\beta|}$ is uniformly selected by G) and elementary manipulations, we get

$$
\begin{aligned}
s_r \overset{\text{def}}{=}\ & \Pr_{\beta,y}[G(\beta, f(r)) = (g(r), \beta)_2] \\
=\ & \sum_{\beta \in B} \frac{1}{|B|} \cdot \Pr_y[G(\beta, f(r)) = (g(r), \beta)_2] \\
=\ & \frac{1}{|B|} \cdot \sum_{\beta \in B} \Pr_y[T(\beta, f(r)) = 1 \oplus (\beta, y)_2 \oplus (g(r), \beta)_2] \\
=\ & \frac{1}{2|B|} \cdot \sum_{\beta \in B} \Pr_y[T(f(r), y) = 1 \mid y \equiv_\beta g(r)] \\
& + \frac{1}{2|B|} \cdot \sum_{\beta \in B} \Pr_y[T(f(r), y) = 0 \mid y \not\equiv_\beta g(r)] \\
=\ & \frac{1}{2} + \frac{1}{2|B|} \cdot \sum_{\beta \in B} \Pr_y[T(f(r), y) = 1 \mid y \equiv_\beta g(r)] \\
& - \frac{1}{2|B|} \cdot \sum_{\beta \in B} \Pr_y[T(f(r), y) = 1 \mid y \not\equiv_\beta g(r)] \\
=\ & \frac{1}{2} + \frac{1}{2|B|} \cdot \frac{1}{2^{n-1}} \cdot \sum_{\beta \in B} \sum_{y \equiv_\beta g(r)} \Pr[T(f(r), y) = 1] \\
& - \frac{1}{2|B|} \cdot \frac{1}{2^{n-1}} \cdot \sum_{\beta \in B} \sum_{y \not\equiv_\beta g(r)} \Pr[T(f(r), y) = 1] \\
=\ & \frac{1}{2} + \frac{1}{2^n \cdot |B|} \cdot \sum_y \sum_{\beta \in B \text{ s.t. } y \equiv_\beta g(r)} \Pr[T(f(r), y) = 1] \\
& - \frac{1}{2^n \cdot |B|} \cdot \sum_y \sum_{\beta \in B \text{ s.t. } y \not\equiv_\beta g(r)} \Pr[T(f(r), y) = 1]
\end{aligned}
$$

Recall that $B = \{0,1\}^n - 0^n$. Now, if $y \neq g(r)$ then the number of $\beta \in B$ for which $y \not\equiv_\beta g(r)$ is 2^{n-1} (and the number of $\beta \in B$ for which $y \equiv_\beta g(r)$ is $2^{n-1}-1$). On the other hand, if $y = g(r)$, then all $\beta \in B$ satisfy $y \equiv_\beta g(r)$. Hence, we get

$$s_r - \frac{1}{2} = \frac{1}{2^n |B|} \cdot \sum_{y \neq g(r)} \left((2^{n-1}-1)\cdot \Pr[T(f(r),y)=1] - 2^{n-1}\cdot \Pr[T(f(r),y)=1]\right)$$

$$+ \frac{1}{2^n |B|} \cdot |B| \cdot \Pr[T(f(r),g(r))=1]$$

$$= -\frac{1}{2^n |B|} \cdot \sum_{y \neq g(r)} \Pr[T(f(r),y)=1] \; + \; \frac{1}{2^n |B|} \cdot |B| \cdot \Pr[T(f(r),g(r))=1]$$

$$= -\frac{1}{|B|} \cdot \sum_{y} \frac{1}{2^n} \cdot \Pr[T(f(r),y)=1]$$

$$+ \frac{1}{2^n |B|} \cdot (|B|+1) \cdot \Pr[T(f(r),g(r))=1]$$

$$= -\frac{1}{|B|} \cdot \Pr_y[T(f(r),y)=1] + \frac{1}{|B|} \cdot \Pr[T(f(r),g(r))=1]$$

Hence, for every r

$$\Pr_{\beta,y}[G(\beta,f(r))=(g(r),\beta)_2] \; = \; \frac{1}{2} + \frac{\Pr[T(f(r),g(r))=1] - \Pr_y[T(f(r),y)=1]}{|B|}$$

and so we have for uniformly chosen r

$$\Pr_{r,\beta,y}[G(\beta,f(r))=(g(r),\beta)_2] \; = \; \frac{1}{2} + \frac{\Pr_r[T(f(r),g(r))=1] - \Pr_{r,y}[T(f(r),y)=1]}{|B|}$$

and the lemma follows.   ∎

2.4 Application to Pseudorandom Generators for Bounded Space

We apply the Computational XOR-Lemma to the construction of pseudorandom generators with linear stretching that withstands tests of linearly bounded space. Namely, on input a random string of length n, the generator outputs a pseudorandom string of length cn withstanding tests of space en (where $e > 0$ is a constant depending on the constant $c > 1$). An alternative construction is immediate from the techniques presented by Nisan in [17].[4] A third alternative construction was suggested by Noam Nisan (private communication) based on the ideas in [3].

The tests (or predictors) that we consider are non-uniform bounded-space machines with one-way access to the input (i.e., the string that they test). Hence, these machines can be represented by finite automata. By an $s(n)$-space bounded

---

[4] Use a constant number of hash functions.

machine we mean a finite automata with $2^{s(n)}$ states that is given an input of length n. For sake of simplicity, we sometimes discuss randomized automata. Clearly, randomness can be eliminated by introducing "more" non-uniformity.

Following is an overview of our construction. We begin by presenting a generator that extends seeds of length n into strings of length cn withstanding tests of space en, for a specific value of $c > 1$ (and $e > 0$). This generator is based on three observations:

1. Given two vectors, their inner-product mod 2 is unpredictable by machines of space significantly smaller than the length of these vectors.
2. With respect to such machines, the exclusive-or of bits resulting from the inner-product mod 2 of one vector and non-cyclic shifts of a second vector is also unpredictable. This holds because a machine predicting this exclusive-or can be transformed into a machine predicting the inner product of two vectors (cf. [10]).
3. Finally, using the computational XOR-Lemma, it follows that the bits resulting from the various inner-products are indistinguishable from random by space bounded machines.

The foregoing steps are detailed in Section 2.4.1. Next, in Section 2.4.2, we use this generator to construct, for every $k > 1$, a generator extending seeds of length n into strings of length $c^k \cdot n$ withstanding tests of space $(e/3)^k \cdot n$.

2.4.1 A construction for a specific expansion constant.
The constants $c_1, \epsilon_1, c_0, \epsilon_0$ in the following construction and analysis will be determined in course of the analysis. In particular, $c_0 = \frac{1}{4}$, $\epsilon_0 = \frac{1}{6}$, $c_1 = 1 + \frac{c_0}{3}$, and $\epsilon_1 = \frac{\epsilon_0}{3}$, will do.

Construction 1: *Using the notation* $p_j(r_1 r_2 \cdots r_{2n}) \overset{\text{def}}{=} r_j r_{j+1} \cdots r_{j+n-1}$ *and* $b(x, s) \overset{\text{def}}{=} \sum_{i=1}^{n} x_i s_i \bmod 2$, *consider the function* $g : \{0,1\}^{3n} \to \{0,1\}^{c_0 n}$ *defined by* $g(x, r) = b(x, p_1(r)) \cdots b(x, p_{c_0 n}(r))$. *Finally, consider the generator*

$$g_1(x, r) = (x, r, g(x, r)). \tag{10}$$

This generator expands seeds of length $3n$ into strings of length $3n + c_0 n = c_1 \cdot 3n$. Clearly, the function g can be computed by an n-space machine. The robustness of the generator against $\epsilon_0 n$-space machines follows from the following three claims.

Claim 1.1: *Let A be an automaton with q states, and x, y be uniformly and independently selected in $\{0, 1\}^n$. Then*

$$\Pr_{x,y}[A(x, y) = b(x, y)] \leq \frac{1}{2} + \sqrt{\frac{2q}{2^n}}.$$

Proof (adapted from [3]): By Lindsey Lemma (see [6, P. 88]), for every $X, Y \subseteq \{0,1\}^n$, it holds that

$$\left| \sum_{x \in X, y \in Y} \frac{b(x,y)}{|X| \cdot |Y|} - \frac{1}{2} \right| \leq \sqrt{\frac{2^n}{|X| \cdot |Y|}} \tag{11}$$

Consider a partition of the set of all possible x's according to the state in which the automaton is after reading x (i.e., the first half of its input), and denote the resulting sets $X_1, X_2, ..., X_q$. Note that for every $x_1, x_2 \in X_j$ and every y, we have $A(x_1, y) = A(x_2, y)$. For each X_i, let Y_i^σ denote the sets of y's for which $A(x,y) = \sigma$ given that $x \in X_i$. It follows that

$$\Delta \overset{\text{def}}{=} \left| \Pr_{x,y}[A(x,y) = b(x,y)] - \frac{1}{2} \right|$$

$$= \sum_{i=1}^{q} \sum_{\sigma \in \{0,1\}} \Pr_{x,y}[x \in X_i \wedge y \in Y_i^\sigma] \cdot \left| \frac{\{(x,y) \in X_i \times Y_i^\sigma : b(x,y) = \sigma\}}{|X_i| \cdot |Y_i^\sigma|} - \frac{1}{2} \right|$$

$$\leq \sum_{i=1}^{q} \sum_{\sigma \in \{0,1\}} \Pr_{x,y}[x \in X_i \wedge y \in Y_i^\sigma] \cdot \sqrt{\frac{2^n}{|X_i| \cdot |Y_i^\sigma|}}$$

$$= 2^{-3n/2} \cdot \sum_{i=1}^{q} \sum_{\sigma \in \{0,1\}} \sqrt{|X_i| \cdot |Y_i^\sigma|}$$

$$\leq 2^{-3n/2} \cdot \sqrt{2q} \cdot 2^n$$

where the first inequality is due to Eq. (11) and the second inequality is due to (a special case of) the Cauchy-Schwartz Inequality.[5] The claim follows. $\square$

Claim 1.2: *Let $S \subseteq \{1, 2, ..., m\}$, where $m < n$. Suppose that automaton A_S has q states and let*

$$p \overset{\text{def}}{=} \Pr_{x,r} \left[A_S(x,r) = \bigoplus_{i \in S} b(x, p_i(r)) \right]$$

where the probability is taken over all random choices of $x \in \{0,1\}^n$ and $r \in \{0,1\}^{2n}$. Then, there exists an automaton A with $q \cdot 2^{2m}$ states satisfying

$$\Pr_{x,y}[A(x,y) = b(x,y)] \geq p$$

where the probability is taken over all random choices of $x, y \in \{0,1\}^n$.

Proof (adapted from [10]): Following is a construction of a randomized automaton A (randomization can be eliminated via non-uniformity). On input x, y, the predictor A produces a random string $r \in \{0,1\}^{2|y|}$ satisfying $y_i =$

---

[5] Specifically, we use $\sum_{j=1}^{m} \sqrt{a_j} \leq \sqrt{m \cdot \sum_{j=1}^{m} a_j}$.

$\sum_{j \in S} r_{i+j-1} \bmod 2$, for every $i \leq n$. This is done by setting the bits of r in increasing order such that r_k is randomly selected if either $k < t \overset{\text{def}}{=} \max(S)$ or $k \geq t+n$, and r_k is set to $y_{k-t+1} - \sum_{j \in S \setminus \{t\}} r_{k-t+j} \bmod 2$ for $k = t, t+1, ..., t+n-1$. Hence, $\bigoplus_{j \in S} p_j(r) = y$, where $\bigoplus_{j \in S} v_j$ denotes the bit-by-bit exclusive or of the vectors v_j (where $j \in S$). The predictor A runs $A_S(x, r)$ and obtains a prediction for $\bigoplus_{j \in S} b(x, p_j(r)) = b(x, \bigoplus_{j \in S} p_j(r)) = b(x, y)$. The predictor uses at most $2m$ more space than G_S (for storing r), and the claim follows. $\square$

Claim 1.3: *For every automaton, T, with q states*

$$|\Pr[T(x, r, g(x, r)) = 1] - \Pr[T(x, r, y) = 1]| < 2^{|g(r)|} \cdot \sqrt{\frac{2q \cdot 2^{2c_0 n}}{2^n}}$$

where (x, r) is selected uniformly in $\{0,1\}^{n+2n}$, the string y is selected uniformly in $\{0,1\}^{|g(x,r)|}$.

Proof: Immediate by combining Claims 1.1 and 1.2 with the Computational XOR-Lemma. $\square$

Setting $c_0 = \frac{1}{4}$ and $\epsilon_1 = \frac{1}{6}$, we conclude that any $\epsilon_1 n$-space bounded machine can distinguish $g_1(x, r)$ (where $xr \in \{0,1\}^{3n}$) from a uniformly chosen string of length $(3 + c_0)n$ with gap at most $2^{-\epsilon_1 n}$. Hence, for constants $c_1 = 1 + \frac{1}{12}$ and $e_1 = \frac{1}{18}$, we have a generator extending strings of length n to strings of length $c_1 n$ so that no $\epsilon_1 n$-space bounded machine can distinguish $g_1(s)$ (where $s = (x, r) \in \{0,1\}^n$) from a uniformly chosen string of length $c_1 n$ with gap greater than $2^{-\epsilon_1 n}$. We say that g_1 has expansion factor c_1 and security constant e_1.

2.4.2 Construction for any expansion constant.

To achieve larger expansion we apply the generator again on small blocks of its output. This idea is taken from [9], but its usage in our context is restricted since in lower level the generator will be applied to shorter strings (and not to strings of the same length as done in [9]). The fact that in lower levels the generator is applied to shorter strings plays a key role in the proof that the resulting generator is indeed pseudorandom with respect to appropriate space-bounded machines.

In the sequel we show how to convert generators with expansion factor c into generators with expansion factor c^2. Larger expansion factors are obtained by repeated application of the construction.

Construction 2: *Let g be a generator with expansion factor c and security constant e. We construct a generator g_2 with expansion factor c^2 and security constant $\frac{e^2}{3}$ as follows: $g_2(s) = g(r_1) \cdots g(r_t)$, where $r_1 \cdots r_t = g(s)$ such that $|r_j| = \frac{e}{2} \cdot |s|$ (for all $1 \leq j \leq t$) and $t = 2c/e$.*

To prove that the generator g_2 has security $\frac{e^2}{3}$ we consider a *hybrid* distribution H that results by selecting at random a string of length cn, partitioning it into t

blocks (each of length $\frac{e}{2}n$), and applying the generator g to each of them. First we show that H is hard to distinguish from random strings of length c^2n. Next, we show that H is hard to distinguish from the strings that g_2 generates on input a random seed of length n.

Claim 2.1 (indistinguishability of H and the uniform distribution): *Suppose that the automaton T has q states. Let $p_H \stackrel{\text{def}}{=} \text{Pr}_{s_1 \cdots s_t}[T(g(s_1) \cdots g(s_t)) = 1]$ and $p_R \stackrel{\text{def}}{=} \text{Pr}_{r_1 \cdots r_t}[T(r_1 \cdots r_t) = 1]$, where the probability is taken over all random choices of $s_1, ..., s_t \in \{0,1\}^{\frac{e}{2}n}$ and $r_1, ..., r_t \in \{0,1\}^{\frac{ce}{2}n}$. Then, there exists an automaton T' with q states satisfying*

$$|\text{Pr}_s[T'(g(s)) = 1] - \text{Pr}_r[T'(r) = 1]| \geq \frac{|p_H - p_R|}{t}$$

where the probability is taken over all random choices of $s \in \{0,1\}^{\frac{e}{2}n}$ and $r \in \{0,1\}^{\frac{ce}{2}n}$. Hence, if $q \leq \frac{e^2}{2}n$, then $|p_R - p_H| < \frac{1}{2} \cdot 2^{-\frac{e^2}{3}n}$.

Proof: For every $0 \leq i \leq t$, define

$$p_i \stackrel{\text{def}}{=} \text{Pr}_{r_1 \cdots r_i s_{i+1} \cdots s_t}[T(r_1 \cdots r_i \, g(s_{i+1}) \cdots g(s_t)) = 1],$$

where the probability is taken over all random choices of $r_1, ..., r_i \in \{0,1\}^{\frac{ce}{2}n}$ and $s_{i+1}, ..., s_t \in \{0,1\}^{\frac{e}{2}n}$. Namely, p_i is the probability that T outputs 1 on input taken from a hybrid distribution consisting of i "random" blocks and $t - i$ "pseudorandom" blocks. Clearly, $p_0 = p_H$ whereas $p_t = p_R$, and there exists $0 \leq i \leq t-1$ such that $|p_i - p_{i+1}| \geq \frac{|p_0 - p_t|}{t}$. The test T' is obtained from T as follows. Fix a sequence $r_1, ..., r_i \in \{0,1\}^{\frac{ce}{2}n}$ and $s_{i+2}, ..., s_t \in \{0,1\}^{\frac{e}{2}n}$ maximizing the distinguishing gap between the ith and $i+1^{\text{st}}$ hybrids. The starting state of test T' is the state to which T arrives on input $r_1, ..., r_i$. The accepting states (i.e. states with output 1) of test T' are the state from which T reaches its accepting state when reading the string $s_{i+2}, ..., s_t$. Clearly, T' has at most q states and distinguishes $r \in \{0,1\}^{\frac{ce}{2}n}$ from $g(s)$ (for $s \in \{0,1\}^{\frac{e}{2}n}$) with gap at least $\frac{|p_H - p_R|}{t}$. Using the security hypothesis for g, it follows that $|p_R - p_H| < t \cdot 2^{-e \cdot \frac{e}{2}n} < \frac{1}{2} 2^{-\frac{e^2}{3}n}$ (for all sufficiently large n). The claim follows. $\square$

Note that the test constructed in the proof of Claim 2.1 examines strings of length $c \cdot \frac{e}{2}n$.

Claim 2.2 (indistinguishability of H and the output of g_2): *Suppose that the automaton T has q states and let $p_G \stackrel{\text{def}}{=} \text{Pr}_s[T(g_2(s)) = 1]$ and $p_H \stackrel{\text{def}}{=} \text{Pr}_{r_1 \cdots r_t}[T(g(r_1) \cdots g(r_t)) = 1]$, where the probability is taken over all random choices of $s \in \{0,1\}^n$ and $r_1, ..., r_t \in \{0,1\}^{\frac{e}{2}n}$. Then, there exists an automaton T' with $q \cdot 2^{\frac{e}{2}n}$ states satisfying $|\text{Pr}_s(T'(g(s)) = 1) - \text{Pr}_r(T'(r) = 1)| \geq p_G - p_H$, where the probability is taken over all random choices of $s \in \{0,1\}^n$ and $r \in \{0,1\}^{cn}$. Hence, if $q \leq \frac{e}{2}n$, then $|p_G - p_H| < \frac{1}{2} 2^{-\frac{e^2}{3}n}$.*

Proof: The test T' is obtained from T as follows. On input $\alpha \in \{0,1\}^{cn}$ (either random or pseudorandom), the test T' breaks α into t blocks, $\alpha_1, ..., \alpha_t$, each of length $\frac{e}{2}n$. Then T' computes $\beta = \beta_1 \cdots \beta_t$ so that $\beta_i = g(\alpha_i)$, and applies T to the string β. (T' accepts α iff T accepts β.) If α is taken from the uniform distribution, then the resulting β is distributed according to H. On the other hand, if α is taken as the output of g on random seed s, then $\beta = g_2(s)$. The test T' distinguishes the above cases with gap $\geq |p_H - p_G|$, and can be implemented using $q \cdot 2^{\frac{e}{2}n}$ states. Using the security hypothesis for g, it follows that $|p_G - p_H| < 2^{-en} < \frac{1}{2}2^{-\frac{e^2}{3}n}$. The claim follows.   □

Note that the test constructed in the proof of Claim 2.2 evaluates g on strings of length $\frac{e}{2}n$. Combining Claims 2.1 and 2.2, we conclude that the generator g_2 has security constant $\frac{e^2}{3}$.

3   A Hard-Core Predicate for All One-Way Functions

A theorem of Goldreich and Levin [10] relates the following two computational tasks (regarding the function f). The first task is inverting a function f; that is, given y, find an x so that $f(x) = y$. The second task is predicting, with non-negligible advatage, the exclusive-or of a subset of the bits of x when only given $f(x)$. More precisely, it has been proved that *if f cannot be efficiently inverted, then given $f(x)$ and r it is infeasible to predict the inner-product mod 2 of x and r better than obvious.*

We present an alternative proof to the original proof as appeared in [10]. The new proof, due to Charlie Rackoff, has two main advantages over the original one: It is simpler to explain and it provides better security (i.e., a more efficient reduction of inverting f to predicting the inner-product). The new proof was inspired by the proof in [1]. (We mention that the original proof provides a better starting point for the generalization presented in [12].)

3.1   Introduction

One-way functions are fundamental to many aspects of theory of computation. Loosely speaking, one-way are those functions that are easy to evaluate but hard to invert. However, many applications such as pseudorandom generators (see [4,23] and [7, Chap. 3]) and secure encryption (see [13] and [8, Chap. 5]) require that the function has a "hard-core" predicate b. This value $b(x)$ should be easy to evaluate on input x, but hard to guess (with a noticeable correlation) when given only the value of $f(x)$. Intuitively, the hard-core predicate "concentrates" the one-wayness of the function in a strong sense. A natural question of practical and theoretical importance is *whether every one-way function has a hard-core predicate.* Prior to [10] only partial answers have been given:

1. Blum and Micali [4] proved that if the discrete exponentiation function is one-way, then it has a hard-core predicate.[6] Analogous results for the RSA

---

[6] This result was generalized to all Abelian groups in [14].

and Rabin functions (i.e. raising to a power and squaring modulo an integer, respectively) were obtained by Alexi, Chor, Goldreich, and Schnorr [1].

2. Yao [23] claimed that any one-way function f can be used to construct another one-way function f^* that has a hard-core predicate. The function f^* partitions its input into many shorter inputs and applies f to each of them in parallel (i.e., $f^*(x_1 \ldots x_{k^3}) = f(x_1) \ldots f(x_{k^3})$, where $|x_i| = k$). This claim was proved in [15] (see also [11]).

The drawback of the first set of results is that they are based on a particular intractability assumption (e.g. the hardness of the discrete logarithm problem). The second result constructs a predicate with security not bounded by a constant power of the security of f.

Goldreich and Levin [10] resolved the foregoing question by providing essentially every one-way function with a hard-core predicate (see Theorem 3 below). More specifically, for any time limit s (e.g. $s(n) = n$, or $s(n) = 2^{\sqrt{n}}$), the following tasks are equivalent for probabilistic algorithms running in time $s(|x|)^{O(1)}$:

1. Given $f(x)$ find x for at least a fraction $s(|x|)^{-O(1)}$ of the x's.
2. Given $f(x)$ and p, $|p| = |x|$, guess the Boolean inner-product of x and p with a correlation (i.e. the difference between the success and failure probabilities) of $s(|x|)^{-O(1)}$.

We mention that, for any polynomial time computable f and b, the smallest (within a polynomial) such s exists and is called the **security** of f and b, respectively. The security is a constructible function, can be computed by trying all small guessing algorithms, and is assumed to grow very fast (at least $n^{1/o(1)}$).

3.2   Definitions

Loosely speaking, a *polynomial-time* function f is called **one-way** if any efficient algorithm can invert it only with negligible success probability. A *polynomial-time* predicate b is called a **hard-core of a function** f if all efficient algorithm, given $f(x)$, can guess $b(x)$ only with success probability that is negligibly better than half.

To simplify our exposition, we associate efficiency with polynomial-time and negligible functions as such decreasing smaller than $1/\text{poly}(n)$. By U_n we denote a random variable uniformly distributed over $\{0, 1\}^n$. For simplicity we consider only *length preserving* functions (i.e., $|f(x)| = |x|$ for every x).

Definition 1 (one-way function): *A* **one-way** *function, f, is a polynomial-time computable function such that for every probabilistic polynomial-time algorithm A', every polynomial $p(\cdot)$, and all sufficiently large n it holds that*

$$\Pr[f(A'(Y_n)) = Y_n] < \frac{1}{2} + \frac{1}{p(n)}$$

where $Y_n = f(U_n)$.

Definition 2 (hard-core predicate): *A polynomial-time computable predicate b : $\{0,1\}^* \to \{0,1\}$ is called a* hard-core *of a function f if for every probabilistic polynomial-time algorithm A', every polynomial $p(\cdot)$, and all sufficiently large n it holds that*

$$\Pr\left[A'(f(U_n)) = b(U_n)\right] < \frac{1}{2} + \frac{1}{p(n)}$$

3.3 The Main Result and Its Proof

The following result asserts that every one-way function has a closely related variant that has a hard-core predicate. The closely related variant (i.e., f' below) is obtained by padding the original function (i.e., f), and the security of f' is closely related to the security of f. Furthermore, the same hard-core predicate is used for all these variants (and is thus "universal" for them).

Theorem 3 (inner product mod 2 is an almost universal hard-core): *Let f be an arbitrary one-way function, and let g be defined by $f'(x,r) \stackrel{\text{def}}{=} (f(x),r)$, where $|x| = |r|$. Let $b(x,r)$ denote the inner-product mod 2 of the binary vectors x and r. Then, the predicate b is a hard-core of the function g.*

In other words, the theorem states that if f is one-way, then it is infeasible to guess the exclusive-or of a random subset of the bits of x when given $f(x)$ and the subset itself. We point out that f' maintains properties of f such as being length-preserving and being one-to-one. Furthermore, an analogous statement holds for collections of one-way functions with/without trapdoor etc. [7, Sec. 2.4].

As stated in Section 3.1, the proof of the foregoing Theorem 3 establishes a tight relation between the security of the one-way functions and the security of the corresponding hard-core predicate. This fact is one major advantage of Theorem 3 over Yao's aforementioned construction [23].

3.3.1 The proof's Basic strategy

The proof uses a "reducibility argument" (see [7, Sec. 2.3.3]). Specifically, we assume (for contradiction) the existence of an efficient algorithm predicting the inner-product with advantage that is not negligible, and derive an algorithm that inverts f with related (i.e. not negligible) success probability. This contradicts the hypothesis that f is a one-way function. Thus, we show that inverting the function f is reduced to predicting $b(x,r)$ from $(f(x),r)$.

Let G be a (probabilistic polynomial-time) algorithm that on input $f(x)$ and r tries to predict the inner-product (mod 2) of x and r. Denote by $\varepsilon_G(n)$ the (overall) advantage of algorithm G in predicting $b(x,r)$ from $f(x)$ and r, where x and r are uniformly chosen in $\{0,1\}^n$. That is,

$$\varepsilon_G(n) \stackrel{\text{def}}{=} \Pr\left[G(f(X_n), R_n) = b(X_n, R_n)\right] - \frac{1}{2}, \tag{12}$$

where here and in the sequel X_n and R_n denote two independent random variables, each uniformly distributed over $\{0,1\}^n$. Assuming, towards the contradiction, that b is not a hard-core of f' means that exists an efficient algorithm G, a polynomial $p(\cdot)$ and an infinite set N so that for every $n \in N$ it holds that $\varepsilon_G(n) > \frac{1}{p(n)}$. We restrict our attention to this algorithm G and to n's in this set N. In the sequel we shorthand ε_G by ε.

Our first observation is that, on at least an $\frac{\varepsilon(n)}{2}$ fraction of the x's of length n, algorithm G has an $\frac{\varepsilon(n)}{2}$ advantage in predicting $b(x, R_n)$ from $f(x)$ and R_n. Namely,

Claim 3.1: *There exists a set $S_n \subseteq \{0,1\}^n$ of cardinality at least $\frac{\varepsilon(n)}{2} \cdot 2^n$ such that for every $x \in S_n$, it holds that*

$$s(x) \overset{\text{def}}{=} \Pr[G(f(x), R_n) = b(x, R_n)] \geq \frac{1}{2} + \frac{\varepsilon(n)}{2}, \tag{13}$$

where here the probability is taken over all possible values of R_n and all internal coin tosses of algorithm G, whereas x is fixed.

Proof: The observation follows by an averaging argument. Namely, write $\text{Exp}[s(X_n)] = \frac{1}{2} + \varepsilon(n)$, and apply Markov Inequality.   □

In the sequel we restrict our attention to x's in S_n. We will show an efficient algorithm that on every input y, with $y = f(x)$ and $x \in S_n$, finds x with very high probability. Contradiction to the one-wayness of f will follow by noting that $\Pr[U_n \in S_n] \geq \frac{\varepsilon(n)}{2}$.

Recall that $b(x, r) = \sum_{i=1}^{n} x_i r_i \bmod 2$, where x_i (resp., r_i) dentoes the ith bit of x (resp., r). We highlight the fact that $b(x, r) \oplus b(x, s) = b(x, r \oplus s)$, which follows by $\sum_{i=1}^{n} x_i r_i + \sum_{i=1}^{n} x_i s_i \equiv \sum_{i=1}^{n} x_i (r_i \oplus + s_i) \pmod{2}$.

3.3.2 A Motivating Discussion

Consider a fixed $x \in S_n$. By definition $s(x) \geq \frac{1}{2} + \frac{\varepsilon(n)}{2} > \frac{1}{2} + \frac{1}{2p(n)}$. Suppose, for a moment, that $s(x) > \frac{3}{4} + \frac{1}{2p(n)}$. In this case (i.e., of $s(x) > \frac{3}{4} + \frac{1}{\text{poly}(|x|)}$) retrieving x from $f(x)$ is quite easy. To retrieve the ith bit of x, denoted x_i, we randomly select $r \in \{0,1\}^n$, and compute $G(f(x), r)$ and $G(f(x), r \oplus e^i)$, where e^i is an n-dimensional binary vector with 1 in the ith component and 0 in all the others, and $v \oplus u$ denotes the addition mod 2 of the binary vectors v and u. Clearly, if both $G(f(x), r) = b(x, r)$ and $G(f(x), r \oplus e^i) = b(x, r \oplus e^i)$, then

$$G(f(x), r) \oplus G(f(x), r \oplus e^i) = b(x, r) \oplus b(x, r \oplus e^i)$$
$$= b(x, e^i)$$
$$= x_i$$

(since $b(x, r) \oplus b(x, s) = b(x, r \oplus s)$). The probability that both equalities hold (i.e., both $G(f(x), r) = b(x, r)$ and $G(f(x), r \oplus e^i) = b(x, r \oplus e^i)$) is at least $1 - 2 \cdot (\frac{1}{4} - \frac{1}{\text{poly}(|x|)}) = \frac{1}{2} - \frac{1}{\text{poly}(|x|)}$. Hence, repeating the above procedure sufficiently many times and ruling by majority, we retrieve x_i with very high probability. Similarly, we can retrieve all the bits of x, and hence invert f on $f(x)$. However, the entire analysis was conducted under (the unjustifiable) assumption that $s(x) > \frac{3}{4} + \frac{1}{2p(|x|)}$, whereas we only know that $s(x) > \frac{1}{2} + \frac{1}{2p(|x|)}$.

The problem with the above procedure is that it doubles the original error probability of algorithm G on inputs of form $(x, \cdot)$. Under the unrealistic assumption, that the G's error on such inputs is significantly smaller than $\frac{1}{4}$, the "error-doubling" phenomenon raises no problems. However, in general (and even in the special case where G's error is exactly $\frac{1}{4}$) the above procedure is unlikely to invert f. Note that the error probability of G can not be decreased by repeating G several times (e.g., G may always answer correctly on three quarters of the inputs, and always err on the remaining quarter). What is required is an *alternative way of using* the algorithm G, a way that does not double the original error probability of G.

The key idea is to generate the r's in a way that requires applying algorithm G only once per each r (and x_i), instead of twice. The good news is that the error probability is no longer doubled, since we only need to use G to get an "estimate" of $b(x, r \oplus e^i)$. The bad news is that we still need to know $b(x, r)$, and it is not clear how we can know $b(x, r)$ without applying G. The answer is that we can guess $b(x, r)$ by ourselves. This is fine if we only need to guess $b(x, r)$ for one r (or logarithmically in $|x|$ many r's), but the problem is that we need to know (and hence guess) $b(x, r)$ for polynomially many r's. An obvious way of guessing these $b(x, r)$'s yields an exponentially vanishing success probability. The solution is to generate these polynomially many r's so that, on one hand they are "sufficiently random" whereas on the other hand we can guess all the $b(x, r)$'s with non-negligible success probability. Specifically, generating the r's in a *particular pairwise independent* manner will satisfy both (seemingly contradictory) requirements. We stress that in case we are successful (in our guesses for the $b(x, r)$'s), we can retrieve x with high probability. Hence, we retrieve x with non-negligible probability.

A word about the way in which the pairwise independent r's are generated (and the corresponding $b(x, r)$'s are guessed) is indeed in place. To generate $m = \text{poly}(n)$ many r's, we uniformly (and independently) select $l \stackrel{\text{def}}{=} \log_2(m + 1)$ strings in $\{0, 1\}^n$. Let us denote these strings by $s^1, ..., s^l$. We then guess $b(x, s^1)$ through $b(x, s^l)$. Let use denote these guesses, which are uniformly (and independently) chosen in $\{0, 1\}$, by σ^1 through σ^l. Hence, the probability that all our guesses for the $b(x, s^i)$'s are correct is $2^{-l} = \frac{1}{\text{poly}(n)}$. The different r's correspond to the different non-empty subsets of $\{1, 2, ..., l\}$; that is, for every (non-empty) $J \subseteq \{1, 2, ..., l\}$, we set $r^J \stackrel{\text{def}}{=} \bigoplus_{j \in J} s^j$. The reader can easily verify that the r^J's are pairwise independent and each is uniformly distributed in $\{0, 1\}^n$ (see details below). The key observation is that

$$b(x, r^J) = b\left(x, \bigoplus_{j \in J} s^j\right) = \bigoplus_{j \in J} b(x, s^j). \tag{14}$$

Hence, our guess for the $b(x, r^J)$'s is $\bigoplus_{j \in J} \sigma^j$, and with non-negligible probability all our guesses are correct.

3.3.3   Back to the Formal Argument

Following is a formal description of the inverting algorithm, denoted A. We assume, for simplicity that f is length preserving (yet this assumption is not essential). On input y (supposedly in the range of f), algorithm A sets $n \overset{\text{def}}{=} |y|$, and $l \overset{\text{def}}{=} \lceil \log_2(2n \cdot p(n)^2 + 1) \rceil$, where $p(\cdot)$ is the polynomial guaranteed above (i.e., $\epsilon(n) > \frac{1}{p(n)}$ for the infinitely many n's in N). Algorithm A uniformly and independently select $s^1, ..., s^l \in \{0,1\}^n$, and $\sigma^1, ..., \sigma^l \in \{0,1\}$. It then computes, for every non-empty set $J \subseteq \{1, 2, ..., l\}$, a string $r^J \leftarrow \bigoplus_{j \in J} s^j$ and a bit $\rho^J \leftarrow \bigoplus_{j \in J} \sigma^j$. Next, for every $i \in \{1, ..., n\}$ and every *non-empty* $J \subseteq \{1, .., l\}$, algorithm A computes $z_i^J \leftarrow \rho^J \oplus G(y, r^J \oplus e^i)$. Finally, algorithm A sets z_i to be the majority of the z_i^J values, and outputs $z = z_1 \cdots z_n$.

(Remark: In an alternative implementation of the foregoing ideas, the inverting algorithm, denoted A', tries all possible values for $\sigma^1, ..., \sigma^l$, and outputs only one of resulting strings z, with an obvious preference to a string z satisfying $f(z) = y$.)

Following is a detailed analysis of the success probability of algorithm A on inputs of the form $f(x)$, for $x \in S_n$, where $n \in N$. We start by showing that if the σ^j's are correct, then, with constant probability, $z_i = x_i$ for all $i \in \{1, ..., n\}$. This is proved by lower bounding the probability that the majority of the z_i^J's equals x_i.

Claim 3.2: *For every $x \in S_n$ and every $i \in \{1, ..., n\}$, it holds that*

$$\Pr\left[|\{J : b(x, r^J) \oplus G(f(x), r^J \oplus e^i) = x_i\}| > \frac{1}{2} \cdot (2^l - 1)\right] > 1 - \frac{1}{2n}$$

where $r^J \overset{\text{def}}{=} \bigoplus_{j \in J} s^j$ and the s^j's are independently and uniformly chosen in $\{0,1\}^n$.

Proof: For every J, define a 0-1 random variable ζ^J, so that ζ^J equals 1 if and only if $b(x, r^J) \oplus G(f(x), r^J \oplus e^i) = x_i$. The reader can easily verify that each r^J is uniformly distributed in $\{0,1\}^n$. It follows that each ζ^J equals 1 with probability $s(x)$, which by $x \in S_n$, is at least $\frac{1}{2} + \frac{1}{2p(n)}$. We show that the ζ^J's are pairwise independent by showing that the r^J's are pairwise independent. For every $J \neq K$ we have, without loss of generality, $j \in J$ and $k \in K \setminus J$. Hence, for every $\alpha, \beta \in \{0,1\}^n$, we have

$$\Pr\left[r^K = \beta \mid r^J = \alpha\right] = \Pr\left[s^k = \beta \mid s^j = \alpha\right]$$
$$= \Pr\left[s^k = \beta\right]$$
$$= \Pr\left[r^K = \beta\right]$$

and pairwise independence of the r^J's follows. Let $m \overset{\text{def}}{=} 2^l - 1$. Using Chebyshev's Inequality, we get

$$\Pr\left[\sum_J \zeta^J \leq \frac{1}{2} \cdot m\right] \leq \Pr\left[\left|\sum_J \zeta^J - \left(\frac{1}{2} + \frac{1}{2p(n)}\right) \cdot m\right| \geq \frac{1}{2p(n)} \cdot m\right]$$

$$< \frac{\text{Var}(\zeta^{\{1\}})}{(\frac{1}{2p(n)})^2 \cdot (2n \cdot p(n)^2)}$$

$$< \frac{\frac{1}{4}}{(\frac{1}{2p(n)})^2 \cdot 2n \cdot p(n)^2}$$

$$= \frac{1}{2n}$$

The claim follows.   □

Recall that if $\sigma^j = b(x, s^j)$, for all j's, then $\rho^J = b(x, r^J)$ for all non-empty J's. In this case z output by algorithm A equals x, with probability at least one half. However, the first event happens with probability $2^{-l} = \frac{1}{2n \cdot p(n)^2}$ independently of the events analyzed in Claim 3.2. Hence, in case $x \in S_n$, algorithm A inverts f on $f(x)$ with probability at least $\frac{1}{4p(|x|)}$ (whereas the modified algorithm, A', succeeds with probability at least $\frac{1}{2}$). Recalling that $|S_n| > \frac{1}{2p(n)} \cdot 2^n$, we conclude that, for every $n \in N$, algorithm A inverts f on $f(U_n)$ with probability at least $\frac{1}{8p(n)^2}$. Noting that A is polynomial-time (i.e., it merely invokes G for $2n \cdot p(n)^2 = \text{poly}(n)$ times in addition to making a polynomial amount of other computations), a contradiction to our hypothesis that f is one-way follows. The theorem follows.   ■

3.3.4   Improving the Efficiency of the Inverting Algorithm

In continuation to the proof of Theorem 3, we present guidelines for a more efficient inverting algorithm. In the sequel it will be more convenient to use the arithmetic of the reals instead of that of the Booleans. Hence, we denote $b'(x, r) = (-1)^{b(r,x)}$ and $G'(y, r) = (-1)^{G(y,r)}$.

1. Prove that, for every x, it holds that $\text{Exp}[b'(x, r) \cdot G'(f(x), r + e^i)] = s'(x) \cdot (-1)^{x_i}$, where $s'(x) \overset{\text{def}}{=} 2 \cdot (s(x) - \frac{1}{2})$.
2. Let v be an l-dimensional Boolean vector, and let R be a uniformly chosen l-by-n Boolean matrix. Prove that for every $v \neq u \in \{0,1\}^l \setminus \{0\}^l$ it holds that vR and uR are pairwise independent and uniformly distributed in $\{0,1\}^n$.
3. Prove that $b'(x, vR) = b'(xR^\top, v)$, for every $x \in \{0,1\}^n$ and $v \in \{0,1\}^l$.
4. Prove that, for every $x \in S_n$, with probability at least $\frac{1}{2}$ (over the choices of R as in Item 2), there exists $\sigma \in \{0,1\}^l$ such that for every $1 \leq i \leq n$ the sign of $\sum_{v \in \{0,1\}^l} b'(\sigma, v) G'(f(x), vR + e^i)$ equals the sign of $(-1)^{x_i}$. (Hint: $\sigma \overset{\text{def}}{=} xR^\top$.)
5. Let B be an 2^l-by-2^l matrix with the (σ, v)-entry being $b'(\sigma, v)$, and let $\bar{g}^i$ be an 2^l-dimensional vector with the vth entry equal $G'(f(x), vR + e^i)$. Consider an inverting algorithm that computes $\bar{z}_i \leftarrow B\bar{g}^i$, for all i's, and forms a matrix Z in which the columns are the $\bar{z}_i$'s. That is, the $(\sigma, i)^{\text{th}}$ entry in Z is $\sum_v b'(\sigma, v) \cdot G'(f(x), vR + e^i)$. The algorithm outputs a row of X such that applying f to it yields $f(x)$, where X is Boolean matrix such that its $(\sigma, i)^{\text{th}}$ entry is 1 iff the $(\sigma, i)^{\text{th}}$ entry in Z is negative.

(a) Evaluate the success probability of this inverting algorithm.

(b) Using the special structure of matrix B, show that the product Bg^i can be computed in time $l \cdot 2^l$.

Hint: B is the Sylvester matrix, which can be written recursively as

$$S_k = \begin{pmatrix} S_{k-1} & S_{k-1} \\ S_{k-1} & \overline{S_{k-1}} \end{pmatrix}$$

where $S_0 = +1$ and $\overline{M}$ means flipping the $+1$ entries of M to -1 and vice versa.

3.4 Hard-Core Functions

We have just seen that every one-way function can be easily modified to have a hard-core predicate. In other words, the result establishes one bit of information about the preimage that is hard to approximate from the value of the function. A stronger result may say that several bits of information about the preimage are hard to approximate. For example, we may want to say that a specific pair of bits is hard to approximate, in the sense that it is infeasible to guess this pair with probability significantly larger than $\frac{1}{4}$. In general, a *polynomial-time* function, h, is called a hard-core of a function f if no efficient algorithm can distinguish $(f(x), h(x))$ from $(f(x), r)$, where r is a random string of length $|h(x)|$. We assume for simplicity that h is length regular (see below).

Definition 4 (hard-core function): *Let $h : \{0,1\}^* \to \{0,1\}^*$ be a polynomial-time computable function, satisfying $|h(x)| = |h(y)|$ for all $|x| = |y|$, and let $l(n) \stackrel{\text{def}}{=} |h(1^n)|$. The function $h : \{0,1\}^* \to \{0,1\}^*$ is called a* hard-core *of a function f if for every probabilistic polynomial-time algorithm D', every polynomial $p(\cdot)$, and all sufficiently large n it holds that*

$$\left| \Pr\left[D'(f(X_n), h(X_n)) = 1 \right] - \Pr\left[D'(f(X_n), R_{l(n)}) = 1 \right] \right| < \frac{1}{p(n)}$$

where X_n and $R_{l(n)}$ are two independent random variables the first uniformly distributed over $\{0,1\}^n$ and the second uniformly distributed over $\{0,1\}^{l(n)}$.

Theorem 5 (almost universal hard-core functions): *Let f be an arbitrary one-way function, and let f_2 be defined by $f_2(x, s) \stackrel{\text{def}}{=} (f(x), s)$, where $|s| = 2|x|$. Let $c > 0$ be a constant, and $l(n) \stackrel{\text{def}}{=} \lceil c \log_2 n \rceil$. Let $b_i(x, s)$ denote the inner-product mod 2 of the binary vectors x and $(s_{i+1}, ..., s_{i+n})$, where $s = (s_1, ..., s_{2n})$. Then the function $h(x, s) \stackrel{\text{def}}{=} b_1(x, s) \cdots b_{l(|x|)}(x, s)$ is a hard-core of the function f_2.*

The proof of the theorem follows by combining a proposition concerning the structure of the specific function h with a general lemma concerning hard-core functions. Loosely speaking, the proposition "reduces" the problem of approximating $b(x, r)$ given $f'(x, r)$ to the problem of approximating the exclusive-or of any non-empty set of the bits of $h(x, s)$ given $f_2(x, s)$, where b and f' are the hard-core and the one-way function presented in Section 3.3. Since we know that the predicate $b(x, r)$ cannot be approximated from $f'(x, r)$, we conclude that no exclusive-or of the bits of $h(x, s)$ can be approximated from $f_2(x, s)$. The general lemma states that, for every "logarithmically shrinking" function h (i.e., h satisfying $|h(x)| = O(\log|x|)$), the function h is a hard-core of a function f if and only if the exclusive-or of any non-empty subset of the bits of h cannot be approximated from the value of f.

Proposition 6 (exclusive-ors of bits of h are hard-core predicates): *Let f, f_2 and b_i's be as above. Let $I(n) \subseteq \{1, 2, ..., l(n)\}$, $n \in \mathbf{N}$, be an arbitrary sequence of non-empty subsets, and let $b_{I(|x|)}(x, s) \overset{\text{def}}{=} \bigoplus_{i \in I(|x|)} b_i(x, s)$. Then, for every probabilistic polynomial-time algorithm A', every polynomial $p(\cdot)$, and all sufficiently large n it holds that*

$$\Pr\left[A'(I(n), f_2(U_{3n})) = b_{I(n)}(U_{3n})\right] < \frac{1}{2} + \frac{1}{p(n)}$$

The proof is analogous to the proof of Claim 1.2 (presented in Section 2.4). Nevertheless, we detail the proof for sake of clarity.

Proof: The proof is by a "reducibility" argument. It is shown that the problem of approximating $b(X_n, R_n)$ given $(f(X_n), R_n)$ is reducible to the problem of approximating $b_{I(n)}(X_n, S_{2n})$ given $(f(X_n), S_{2n})$, where X_n, R_n and S_{2n} are independent random variable and the last is uniformly distributed over $\{0, 1\}^{2n}$. The underlying observation is that, for every $|s| = 2 \cdot |x|$,

$$b_I(x, s) = \bigoplus_{i \in I} b_i(x, s) = \bigoplus_{i \in I} b(x, \text{sub}_i(s)) = b(x, \bigoplus_{i \in I} \text{sub}_i(s))$$

where $\text{sub}_i(s_1, ..., s_{2n}) \overset{\text{def}}{=} (s_{i+1}, ..., s_{i+n})$. Furthermore, the reader can verify[7] that for every non-empty $I \subseteq \{1, ..., n\}$, the random variable $\bigoplus_{i \in I} \text{sub}_i(S_{2n})$ is uniformly distributed over $\{0, 1\}^n$, and that given a string $r \in \{0, 1\}^n$ and such a set I one can efficiently select a string uniformly in the set $\{s : \bigoplus_{i \in I} \text{sub}_i(s) = r\}$.

Now, assume to the contradiction, that there exists an efficient algorithm A', a polynomial $p(\cdot)$, and an infinite sequence of sets (i.e., $I(n)$'s) and n's such that

$$\Pr\left[A'(I(n), f_2(U_{3n})) = b_{I(n)}(U_{3n})\right] \geq \frac{1}{2} + \frac{1}{p(n)}$$

We first observe that for n's satisfying the above inequality we can find in probabilistic polynomial time (in n) a set I satisfying

$$\Pr\left[A'(I, f_2(U_{3n})) = b_I(U_{3n})\right] \geq \frac{1}{2} + \frac{1}{2p(n)}$$

(i.e., by going over all possible I's and experimenting with algorithm A' on each of them). Of course we may be wrong in these experiments, but the error probability can be made exponentially small.

---

[7] Indeed, see the proof of Claim 1.2.

We now present an algorithm for approximating $b(x, r)$, from $y \stackrel{\text{def}}{=} f(x)$ and r. On input y and r, the algorithm first finds a set I as described above (this stage depends only on $|x|$, which equals $|r|$). Once I is found, the algorithm uniformly select a string s so that $\bigoplus_{i \in I} \text{sub}_i(s) = r$, and return $A'(y, s)$. Evaluation of the success probability of this algorithm is left as an exercise. ∎

Lemma 7 (Computational XOR Lemma, revisited): *Let f and h be arbitrary length regular functions, and let $l(n) \stackrel{\text{def}}{=} |h(1^n)|$. Let D be an algorithm. Denote*

$$p \stackrel{\text{def}}{=} \Pr\left[D(f(X_n), h(X_n)) = 1\right] \quad and \quad q \stackrel{\text{def}}{=} \Pr\left[D(f(X_n), R_{l(n)}) = 1\right]$$

where X_n and R_l are as above. Let G be an algorithm that on input y and S (and $l(n)$), selects r uniformly in $\{0,1\}^{l(n)}$, and outputs $D(y, r) \oplus 1 \oplus (\bigoplus_{i \in S} r_i)$, where $r = r_1 \cdots r_l$ and $r_i \in \{0,1\}$. Then,

$$\Pr\left[G(f(X_n), I_l, l(n)) = \bigoplus_{i \in I_l} h_i(X_n)\right] = \frac{1}{2} + \frac{p - q}{2^{l(n)} - 1}$$

where I_l is a randomly chosen non-empty subset of $\{1, ..., l(n)\}$ and $h_i(x)$ denotes the ith bit of $h(x)$.

Proof: See Section 2. ∎

It follows that, for logarithmically shrinking h's, the existence of an efficient algorithm that distinguishes (with a gap that is not negligible in n) the random variables $(f(X_n), h(X_n))$ and $(f(X_n), R_{l(n)})$ implies the existence of an efficient algorithm that approximates the exclusive-or of a random non-empty subset of the bits of $h(X_n)$ from the value of $f(X_n)$ with an advantage that is not negligible.

References

1. Alexi, W., Chor, B., Goldreich, O., Schnorr, C.P.: RSA and Rabin Functions: Certain Parts Are As Hard As the Whole. SIAM Journ. on Computing 1988, 194–209 (1984)
2. Alon, N., Goldreich, O., Håstad, J., Peralta, R.: Simple Constructions of Almost k-wise Independent Random Variables. Journal of Random Structures and Algorithms 3(3), 289–304 (1992)
3. Babai, L., Nisan, N., Szegedy, M.: Multiparty protocols and logspace-hard pseudorandom sequences. In: 21st STOC, pp. 1–11 (1989)
4. Blum, M., Micali, S.: How to Generate Cryptographically Strong Sequences of Pseudo-Random Bits. SIAM Journ. on Computing 1984, 850–864 (1982); Preliminary version in 23rd FOCS 1982
5. Chor, B., Friedmann, J., Goldreich, O., Hastad, J., Rudich, S., Smolansky, R.: The Bit Extraction Problem or t-Resilient Functions. In: Proc. of the 26th IEEE Symp. on Foundation Of Computer Science (FOCS), pp. 396–407 (1985)
6. Erdos, P., Spenser, J.: Probabilistic Methods in Combinatorics. Academic Press, New York (1974)

7. Goldreich, O.: Foundation of Cryptography: Basic Tools. Cambridge University Press, Cambridge (2001)
8. Goldreich, O.: Foundation of Cryptography: Basic Applications. Cambridge University Press, Cambridge (2004)
9. Goldreich, O., Goldwasser, S., Micali, S.: How to Construct Random Functions. Jour. of the ACM 33(4), 792–807 (1986)
10. Goldreich, O., Levin, L.A.: Hard-core Predicates for any One-Way Function. In: 21st STOC, pp. 25–32 (1989)
11. Goldreich, O., Nisan, N., Wigderson, A.: On Yao's XOR-Lemma. In: Goldreich, O., et al.: Studies in Complexity and Cryptography. LNCS, vol. 6650, pp. 273–301. Springer, Heidelberg (2011)
12. Goldreich, O., Rubinfeld, R., Sudan, M.: Learning polynomials with queries: the highly noisy case. SIAM J. Discrete Math. 13(4), 535–570 (2000)
13. Goldwasser, S., Micali, S.: Probabilistic Encryption. JCSS 28(2), 270–299 (1982); Preliminary version in 14th STOC 1982
14. Kaliski Jr., B.S.: Elliptic Curves and Cryptography: A Pseudorandom Bit Generator and Other Tools, Ph.D. Thesis, LCS, MIT (1988)
15. Levin, L.A.: One-Way Function and Pseudorandom Generators. Combinatorica 7(4), 357–363 (1987); A preliminary version in 19th STOC 1985
16. Naor, J., Naor, M.: Small-bias Probability Spaces: Efficient Constructions and Applications. In: 22nd STOC, pp. 213–223 (1990)
17. Nisan, N.: Pseudorandom Generators for Space-Bounded Computations. In: 22nd STOC, pp. 204–212 (1990)
18. Rabin, M.O.: Digitalized Signatures and Public Key Functions as Intractable as Factoring, MIT/LCS/TR-212 (1979)
19. Rivest, R., Shamir, A., Adleman, L.: A Method for Obtaining Digital Signatures and Public Key Cryptosystems. Comm. ACM 21, 120–126 (1978)
20. Vazirani, U.V.: Randomness, Adversaries and Computation, Ph.D. Thesis, EECS, UC Berkeley (1986)
21. Vazirani, U.V.: Efficiency Considerations in Using Semi-random Sources. In: Proc. 19th ACM Symp. on Theory of Computing, pp. 160–168 (1987)
22. Vazirani, U.V., Vazirani, V.V.: Efficient and Secure Pseudo-Random Number Generation. In: Proc. 25th IEEE Symp. on Foundation of Computer Science, pp. 458–463 (1984)
23. Yao, A.C.: Theory and Applications of Trapdoor Functions. In: Proc. of the 23rd IEEE Symp. on Foundation of Computer Science, pp. 80–91 (1982)

On Yao's XOR-Lemma

Oded Goldreich, Noam Nisan, and Avi Wigderson

Abstract. A fundamental lemma of Yao states that computational weak-unpredictability of Boolean predicates is amplified when the results of several independent instances are XOR together. We survey two known proofs of Yao's Lemma and present a third alternative proof. The third proof proceeds by first proving that a function constructed by *concatenating* the values of the original function on several independent instances is much more unpredictable, with respect to specified complexity bounds, than the original function. This statement turns out to be easier to prove than the XOR-Lemma. Using a result of Goldreich and Levin (1989) and some elementary observation, we derive the XOR-Lemma.

Keywords: Yao's XOR Lemma, Direct Product Lemma, One-Way Functions, Hard-Core Predicates, Hard-Core Regions.

An early version of this survey appeared as TR95-050 of *ECCC*, and was revised several times (with the latest revision posted in January 1999). Since the first publication of this survey, Yao's XOR Lemma has been the subject of intensive research. The current revision contains a short review of this research (see Section 7), but the main text (i.e., Sections 1–6) is *not* updated according to these subsequent discoveries. The current version also include a new appendix (Appendix B), which discusses a variant of the XOR Lemma, called the Selective XOR Lemma.

1 Introduction

A fundamental lemma of Yao states that computational weak-unpredictability of Boolean predicates is amplified when the results of several independent instances are XOR together. Indeed, this is analogously to the information theoretic wire-tape channel Theorem (cf., Wyner), but the computational analogue is significanly more complex.

Loosely speaking, by weak-unpredictability we mean that any efficient algorithm will fail to guess the value of the function with probability beyond a stated bound, where the probability is taken over all possible inputs (say, with uniform probability distribution). In particular, the lemma known as Yao's XOR Lemma asserts that if the predicate f is weakly-unpredictable (within some complexity bound), then for sufficiently large t (which depends on the bound) the predicate $F(x_1, ..., x_t) \stackrel{\text{def}}{=} \oplus_{i=1}^{t} f(x_i)$ is almost unpredictable within a related complexity bound (i.e., algorithms of this complexity cannot do much better than flip a coin for the answer).

O. Goldreich et al.: Studies in Complexity and Cryptography, LNCS 6650, pp. 273–301, 2011.
© Springer-Verlag Berlin Heidelberg 2011

Yao stated the XOR Lemma in the context of one-way functions, where the predicate f is the composition of an easy to compute Boolean predicate and the inverse of the one-way function (i.e., $f(x) = b(g^{-1}(x))$, where g is a 1-1 one-way function and b is an easy to compute predicate). Clearly, this is a special case of the setting described above. Yet, the XOR Lemma is sometimes used within the more general setting (under the false assumption that proofs for this setting have appeared in the literature). Furthermore, in contrary to common beliefs, the lemma itself has not appeared in Yao's original paper "Theory and Applications of Trapdoor Functions" [17] (but rather in oral presentations of his work).

A proof of Yao's XOR Lemma has first appeared in Levin's paper [12]. Levin's proof is for the context of one-way functions and is carried through in a uniform model of complexity. The presentation of this proof in [12] is very succinct and does not decouple the basic approach from difficulties arising from the uniform-complexity model. In Section 3, we show that Levin's basic approach suffices for the general case (mentioned above) provided it is stated in terms of non-uniform complexity. The proof also extends to a uniform-complexity setting, provided that some sampling condition (which is satisfied in the context of one-way functions) holds. We do not know whether the XOR Lemma holds in the uniform-complexity model in case this sampling condition is not satisfied.

Recently, Impagliazzo has shown that, in the non-uniform model, any weakly-unpredictable predicate has a "hard-core"[1] on which it is almost unpredictable [7]. Using this result, Impagliazzo has presented an alternative proof for the general case of the XOR-Lemma (within the non-uniform model). We present this proof in Section 4.

A third proof for the general case of the XOR-Lemma is presented in Section 5. This proof proceeds by first proving that a function constructed by *concatenating* the values of the predicate on several independent instances is much more unpredictable, with respect to specified complexity bounds, than the original predicate. Loosely speaking, it is hard to predict the value of the function with probability substantially higher than δ^t, where δ is a bound on the probability of predicting the predicate and t is the number of instances concatenated. Not surprisingly, this statement turns out to be easier to prove than the XOR-Lemma. Using a result of Goldreich and Levin [5] and some elementary observation, we derive the XOR-Lemma.

We remark that Levin's proof yields a stronger quantitative statement of the XOR Lemma than the other two proofs. In fact, the quantitative statement provided by Levin's proof is almost optimal. Both Levin's proof and our proof can be transformed to the uniform-complexity provided some natural sampling condition holds. We do not know how to transform Impagliazzo's proof to the uniform-complexity setting, even under this condition.

---

[1] Here the term 'hard-core' means a subset of the predicate's domain. This meaning is certainly different from the usage of the term 'hard-core' in [5], where it means a strongly-unpredicatable predicate associated with a one-way function.

A different perspective on the concatenating problem considered above is presented in Section 6, where we consider the conditional entropy of the function's value given the result of a computation (rather than the probability that the two agree).

2   Formal Setting

We present a general framework, and view the context of one-way functions as a specail case. The general framework is presented in term of non-uniform complexity, but uniformity conditions can be added in.

2.1   The Basic Setting

The basic framework consists of a Boolean predicate $f : \{0,1\}^* \to \{0,1\}$ and a non-uniform complexity class such as $\mathcal{P}/\text{poly}$. Specifically, we consider all families of polynomial-size circuits and for each family, $\{C_n\}$, we consider the probability that it correctly computes f, where the probability is taken over all n-bit inputs with uniform probability distribution. Alternatively, one may consider the most successful n-bit input circuit among all circuits of a given size. This way we obtain a bound on unpredictability of f with respect to a specific complexity class.

In the sequel, it will be more convenient to redefine f as mapping bit string into $\{\pm 1\}$ and to consider the correlation of a circuit (outputting a value in $\{\pm 1\}$) with the value of the function (i.e., redefine $f(x) \stackrel{\text{def}}{=} (-1)^{f(x)}$).[2] Using this notation allows to replace $\text{Prob}[C(X) = f(X)]$ by $(1 + \text{E}[C(X) \cdot f(X)])/2$, by noting that $\text{E}[C(X) \cdot f(X)] = \text{Prob}[C(X) = f(X)] - \text{Prob}[C(X) \neq f(X)]$.

We also generalize the treatment to arbitrary distributions over the set of n-bit long inputs (rather than uniform ones) and to "probabilistic" predicates (or processes) that on input x return some distribution on $\{\pm 1\}$; that is, for a fixed x, we let $f(x)$ be a random variable distributed over $\{\pm 1\}$ (rather than a fixed value). One motivation for this generalization is that it allows us to treat as a special case 'hard predicates' of one-way functions, when the functions are not necessarily 1-1.

Definition 1 (algorithmic correlation): *Let P be a randomized process/ algorithm that maps bit strings into values in $\{\pm 1\}$ and let $\mathbf{X} \stackrel{\text{def}}{=} \{X_n\}$ be a probability ensemble such that, for each n, the random variable X_n is distributed over $\{0,1\}^n$. The* correlation *of a circuit family $\mathbf{C} = \{C_n\}$ with P over $\mathbf{X}$ is defined as $c:\mathbb{N} \to \mathbb{R}$ such that*

$$c(n) \stackrel{\text{def}}{=} \text{E}[C_n(X_n) \cdot P(X_n)],$$

---

[2] This suggestion, of replacing the standard $\{0,1\}$ by $\{\pm 1\}$ and using correlations rather than probabilities, is due to Levin. It is indeed amazing how this simple change of notation simplifies both the statements and the proofs.

where the expectation is taken over the random variable X_n (and the process P).
We say that a complexity class (i.e., a set of circuit families) *has* correlation *at*
most $c(\cdot)$ with P over $\mathbf{X}$ if, for every circuit family $\mathbf{C}$ in this class, the correlation
of $\mathbf{C}$ with P over $\mathbf{X}$ is bounded by $c(\cdot)$.

The foregoing definition may be used to discuss both uniform and non-uniform
complexity classes. In the next subsection we relate the Definition 1 to the standard treatment of unpredictability within the context of one-way functions.

2.2   The Context of One-Way Functions

For sake of simplicity, we consider only length-preserving functions (i.e., func-
tions $f : \{0,1\}^* \to \{0,1\}^*$ satisfying $|f(x)| = |x|$ for all x). A one-way function
$f : \{0,1\}^* \to \{0,1\}^*$ is a function that is easy to compute but hard to invert.
Namely, there exists a polynomial-time algorithm for computing f, but for any
probabilistic polynomial-time[3] algorithm A, the probability that $A(f(x))$ is a
preimage of $f(x)$ is negligible (i.e., smaller than $1/p(|x|)$ for any positive poly-
nomial p), where the probability is taken uniformly over all $x \in \{0,1\}^n$ and all
possible internal coin tosses of algorithm A.

Let $b : \{0,1\}^* \to \{\pm 1\}$ be an easy to compute predicate and let $\delta : \mathbb{N} \to \mathbb{R}$.
The predicate b is said to be at most δ-correlated to f in polynomial-time if
for any probabilistic polynomial-time algorithm G, the expected correlation of
$G(f(x))$ and $b(x)$, is at most $\delta(n)$ (for all but finitely many n's). (Again, the
probability space is uniform over all $x \in \{0,1\}^n$ and all possible internal coin
tosses of the algorithm.) Thus, although b is easy to evaluate (i.e., the mapping
$x \mapsto b(x)$ is polynomial-time computable), it is hard to predict $b(x)$ from $f(x)$,
for a random x.

Let us relate the latter notion to Definition 1. Suppose, first, that f is 1-1.
Then, saying that b is at most δ-correlated to f in polynomial-time is equivalent to saying that the class of (probabilistic) polynomial-time algorithms has
correlation at most $\delta(\cdot)$ with the predicate $P(x) \stackrel{\text{def}}{=} b(f^{-1}(x))$, over the uniform
distribution. Note that if f is polynomial-time computable and b is at most
$(1 - (1/\text{poly}))$-correlated to f in polynomial-time, then f must be one-way (because otherwise $b(x)$ can be correlated too well by first obtaining $f^{-1}(x)$ and
then evaluating b),

The treatment can be extended to arbitrary one-way functions, which are not
necessarily 1-1. Let f be such a function and b a predicate that is at most δ-
correlated to f (by polynomial-time algorithms). Define the probability ensemble
$\mathbf{X} = \{X_n\}$ by letting $X_n = f(r)$, where r is uniformly selected in $\{0,1\}^n$,
and define the randomized process $P(x)$ by uniformly selecting $r \in f^{-1}(x)$ and
outputting $b(r)$. Now, it follows that the class of (probabilistic) polynomial-time
algorithms has correlation at most $\delta(\cdot)$ with the predicate P over $\mathbf{X}$.

---

[3] Here we adopt the standard definition of one-way function; however, our treatment
applies also to the general definition where inverting is infeasible with respect to a
specified time bound and success probability.

2.3   Getting Random Examples

An important issue regarding the general setting, is whether it is possible to obtain random examples of the distribution $(X_n, P(X_n))$. Indeed, random examples are needed in all known proofs of the XOR Lemma (i.e., they are used in the algorithms deriving a contradiction to the difficulty of correlating the basic predicate).[4] Other than this aspect (i.e., the use of random examples), two of the three proofs can be adapted to the uniform-complexity setting (see Section 2.5).

Note that in the context of one-way functions such random examples can be generated by a probabilistic polynomial-time algorithm. Specifically, although the corresponding P is assumed not to be polynomial-time computable, it is easy to generate randomly pairs $(x, P(x))$ for $x \leftarrow X_n$. (This is done, by uniformly selecting $r \in \{0,1\}^n$, and outputting the pair $(f(r), b(r)) = (f(r), P(f(r)))$.) Thus, we can prove the XOR Lemma in the (uniform-complexity) context of one-way functions.

We also note that the effect of random examples can be easily simulated by non-uniform polynomial-size circuits (i.e., random examples can be hard-wired into the circuit). Thus, we can prove the XOR Lemma in the general non-uniform complexity setting.

2.4   Three (Non-uniform) Forms of the XOR Lemma

Following the description in the introduction (and Yao's expositions), the basic form of the XOR Lemma states that the tractable algorithmic correlation of the XOR-predicate $P^{(t)}(x_1, ..., x_t) \stackrel{\text{def}}{=} \prod_{i=1}^{t} P(x_i)$ decays exponentially with t (upto a negligible fraction). Namely:

Lemma 1 (XOR Lemma – Yao's version): *Let P and $\mathbf{X} = \{X_n\}$ be as in Definition 1. For every function $t : \mathbb{N} \to \mathbb{N}$, define the predicate*

$$P^{(t)}(x_1, ..., x_{t(n)}) \stackrel{\text{def}}{=} \prod_{i=1}^{t(n)} P(x_i),$$

where $x_1, ..., x_{t(n)} \in \{0,1\}^n$, and let $\mathbf{X}^{(t)} \stackrel{\text{def}}{=} \{X_n^{(t)}\}$ be a probability ensemble such that $X_n^{(t)}$ consists of $t(n)$ independent copies of X_n.

(hypothesis) *Let $s : \mathbb{N} \to \mathbb{N}$ be a size function, and $\delta : \mathbb{N} \to [-1, +1]$ be a function that is* bounded-away-from-1 *(i.e., $|\delta(n)| < 1 - \frac{1}{p(n)}$, for some polynomial p and all sufficiently large n's). Suppose that δ is an upper bound on the correlation of families of $s(\cdot)$-size circuits with P over $\mathbf{X}$.*

(conclusion) *Then, there exists a bounded-away-from-1 function $\delta' : \mathbb{N} \to [-1, +1]$ and a polynomial p such that, for every function $t : \mathbb{N} \to \mathbb{N}$ and every function $\epsilon : \mathbb{N} \to [0, 1]$, the function*

$$\delta^{(t)}(n) \stackrel{\text{def}}{=} p(n) \cdot \delta'(n)^{t(n)} + \epsilon(n)$$

---

[4] This assertion refers to what was known at the time this survey was written. As noted in Section 7, the situation regarding this issue has changed recently.

is an upper bound on the correlation of families of $s'(\cdot)$-size circuits with $P^{(t)}$ over $\mathbf{X}^{(t)}$, where

$$s'(t(n) \cdot n) \stackrel{\text{def}}{=} \text{poly}\left(\frac{\epsilon(n)}{n}\right) \cdot s(n) - \text{poly}(n \cdot t(n)).$$

All three proofs presented below establish Lemma 1. The later two proofs do so for various values of δ' and p; that is, in Impagliazzo's proof (see Section 4) $\delta'(n) = \frac{1+\delta(n)}{2} + o(1 - \delta(n))$ and $p(n) = 2$, whereas in our proof (see Section 5) $\delta'(n) = \sqrt[3]{\frac{1+\delta(n)}{2}}$ and $p(n) = o(n)$. Levin's proof (see Section 3) does even better; it establishes the following:

Lemma 2 (XOR Lemma – Levin's version): *Yao's version holds with $\delta' = \delta$ and $p = 1$.*

Lemma 2 still contains some slackness; specifically, the closest one wants to get to the "obvious" bound of $\delta^{(t)}(n) = \delta(n)^{t(n)}$, the more one losses in terms of the complexity bounds (i.e., bounds on circuit size).[5] In particular, if one wishes to have $s'(t(n) \cdot n) = \frac{s(n)}{\text{poly}(n)}$, then one can only get a result for $\epsilon(n) = 1/\text{poly}(n)$ (i.e., get $\delta^{(t)}(n) = \delta(n)^{t(n)} + 1/p(n)$, for any polynomial p). We do not know how to remove this slackness. We even do not know if it can be reduced "a little" as follows.

Lemma 3 (XOR Lemma – dream version – a conjecture): *For some fixed negligible function μ (e.g., $\mu(n) \stackrel{\text{def}}{=} 2^{-n}$ or even $\mu(n) \stackrel{\text{def}}{=} 2^{-(\log_2 n)^2}$), Yao's version holds with $\delta^{(t)}(n) = \delta'(n)^{t(n)} + \mu(n)$, and $s'(t(n) \cdot n) = \frac{s(n)}{\text{poly}(n)}$.*

Steven Rudich has observed that the Dream Version does not hold in a relativized world. Specifically, his argument proceeds as follows. Fix μ as in the Dream Version and set t such that $\delta^{(t)} < 2\mu(n)$. Consider an oracle that for every $(x_1, ..., x_{t(n)}) \in (\{0,1\}^n)^{t(n)}$ and for a $2\mu(n)$ fraction of the r's in $\{0,1\}^n$, answers the query $(x_1, ..., x_{t(n)}, r)$ with $(P(x_1), ..., P(x_t))$, otherwise the oracle answers with a special symbol. These r's may be selected at random (thus constructing a random oracle). The hypothesis of the lemma may hold relative to this oracle, but the conclusion cannot possibly hold. Put differently, one can argue that there is no (polynomial-time) "black-box" reduction of the task of correlating P (by at least δ) to the task of correlating $P^{(t)}$ (by at least μ). The reason being that the polynomial-time machine (effecting this reduction) cannot distinguish a black-box of negligible correlation (i.e., correlation 2μ) from a black-box of zero correlation.

2.5 Uniform Forms of the XOR Lemma

So far, we have stated three forms of the XOR Lemma in terms of non-uniform complexity. Analogous statements in terms of uniform complexity can be made

---

[5] I.e., $\delta^{(t)}(n) = \delta'(n)^{t(n)} + \epsilon(n)$ is achieved for $s'(t(n) \cdot n) = \text{poly}(\epsilon(n)/n) \cdot s(n)$.

as well. These statements relate to the time required to construct the circuits in the hypothesis and those in the conclusion. For example, one may refer to circuit families, $\{C_n\}$, for which, given n, the circuit C_n can be constructed in $\text{poly}(|C_n|)$-time. In addition, all functions referred to in the statement of the lemma (i.e., $s, t : \mathbb{N} \to \mathbb{N}$, $\delta : \mathbb{N} \to [-1, +1]$ and $\epsilon : \mathbb{N} \to [-1, +1]$) need to be computable within corresponding time bounds. Such analogues of the two first versions can be proven, provided that one can construct random examples of the distribution $(X_n, P(X_n))$ within the stated (uniform) complexity bounds (and in particular in polynomial-time). See Section 2.3 as well as comments in the subsequent sections.

3   Levin's Proof

The key ingredient in Levin's proof is the following lemma, which provides an accurate account of the decrease of the computational correlation in the case that two predicates are xor-ed together. It should be stressed that the statement of the lemma is intentionally asymmetric with respect to the two predicates.

Lemma 4 (Isolation Lemma): *Let P_1 and P_2 be two predicates, $l : \mathbb{N} \to \mathbb{N}$ be a length function, and $P(x) \overset{\text{def}}{=} P_1(y) \cdot P_2(z)$ where $x = yz$ and $|y| = l(|x|)$. Let $\mathbf{X} = \{X_n\}$ be a probability ensemble such that the first $l(n)$ bits of X_n are statistically independent of the rest, and let $\mathbf{Y} = \{Y_{l(n)}\}$ (resp., $\mathbf{Z} = \{Z_{n-l(n)}\}$) denote the projection of $\mathbf{X}$ on the first $l(\cdot)$ bits (resp., last $n - l(n)$ bits).*

(hypothesis) *Suppose that $\delta_1(\cdot)$ is an upper bound on the correlation of families of $s_1(\cdot)$-size circuits with P_1 over $\mathbf{Y}$, and that $\delta_2(\cdot)$ is an upper bound on the correlation of families of $s_2(\cdot)$-size circuits with P_2 over $\mathbf{Z}$.*
(conclusion) *Then, for every function $\epsilon : \mathbb{N} \to \mathbb{R}$, the function*

$$\delta(n) \overset{\text{def}}{=} \delta_1(l(n)) \cdot \delta_2(n - l(n)) + \epsilon(n)$$

is an upper bound on the correlation of families of $s(\cdot)$-size circuits with P over $\mathbf{X}$, where

$$s(n) \overset{\text{def}}{=} \min\left\{ \frac{s_1(l(n))}{\text{poly}(n/\epsilon(n))} \ , \ s_2(n - l(n)) - n \right\}$$

The lemma is asymmetric with respect to the dependency of $s(\cdot)$ on the s_i's. The fact that $s(\cdot)$ maybe almost equal to $s_2(\cdot)$ plays a central role in deriving the XOR Lemma from the Isolation Lemma.

3.1   Proof of the Isolation Lemma

Assume, towards the contradiction, that a circuit family $\mathbf{C}$ (of size $s(\cdot)$) has correlation greater than $\delta(\cdot)$ with P over $\mathbf{X}$. Thus, denoting by Y_l (resp., Z_m)

the projection of X_n on the first $l \stackrel{\text{def}}{=} l(n)$ bits (resp., last $m \stackrel{\text{def}}{=} n - l(n)$ bits), we get

$$
\begin{aligned}
\delta(n) &< \mathrm{E}[C_n(X_n) \cdot P(X_n)] \\
&= \mathrm{E}[C_n(Y_l, Z_m) \cdot P_1(Y_l) \cdot P_2(Z_m)] \\
&= \mathrm{E}[P_1(Y_l) \cdot \mathrm{E}[C_n(Y_l, Z_m) \cdot P_2(Z_m)]]
\end{aligned}
$$

where, in the last expression, the outer expectation is over Y_l and the inner one is over Z_m. For every fixed $y \in \{0,1\}^l$, let

$$T(y) \stackrel{\text{def}}{=} \mathrm{E}[C_n(y, Z_m) \cdot P_2(Z_m)]. \tag{1}$$

Then, by the foregoing,

$$\mathrm{E}[T(Y_l) \cdot P_1(Y_l)] > \delta(n). \tag{2}$$

We shall see that Eq. (2) either contradicts the hypothesis concerning P_2 (see Claim 4.1) or contradicts the hypothesis concerning P_1 (by a slightly more involved argument).

Claim 4.1: For all but finitely many n's and every $y \in \{0,1\}^l$

$$|T(y)| \le \delta_2(m).$$

Proof: Otherwise, fixing a y contradicting the claim, we get a circuit $C'_m(z) \stackrel{\text{def}}{=} C_n(y, z)$ of size $s(n) + l < s_2(m)$, having greater correlation with P_2 than that allowed by the lemma's hypothesis. □

By Claim 4.1, the value $T(y)/\delta_2(m)$ lies in the interval $[-1, +1]$; while, on the other hand (by Eq. (2)), it (i.e., $T(\cdot)/\delta_2(m)$) has good correlation with P_1. In the rest of the argument we "transform" the function T into a circuit which contradicts the hypothesis concerning P_1. Suppose for a moment, that one could compute $T(y)$, on input y. Then, one would get an algorithm with output in $[-1, +1]$ that has correlation at least $\delta(n)/\delta_2(m) > \delta_1(l)$ with P_1 over Y_l, which is almost in contradiction to the hypothesis of the lemma.[6] The same holds if one can approximate $T(y)$ "well enough" using circuits of size $s_1(l)$. Indeed, the lemma follows by observing that such an approximation is possible. Namely:

Claim 4.2: For every n, $l = l(n)$, $m = n - l$, $q = \text{poly}(n/\epsilon(n))$ and $y \in \{0,1\}^l$, let

$$\tilde{T}(y) \stackrel{\text{def}}{=} \frac{1}{q} \sum_{i=1}^{q} C_n(y, z_i) \cdot \sigma_i$$

where $(z_1, \sigma_1), ..., (z_q, \sigma_q)$ is a sequence of q independent samples from the distribution $(Z_m, P_2(Z_m))$. Then,

$$\mathrm{Prob}[|T(y) - \tilde{T}(y)| > \epsilon(n)] < 2^{-l(n)}$$

---

[6] See discussion below; the issue is that the output is in the interval $[-1, +1]$ rather than being a binary value in $\{\pm 1\}$.

Proof: Immediate by the definition of $T(y)$ and application of Chernoff bound.□

Claim 4.2 suggests an approximation algorithm (for the function T), where we assume that the algorithm is given as auxiliary input a sequence of samples from the distribution $(Z_m, P_2(Z_m))$. (The algorithm merely computes the average of $C_n(y, z_i) \cdot \sigma_i$ over the sample sequence $(z_1, \sigma_1), ..., (z_q, \sigma_q)$.)

If such a sample sequence can be generated efficiently, by a uniform algorithm (as in the context of one-way functions), then we are done. Otherwise, we use non-uniformity to obtain a fixed sequence that is good for all possible y's. (Such a sequence does exist since with positive probability, a randomly selected sequence, from the above distribution, is good for all $2^{l(n)}$ possible y's.) Thus, there *exists* a circuit of size poly$(n/\epsilon(n)) \cdot s(n)$ that, on input $y \in \{0,1\}^{l(n)}$, outputs a value $(T(y) \pm \epsilon(n))/\delta_2(m)$.

We note that this output is at least $\frac{\delta(n)}{\delta_2(m)} - \frac{\epsilon(n)}{\delta_2(m)} = \delta_1(l)$ correlated with P_1, which almost contradicts the hypothesis of the lemma. The only problem is that the resulting circuit has output in the interval $[-1, +1]$ instead of a binary output in $\{\pm 1\}$. This problem is easily corrected by modifying the circuit so that on output $r \in [-1, +1]$ it outputs $+1$ with probability $(1 + r)/2$ and -1 otherwise. Noting that this modification preserves the correlation of the circuit, we derive a contradiction to the hypothesis concerning P_1. ∎

3.2 Proof of Lemma 2

The stronger version of the XOR Lemma (i.e., Lemma 2) follows by a (careful) successive application of the Isolation Lemma. Loosely speaking, we write $P^{(t)}(x_1, x_2, ..., x_{t(n)}) = P(x_1) \cdot P^{(t-1)}(x_2, ..., x_{t(n)})$, assume that $P^{(t-1)}$ is hard to correlate as claimed, and apply the Isolation Lemma to $P \cdot P^{(t-1)}$. This way, the lower bound on the size of circuits correlating $P^{(t)}$ is related to the lower bound assumed for circuits correlating the original P, since the lower bound derived for $P^{(t-1)}$ is larger and is almost preserved by the Isolation Lemma (losing only an additive term!).

3.3 Remarks Concerning the Uniform Complexity Setting

A uniform-complexity analogue of Lemma 2 can be proven provided that one can construct random examples of the distribution $(X_n, P(X_n))$ within the stated (uniform) complexity bounds. To this end, one should state and prove a uniform-complexity version of the Isolation Lemma, which also assumes that example from both distributions (i.e., $(Y_l, P_1(Y_l))$ and $(Z_m, P_2(Z_m)))$[7] can be generated within the relevant time complexity; certainly, sampleability in probabilistic polynomial-time suffices. Furthermore, in order to derive the XOR Lemma it is important to prove a strong statement regarding the relationship between the time required to construct the circuits referred to in the lemma. Namely:

Lemma 5 (Isolation Lemma – uniform complexity version): *Let* $P_1, P_2, l, P,$ **X, Y** *and* **Z** *be as in Lemma 4.*

---

[7] Actually, it suffices to be able to sample the distributions Y_l and $(Z_m, P_2(Z_m))$.

(hypothesis) *Suppose that $\delta_1(\cdot)$ (resp., δ_2) is an upper bound on the correlation of $t_1(\cdot)$-time-constructible families of $s_1(\cdot)$-size (resp., $t_2(\cdot)$-time-constructible families of $s_2(\cdot)$-size) circuits with P_1 over $\mathbf{Y}$ (resp., P_2 over $\mathbf{Z}$). Furthermore, suppose that one can generate in polynomial-time a random sample from the distribution $(Y_l, Z_m, P_2(Z_m))$.*
(conclusion) *Then, for every function $\epsilon : \mathbb{N} \to \mathbb{R}$, the function*

$$\delta(n) \overset{\text{def}}{=} \delta_1(l(n)) \cdot \delta_2(n - l(n)) + \epsilon(n)$$

is an upper bound on the correlation of $t(\cdot)$-time-constructible families of $s(\cdot)$-size circuits with P over $\mathbf{X}$, where

$$s(n) \overset{\text{def}}{=} \min \left\{ \frac{s_1(l(n))}{\text{poly}(n/\epsilon(n))} \ , \ s_2(n - l(n)) - n \right\}$$

$$t(n) \overset{\text{def}}{=} \min \{ t_1(l(n)) \ , \ t_2(n - l(n)) \} - \text{poly}(n/\epsilon(n)) \cdot s(n).$$

The uniform-complexity version of the Isolation Lemma is proven by adapting the proof of Lemma 4 as follows. First, a weaker version of Claim 4.1 is stated, asserting that (for all but finitely many n's) it holds that

$$\text{Prob}[|T(Y_l)| > \delta_2(m) + \epsilon'(n)] < \epsilon'(n),$$

where $\epsilon'(n) \overset{\text{def}}{=} \epsilon(n)/3$. The new claim is valid, since otherwise, one can find in poly$(n/\epsilon(n))$-time a y violating it; to this end we need to sample Y_l and, for each sample y, approximate the value of $T(y)$ (by using poly$(n/\epsilon(n))$ samples of $(Z_m, P_2(Z_m))$). Once a good y is found, we incorporate it in the construction of C_n, obtaining a circuit that contradicts the hypothesis concerning P_2. (We stress that we have presented an efficient algorithm for constructing a circuit for P_2, given an algorithm that constructs the circuit C_n. Furthermore, the running time of our algorithm is the sum of the time required to construct C_n and the time required for sampling $(Z_m, P_2(Z_m))$ sufficiently many times and for evaluating C_n on sufficiently many instances.)

Clearly, Claim 4.2 remains unchanged (except for the replacing $\epsilon(n)$ by ϵ'). Using the hypothesis that samples from $(Z_m, P_2(Z_m))$ can be efficiently generated, we can construct a circuit for correlating P_1 within time $t(n) + \text{poly}(n/\epsilon(n)) \cdot (n + s(n))$. This circuit is merely an approximater of the function T, which operates by averaging (as in Claim 4.2); this circuit is constructed by first constructing C_n, generating poly$(n/\epsilon(n))$ samples of $(Z_m, P_2(Z_m))$ and incorporating them in corresponding copies of C_n – thus justifying the above time and size bounds. However, unlike in the non-uniform case, we are not guaranteed that $|T(y)|$ is bounded above (by $\delta_2(m) + \epsilon'(n)$) for all y's. Yet, if we modify our circuit to do nothing whenever its estimate violates the bound, we loss at most $\epsilon'(n)$ of the correlation and we can proceed as in the non-uniform case.

Proving a uniform complexity version of Lemma 2: As in the non-uniform case, the (strong form of the) XOR Lemma follows by a (careful) successive application of the Isolation Lemma. Again, we write $P^{(\tau)}(x_1, x_2, ..., x_{\tau(n)}) = P(x_1) \cdot P^{(\tau-1)}(x_1, ..., x_{\tau(n)-1})$, assume that $P^{(\tau-1)}$ is hard to correlate as claimed, and apply the Isolation Lemma to $P \cdot P^{(\tau-1)}$. This way, the lower bounds on circuits correlating $P^{(\tau)}$ is related to the lower bound assumed for circuits correlating the original P and is almost the bound derived for $P^{(\tau-1)}$ (losing only an additive terms!). This almost concludes the proof, except that we have implicitly assumed that we know the value of τ for which the XOR Lemma first fails; this value is needed in order to construct the circuit violating the hypothesis for the original P. In the non-uniform case this value of τ can be incorporated into the circuit, but in the uniform-complexity case we need to find it. This is not a big problem as they are only polynomially many possible values and we can test each of them within the allowed time complexity.

4   Impagliazzo's Proof

The key ingredient in Impagliazzo's proof is the notion of a hard-core region of a weakly-unpredictable predicate and a lemma that asserts that every weakly-unpredictable predicate has a hard-core region of substantial size.

Definition 2 (hard-core region of a predicate): *Let $f : \{0,1\}^* \to \{0,1\}$ be a Boolean predicate, $s : \mathbb{N} \to \mathbb{N}$ be a size function, and $\epsilon : \mathbb{N} \to [0,1]$ be a function.*

- *We say that a sequence of sets, $\mathbf{S} = \{S_n \subseteq \{0,1\}^n\}$, is a hard-core (region) of f with respect to $s(\cdot)$-size circuits families and advantage $\epsilon(\cdot)$ if for every n and every circuit C_n of size at most $s(n)$, it holds that*

$$\text{Prob}[C_n(X_n) = f(X_n)] \leq \frac{1}{2} + \epsilon(n)$$

  where X_n is a random variable uniformly distributed on S_n.
- *We say that f has a hard-core (region) of density $\rho(\cdot)$ with respect to $s(\cdot)$-size circuits families and advantage $\epsilon(\cdot)$ if there exists a sequence of sets $\mathbf{S} = \{S_n \subseteq \{0,1\}^n\}$ such that $\mathbf{S}$ is a hard-core of f with respect to the above and $|S_n| \geq \rho(n) \cdot 2^n$.*

We stress that the usage of the term 'hard-core' in the above definition (and in the rest of this section) is different from the usage of this term in [5]. Observe that every strongly-unpredictable predicate has a hard-core of density 1 (i.e., the entire domain itself). Impagliazzo proves that also weakly-unpredicatabe predicates have hard-core sets that have density related to the amount of unpredictability. Namely:

Lemma 6 (existence of hard-core regions for unpredictable predicates): *Let $f : \{0,1\}^* \to \{0,1\}$ be a Boolean predicate, $s : \mathbb{N} \to \mathbb{N}$ be a size function, and $\rho : \mathbb{N} \to [0,1]$ be a noticeable function (i.e., $\rho(n) > 1/\text{poly}(n)$), such that for every n and every circuit C_n of size at most $s(n)$ it holds that*

$$\text{Prob}[C_n(U_n) = f(U_n)] \leq 1 - \rho(n),$$

where U_n is a random variable uniformly distributed on $\{0,1\}^n$. Then, for every function $\epsilon : \mathbb{N} \to [0,1]$, the function f has a hard-core of density $\rho'(\cdot)$ with respect to $s'(\cdot)$-size circuits families and advantage $\epsilon(\cdot)$, where $\rho'(n) \stackrel{\text{def}}{=} (1 - o(1)) \cdot \rho(n)$ and $s'(n) \stackrel{\text{def}}{=} s(n)/\text{poly}(n/\epsilon(n))$.

The proof of Lemma 6 is given in Appendix A. Using Lemma 6, we derive a proof of the XOR-Lemma, for the special case of uniform distribution.

Suppose that $\delta(\cdot)$ is a bound on the correlation of $s(\cdot)$-circuits with f over the uniform distribution. Then, it follows that such circuits cannot guess the value of f better than with probability $p(n) \stackrel{\text{def}}{=} \frac{1+\delta(n)}{2}$ and the existence of a hard-core $\mathbf{S} = \{S_n\}$ (w.r.t. $s'(n)$-circuits and $\epsilon(n)$-advantage) with density $\rho'(n) \stackrel{\text{def}}{=} (1 - o(1)) \cdot (1 - p(n))$ follows. Clearly,

$$\rho'(n) = (1 - o(1)) \cdot \frac{1 - \delta(n)}{2} > \frac{1}{3} \cdot (1 - \delta(n)).$$

Now, suppose that in contradiction to the XOR Lemma, the predicate $F^{(t)}$ defined as $F^{(t)}(x_1, ..., x_t) \stackrel{\text{def}}{=} \oplus_i f(x_i)$ can be correlated by "small" circuits with correlation greater than $c'(n) \stackrel{\text{def}}{=} 2 \cdot (\frac{2+\delta(n)}{3})^t + \epsilon(n)$. In other words, such circuits can guess $F^{(t)}$ with success probability at least $\frac{1}{2} + \frac{1}{2} \cdot c'(n)$. However, the probability that none of the t arguments to $F^{(t)}$ falls in the hard-core is at most $(1 - \rho'(n))^t$. Thus, conditioned on the event that at least one argument falls in the hard-core $\mathbf{S}$, the circuit guess $F^{(t)}$ correctly with probability at least

$$\frac{1}{2} + \frac{1}{2} \cdot c'(n) - (1 - \rho'(n))^t > \frac{1}{2} + \frac{\epsilon(n)}{2}.$$

Note, however, that this does not seem to yield an immediate contradition to the definition of a hard-core of f, yet we shall see that such a contradiction can be derived.

For every non-empty $I \subseteq \{1, ..., t\}$, we consider the event, denoted E_I, that represents the case that the arguments to $F^{(t)}$ that fall in the hard-core of f are exactly those with index in I. We have just shown that, conditioned on the union of these events, the circuit guesses the predicate $F^{(t)}$ correctly with probability at least $\frac{1}{2} + \frac{\epsilon(n)}{2}$. Thus, there exists an (non-empty) I such that, conditioned on E_I, the circuit guesses $F^{(t)}$ correctly with probability at least $\frac{1}{2} + \frac{\epsilon(n)}{2}$. Let $i \in I$ be arbitrary. By another averaging argument, we fix all inputs to the circuit except the ith input and obtain a circuit that guesses f correctly with probability at least $\frac{1}{2} + \frac{\epsilon(n)}{2}$. (For these fixed x_j's, $j \neq i$, the circuit incorporates also the value of $\oplus_{j \neq i} f(x_j)$.) This contradicts the hypothesis that $\mathbf{S}$ is a hard-core.

Generalization. We have just established the validity of the Lemma 1 for the case of the uniform probability ensemble and parameters $p(n) = 2$ and $\delta'(n) = \frac{2+\delta(n)}{3}$. The bound for δ' can be improved to $\delta'(n) = \frac{1+\delta(n)}{2} + o(1 - \delta(n))$. The argument extends to arbitrary probability ensembles. To this end one needs to properly generalize Definition 2 and prove a generalization of Lemma 6; for details the interested reader is referred to Appendix A.

5   Going through the Direct Product Problem

The third proof of the XOR Lemma proceeds in two steps. First it is shown that the success probability of feasible algorithms that try to predict the values of a predicate on several unrelated arguments decreases exponentially with the number of arguments. This statement is a generalization of another theorem due to Yao [17], hereafter called the *Concatenation Lemma*. Invoking a result of Goldreich and Levin [5], the XOR-Lemma follows.

5.1   The Concatenation Lemma

(This lemma is currently called the *Direct Product Theorem*.)

Lemma 7 (concatenation lemma): *Let* P, $\mathbf{X} = \{X_n\}$, $s : \mathbb{N} \to \mathbb{N}$, *and* $\delta : \mathbb{N} \to [-1, +1]$ *be as in Lemma 1. For every function* $t : \mathbb{N} \to \mathbb{N}$, *define the function* $F^{(t)}(x_1, ..., x_{t(n)}) \stackrel{\text{def}}{=} (P(x_1), ..., P(x_{t(n)}))$, *where* $x_1, ..., x_{t(n)} \in \{0, 1\}^n$, *and the probability ensemble* $\mathbf{X}^{(t)} = \{X_n^{(t)}\}$, *where* $X_n^{(t)}$ *consists of* $t(n)$ *independent copies of* X_n.

(hypothesis) *Suppose that* δ *is an upper bound on the correlation of families of* $s(\cdot)$-*size circuits with* P *over* $\mathbf{X}$. *Namely, suppose that for every* n *and for every* $s(n)$-*size circuit* C, *it holds that*

$$\mathrm{Prob}[C(X_n) = P(X_n)] \leq p(n) \stackrel{\text{def}}{=} \frac{1 + \delta(n)}{2} .$$

(conclusion) *Then, for every function* $\epsilon : \mathbb{N} \to [0, +1]$, *for every* n *and for every* $\mathrm{poly}(\frac{\epsilon(n)}{n}) \cdot s(n)$-*size circuit* C', *it holds that*

$$\mathrm{Prob}[C'(X_n^{(t)}) = F^{(t)}(X_n^{(t)})] \leq p(n)^{t(n)} + \epsilon(n).$$

Remark. Nisan et. al. [14] have used the XOR-Lemma in order to derive the Concatenation Lemma. Our feeling is that the Concatenation Lemma is more "basic" than the XOR Lemma, and thus that their strategy is not very natural.[8] In fact, this feeling was our motivation for trying to find a "direct" proof for the Concatenation Lemma. Extrapolating from the situation regarding the two original lemmata of Yao (i.e., the XOR Lemma and the Concatenation Lemma w.r.t. one-way functions),[9] we believed that such a proof (for the Concatenation

---

[8] This assertion is supported by a recent work of Viola and Wigderson, which provides a very simple proof that, in the general setting, the XOR Lemma implies the Concatenation Lemma [16, Prop. 1.4].

[9] Yao's original XOR Lemma (resp., Concatenation Lemma) refers to the setting of one-way functions. In this setting, the basic predicate P is a composition of an easy to compute predicate b and the inverse of a 1-1 one-way function f; i.e., $P(x) \stackrel{\text{def}}{=} b(f^{-1}(x))$. For years, the first author has considered the proof of the XOR Lemma (even for this setting) too complicated to be presented in class; whereas, a proof of the Concatenation Lemma (for this setting) has appeared in his classnotes [1] (see also [2]).

Lemma) should be easy to find. Indeed, we consider the following proof of Concatenation Lemma much simpler than the proofs of the XOR Lemma (given in previous sections).

A tight two-argument version. Lemma 7 is derived from the following Lemma 8 (which is a tight two-argument version of Lemma 7) analogously to the way that Lemma 2 was derived from Lemma 4; that is, we write $F^{(t)}(x_1, x_2, ..., x_{t(n)}) = (P(x_1), F^{(t-1)}(x_2, ..., x_{t(n)}))$, assume that $F^{(t-1)}$ is hard to guess as claimed, and apply the Concatenation Lemma to $(P, F^{(t-1)})$. This way, the lower bound on circuits guessing $F^{(t)}$ is related to the lower bound assumed for circuits guessing the original P and is almost the bound derived for $F^{(t-1)}$ (losing only an additive term!). It is thus left to prove the following two-argument version.

Lemma 8 (two argument version of concatenation lemma): *Let F_1 and F_2 be two functions, $l : \mathbb{N} \to \mathbb{N}$ be a length function, and $F(x) \stackrel{\text{def}}{=} (F_1(y), F_2(z))$ where $x = yz$ and $|y| = l(|x|)$. Let $\mathbf{X} = \{X_n\}$, $\mathbf{Y} = \{Y_{l(n)}\}$ and $\mathbf{Z} = \{Z_{n-l(n)}\}$ be probability ensembles as in Lemma 4 (i.e., $X_n = (Y_{l(n)}, Z_{n-l(n)})$).*

(hypothesis) *Suppose that $p_1(\cdot)$ is an upper bound on the probability that families of $s_1(\cdot)$-size circuits guess F_1 over $\mathbf{Y}$. Namely, for every such circuit family $\mathbf{C} = \{C_l\}$ it holds that*

$$\text{Prob}[C_l(Y_l) = F_1(Y_l)] \le p_1(l).$$

Likewise, suppose that $p_2(\cdot)$ is an upper bound on the probability that families of $s_2(\cdot)$-size circuits guess F_2 over $\mathbf{Z}$.

(conclusion) *Then, for every function $\epsilon : \mathbb{N} \to \mathbb{R}$, the function $p(n) \stackrel{\text{def}}{=} p_1(l(n)) \cdot p_2(n - l(n)) + \epsilon(n)$ is an upper bound on the probability that families of $s(\cdot)$-size circuits guess F over $\mathbf{X}$, where*

$$s(n) \stackrel{\text{def}}{=} \min \left\{ \frac{s_1(l(n))}{\text{poly}(n/\epsilon(n))} \ , \ s_2(n - l(n)) - n \right\}.$$

Proof: Let $\mathbf{C} = \{C_n\}$ be a family of $s(\cdot)$-size circuits. Fix an arbitrary n, and write $C = C_n$, $\epsilon = \epsilon(n)$, $l = l(n)$, $m = n - l(n)$, $Y = Y_l$ and $Z = Z_m$. Abusing notation, we let $C_1(x, y)$ denote the first component of $C(x, y)$ (i.e., the guess for $F_1(x)$) and likewise $C_2(x, y)$ is C's guess for $F_2(y)$. It is instructive to write the success probability of C as follows:

$$\text{Prob}[C(Y, Z) = F(Y, Z)] = \text{Prob}[C_2(Y, Z) = F_2(Z)]$$
$$\cdot \text{Prob}[C_1(Y, Z) = F_1(Y) \mid C_2(Y, Z) = F_2(Z)]$$

The basic idea is that using the hypothesis regarding F_2 allows to bound the first factor by $p_2(m)$, whereas the hypothesis regarding F_1 allows to bound the second factor by approximately $p_1(l)$. The basic idea for the latter step is that a sufficiently large sample of $(Z, F_2(Z))$, which may be hard-wired into the circuit, allows to use the conditional probability space (in such a circuit), provided the condition holds with noticeable probability. The last caveat motivates a separate treatment for y's with noticeable $\text{Prob}[C_2(y, Z) = F_2(Z)]$ and for the rest.

We call y **good** if $\text{Prob}[C_2(y, Z) = F_2(Z)] \geq \epsilon/2$ and **bad** otherwise. Let G be the set of good y's. Then, using $\text{Prob}[C(Y, Z) = F(Y, Z)] < \epsilon/2$ for every bad y, we upper bound the success probability of C as follows

$$
\begin{aligned}
\text{Prob}[C(Y, Z) = F(Y, Z)] &= \text{Prob}[C(Y, Z) = F(Y, Z) \,\&\, Y \in G] \\
&\quad + \text{Prob}[C(Y, Z) = F(Y, Z) \,\&\, Y \notin G] \\
&< \text{Prob}[C(Y, Z) = F(Y, Z) \,\&\, Y \in G] + \frac{\epsilon}{2}.
\end{aligned}
$$

Thus, using $p(n) = p_1(l) \cdot p_2(m) + \epsilon$, it remains to prove that

$$
\text{Prob}[C(Y, Z) = F(Y, Z) \,\&\, Y \in G] \leq p_1(l) \cdot p_2(m) + \epsilon/2. \tag{3}
$$

We proceed according to the foregoing outline. We first show that $\text{Prob}[C_2(Y, Z) = F_2(Z)]$ cannot be too large, as otherwise the hypothesis concerning F_2 is violate. Actually, we prove the following

Claim 8.1: For every y, it holds that

$$
\text{Prob}[C_2(y, Z) = F_2(Z)] \leq p_2(m).
$$

Proof: Otherwise, using any $y \in \{0, 1\}^l$ such that $\text{Prob}[C_2(y, Z) = F_2(Z)] > p_2(m)$, we get a circuit $C'(z) \stackrel{\text{def}}{=} C_2(y, z)$ that contradicts the lemma's hypothesis concerning F_2. $\qquad \square$

Next, we use Claim 8.1 in order to relate the success probability of C to the success probability of small circuits for F_1.

Claim 8.2: There exists a circuit C' of size $s_1(l)$ such that

$$
\text{Prob}[C'(Y) = F_1(Y)] \geq \frac{\text{Prob}[C(Y, Z) = F(Y, Z) \,\&\, Y \in G]}{p_2(m)} - \frac{\epsilon}{2}.
$$

Proof: The circuit C' is constructed as suggested in the foregoing outline. Specifically, we take a $\text{poly}(n/\epsilon)$-large sample, denoted S, from the distribution $(Z, F_2(Z))$ and let $C'(y) \stackrel{\text{def}}{=} C_1(y, z)$, where (z, β) is a uniformly selected among the elements of S for which $C_2(y, z) = \beta$ holds. Details follow.

Let S be a sequence of $t \stackrel{\text{def}}{=} \text{poly}(n/\epsilon)$ pairs, generated by taking t independent samples from the distribution $(Z, F_2(Z))$. We stress that we do not assume here that such a sample can be produced by an efficient (uniform) algorithm (but, jumping ahead, we remark that such a sequence can be fixed non-uniformly). For each $y \in G \subseteq \{0, 1\}^l$, we denote by S_y the set of pairs $(z, \beta) \in S$ for which $C_2(y, z) = \beta$. Note that S_y is a random sample for the residual probability space defined by $(Z, F_2(Z))$ conditioned on $C_2(y, Z) = F_2(Z)$. Also, with overwhelmingly high probability, $|S_y| = \Omega(l/\epsilon^2)$ (since $y \in G$ implies

$\text{Prob}[C_2(y, Z) = F_2(Z)] \geq \epsilon/2)$. Thus, with overwhelming probability (i.e., probability greater than $1 - 2^{-l}$), taken over the choices of S, the sample S_y provides a good approximation to the conditional probability space, and in particular

$$\frac{|\{(z,\beta) \in S_y : C_1(y, z) = F_1(y)\}|}{|S_y|} \geq \text{Prob}[C_1(y, Z) = F_1(y) \,|\, C_2(y, Z) = F_2(Z)] - \frac{\epsilon}{2} \quad (4)$$

Thus, with positive probability, Eq. (4) holds for all $y \in G \subseteq \{0,1\}^l$. The circuit C' guessing F_1 is now defined as follows. A set $S = \{z_i, \beta_i\}$ satisfying Eq. (4) for all good y's is "hard-wired" into the circuit C'. (In particular, S_y is not empty for any good y.) On input y, the circuit C' first determines the set S_y, by running C for t times and checking, for each $i = 1, ..., t$, whether $C_2(y, z_i) = \beta_i$. In case S_y is empty, the circuit returns an arbitrary value. Otherwise, the circuit selects uniformly a pair $(z, \beta) \in S_y$ and outputs $C_1(y, z)$. (This latter random choice can be eliminated by a standard averaging argument.) Using the definition of C' and Eq. (4), we get

$$\begin{aligned}
&\text{Prob}[C'(Y) = F_1(Y)] \\
&\geq \sum_{y \in G} \text{Prob}[Y = y] \cdot \text{Prob}[C'(y) = F_1(y)] \\
&= \sum_{y \in G} \text{Prob}[Y = y] \cdot \frac{|\{(z, \beta) \in S_y : C_1(y, z) = F_1(y)\}|}{|S_y|} \\
&\geq \sum_{y \in G} \text{Prob}[Y = y] \cdot \left(\text{Prob}[C_1(y, Z) = F_1(y) \,|\, C_2(y, Z) = F_2(Z)] - \frac{\epsilon}{2} \right) \\
&\geq \left(\sum_{y \in G} \text{Prob}[Y = y] \cdot \frac{\text{Prob}[C(y, Z) = F(y, Z)]}{\text{Prob}[C_2(y, Z) = F_2(Z)]} \right) - \frac{\epsilon}{2}.
\end{aligned}$$

Next, using Claim 8.1, we get

$$\text{Prob}[C'(Y) = F_1(Y)] \geq \left(\sum_{y \in G} \text{Prob}[Y = y] \cdot \frac{\text{Prob}[C(y, Z) = F(y, Z)]}{p_2(m)} \right) - \frac{\epsilon}{2}$$

and the claim follows.   □

Now, by the lemma's hypothesis concerning F_1, we have $\text{Prob}[C'(Y) = F_1(Y)] \leq p_1(l)$, and so using Claim 8.2 we get

$$\begin{aligned}
\text{Prob}[Y \in G \,\&\, C(Y, Z) = F(Y, Z)] &\leq (p_1(l) + \epsilon/2) \cdot p_2(m) \\
&\leq p_1(l) \cdot p_2(m) + \epsilon/2.
\end{aligned}$$

This proves Eq. (3) and the lemma follows.   ∎

5.2 Deriving the XOR Lemma from the Concatenation Lemma

Using the techniques of Goldreich and Levin [5], we obtain the following result.

Lemma 9 (hard-core predicate of unpredictable functions): *Let $F : \{0,1\}^* \to \{0,1\}^*$, $p : \mathbb{N} \to [0,1]$, and $s : \mathbb{N} \to \mathbb{N}$, and let $\mathbf{X} = \{X_n\}$ be as in Definition 1. For $\alpha, \beta \in \{0,1\}^\ell$, we denote by $IP_2(\alpha, \beta)$ the inner-product mod 2 of α and β, viewed as binary vectors of length ℓ.*

(hypothesis) *Suppose that, for every n and for every $s(n)$-size circuit C, it holds that*

$$\mathrm{Prob}[C(X_n) = F(X_n)] \leq p(n).$$

(conclusion) *Then, for some constant $c > 0$, for every n and for every $\mathrm{poly}(\frac{p(n)}{n})$· $s(n)$-size circuit C', it holds that*

$$\mathrm{Prob}[C'(X_n, U_\ell) = IP_2(F(X_n), U_\ell)] \leq \frac{1}{2} + c \cdot \sqrt[3]{n^2 \cdot p(n)},$$

where U_ℓ denotes the uniform distribution over $\{0,1\}^\ell$, with $\ell \stackrel{\mathrm{def}}{=} |F(X_n)|$. (That is, C' has correlation at most $2c\sqrt[3]{n^2 p(n)}$ with IP_2 over $(F(X_n), U_\ell)$.)

Proof Sketch: Let $q(n) \stackrel{\mathrm{def}}{=} c\sqrt[3]{n^2 p(n)}$. Suppose that C' contradicts the conclusion of the lemma. Then, there exists a set S such that $\mathrm{Prob}[X_n \in S] \geq q(n)$ and for every $x \in S$ the probability that $C'(x, U_\ell) = IP_2(F(x), U_\ell)$ is at least $\frac{1}{2} + \frac{q(n)}{2}$, where the probability is taken over U_ℓ (while x is fixed). Employing the techniques of [5][10], we obtain a randomized circuit C (of size at most a $\mathrm{poly}(n/p(n))$ factor larger than C') such that, for every $x \in S$, it holds that $\mathrm{Prob}[C(X_n) = F(X_n)] \geq c' \cdot (q(n)/n)^2$ (where the constant $c' > 0$ is determined in the proof of [5] according to Chebishev's Inequality).[11] Thus, C satisfies

$$\mathrm{Prob}[C(X_n) = F(X_n)] \geq \mathrm{Prob}[C(X_n) = F(X_n) \wedge X_n \in S]$$
$$= \mathrm{Prob}[X_n \in S] \cdot \mathrm{Prob}[C(X_n) = F(X_n) | X_n \in S]$$
$$\geq q(n) \cdot \left(c' \cdot (q(n)/n)^2\right) = p(n)$$

in contradiction to the hypothesis. The lemma follows. ∎

Conclusion. Combining the Concatenation Lemma (Lemma 7) with Lemma 9 we establish the validity of Lemma 1 for the third time; this time with respect to the parameters $p(n) = cn^{2/3} = o(n)$ and $\delta'(n) = \sqrt[3]{\frac{1+\delta(n)}{2}}$. Details follow.

Starting with a predicate for which δ is a correlation bound and using Lemma 7, we get a function that is hard to guess with probability substantially higher than

---

[10] See alternative expositions in either [4, Sec. 7.1.3] or [3, Sec. 2.5.2].

[11] The algorithm in [5] will actually retrieve all values $\alpha \in \{0,1\}^\ell$ for which the correlation of $C'(x, U_\ell)$ and $IP_2(\alpha, U_\ell)$ is at least $q(n)$. With overwhelming probability it outputs a list of $O((n/q(n))^2)$ strings containing all the values just mentioned and thus uniformly selecting one of the values in the list yields $F(x)$ with probability at least $1/O((n/q(n))^2)$.

$(\frac{1+\delta(n)}{2})^{t(n)}$. Applying Lemma 9 establishes that given $(x_1, ..., x_{t(n)})$ and a uniformly chosen subset $S \subseteq \{1, 2, ..., t(n)\}$ it is hard to correlate $\oplus_{i \in S} P(x_i)$ better than with correlation

$$O\left(\sqrt[3]{n^2 \cdot \left(\frac{1+\delta(n)}{2}\right)^{t(n)}}\right) = o(n) \cdot \left(\sqrt[3]{\frac{1+\delta(n)}{2}}\right)^{t(n)}.$$

This is almost what we need, but not quite (what we need is a statement concerning $S = \{1, ..., t(n)\}$). The gap is easily bridged by some standard "padding" trick. For example, by using a sequence of fixed pairs (z_i, σ_i), such that $\sigma_i = P(z_i)$, we reduce the computation of $\oplus_{i \in S} P(x_i)$ to the computation of $\oplus_{i=1}^{t(n)} P(y_i)$ by setting $y_i = x_i$ if $i \in S$ and $y_i = z_i$ otherwise. (See Appendix B for more details.) Thus, Lemma 1 follows (with the stated parameters).

5.3   Remarks Concerning the Uniform Complexity Setting

A uniform-complexity analogue of the foregoing proof can be carried out provided that one can construct random examples of the distribution $(X_n, P(X_n))$ within the stated (uniform) complexity bounds (and in particular in polynomial-time). Actually, this condition is required only for the proof of the Concatenation Lemma. Thus we confine ourselves to presenting a uniform-complexity version of the Concatenation Lemma.

Lemma 10 (Concatenation Lemma – uniform complexity version): *Let* $P, \mathbf{X}$, s, δ, t *and* $F^{(t)}$ *be as in Lemma 7.*

(hypothesis) *Suppose that* $\delta(\cdot)$ *is an upper bound on the correlation of* $T(\cdot)$-*time-constructible families of* $s(\cdot)$-*size circuits with* P *over* $\mathbf{X}$. *Furthermore, suppose that one can generate in polynomial-time a random sample from the distribution* $(X_n, P(X_n))$.

(conclusion) *Then, for every function* $\epsilon : \mathbb{N} \to [0, +1]$, *the function* $q(n) \overset{\text{def}}{=} p(n)^{t(n)} + \epsilon(n)$ *is an upper bound on the correlation of* $T'(\cdot)$-*time-constructible families of* $s'(\cdot)$-*size circuits with* F *over* $\mathbf{X}^{(t)}$, *where* $T'(t(n) \cdot n) = \text{poly}$ $(\epsilon(n)/n) \cdot T(n)$ *and* $s'(t(n) \cdot n) = \text{poly}(\epsilon(n)/n) \cdot s(n)$.

The uniform-complexity version of the Concatenation Lemma is proven by adapting the proof of Lemma 7 as follows. Firstly, we observe that it suffices to prove an appropriate (uniform-complexity) version of Lemma 8. This is done by first proving a weaker version of Claim 8.1 that asserts that for all but at most an $\epsilon(n)/8$ measure of the y's (under Y), it holds that

$$\text{Prob}[C_2(y, Z) = F_2(Z)] \le p_2(m) + \epsilon(n)/8.$$

This holds because otherwise one may sample Y with the aim of finding a y such that $\text{Prob}[C_2(y, Z) = F_2(Z)] > p_2(m)$ holds, and then use this y to construct (uniformly!) a circuit that contradicts the hypothesis concerning F_2. Next, we prove a weaker version of Claim 8.2 by observing that, for a uniformly selected pair sequence S, with overwhelmingly high probability (and not only with positive probability), Eq. (4) holds for all good $y \in \{0,1\}^l$. Thus, if we generate S by taking random samples from the distribution $(Z_m, F_2(Z_m))$, then with overwhelmingly high probability we end-up with a circuit as required by the modified claim. (The modified claim has $p_2(m) + \epsilon/8$ in the denominator (rather than $p_2(m)$) as well as an extra additive term of $\epsilon/8$.) Using the hypothesis concerning F_1, we are done as in the non-uniform case.

6   A Different Perspective: The Entropy Angle

The XOR Lemma and the Concatenation Lemma are special cases of the so-called "direct sum conjecture" asserting that computational difficulty increases when many independent instances of the problem are to be solved. In both cases the "direct sum conjecture" is postulated by considering insufficient resources and bounding the probability that these tasks can be performed within these resources, as a function of the number of instances. In this section we suggest an analogous analysis based on entropy rather than probability. Specifically, we consider the amount of information remaining in the task (e.g., of computing $f(x)$) when given the result of a computation (e.g., $C(x)$). This analysis turns out to be much easier.

Proposition 11. *Let f be a predicate, X be a random variable and C be a class of circuits so that for every circuit $C \in \mathcal{C}$*

$$\mathrm{H}(f(X)|C(X)) \geq \epsilon,$$

where H *denotes the* (conditional) *binary entropy function. Furthermore, suppose that, for every circuit $C \in \mathcal{C}$, fixing any of the inputs of C yields a circuit also in $\mathcal{C}$. Then, for every circuit $C \in \mathcal{C}$, it holds that*

$$\mathrm{H}(f(X^{(1)}), ..., f(X^{(t)})|C(X^{(1)}, ..., X^{(t)})) \geq t \cdot \epsilon,$$

where the $X^{(i)}$'s are independently distributed copies of X.

We stress that the class $\mathcal{C}$ in Proposition 11 may contain circuits with several Boolean outputs. Furthermore, for a meaningful conclusion, the class $\mathcal{C}$ must contain circuits with t outputs (otherwise, for a circuit C with much fewer outputs, the conditional entropy $\mathrm{H}(f(x_1), ..., f(x_t)|C(x_1, ..., x_t))$ is large merely due to information theoretical reasons). On the other hand, the more outputs the circuits in $\mathcal{C}$ have, the stronger the hypothesis of Proposition 11 is. In particular, the number of outputs must be smaller that $|X|$ otherwise the value of the circuit $C(x) = x$ determines $f(x)$ (i.e., $\mathrm{H}(f(x)|x) = 0$). Thus, a natural instantiation of Proposition 11 is for a family of small (e.g., poly-size) circuits each having t outputs.

Proof: By definition of conditional entropy, we have for every $C \in \mathcal{C}$,

$$\mathrm{H}(f(X^{(1)}), ..., f(X^{(t)})|C(X^{(1)}, ..., X^{(t)}))$$

$$= \sum_{i=1}^{t} \mathrm{H}(f(X^{(i)})|C(X^{(1)}, ..., X^{(t)}), f(X^{(1)}), ..., f(X^{(i-1)}))$$

$$\geq \sum_{i=1}^{t} \mathrm{H}(f(X^{(i)})|C(X^{(1)}, ..., X^{(t)}), X^{(1)}, ..., X^{(i-1)}).$$

Now, for each i, we show that

$$\mathrm{H}(f(X^{(i)})|C(X^{(1)}, ..., X^{(t)}), X^{(1)}, ..., X^{(i-1)}) \geq \epsilon.$$

We consider all possible settings of all variables, except $X^{(i)}$, and bound the conditional entropy under this setting (which does not effect $X^{(i)}$). The fixed values $X^{(j)} = x_j$ can be eliminated from the entropy condition and incorporated into the circuit. However, fixing some of the inputs in the circuit C yields a circuit also in $\mathcal{C}$ and so we can apply the proposition's hypothesis and get

$$\mathrm{H}(f(X^{(i)})|C(x_1, ..., x_{i-1}, X^{(i)}, x_{i+1}, ..., x_t)) \geq \epsilon.$$

The proposition follows.   ∎

Proposition 11 vs the Concatenation Lemma. We compare the hypotheses and conclusions of these two results.

The hypotheses. The hypothesis in Proposition 11 is related to the hypotheses in the Concatenation Lemma. Clearly, an entropy lower bound (on a single bit) translates to some unpredictability bound on this bit. (This does not hold for many bits as can be seen below.) The other direction (i.e., unpredictability implies a lower bound on the conditional entropy) is obvious for a single bit.

The conclusions. For $t = O(\log n)$ the conclusion of Proposition 11 is implied by the conclusion of the Concatenation Lemma, but for sufficiently large t the conclusion of Proposition 11 does not imply the conclusion of Concatenation Lemma. Details follow.

1. To show that, for $t = O(\log n)$, the conclusion of the Concatenation Lemma implies the conclusion of Proposition 11, suppose that for a small circuit C it holds that $h \stackrel{\text{def}}{=} \mathrm{H}(f(X^{(1)}), ..., f(X^{(t)})|C(X^{(1)}, ..., X^{(t)})) = o(t)$. Then, for every value of C, denoted v, there exists a string $w = w(v)$ such that $\mathrm{Prob}[f(X^{(1)}), ..., f(X^{(t)}) = w|C(X^{(1)}, ..., X^{(t)}) = v] \geq 2^{-h}$. Hardwiring these 2^t strings $w(\cdot)$ into C, we obtain a small circuit that predicts $f(X^{(1)}), ..., f(X^{(t)})$ with probability at least $2^{-h} = 2^{-o(t)}$, in contradiction to the conclusion of the Concatenation Lemma.

2. To show that the conclusion of Proposition 11 does not imply the conclusion of the Concatenation Lemma, consider the possibility of a small

(randomized) circuit C that with probability $1-\epsilon$ correctly determines all the f values (i.e., $\mathrm{Prob}[C(X^{(1)}, ..., X^{(t)}) = f(X^{(1)}), ..., f(X^{(t)})] = 1 - \epsilon$), and yields no information (e.g., outputs a special fail symbol) otherwise. Then, although C has success probability $1 - \epsilon$, the conditional entropy is $(1 - \epsilon) \cdot 0 + \epsilon \cdot t$ (assuming that $\mathrm{Prob}[f(X) = 1] = 1/2$).

7   Subsequent Work

Since the first publication of this survey, Yao's XOR Lemma has been the subject of intensive research. Here we only outline three themes that were pursued, while referring the interested reader to [10] and the references therein.

Derandomization. A central motivation for Impagliazzo's work [7,8] has been the desire to present "derandomized versions" of the XOR Lemma; that is, predicates that use their input in order to define a sequence of related instances, and take the XOR of the original predicate on these instances.[12] The potential benefit in such a construction is that the hardness of the resulting predicate is related to shorter inputs (i.e., the seed of a generator of a t-long sequence of n-bit long strings, rather than the tn-bit long sequence itself). Indeed, Impagliazzo's work [7,8] presented such a construction (based on a pairwise independent generator), and left the question of providing a "full derandomization" (that uses a seed of length $O(n)$ to generate t instances) to subsequent work. The goal was achieved by Impagliazzo and Wigderson [11] by using a generator that combines Impagliazzo's generator [7,8] with a new generator, which in turn combines an expander walk generator with the Nisan-Wigderson generator [15].

Avoiding the use of random examples. As pointed out in Section 2.3, all proofs presented in this survey make an essential use of random examples. For more than a decade, this feature stood in the way of a general uniform version of the XOR Lemma (i.e., all uniform proofs assumed access to such random examples). This barrier was lifted by Impagliazzo, Jaiswal, and Kabanets [9], which culminated in comprehensive treatment of [10]. The latter work provides simplified, optimized, and derandomized versions of the XOR and Concatenation Lemmas.[13] The key idea is to use the hypothetical solver of the concatenated problem in order to obtain a sequence of random examples that are all good with noticeable probability. An instance of the original problem is then solved by hiding it in a random sequence that has a fair intersection with the initial sequence of random examples. The interested reader is referred to [10] for a

---

[12] That is, the predicate consists of an "instance generator" and multiple applications of the original predicate, P. Specifically, on input an s-bit long seed, denoted y, the generator produces a t-long sequence of n-bit long strings (i.e., $(x_1, ..., x_t) \leftarrow G(y)$), and the value of the new predicate is defined as the XOR of the values of P on these t strings (i.e., $\oplus_{i=1}^{t} P(x_i)$).

[13] The focus of [10] is actually on the Concatenation Lemma, which is currently called the Direct Product Theorem. See next paragraph regarding the relation to the XOR Lemma.

mature description of this idea (and its sources of inspirarion) as well as for a discussion of the relation this problem (i.e., proofs of the Concatenation Lemma) and list-decoding of the direct product code.

The relation between the XOR and Concatenation Lemmas. In Section 5 we advocated deriving the XOR Lemma from the Concatenation Lemma, and this suggestion was adopted in several works (including [9,10]). Our intuition that the Concatenation Lemma is simpler than the XOR Lemma is supported by a recent work of Viola and Wigderson, which provides a very simple proof that, in the general setting, the XOR Lemma implies the Concatenation Lemma [16, Prop. 1.4]. We mention that the both directions of the equivalence between the Concatenation Lemma and the XOR Lemma pass through an intermediate lemma called the Selective XOR Lemma (see [4, Exer. 7.17]). For further discussion see Appendix B.

Acknowledgement. We wish to thank Mike Saks for useful discussions regarding Levin's proof of the XOR Lemma. We also thank Salil Vadhan and Ronen Shaltiel for pointing out errors in previous versions, and for suggesting ways to fix these errors.

References

1. Goldreich, O.: Foundation of Cryptography – Class Notes. Computer Science Department, Technion, Haifa, Israel (Spring 1989)
2. Goldreich, O.: Foundation of Cryptography – Fragments of a Book, Available from ECCC (February 1995)
3. Goldreich, O.: Foundation of Cryptography: Basic Tools. Cambridge University Press, Cambridge (2001)
4. Goldreich, O.: Computational Complexity: A Conceptual Perspective. Cambridge University Press, Cambridge (2008)
5. Goldreich, O., Levin, L.A.: A Hard-Core Predicate for all One-Way Functions. In: 21st STOC, pp. 25–32 (1989)
6. Håstad, J., Impagliazzo, R., Levin, L.A., Luby, M.: A Pseudorandom Generator from any One-way Function. SICOMP 28(4), 1364–1396 (1999); Combines papers of Impagliazzo et al. (21st STOC, 1989) and Håstad (22nd STOC, 1990)
7. Impagliazzo, R.: See [8], which appeared after our first posting (1994) (manuscript)
8. Impagliazzo, R.: Hard-core Distributions for Somewhat Hard Problems. In: 36th FOCS, pp. 538–545 (1995); This is a later version of [7]
9. Impagliazzo, R., Jaiswal, R., Kabanets, V.: Approximately List-Decoding Direct Product Codes and Uniform Hardness Amplification. In: 47th FOCS, pp. 187–196 (2006)
10. Impagliazzo, R., Jaiswal, R., Kabanets, V., Wigderson, A.: Uniform Direct Product Theorems: Simplified, Optimized, and Derandomized. SIAM J. Comput. 39(4), 1637–1665 (2010); Preliminary version in 40th STOC (2008)
11. Impagliazzo, R., Wigderson, A.: P=BPP if E requires exponential circuits: Derandomizing the XOR Lemma. In: 29th STOC, pp. 220–229 (1997)
12. Levin, L.A.: One-Way Functions and Pseudorandom Generators. Combinatorica 7(4), 357–363 (1987)
13. Levin, L.A.: Average Case Complete Problems. SICOMP 15, 285–286 (1986)

14. Nisan, N., Rudich, S., Saks, M.: Products and Help Bits in Decision Trees. In: 35th FOCS, pp. 318–329 (1994)
15. Nisan, N., Wigderson, A.: Hardness vs Randomness. JCSS 49(2), 149–167 (1994)
16. Viola, E., Wigderson, A.: Norms, XOR Lemmas, and Lower Bounds for Polynomials and Protocols. Theory of Computing 4(1), 137–168 (2008); Preliminary version in IEEE Conf. on Comput. Complex. (2007)
17. Yao, A.C.: Theory and Application of Trapdoor Functions. In: 23rd FOCS, pp. 80–91 (1982)

Appendix A: Proof of a Generalization of Lemma 6

We first generalize Impagliazzo's treatment to the case of non-uniform distributions; Impagliazzo's treatment is regained by letting $\mathbf{X}$ be the uniform probability ensemble.

Definition 3 (hard-core of a predicate relative to a distribution): *Let* $f:\{0,1\}^* \to \{0,1\}$ *be a Boolean predicate,* $s:\mathbb{N} \to \mathbb{N}$ *be a size function,* $\epsilon:\mathbb{N} \to [0,1]$ *be a function, and* $\mathbf{X} = \{X_n\}$ *be a probability ensemble.*

- *We say that a sequence of sets,* $\mathbf{S} = \{S_n \subseteq \{0,1\}^n\}$, *is a* hard-core *of* f *relative to* $\mathbf{X}$ *with respect to* $s(\cdot)$-*size circuits families and advantage* $\epsilon(\cdot)$ *if for every* n *and every circuit* C_n *of size at most* $s(n)$, *it holds that*

$$\mathrm{Prob}[C_n(X_n) = f(X_n) | X_n \in S_n] \leq \frac{1}{2} + \epsilon(n).$$

- *We say that* f *has a* hard-core *of density* $\rho(\cdot)$ *relative to* $\mathbf{X}$ *with respect to* $s(\cdot)$-*size circuits families and advantage* $\epsilon(\cdot)$ *if there exists a sequence of sets* $\mathbf{S} = \{S_n \subseteq \{0,1\}^n\}$ *such that* $\mathbf{S}$ *is a hard-core of* f *relative to* $\mathbf{X}$ *with respect to the above and* $\mathrm{Prob}[X_n \in S_n] \geq \rho(n)$.

Lemma 12 (generalization of Lemma 6): *Let* $f:\{0,1\}^* \to \{0,1\}$ *be a Boolean predicate,* $s:\mathbb{N} \to \mathbb{N}$ *be a size function,* $\mathbf{X} = \{X_n\}$ *be a probability ensemble, and* $\rho:\mathbb{N} \to [0,1]$ *be a noticeable function such that for every* n *and every circuit* C_n *of size at most* $s(n)$, *it holds that*

$$\mathrm{Prob}[C_n(X_n) = f(X_n)] \leq 1 - \rho(n).$$

Then, for every function $\epsilon:\mathbb{N} \to [0,1]$, *the function* f *has a hard-core of density* $\rho'(\cdot)$ *relative to* $\mathbf{X}$ *with respect to* $s'(\cdot)$-*size circuits families and advantage* $\epsilon(\cdot)$, *where* $\rho'(n) \stackrel{\text{def}}{=} (1 - o(1)) \cdot \rho(n)$ *and* $s'(n) \stackrel{\text{def}}{=} s(n)/\mathrm{poly}(n/\epsilon(n))$.

Proof: We start by proving a weaker statement; namely, that $\mathbf{X}$ "dominates" an ensemble $\mathbf{Y}$ under which the function f is strongly unpredictable. Our notion of domination originates in a different work of Levin [13]. Specifically, referring to a fixed function ρ, we define domination as assigning probability mass that is at least a ρ fraction of the mass assigned by the dominated ensemble; namely:

Definition: *Fixing the function ρ for the rest of the proof*, we say that the ensemble $\mathbf{X} = \{X_n\}$ dominates the ensemble $\mathbf{Y} = \{Y_n\}$ if for every string α,

$$\text{Prob}[X_n = \alpha] \geq \rho(|\alpha|) \cdot \text{Prob}[Y_n = \alpha].$$

In this case we also say that $\mathbf{Y}$ is dominated by $\mathbf{X}$. We say that $\mathbf{Y}$ is critically dominated by $\mathbf{X}$ if for every string α either $\text{Prob}[Y_n = \alpha] = (1/\rho(|\alpha|)) \cdot \text{Prob}[X_n = \alpha]$ or $\text{Prob}[Y_n = \alpha] = 0$. (Actually, to avoid trivial difficulties, we allow at most one string $\alpha \in \{0,1\}^n$ such that $0 < \text{Prob}[Y_n = \alpha] < (1/\rho(|\alpha|)) \cdot \text{Prob}[X_n = \alpha]$.)

The notions of domination and critical domination play central roles in the following proof, which consists of two parts. In the first part (cf., Claim 12.1), we prove the existence of a ensemble dominated by $\mathbf{X}$ such that f is strongly unpredictable under this ensemble. In the second part (cf., Claims 12.2 and 12.3), we essentially prove that the existence of such a dominated ensemble implies the existence of an ensemble that is *critically* dominated by $\mathbf{X}$ such that f is strongly unpredictable under the latter ensemble. However, such a critically dominated ensemble defines a hard-core of f relative to $\mathbf{X}$, and the lemma follows. Before starting, we make the following simplifying assumptions (used in Claim 12.3).

Simplifying assumptions: Without loss of generality, the following two conditions hold:

1. $\log_2 s(n) \leq n$.
   (Otherwise the hypothesis of the lemma cannot hold.)
2. $\text{Prob}[X_n = x] < \text{poly}(n)/s(n)$, for all x's.
   (This assumption is justified since x's violating this condition cannot contribute to the hardness of f with respect to X_n, because one can incorporate all these $s(n)/poly(n)$ many violating x's with their corresponding $f(x)$'s into the circuit).

Claim 12.1: Under the hypothesis of the lemma it holds that there exists a probability ensemble $\mathbf{Y} = \{Y_n\}$ such that $\mathbf{Y}$ is dominated by $\mathbf{X}$ and, for every $s'(n)$-circuit C_n, it holds that

$$\text{Prob}[C_n(Y_n) = f(Y_n)] \leq \frac{1}{2} + \frac{\epsilon(n)}{2}. \tag{5}$$

Proof:[14] We start by assuming, towards the contradiction, that for every distribution Y_n that is dominated by X_n there exists an $s'(n)$-size circuits C_n such that $\text{Prob}[C_n(Y_n) = f(Y_n)] > 0.5 + \epsilon'(n)$, where $\epsilon'(n) = \epsilon(n)/2$. One key observation is that there is a correspondence between the set of all distributions that are each dominated by X_n and the set of all the convex combinations of critically dominated (by X_n) distributions; that is, each dominated distribution is a convex combinations of critically dominated distributions and vice versa. Thus, considering an enumeration $Y_n^{(1)}, ..., Y_n^{(t)}$ of all the critically dominated (by X_n)

---

[14] The current text was revised following the revision in [4, Sec. 7.2.2.1].

distributions, we conclude that, for every distribution (or convex combination) π on $[t]$, there exists an $s'(n)$-size circuits C_n such that

$$\sum_{i=1}^{t} \pi(i) \cdot \mathrm{Prob}[C_n(Y_n^{(i)}) = f(Y_n^{(i)})] > 0.5 + \epsilon'(n). \tag{6}$$

Now, consider a finite game between two players, where the first player selects a critically dominated (by X_n) distribution, and the second player selects an $s'(n)$-size circuit and obtains a payoff as determined by the corresponding success probability; that is, if the first player selects the ith critically dominated distribution and the second player selects the circuit C, then the payoff equals $\mathrm{Prob}[C(Y_n^{(i)}) = f(Y_n^{(i)})]$. Taking this perspective Eq. (6) means that, for any randomized strategy for the first player, there exists a deterministic strategy for the second player yielding average payoff greater than $0.5 + \epsilon'(n)$. The Min-Max Principle asserts that, in such a case, there exists a randomized strategy for the second player that yields average payoff greater than $0.5 + \epsilon'(n)$ no matter what strategy is employed by the first player. This means that there exists a distribution, denoted D_n, on $s'(n)$-size circuits such that for every i it holds that

$$\mathrm{Prob}[D_n(Y_n^{(i)}) = f(Y_n^{(i)})] > 0.5 + \epsilon'(n), \tag{7}$$

where the probability refers both to the choice of the circuit D_n and to the random variable $Y_n^{(i)}$. Let $B_n = \{x : \mathrm{Prob}[D_n(x) = f(x)] \leq 0.5 + \epsilon'(n)\}$. Then, $\mathrm{Prob}[X_n \in B_n] < \rho(n)$, because otherwise we reach a contradiction to Eq. (7) by defining Y_n such that $\mathrm{Prob}[Y_n = x] = \mathrm{Prob}[X_n = x]/\mathrm{Prob}[X_n \in B_n]$ if $x \in B_n$ and $\mathrm{Prob}[Y_n = x] = 0$ otherwise.[15] By employing standard amplification to D_n, we obtain a distribution D'_n over $\mathrm{poly}(n/\epsilon'(n)) \cdot s'(n)$-size circuits such that for every $x \in \{0,1\}^n \setminus B_n$ it holds that $\mathrm{Prob}[D'_n(x) = f(x)] > 1 - 2^{-n}$. It follows that there exists an $s(n)$-sized circuit C_n such that $C_n(x) = f(x)$ for every $x \in \{0,1\}^n \setminus B_n$, which implies that $\mathrm{Prob}[C_n(X_n) = f(X_n)] \geq \mathrm{Prob}[X_n \in \{0,1\}^n \setminus B_n] > 1 - \rho(n)$, in contradiction to the theorem's hypothesis. The claim follows.   □

From a dominated ensemble to a hard-core. In the rest of the proof, we fix an arbitrary ensemble, denoted $\mathbf{Y} = \{Y_n\}$ satisfying Claim 12.1. Using this ensemble, which is dominated by $\mathbf{X}$, we prove the validity of the lemma (i.e., the existence of a hard-core) by a probabilistic argument. Specifically, we consider the following probabilistic construction.

Probabilistic construction: We define a random set $R_n \subseteq \{0,1\}^n$ by selecting each string $x \in \{0,1\}^n$ to be in R_n with probability

$$p(x) \overset{\mathrm{def}}{=} \frac{\rho(n) \cdot \mathrm{Prob}[Y_n = x]}{\mathrm{Prob}[X_n = x]} \leq 1 \tag{8}$$

---

[15] Note that Y_n is dominated by X_n, whereas by the hypothesis $\mathrm{Prob}[D_n(Y_n) = f(Y_n)] \leq 0.5 + \epsilon'(n)$. Using the fact that any dominated distribution is a convex combination of critically dominated distributions, it follows that $\mathrm{Prob}[D_n(Y_n^{(i)}) = f(Y_n^{(i)})] \leq 0.5 + \epsilon'(n)$ holds for some critically dominated $Y_n^{(i)}$.

independently of the choices made for all other strings. Note that the inequality holds because **X** dominates **Y**.

First we show that, with overwhelmingly high probability over the choive of R_n, it holds that $\text{Prob}[X_n \in R_n] \approx \rho(n)$.

Claim 12.2: Let $\alpha > 0$ and suppose that $\text{Prob}[X_n = x] \le \rho(n) \cdot \alpha^2/\text{poly}(n)$, for every x. Then, for all but at most a $2^{-\text{poly}(n)}$ measure of the choices of R_n, it holds that

$$|\text{Prob}[X_n \in R_n] - \rho(n)| < \alpha \cdot \rho(n).$$

Proof: For every $x \in \{0,1\}^n$, let $w_x \overset{\text{def}}{=} \text{Prob}[X_n = x]$. We define random variables $\zeta_x = \zeta_x(R_n)$, over the probability space defined by the random choices of R_n, such that ζ_x indicate whether $x \in R_n$; that is, the ζ_x's are independent of one another, and $\text{Prob}[\zeta_x = 1] = p(x)$ (and $\zeta_x = 0$ otherwise). Thus, for every possible choice of R_n, it holds that

$$\text{Prob}[X_n \in R_n] = \sum_x \zeta_x(R_n) \cdot w_x$$

and consequently we are interested in the behaviour of the sum $\sum_x w_x \zeta_x$ as a random variable (over the probability space of all possible choices of R_n). Taking expactation (over the possible choices of R_n), we get

$$\text{E}\left[\sum_x w_x \zeta_x\right] = \sum_x p(x) \cdot w_x$$
$$= \sum_x \frac{\rho(n) \cdot \text{Prob}[Y_n = x]}{\text{Prob}[X_n = x]} \cdot \text{Prob}[X_n = x]$$
$$= \rho(n).$$

Now, using Chernoff bound, we get

$$\text{Prob}\left[\left|\sum_x w_x \zeta_x - \rho(n)\right| > \alpha \cdot \rho(n)\right] < \exp\left(-\Omega\left(\frac{\alpha^2 \rho(n)}{\max_x\{w_x\}}\right)\right).$$

Finally, using the claim's hypotheses $w_x \le \alpha^2 \cdot \rho(n)/\text{poly}(n)$ (for all x's), the latter expression is bounded by $\exp(-\text{poly}(n))$, and the claim follows. □

Finally, we show that R_n is likely to be a hard-core of f realtive to **X** (w.r.t. sufficiently small circuits).

Claim 12.3:[16] For all but at most a $2^{-\text{poly}(n)}$ measure of the choices of R_n, it holds that every circuit C_n of size $s'(n)$ satisfies

$$\text{Prob}[C_n(X_n) = f(X_n)|X_n \in R_n] < \frac{1}{2} + \epsilon(n).$$

Proof: We define the same random variables $\zeta_x = \zeta_x(R_n)$ as in the proof of Claim 12.2; that is, $\zeta_x(R_n) = 1$ if $x \in R_n$ and $\zeta_x(R_n) = 0$ otherwise. Also, as

---

[16] The current statement and its proof were somewhat revised.

before, $w_x \stackrel{\text{def}}{=} \text{Prob}[X_n = x]$, for every $x \in \{0,1\}^n$. Fixing any circuit C_n, let C be the set of inputs on which C_n correctly computes f; namely,

$$C \stackrel{\text{def}}{=} \{x : C_n(x) = f(x)\}. \tag{9}$$

For every choice of R_n, we are interested in the probability

$$\text{Prob}[X_n \in C | X_n \in R_n] = \frac{\text{Prob}[X_n \in C \wedge X_n \in R_n]}{\text{Prob}[X_n \in R_n]} \tag{10}$$

We first determine the expected value of the numerator of Eq. (10), where the expactation is taken over the possible choices of R_n. We rewrite the numerator as $\sum_{x \in C} \zeta_x(R_n) \cdot w_x$, and lower bound it as follows

$$
\begin{aligned}
\text{E}\left[\sum_{x \in C} \zeta_x \cdot w_x\right] &= \sum_{x \in C} p(x) \cdot w_x \\
&= \sum_{x \in C} \frac{\rho(n) \cdot \text{Prob}[Y_n = x]}{\text{Prob}[X_n = x]} \cdot \text{Prob}[X_n = x] \\
&= \rho(n) \cdot \text{Prob}[Y_n \in C] \\
&\leq \rho(n) \cdot \left(\frac{1}{2} + \frac{\epsilon(n)}{2}\right),
\end{aligned}
$$

where the last inequality is due to the hypothesis regarding Y_n. Next, we use a (multiplicative) Chernoff bound, and get

$$
\begin{aligned}
\text{Prob}\left[\sum_{x \in C} w_x \zeta_x > \left(\frac{1}{2} + \frac{2\epsilon(n)}{3}\right) \cdot \rho(n)\right] &< \exp\left(-\Omega\left(\frac{\epsilon(n)^2 \rho(n)}{\max_x\{w_x\}}\right)\right) \\
&< \exp\left(-\Omega\left(\frac{\epsilon(n)^2 s(n) \log_2 s(n)}{\text{poly}(n)}\right)\right),
\end{aligned}
$$

where the last inequality uses the simplifying assumptions regarding the w_x's and $s(n)$ (i.e., $w_x < \text{poly}(n)/s(n)$ and $\log_2 s(n) \leq n$). Thus, for all but at most a $\exp(-\text{poly}(n) \cdot s'(n) \log_2 s'(n))$ measure of the R_n's, the numerator of Eq. (10) is at most $(\frac{1}{2} + \frac{2\epsilon(n)}{3}) \cdot \rho(n)$. This holds for each possible circuit of size $s'(n)$. Applying the union bound to the set of all $2^{s'(n)(O(1) + 2 \log_2 s'(n))}$ possible circuits of size $s'(n)$, we conclude that the probability that for some of these circuits the numerator of Eq. (10) is greater than $(\frac{1}{2} + \frac{2\epsilon(n)}{3}) \cdot \rho(n)$ is at most $\exp(-\text{poly}(n))$, where the probability is taken over the choice of R_n. Using Claim 12.2, we conclude that, for a similar measure of R_n's, the denumerator of Eq. (10) is at least $(1 - \frac{\epsilon(n)}{3}) \cdot \rho(n)$. The claim follows. $\qquad \square$

Conclusion. The lemma now follows by combining the foregoing three claims. Claim 12.1 provides us with a suitable **Y** for which we apply the probabilistic construction, whereas Claims 12.2 and 12.3 establish the existence of a set R_n such that both

$$\text{Prob}[X_n \in R_n] > (1 - o(1)) \cdot \rho(n)$$

and

$$\text{Prob}[C_n(X_n) = f(X_n) | X_n \in R_n] < \frac{1}{2} + \epsilon(n)$$

holds for all possible circuits, C_n, of size $s'(n)$. The lemma follows. ∎

Appendix B: On the Selective XOR Lemma

Following [4, Exer. 7.17], we explicitly introduce a variant of the XOR Lemma, called the *Selective XOR Lemma*. Recall that the standard XOR Lemma refers to the predicate $P^{(t)}(x_1, ..., x_{t(n)}) \stackrel{\text{def}}{=} \prod_{i=1}^{t(n)} P(x_i)$, where P is the original predicate and $x_i \in \{0,1\}^n$ for every i. Instead, the Selective XOR Lemma refers to the predicate $Q^{(t)}(x_1, ..., x_{t(n)}, S) \stackrel{\text{def}}{=} \prod_{i \in S} P(x_i)$, where the x_i's are as before and $S \subseteq \{1, ..., t(n)\}$ is represented as an $t(n)$-bit long string. Thus, we have the following variant of Lemma 1.

Lemma 13 (Selective XOR Lemma): *Let P and $\mathbf{X} = \{X_n\}$ be as in Definition 1. For every function $t : \mathbb{N} \to \mathbb{N}$, define the predicate*

$$Q^{(t)}(x_1, ..., x_{t(n)}, S) \stackrel{\text{def}}{=} \prod_{i \in S} P(x_i) ,$$

where $x_1, ..., x_{t(n)} \in \{0,1\}^n$ and $S \subseteq \{1, ..., t(n)\}$. Let $\mathbf{Y}^{(t)} \stackrel{\text{def}}{=} \{(X_n^{(t)}, U_{t(n)})\}$, where $X_n^{(t)}$ is as in Lemma 1 and $U_{t(n)}$ be a random variable that is independently and uniformly distributed over $\{0,1\}^{t(n)}$.

(hypothesis) *As in Lemma 1; that is, suppose that for some function $s : \mathbb{N} \to \mathbb{N}$ and some bounded-away-from-1 function $\delta : \mathbb{N} \to [-1, +1]$, it holds that δ is an upper bound on the correlation of families of $s(\cdot)$-size circuits with P over $\mathbf{X}$.*

(conclusion) *Analogously to Lemma 1, there exists a bounded-away-from-1 function $\delta' : \mathbb{N} \to [-1, +1]$ and a polynomial p such that, for every function $t : \mathbb{N} \to \mathbb{N}$ and every function $\epsilon : \mathbb{N} \to [0, 1]$, the function*

$$\delta^{(t)}(n) \stackrel{\text{def}}{=} p(n) \cdot \delta'(n)^{t(n)} + \epsilon(n)$$

is an upper bound on the correlation of families of $s'(\cdot)$-size circuits with $Q^{(t)}$ over $\mathbf{Y}^{(t)}$, where

$$s'(t(n) \cdot n) \stackrel{\text{def}}{=} \text{poly}\left(\frac{\epsilon(n)}{n}\right) \cdot s(n) - \text{poly}(n \cdot t(n)).$$

In this appendix we discuss the relation of the Selective XOR Lemma to the XOR Lemma and to the Concatenation Lemma.

The Selective XOR Lemma vs the Concatenation Lemma. As shown in Section 5.2, the Concatenation Lemma implies the Selective XOR Lemma (by using Lemma 9). The opposite implication was recently shown in [16, Prop. 1.4]. The proof boils down to showing that any algorithm that computes the concatenation of the t values, can be used to correlate the selective XOR as follows: On input $(x_1, ..., x_t, S)$, we obtain (from the algorithm) a guess $(b_1, ..., b_t)$ for $(P(x_1), ..., P(x_t))$, and output $b(S) \stackrel{\text{def}}{=} \prod_{i \in S} b_i$. Note that if $(b_1, ..., b_t) = (P(x_1), ..., P(x_t))$, then our answer $b(S)$ is correct for any S, whereas if $(b_1, ..., b_t) \neq (P(x_1), ..., P(x_t))$, then $\text{Prob}_S[b(S) = \prod_{i \in S} P(x_i)] = 1/2$. Thus, if the algorithm is correct with probability p, then our answer has correlation p with $Q^{(t)}$.

The Selective XOR Lemma implies the XOR Lemma. This implication was sketched in Section 5.2, and we provide more details next. We show how to use an algorithm that correlates $P^{(t)}$ in order to correlate $Q^{(t)}$. We shall use t random examples, denoted $(z_1, P(z_1)), ..., (z_t, P(z_t))$. On input $(x_1, ..., x_t, S)$, we set $x'_i = x_i$ if $i \in S$ and $x'_i = z_i$ otherwise, obtain (from the algorithm) a guess b for $P^{(t)}(x'_1, ..., x'_t)$, and output $b \cdot \prod_{i \in [t] \setminus S} P(z_i)$. Thus, our answer is correct if and only if $b = P^{(t)}(x'_1, ..., x'_t)$, because $P^{(t)}(x'_1, ..., x'_t)$ equals $Q^{(t)}(x_1, ..., x_t, S) \cdot \prod_{i \in [t] \setminus S} P(z_i)$.

The XOR Lemma implies the Selective XOR Lemma. Following [16, Prop. 1.4], we show how to use an algorithm that correlates $Q^{(3t)}$ in order to correlate $P^{(t)}$. Here we shall use $3t$ random examples, denoted $(z_1, P(z_1)), ..., (z_{3t}, P(z_{3t}))$. On input $(x_1, ..., x_t)$, we select at random a subset $S \subseteq \{1, ..., 3t\}$, and let $i_1, ..., i_t$ be arbitrary t distinct elements of S (assuming that $|S| \geq t$). Next, we set $x'_{i_j} = x_j$ for every $j = 1, .., t$, and set $x'_i = z_i$ for every $i \in S'$, where $S' \stackrel{\text{def}}{=} \{1, ..., 3t\} \setminus \{i_j : j = 1, ..., t\}$. We obtain (from the algorithm) a guess b for $Q^{(3t)}(x'_1, ..., x'_{3t}, S)$, and output $b \cdot \prod_{i \in S \setminus S'} P(z_i)$. Thus, our answer is correct if and only if $b = Q^{(3t)}(x'_1, ..., x'_{3t}, S)$, because $Q^{(3t)}(x'_1, ..., x'_{3t}, S)$ equals $P^{(t)}(x_1, ..., x_t) \cdot \prod_{i \in S \setminus S'} P(z_i)$. Note that this works assuming that $|S| \geq t$, which holds with probability $1 - 2^{-\Omega(t)}$. Thus, our correlation with $P^{(t)}$ is lower bounded by $p - 2^{-\Omega(t)}$, where p is the correlation of the given algorithm with $Q^{(3t)}$.

A Sample of Samplers: A Computational Perspective on Sampling

Oded Goldreich

Abstract. We consider the problem of estimating the average of a huge set of values. That is, given oracle access to an arbitrary function $f : \{0,1\}^n \to [0,1]$, we wish to estimate $2^{-n} \sum_{x \in \{0,1\}^n} f(x)$ upto an additive error of ϵ. We are allowed to employ a randomized algorithm that may err with probability at most δ.

We survey known algorithms for this problem and focus on the ideas underlying their construction. In particular, we present an algorithm that makes $O(\epsilon^{-2} \cdot \log(1/\delta))$ queries and uses $n + O(\log(1/\epsilon)) + O(\log(1/\delta))$ coin tosses, both complexities being very close to the corresponding lower bounds.

Keywords: Sampling, randomness complexity, saving randomness, pairwise independent random variables, Expander graphs, random walks on graphs, information theoretic lower bounds.

An earlier version of this survey appeared as TR97-020 of *ECCC*. The current version includes a quantitative improvement in Theorem 6.1, which is obtained by the subsequent work of [26].

Preface. The idea of writing this survey occurred to me when finding out that a brilliant, young researcher who has worked in very related areas was unaware of the Median-of-Averages Sampler (of [7]). It then occurred to me that many of the results surveyed here have appeared in papers devoted to other subjects (indeed, the Median-of-Averages Sampler is an excellent example), and have thus escaped the attention of a wider community, which might have cared to know about them. I thus decided to write a survey that focuses on these very basics.

1 Introduction

In many settings repeated sampling is used to estimate the average value of a huge set of values. Namely, one has access to a value function ν, which is defined over a huge space (say, $\nu : \{0,1\}^n \to [0,1]$), and wishes to approximate $\bar{\nu} \overset{\text{def}}{=} \frac{1}{2^n} \sum_{x \in \{0,1\}^n} \nu(x)$ without having to inspect the value of ν on the entire domain. It is well-known that sampling ν at sufficiently many (random) points yields such an approximation, but we are interested in the complexity of the approximation. Specifically, (1) how many samples are required? (2) how much randomness is required to generate these samples? and (3) is this generation procedure efficient?

O. Goldreich et al.: Studies in Complexity and Cryptography, LNCS 6650, pp. 302–332, 2011.

We comment that it is essential to have the range of ν be bounded (or else no reasonable approximation may be possible). Our convention of having $[0, 1]$ be the range of ν is adopted for simplicity, and the problem for other (predetermined) ranges can be treated analogously.

1.1 Formal Setting

Our notion of approximation depends on two parameters: *accuracy* (denoted ϵ) and *error probability* (denoted δ). We wish to have an algorithm that, with probability at least $1 - \delta$, gets within ϵ of the correct value. This leads to the following definition.

Definition 1.1 (sampler): *A sampler is a randomized algorithm that on input parameters n (length), ϵ (accuracy) and δ (error), and oracle access to any function $\nu : \{0, 1\}^n \to [0, 1]$, outputs, with probability at least $1 - \delta$, a value that is at most ϵ away from $\bar{\nu} \stackrel{\text{def}}{=} \frac{1}{2^n} \sum_{x \in \{0,1\}^n} \nu(x)$. Namely,*

$$\Pr\left[|\text{sampler}^{\nu}(n, \epsilon, \delta) - \bar{\nu}| > \epsilon\right] < \delta, \tag{1}$$

where the probability is taken over the internal coin tosses of the sampler.

We are interested in "the complexity of sampling" quantified as a function of the parameters n, ϵ and δ. Specifically, we will consider three complexity measures:

1. Sample Complexity: The number of oracle queries made by the sampler.
2. Randomness Complexity: The number of (unbiased) coin tosses performed by the sampler.
3. Computational Complexity: The running-time of the sampler.
   We say that a sample is efficient if its running-time is polynomial in the total length of its queries (i.e., polynomial in both its sample complexity and in the length parameter, n).

We will focus on efficient samplers. Furthermore, we will focus on efficient samplers that have optimal (upto a constant factor) sample complexity, and will be interested in having the randomness complexity be as low as possible.

1.2 Overview

The straightforward method (or the naive sampler) consists of *uniformly and independently* selecting sufficiently many sample points (queries), and outputting the average value of the function on these points. Using Chernoff Bound one can easily show that $O(\frac{\log(1/\delta)}{\epsilon^2})$ sample points suffice. The naive sampler is optimal (upto a constant factor) in its sample complexity, but is quite wasteful in randomness. In Section 2, we discuss the naive sampler and present lower (and upper) bounds on the sample and randomness complexities of samplers. These bounds will guide our quest for improvements.

Pairwise-independent sampling yields a great saving in the randomness complexity. In Section 3 we present the Pairwise-Independent Sampler, and discuss its advantages and disadvantages. Specifically, for constant $\delta > 0$, the Pairwise-Independent Sampler is optimal upto a constant factor in both its sample and randomness complexities. However, for small δ (i.e., $\delta = o(1)$), its sample complexity is wasteful.

An additional idea is required for going further, and a relevant tool – random walks on expander graphs (see Appendix A) – is also used. In Section 4, we combine the Pairwise-Independent Sampler with the Expander Random Walk Technique to obtain a new sampler. Loosely speaking, the new sampler uses a random walk on an expander to generate a sequence of $\ell \stackrel{\text{def}}{=} O(\log(1/\delta))$ (related) random pads for ℓ invocations of the Pairwise-Independent Sampler. Each of these invocations returns an ϵ-close approximation with probability at least 0.99. The expander walk technique yields that, with probability at least $1 - \exp(-\ell) = 1 - \delta$, most of these ℓ invocations return an ϵ-close approximation. Thus, the median value is an (ϵ, δ)-approximation to the correct value (i.e., an approximation that, with probability at least $1 - \delta$, is within an additive term of ϵ of the correct value). The resulting sampler, called the Median-of-Averages Sampler, has sample complexity $O(\frac{\log(1/\delta)}{\epsilon^2})$ and randomness complexity $2n + O(\log(1/\delta))$.

In Section 5 we present an alternative sampler that improves over the pairwise-independent sampler. Maintaining the sample complexity of the latter (i.e., $O(1/\delta\epsilon^2)$), the new sampler has randomness complexity $n + O(\log(1/\delta\epsilon))$ (rather than $2n$). Combining this new sampler with the Expander Random Walk Technique, we obtain sample complexity $O(\frac{\log(1/\delta)}{\epsilon^2})$ and randomness complexity $n + O(\log(1/\delta)) + O(\log(1/\epsilon))$. Better bounds are obtained for the case of "Boolean samplers" (i.e., algorithms that must only well-approximate Boolean functions). In addition, in Section 5 we present two general techniques for improving existing samplers.

We conclude with some open problems (see Section 6). In particular, we discuss the notion of "oblivious" (or "averaging") samplers, which is closely related to the notion of randomness extractors (see Section 7.2 and more details in [28]).[1] Section 7 sketches the outline of an alternative survey that focuses on the notion of "averaging" samplers and on their relation to general samplers, on the one hand, and to randomness extractors, on the other hand.

The Hitting Problem. In order to distinguish the all-zero function from a function having at least an ϵ fraction of non-zero values, the sampler must query the function at a non-zero value (or "hit" some non-zero value). Thus, any sampler solves the *hitting problem*, as surveyed in Appendix C. That is, given an oracle to a Boolean function having at least an ϵ fraction of 1's, the "hitter" is required to

---

[1] Indeed, the current text focuses on general samplers, which are not necessarily of the "averaging" type (e.g., the aforementioned Median-of-Averages Sampler). Thus, this survey barely mentions the vast body of work that focuses on randomness extractors, and the interested reader is indeed referred to [28].

find an input that evaluates to 1. As noted above, each sampler can be used for this purpose, but this is an over-kill. Indeed, all results and techniques regarding samplers (presented in the main text of this survey) have simpler analogues for the hitting problem. Thus, Appendix C can be read as a warm-up towards the rest of the survey.

2   The Information Theoretic Perspective

The Naive Sampler, presented below, corresponds to the information theoretical (or statistician) perspective of the problem. We augment it by a lower bound on the *sample complexity* of samplers, which is in the spirit of these areas. We conclude with lower and upper bounds on the *randomness complexity* of samplers. The latter lower bound is also information theoretic in nature, but it refers to a concern that is more common in computer science.

2.1   The Naive Sampler

The straightforward sampling method consists of randomly selecting a small sample set S and outputting $\frac{1}{|S|} \sum_{x \in S} \nu(x)$ as an estimate to $\bar{\nu}$. More accurately, we select m *independently and uniformly distributed* strings in $\{0, 1\}^n$, denoted $s_1, ..., s_m$, and output $\frac{1}{m} \sum_{i=1}^{m} \nu(s_i)$ as our estimate. Setting $m = \frac{\ln(2/\delta)}{2\epsilon^2}$, we refer to this procedure as to the Naive Sampler.

To analyze the performance of the Naive Sampler, we use the Chernoff Bound. Specifically, we define m independent random variables, denoted $\zeta_1, ..., \zeta_m$, such that $\zeta_i \stackrel{\text{def}}{=} \nu(s_i)$, where the s_i's are independently and uniformly distributed in $\{0, 1\}^n$. By Chernoff Bound:

$$\Pr \left[\left| \bar{\nu} - \frac{1}{m} \sum_{i=1}^{m} \zeta_i \right| > \epsilon \right] \leq 2 \exp \left(-2\epsilon^2 m \right) \tag{2}$$

$$< \delta \tag{3}$$

where Eq. (3) is due to $m = \ln(2/\delta)/2\epsilon^2$. Observing that $\frac{1}{m} \sum_{i=1}^{m} \zeta_i$ represents the estimate output by the Naive Sampler, we have established that the Naive Sampler indeed satisfies Definition 1.1 (i.e., is indeed a sampler). We now consider the complexity of the Naive Sampler

- Sample Complexity: $m \stackrel{\text{def}}{=} \frac{\ln(2/\delta)}{2\epsilon^2} = \Theta(\frac{\log(1/\delta)}{\epsilon^2})$.
- Randomness Complexity: $m \cdot n = \Theta(\frac{\log(1/\delta)}{\epsilon^2} \cdot n)$.
- Computational Complexity: indeed efficient.

In light of Theorem 2.1 (below), the sample complexity of the Naive Sampler is optimal upto a constant factor. However, as we will shortly see, it is extremely wasteful in its usage of randomness. In fact, the rest of this survey is devoted to presenting ways for redeeming the latter aspect.

2.2   A Sample Complexity Lower Bound

We first assert that the Naive Sampler is quite good as far as sample complexity is concerned. The following theorem is analogous to many results known in statistics, though we are not aware of a reference prior to [10] where it can be found.

Theorem 2.1 [10]: *Any sampler has sample complexity bounded below by*

$$\min\left\{2^{(n-4)/2}, \frac{\ln(1/O(\delta))}{4\epsilon^2}\right\}$$

provided $\epsilon \leq \frac{1}{8}$ and $\delta \leq \frac{1}{6}$.

Note that a (constant factor) gap remains between the lower bound asserted here and the upper bound established by the Naive Sampler. We conjecture that the lower bound can be improved. Motivated by the lower bound, we say that a sampler is **sample-optimal** if its sample complexity is $O(\frac{\log(1/\delta)}{\epsilon^2})$.

2.3   Randomness Complexity Lower and Upper Bounds

We first assert that the Naive Sampler is quite bad as far as randomness complexity is concerned. First evidence towards our claim is provided by a non-explicit (and so inefficient) sampler:

Theorem 2.2 [10]: *There exists a* (non-efficient) *sampler with sample complexity $\frac{2\ln(4/\delta)}{\epsilon^2}$ and randomness complexity $n + 2\log_2(2/\delta) + \log_2\log_2(1/\epsilon)$.*

The proof is by a probabilistic argument that, given the Naive Sampler, asserts the existence of a relatively small set of possible coin tosses under which this sampler behaves almost as under all possible coin tosses (with respect to any possible function ν). Actually, the randomness bound can be improved to $n + \log_2(1/\delta) - \log_2\log_2(1/\delta)$ while using a constant factor larger sample complexity and more sophisticated techniques [30]. More generally:

Theorem 2.3 [30]: *For every function $s : [0,1]^2 \to \mathbb{R}$ such that $s(\epsilon, \delta) \geq \frac{2\log_2(1/\delta)}{\epsilon^2}$, there exists a* (non-efficient) *sampler with sample complexity $s(\epsilon, \delta)$ and randomness complexity*

$$n + \log_2(1/\delta) + 2\log_2(4/\epsilon) - \log_2 s(\epsilon, \delta)$$

This gets us very close to the following lower bound.

Theorem 2.4 [10]: *Let $s : \mathbb{N} \times [0,1]^2 \to \mathbb{R}$. Any sampler that has sample complexity at most $s(n, \epsilon, \delta)$, has randomness complexity at least*

$$n + \log_2(1/\delta) - \log_2 s(n, \epsilon, \delta) - \log_2(1 - 2\epsilon)^{-1} - 2,$$

provided $\epsilon, \delta < 0.5$ and $s(n, \epsilon, \delta) \leq 2^{n-1}$.

The dependency of the lower bound on the sample complexity should not come as a surprise. After all, there exists a deterministic sampler that queries the function on the entire domain. Furthermore, the upper bound of Theorem 2.3 does express a similar trade-off between randomness complexity and sample complexity. Similarly, one should not be surprised at the effect of $1 - 2\epsilon$ on the bound: For example, when $\epsilon = 0.5$, a sample may merely output $\widetilde{\nu} = \frac{1}{2}$ as its estimate and always be within ϵ of the average of any function $\nu : \{0,1\}^n \to [0,1]$.

Using Theorem 2.4, we obtain a lower bound on the randomness complexity of any *sample-optimal* sampler:

Corollary 2.5 [10]: *Any sampler that has sample complexity* $O(\frac{\log(1/\delta)}{\epsilon^2})$, *has randomness complexity at least[2]*

$$n + (1 - o(1)) \cdot \log_2(1/\delta) - 2\log_2(1/\epsilon),$$

provided $\epsilon, \delta < 0.4$ *and* $\frac{\log(1/\delta)}{\epsilon^2} = o(2^n)$.

3 The Pairwise-Independent Sampler

To motivate the Pairwise-Independent Sampler, let us confront two well-known central limit theorems: Chernoff Bound, which refers to *totally independent* random variables, and Chebyshev's Inequality, which refers to *pairwise-independent* random variables

Chernoff Bound: Let $\zeta_1, ..., \zeta_m$ be *totally independent* random variables, each ranging in $[0,1]$ and having expected value μ. Then,

$$\Pr\left[\left|\mu - \frac{1}{m}\sum_{i=1}^{m}\zeta_i\right| > \epsilon\right] \leq 2\exp\left(-2\epsilon^2 m\right)$$

Chebyshev's Inequality: Let $\zeta_1, ..., \zeta_m$ be *pairwise-independent* random variables, each ranging in $[0,1]$ and having expected value μ. Then,

$$\Pr\left[\left|\mu - \frac{1}{m}\sum_{i=1}^{m}\zeta_i\right| > \epsilon\right] \leq \frac{1}{4\epsilon^2 m}$$

Our conclusion is that these two bounds essentially agree when $m = O(1/\epsilon^2)$. That is, in both cases $\Theta(1/\epsilon^2)$ identical random variables are necessary and sufficient to guarantee a concentration within ϵ with constant probability. Thus, if this is what we want, then there is no point in using the more sophisticated Chernoff Bound, which requires more of the random variables.

In the context of sampling, our conclusion is that for achieving an approximation to within ϵ accuracy with constant error probability, using $O(1/\epsilon^2)$ pairwise-independent random sample points is as good as using $O(1/\epsilon^2)$ totally independent random sample points. Furthermore, in the first case we may be save a lot in terms of randomness.

---

[2] The $o(1)$ term is actually $\frac{\log_2 O(\log(1/\delta))}{\log_2(1/\delta)}$.

The Pairwise-Independent Sampler [12]: On input parameters n, ϵ and δ, set $m \overset{\text{def}}{=} \frac{1}{4\epsilon^2\delta}$ and generate a sequence of m *pairwise-independently and uniformly distributed* strings in $\{0,1\}^n$, denoted $s_1, ..., s_m$. Using the oracle access to ν, output $\frac{1}{m}\sum_{i=1}^{m} \nu(s_i)$ as the estimate to $\bar{\nu}$. Using Chebyshev's Inequality, one can easily see that the Pairwise-Independent Sampler indeed satisfies Definition 1.1 (i.e., is indeed a sampler).

There are two differences between the Naive Sampler and the Pairwise-Independent Sampler. Whereas the former uses independently selected sample points, the latter uses a sequence of pairwise independent sample points. As we shall see, this allows the latter sampler to use much less randomness. On the other hand, the Naive Sampler uses $O(\frac{\log(1/\delta)}{\epsilon^2})$ samples (which is optimal upto a constant factor), whereas the Pairwise-Independent Sampler uses $O(\frac{1}{\epsilon^2\delta})$ samples. However, for constant δ, both samplers use essentially the same number of sample points. Thus, for constant δ, the Pairwise-Independent Sampler offers a saving in randomness while being sample-optimal.

Generating a Pairwise-Independent sequence: Whereas generating m totally independent random points in $\{0,1\}^n$ requires $m \cdot n$ unbiased coin flips, one can generate m ($m \le 2^n$) pairwise-independent random points using only $O(n)$ unbiased coin flips. We present two well-known ways of doing this.

1. **Linear functions over finite fields:** We associate $\{0,1\}^n$ with the finite field $F \overset{\text{def}}{=} \mathrm{GF}(2^n)$. Let $\alpha_1, ..., \alpha_m$ be $m \le |F|$ distinct elements of F. To generate a (pairwise-independent) sequence of length m, we uniformly and independently select $s, r \in F$, and let the ith element in the sequence be $e_i \overset{\text{def}}{=} r + \alpha_i s$ (where the arithmetic is that of F). The analysis of this construction "reduces" the stochastic independence of e_i and e_j to the linear independence of the vectors $(1 , \alpha_i)$ and $(1 , \alpha_j)$: For every $i \ne j$ and every $a, b \in F$, we have

$$\Pr_{r,s}[e_i = a \wedge e_j = b] = \Pr_{r,s}\left[\begin{pmatrix} 1 & \alpha_i \\ 1 & \alpha_j \end{pmatrix}\begin{pmatrix} r \\ s \end{pmatrix} = \begin{pmatrix} a \\ b \end{pmatrix}\right]$$

$$= \Pr_{r,s}\left[\begin{pmatrix} r \\ s \end{pmatrix} = \begin{pmatrix} 1 & \alpha_i \\ 1 & \alpha_j \end{pmatrix}^{-1}\begin{pmatrix} a \\ b \end{pmatrix}\right]$$

$$= \frac{1}{|F|^2}.$$

Only $2n$ random coins are required in this construction, but the drawback is that we need a representation of the field F (i.e., an irreducible polynomial of degree n over $\mathrm{GF}(2)$) which may not be easy to find in general.[3] Still, for specific values of n a good representation exists: Specifically, for $n = 2 \cdot 3^\ell$ (with ℓ integer), the polynomial $x^n + x^{n/2} + 1$ is irreducible [17, p. 96], and so we obtain a representation of $\mathrm{GF}(2^n)$ for such n's.

---

[3] Things are not better if we wish to work with a large field of prime cardinality; since we need to find such a prime.

2. Toeplitz matrices: To avoid problems with non-trivial representation, one may use the following construction. We associate $\{0,1\}^n$ with the n-dimensional vector space over GF(2). Let $v_1, ..., v_m$ be $m \leq 2^n$ distinct vectors in this vector space. A Toeplitz matrix is a matrix with all diagonals being homogeneous; that is, $T = (t_{i,j})$ is a Toeplitz matrix if $t_{i,j} = t_{i+1,j+1}$, for all i, j. Note that a Toeplitz matrix is determined by its first row and first column (i.e., the values of $t_{1,j}$'s and $t_{i,1}$'s). To generate a (pairwise-independent) sequence of length m, we uniformly and independently select an n-by-n Boolean Toeplitz matrix, T, and an n-dimensional Boolean vector u. We let the ith element in the sequence be $e_i \overset{\text{def}}{=} T v_i + u$ (where the arithmetic is that of the vector space). The analysis of this construction is given in Appendix B. Here, we merely note that $3n - 1$ random coins suffice for this construction,

Plugging-in either of these constructions, we obtain the following complexities for the Pairwise-Independent Sampler

- Sample Complexity: $\frac{1}{4\delta\epsilon^2}$.
- Randomness Complexity: $2n$ or $3n-1$, depending on which of the constructions is used.
- Computational Complexity: Indeed efficient.

We note that for constant δ, the sample and randomness complexities match the lower bounds upto a constant factor. However, as δ decreases, the sample complexity of the Pairwise-Independent Sampler increases faster than the corresponding complexity of the Naive Sampler. Redeeming this state of affairs is our next goal.

4   The (Combined) Median-of-Averages Sampler

Our goal here is to decrease the sample complexity of the Pairwise-Independent Sampler while essentially maintaining its randomness complexity. To motivate the new construction we first consider an oversimplified version of it.

Median-of-Averages Sampler (oversimplified): On input parameters n, ϵ and δ, set $m \overset{\text{def}}{=} \Theta(\frac{1}{\epsilon^2})$ and $\ell \overset{\text{def}}{=} \Theta(\log(1/\delta))$, generate ℓ independent m-element sequences, each being a sequence of m *pairwise-independently and uniformly distributed* strings in $\{0,1\}^n$. Denote the sample points in the ith sequence by $s_1^i, ..., s_m^i$. Using the oracle access to ν, compute $\widetilde{\nu}^i \overset{\text{def}}{=} \frac{1}{m} \sum_{j=1}^m \nu(s_j^i)$, for $i = 1, ..., \ell$, and output the *median value* among these $\widetilde{\nu}^i$'s. Using Chebyshev's Inequality (as in previous section), for each i, it holds that

$$\Pr[|\widetilde{\nu}^i - \bar{\nu}| > \epsilon] < 0.1$$

and so

$$\Pr\left[|\{i : |\widetilde{\nu}^i - \bar{\nu}| > \epsilon\}| \geq \frac{\ell}{2}\right] < \sum_{j=\ell/2}^{\ell} \binom{\ell}{j} \cdot 0.1^j \cdot 0.9^{\ell-j}$$
$$< 2^\ell \cdot 0.1^{\ell/2}$$
$$\leq \delta,$$

where the last inequality is due to the choice of ℓ. Thus, the oversimplified version described above is indeed a sampler and has the following complexities

- Sample Complexity: $\ell \cdot m = O(\frac{\log(1/\delta)}{\epsilon^2})$.
- Randomness Complexity: $\ell \cdot O(n) = O(n \cdot \log(1/\delta))$.
- Computational Complexity: Indeed efficient.

Thus, the sample complexity is optimal (upto a constant factor), but the randomness complexity is higher than what we aim for. To reduce the randomness complexity, we use the same approach as above, but take dependent sequences rather than independent ones. The dependency we use is such that essentially preserves the probabilistic behavior of independent choices. Specifically, we use random walks on expander graphs (cf., Appendix A) to generate a sequence of ℓ "seeds" each of length $O(n)$. Each seed is used to generate a sequence of m pairwise independent elements in $\{0,1\}^n$, as above. Let us generalize this construction as follows.

Theorem 4.1 (general median-composition [7]): *Suppose we are given an efficient sampler of sample complexity $s(n,\epsilon,\delta)$ and randomness complexity $r(n,\epsilon,\delta)$. Then:*

1. *There exists an efficient sampler with sample complexity $O(s(n,\epsilon,0.01) \cdot \log(1/\delta))$ and randomness complexity $r(n,\epsilon,0.01) + O(\log(1/\delta))$.*
2. *For any $c > 4$, there exists an $\alpha > 0$ and an efficient sampler with sample complexity $O(s(n,\epsilon,\alpha) \cdot \log(1/\delta))$ and randomness complexity $r(n,\epsilon,\alpha) + c \cdot \log_2(1/\delta)$.*

Proof: For Item 1, let $r \stackrel{\text{def}}{=} r(n,\epsilon,0.01)$. We use an explicit construction of expander graphs with vertex set $\{0,1\}^r$, degree d and second eigenvalue λ so that $\lambda/d < 0.1$. We consider a random walk of (edge) length $\ell - 1 = O(\log(1/\delta))$ on this expander, and use each of the ℓ vertices along the path as random coins for the given sampler. Thus, we obtain ℓ estimates to $\bar{\nu}$ and output the median value as the estimate of the new sampler. To analyze the performance of the resulting sampler, we let W denote the set of coin tosses (for the basic sampler) that make the basic sampler output an estimate that is ϵ-far from the correct value (i.e., $\bar{\nu}$). Thus, W denotes the set of coin tosses that are bad for the basic sampler, and by the hypothesis $\frac{|W|}{2^r} \leq 0.01$. Using Theorem A.4 (with some W_i's set to W and the others set to $\{0,1\}^r$), we infer that the probability that at least $\ell/2$ vertices of the path reside in W is smaller than

$$\sum_{j=\ell/2}^{\ell} \binom{\ell}{j} \cdot 0.02^{j/2} < 2^\ell \cdot 0.02^{\ell/4}$$

$$\leq \delta.$$

Note that we have used $\ell \cdot s(n,\epsilon,0.01)$ samples and $r + (\ell - 1) \cdot \log_2 d = r + O(\log(1/\delta))$ coin tosses. Item 1 follows.

Item 2 is proved using the same argument but using Ramanujan Graphs (and slightly more care). Specifically, we use Ramanujan graphs (i.e., expanders with $\lambda \leq 2\sqrt{d-1}$) with vertex set $\{0,1\}^r$, where $r \stackrel{\text{def}}{=} r(n,\epsilon,\alpha)$ and $\alpha = (\frac{\lambda}{d})^2$. Repeating the foregoing argument, with $\ell - 1 = \frac{2\log_2(1/\delta)}{\log_2(\alpha/8)}$, we obtain an efficient sampler that uses $\ell \cdot s(n,\epsilon,\alpha)$ samples and $r + (\ell-1) \cdot \log_2 d = r + (4 + \frac{16}{(\log_2 d)-8}) \cdot \log_2(1/\delta)$ coin tosses. Since this can be done with a sufficiently large d, Item 2 follows. ∎

Combining the Pairwise-Independent Sampler with Theorem 4.1, we get

Corollary 4.2 (The Median-of-Averages Sampler [7]): *There exists an efficient sampler with*

- Sample Complexity: $O(\frac{\log(1/\delta)}{\epsilon^2})$.
- Randomness Complexity: $O(n + \log(1/\delta))$.

Furthermore, we can obtain randomness complexity $2n + (4 + o(1)) \cdot \log_2(1/\delta)$.

In the next section, we further reduce the randomness complexity of samplers (from $2n + O(\log(1/\delta))$) to $n + O(\log(1/\epsilon) + \log(1/\delta))$, while maintaining the sample complexity (up-to a multiplicative constant).

Generalizing Theorem 4.1. A close look at the proof of Theorem 4.1 reveals the fact that the median value obtained via an expander random walk (on the vertex set $\{0,1\}^r$) is used as a sampler of accuracy 0.49 and error probability δ. This suggests the following generalization of Theorem 4.1: *Suppose we are given two efficient samplers such that the ith sampler has sample complexity $s_i(n,\epsilon,\delta)$ and randomness complexity $r_i(n,\epsilon,\delta)$. Then, for every $\delta_0 \in (0,0.5)$, there exists an efficient sampler of sample complexity $s_2(r, 0.5-\delta_0, \delta) \cdot s_1(n,\epsilon,\delta_0)$ and randomness complexity $r_2(r, 0.5-\delta_0, \delta)$, where $r \stackrel{\text{def}}{=} r_1(n,\epsilon,\delta_0)$.* Theorem 4.1 is derived as a special case, when using the expander random walk as the second sampler and setting $\delta_0 = 0.01$.

5 The Expander Sampler and Two Generic Transformations

The main result of this section is the following:

Theorem 5.1 [7,16]: *There exists an efficient sampler that has*

- Sample Complexity: $O(\frac{\log(1/\delta)}{\epsilon^2})$.
- Randomness Complexity: $n + \log_2(1/\epsilon) + O(\log(1/\delta))$.

The theorem is proved by applying Theorem 4.1 to a new efficient sampler that makes $O(\frac{1}{\delta\epsilon^2})$ oracle queries and tosses $n+\log_2(1/\epsilon)$ coins. We start by presenting a sampler for the special case of Boolean functions.

Definition 5.2 (Boolean sampler): *A* Boolean sampler *is a randomized algorithm that on input parameters* n, ϵ *and* δ, *and oracle access to any Boolean function* $\nu : \{0,1\}^n \to \{0,1\}$, *outputs, with probability at least* $1 - \delta$, *a value that is at most* ϵ *away from* $\bar{\nu} \stackrel{\text{def}}{=} \frac{1}{2^n} \sum_{x \in \{0,1\}^n} \nu(x)$. *Namely,*

$$\Pr[|\text{sampler}^\nu(n, \epsilon, \delta) - \bar{\nu}| > \epsilon] < \delta$$

where the probability is taken over the internal coin tosses of the sampler.

That is, unlike (general) samplers, a *Boolean sampler* is required to work well only when given access to a *Boolean* function. The rest of this section is organized as follows:

In Section 5.1 we present the Expander Sampler, which is a Boolean sampler of sample complexity $O(1/\delta\epsilon^2)$ and randomness complexity n. This sample complexity is obtained by using Ramanujan Graphs (rather than arbitrary expanders).

In Section 5.2 we present a (general) transformation of Boolean samplers to general ones.

In Section 5.3 we revisit the Expander Sampler, while using an arbitrary expander. More importantly, we present another generic composition of samplers, and obtain an alternative construction by using this composition in conjunction with the aforementioned sampler. Unlike the composition method that underlies Theorem 4.1, which reduces the error complexity (in an efficient manner), the current composition reduces the sample complexity.

Theorem 5.1 is proved by combining the ideas of Sections 5.1 and 5.2. An alternative proof of a somewhat weaker result is obtained by combining the ideas of Sections 5.1 and 5.3.

5.1   A Sampler for the Boolean Case

We start by presenting a sampler for the special case of Boolean functions. Our sampling procedure is exactly the one suggested by Karp, Pippinger and Sipser for hitting a witness set [22] (cf., Appendix C), yet the analysis is somewhat more involved. Furthermore, to get an algorithm that samples the universe only on $O(1/\delta\epsilon^2)$ points, it is crucial to use a Ramanujan graph in role of the expander in the Karp-Pippinger-Sipser method.

The sampler. We use an expander of degree $d = 4/\delta\epsilon^2$ second eigenvalue bounded by λ and associate the vertex set of the expander with $\{0,1\}^n$. The sampler consists of uniformly selecting a vertex, v, (of the expander) and averaging over the values assigned (by ν) to all the neighbors of v; that is, *the algorithm outputs the estimate*

$$\widetilde{\nu} \stackrel{\text{def}}{=} \frac{1}{d} \sum_{u \in N(v)} \nu(u), \tag{4}$$

where $N(v)$ *denotes the set of neighbors of vertex* v.

This algorithm has

- Sample Complexity: $O(\frac{1}{\delta \epsilon^2})$.
- Randomness Complexity: n.
- Computational Complexity: Indeed efficient; that is, polynomial in n, ϵ^{-1} and δ^{-1}.

Lemma 5.3 [16]: *The foregoing algorithm constitutes an efficient Boolean sampler.*

Proof: We denote by B the set of *bad* choices for the algorithm; namely, the set of vertices that once selected by the algorithm yield a wrong estimate. That is, $v \in B$ if

$$\left| \frac{1}{d} \sum_{u \in N(v)} \nu(u) - \bar{\nu} \right| > \epsilon. \tag{5}$$

Denote by B' the subset of $v \in B$ for which

$$\frac{1}{d} \sum_{u \in N(v)} \nu(u) > \bar{\nu} + \epsilon. \tag{6}$$

It follows that each $v \in B'$ has ϵd too many neighbors in the set $A \stackrel{\text{def}}{=} \{u : \nu(u) = 1\}$; namely,

$$|\{u \in N(v) : u \in A\}| > (\rho(A) + \epsilon) \cdot d, \tag{7}$$

where $\rho(A) \stackrel{\text{def}}{=} \frac{|A|}{N}$ and $N \stackrel{\text{def}}{=} 2^n$. Using the Expander Mixing Lemma (i.e., Lemma A.2), we get that

$$\begin{aligned}
\epsilon \cdot \rho(B') &= \left| \frac{|B'| \cdot (\rho(A) + \epsilon)d}{dN} - \rho(B') \cdot \rho(A) \right| \\
&\leq \left| \frac{|(B' \times A) \cap E|}{|E|} - \frac{|A|}{|V|} \cdot \frac{|B'|}{|V|} \right| \\
&\leq \frac{\lambda}{d} \cdot \sqrt{\rho(A) \cdot \rho(B')}.
\end{aligned}$$

Thus,

$$\rho(B') \leq \left(\frac{\lambda}{d\epsilon} \right)^2 \cdot \rho(A). \tag{8}$$

Using $\lambda \le 2\sqrt{d}$ and $d = \frac{4}{\delta\epsilon^2}$, we get $\rho(B') \le \delta \cdot \rho(A)$. Using a similar argument,[4] we can show that $\rho(B \setminus B') \le \delta \cdot (1 - \rho(A))$. Thus, $\rho(B) \le \delta$, and the claim follows. ∎

Comment 5.4 [16]: *Observe that if we were to use an arbitrary d-regular graph with second eigenvalue λ, then the foregoing proof would hold provided that*

$$\frac{\lambda}{d} \le \sqrt{\delta\epsilon^2}. \tag{9}$$

This yields, for any such d-regular graph, an efficient Boolean sampler with sam-ple complexity d and randomness complexity n.

5.2   From Boolean Samplers to General Samplers

The following generic transformation was suggested to us by Luca Trevisan.

Theorem 5.5 (Boolean samplers imply general ones): *Suppose we are given an efficient Boolean sampler of sample complexity $s(n, \epsilon, \delta)$ and randomness com-plexity $r(n, \epsilon, \delta)$. Then, there exists an efficient sampler with sample complexity $s(n + \log_2(1/\epsilon), \epsilon/2, \delta)$ and randomness complexity $r(n + \log_2(1/\epsilon), \epsilon/2, \delta)$.*

Proof: As a mental experiment, given an arbitrary function $\nu : \{0,1\}^n \to [0,1]$, we define a Boolean function $\mu : \{0,1\}^{n+\ell} \to \{0,1\}$, where $\ell \stackrel{\text{def}}{=} \log_2(1/\epsilon)$, as follows: For every x and $i = 1, ..., \epsilon^{-1}$, we set $\mu(x, i) \stackrel{\text{def}}{=} 1$ if and only if $\nu(x) > (i - 0.5) \cdot \epsilon$ (i.e., iff $i < \epsilon^{-1}\nu(x) + 0.5$). Then, for every x, it holds that $|\nu(x) - \epsilon \cdot \sum_{i=1}^{1/\epsilon} \mu(x, i)| \le \epsilon/2$. Thus, if we were to sample μ and obtain an $\epsilon/2$-approximation of $\bar{\mu}$ then we get an ϵ-approximation of $\bar{\nu}$. Now, although we don't have actual access to μ we can emulate its answers given an oracle to ν.

Given a Boolean sampler, B, we construct a general sampler, A, as follows. On input n, ϵ, δ and access to an arbitrary ν as above, algorithm A sets $n' = n + \ell$, $\epsilon' = \epsilon/2$, and $\delta' = \delta$, and invoke B on input n', ϵ', δ'. When B makes a query $(x, i) \in \{0,1\}^n \times \{0,1\}^\ell$, algorithm A queries for $\nu(x)$ and returns 1 if and only if $\nu(x) > (i - 0.5) \cdot \epsilon$. When B halts with output v, algorithm A does the same. The theorem follows. ∎

Combining the sampler of Section 5.1 with Theorem 5.5, we get

Corollary 5.6 (The Expander Sampler, revisited): *There exists an efficient sampler that has*

- Sample Complexity: $O(\frac{1}{\delta\epsilon^2})$.
- Randomness Complexity: $n + \log_2(1/\epsilon)$.

Theorem 5.1 follows by combining Corollary 5.6 with Theorem 4.1.

---

[4] That is, we consider the set $B'' \stackrel{\text{def}}{=} B \setminus B'$, and observe that every $v \in B''$ has ϵd too many neighbours in $A'' \stackrel{\text{def}}{=} \{0,1\}^n \setminus A$. Hence, we conclude that $\rho(B'') \le \delta \cdot \rho(A'')$.

5.3   An Alternative Construction

Using an arbitrary expander graph (with $d = \text{poly}(1/\epsilon\delta)$ and $\frac{\lambda}{d} < \sqrt{\delta\epsilon^2}$) and invoking Comment 5.4, we have an efficient Boolean sampler with sample complexity $\text{poly}(1/\epsilon\delta)$ and randomness complexity n. Using Theorem 5.5, we get

Corollary 5.7 (The Expander Sampler, revisited again): *There exists an efficient sampler with* sample complexity $\text{poly}(1/\epsilon\delta)$ *and* randomness complexity $n + \log_2(1/\epsilon)$.

To derive (a weaker form of) Theorem 5.1 via the foregoing sampler, we first need to reduce its sample complexity. This is done via the following general transformation. We say that a sampler is of the **averaging type** if its output is the average value obtained on its queries, which in turn are determined as a function of its own coin tosses (independently of the answers obtained on previous queries).

Theorem 5.8 (reducing sample complexity (or "sampling the sample")): *Suppose we are given two efficient samplers such that the ith sampler has sample complexity $s_i(n, \epsilon, \delta)$ and randomness complexity $r_i(n, \epsilon, \delta)$. Further suppose that the first sampler is of the averaging type. Then, there exists an efficient sampler of sample complexity $s_2(\log_2 s_1(n, \epsilon/2, \delta/2), \epsilon/2, \delta/2)$ and randomness complexity $r_1(n, \epsilon/2, \delta/2) + r_2(\log_2 s_1(n, \epsilon/2, \delta/2), \epsilon/2, \delta/2)$. Furthermore, if also the second sampler is of the averaging type, then so is the resulting sampler.*

Proof: We compose the two samplers as follows. Setting $m \overset{\text{def}}{=} s_1(n, \epsilon/2, \delta/2)$, we invoke the first sampler and determine the m queries it would have asked (given a particular choice of its coins).[5] We then use the second sampler to sample these m queries (invoking it with parameters $\log_2 m, \epsilon/2$ and $\delta/2$). Specifically, we let the second sampler make virtual queries into the domain $[m] \overset{\text{def}}{=} \{1, ..., m\}$ and answer a query $q \in [m]$ by the value of the function at the ith query specified by the first sampler. That is, given access to a function $\nu : \{0,1\}^n \to [0,1]$, and determining a sequence r of coins for the first sampler, we consider the function $\nu_r : [m] \to [0,1]$ defined by letting $\nu_r(i) = \nu(q_{r,i})$ where $q_{r,i}$ is the ith query made by the first sampler on coins r. We run the second sampler providing it virtual access to the function ν_r in the obvious manner, and output its output. Thus, the complexities are as claimed and the combined sampler errs if either $|\bar{\nu} - \frac{1}{m}\sum_{i=1}^m \nu(q_{r,i})| > \frac{\epsilon}{2}$ or $|\frac{1}{m}\sum_{i=1}^m \nu(q_{r,i}) - \tilde{\nu}_r| > \epsilon/2$, where $\tilde{\nu}_r$ is the estimate output by the second sampler when given virtual access to ν_r. Observing that the first event means that the first sampler errs (here we use the hypothesis that this sampler is averaging) and that the second event means that the second sampler errs (here we use $\sum_{i=1}^m \nu(q_{r,i}) = \bar{\nu}_r$), we are done.   ∎

---

[5] Here we use the hypothesis that the first sampler is non-adaptive; that is, its queries are determined by its coin tosses (independently of the answers obtained on previous queries).

It is tempting to try to improve the sample complexity of the sampler asserted in Corollary 5.7 by combining it with the Pairwise-Independent Sampler, via Theorem 5.8. The problem is that the former sampler, which we wish to use in the role of the outer sampler, is not of the averaging type. Indeed, the expander sampler (of Comment 5.4) is of the averaging type, but the proof of Theorem 5.5 does not preserve this feature. Instead, as shown in Theorem 5.10 (below), any Boolean sampler of the averaging type is a general sampler of the averaging time, except that its accuracy and error probability may increase by a constant factor. Thus, combining the sampler of Comment 5.4 with the Pairwise-Independent Sampler, via Theorem 5.8, we obtain:

Corollary 5.9 (sampling the Expander Sampler): *There exists an efficient sampler that has*

- Sample Complexity: $O(\frac{1}{\delta \epsilon^2})$.
- Randomness Complexity: $n + O(\log(1/\epsilon)) + O(\log(1/\delta))$.

Indeed, the sampler asserted in Corollary 5.9 operates by selecting a random vertex in an expander and taking a pairwise-independent sample of its neighbor set. A weaker form of Theorem 5.1 (i.e., with an $O(\log(1/\epsilon)$ term rather than with a $\log_2(1/\epsilon)$ term) follows by combining Corollary 5.9 with Theorem 4.1.

It is left to establish the aforementioned claim by which any Boolean sampler of the averaging type is a general sampler (of the averaging time), except that its accuracy and error probability may increase by a constant factor. (A similar statement was proved in [30].)

Theorem 5.10 (Boolean vs general samplers of the averaging type): *Every Boolean sampler of the averaging type, having sample complexity $s(n, \epsilon, \delta)$ and randomness complexity $r(n, \epsilon, \delta)$, is a general sampler (of the averaging type) with sample complexity $s(n, \epsilon/4, \delta/3)$ and randomness complexity $r(n, \epsilon/4, \delta/3)$.*

Proof: For any function $\nu : \{0, 1\}^n \to [0, 1]$, we consider a random function $\rho : \{0, 1\}^n \to \{0, 1\}$ such that, for every x, we set $\rho(x) = 1$ with probability $\nu(x)$, independently of the setting of all other arguments. Clearly, with probability $1 - \exp(-2\epsilon^2 2^n) > 1 - \delta$, it holds that $|\bar\nu - \bar\rho| < \epsilon$. Furthermore, fixing any possible outcome of the sampler's coins, with probability at least $1 - \exp(-8\epsilon^2 s)$ over the choice of ρ, the average of the ρ-values queried by the sampler is 2ϵ-close to the average of the ν-values, where s denotes the number of queries. Since (by Theorem 2.1) $s > \epsilon^{-2} \log(1/\delta)/8$, with probability at least $1 - \delta$ over the choice of ρ, the average that the Boolean sampler outputs when given access to ν is 2ϵ-close to the average it would have output on a random ρ, which in turn (with probability at least $1 - \delta$ over the sampler's coins) is ϵ-close to $\bar\rho$. Thus, with probability at least $1 - 3\delta$ (over the sampler's coins), the Boolean sampler outputs a value that is 4ϵ-close to $\bar\nu$, ∎

6   Conclusions and Open Problems

The main results surveyed in the text are summarized in Figure 1. The first row tabulates $\Omega(\epsilon^{-2}\log(1/\delta))$ as a lower bound on sample complexity and the subsequent three rows refer to sample-optimal samplers (i.e., samplers of sample complexity $O(\epsilon^{-2}\log(1/\delta))$). The last row refers to a sampler (cf., Thm. 6.1 below) that has randomness complexity closer to the lower bound. However, this sampler is not sample-optimal.

	sample complexity	randomness complexity	pointer
lower bound	$\Omega(\frac{\log(1/\delta)}{\epsilon^2})$		Thm. 2.1
lower bound	for $O(\frac{\log(1/\delta)}{\epsilon^2})$	$n + (1 - o(1)) \cdot \log_2(1/\delta) - 2\log_2(1/\epsilon)$	Cor. 2.5
upper bound	$O(\frac{\log(1/\delta)}{\epsilon^2})$	$n + \log_2(1/\delta)$	Thm. 2.3
algorithm	$O(\frac{\log(1/\delta)}{\epsilon^2})$	$n + O(\log(1/\delta)) + \log_2(1/\epsilon)$	Thm. 5.1
algorithm	$\mathrm{poly}(\epsilon^{-1}, \log(1/\delta))$	$n + (1 + \alpha) \cdot \log_2(1/\delta), \forall \alpha > 0$	Thm. 6.1

Fig. 1. Summary of main results

The randomness complexity of sample-optimal samplers. A closer look at the randomness complexity of sample-optimal samplers is provided in Figure 2. The first two rows tabulate lower and upper bounds, which are $2\log_2(1/\epsilon) + O(1)$ apart. Our conjecture is that the lower bound can be improved to match the upper bound.[6] The efficient samplers use somewhat more than $n + 4 \cdot \log_2(1/\delta)$ coins, where one factor of 2 is due to the use of expanders and the other to the "median-of-averages paradigm". As long as we stick to using expanders in the Median-of-Averages Sampler, there is no hope to reduce the first factor, which is due to the relation between the expander degree and its second eigenvalue. In fact, achieving a factor of 4 rather than a bigger factor is due to the use of Ramanujan Graphs (which have the best possible such relation).

Boolean samplers vs general ones. Another fact presented in Figure 2 is that we can currently do better if we are guaranteed that the oracle function is Boolean (rather than mapping to the interval [0, 1]). We stress that the lower bound holds also with respect to samplers that need only to work for Boolean functions.

Adaptive vs non-adaptive. All known samplers are non-adaptive; that it, they determine the sample points (queries) solely as a function of their coin tosses. In contrast, *adaptive* samplers may determine the next query depending on the value of the function on previous queries. Intuitively, adaptivity should not help the sampler. Indeed, all lower bounds refer also to adaptive samplers, whereas all

---

[6] Partial support for this conjecture was offered to us recently by Ronen Shaltiel (priv. comm., 2010). He observed that one $\log_2(1/\epsilon)$ term can be shaved off the lower bound in the special case of averaging samplers (see below), by using the connection to randomness extractors and a lower bound on entropy loss due to [25].

lower bound (even for Boolean)	$n + \log_2(1/\delta) - 2\log_2(1/\epsilon) - \log_2\log_2(1/\delta) - O(1)$
upper bound	$n + \log_2(1/\delta) - \log_2\log_2(1/\delta)$
efficient samplers	$n + (4+\alpha)\log_2(1/\delta) + \log_2(1/\epsilon)$, for any $\alpha > 0$
efficient Boolean samplers	$n + (4+\alpha)\log_2(1/\delta)$, for any $\alpha > 0$

Fig. 2. The randomness complexity of samplers that make $\Theta(\frac{\log(1/\delta)}{\epsilon^2})$ queries

upper bound only utilizes non-adaptive samplers. This indicates that the difference between adaptive samplers and non-adaptive ones can not be significant. In a preliminary version of this survey we advocated providing a direct and more tight proof of the foregoing intuition. When referring to the sample complexity, such a simple proof was provided in [6, Lem. 9]: It amounts to observing that adapting queries made to a random isomorphic copy of a function f are equivalent to uniformly and independently distributed queries made to f. Thus, adaptivity offers no advantage in this setting.

Averaging (or oblivious) *samplers.* A special type of non-adaptive samplers are ones that output the average value of the function over their sample points. Such samplers were first defined in [9], where they were called "oblivious", but we prefer the term *averaging*. (Recall that we have already defined and used such samplers in Section 5.3.) We mention that averaging samplers have some applications not offered by arbitrary non-adaptive samplers (cf., [9] and [29]). More importantly, averaging samplers are very appealing, since averaging over a sample seem *the natural thing to do*. Furthermore, as pointed out in [30], averaging samplers are closely related to randomness extractors (see Section 7 and more details in [28]). Note that the Naive Sampler, the Pairwise-Independent Sampler, and the Expander Sampler are all averaging samplers, although they differ in the way they generate their sample. However, the Median-of-Averages Sampler, as its name indicates, is not an averaging sampler. Thus, obtaining an averaging sampler of relatively low sample and randomness complexities requires an alternative approach. The best results are obtained via the connection to randomness extractors, and are summarized below.

Theorem 6.1 (efficient averaging samplers [26, Cor. 7.3]):[7] *For every constant $\alpha > 0$, there exists an efficient averaging sampler with*

- Sample Complexity: $\mathrm{poly}(\epsilon^{-1}, \log(1/\delta))$.
- Randomness Complexity: $n + (1+\alpha) \cdot \log_2(1/\delta)$.

We stress that this sampler is not sample-optimal (i.e., the polynomial in ϵ^{-1} is not quadratic). It would be interesting to obtain an efficient sample-optimal *averaging* sampler of low randomness complexity, say, one that uses $O(n+\log(1/\delta))$

---

[7] The result builds on [30], and uses [18, Thm. 1.5] in order to remove a mild restriction on the value of ϵ.

coins. We mention that non-explicit sample-optimal averaging samplers of low randomness complexity do exist; specifically, Theorems 2.2 and 2.3 holds with averaging-samplers (see [10,30], resp.).

7 Postscript: A Different Perspective

As stated in the introduction, the intention of the current survey was to provide a wide audience of theoretical computer scientists with a basic tutorial regarding samplers. The focus of this tutorial was on the complexity of sampling, and our aim was to *simultaneously* minimize three complexity measures: (1) the sample complexity, (2) the randomness complexity, and (3) the computational complexity. We actually focused on the minimization of the first two, while requiring that a minimal level of computational efficiency is maintained (i.e., that the sampler works in time that is polynomial in the total length of the queries made).

From our perspective, averaging samplers are of no special interest, except maybe for their natural appeal. An alternative perspective, strongly advocated by Ronen Shaltiel and Amnon Ta-Shma, may put averaging samplers and their relation to general samplers at the main focus. This is likely to yield a very interesting survey, which we outline in the rest of this section, but it is not the one we set out to write...

7.1 Average Samplers versus General Samplers

The alternative survey will focus on the question of whether non-averaging samplers can outperform averaging samplers. As noted by Amnon and Ronen, a good starting point for such a survey is the observation that the median of averages operation can be used for improving the performance of samplers, but it yields non-averaging samplers. Specifically, the median of averages operation can be combined with simple averaging samplers (e.g., the pairwise independent ones) to yield very strong and simple non-averaging samplers. Another interesting observation is that the currently known lower bound on the randomness complexity of sample-optimal averaging samplers is higher than the currently know bound for general samplers (see Footnote 6). Finally, when viewing the minimization of sample complexity as the primary goal and the minimization of the randomness complexity as the secondary goal, the median of averages operation enables constructing *efficient* samplers that are by far better (and also much simpler) than the currently known *efficient* averaging samplers.

Another interesting parameter is the Boolean versus general distinction, which was discussed in prior sections. Recall that in the case of averaging samplers, the two notions are almost identical (see Theorem 5.10), whereas for general sampler we currently lose a $\log_2(1/\epsilon)$ term in the randomness complexity (see Theorem 5.5). Focusing on sample-optimal samplers, we summarize the currently

lower bound (even for Boolean)	$n + \log_2(1/\delta) - 2\log_2(1/\epsilon) - \ell - O(1)$
lower bound for averaging samplers	$n + \log_2(1/\delta) - \log_2(1/\epsilon) - \ell - O(1)$
upper bound (by averaging samplers)	$n + \log_2(1/\delta) - \ell$
efficient samplers	$n + (4 + \alpha) \cdot \log_2(1/\delta) + \log_2(1/\epsilon), \forall \alpha > 0$
efficient averaging samplers	$n + (1 + \alpha) \cdot \log_2(1/\delta) + \widetilde{O}(s), \forall \alpha > 0$

Fig. 3. The randomness complexity of samplers that make $s \overset{\text{def}}{=} \Theta(\frac{\log(1/\delta)}{\epsilon^2})$ queries, where ℓ denotes $\log_2 \log_2(1/\delta)$

known results in Figure 3, where the three first rows ignore the question of efficiency (and the last row of Figure 3 is justified by combining Theorems 6.1 and 5.8).[8]

7.2 Average Samplers versus Randomness Extractors

We start by recalling the basic definition of randomness extractors, while (slightly) changing some common conventions to better fit our discussion.[9] Loosely speaking, a randomness extractor is a function Ext : $\{0,1\}^r \times [s] \to \{0,1\}^n$ that uses an $(\log_2 s)$-bit long random seed in order to transform an r-bit long (outcome of a) weak source of randomness into an n-bit long string that is almost uniformly distributed in $\{0,1\}^n$. Specifically, we consider arbitrary weak sources that are restricted (only) in the sense that, for a parameter k, no string appears as the source outcome with probability that exceeds 2^{-k}. Such sources are called (r,k)-sources (and k is called the min-entropy). A special type of (r,k)-sources are (r,k)-flat sources, which are sources in which each string appears with probability that equals either 2^{-k} or 0. We say that two distributions are ϵ-close if the statistical difference (a.k.a variation distance) between them is at most ϵ. Now, Ext is called a (k,ϵ)-extractor if for any (r,k)-source X it holds that $\text{Ext}(X, U_s)$ is ϵ-close to the uniform distribution over n-bit strings, where U_s denotes the uniform distribution over $[s]$.

There is a close relationship between extractors and averaging samplers. In order to discuss this relationship, it will be more convenient to state the performance guarantees of the sampler (i.e., ϵ and δ) in terms of its complexities (i.e., s and r), rather than the other way around (as done in the rest of this survey). Thus, we may say that a certain oracle machine (which has certain sample and randomness complexities) is an (ϵ, δ)-sampler if it satisfies Eq. (1) for these particular values of ϵ and δ.

We shall first show that any averaging sampler gives rise to an extractor. Let $G : \{0,1\}^r \to (\{0,1\}^n)^s$ be the sample generating algorithm of an averaging

---

[8] Specifically, we invoke Theorem 5.8 when using the sampler of Theorem 6.1 as the first (i.e., "outer") sampler, and the Naive Sampler as the second (i.e., "inner") sampler.

[9] Typically, extractors are defined as mapping $\{0,1\}^n \times \{0,1\}^s$ to $\{0,1\}^m$.

(ϵ, δ)-sampler. That is, G uses r bits of randomness and generates s sample points in $\{0,1\}^n$ such that, for every $f : \{0,1\}^n \to [0,1]$ with probability at least $1 - \delta$, the average of the f-values of these s pseudorandom points resides in the interval $[\overline{f} \pm \epsilon]$, where $\overline{f} \stackrel{\text{def}}{=} \sum_{x \in \{0,1\}^n} f(x)/2^n$. Define $\text{Ext} : \{0,1\}^r \times [s] \to \{0,1\}^n$ such that $\text{Ext}(\omega, i)$ is the ith sample generated by $G(\omega)$. We shall prove that Ext is a $(k, 2\epsilon)$-extractor, for $k = r - \log_2(\epsilon/\delta)$.

Suppose towards the contradiction that there exists a (r, k)-source X such that for some $S \subset \{0,1\}^n$ it is the case that $\Pr[\text{Ext}(X, U_s) \in S] > 2^{-n} \cdot |S| + 2\epsilon$. Then, without loss of generality, X is (r, k)-flat, and we consider the set

$$B = \{x \in \{0,1\}^r : \Pr[\text{Ext}(x, U_s) \in S] > 2^{-n} \cdot |S| + \epsilon\}.$$

Then, $|B| > \epsilon \cdot 2^k = \delta \cdot 2^r$, where the inequality holds since $\Pr[\text{Ext}(X, U_s) \in S] \leq \Pr[X \in B] + 2^{-n} \cdot |S| + \epsilon$. Defining $f(z) = 1$ if $z \in S$ and $f(z) = 0$ otherwise, it holds that $\overline{f} = |S|/2^m$. But, for every $\omega \in B$, the f-average of the sample $G(\omega)$ is greater than $\overline{f} + \epsilon$, in contradiction to the hypothesis that the sampler has error probability δ (with respect to accuracy ϵ).

We now turn to show that extractors give rise to averaging samplers. Let $\text{Ext} : \{0,1\}^r \times [s] \to \{0,1\}^n$ be a (k, ϵ)-extractor. Consider the sample generation algorithm $G : \{0,1\}^r \to (\{0,1\}^n)^s$ defined by $G(\omega) = (\text{Ext}(\omega, i))_{i \in [s]}$. We prove that G corresponds to an averaging (ϵ, δ)-sampler, for $\delta = 2^{-(r-k-1)}$.

Suppose towards the contradiction that there exists a function $f : \{0,1\}^n \to [0,1]$ such that for $\delta 2^r = 2^{k+1}$ strings $\omega \in \{0,1\}^r$ the average f-value of the sample $G(\omega)$ deviates from $\overline{f}$ by more than ϵ. Suppose, without loss of generality, that for at least half of these ω's the average is greater than $\overline{f} + \epsilon$, and let B denote the set of these ω's. Then, for X that is uniformly distributed on B (and is thus a (r, k)-source), we have

$$\text{Exp}[f(\text{Ext}(X, U_s))] > \text{Exp}[f(U'_n)] + \epsilon,$$

where U'_n denotes the uniform distribution on n-bit long strings. But, since $|f(z)| \leq 1$ for every z, this contradicts the hypothesis that $\text{Ext}(X, U_s)$ is ϵ-close to U'_n, because $|\text{Exp}[f(Y)] - \text{Exp}[f(Z)]|$ is upper bounded by the statistical difference between Y and Z (times $\max_z\{|f(z)|\}$). Summarizing the foregoing discussion, we obtain:

Theorem 7.1 (averaging samplers vs randomness extractors): *Let $r, s, k \in \mathbb{N}$ and $\epsilon, \delta \in [0, 1]$. Then:*

1. *If $\text{Ext} : \{0,1\}^r \times [s] \to \{0,1\}^n$ is a (k, ϵ)-extractor, then the sample generating algorithm $G : \{0,1\}^r \to (\{0,1\}^n)^s$ defined by $G(\omega) = (\text{Ext}(\omega, i))_{i \in [s]}$ yields an averaging (ϵ, δ)-sampler for $\delta = 2^{-(r-k-1)}$ (i.e., $r - k = \log_2(1/\delta) + 1$).*

2. *If $G : \{0,1\}^r \to (\{0,1\}^n)^s$ is the sample generating algorithm of an averaging (ϵ, δ)-sampler, then the algorithm $\text{Ext} : \{0,1\}^r \times [s] \to \{0,1\}^n$ defined by $\text{Ext}(\omega, i) = G(\omega)_i$ is a $(k, 2\epsilon)$-extractor, for $k = r - \log_2(\epsilon/\delta)$ (i.e., $r - k = \log_2(1/\delta) - \log_2(1/\epsilon)$).*

Note that starting with a $(k, 2\epsilon)$-extractor and applying both parts of Theorem 7.1, we obtain a $(k', 2\epsilon)$-extractor for $k' = k + 1 + \log_2(1/\epsilon)$. Thus, the translation offered by Theorem 7.1 is not optimal, yet the bounds provided in both directions are (in general) tight.[10]

The connection to averaging samplers and the desire to have averaging samplers of optimal sample and randomness complexities calls attention to a research direction regarding extractors that did not receive much attention. We refer to the construction of extractors with strongly optimal seed length and almost optimal extraction rate. That is, the seed length, which is $\log_2 s$ in terms of this section, should be optimal up to a *constant additive term*, whereas the extraction rate (i.e., n/k) (or rather the inverse loss rate (i.e., $(r-k)/(n-k)$)) should be close to 1.

Acknowledgments. I would like to thank Noga Alon, Nabil Kahale, Ronen Shaltiel, Amnon Ta-Shma, Luca Trevisan, and Salil Vadhan for useful discussions.

References

1. Ajtai, M., Komlos, J., Szemerédi, E.: Deterministic Simulation in LogSpace. In: Proc. 19th STOC, pp. 132–140 (1987)
2. Alon, N.: Eigenvalues, Geometric Expanders, Sorting in Rounds and Ramsey Theory. Combinatorica 6, 231–243 (1986)
3. Alon, N., Bruck, J., Naor, J., Naor, M., Roth, R.: Construction of Asymptotically Good, Low-Rate Error-Correcting Codes through Pseudo-Random Graphs. IEEE Transactions on Information Theory 38, 509–516 (1992)
4. Alon, N., Milman, V.D.: λ_1, Isoperimetric Inequalities for Graphs and Superconcentrators. J. Combinatorial Theory, Ser. B 38, 73–88 (1985)
5. Alon, N., Spencer, J.H.: The Probabilistic Method. John Wiley & Sons, Inc., Chichester (1992)

---

[10] To see the tightness of Part 1, consider an arbitrary (k, ϵ)-extractor, $\mathrm{Ext} : \{0,1\}^r \times [s] \to \{0,1\}^n$, and modify it such that, for every $x' \in \{0,1\}^k$ and $i \in [3\epsilon \cdot s]$, it holds that $\mathrm{Ext}(0^{r-k}x', i) = 0^n$. Then, the modified extractor is a $(k+2, 2\epsilon)$-extractor, but the resulting averaging sampler has error probability *at least* 2^{-r+k} with respect to deviation 2ϵ. (Recall that Part 1 asserts that the resulting averaging sampler has error probability at most $2^{-(r-k-3)}$ with respect to deviation 2ϵ.) To see the tightness of Part 2, consider an arbitrary avearging (ϵ, δ)-sampler with a sample generating algorithm $G : \{0,1\}^r \to (\{0,1\}^n)^s$, and modify the latter to be identically zero on $\delta 2^r$ seeds; that is, for an arbitrary $B \subset \{0,1\}^r$ of size $\delta 2^r$, redefine G such that for every $x \in B$ it holds that $G(x) = (0^n)^s$. Then, the modified averaging sampler is an $(\epsilon, 2\delta)$-sampler, but (as shown next) the resulting extractor can be a $(k', c\epsilon)$-extractor only if $k' > k + \log_2(1/\epsilon) - c'$, where $k \stackrel{\text{def}}{=} r - \log_2(1/\delta)$ and $c' = \log_2(c+1)$. The lower bound on k' holds because a (k', r)-source may assign B probability $2^{k-k'}$, whereas 0^n should be assigned probability at most $c\epsilon + 2^{-n}$. Thus, $2^{k-k'} \le c\epsilon + 2^{-n}$, which implies $k' - k > \log_2(1/\epsilon) - c'$. (Recall that Part 2 asserets that the resulting construct is a $(k', 2\epsilon)$-extractor for $k' = k + \log_2(1/\epsilon) + 1$.)

6. Bar-Yossef, Z., Kumar, R., Sivakumar, D.: Sampling Algorithms: Lower Bounds and Applications. In: 33rd STOC, pp. 266–275 (2001)
7. Bellare, M., Goldreich, O., Goldwasser, S.: Randomness in Interactive Proofs. Computational Complexity 4(4), 319–354 (1993); Extended abstract in 31st FOCS, pp. 318–326 (1990)
8. Bellare, M., Goldreich, O., Goldwasser, S.: Addendum to [7]. (May 1997), http://theory.lcs.mit.edu/~oded/papers.html
9. Bellare, M., Rompel, J.: Randomness-efficient oblivious sampling. In: 35th FOCS (1994)
10. Canetti, R., Even, G., Goldreich, O.: Lower Bounds for Sampling Algorithms for Estimating the Average. In: IPL, vol. 53, pp. 17–25 (1995)
11. Carter, L., Wegman, M.: Universal Classes of Hash Functions. J. Computer and System Sciences 18, 143–154 (1979)
12. Chor, B., Goldreich, O.: On the Power of Two–Point Based Sampling. Jour. of Complexity 5, 96–106 (1989)
13. Cohen, A., Wigderson, A.: Dispensers, Deterministic Amplification, and Weak Random Sources. In: 30th FOCS, pp. 14–19 (1989)
14. Gaber, O., Galil, Z.: Explicit Constructions of Linear Size Superconcentrators. JCSS 22, 407–420 (1981)
15. Goldreich, O., Impagliazzo, R., Levin, L.A., Venkatesan, R., Zuckerman, D.: Security Preserving Amplification of Hardness. In: 31st FOCS, pp. 318–326 (1990)
16. Goldreich, O., Wigderson, A.: Tiny Families of Functions with Random Properties: A Quality–Size Trade–off for Hashing. Journal of Random structures and Algorithms 11(4), 315–343 (1997)
17. Golomb, S.W.: Shift Register Sequences. Aegean Park Press, (1982) (revised edition)
18. Guruswami, V., Umans, C., Vadhan, S.: Unbalanced Expanders and Randomness Extractors from Parvaresh-Vardy Codes. JACM 56(4) (2009); Preliminary version in 22nd CCC 2007
19. Hoory, S., Linial, N., Wigderson, A.: Expander Graphs and their Applications. Bull. AMS 43(4), 439–561 (2006)
20. Impagliazzo, R., Zuckerman, D.: How to Recycle Random Bits. In: 30th FOCS, pp. 248–253 (1989)
21. Kahale, N.: Eigenvalues and Expansion of Regular Graphs. Journal of the ACM 42(5), 1091–1106 (1995)
22. Karp, R.M., Pippinger, N., Sipser, M.: A Time-Randomness Tradeoff. In: AMS Conference on Probabilistic Computational Complexity, Durham, New Hampshire (1985)
23. Lubotzky, A., Phillips, R., Sarnak, P.: Explicit Expanders and the Ramanujan Conjectures. In: Proc. 18th STOC, pp. 240–246 (1986)
24. Margulis, G.A.: Explicit Construction of Concentrators. Prob. Per. Infor. 9(4), 71–80 (1973); In Russian, English translation in Problems of Infor. Trans., 325–332 (1975)
25. Radhakrishnan, J., Ta-Shma, A.: Bounds for Dispersers, Extractors, and Depth-Two Superconcentrators. SIAM J. Discrete Math. 13(1), 2–24 (2000)
26. Reingold, O., Vadhan, S., Wigderson, A.: Entropy Waves, the Zig-Zag Graph Product, and New Constant-Degree Expanders and Extractors. ECCC, TR01-018, 2001; Preliminary version in 41st FOCS, pp. 3–13 (2000)
27. Sipser, M.: Expanders, Randomness or Time vs Space, Proceedings of the Structure in Complexity Theory (1986)

28. Shaltiel, R.: Recent Developments in Explicit Constructions of Extractors. In: Paun, G., Rozenberg, G., Salomaa, A. (eds.) Current Trends in Theoretical Computer Science: The Challenge of the New Century. Algorithms and Complexity, vol. 1, pp. 67–95. World scietific (2004); Preliminary version in Bulletin of the EATCS 77, pages 67–95, 2002
29. Trevisan, L.: When Hamming meets Euclid: The Approximability of Geometric TSP and MST. In: 29th STOC, pp. 21–29 (1997)
30. Zuckerman, D.: Randomness-Optimal Oblivious Sampling. In: Journal of Random Structures and Algorithms, vol. 11(4), pp. 345–367 (1997); Preliminary version in 28th STOC, pages 286–295, 1996

Appendix A: Expanders and Random Walks

This appendix provides more background on expanders than the very minimum that is needed for the main text. On the other hand, there is much more to be learned about this subject (see, e.g., [19]).

A.1 Expanders

An (N, d, λ)-expander is a d-regular graph with N vertices such that the absolute value of all eigenvalues (except the biggest one) of its adjacency matrix is bounded by λ. A (d, λ)-family is an infinite sequence of graphs so that the nth graph is a $(2^n, d, \lambda)$-expander. We say that such a family is *efficiently constructible* if there exists a polynomial-time algorithm that given a vertex, v, in the expander and an index $i \in [d] \stackrel{\text{def}}{=} \{1, ..., d\}$, returns the ith neighbor of v. We first recall that for $d = 16$ and some $\lambda < 16$, efficiently constructible $(16, \lambda)$-families do exist (cf., [14]).[11]

In our applications we use (parameterized) expanders satisfying $\frac{\lambda}{d} < \alpha$ and $d = \text{poly}(1/\alpha)$, where α is an application-specific parameter. Such (parameterized) expanders are also efficiently constructible. For example, we may obtain them by taking paths of length $O(\log(1/\alpha))$ on an expander as above. Specifically, given a parameter $\alpha > 0$, we obtain an efficiently constructible (D, Λ)-family satisfying $\frac{\Lambda}{D} < \alpha$ and $D = \text{poly}(1/\alpha)$ as follows. We start with a constructible $(16, \lambda)$-family, set $k \stackrel{\text{def}}{=} \log_{16/\lambda}(1/\alpha) = O(\log 1/\alpha)$ and consider the paths of length k in each graph. This yields a constructible $(16^k, \lambda^k)$-family, and indeed both $\frac{\lambda^k}{16^k} < \alpha$ and $16^k = \text{poly}(1/\alpha)$ hold.

Comment: To obtain the best constants in Sections 4 and 5, one may use efficiently constructible Ramanujan Graphs [23]. Furthermore, using Ramanujan

---

[11] These are minor technicalities, which can be easily fixed. Firstly, the Gaber–Galil expanders are defined (only) for graph sizes that are perfect squares [14]. This suffices for even n's. For odd n's, we may use a trivial modification such as taking two copies of the graph of size 2^{n-1} and connecting each pair of corresponding vertices. Finally, we add multiple edges so that the degree becomes 16, rather than being 14 for even n's and 15 for odd n's.

Graphs is essential for our proof of the second item of Theorem 4.1 as well as of Lemma 5.3. Ramanujan Graphs satisfy $\lambda \leq 2\sqrt{d-1}$ and so, setting $d = 4/\alpha$, we obtain $\frac{\lambda}{d} < \alpha$, where α is an application-specific parameter. Here some minor technicalities arise since these graphs are given only for certain degrees and certain sizes. Specifically, they can be efficiently constructed for $\frac{1}{2} \cdot q^k \cdot (q^{2k} - 1)$ vertices, where q is a prime such that $q \equiv d - 1 \equiv 1 \bmod 4$ and $d - 1$ is a prime that is a quadratic residue modulo q (cf., [3, Sec. II]). This technical difficulty may be resolved in two ways:

1. Fixing d and ϵ, N, we may find q and k satisfying the foregoing conditions with $\frac{1}{2} \cdot q^k \cdot (q^{2k} - 1) \in [(1-\epsilon) \cdot N, N]$, in time polynomial in $1/\epsilon$ (and in $\log N$). This defines a Ramanujan Graph that is adequate for all our applications (since it biases the desired sample in $[N]$ only by ϵ).
2. Fixing d and ϵ, N, we may find q and k satisfying the foregoing conditions with $\frac{1}{2} \cdot q^k \cdot (q^{2k} - 1) \in [N, 2N]$, in time polynomial in $\log N$. We may easily modify our applications so that whenever we obtain a vertex not in $[N]$ we just ignore it. One can easily verify that the analysis of the application remains valid.

A.2 The Expander Mixing Lemma

The following lemma is folklore and has appeared in many papers. Loosely speaking, the lemma asserts that expander graphs (for which $d \gg \lambda$) have the property that the fraction of edges between two large sets of vertices approximately equals the product of the densities of these sets. This property is called *mixing*.

Lemma A.2 (Expander Mixing Lemma): *Let $G = (V, E)$ be an expander graph of degree d and λ be an upper bound on the absolute value of all eigenvalues, except the biggest one, of the adjacency matrix of the graph. Then, for every two subsets, $A, B \subseteq V$, it holds*

$$\left| \frac{|(A \times B) \cap E|}{|E|} - \frac{|A|}{|V|} \cdot \frac{|B|}{|V|} \right| \leq \frac{\lambda \sqrt{|A| \cdot |B|}}{d \cdot |V|} < \frac{\lambda}{d}.$$

The lemma (and a proof) appears as Corollary 2.5 in [5, Chap. 9].

A.3 Random walks on Expanders

A fundamental discovery of Ajtai, Komlos, and Szemerédi [1] is that random walks on expander graphs provide a good approximation to repeated independent attempts to hit any arbitrary fixed subset of sufficient density (within the vertex set). The importance of this discovery stems from the fact that a random walk on an expander can be generated using much fewer random coins than required for generating independent samples in the vertex set. Precise formulations of the foregoing discovery were given in [1,13,15] culminating in Kahale's optimal analysis [21, Sec. 6].

Theorem A.3 (Expander Random Walk Theorem [21, Cor. 6.1]): *Let $G = (V, E)$ be an expander graph of degree d and λ be an upper bound on the absolute value of all eigenvalues, except the biggest one, of the adjacency matrix of the graph. Let W be a subset of V and $\rho \overset{\text{def}}{=} |W|/|V|$. Then, the fraction of random walks (in G) of (edge) length ℓ that stay within W is at most*

$$\rho \cdot \left(\rho + (1 - \rho) \cdot \frac{\lambda}{d} \right)^{\ell} \tag{10}$$

A more general bound (which is weaker for the above special case) was pointed out to us by Nabil Kahale (personal communication, April 1997):

Theorem A.4 (Expander Random Walk Theorem – general case): *Let $G = (V, E)$, d and λ be as in Theorem A.3. Let $W_0, W_1, ..., W_\ell$ be subsets of V with densities $\rho_0, ..., \rho_\ell$, respectively. Then the fraction of random walks (in G) of (edge) length ℓ that intersect $W_0 \times W_1 \times \cdots \times W_\ell$ is at most*

$$\sqrt{\rho_0 \rho_\ell} \cdot \prod_{i=1}^{\ell} \sqrt{\rho_i + (1 - \rho_i) \cdot \left(\frac{\lambda}{d} \right)^2} \tag{11}$$

Theorem A.4 improves over a previous bound of [7] (see [8]). Comments regarding the proofs of both theorems follow.

On the proofs of Theorems A.3 and A.4. The basic idea is viewing events occuring during the random walk as an evolution of a corresponding probability vector under suitable transformations. The transformations correspond to taking a random step in G and to passing through a "sieve" that keeps only the entries that correspond to the current set W. The key observation is that the first transformation shrinks the component that is orthogonal to the uniform distribution, whereas the second transformation shrinks the component that is in the direction of the uniform distribution. Details follow.

Let A be a matrix representing the random walk on G (i.e., A is the adjacency matrix of G divided by the degree, d). Let $\bar{\lambda}$ denote the absolute value of the second largest eigenvalue of A (i.e., $\bar{\lambda} \overset{\text{def}}{=} \lambda/d$). Let P (resp., P_i) be a 0-1 matrix that has 1-entries only on its diagonal such that entry (j, j) is set to 1 if and only if $j \in W$ (resp., $j \in W_i$). Then, we are interested in the vector obtained when applying $(PA)^\ell$ (resp., $P_\ell A \cdots P_1 A$) to the vector representing the uniform distribution; that is, the probability that we are interested in is the sum of the component of the resulting vector.

The best bounds are obtained by applying the following technical lemma, which refer to the effect of a single PA application. For any n-by-n stochastic matrix M, we let $\|M\|$ denote the norm of M defined as the maximum of $\|Mx\|$ taken over all normal vectors x (i.e., $x \in \mathbb{R}^n$ with $\|x\| = 1$), where $\|x\|$ denote the Euclidean norm of $x \in \mathbb{R}^n$.

Lemma A.5 ([21, Lem. 3.2] restated): *Let M be a symmetric stochastic matrix and let δ denote the absolute value of the second largest eigenvalue of M. Let P be a 0-1 matrix that has 1's only on the diagonal and let ρ be the fraction of 1's on the diagonal. Then, $\|PMP\| \le \rho + (1-\rho) \cdot \delta$.*

A proof of a weaker bound is presented below.

Proof of Theorem A.3: Let $u \in \mathbb{R}^n$ be the vector representing the uniform distribution over $V \equiv \{1, ..., n\}$ (i.e., $u = (n^{-1}, ..., n^{-1})$). Let P be a 0-1 matrix such that the only 1-entries are in entries (i, i) with $i \in W$. Thus, the probability that a random walk of length ℓ stays within W is the sum of the entries of the vector

$$x \stackrel{\text{def}}{=} (PA)^\ell Pu. \tag{12}$$

In other words, denoting by $\|x\|_1$ the L_1 norm of x, we are interested in an upper bound on $\|x\|_1$. Since x has at most ρn non-zero entries (i.e., $x = Px'$ for some x'), we have $\|x\|_1 \le \sqrt{\rho n} \cdot \|x\|$. Invoking Lemma A.5 we get

$$\|x\|_1 \le \sqrt{\rho n} \cdot \|(PA)^\ell Pu\|$$
$$\le \sqrt{\rho n} \cdot \|PAP\|^\ell \cdot \|Pu\|$$
$$\le \sqrt{\rho n} \cdot \left(\rho + (1-\rho) \cdot \bar{\lambda}\right)^\ell \cdot \sqrt{\rho/n}$$

and the theorem follows. ∎

Proof of Theorem A.4: Using the same argument, we need to upper bound the L_1 norm of x given by

$$x \stackrel{\text{def}}{=} P_\ell A \cdots P_1 A P_0 u. \tag{13}$$

We observe that $\|P_j A\| = \sqrt{\|P_j A^2 P_j\|}$ and use Lemma A.5 to obtain $\|P_j A^2 P_j\| \le \rho_j + (1-\rho_j) \cdot \bar{\lambda}^2$. Thus, we have

$$\|x\|_1 \le \sqrt{\rho_\ell n} \cdot \|P_\ell A \cdots P_1 A P_0 u\|$$
$$\le \sqrt{\rho_\ell n} \cdot \prod_{j=1}^{\ell} \|P_j A\| \cdot \|P_0 u\|$$
$$\le \sqrt{\rho_\ell n} \cdot \prod_{j=1}^{\ell} \sqrt{\rho_j + (1-\rho_j) \cdot \bar{\lambda}^2} \cdot \sqrt{\rho_0/n}$$

and the theorem follows. ∎

Proof of a weak version of Lemma A.5. Rather than proving that $\|PMP\| \le \rho + (1-\rho) \cdot \delta$, we shall only prove that $\|PMP\| \le \|PM\| \le \sqrt{\rho + \delta^2}$. That is, we shall prove that, for every z, it holds that $\|PMz\| \le (\rho + \delta^2)^{1/2} \cdot \|z\|$.

Intuitively, M shrinks the component of z that is orthogonal to the uniform vector u, whereas P shrinks the component of z that is in the direction of u. Specifically, we decompose $z = z_1 + z_2$ such that z_1 is the projection of z on u and z_2 is the component orthogonal to u. Then, using the triangle inequality and other obvious facts (which imply $\|PM z_1\| = \|P z_1\|$ and $\|PM z_2\| \leq \|M z_2\|$), we have

$$
\begin{aligned}
\|PM z_1 + PM z_2\| &\leq \|PM z_1\| + \|PM z_2\| \\
&\leq \|P z_1\| + \|M z_2\| \\
&\leq \sqrt{\rho_i} \cdot \|z_1\| + \delta \cdot \|z_2\|
\end{aligned}
$$

where the last inequality uses the fact that P shrinks any uniform vector by eliminating $1 - \rho_i$ of its elements, whereas M shrinks the length of any eigenvector except u by a factor of at least δ. Using the Cauchy-Schwartz inequality[12], we get

$$
\begin{aligned}
\|PM z\| &\leq \sqrt{\rho_i + \delta^2} \cdot \sqrt{\|z_1\|^2 + \|z_2\|^2} \\
&= \sqrt{\rho_i + \delta^2} \cdot \|z\|,
\end{aligned}
$$

where the equality is due to the fact that z_1 is orthogonal to z_2.   ∎

Appendix B: Analyzing the Toeplitz Matrix Construction

For every $i \neq j$ and $a, b \in \mathrm{GF}(2)^n$, we have

$$
\begin{aligned}
\mathrm{Pr}_{T,u} \begin{bmatrix} e_i = a \\ e_j = b \end{bmatrix} &= \mathrm{Pr}_{T,u}\,[e_i = a | e_i \oplus e_j = a \oplus b] \cdot \mathrm{Pr}_{T,u}\,[e_i \oplus e_j = a \oplus b] \\
&= \mathrm{Pr}_{T,u}\,[T v_i + u = a | T w = c] \cdot \mathrm{Pr}_T\,[T w = c],
\end{aligned}
$$

where $w = v_i \oplus v_j \neq 0^n$ and $c = a \oplus b$. Clearly, for any $c \in \mathrm{GF}(2)^n$ and any T':

$$
\begin{aligned}
\mathrm{Pr}_{T,u}[T v_i + u = a | T w = c] &= \mathrm{Pr}_u[T' v_i + u = a] \\
&= 2^{-n}
\end{aligned}
$$

It is thus left to show that, for any $w \neq 0^n$, when T is a uniformly chosen Toeplitz matrix, the vector $T w$ is uniformly distributed over $\mathrm{GF}(2)^n$. It may help to consider first the distribution of $M w$, where M is a uniformly distributed n-by-n matrix. In this case $M w$ is merely the sum of several (not zero) uniformly and independently chosen column vectors, and so is uniformly distributed over $\mathrm{GF}(2)^n$. The argument regarding a uniformly chosen Toeplitz matrix is slightly more involved.

Let f be the first non-zero entry of $w = (w_1, ..., w_n) \neq 0^n$ (i.e., $w_1 = \cdots = w_{f-1} = 0$ and $w_f = 1$). We make the mental experiment of selecting $T = (t_{i,j})$,

---

[12] That is, we get $\sqrt{\rho_i}\|z_1\| + \delta\|z_2\| \leq \sqrt{\rho_i + \delta^2} \cdot \sqrt{\|z_1\|^2 + \|z_2\|^2}$, by using $\sum_{i=1}^n a_i \cdot b_i \leq \left(\sum_{i=1}^n a_i^2\right)^{1/2} \cdot \left(\sum_{i=1}^n b_i^2\right)^{1/2}$, with $n = 2$, $a_1 = \sqrt{\rho_i}$, $b_1 = \|z_1\|$, etc.

by uniformly selecting elements determining T as follows. First we uniformly and independently select $t_{1,n}, ..., t_{1,f}$. Next, we select $t_{2,f}, ..., t_{n,f}$ (here it is important to select $t_{j,f}$ before $t_{j+1,f}$). Finally, we select $t_{n,f-1}, ..., t_{n,1}$. Clearly, this determines a uniformly chosen Toeplitz matrix, denoted T. We conclude by showing that each of the bits of Tw is uniformly distributed given the previous bits. To prove the claim for the jth bit of Tw, consider the time by which $t_{1,n}, ..., t_{1,f}, ..., t_{j-1,f}$ were determined. Note that these determine the first $j-1$ bits of Tw. The key observation is that the value of the jth bit of Tw is a linear combination of the above determined values XORed with the still undetermined $t_{j,f}$. (Here we use the hypothesis that $w_1 = \cdots = w_{f-1} = 0$ and $w_f = 1$.) Thus, uniformly selecting $t_{j,f}$ makes the jth bit of Tw be uniformly distributed given the past. ∎

Appendix C: The Hitting problem

The hitting problem is a one-sided version of the Boolean sampling problem. Given parameters n (length), ϵ (density) and δ (error), and oracle access to any function $\sigma : \{0,1\}^n \to \{0,1\}$ such that $|\{x : f(x) = 1\}| \geq \epsilon 2^n$, the task to find a string that is mapped to 1. That is:

Definition C.1 (hitter): *A hitter is a randomized algorithm that on input parameters n, ϵ and δ, and oracle access to any function $\sigma : \{0,1\}^n \to \{0,1\}$, such that $|\{x : f(x) = 1\}| \geq \epsilon 2^n$, satisfies*

$$\Pr[\sigma(\text{hitter}^\sigma(n, \epsilon, \delta)) = 1] > 1 - \delta.$$

Observe that, on input parameters n, ϵ and δ, any sampler must be able to distinguish the all-zero function from any function $\sigma : \{0,1\}^n \to \{0,1\}$ such that $|\{x : f(x) = 1\}| \geq 2\epsilon 2^n$. Thus, in the latter case, the sampler must obtain (with probability at least $1-\delta$) the value 1 for at least one of its queries, and outputting such a query satisfies the requirement for a hitter (w.r.t parameters n, 2ϵ and δ).

We note that all results and techniques regarding sampling (presented in the main text), have simpler analogous with respect to the hitting problem. In fact, this appendix may be read as a warm-up towards the main text.

C.1   The Information Theoretic Perspective

Analogously to the Naive Sampler, we have the Naive Hitter that *independently* selects $m \stackrel{\text{def}}{=} \frac{\ln(1/\delta)}{\epsilon}$ uniformly distributed sample points and queries the oracle on each. Clearly, the probability that the hitter fails to sample a point of value 1 is at most $(1 - \epsilon)^m = \delta$. The complexities of this hitter are as follows

- Sample Complexity: $m \stackrel{\text{def}}{=} \frac{\ln(1/\delta)}{\epsilon} = \Theta(\frac{\log(1/\delta)}{\epsilon})$.
- Randomness Complexity: $m \cdot n = \Theta(\frac{\log(1/\delta)}{\epsilon} \cdot n)$.
- Computational Complexity: Indeed efficient.

It is easy to prove that the Naive Hitter is sample-optimal. That is:

Theorem C.2 (sample complexity lower bound): *Any hitter has sample complexity bounded below by*

$$\min\left\{ 2^{n-O(1)}, \frac{\ln(1/2\delta)}{2\epsilon} \right\}$$

provided $\epsilon \le \frac{1}{8}$.

Proof Sketch: Let A be a hitter with sample complexity $m = m(n, \epsilon, \delta)$ and let σ be a function selected at random by setting its value independently on each argument such that $\Pr(\sigma(x)=1) = 1.5\epsilon$. Then,

$$\Pr_\sigma[\sigma(A^\sigma(n, \epsilon, \delta)) \ne 1] = (1 - 1.5\epsilon)^m,$$

where the probability is taken over the choice of σ and the internal coin tosses of A. On the other hand, using a Multiplicative Chernoff Bound:

$$\Pr_\sigma[|\{x : \sigma(x)=1\}| < \epsilon 2^n] = 2\exp(-\Omega(\epsilon 2^n)).$$

We may assume that $\Omega(\epsilon 2^n) > \log_2(1/\delta)$ and so the probability that σ has at least ϵ fraction of 1's and yet algorithm A fails is at least $(1 - 1.5\epsilon)^m - \delta > \delta$, unless $m > \frac{\ln(1/2\delta)}{\ln(1-1.5\epsilon)} > \frac{\ln(1/2\delta)}{2\epsilon}$. ∎

Theorem C.3 (randomness complexity lower bound): *Let* $s : \mathbb{N} \times [0,1]^2 \to \mathbb{R}$. *Any sampler that has sample complexity at most* $s(n, \epsilon, \delta)$, *has randomness complexity at least*

$$r > n - \log_2 s(n, \epsilon, \delta) + \log_2((1 - \epsilon)/\delta).$$

Proof Sketch: Let A be a hitter with sample complexity $s = s(n, \epsilon, \delta)$, and randomness complexity $r = r(n, \epsilon, \delta)$. Consider any subset of $\delta 2^r$ possible sequence of coin tosses for A and all $\delta 2^r \cdot s$ points that are queried at any of these coin-sequences. We argue that $\delta 2^r \cdot s > (1 - \epsilon)2^n$ must hold, or else there exists a function σ that evaluates to 0 on each of these points and to 1 otherwise (contradicting the requirement that this function be "hit" with probability at least $1 - \delta$). Thus, $r > n + \log_2(1 - \epsilon) - \log_2 s + \log_2(1/\delta)$. ∎

C.2   The Pairwise-Independent Hitter

Using a pairwise-independent sequence of uniformly distributed sample points rather than a totally independent one, we obtain the pairwise-independent hitter. Here we set $m \stackrel{\text{def}}{=} \frac{1-\epsilon}{\delta\epsilon}$. Letting ζ_i represent the σ-value of the ith sample point, considering only σ's with an ϵ-fraction of 1-values,[13] and using Chebyshev's

---

[13] Considering only σ's with *exactly* an ϵ-fraction of 1-values implies that $\text{Var}[\zeta_i] = (1-\epsilon)\epsilon$. Needless to say, if the hitter works well for all these functions, then it works well for all functions having *at least* an ϵ-fraction of 1-values.

Inequality we have

$$\Pr\left[\sum_{i=1}^{m} \zeta_i = 0\right] \leq \Pr\left[\left|m\epsilon - \sum_{i=1}^{m} \zeta_i\right| \geq \epsilon m\right]$$
$$\leq \frac{m \cdot (1 - \epsilon)\epsilon}{(\epsilon m)^2}$$
$$= \delta.$$

Recalling that we can generate $2^n - 1$ pairwise-independent samples using $2n$ coins, the pairwise-independent hitter achieves

- Sample Complexity: $\frac{1}{\delta\epsilon}$ (reasonable for constant δ).
- Randomness Complexity: $2n$
- Computational Complexity: Indeed efficient.

C.3   The Combined Hitter

Our goal here is to decrease the sample complexity of the Pairwise-Independent Hitter while essentially maintaining its randomness complexity. To motivate the new construction we first consider an oversimplified version of it.

Combined Hitter (oversimplified): On input parameters n, ϵ and δ, set $m \overset{\text{def}}{=} \frac{2}{\epsilon}$ and $\ell \overset{\text{def}}{=} \log_2(1/\delta)$, generate ℓ independent m-element sequences, each being a sequence of m *pairwise-independently and uniformly distributed* strings in $\{0,1\}^n$. Denote the sample points in the ith sequence by $s_1^i, ..., s_m^i$. We merely try all these $\ell \cdot m$ samples as hitting points. Clearly, for each $i = 1, ..., \ell$,

$$\Pr[(\forall j \in \{1, .., m\}) \, \sigma(s_j^i) = 0] < \frac{1}{2}$$

and so the probability that none of these s_j^i "hits σ" is at most $0.5^\ell = \delta$. Thus, the oversimplified version described above is indeed a hitter and has the following complexities:

- Sample Complexity: $\ell \cdot m = O(\frac{\log(1/\delta)}{\epsilon})$.
- Randomness Complexity: $\ell \cdot O(n) = O(n \cdot \log(1/\delta))$.
- Computational Complexity: Indeed efficient.

Thus, the sample complexity is optimal (upto a constant factor), but the randomness complexity is higher than what we aim for. To reduce the randomness complexity, we use the same approach as above, but take dependent sequences rather than independent ones. The dependency we use is such that essentially preserves the probabilistic behavior of independent choices. Specifically, we use random walks on expander graphs (cf., Appendix A) to generate a sequence of ℓ "seeds" each of length $O(n)$. Each seed is used to generate a sequence of m pairwise independent elements in $\{0,1\}^n$, as above. Thus, we obtain:

Corollary C.4 (The Combined Hitter): *There exists an efficient hitter with*

- Sample Complexity: $O(\frac{\log(1/\delta)}{\epsilon})$.
- Randomness Complexity: $2n + O(\log(1/\delta))$.

Furthermore, we can obtain randomness complexity $2n + (2 + o(1)) \cdot \log_2(1/\delta))$.

Proof Sketch: We use an explicit construction of expander graphs with vertex set $\{0,1\}^{2n}$, degree d and second eigenvalue λ so that $\lambda/d < 0.1$. We consider a random walk of (edge) length $\ell - 1 = \log_2(1/\delta)$ on this expander, and use each of the ℓ vertices along the path as random coins for the Pairwise-Independent Hitter, which in turn makes $m \overset{\text{def}}{=} \epsilon/3$ trials. To analyze the performance of the resulting algorithm, we let W denote the set of coin tosses (for the basic hitter) on which the basic hitter fails to output a point that evaluates to 1. By the hypothesis, $\frac{|W|}{2^{2n}} \leq 1/3$, and using Theorem A.3, the probability that all vertices of a random path reside in W is bounded above by $(0.34 + 0.1)^\ell < \delta$. The furthermore clause follows by using a Ramanujan Graph and an argument as in the proof of Item 2 of Theorem 4.1.   ∎

C.4   The Expander Hitter

Our goal here is to decrease the randomness complexity of hitters from $2n + O(\log(1/\delta))$ to $n + O(\log(1/\delta))$, while preserving the sample complexity of $O(\epsilon^{-1} \log(1/\delta))$. The first step is to get an analogous improvement with respect to the Pairwise-Independent Hitter (which has sample complexity $O(1/\delta\epsilon)$).

We use a Ramanujan Graph of degree $d = O(1/\epsilon\delta)$ and vertex-set $\{0,1\}^n$. The hitter uniformly selects a vertex in the graph and use its neighbors as a sample. Suppose we try to hit a 1-value of a function σ and let $S \overset{\text{def}}{=} \{u : \sigma(u) = 1\}$. Let $B \overset{\text{def}}{=} \{v : \mathrm{N}(v) \cap S = \emptyset\}$ be the set of bad vertices (i.e., choosing any of these results in not finding a preimage of 1). Using the Expander Mixing Lemma we have

$$\rho(B)\rho(S) = \left| \frac{|(B \times S) \cap E|}{|E|} - \rho(B)\rho(S) \right|$$
$$\leq \frac{\lambda}{d} \cdot \sqrt{\rho(B)\rho(S)}$$

Hence, $\rho(B)\rho(S) \leq (\lambda/d)^2 = \epsilon\delta$ and using $\rho(S) \geq \epsilon$ we get $\rho(B) \leq \delta$. The complexities of this hitter are as follows:

- Sample Complexity: $O(\frac{1}{\delta\epsilon})$
- Randomness Complexity: n
- Computational Complexity: Indeed efficient.

Adapting the argument in the proof of Corollary C.4, we obtain

Corollary C.5 (The Combined Hitter, revisited): *There exists an efficient hitter with*

- Sample Complexity: $O(\frac{\log(1/\delta)}{\epsilon})$.
- Randomness Complexity: $n + (2 + o(1)) \cdot \log_2(1/\delta)$.

Short Locally Testable Codes and Proofs

Oded Goldreich

Abstract. We survey known results regarding locally testable codes and locally testable proofs (known as PCPs), with emphasis on the length of these constructs. Local testability refers to approximately testing large objects based on a very small number of probes, each retrieving a single bit in the representation of the object. This yields super-fast approximate-testing of the corresponding property (i.e., be a codeword or a valid proof). We also review the related concept of local decodable codes.

The survey consists of two independent (i.e., self-contained) parts that cover the same material at different levels of rigor and detail. Still, in spite of the repetitions, there may be a benefit in reading both parts.

Keywords: Error Correcting Codes, Property Testing, Probabilistically Checkable Proofs (PCP), Locally Testable Codes, Locally Decodable Codes, Self-Correction, Low-Degree Tests, Derandomization, Private Information Retrieval.

A previous version of this survey appeared as TR05-014 of *ECCC*; in fact, this earlier version [36] is cited in the text, when reporting of subsequent developments. The current version also appeared in [38].

PART I: A HIGH-LEVEL OVERVIEW

The title of this survey refers to two types of objects (i.e., codes and proofs) and two adjectives (i.e., *local testability* and *short*). A clarification of these terms is in place.

Codes, proofs and their length. Codes are sets of strings (of equal length), typically, having a large pairwise distance. Equivalently, codes are viewed as mappings from short (k-bit) strings to longer (n-bit) strings, called codewords, such that the codewords are distant from one another. We will focus on *codes with relative constant distance*; that is, every two n-bit codewords are at distance $\Omega(n)$ apart. The length of the code is measured in terms of the length of the pre-image (i.e., we are interested in the growth of n as a function of k). Turning to proofs, these are defined with respect to a verification procedure for assertions of a certain length, and their length is measured in terms of the length of the assertion. The verification procedure must satisfy the natural completeness and soundness properties: For valid assertions there should be strings, called proofs, that are accepted (in conjunction with the assertion) by the verification procedures, whereas for false assertions no such strings may exist. The reader may envision proof systems for the set of satisfiable propositional formulae (i.e., assertions of satisfiability of given formulae).

O. Goldreich et al.: Studies in Complexity and Cryptography, LNCS 6650, pp. 333–372, 2011.

Local testability. By local testability we mean that the object can be tested for the natural property (i.e., being a codeword or a valid proof) using a small (typically constant)[1] number of probes, each recovering individual bits in a standard representation of the object. Thus, local testability allows for super-fast testing of the corresponding objects. The tests are probabilistic and hence the result is correct only with high probability.[2] Furthermore, correctness refers to a *relaxed notion of deciding* (which was formulated, in general terms, in the context of property testing [58, 39]): It is required that valid objects be accepted with high probability, whereas objects that are "far" from being valid should be rejected with high probability. Specifically, in the case of codes, codewords should be accepted (with high probability), whereas strings that are "far" from the code should be rejected (with high probability). In the case of proofs, valid proofs (which exist for correct assertions) should be accepted (with high probability), whereas strings that are "far" from being valid proofs (and, in particular, all strings in case no valid proofs exist) should be rejected (with high probability).[3]

Our notion of locally testable proofs is closely related to the notion of a PCP (i.e., probabilistically checkable proof), and we will ignore the difference in the sequel. The difference is that in the definition of locally testable proofs we required rejection of strings that are far from any valid proof, also in the case that valid proofs exists (i.e., the assertion is valid). In contrast, the standard rejection criteria of PCPs refers only to false assertions. Still, all known PCP constructions actually satisfy the stronger definition.

The very possibility of local testability. Indeed, local testability of either codes or proofs is quite challenging, regardless of the issue of length:

- For codes, the simplest example of a locally testable code (of constant relative distance) is the Hadamard code and testing it amounts to linearity testing. However, the exact analysis of the natural linearity tester (of Blum, Luby and Rubinfeld [22]) turned out to be highly complex (cf. [22, 6, 31, 12, 13, 10, 47]).
- For proofs, the simplest example of a locally testable proof is the "inner verifier" of the PCP construction of Arora, Lund, Motwani, Sudan and Szegedy [4], which in turn is based on the Hadamard code.

In both cases, the constructed object has exponential length in terms of the relevant parameter (i.e., the amount of information being encoded in the code or the length of the assertion being proved).

Local testability at a polynomial blow-up. Achieving local testability by codes and proofs that have polynomial length turns out to be even more challenging.

---

[1] In this part, we associate local testability with tests that perform a constant number of probes.

[2] It is easy to see that deterministic tests will perform very poorly, and the same holds with respect to probabilistic tests that make no error.

[3] Indeed, in the case the assertion is false, there exist no valid proofs. In this case all strings are defined to be far from a valid proof.

– In the case of codes, a direct interpretation of *low-degree tests* (cf. [6, 7, 35, 58, 34]), proposed in [34, 58], yields a locally testable code of quadratic length over a *sufficiently large alphabet*. Similar (and actually better) results for *binary* codes required additional ideas, and have appeared only later (cf. [42]).

– The case of proofs is far more complex: Achieving locally testable proof of polynomial length is essentially the contents of the celebrated PCP Theorem of Arora, Lund, Motwani, Safra, Sudan and Szegedy [5, 4].

We focus on even *shorter* codes and proofs; specifically, codes and proofs of *nearly linear length*. The latter term has been given quite different interpretations, and here we adopt the most strict interpretation by which nearly linear means linear up to polylogarithmic factors.

Local testability with a polylogarithmic (length) overhead: The ultimate goal is to obtain locally testable codes and proofs of minimal length. The currently known results get very close to obtaining this goal.

Theorem 1 (Dinur [26], building on [20]): *There exist locally testable codes and proofs of length that is only a polylogarithmic factor larger than the relevant parameter. That is, the length function $\ell : \mathbb{N} \to \mathbb{N}$ satisfies $\ell(k) = \widetilde{O}(k) = k \cdot \mathrm{poly}(\log k)$.*

One may wonder whether or not a polylogarithmic overhead in inherent to local testability of codes and proofs. This is indeed a fundamental open problem.

Open Problem 2. *Do there exist locally testable codes and proofs of linear length?*

In the rest of this part of the survey, we motivate the study of short locally testable objects, comment on the relation between such codes and proofs, and discuss a somewhat related coding problem.

Motivation for the study of short locally testable codes and proofs

Local testability offers an extremely strong notion of efficient testing: The tester makes only a constant number of bit probes, and determining the probed locations (as well as the final decision) is typically done in time that is poly-logarithmic in the length of the probed object.

The length of an error-correcting code is widely recognized as one of the two most fundamental parameters of the code (the second one being its distance). In particular, the length of the code is of major importance in applications, because it determines the overhead involved in encoding information.

The same considerations apply also to proofs. However, in the case of proofs, this obvious point was blurred by the indirect, unexpected and highly influential applications of locally testable proofs (known as PCPs) to the theory of approximation algorithms. In our view, the significance of locally testable proofs (i.e., PCPs) extends far beyond their applicability to deriving non-approximability results. The mere fact that proofs can be transformed into a format that supports super-fast probabilistic verification is remarkable. From this perspective, the question of how much redundancy is introduced by such a transformation is a fundamental one. Furthermore, locally testable proofs (i.e., PCPs) have been used not only to derive non-approximability results but also for obtaining positive results (e.g., CS-proofs [49,54] and their applications [8,24]), and the length of the PCP affects the complexity of those applications.

Turning back to the celebrated application of PCP to the study of approximation algorithms, we note that the length of PCPs is also relevant to non-approximability results; specifically, the length of PCPs affects the *tightness with respect to the running time* of the non-approximability results derived. For example, suppose (exact) SAT has complexity $2^{\Omega(n)}$. The original PCP Theorem [5,4] only implies that approximating MaxSAT requires time 2^{n^α}, for some (small) $\alpha > 0$. The work of [56] makes α arbitrarily close to 1, whereas the results of [42,21] further improve the lower bound to $2^{n^{1-o(1)}}$ and the results of [20,26] yields a lower bound of $2^{n/\mathrm{poly}(\log n)}$.[4]

On the relation between locally testable codes and proofs

Locally testable codes seem related to locally testable proofs (PCPs). In fact, the use of codes with some "local testability" features is implicit in known PCP constructions. Furthermore, the known constructions of locally testable proofs (PCPs) provides a transformation of *standard proofs* (for say SAT) to *locally testable proofs* (i.e., PCP-oracles) such that transformed strings are accepted with probability one by the PCP verifier. Moreover, starting from different standard proofs, one obtains locally testable proofs that are far apart, and hence constitute a good code. It is tempting to think that the PCP verifier yields a codeword tester, but this is not really the case. Note that our definition of a locally testable proof requires rejection of strings that are far from any valid proof, but it is not clear that the only valid proofs (w.r.t the constructed PCP verifier) are those that are obtained by the aforementioned transformation of standard proofs to locally testable ones.[5] In fact, the standard PCP constructions accept also valid proofs that are not in the range of the corresponding transformation.

In spite of the above, locally testable codes and proofs are related, and the feeling is that locally testable codes are the combinatorial counterparts of locally testable proofs (PCPs), which are complexity theoretic in nature. From that perspective, one should expect (or hope) that it would be easier to construct locally testable codes than it is to construct PCPs. This feeling was among the main motivations of Goldreich and Sudan, and indeed their first result was along this vein: They showed a relatively simple construction (i.e., simple in comparison to PCP constructions) of a locally testable code of length $\ell(k) = k^c$

---

[4] Using [55] (or [27]) allows to achieve the lower bound of $2^{n^{1-o(1)}}$ simultaneously with optimal approximation ratios, but this is currently unknown for the better lower bound of $2^{n/\mathrm{poly}(\log n)}$.

[5] Let alone that the standard definition of PCP refers only to the case of false assertions, in which case all strings are far from a valid proof (which does not exist).

for any constant $c > 1$ [42, Sec. 3]. Unfortunately, their stronger result, providing a locally testable code of shorter length (i.e., length $\ell(k) = k^{1+o(1)}$) is obtained by constructing and using a corresponding locally testable proof (i.e., PCP). Subsequent works have mostly followed this route, with the notable exception of Meir's work [52].

Locally Decodable Codes

Locally *decodable* codes are in some sense complimentary to local *testable* codes. Here, one is given a slightly corrupted codeword (i.e., a string close to some unique codeword), and is required to recover individual bits of the encoded information based on a constant number of probes (per recovered bit). That is, a code is said to be locally decodable if whenever relatively few location are corrupted, the decoder is able to recover each information-bit, with high probability, based on a constant number of probes to the (corrupted) codeword.

The best known locally decodable codes are of strictly sub-exponential length. Specifically, k information bits can be encoded by codewords of length $n = \exp(k^{o(1)})$ that are locally decodable using three bit-probes (cf. [29], building over [62]). The problem is related to the construction of (information theoretic secure) Private Information Retrieval schemes, introduced in [25].

A natural relaxation of the definition of locally decodable codes requires that, whenever few location are corrupted, the decoder should be able to recover most of the individual information-bits (based on a constant number of queries), and for the rest of the locations the decoder may output a fail symbol (but not the wrong value). That is, the decoder must still avoid errors (with high probability), but on a few bit-locations it is allowed to sometimes say "don't know". This relaxed notion of local decodability can be supported by codes that have length $\ell(k) = k^c$ for any constant $c > 1$ (cf. [15]).

An obvious open problem is to separate locally decodable codes from relaxed locally decodable codes. This may follow by either improving the $\Omega(k^{1+\frac{1}{q-1}})$ lower bound on the length of q-query locally decodable codes (of [46]), or by providing relaxed locally decodable codes of length $\ell(k) = k^{1+o(1)}$.

PART II: A MORE DETAILED AND RIGOROUS ACCOUNT

In this part we provide a general treatment of local testability. In contrast to Part I, here we allow the tester to use a number of queries that is a (typically small) predetermined function of the length parameter, rather than insisting on a constant number of queries. The latter special case is indeed an important one.

1 Introduction

Codes (i.e., error correcting codes) and proofs (i.e., automatically verifiable proofs) are fundamental to computer science as well as to related disciplines such as mathematics and computer engineering. Redundancy is inherent to error-correcting codes, whereas testing validity is inherent to proofs. In this survey we also consider less traditional combinations such as testing validity of codewords and the use of proofs that contain redundancy. The reader may wonder why we explore these non-traditional possibilities, and the answer is that they offer various advantages (as will be elaborated next).

Testing the validity of codewords is natural in settings in which one may want to take an action in case the codeword is corrupted. For example, when storing data in an error correcting format, one may want to recover the data and re-encode it whenever one finds that the current encoding is corrupted. Doing so may allow to maintain the data integrity over eternity, although the encoded bits may all get corrupted in the course of time. Of course, one can use the error-correcting decoding procedure associated with the code in order to check whether the current encoding is corrupted, but the question is whether one can check (or just approximately check) this property *much faster*.

Loosely speaking, locally testable codes are error correcting codes that allow for a super-fast testing of whether or not a give string is a valid codeword. In particular, the tester works in sub-linear time and reads very few of the bits of the tested object. Needless to say, the answer provided by such a tester can only be approximately correct, but this would suffice in many applications (including the one outlined above).

Similarly, locally testable proofs are proofs that allow for a super-fast probabilistic verification. Again, the tester works in sub-linear time and reads very few of the bits of the tested object. The tester's (a.k.a. verifier's) verdict is only correct with high probability, but this may suffice for many applications, where the assertion is rather mundane but of great practical importance. In particular, it suffices in applications in which proofs are used for establishing the correctness of *specific* computations of practical interest. Lastly, we comment that such *locally testable proofs must be redundant* (or else there would be no chance for verifying them based on inspecting only a small portion of them).

Our focus is on relatively *short* locally testable codes and proofs, which is not surprising in view of the fact that *we envision such objects being actually used in practice*. Of course, we do not mean to suggest that one may use in practice any of the constructions surveyed here (especially not the ones that provide the stronger bounds). We rather argue that this direction of research may find applications in practice. Furthermore, it may even be the case that some of the current concepts and techniques may lead to such applications.

Organization: In Section 2 we provide a quite comprehensive definitional treatment of locally testable codes and proofs, while relating them to PCPs, PCPs of proximity, and property testing. In Section 3, we survey the main results regarding locally testable codes and proofs as well as many of the underlying ideas. In Section 4 we consider locally decodable codes, which are somewhat complementary to locally testable codes.

Caveat: Our exposition of locally testable/decodable codes is aimed at achieving the best possible length, regardless of whether or not the code is popular (i.e., used in practice). Thus, we do not survey here results that refer to the testing (and decoding) features of various popular codes, unless these features are instructive for our aim.

2 Definitions

Local testability is formulated by considering oracle machines. That is, the tester is an oracle machine, and the object that it tests is viewed as an oracle. For simplicity, we confine ourselves to *non-adaptive* probabilistic oracle machines; that is, machines that determine their queries based on their explicit input (which in case of codes is merely a length parameter) and their internal coin tosses (but not depending on previous oracle answers). When talking about oracle access to a string $w \in \{0,1\}^n$ we viewed w as a function $w : \{1, ..., n\} \to \{0,1\}$.

2.1 Codeword Testers

We consider codes mapping sequences of k (input) bits into sequences of $n \geq k$ (output) bits. Such a generic code is denoted by $C : \{0,1\}^k \to \{0,1\}^n$, and the elements of $\{C(x) : x \in \{0,1\}^k\} \subseteq \{0,1\}^n$ are called codewords (of C).

The distance of a code $C : \{0,1\}^k \to \{0,1\}^n$ is the minimum (Hamming) distance between its codewords; that is, $\min_{x \neq y}\{\Delta(C(x), C(y))\}$, where $\Delta(u,v)$ denotes the number of bit-locations on which u and v differ. Throughout this work, *we focus on codes of linear distance*; that is, codes $C : \{0,1\}^k \to \{0,1\}^n$ of distance $\Omega(n)$.

The distance of $w \in \{0,1\}^n$ from a code $C : \{0,1\}^k \to \{0,1\}^n$, denoted $\Delta_C(w)$, is the minimum distance between w and the codewords; that is, $\Delta_C(w) \overset{\text{def}}{=} \min_x\{\Delta(w, C(x))\}$. For $\delta \in [0,1]$, the n-bit long strings u and v are said to be δ-far (resp., δ-close) if $\Delta(u,v) > \delta \cdot n$ (resp., $\Delta(u,v) \leq \delta \cdot n$). Similarly, w is δ-far from C (resp., δ-close to C) if $\Delta_C(w) > \delta \cdot n$ (resp., $\Delta_C(w) \leq \delta \cdot n$).

Definition 2.1 (codeword tests, basic version): *Let* $C : \{0,1\}^k \to \{0,1\}^n$ *be a code of distance d, and let $q \in \mathbb{N}$ and $\delta \in (0,1)$. A q-local (codeword) δ-tester for C is a probabilistic (non-adaptive) oracle machine M that makes at most q queries and satisfies the following two conditions:*

Accepting codewords (a.k.a. completeness): *For any* $x \in \{0,1\}^k$, *given oracle access to* $w = C(x)$, *machine M accepts with probability 1. That is,* $\Pr[M^{C(x)}(1^k) = 1] = 1$, *for any* $x \in \{0,1\}^k$.

Rejection of non-codeword (a.k.a. soundness): *For any* $w \in \{0,1\}^n$ *that is δ-far from C, given oracle access to w, machine M rejects with probability at least $1/2$. That is,* $\Pr[M^w(1^k) = 1] \leq 1/2$, *for any* $w \in \{0,1\}^n$ *that is δ-far from C.*

We call q the query complexity *of M, and δ the* proximity parameter.

The above definition is interesting only in case δn is smaller than the covering radius of C (i.e., the smallest r such that for every $w \in \{0,1\}^n$ it holds that $\Delta_{\mathrm{C}}(w) \leq r$). Clearly, $r \geq d/2$, and so the definition is certainly interesting in the case that $\delta < d/2n$, and indeed we will focus on this case. On the other hand, observe that $q = \Omega(1/\delta)$ must hold, which means that we focus on the case that $d = \Omega(n/q)$.

We next consider families of codes $\mathrm{C} = \{\mathrm{C}_k : \{0,1\}^k \to \{0,1\}^{n(k)}\}_{k \in K}$, where $n, d : \mathbb{N} \to \mathbb{N}$ and $K \subseteq \mathbb{N}$, such that C_k has distance $d(k)$. In accordance with the above, our main interest is in the case that $\delta(k) < d(k)/2n(k)$. Furthermore, seeking constant query complexity, we focus on the case $d = \Omega(n)$.

Definition 2.2 (codeword tests, asymptotic version): *For functions $n, d : \mathbb{N} \to \mathbb{N}$, let $\mathrm{C} = \{\mathrm{C}_k : \{0,1\}^k \to \{0,1\}^{n(k)}\}_{k \in K}$ be such that C_k is a code of distance $d(k)$. For functions $q : \mathbb{N} \to \mathbb{N}$ and $\delta : \mathbb{N} \to (0,1)$, we say that a machine M is a q-local (codeword) δ-tester for $\mathrm{C} = \{\mathrm{C}_k\}_{k \in K}$ if, for every $k \in K$, machine M is a $q(k)$-local $\delta(k)$-tester for C_k. Again, q is called the* query complexity *of M, and δ the* proximity parameter.

Recall that being particularly interested in constant query complexity (and recalling that $d(k)/n(k) \geq 2\delta(k) = \Omega(1/q(k))$), we focus on the case that $d = \Omega(n)$ and δ is a constant smaller than $d/2n$. In this case, we may consider a stronger definition.

Definition 2.3 (locally testable codes): *Let n, d and C be as in Definition 2.2 and suppose that $d = \Omega(n)$. We say that C is* locally testable *if for every constant $\delta > 0$ there exists a constant q and a probabilistic polynomial-time oracle machine M such that M is a q-local δ-tester for C.*

We will be concerned of the growth rate of n as a function of k, for locally testable codes $\mathrm{C} = \{\mathrm{C}_k : \{0,1\}^k \to \{0,1\}^{n(k)}\}_{k \in K}$ of distance $d = \Omega(n)$. More generally, for $d = \Omega(n)$, we will be interested in the trade-off between n, the proximity parameter δ, and the query complexity q.

2.2 Proof Testers

We start by recalling the standard definition of PCP. (For an introduction to the subject as well as a wider perspective, see [37, Chap. 9]).

Definition 2.4 (PCP, standard definition): *A* probabilistically checkable proof (PCP) system *for a set S is a probabilistic* (non-adaptive) *polynomial-time oracle machine* (called a verifier), *denoted V, satisfying*

Completeness: *For every $x \in S$ there exists an oracle π_x such that V, on input x and access to oracle π_x, always accepts x; that is, $\Pr[V^{\pi_x}(x)=1] = 1$.*

Soundness: *For every $x \notin S$ and every oracle π, machine V, on input x and access to oracle π, rejects x with probability at least $\frac{1}{2}$; that is, $\Pr[V^{\pi}(x) = 1] \leq 1/2$,*

Let $Q_x(r)$ denote the set of oracle positions inspected by V on input x and random-tape $r \in \{0,1\}^{\mathrm{poly}(|x|)}$. The query complexity *of V is defined as* $q(n) \stackrel{\mathrm{def}}{=} \max_{x \in \{0,1\}^n, r \in \{0,1\}^{\mathrm{poly}(n)}} \{|Q_x(r)|\}$. The proof complexity *of V is defined as* $p(n) \stackrel{\mathrm{def}}{=} \max_{x \in \{0,1\}^n} \{|\bigcup_{r \in \{0,1\}^{\mathrm{poly}(n)}} Q_x(r)|\}$.

Note that in the case that the verifier V uses a logarithmic number of coin tosses, its proof complexity is polynomial. In general, the proof complexity is upper-bounded by $2^r \cdot q$, where r and q are the randomness complexity and the query complexity of the proof tester. Thus, the trade-off between the query complexity and the proof complexity is typically captured by the trade-off between the query complexity and the randomness complexity. Furthermore, focusing on the randomness complexity allows for better bounds when composing proofs (cf. §3.2.2).

All known PCP constructions can be easily modified such that the oracle locations accessed by V are a prefix of the oracle (i.e., $\bigcup_{r \in \{0,1\}^{\mathrm{poly}(|x|)}} Q_x(r) \subseteq \{1, ..., p(|x|)\}$, for every x).[6] (For simplicity, the reader may assume that this is the case throughout the rest of this exposition.) More importantly, all known PCP constructions can be easily modified to satisfy the following definition, which is closer in spirit to the definition of locally testable codes.

Definition 2.5 (PCP, augmented): *For functions $q : \mathbb{N} \to \mathbb{N}$ and $\delta : \mathbb{N} \to (0,1)$, we say that a PCP system V for a set S is a q-locally δ-testable proof system if it has query complexity q and satisfies the following condition, which augments the standard soundness condition.[7]*

Rejecting invalid proofs: *For every $x \in \{0,1\}^*$ and every oracle π that is δ-far from $\Pi_x \stackrel{\mathrm{def}}{=} \{w : \Pr[V^w(x) = 1] = 1\}$, machine V, on input x and access to oracle π, rejects x with probability at least $\frac{1}{2}$.*

The proof complexity of V is defined as in Definition 2.4.

Note that Definition 2.5 uses the tester V itself in order to define the set (denoted Π_x) of valid proofs (for $x \in S$). That is, V is used both to define the set of valid proofs and to test for the proximity of a given oracle to this set. A more general definition (presented next), refers to an arbitrary proof system, and lets Π_x equal the set of valid proofs (in that system) for $x \in S$. Obviously, it must hold that $\Pi_x \neq \emptyset$ if and only if $x \in S$. Typically, one also requires the existence of

---

[6] Recall that p denotes the proof complexity of the system. In fact, for every $x \in \{0,1\}^n$, it holds that $\bigcup_{r \in \{0,1\}^{\mathrm{poly}(n)}} Q_x(r) = \{1, ..., p(n)\}$.

[7] Definition 2.5 relies on two natural conventions:

1. All strings in Π_x are of the same length, which equals $|\bigcup_{r \in \{0,1\}^{\mathrm{poly}(n)}} Q_x(r)|$, where $Q_x(r)$ is as in Definition 2.4. Furthermore, we consider only π's of this length.
2. If $\Pi_x = \emptyset$ (which happens if and only if $x \notin S$), then every π is considered δ-far from Π_x.

These conventions will also be used in Definition 2.6.

a polynomial-time procedure that, on input a pair (x, π), determines whether or not $\pi \in \Pi_x$.[8] For simplicity we assume that, for some function $p : \mathbb{N} \to \mathbb{N}$ and every $x \in \{0,1\}^*$, it holds that $\Pi_x \subseteq \{0,1\}^{p(|x|)}$. The resulting definition follows.

Definition 2.6 (locally testable proofs): *Suppose that, for some function $p :$ $\mathbb{N} \to \mathbb{N}$ and every $x \in \{0,1\}^*$, it holds that $\Pi_x \subseteq \{0,1\}^{p(|x|)}$. For functions $q :$ $\mathbb{N} \to \mathbb{N}$ and $\delta : \mathbb{N} \to (0,1)$, we say that a probabilistic (non-adaptive) polynomial-time oracle machine V is a q-locally δ-tester for proofs in $\{\Pi_x\}_{x \in \{0,1\}^*}$ if V has query complexity q and satisfies the following conditions:*

Technical condition: *On input x, machine V issues queries in $\{1, ..., p(|x|)\}$.*
Accepting valid proofs: *For every $x \in \{0,1\}^*$ and every oracle $\pi \in \Pi_x$, machine V, on input x and access to oracle π, accepts x with probability 1.*
Rejecting invalid proofs: *For every $x \in \{0,1\}^*$ and every oracle π that is δ-far from Π_x, machine V, on input x and access to oracle π, rejects x with probability at least $\frac{1}{2}$.*

The proof complexity *of V is defined as p,[9] and δ is called the* proximity parameter. *In such a case, we say that $\Pi = \{\Pi_x\}_{x \in \{0,1\}^*}$ is q-locally δ-testable, and that $S = \{x \in \{0,1\}^* : \Pi_x \neq \emptyset\}$ has q-locally δ-testable proofs of length p.*
We say that Π is locally testable *if for every constant $\delta > 0$ there exists a constant q such that Π is q-locally δ-testable. In such a case, we say that S has* locally testable proofs of length p.*

This notion of locally testable proofs is closely related to the notion of probabilistically checkable proofs (i.e., PCPs). The difference is that in the definition of locally testable proofs we required rejection of strings that are far from any valid proof, also in the case that valid proofs exists (i.e., the assertion is valid). In contrast, the standard rejection criteria of PCPs refers only to false assertions. Still, all known PCP constructions actually satisfy the stronger definition.[10]

Needless to say, the new term "locally testable proof" was introduced to match the term "locally testable codes". In retrospect, "locally testable proofs" seems a more fitting term than "probabilistically checkable proofs", because it stresses the positive aspect (of locality) rather than the negative aspect (of being probabilistic). The latter perspective has been frequently advocated by Leonid Levin.

---

[8] Recall that in the case that the verifier V uses a logarithmic number of coin tosses, its proof complexity is polynomial (and so the "effective length" of the strings in Π_x must be polynomial in $|x|$). Furthermore, if in addition it holds that $\Pi_x = \{w : \Pr[V^w(x) = 1] = 1\}$, then (scanning all possible coin tosses of) V yields a polynomial-time procedure for determining whether a given pair (x, π) satisfies $\pi \in \Pi_x$.

[9] Note that by the technical condition, the current definition of the proof complexity of V is lower-bounded by the definition used in Definition 2.4.

[10] In some cases this holds only under a weighted version of the Hamming distance, rather under the standard Hamming distance. Alternatively, these constructions can be easily modified to work under the standard Hamming distance.

2.3 Discussion

We first comment about a few definitional choices made above. Firstly, we chose
to present testers that always accept valid objects (i.e., accept valid codewords
(resp., valid proofs) with probability 1). This is more appealing than allowing
two-sided error, but the latter weaker notion is meaningful too. A second choice
was to fix the error probability (i.e., probability of accepting far from valid
objects), rather than introducing yet another parameter. Needless to say, the
error probability can be reduced by sequential applications of the tester.

In the rest of this section, we consider an array of definitional issues. First,
we consider two natural strengthenings of the definition of local testability
(cf. §2.3.1). We next discuss the relation of local testability to property testing (cf. §2.3.2), and the relation of locally testable proofs to PCP of proximity
(as defined in [15], cf. §2.3.3). Finally, we discuss the relation between locally
testable codes and proofs (cf. §2.3.4), and the motivation for the study of *short*
local testable codes and proofs (cf. §2.3.5).[11] Finally (in §2.3.6), we mention a
weaker definition, which seem natural only in the context of codes.

2.3.1 Stronger Definitions

The definitions of testers presented so far, allow for the construction of a different
tester for each relevant value of the proximity parameter. However, whenever
such testers are actually constructed, they tend to be "uniform" over all relevant
values of the proximity parameter. Thus, it is natural to present a single tester for
all relevant values of the proximity parameter, provide this tester with the said
parameter, allow it to behave accordingly, and measure its query complexity as
a function of that parameter. For example, we may strengthen Definition 2.3, by
requiring the existence of a function $q : (0,1) \to \mathbb{N}$ and an oracle machine M such
that, for every constant $\delta > 0$, all (sufficiently large) k and all $w \in \{0,1\}^{n(k)}$,
the following conditions hold:

1. On input $(1^k, \delta)$, machine M makes $q(\delta)$ queries.
2. If w is a codeword of C then $\Pr[M^w(1^k, \delta) = 1] = 1$.
3. If w is δ-far from $\{C(x) : x \in \{0,1\}^k\}$ then $\Pr[M^w(1^k, \delta) = 1] \leq 1/2$.

An analogous strengthening applies to Definition 2.6. A special case of interest
is when $q(\delta) = O(1/\delta)$. In this case, it makes sense to ask whether or not an even
stronger "uniformity" condition may hold. Like in Definitions 2.1 and 2.2 (resp.,
Definitions 2.5 and 2.6), the tester M is not given the proximity parameter (and
so its query complexity cannot depend on it), but we only require it to reject
with probability proportional to the distance of the oracle from the relevant set.
For example, we may strengthen Definition 2.3, by requiring the existence of an
oracle machine M and a *constant* q such that, for every constant $\delta > 0$, every
(sufficiently large) k and $w \in \{0,1\}^{n(k)}$, the following conditions hold:

---

[11] The text of §2.3.5 is almost identical to a corresponding motivational text that
appears in Part I.

1. On input 1^k, machine M makes q queries.
2. If w is a codeword of C then $\Pr[M^w(1^k, \delta) = 1] = 1$.
3. If w is δ-far from $\{C(x) : x \in \{0,1\}^k\}$ then $\Pr[M^w(1^k, \delta) = 1] < 1 - \Omega(\delta)$.

2.3.2 Relation to Property Testing

Locally testable codes (and their corresponding testers) are essentially special cases of property testing algorithms, as defined in [58,39]. Specifically, the property being tested is membership in a predetermined code. The only difference between the definitions presented in Section 2.1 and the formulation that is standard in the property testing literature is that in the latter the tester is given the proximity parameter as input and determines its behavior (and in particular the number of queries) accordingly. This difference is eliminated in the first strengthening outlined in §2.3.1, while the second strengthening is related to the notion of proximity oblivious testing (cf. [40]). We note, however, that most of the property testing literature is concerned with "natural" objects (e.g., graphs, sets of points, functions) presented in a "natural" form rather than with objects designed artificially to withstand errors (i.e., codewords of error correcting codes).

Our general formulation of proof testing (i.e., Definition 2.6) can be viewed as a generalization of property testing. That is, we view the set Π_x as a set of objects having a certain x-dependent property (rather than as a set of valid proofs for some property of x). In other words, Definition 2.6 allows to consider properties that are parameterized by auxiliary information (i.e., x), whereas traditional property testing may be viewed as referring to the case that x only determines the length of strings in Π_x (e.g., $\Pi_x = \emptyset$ for every $x \notin \{1\}^*$ or, equivalently, $\Pi_x = \Pi_y$ for every $|x| = |y|$).[12]

2.3.3 Relation to PCPs of Proximity

Our definition of a locally testable proof is related but different from the definition of a PCP of proximity (appearing in [15]).[13] We start by reviewing the definition of a PCP of proximity.

Definition 2.7 (PCPs of Proximity): *A PCP of proximity for a set S with proximity parameter δ is a probabilistic* (non-adaptive) *polynomial-time oracle machine, denoted V, satisfying*

Completeness: *For every $x \in S$ there exists a string π_x such that V always accepts when given access to the oracle (x, π_x); that is, $\Pr[V^{x,\pi_x}(1^{|x|}) = 1] = 1$.*

---

[12] In fact, in the context of property testing, the length of the oracle must always be given to the tester (although some sources neglect to state this fact).

[13] We mention that PCPs of proximity are almost identical to Assignment Testers, defined independently by Dinur and Reingold [28]. Both notions are (important) special cases of the general definition of a "PCP spot-checker" formulated before in [30].

Soundness: *For every x that is δ-far from $S \cap \{0,1\}^{|x|}$ and for every string π, machine V rejects with probability at least $\frac{1}{2}$ when given access to the oracle (x, π); that is, $\Pr[M^{x,\pi}(1^{|x|}) = 1] \leq 1/2$.*

The query complexity of V is defined as in case of PCP, but here also queries to the x-part are counted.

The oracle (x, π) is actually a concatenation of two oracles: the input-oracle x (which replaces an explicitly given input in the definitions of PCPs and locally testable proofs), and a proof-oracle π (exactly as in the prior definitions). Note that Definition 2.7 refers to the distance of the input-oracle to S, whereas locally testable proofs refer to the distance of the proof-oracle from the set Π_x of valid proofs of membership of $x \in S$.

Still, PCPs of proximity can be defined within the framework of locally testable proofs. Specifically, consider an extension of Definition 2.6, where (relative) distances are measured according to a weighted Hamming distance; that is, for a weight function $\omega : \{1, ..., n\} \to [0,1]$ and $u, v \in \{0,1\}^n$, we let $\delta_\omega(u, v) = \sum_{i=1}^{n} \omega(i) \cdot \Delta(u_i, v_i)$. (Indeed, the standard notion of relative distance between $u, v \in \{0,1\}^n$ is obtained by $\delta_\omega(u, v)$ when using the uniform weighting function (i.e., $\omega(i) = 1/n$ for every $i \in \{1, ..., n\}$).) Now, Definition 2.7 can be viewed as a special case of (the extended) Definition 2.6 when applied to the (rather artificial) set of proofs $\Pi_{1^n} = \{(x, \pi) : x \in S \cap \{0,1\}^n \wedge \pi \in \Pi'_x\}$, where $\Pi'_x = \{\pi : \Pr[V^{x,\pi}(1^{|x|}) = 1] = 1\}$, by using the weighted Hamming distance δ_ω for ω that is uniform on the input-part of the oracle; that is, for $(x, \pi), (x', \pi') \in \{0,1\}^{n+p}$, we use $\delta_\omega((x, \pi), (x', \pi')) \stackrel{\text{def}}{=} \Delta(x, x')/n$, which corresponds to $\omega(i) = 1/n$ if $i \in \{1, ..., n\}$ and $\omega(i) = 0$ otherwise. Alternatively, weights can be approximately replaced by repetitions (provided that the tester checks the consistency of the repetitions).[14]

We mention that PCPs of proximity (of constant query complexity) yield a simple way of obtaining locally testable codes. More generally, we can combine any code C_0 with any PCP of proximity V, and obtain a q-locally testable code with distance essentially determined by C_0 and rate determined by V, where q is the query complexity of V. Specifically, x will be encoded by appending $c = C_0(x)$ by a proof that c is a codeword of C_0, and distances will be determined by the weighted Hamming distance that assigns uniform weights to the first part of the new code. As in the previous paragraph, these weights can be implemented by making suitable repetitions.

---

[14] That is, given a verifier V as in Definition 2.7, and denoting by n and $p = p(n)$ the sizes of the two parts of its oracle, we consider proofs of length $t \cdot n + p$, where $t = p/o(n)$ (e.g., $t = (p/n) \cdot \log n$). We consider a verifier V' with syntax as in Definition 2.6 that, on input 1^n and oracle access to $w = (u_1, ..., u_t, v) \in \{0,1\}^{t \cdot n + p}$, where $u_i \in \{0,1\}^n$ and $v \in \{0,1\}^p$, selects uniformly $i \in \{1, ..., t\}$ and invokes $V^{u_i, v}(1^n)$. In addition, V' performs a number of repetition tests that is inversely proportional to the proximity parameter, where in each test V' selects uniformly $i, i' \in \{1, ..., t\}$ and $j \in \{1, ..., n\}$ and checks that u_i and $u_{i'}$ agree on their j-th bit. Thus, V' essentially emulates the PCP of proximity V, and the fact that V satisfies Definition 2.7 can be captured by saying that V' satisfies Definition 2.6.

Finally, we comment that the definition of a PCP of proximity can be extended by providing the verifier with part of the input in an explicit form. That is, referring to Definition 2.7, we let $x = (x', x'')$, and provide V with explicit input $(x', 1^{|x|})$ and input-oracle x'' (rather than with explicit input $1^{|x|}$ and input-oracle x). Clearly, the extended formulation implies PCP as a special case (i.e., $x'' = \lambda$). More interestingly, an extended PCP of proximity for a set of pairs R (e.g., the witness relation of an NP-set), yields a PCP for the set $S \stackrel{\text{def}}{=} \{x' : \exists x'' \text{ s.t. } (x', x'') \in R\}$.

2.3.4 Relating Locally Testable Codes and Proofs

Locally testable codes can be thought of as the combinatorial counterparts of the complexity theoretic notion of locally testable proofs (PCPs). This perspective raises the question of whether one of these notions implies (or is useful towards the understanding of) the other.

Do PCPs imply locally testable codes? The use of codes with features related to local testability is implicit in known PCP constructions. Furthermore, the known constructions of locally testable proofs (PCPs) provides a transformation of *standard proofs* (for say SAT) to *locally testable proofs* (i.e., PCP-oracles), such that transformed strings are accepted with probability one by the PCP verifier. Specifically, denoting by S_x the set of standard proofs referring to an assertion x, there exists a polynomial-time mapping f_x of S_x to $R_x \stackrel{\text{def}}{=} \{f_x(y) : y \in S_x\}$ such that for every $\pi \in R_x$ it holds that $\Pr[V^\pi(x) = 1] = 1$, where V is the PCP verifier. Moreover, starting from different standard proofs, one obtains locally testable proofs that are far apart, and hence constitute a good code (i.e., for every x and every $y \neq y' \in S_x$, it holds that $\Delta(f_x(y), f_x(y')) \geq \Omega(|f_x(y)|)$). It is tempting to think that the PCP verifier yields a codeword tester, but this is not really the case. Note that Definition 2.5 requires rejection of strings that are far from any valid proof (i.e., any string far from Π_x), but it is not clear that the only valid proofs (w.r.t V) are those in R_x (i.e., the proofs obtained by the transformation f_x of standard proofs (in S_x) to locally testable ones).[15] In fact, the standard PCP constructions accept also valid proofs that are not in the range of the corresponding transformation (i.e., f_x); that is, Π_x as in Definition 2.5 is a strict subset of R_x (rather than $\Pi_x = R_x$). We comment that most known PCP constructions can be (non-trivially)[16] modified to yield $\Pi_x = R_x$, and thus to yield a locally testable code (but this is not necessarily the best way to design locally testable codes, see one alternative in §2.3.3).

Do locally testable codes imply PCPs? Saying that locally testable codes are the combinatorial counterparts of locally testable proofs (PCPs), raises the expectation (or hope) that it would be easier to construct locally testable codes than it

---

[15] Let alone that Definition 2.4 refers only to the case of false assertions, in which case all strings are far from a valid proof (which does not exist).

[16] The interested reader is referred to [42, Sec. 5.2] for a discussion of typical problems that arise.

is to construct PCPs. The reason being that combinatorial objects (e.g., codes) should be easier to understand than complexity theoretic ones (e.g., PCPs). Indeed, this feeling was among the main motivations of Goldreich and Sudan, and their first result (cf. [42, Sec. 3]) was along this vein: They showed a relatively simple construction (i.e., simple in comparison to PCP constructions) of a locally testable code of length $\ell(k) = k^c$ for any constant $c > 1$. Unfortunately, their stronger result, providing a locally testable code of shorter length (i.e., length $\ell(k) = k^{1+o(1)}$) is obtained by constructing (cf. [42, Sec. 4]) and using (cf. [42, Sec. 5]) a corresponding locally testable proof (i.e., PCP). Subsequent works have mostly followed this route, with the notable exception of Meir's work [52], which provides a combinatorial construction of a locally testable code that does not seem to yield a corresponding locally testable proof.[17]

2.3.5 Motivation for the Study of Short Locally Testable Codes and Proofs

Local testability offers an extremely strong notion of efficient testing: The tester makes only a constant number of bit probes, and determining the probed locations (as well as the final decision) is typically done in time that is polylogarithmic in the length of the probed object. Recall that the tested object is supposed to be related to some primal object; in the case of codes, the probed object is supposed to encode the primal object, whereas in the case of proofs the probed object is supposed to help verify some property of the primal object. In both cases, the length of the secondary (probed) object is of natural concern, and this length is stated in terms of the length of the primary object.

The length of codewords in an error-correcting code is widely recognized as one of the two most fundamental parameters of the code (the second one being the code's distance). In particular, the length of the code is of major importance in applications, because it determines the overhead involved in encoding information.

As argued in Section 1, the same considerations apply also to proofs. However, in the case of proofs, this obvious point was blurred by the indirect, unexpected and highly influential applications of PCPs to the theory of approximation algorithms. In our view, the significance of locally testable proofs (or PCPs) extends far beyond their applicability to deriving non-approximability results. The mere fact that proofs can be transformed into a format that supports super-fast probabilistic verification is remarkable. From this perspective, the question of how much redundancy is introduced by such a transformation is a fundamental one. Furthermore, locally testable proofs (i.e., PCPs) have been used not only to derive non-approximability results but also for obtaining positive results (e.g., CS-proofs [49,54] and their applications [8,24]), and the length of the PCP affects the complexity of those applications.

---

[17] We mention that the prior work of Ben-Sasson and Sudan [20] also shows some deviation from this route (i.e., it reversed the course to the "right one"): First codes are constructed, and next they are used towards the construction of proofs (rather than the other way around).

Turning back to the celebrated application of PCP to the study of approximation algorithms, we note that the length of PCPs is also relevant to non-approximability results; specifically, the length of PCPs affects the *tightness with respect to the running time* of the non-approximability results derived from these PCPs. For example, suppose (exact) SAT has complexity $2^{\Omega(n)}$. The original PCP Theorem [5,4] only implies that approximating MaxSAT requires time $2^{n^{\alpha}}$, for some (small) $\alpha > 0$. The work of [56] makes α arbitrarily close to 1, whereas the results of [42,21] further improve the lower bound to $2^{n^{1-o(1)}}$ and the results of [20,26] yields a lower bound of $2^{n/\text{poly}(\log n)}$. We mention that the result of [55] (cf. [27]) allows to achieve the lower bound of $2^{n^{1-o(1)}}$ simultaneously with optimal approximation ratios, but this is currently unknown for the better lower bound of $2^{n/\text{poly}(\log n)}$.

2.3.6 A Weaker Definition

One of the concrete motivations for local testable codes refers to settings in which one may want to re-encode the information when discovering that the codeword is corrupted. In such a case, assuming that re-encoding is based solely on the corrupted codeword, one may assume (or rather needs to assume) that the corrupted codeword is not too far from the code. Thus, the following version of Definition 2.1 may suffice for various applications.

Definition 2.8 (weak codeword tests): *Let* C $: \{0,1\}^k \to \{0,1\}^n$ *be a code of distance* d, *and let* $q \in \mathbb{N}$ *and* $\delta_1, \delta_2 \in (0,1)$ *be such that* $\delta_1 < \delta_2$. *A weak* q-local (codeword) (δ_1, δ_2)-tester *for* C *is a probabilistic* (non-adaptive) *oracle machine* M *that makes at most* q *queries, accepts any codeword, and rejects non-codewords that are both* δ_1-far *and* δ_2-close *to* C. *That is, the rejection condition of Definition 2.1 is modified as follows.*

Rejection of non-codeword (weak version): *For any* $w \in \{0,1\}^n$ *such that* $\Delta_C(w) \in [\delta_1 n, \delta_2 n]$, *given oracle access to* w, *machine* M *rejects with probability at least* $1/2$.

Needless to say, there is something highly non-intuitive in this definition: It requires rejection of non-codewords that are somewhat far from the code, but not the rejection of codewords that are very far from the code. Still, such weak codeword testers may suffice in some applications. Interestingly, such weak codeword testers do exist and even achieve linear length (cf. [59, Chap. 5]). We note that the non-monotonicity of the rejection probability of testers has been observed before, the most famous example being linearity testing (cf. [22] and [10]).

2.4 A Confused History

There is a fair amount of confusion regarding credits for some of the definitions presented in this section.[18] We refer mainly to the definition of locally testable

---

[18] Some confusion exists also with respect to some of the results and constructions described in Section 3, but in comparison to what will be discussed here the latter confusion is minor.

codes. This definition (or at least a related notion)[19] is arguably implicit in [7] as well as in subsequent works on PCP (see §2.3.4). Furthermore, the definition of locally testable codes has appeared independently in the works of Friedl and Sudan [34] and Rubinfeld and Sudan [58] as well as in the PhD Thesis of Arora [3].

3 Results and Ideas

We review the known constructions of locally testable codes and proofs, starting from codes and proofs of exponential length and concluding with codes and proofs of nearly linear length. We mention that random linear codes (of linear length) require any codeword tester to read a linear number of bits of the codeword [18], providing an indication to the non-triviality of local testability.

3.1 The Mere Existence of Locally Testable Codes and Proofs

The mere existence of locally testable codes and proofs, regardless of their length, is non-obvious. Thus, we start by recalling the simplest constructions known.

3.1.1 The Hadamard Code Is Locally Testable

The simplest example of a locally testable code (of constant relative distance) is the Hadamard code. This code, denoted C_{Had}, maps $x \in \{0,1\}^k$ to a string, of length $n = 2^k$, that provides the evaluation of all GF(2)-linear functions at x; that is, the coordinates of the codeword are associated with linear functions $\ell(z) = \sum_{i=1}^k \ell_i z_i$ and so $C_{\text{Had}}(x)_\ell = \ell(x) = \sum_{i=1}^k \ell_i x_i$. Testing whether a string $w \in \{0,1\}^{2^k}$ is a codeword amounts to linearity testing. This is the case because w is a codeword of C_{Had} if and only if, when viewed as a function $w : \{0,1\}^k \to \{0,1\}$, it is linear (i.e., $w(z) = \sum_{i=1}^k c_i z_i$ for some c_i's, or equivalently $w(y+z) = w(y) + w(z)$ for all y, z). Specifically, local testability is achieved by uniformly selecting $y, z \in \{0,1\}^k$ and checking whether $w(y+z) = w(y) + w(z)$. The exact analysis of this natural tester, due to Blum, Luby and Rubinfeld [22], turned out to be highly complex (cf. [22, 6, 31, 12, 13, 10, 47]). Denoting by $\text{rej}(w)$ the probability that the test rejects the string w and by $R(\delta)$ be the minimum of $\text{rej}(w)$ taken over all strings that are at distance $\delta \cdot |w|$ from C_{Had}, it is known that $R(\delta) \geq \Gamma(\delta)$, where the function $\Gamma : [0, 0.5] \to [0,1]$ is defined as follows:

$$\Gamma(x) \stackrel{\text{def}}{=} \begin{cases} 3x - 6x^2 & 0 \leq x \leq 5/16 \\ 45/128 & 5/16 \leq x \leq \tau_2 \text{ where } \tau_2 \approx 44.9962/128 \\ x + \delta(x) & \tau_2 \leq x \leq 1/2, \\ & \text{where } \delta(x) \stackrel{\text{def}}{=} 1376x^3(1-2x)^{12}. \end{cases} \tag{1}$$

---

[19] The related notion refers to the following relaxed notion of codeword testing: For two fixed good codes $C_1 \subseteq C_2 \subset \{0,1\}^n$, one has to accept (with high probability) every codeword of C_1, but reject (with high probability) every string that is far from being a codeword of C_2. Indeed, our definitions refer to the special (natural) case that $C_2 = C_1$, but the more general case suffices for the construction of PCPs (and is implicitly achieved in most of them).

The lower bound Γ is composed of three different bounds with "phase transitions" at $x = \frac{5}{16}$ and at $x = \tau_2$ (where $\tau_2 \approx \frac{44,9962}{128}$ is the solution to $x + \delta(x) = 45/128$).[20] It was shown in [10] that the first segment of this bound (i.e., for $x \in [0, 5/16]$) is the best possible, and that the first "phase transitions" (i.e., at $x = \frac{5}{16}$) is indeed a reality; in other words, $R = \Gamma$ in the interval $[0, 5/16]$.[21] We highlight the fact that the detection probability of the aforementioned test does not increase monotonically with the distance (of the string from the code), since Γ decreases in the interval $[1/4, 5/16]$ (while equaling R in this interval).

Other codes. We mention that Reed-Muller Codes of constant order are also locally testable [1]. These codes have sub-exponential length, but are quite popular in practice. The Long Code is also locally testable [11], but this code has double-exponential length (and was introduced merely for the design of PCPs).[22]

3.1.2   The Hadamard-Based PCP of [4]

The simplest example of a locally testable proof (for a set not known to be in $\mathcal{BPP}$) is the "inner verifier" of the PCP construction of Arora, Lund, Motwani, Sudan and Szegedy [4], which in turn is based on the Hadamard code. Specifically, proofs of the satisfiability of a given system of quadratic equations over $GF(2)$ are presented by providing a Hadamard encoding of the outer-product of a satisfying assignment with itself (i.e., a satisfying assignment $\alpha \in \{0,1\}^n$ is presented by $C_{\mathrm{Had}}(\beta)$, where $\beta = (\beta_{i,j})_{i,j \in [n]}$ and $\beta_{i,j} = \alpha_i \alpha_j$). Given an alleged proof $\pi \in \{0,1\}^{2^{n^2}}$, the proof-tester proceeds as follows:

1. Tests that π is indeed a codeword of the Hadamard Code. If the test passes then w is close to some $C_{\mathrm{Had}}(\beta)$, for an arbitrary $\beta = (\beta_{i,j})_{i,j \in [n]}$.
2. Tests that the aforementioned β is indeed an outer-product of some $\alpha \in \{0,1\}^n$ with itself. Note that the Hadamard encoding of α is supposed to be part of the Hadamard encoding of β (because $\sum_{i=1}^n c_i \alpha_i = \sum_{i=1}^n c_i \alpha_i^2$ is supposed to equal $\sum_{i=1}^n c_i \beta_{i,i}$). So we would like to test that the latter codeword matches the former one. Specifically, we wish to test whether $(\beta_{i,j})_{i,j \in [n]}$ equals $(\alpha_i \alpha_j)_{i,j \in [n]}$ (i.e., the equality of two matrices). This can be done by uniformly selecting $(r_1, ..., r_n), (s_1, ..., s_n) \in \{0,1\}^n$, and comparing $\sum_{i,j} r_i s_j \beta_{i,j}$ and $\sum_{i,j} r_i s_j \alpha_i \alpha_j = (\sum_i r_i \alpha_i)(\sum_j s_j \alpha_j)$.
   The above would have been fine if $w = C_{\mathrm{Had}}(\beta)$, but we only know that w is close to $C_{\mathrm{Had}}(\beta)$. The Hadamard encoding of α is a tiny part of the latter, and so we should not try to retrieve the latter directly (because this tiny part

---

[20] The third segment is due to [47], which improves over the prior bound of [10] that asserted $R(x) \geq \max(45/128, x)$ for every $x \in [5/16, 1/2]$.

[21] In contrast, the lower bound provided by the other two segments (i.e., for $x \in [5/16, 1/2]$) is unlikely to be tight, and in particular it is unlikely that the "phase transitions" at $x = \tau_2$ represents the behavior of R itself. Also note that $\delta(x) > 59(1 - 2x)^{12}$ for every $x > \tau_2$, but $\delta(x) < 0.0001$ for every $x < 1/2$.

[22] Interestingly, the best results are obtained by using a relaxed notion of local testability [44, 45].

may be totally corrupted). Instead, we use the paradigm of self-correction (cf. [22]): In general, for any fixed $c = (c_{i,j})_{i,j \in [n]}$, whenever we wish to retrieve $\sum_{i=1}^{n} c_{i,j} \beta_{i,j}$, we uniformly select $r = (r_{i,j})_{i,j \in [n]}$ and retrieve both $w(r)$ and $w(r + c)$. Thus, we obtain a self-corrected value of $w(c)$; that is, if w is δ-close to $C_{\mathrm{Had}}(\beta)$ then $w(r + c) - w(r) = \sum_{i=1}^{n} c_{i,j} \beta_{i,j}$ with probability at least $1 - 2\delta$ (over the choice of r).

Using self-correction, we indirectly obtain bits in $C_{\mathrm{Had}}(\alpha)$, for $\alpha = (\alpha_i)_{i \in [n]} = (\beta_{i,i})_{i \in [n]}$. Similarly, we can obtain any other desired bit in $C_{\mathrm{Had}}(\beta)$, which in turn allows us to test whether $(\beta_{i,j})_{i,j \in [n]} = (\alpha_i \alpha_j)_{i,j \in [n]}$. In fact, we are checking whether $(\beta_{i,j})_{i,j \in [n]} = (\beta_{i,i} \beta_{j,j})_{i,j \in [n]}$, by comparing $\sum_{i,j} r_i s_j \beta_{i,j}$ and $(\sum_i r_i \beta_{i,i})(\sum_j s_j \beta_{j,j})$, for randomly selected $(r_1, ..., r_n), (s_1, ..., s_n) \in \{0, 1\}^n$.

3. Finally, we need to check whether the aforementioned α satisfies the given system of equations. Towards this end, we uniformly selects a linear combination of the equations, and check whether α satisfies the resulting (single) equation. Note that the value of the corresponding linear expression (in quadratic (and linear) forms) appears as a bit of the Hadamard encoding of β, but again we retrieve it from w by using self correction.

One key observation underlying the analysis of Steps 2 and 3 is that for $(u_1, ..., u_n) \neq (v_1, ..., v_n) \in \{0, 1\}^n$, if we uniformly select $(r_1, ..., r_n) \in \{0, 1\}^n$ then $\Pr[\sum_i r_i u_i = \sum_i r_i v_i] = 1/2$. Similarly, for n-by-n matrices $A \neq B$, when $r, s \in \{0, 1\}^n$ are uniformly selected (vectors), it holds that $\Pr[As = Bs] = 2^{-\mathrm{rank}(A-B)}$ and it follows that $\Pr[rAs = rBs] \leq 3/4$.

3.2 Locally Testable Codes and Proofs of Polynomial Length

The constructions presented in Section 3.1 have exponential length in terms of the relevant parameter (i.e., the amount of information being encoded in the code or the length of the assertion being proved). Achieving local testability by codes and proofs that have polynomial length turns out to be more challenging.

3.2.1 Locally Testable Codes of Quadratic Length

A direct interpretation of *low-degree tests* (cf. [6,7,35,58,34]), proposed by Friedl and Sudan [34] and Rubinfeld and Sudan [58], yields a locally testable code of quadratic length over a *sufficiently large alphabet*. Similar (and actually better) results for *binary* codes required additional ideas, and have appeared only later (cf. [42]). We sketch both constructions below, starting with locally testable codes over very large alphabets (which are defined analogously to the binary case).

We will consider a code $C : \Sigma^k \to \Sigma^n$ of linear distance, with $|\Sigma| \gg k$ and $n > k^2$. For parameters $m \ll d < \log k$ (such that $k < d^m$), consider a finite field F of size $O(d)$ and an alphabet $\Sigma = F^{d+1}$ (see below).[23] Viewing the

---

[23] Indeed, it would have been more natural to present the code as a mapping from sequences over F to sequences over $\Sigma = F^{d+1}$. Following the convention of using the same alphabet for both the information and the codeword, we just pack every $d + 1$ elements of F as an element of Σ.

information as an m-variant polynomial p of total degree d over F, we encode it by providing its value on all possible lines over F^m, where each such line is defined by two points in F^m. Actually, the value of p on such a line can be represented by a univariate polynomial of degree d. Thus, the code maps $\log_2 |F|^{\binom{m+d}{d}} > (d/m)^m \log |F|$ bits of information (which may be viewed as $k \stackrel{\text{def}}{=} (d/m)^m / (d+1) \approx d^{m-1}/m^m$ long sequences over $\Sigma = F^{d+1}$) to sequences of length $n \stackrel{\text{def}}{=} |F|^{2m} = O(d)^{2m}$ over Σ. Note that the smaller m, the better the rate (i.e., relation of n to k) is, but this comes at the expense of using a larger alphabet. In particular, we consider two instantiations:

1. Using $d = m^m$, we get $k \approx m^{m^2-2m}$ and $n = m^{2m^2+o(m)}$, which yields $n \approx \exp(\sqrt{\log k}) \cdot k^2$ and $\log |\Sigma| = \log |F|^{d+1} \approx d \log d \approx \exp(\sqrt{\log k})$.
2. Letting $d = m^c$ for any constant $c > 1$, we get $k \approx m^{(c-1)m}$ and $n = m^{2cm+o(m)}$, which yields $n \approx k^{2c/(c-1)}$ and $\log |\Sigma| \approx d \log d \approx (\log k)^c$.

As for the codeword tester, it uniformly selects two intersecting lines and checks that the corresponding univariate polynomials agree on the point of intersection. Thus, this tester makes two queries (to an oracle over the alphabet Σ). The analysis of this tester reduces to the analysis of the corresponding low degree test, undertaken in [4, 56].

The above tester uses only two queries, but the entire description (which refers to codes over a large alphabet) deviates from the bulk of our treatment, which has focused on a binary alphabet. We comment that 2-query locally testable *binary* codes are essentially impossible (cf., [14]), but we have already seem that 3-query tests are possible. A natural way of reducing the alphabet size of codes is via the well-known paradigm of *concatenated codes* [32].[24] However, local testability can be maintained only in special cases. In particular, observe that, for each of the two queries made by the tester of C, the tester does not need the entire polynomial represented in $\Sigma = F^{d+1}$, but rather only its value at a specific point. Thus, encoding Σ by an error correcting code that supports recovery of the said value while using a constant number of probes will do.[25] In particular, for integers h, e such that $d + 1 = h^e$, Goldreich and Sudan used an encoding of $F^{d+1} = F^{h^e}$ by sequences of length $|F|^{eh}$ over F, and provided a testing and

---

[24] A concatenated code is obtained by encoding the symbols of an "outer code" (using the coding method of the "inner code"). Specifically, let $C_1 : \Sigma_1^{k_1} \to \Sigma_1^{n_1}$ be the outer code and $C_2 : \Sigma_2^{k_2} \to \Sigma_2^{n_2}$ be the inner code, where $\Sigma_1 \equiv \Sigma_2^{k_2}$. Then, the concatenated code $C : \Sigma_2^{k_1 k_2} \to \Sigma_2^{n_1 n_2}$ is obtained by $C(x_1, ..., x_{k_1}) = (C_2(y_1), ..., C_2(y_{n_1}))$, where $x_i \in \Sigma_2^{k_2} \equiv \Sigma_1$ and $(y_1, ..., y_{n_1}) = C_1(x_1, ..., x_{k_1})$. Using a good inner code for relatively short sequences, allows to transform good codes for a large alphabet into good codes for a smaller alphabet.

[25] Indeed, this property is related to locally decodable codes, to be discussed in Section 4. Here we need to recover one out of $|F|$ specific linear combinations of the encoded $(d+1)$-long sequence of F-symbols. In contrast, locally decodable refers to recovering one out of the original F-symbols of the $(d+1)$-long sequence.

recovery procedure that makes $O(e)$ queries [42, Sec. 3.3]. We mention that the case of $e = 1$ and $|F| = 2$ corresponds to the Hadamard code, and that a bigger constant e allow for shorter codes. The resulting concatenated code, C', is a locally testable code over F, and has length $n \cdot O(d)^{eh} = n \cdot \exp((e \log d) \cdot d^{1/e})$. Using constant $e = 2c$ and setting $d = m^c \approx (\log k)^c$, we get $n \approx k^{2c/(c-1)} \cdot \exp(\widetilde{O}(\log k)^{1/2})$ and $|F| = \text{poly}(\log k)$. Finally, a *binary* locally testable code is obtained by concatenating C' with the Hadamard code, while noting that the latter supports a "local recovery" property that suffices to emulate the tester for C'. In particular, the tester of C' merely checks a linear (over F) equation referring to a constant number of F-elements, and for $F = GF(2^\ell)$, this can be emulated by checking *related* random linear combinations of the bits representing these elements, which in turn can be locally recovered (or rather self-corrected) from the Hadamard code. The final result is a locally testable (binary) code of nearly quadratic length.[26]

3.2.2   Locally Testable Proofs of Polynomial Length: The PCP Theorem

The case of proofs is far more complex: Achieving locally testable proofs of polynomial length is essentially the contents of the celebrated PCP Theorem of Arora, Lund, Motwani, Safra, Sudan and Szegedy [5, 4]. The construction is analogous to (but far more complex than) the one presented in the case of codes:[27] First one constructs proofs over a large alphabet, and next one composes such proofs with corresponding "inner" proofs (over a smaller alphabet, and finally a binary one). Our exposition focuses on the construction of these proof systems and blurs the issues involved in their composition.[28]

The first step is to introduce the following NP-complete problem. The input to the problem consists of a finite field F, a subset $H \subset F$ of size $\lfloor |F|^{1/15} \rfloor$, an integer $m < |H|$, and a $(3m + 4)$-variant polynomial $P : F^{3m+4} \to F$ of total degree $3m|H| + O(1)$. The problem is to determine whether there exists an m-variant ("assignment") polynomial $A : F^m \to F$ of total degree $m|H|$ such that $P(x, z, y, \tau, A(x), A(y), A(z)) = 0$ for every $x, y, z \in H^m$ and $\tau \in \{0, 1\}^3 \subset H$. Note that the problem-instance can be explicitly described by a sequence of $|F|^{3m+4} \log_2 |F|$ bits, whereas the solution sought can be explicitly described by a sequence of $|F|^m \log_2 |F|$ bits. We comment that the NP-completeness of the aforementioned problem can be proved via a reduction from 3SAT, by identifying the variables of the formula with H^m and essentially letting P be a low-degree extension of a function $f : H^{3m} \times \{0, 1\}^3 \to \{0, 1\}$ that encodes the structure of

---

[26] Actually, the aforementioned result is only implicit in [42], because Goldreich and Sudan apply these ideas directly to a truncated version of the low-degree based code.

[27] Our presentation reverses the historical order in which the corresponding results (for codes and proofs) were achieved. That is, the constructions of locally testable proofs of polynomial length predated the coding counterparts.

[28] This section is significantly more complex than the rest of this article, and some readers may prefer to skip it and proceed directly to Section 3.3. For further details regarding the proof composition paradigm, the reader is referred to [37, Sec. 9.3.2].

the formula (by considering all possible 3-clauses). In fact, the resulting P has degree $|H|$ in each of the first $3m$ variables and constant degree in each of the other variables, and this fact can be used to improve the parameters below (but not in a fundamental way).

The proof that a given input P satisfies the aforementioned condition consists of an m-variant polynomial $A : F^m \to F$ (which is supposed to be of total degree $m|H|$) as well as $3m + 4$ auxiliary polynomials $A_i : F^{3m+1} \to F$, for $i = 1, ..., 3m + 1$ (each supposedly of degree $(3m|H| + O(1)) \cdot m|H|$). The polynomial A is supposed to satisfy the conditions of the problem, and in particular $P(x, z, y, \tau, A(x), A(y), A(z)) = 0$ should hold for every $x, y, z \in H^m$ and $\tau \in \{0,1\}^3 \subset H$. Furthermore, $A_0(x, z, z, \tau) \overset{\text{def}}{=} P(x, z, y, \tau, A(x), A(y), A(z))$ should vanish on H^{3m+1}. The auxiliary polynomials are given to assist the verification of the latter condition. In particular, it should be the case that A_i vanishes on $F^i H^{3m+1-i}$, a condition that is easy to test for A_{3m+1} (assuming that A_{3m+1} is a low degree polynomial). Checking that A_{i-1} agrees with A_i on $F^{i-1} H^{3m+1-(i-1)}$, for $i = 1, ..., 3m+1$, and that all A_i's are low degree polynomials, establishes the claim for A_0. Thus, testing an alleged proof $(A, A_1, ..., A_{3m+1})$ is performed as follows:

1. Testing that A is a polynomial of total degree $m|H|$. This is done by selecting a random line through F^m, and testing whether A restricted to this line agrees with a degree $m|H|$ univariate polynomial.

2. Testing that, for $i = 1, ..., 3m + 1$, the polynomial A_i is of total degree $d \overset{\text{def}}{=} (3m|H| + O(1)) \cdot m|H|$. Here we select a random line through F^{3m+1}, and test whether A_i restricted to this line agrees with a degree d univariate polynomial.

3. Testing that, for $i = 1, ..., 3m + 1$, the polynomial A_i agrees with A_{i-1} on $F^{i-1} H^{3m+1-(i-1)}$. This is done by uniformly selecting $r' = (r_1, ..., r_{i-1}) \in F^{i-1}$ and $r'' = (r_{i+1}, ..., r_{3m+1}) \in F^{3m+1-i}$, and comparing $A_{i-1}(r', e, r'')$ to $A_i(r', e, r'')$, for every $e \in H$. In addition, we check that both functions when restricted to the axis-parallel line $(r', \cdot, r'')$ agree with a univariate polynomial of degree d.[29] We stress that the values of A_0 are computed according to the given polynomial P by accessing A at the appropriate locations (i.e., by definition $A_0(x, z, z, \tau) = P(x, z, y, \tau, A(x), A(y), A(z))$).

4. Testing that A_{3m+1} vanishes on F^{3m+1}. This is done by uniformly selecting $r \in F^{3m+1}$, and testing whether $F(r) = 0$.

The above description (which follows [60, Apdx. C]) is somewhat different than the original presentation in [4], which in turn follows [6,7,31].[30] The above tester may be viewed as making $O(m|F|)$ queries to an oracle over the alphabet F, or

---

[29] Thus, effectively, we are self-correcting the values at H (on the said line), based on the values at F (on that line).

[30] The point is that the sum-check, which originates in [51], is replaced by an analogous process (which happens to be non-adaptive).

alternatively, as making $O(m|F|\log|F|)$ binary queries.[31] Note that we have already obtained a highly non-trivial tester. It makes $O(m|F|\log|F|)$ queries in order to verify a claim regarding an input of length $n \stackrel{\text{def}}{=} |F|^{3m+4}\log_2|F|$. Using $m = \log n/\log\log n$, $|H| = \log n$ and $|F| = \text{poly}(\log n)$, we have obtained a tester of poly-logarithmic query complexity.

To further reduce the query complexity, one invokes the "proof composition" paradigm, introduced by Arora and Safra [5]. Specifically, one composes an "outer" tester (as described above) with an "inner" tester that checks the residual condition that the "outer" tester determines for the answers it obtains. This composition is more problematic than one suspects, because we wish the "inner" tester to perform its task without reading its entire input (i.e., the answers to the "outer" tester). This seems quite paradoxical, since it is not clear how the "inner" tester can operate without reading its entire input. The problem can be resolved by using a "proximity tester" (i.e., a PCP of proximity) as an "inner" tester, provided that it suffices to have such a proximity test (for the answers to the "outer" tester). Thus, the challenge is to reach a situation in which the "outer" tester is robust in the sense that, when the assertion is false, the answers obtained by this tester are far from being convincing (i.e., they are far from any sequence of answers that is accepted by this tester). Two approaches towards obtaining such robust testers are known.

– One approach, introduced in [4], is to convert the "outer" tester into one that makes a constant number of queries over some larger alphabet, and furthermore have the answer be presented in an error correcting format. Thus, robustness is guaranteed by the fact that the answers correspond to a constant-length sequence of codewords, and so any two (properly formatted) sequences are at constant relative distance of one another.

The implementation of this approach consists of two steps (and is based on some specifics). The first step is to convert the "outer" tester into one that makes a constant number of queries over some larger alphabet. This step uses the so-called parallelization technique (cf. [50,4]). Next, one applies an error correcting code to these $O(1)$ longer answers, and assumes that the "proximity tester" can handle inputs presented in this format (i.e., that it can test an input that is presented by an encoding of a constant number of its parts).[32]

– An alternative approach, pursued and advocated in [15], is to take advantage of the specific structure of the queries, "bundle" the answers together and furthermore show that the "bundled" answers are "robust" in a sense that fits

---

[31] Another alternative perspective is obtained by applying so-called parallelization (cf. [50,4]). The result is a test making a constant number of queries that are each answered by strings of length $\text{poly}(|F|)$.

[32] The aforementioned assumption holds trivially in case one uses a generic "proximity tester" (i.e., a PCP of proximity or an Assignment Tester) as done in [28]. But the aforementioned approach can be (and was in fact originally) applied with a specific "proximity tester" that can only handle inputs presented in one specific format (cf. [4]).

proximity testing. In particular, the (generic) parallelization step is avoided, and is replaced by a closer analysis of the specific (outer) tester. We will demonstrate this approach next.

First, we show how the queries of the aforementioned tester can be "bundled" (into a constant number of bundles). In particular, we consider the following "bundling" that accommodates all types of tests (and in particular the $m + 1$ different sub-tests performed in Steps 2 and 3). Consider

$$B(x_1, ..., x_{3m+1}) = (A_1(x_1, x_2, ..., x_{3m+1}), A_2(x_2, ..., x_{3m+1}, x_1), ..., A_{3m+1}(x_{3m+1}, x_1, ..., x_{3m}))$$

and perform all $3m + 1$ tests of Step (3) by selecting uniformly $(r_2, ..., r_{3m+1}) \in F^{3m}$ and querying B at $(e, r_2, ..., r_{3m+1})$ and $(r_{3m+1}, e, ..., r_{3m})$ for all $e \in F$. Thus, all $3m + 1$ tests of Step (3) can be performed by retrieving the values of B on a single *axis parallel* random line through F^{3m+1}. Furthermore, note that all $3m + 1$ tests of Step (2) can be performed by retrieving the values of B on a single (arbitrary) random line through F^{3m+1}. Finally, observe that these tests are "robust" in the sense that if, for some i, the function A_i is (say) 0.01-far from satisfying the condition (i.e., being low-degree or agreeing with A_{i-1}) then with constant probability many of the values of A_i on an appropriate random line will not fit to what is needed. This robustness property is inherited by B, as well as by B' (resp., A') that is obtained by applying a good binary error-correcting code on B (resp., on A). Thus, we may replace A and the A_i's by A' and B', and conduct all all tests by making $O(m^2|F| \log |F|)$ queries to $A' : F^m \times [O(\log |F|)] \rightarrow \{0, 1\}$ and $B' : F^{3m+1} \times [O(\log |F|^{3m+1})] \rightarrow \{0, 1\}$. The *robustness property* asserts that if the original polynomial P had no solution (i.e., an A as above) then the answers obtained by the tester will be far from satisfying the residual decision predicate of the tester.

Once the robustness property of the resulting ("outer") tester fits the proximity testing feature of the "inner tester", composition is possible. Indeed, we compose the "outer" tester with an "inner tester" that checks whether the residual decision predicate of the "outer tester" is satisfies. The benefit of this composition is that query complexity is reduced from poly-logarithmic to polynomial in a double-logarithm. At this point we can afford the Hadamard-Based proof tester (because the overhead in the proof complexity will only be exponential in a polynomial in a double-logarithmic function), and obtain a locally testable proof of polynomial length. That is, we compose the poly(log log)-query tester (acting as an outer tester) with the Hadamard-Based tester (acting as an inner tester), and obtain a locally testable proof of polynomial length (as asserted by the PCP Theorem).

Digest: the proof composition paradigm. The PCP Theorem asserts a PCP system that obtains simultaneously the minimal possible randomness and query complexity (up to a multiplicative factor, assuming that $\mathcal{P} \neq \mathcal{NP}$). The foregoing construction obtains this remarkable result by combining two different PCPs: the first PCP obtains logarithmic randomness but uses poly-logarithmically many queries, whereas the second PCP uses a constant number of queries but has polynomial randomness complexity. We stress that *each of these two PCP systems is highly non-trivial and very interesting by itself.* We also highlight the fact that these PCPs are combined using a very simple composition method (which refers to auxiliary properties such as robustness and proximity testing). Details follow.[33]

Loosely speaking, the proof composition paradigm refers to composing two proof systems such that the "inner" verifier is used for probabilistically verifying the acceptance criteria of the "outer" verifier. That is, the combined verifier selects coins for the "outer" verifier, determines the corresponding locations that the "outer" verifier wishes to inspect (in the proof), and verifies that the "outer" verifier would have accepted the values that reside in these locations. The latter verification is performed by invoking the "inner" verifier, *without reading the values residing in all the aforementioned locations*. Indeed, the aim is to conduct this ("composed") verification while using much fewer queries than the query complexity of the "outer" proof system. In particular, the inner verifier cannot afford to read its input, which makes the composition more subtle than the term suggests.

In order for the proof composition to work, the combined verifiers should satisfy some auxiliary conditions. Specifically, the *outer* verifier should be robust in the sense that its soundness condition guarantee that, with high probability, the oracle answers are "far" from satisfying the residual decision predicate (rather than merely not satisfying it).[34] The *inner* verifier is given oracle access to its input and is charged for each query made to it, but is only required to reject (with high probability) inputs that are far from being valid (and, as usual, accept inputs that are valid). That is, the inner verifier is actually a verifier of proximity.

Composing two such PCPs yields a new PCP, where the new proof oracle consists of the proof oracle of the "outer" system and a sequence of proof oracles for the "inner" system (one "inner" proof per each possible random-tape of the "outer" verifier). The resulting verifier selects coins for the outer-verifier and uses the corresponding "inner" proof in order to verify that the outer-verifier would have accepted under this choice of coins. Note that such a choice of coins determines locations in the "outer" proof that the outer-verifier would have inspected, and the combined verifier provides the inner-verifier with oracle access to these locations (which the inner-verifier considers as its input) as well as with oracle access to the corresponding "inner" proof (which the inner-verifier considers as its proof-oracle).

The quantitative effect of such a composition is easy to analyze. Specifically, composing an outer-verifier of randomness-complexity r' and query-complexity q' with an inner-verifier of randomness-complexity r'' and query-complexity q'' yields a PCP of randomness-complexity $r(n) = r'(n) + r''(q'(n))$ and query-complexity $q(n) = q''(q'(n))$, because $q'(n)$ represents the length of the input (or-

---

[33] Our presentation of the composition paradigm follows [15], rather than the original presentation of [5, 4].

[34] Furthermore, the latter predicate, which is well-defined by the non-adaptive nature of the outer verifier, must have a circuit of size bounded by a polynomial in the number of queries.

acle) that is accessed by the inner-verifier. Thus, assuming $q''(m) \ll m$, the query complexity is significantly decreased (from $q'(n)$ to $q''(q'(n))$), while the increase in the randomness complexity is moderate provided that $r''(q'(n)) \ll r'(n)$. Furthermore, the verifier resulting from the composition inherits the robustness features of the composed verifier, which is important in case we wish to compose the resulting verifier with another inner-verifier.

3.3 Locally Testable Codes and Proofs of Nearly Linear Length

We now move on to even *shorter* codes and proofs; specifically, codes and proofs of *nearly linear length*. The latter term has been given quite different interpretations, and we start by sorting these out. Currently, this taxonomy is relevant mainly for second-level discussions and review of some past works.[35]

3.3.1 Types of Nearly Linear Functions
A few common interpretations of this term are listed below (going from the most liberal to the most strict one).

T1-nearly linear: A very liberal notion, which seems at the verge of an abuse of the term, refers to a sequence of functions $f_\epsilon : \mathbb{N} \to \mathbb{N}$ such that, for every $\epsilon > 0$, it holds that $f_\epsilon(n) \le n^{1+\epsilon}$. That is, each function is actually of the form $n \mapsto n^c$, for some constant $c > 1$, but the sequence as a whole can be viewed as approaching linearity.

The PCP of Polishchuk and Spielman [56] and the simpler locally testable code of Goldreich and Sudan [42, Thm. 2.4] have nearly linear length in this sense.

T2-nearly linear: A more reasonable notion of nearly linear functions refers to individual functions f such that $f(n) = n^{1+o(1)}$. Specifically, for some function $\epsilon : \mathbb{N} \to [0,1]$ that goes to zero, it holds that $f(n) \le n^{1+\epsilon(n)}$. Common sub-types include the following:
1. $\epsilon(n) = 1/\log\log n$.
2. $\epsilon(n) = 1/(\log n)^c$ for some constant $c \in (0,1)$.
   The locally testable codes and proofs of [42, 21, 15] have nearly linear length in this sense. Specifically, in [42, Sec. 4-5] and [21] any $c > 1/2$ will do, whereas in [15] any $c > 0$ will do.
3. $\epsilon(n) = \frac{\exp((\log\log n)^c)}{\log n}$ for some constant $c \in (0,1)$.
   Note that $\mathrm{poly}(\log\log n) < \exp((\log\log n)^c) < (\log n)^{o(1)}$, for any constant $c \in (0,1)$.

Indeed, the case in which $\epsilon(n) = \frac{O(\log\log n)}{\log n}$ (or so) deserves a special category, presented next.

---

[35] Things were different when the original version of this text [36] was written. At that time, only T2-nearly linear length was know for $O(1)$-local testability, and the T3-nearly linear result achieved by Dinur [26] seemed a daring conjecture (which was, nevertheless, stated in [36, Conj. 3.3]).

T3-nearly linear: The strongest notion interprets near-linearity as linearity up to a poly-logarithmic (or quasi-poly-logarithmic) factor. In the former case $f(n) = \widetilde{O}(n) \stackrel{\text{def}}{=} \text{poly}(\log n) \cdot n$, which corresponds to the case of $f(n) \leq n^{1+\epsilon(n)}$ with $\epsilon(n) = O(\log \log n)/\log n$, whereas the latter case corresponds to $\epsilon(n) = \text{poly}(\log \log n)/\log n$ (i.e., in which case $f(n) \leq (\log n)^{\text{poly}(\log \log n)} \cdot n$).

The recent results of [20, 26] refer to this notion.

We note that while [20, 26] achieve T3-nearly linear length, the low-error results of [55, 27] only achieve T2-nearly linear length.

3.3.2 Local Testability with Nearly Linear Length

The celebrated gap amplification technique of Dinur [26] is best known for providing an alternative proof of the PCP Theorem. However, applying this technique to a PCP that was (previously) provided by Ben-Sasson and Sudan [20] yields locally testable codes and proofs of T3-nearly linear length. In particular, the overhead in the code and proof length is only polylogarithmic in the length of the primal object (which establishes [36, Conj. 3.3]).

Theorem 3.1 (Dinur [26], building on [20]): *There exists a constant q and a poly-logarithmic function $f : \mathbb{N} \rightarrow \mathbb{N}$ such that there exist q-locally testable codes and proofs of length $f(k) \cdot k$, where k denotes the length of the actual information (i.e., the assertion in case of proofs and the encoded information in case of codes).*

The proof of Theorem 3.1 combines the PCP system of Ben-Sasson and Sudan [20] with the gap amplification method of Dinur [26]. The latter is reviewed in §3.3.3. We mention that the PCP system of [20] is based on the NP-completeness of a certain code (of length $n = \widetilde{O}(k)$), and on a randomized reduction of testing whether a given n-bit long string is a codeword to a constant number of similar tests that refer to $\sqrt{n}$-bit long strings. Applying this reduction $\log \log n$ times yields a PCP of query complexity $\text{poly}(\log n)$ and length $\widetilde{O}(n)$, which in turn yields a 3-query "PCP with soundness error $1 - 1/\text{poly}(\log n)$".

We mention that in the original version of this survey [36], we conjectured that a polylogarithmic (length) overhead is inherent to local testability (or, at least, that linear length $O(1)$-local testability is impossible). We currently have mixed feelings with respect to this conjecture (even when confined to proofs), and thus rephrase it as an open problem.

Open Problem 3.2. *Determine whether there exist locally testable codes and proofs of linear length.*

3.3.3 The Gap Amplification Method

Essentially, Theorem 3.1 is proved by applying the gap amplification method (of Dinur [26]) to the (weak) PCP system constructed by Ben-Sasson and Sudan [20]. The latter PCP system has length $\ell(k) = \widetilde{O}(k)$, but its soundness error is $1 - 1/\text{poly}(\log k)$ (i.e., its rejection probability is at least $1/\text{poly}(\log k)$). Each application of the gap amplification step *doubles the rejection probability while essentially maintaining the initial complexities.* That is, in each step, the constant query complexity of the verifier is preserved and its randomness complexity is increased only by a constant term (and so the length of the PCP oracle is increased only by a constant factor). Thus, starting from the system of [20] and applying $O(\log \log k)$ amplification steps, we essentially obtain Theorem 3.1. (Note that a PCP system of polynomial length can be obtained by starting from a trivial "PCP" system that has rejection probability $1/\text{poly}(k)$, and applying $O(\log k)$ amplification steps.)

In order to describe the aforementioned process we need to *redefine PCP systems so as to allow arbitrary soundness error.* In fact, for technical reasons, it is more convenient to describe the process as an iterated reduction of a "constraint satisfaction" problem to itself. Specifically, we refer to systems of 2-variable constraints, which are readily represented by (labeled) graphs such that the vertices correspond to (non-Boolean) variables and the edges are associated with constraints.

Definition 3.3 (CSP with 2-variable constraints): *For a fixed finite set Σ, an instance of CSP consists of a graph $G = (V, E)$ (which may have parallel edges and self-loops) and a sequence of 2-variable constraints $\Phi = (\phi_e)_{e \in E}$ associated with the edges, where each constraint has the form $\phi_e : \Sigma^2 \to \{0, 1\}$. The value of an assignment $\alpha : V \to \Sigma$ is the number of constraints satisfied by α; that is, the value of α is $|\{(u, v) \in E : \phi_{(u,v)}(\alpha(u), \alpha(v)) = 1\}|$. We denote by $\text{vlt}(G, \Phi)$ (standing for violation) the fraction of unsatisfied constraints under the best possible assignment; that is,*

$$\text{vlt}(G, \Phi) = \min_{\alpha:V \to \Sigma} \left\{ \frac{|\{(u, v) \in E : \phi_{(u,v)}(\alpha(u), \alpha(v)) = 0\}|}{|E|} \right\} \qquad (2)$$

For various functions $\tau : \mathbb{N} \to (0, 1]$, we will consider the promise problem $\text{gapCSP}_\tau^\Sigma$, having instances as above, such that the YES-instances are fully satisfiable instances (i.e., $\text{vlt} = 0$) and the NO-instances are pairs (G, Φ) for which $\text{vlt}(G, \Phi) \geq \tau(|G|)$ holds, where $|G|$ denotes the number of edges in G.

Note that 3SAT is reducible to $\text{gapCSP}_{\tau_0}^{\Sigma_0}$ for $\Sigma_0 = \{\text{F}, \text{T}\}^3$ and $\tau_0(m) = 1/m$ (e.g., replace each clause by a vertex, and use edge constraints that enforce mutually consistent and satisfying assignments to each pair of clauses). Furthermore, the PCP system of [20] yields a reduction of 3SAT to $\text{gapCSP}_{\tau_1}^{\Sigma_0}$ for $\tau_1(m) = 1/\text{poly}(\log m)$ where the size of the graph is nearly linear in the length of the input formula. Our goal is to reduce $\text{gapCSP}_{\tau_0}^{\Sigma_0}$ (or rather $\text{gapCSP}_{\tau_1}^{\Sigma_0}$) to gapCSP_c^Σ, for some fixed finite Σ and constant $c > 0$, where in the case of $\text{gapCSP}_{\tau_1}^{\Sigma_0}$ we wish the reduction to preserve the length of the instance up to a polylogarithmic factor. The PCP Theorem (resp., a PCP of nearly linear length) follows by showing a simple PCP system for gapCSP_c^Σ. As noted above, the reduction is obtained by repeated applications of an amplification step that is captured by the following lemma.

Lemma 3.4 (amplifying reduction of gapCSP to itself): *For some finite Σ and constant $c > 0$, there exists a polynomial-time computable function f such that, for every instance (G, Φ) of gapCSP^Σ, it holds that $(G', \Phi') = f(G, \Phi)$ is an instance of gapCSP^Σ and the two instances are related as follows:*

1. *If $\text{vlt}(G, \Phi) = 0$ then $\text{vlt}(G', \Phi') = 0$.*
2. *$\text{vlt}(G', \Phi') \geq \min(2 \cdot \text{vlt}(G, \Phi), c)$.*
3. *$|G'| = O(|G|)$.*

That is, satisfiable instances are mapped to satisfiable instances, whereas instances that violate a ν fraction of the constraints are mapped to instances that violate at least a $\min(2\nu, c)$ fraction of the constraints. Furthermore, the mapping increases the number of edges (in the instance) by at most a constant factor. We stress that both Φ and Φ' consists of Boolean constraints defined over Σ^2. Thus, by iteratively applying Lemma 3.4 for a logarithmic (resp., double-logarithmic) number of times, we reduce $\text{gapCSP}^\Sigma_{\tau_0}$ (resp., $\text{gapCSP}^\Sigma_{\tau_1}$) to gapCSP^Σ_c.

Outline of the proof of Lemma 3.4: Before turning to the proof, let us highlight the difficulty that it needs to address. Specifically, the lemma asserts a "violation amplifying effect" (i.e., Items 1 and 2), while maintaining the alphabet Σ and allowing only a moderate increase in the size of the graph (i.e., Item 3). Waiving the latter requirements allows a relatively simple proof that mimics (an augmented version of) the "parallel repetition" of the corresponding PCP. Thus, the challenge is significantly decreasing the "size blow-up" that arises from parallel repetition and maintaining a fixed alphabet. The first goal (i.e., Item 3) calls for a suitable derandomization, and indeed we shall use a "pseudorandom" generator based on random walks on expander graphs. The second goal (i.e., fixed alphabet) can be handled by using the proof composition paradigm, which was outlined in §3.2.2.

The lemma is proved by presenting a three-step reduction. The first step is a pre-processing step that makes the underlying graph suitable for further analysis (e.g., the resulting graph will be an expander). The value of vlt may decrease during this step by a constant factor. The heart of the reduction is the second step in which we increase vlt by any desired constant factor. This is done by a construction that corresponds to taking a random walk of constant length on the current graph. The latter step also increases the alphabet Σ, and thus a post-processing step is employed to regain the original alphabet (by using any inner PCP systems; e.g., the one presented in §3.1.2). Details follow.

We first stress that the aforementioned Σ and c, as well as the auxiliary parameters d and t (to be introduced in the following two paragraphs), are fixed constants that will be determined such that various conditions (which arise in the course of our argument) are satisfied. Specifically, t will be the last parameter to be determined (and it will be made greater than a constant that is determined by all the other parameters).

We start with the pre-processing step. Our aim in this step is to reduce the input (G, Φ) of gapCSP^Σ to an instance (G_1, Φ_1) such that G_1 is a d-regular expander graph.[36] Furthermore, each vertex in G_1 will have at least $d/2$ self-loops, the number of edges will be preserved up to a constant factor (i.e., $|G_1| = O(|G|)$), and $\mathtt{vlt}(G_1, \Phi_1) = \Theta(\mathtt{vlt}(G, \Phi))$. This step is quite simple: essentially, the original vertices are replaced by expanders of size proportional to their degree, and a big (dummy) expander is "superimposed" on the resulting graph.

The main step is aimed at increasing the fraction of violated constraints by a sufficiently large constant factor. The intuition underlying this step is that the probability that a random (t-edge long) walk on the expander G_1 intersects a fixed set of edges is closely related to the probability that a random sample of (t) edges intersects this set. Thus, we may expect such walks to hit a violated edge with probability that is $\min(\Theta(t \cdot \nu), c)$, where ν is the fraction of violated edges. Indeed, the current step consists of reducing the instance (G_1, Φ_1) of $\mathtt{gapCSP}^\Sigma$ to an instance (G_2, Φ_2) of $\mathtt{gapCSP}^{\Sigma'}$ such that $\Sigma' = \Sigma^{d^t}$ and the following holds:

1. The vertex set of G_2 is identical to the vertex set of G_1, and each t-edge long path in G_1 is replaced by a corresponding edge in G_2, which is thus a d^t-regular graph.
2. The constraints in Φ_2 refer to each element of Σ' as a Σ-labeling of the ("distance $\leq t$") neighborhood of a vertex, and mandates that the two corresponding labelings (of the endpoints of the G_2-edge) are consistent as well as satisfy Φ_1. That is, the following two types of conditions are enforced by the constraints of Φ_2:
   (consistency): If vertices u and w are connected in G_1 by a path of length at most t and vertex v resides on this path, then the Φ_2-constraint associated with the G_2-edge between u and w mandates the equality of the entries corresponding to vertex v in the Σ'-labeling of vertices u and w.
   (satisfying Φ_1): If the G_1-edge (v, v') is on a path of length at most t starting at u, then the Φ_2-constraint associated with the G_2-edge that corresponds to this path enforces the Φ_1-constraint that is associated with (v, v').

Clearly, $|G_2| = d^{t-1} \cdot |G_1| = O(|G_1|)$, because d is a constant and t will be set to a constant. (Indeed, the relatively moderate increase in the size of the graph corresponds to the low randomness-complexity of selecting a random walk of length t in G_1.)

Turning to the analysis of this step, we note that $\mathtt{vlt}(G_1, \Phi_1) = 0$ implies $\mathtt{vlt}(G_2, \Phi_2) = 0$. The interesting fact is that the fraction of violated constraints increases by a factor of $\Omega(\sqrt{t})$; that is, $\mathtt{vlt}(G_2, \Phi_2) \geq \min(\Omega(\sqrt{t} \cdot \mathtt{vlt}(G_1, \Phi_1)), c)$. Here we merely provide a rough intuition and refer the interested reader to [26].

---

[36] A d-regular graph is a graph in which each vertex is incident to exactly d edges. Loosely speaking, an expander graph has the property that each moderately balanced cut (i.e., partition of its vertex set) has relatively many edges crossing it. An equivalent definition, also used in the actual analysis, is that, except for the largest eigenvalue (which equals d), all the eigenvalues of the corresponding adjacency matrix have absolute value that is bounded away from d.

We may focus on any Σ'-labeling of the vertices of G_2 that is consistent with some Σ-labeling of G_1, because relatively few inconsistencies (among the Σ-values assigned to a vertex by the Σ'-labeling of other vertices) can be ignored, while relatively many such inconsistencies yield violation of the "equality constraints" of many edges in G_2. Intuitively, relying on the hypothesis that G_1 is an expander, it follows that the set of violated edge-constraints (of Φ_1) with respect to the aforementioned Σ-labeling causes many more edge-constraints of Φ_2 to be violated (because each edge-constraint of Φ_1 is enforced by many edge-constraints of Φ_2). The point is that *any set F of edges of G_1 is likely to appear on a $\min(\Omega(t) \cdot |F|/|G_1|, \Omega(1))$ fraction of the edges of G_2* (i.e., t-paths of G_1). (Note that the claim would have been obvious if G_1 were a complete graph, but it also holds for an expander.)[37]

The factor of $\Omega(\sqrt{t})$ gained in the second step makes up for the constant factor lost in the first step (as well as the constant factor to be lost in the last step). Furthermore, for a suitable choice of the constant t, the aforementioned gain yields an overall constant factor amplification (of vlt). However, so far we obtained an instance of gapCSP$^{\Sigma'}$ rather than an instance of gapCSP$^{\Sigma}$, where $\Sigma' = \Sigma^{d^t}$. The purpose of the last step is to reduce the latter instance to an instance of gapCSP$^{\Sigma}$. This is done by viewing the instance of gapCSP$^{\Sigma'}$ as a PCP-system,[38] and composing it with an inner-verifier using the proof composition paradigm outlined in §3.2.2. We stress that the inner-verifier used here needs only handle instances of constant size (i.e., having description length $O(d^t \log |\Sigma|)$), and so the verifier presented in §3.1.2 will do. The resulting PCP-system uses randomness $r \overset{\text{def}}{=} \log_2 |G_2| + O(d^t \log |\Sigma|)^2$ and a constant number of binary queries, and has rejection probability $\Omega(\text{vlt}(G_2, \Phi_2))$, which is independent of the choice of the constant t. For $\Sigma = \{0,1\}^{O(1)}$, we can obtain an instance of gapCSP$^{\Sigma}$ that has a $\Omega(\text{vlt}(G_2, \Phi_2))$ fraction of violated constraints. Furthermore, the size of the resulting instance (which is used as the output (G', Φ') of the three-step reduction) is $O(2^r) = O(|G_2|)$, where the equality uses the fact that d and t are constants. Recalling that $\text{vlt}(G_2, \Phi_2) \geq \min(\Omega(\sqrt{t} \cdot \text{vlt}(G_1, \Phi_1)), c)$ and $\text{vlt}(G_1, \Phi_1) = \Omega(\text{vlt}(G, \Phi))$, this completes the (outline of the) proof of the entire lemma.   □

Reflection. In contrast to the proof outlined in §3.2.2. which combines two remarkable constructs by using a simple composition method, the current proof of the PCP Theorem is based on developing a powerful "combining method" that improves the quality of the main system to which it is applied. This new method, captured by the amplification step (Lemma 3.4), does not merely obtain the best of the combined systems, but rather obtains a better system than the one given. However, the quality-amplification offered by Lemma 3.4 is rather moderate, and thus many applications are required in order to derive the desired

---

[37] We mention that, due to a technical difficulty, it is easier to establish the claimed bound of $\Omega(\sqrt{t} \cdot \text{vlt}(G_1, \Phi_1))$ rather than $\Omega(t \cdot \text{vlt}(G_1, \Phi_1))$.

[38] The PCP-system referred to here has arbitrary soundness error (i.e., it rejects the instance (G_2, Φ_2) with probability $\text{vlt}(G_2, \Phi_2) \in [0, 1]$).

result. Taking the opposite perspective, one may say that remarkable results are obtained by a gradual process of many moderate amplification steps.

3.4   Additional Considerations

Our motivation for studying locally testable codes and proofs referred to super-fast testing, but our actual definitions have focused on the query complexity of these testers. While the query complexity of testing has a natural appeal, the hope is that low query complexity testers would also yield super-fast testing. Indeed, in the case of codes, it is typically the case that the testing time is related to the query complexity. However, in the case of proofs there is a seemingly unavoidable (linear) dependence of the verification time on the input length. This (linear) dependence can be avoided if one considers PCP-of-Proximity (see Section 2.3.3) rather than standard PCP. But even in this case, additional work is needed in order to derive testers that work is sub-linear time. The interested reader is referred to [16, 53].

4   Locally Decodable Codes

Locally *decodable* codes are complimentary to local *testable* codes. Recall that the latter are required to allow for super-fast rejection of strings that are far from being codewords (while accepting all codewords). In contrast, in case of locally decodable codes, we are guaranteed that the input is close to a codeword, and are required to recover individual bits of the encoded information based on a *small number of probes* (per recovered bit). As in case of local testability, the case when the operation (in this case decoding) is performed based on a *constant number of probes* is of special interest.

Local decodability is of natural practical appeal, which in turn provides additional motivation for local testability. The point is that it makes little sense to try to recover part of the data when the codeword is too corrupt. Thus, one should first apply local testability to check that the received codeword is not too corrupt, and apply local decodability only in case the codeword test passes.

4.1   Definitions

We follow the conventions of Section 2.1, but extend the treatment to codes over any finite alphabet Σ (rather than insisting on $\Sigma = \{0, 1\}$).

Definition 4.1 (locally decodable codes, basic version): *Let* $C : \Sigma^k \to \Sigma^n$ *be a code, and let* $q \in \mathbb{N}$ *and* $\delta \in (0, 1)$. *A* q-local δ-decoder for C *is a probabilistic (non-adaptive) oracle machine* M *that makes at most* q *queries and satisfies the following condition:*

Local recovery from somewhat corrupted codewords: *For every* $i \in [k]$ *and* $x = (x_1, ..., x_k) \in \Sigma^k$, *and any* $w \in \Sigma^n$ *that is* δ-close *to* $C(x)$, *on input* i *and oracle access to* w, *machine* M *outputs* x_i *with probability at least* $2/3$. *That is,* $\Pr[M^w(1^k, i) = x_i] > 2/3$, *for any* $w \in \Sigma^n$ *that is* δ-far *from* $C(x)$.

We call q the query complexity *of M, and δ the* proximity parameter.

Note that the proximity parameter must be smaller than the covering radius of the code (as otherwise the definition cannot possibly be satisfied (at least for some w and i)). One may strengthen Definition 4.1 by requiring that the bits of an uncorrupted codeword be always recovered correctly (rather than with high probability); that is, for every $i \in [k]$ and $x = (x_1, ..., x_k) \in \Sigma^k$, it must hold that $\Pr[M^{C(x)}(1^k, i) = x_i] = 1$. Turning to families of codes, we present the following definition (which potentially allows the alphabet to grow with k).

Definition 4.2 (locally decodable codes, asymptotic version): *For functions $n, \sigma : \mathbb{N} \to \mathbb{N}$, let $C = \{C_k : [\sigma(k)]^k \to [\sigma(k)]^{n(k)}\}_{k \in K}$. We say that C is a* local decodable code *if there exist constants $\delta > 0$ and q and a machine M that is a q-local δ-decoder for C_k, for every $k \in K$.*

We mention that locally decodable codes are related to (information theoretic secure) Private Information Retrieval (PIR) schemes, introduced in [25]. In the latter a user wishes to recover a bit of data from a k-bit long database, copies of which are held by s servers, without revealing any information to any single server. To that end, the user (secretly) communicates with each of the servers, and the issue is to minimize the total amount of communication. As we shall see, certain s-server PIR schemes yield $2s$-locally decodable codes of length exponential in the communication complexity of the PIR.

Related notions of local recovery. The notion of local decodability is a special case of a general notion of local recovery, where one may be required to recover an arbitrary function $f : \Sigma^k \to \{0,1\}^*$ of the original information based on a constant number of probes to the (corrupted) codeword. The function f must be restricted in two ways: Firstly, it should have a small range (e.g., its range may be Σ), and secondly it should come from a small predetermined set $\mathcal{F}$ of functions. Definition 4.1 may be recast in these terms, by considering the set of projection functions (i.e., $\{f_i : \Sigma^k \to \Sigma\}$ where $f_i(x_1, ..., x_k) = x_i$). We believe that this is the most natural special case of the general notion of local recovery. In §3.2.1 we referred to another special case, where the alphabet is associated with a finite field F and the recovery function $f_e : F^k \to F$ is one out of $|F|$ possible linear functions (specifically, $f_e(x_1, ..., x_k) = \sum_{i=1}^k e^{i-1} x_i$, for $e \in F$).[39] Another natural case (also used in §3.2.1) is that of the recovery of (correct) symbols of the codeword, which may be viewed as self-correction. (In this case each admissible function determines one codeword symbol as a function of the encoded message.)

4.2 Results

The best known locally decodable codes are of strictly sub-exponential length; that is, k information bits can be encoded by codewords of length $n = \exp(k^{o(1)})$

---

[39] Indeed, the value $f_e(x_1, ..., x_k)$ is the evaluation at e of the polynomial $p(\zeta) = \sum_{i=1}^k x_i \zeta^{i-1}$ represented by the coefficients $(x_1, ..., x_k)$.

that are locally decodable (cf. [29], building on [62]). This result disproves [36, Conj. 4.4],

Theorem 4.3 (Efremenko [29], building on Yekhanin [62]): *For some $\delta > 0$ there exists a code* $C : \{0,1\}^k \to \{0,1\}^n$ *that has a 3-local δ-decoder such that* $n = \exp(2^{\tilde{O}(\sqrt{\log k})}) = \exp(k^{o(1)})$. *Furthermore, 2^d-local decodability can be obtained with* $n = \exp(2^{\tilde{O}(\sqrt[d]{\log k})})$.

In this section we only outline a couple of codes of lesser performance. Specifically, we will present longer codes that are $O(1)$-locally decodable as well as shorter codes that are poly($\log k$)-locally decodable.

4.2.1   Locally Decodable Codes of Sub-Exponential Length

For any $d \geq 1$, there is a simple construction of a 2^d-locally 2^{-d-2}-decodable binary code of length $n = 2^{d \cdot k^{1/d}}$. For $h = k^{1/d}$, we identify $[k]$ with $[h]^d$, and view $x \in \{0,1\}^k$ as $(x_{i_1,\ldots,i_d})_{i_1,\ldots,i_d \in [h]}$. We encode x by providing the parity of all $x_{i_1,\ldots,i_d}$ residing in each of the $(2^h)^d$ sub-cubes of $[h]^d$; that is, for every $(S_1,\ldots,S_d) \in 2^{[h]} \times \cdots \times 2^{[h]}$, we provide

$$C(x)_{S_1,\ldots,S_d} = \oplus_{i_1 \in S_1,\ldots,i_d \in S_d} x_{i_1,\ldots,i_d}. \tag{3}$$

Indeed, the Hadamard code is the special case in which $d = 1$. To recover the value of $x_{i_1,\ldots,i_d}$, at any desired $(i_1,\ldots,i_d) \in [h]^d$, the decoder uniformly selects $(R_1,\ldots,R_d) \in 2^{[h]} \times \cdots \times 2^{[h]}$, and recovers the (possibly corrupted) values $C(x)_{S_1,\ldots,S_d}$, where each S_j either equals R_j or equals $R_j \triangle \{i_j\}$, where $R\triangle\{i\} = R \setminus \{i\}$ if $i \in R$ and $R\triangle\{i\} = R \cup \{i\}$ otherwise. The key observation is that each of the decoder's queries is uniformly distributed. Thus, with probability at least $3/4$, XORing the 2^d answers, yields the desired result (because $\oplus_{S_1 \in \{R_1, R_1 \triangle \{i_1\}\},\ldots,S_d \in \{R_d, R_d \triangle \{i_d\}\}} C(x)_{S_1,\ldots,S_d}$ equals $C(x)_{\{i_1\},\ldots,\{i_d\}} = x_{i_1,\ldots,i_d}$).

We comment that a related code (of length $n = 2^{d^d \cdot k^{1/d}}$) allows for recovery based on $d + 1$ (rather 2^d) queries. The original presentation, due to [2] (building on [25]), is in terms of PIR schemes (with $s = (d+1)/2$ servers and overall communication $d^d \cdot k^{1/d} = \exp(\tilde{O}(s)) \cdot k^{1/(2s-1)}$). In particular, in the case that $d = 2$, we use two servers, sending (R_1, R_2, R_3) to one server and $(R_1 \triangle \{i_1\}, R_2 \triangle \{i_2\}, R_3 \triangle \{i_3\})$ to the other server. Upon receiving (S_1, S_2, S_3), each server replies with the bit $C(x)_{S_1,S_2,S_3}$ as well as the three $k^{1/3}$-bit long sequences $(C(x)_{S_1 \triangle \{i\}, S_2, S_3})_{i \in [k^{1/3}]}$, $(C(x)_{S_1, S_2 \triangle \{i\}, S_3})_{i \in [k^{1/3}]}$, and $(C(x)_{S_1,S_2,S_3 \triangle \{i\}})_{i \in [k^{1/3}]}$, which contain the bits $C(x)_{S_1 \triangle \{i_1\}, S_2, S_3}$, $C(x)_{S_1, S_2 \triangle \{i_2\}, S_3}$, and $C(x)_{S_1, S_2, S_3 \triangle \{i_3\}}$. Thus, the user obtains the bits $C(x)_{R_1,R_2,R_3}$, $C(x)_{R_1 \triangle \{i_1\}, R_2, R_3}$, $C(x)_{R_1, R_2 \triangle \{i_2\}, R_3}$, and $C(x)_{R_1, R_2, R_3 \triangle \{i_3\}}$ from the first server, and the bits $C_{R_1 \triangle \{i_1\}, R_2 \{i_2\}, R_3 \triangle \{i_3\}}$, $C_{R_1, R_2 \{i_2\}, R_3 \triangle \{i_3\}}$, $C_{R_1 \triangle \{i_1\}, R_2, R_3 \triangle \{i_3\}}$, $C_{R_1 \triangle \{i_1\}, R_2 \triangle \{i_2\}, R_3}$ from the second server.

The corresponding locally decodable code is obtained by a generic transformation that applies to any PIR scheme with s servers, in which the user makes uniformly distributed queries of length $\mathsf{qst}(k)$, gets answers of length $\mathsf{ans}(k)$, and recovers the desired value by XORing some predetermined bits contained in the answers. In this case, the resulting code will contain the Hadamard encoding of each of the possible answers provided by each of the servers; that is, if the j-th server answers according to $A_j(x,q) \in \{0,1\}^{\mathsf{ans}(k)}$, where $x \in \{0,1\}^k$ and $q \in \{0,1\}^{\mathsf{qst}(k)}$, then $C(x)_{j,q,\ell} = C_{\mathrm{Had}}(A_j(x,q))_\ell$, for every $\ell \in \{0,1\}^{\mathsf{ans}(k)}$. Thus, the length of the code is $s \cdot 2^{\mathsf{qst}(k)} \cdot 2^{\mathsf{ans}(k)}$. Now, on input $i \in [k]$, the decoder emulates the PIR user, obtaining the query sequence $(q_1, ..., q_s)$ and the desired linear combinations $(\ell_1, ..., \ell_s)$. It uniformly selects $r_1, ..., r_s \in \{0,1\}^{\mathsf{ans}(k)}$, queries the (possibly corrupted) codeword at locations $(1, q_1, r_1), (1, q_1, r_1 \oplus \ell_1), ..., (s, q_s, r_s), (s, q_s, r_s \oplus \ell_s)$, and XORs the corresponding $2s$ answers. Note that each of these queries is uniformly distributed in $\{j\} \times \{0,1\}^{\mathsf{qst}(k)} \times \{0,1\}^{\mathsf{ans}(k)}$, for some $j \in [s]$, and that $C(x)_{j,q_j,r_j} \oplus C(x)_{j,q_j,r_j \oplus \ell_j} = C_{\mathrm{Had}}(A_j(x,q_j))_{\ell_j}$.

4.2.2 Polylog-local Decoding for Codes of Nearly Linear Length

We will consider a code $C : \Sigma^k \to \Sigma^n$ of linear distance, while identifying Σ with a finite field (denoted F). For parameters h and $m = \log_h k$, consider a finite field F of size $O(m \cdot h)$, and a subset $H \subset F$ of size h. Viewing the information as a function $f : H^m \to F$, we encode it by providing the values of its low-degree extension $\widehat{f} : F^m \to F$ on all points in F, where $\widehat{f}$ is an m-variant polynomial of degree $|H| - 1$ in each variable. Thus, the code maps $k = h^m$ long sequences over F (which may be viewed as $h^m \log|F|$ bits of information) to sequences of length $n \stackrel{\text{def}}{=} |F|^m = O(mh)^m = O(m)^m \cdot k$ over F. This code has relative distance $mh/|F|$. Note that the smaller m, the better the rate (i.e., relation of n to k is), but this comes at the expense of using a larger alphabet F (as well as larger query complexity of the decoder presented below).

The decoder works by applying the self-correction paradigm. Given a point $x \in H^m$ and access to an oracle $w : F^m \to F$ that is $1/2$-close to $\widehat{f}$, the value of $f(x)$ is recovered by uniformly selecting a line through x, querying for the $|F|$ values of w along the line, finding the degree mh univariate polynomial with the greatest agreement with these values, and evaluating it at the appropriate point. Thus, we obtain an $|F|$-local decoder.

Using a constant m, we obtain an $O(k^{1/m})$-locally decodable code of constant rate (i.e., $n = O(k)$), over an alphabet of size $O(k^{1/m})$. On the other hand, using $m = \epsilon \log k / \log \log k$ (for any constant $\epsilon > 0$), we obtain a poly$(\log k)$-locally decodable code of length $n = k^{1+\epsilon}$, over an alphabet of size poly$(\log k)$. Concatenation with any reasonable[40] binary code (coupled with a trivial decoder that reads the entire codeword), yields a binary poly$(\log k)$-locally decodable code of length $n = k^{1+\epsilon}$.

---

[40] Indeed, we may use any good code (i.e., linear length and linear distance), as such can be easily constructed for block length $O(\log \log k)$. But we can even use the Hadamard code, because the length overhead caused by it in this setting is negligible.

4.2.3   Lower bounds

It is known that locally decodable codes cannot be T2-nearly linear:[41] Specifically, any q-locally decodable code $C : \Sigma^k \to \Sigma^n$ must satisfy $n = \Omega(k^{1 + \frac{1}{q-1}})$ (cf. [46]). For $q = 2$ and $\Sigma = \{0,1\}$, an exponential lower bound is known (cf. [48], following [41]).

We mention that our past conjectures regarding lower bounds for locally decodable (binary) codes were disproved twice. Our conjectured lower bound of $n > \exp(k^{\Omega(1/q)})$ for q-locally decodable codes was disproved by [9], and our conjectured lower bound of $n > \exp(k^{\Omega(1)})$ for any locally decodable code was disproved by [29] (after being vastly shaken by [62]). Given this history, we dare not make any further conjectures, but instead pose the following open problem.

Open Problem 4.4. *Determine whether there exist locally decodable codes of polynomial length.*

Recall that we know, for a fact, that T2-nearly linear length is impossible, and it is very tempting to conjecture that T1-nearly linear length is impossible too (i.e., any locally decodable code $C : \Sigma^k \to \Sigma^n$ requires $n > k^{1+\Omega(1)}$). Still, let us pose this too as an open problem.

4.3   Relaxations

In light of the fact that locally decodable codes cannot be T2-nearly linear, it is natural to seek relaxations to the notion of locally decodable codes. One natural relaxation requires local recovery of most individual information-bits, allowing for recovery-failure (but not error) on the rest [15]: That is, it is requires that, whenever few location are corrupted, the decoder should be able to recover most of the individual information-bits, based on a constant number of queries, and for the rest of the locations the decoder may output a fail symbol (but not the wrong value). Augmenting these requirements by the requirement that whenever the codeword is not corrupted – all bits are recovered correctly (with high probability), yields the following definition.

Definition 4.5 (locally decodable codes, relaxed): *For functions $n, \sigma : \mathbb{N} \to \mathbb{N}$, let $C = \{C_k : \{0,1\}^k \to \{0,1\}^{n(k)}\}_{k \in K}$. For $q \in \mathbb{N}$ and $\delta, \rho \in (0,1)$, a q-local relaxed (δ, ρ)-decoder for C is a probabilistic (non-adaptive) oracle machine M that makes at most q queries and satisfies the following conditions:*

Local recovery from uncorrupted codewords: For every $i \in [k]$ and $x = (x_1, ..., x_k) \in \Sigma^k$, it holds that $\Pr[M^{C(x)}(1^k, i) = x_i] > 2/3$,

Relaxed local recovery from somewhat corrupted codewords: For every $x = (x_1, ..., x_k) \in \Sigma^k$, and any $w \in \Sigma^n$ that is δ-close to $C(x)$, the following two conditions hold:

---

[41] See terminology in §3.3.1.

1. *For every $i \in [k]$, it holds that $\Pr[M^{C(x)}(1^k, i) \in \{x_i, \perp\}] > 2/3$, where $\perp$ is a special ("failure") symbol.*
2. *There exists a set $I_w \subseteq [k]$ of size at least ρk such that, for every $i \in I_w$, it holds that $\Pr[M^{C(x)}(1^k, i) = x_i] > 2/3$.*[42]

In such a case, C is said to be locally relaxed-decodable.

It turns out (cf. [15]) that Condition 2, in the relaxed recovery requirement, essentially follows from the other requirements. That is, codes satisfying the other requirements can be transformed into locally relaxed-decodable codes, while essentially preserving their rate (and distance). Furthermore, the resulting codes satisfy the following stronger form of Condition 2: *There exists a set $I_w \subseteq [k]$ of density at least $1 - O(\Delta(w, C(x))/n)$ such that for every $i \in I_w$ it holds that $\Pr[M^{C(x)}(1^k, i) = x_i] > 2/3$.*

Theorem 4.6 [15]: *There exist locally relaxed-decodable codes of T1-nearly linear length. Specifically, for every $\epsilon > 0$, there exists codes of length $n = k^{1+\epsilon}$ that have a $O(1/\epsilon^2)$-local relaxed $(\Omega(\epsilon), 1 - O(\epsilon))$-decoder.*

An obvious open problem is to separate locally decodable codes from relaxed ones. This may follow by either improving the aforementioned lower bound on the length of locally decodable codes or by providing relaxed locally decodable codes of T2-nearly linear length.

Acknowledgments. We are grateful to Madhu Sudan, Luca Trevisan and Salil Vadhan for related discussions. We are also grateful to Omer Tamuz for useful comments and suggestions regarding this article.

References

1. Alon, N., Kaufman, T., Krivelevich, M., Litsyn, S.N., Ron, D.: Testing low-degree polynomials over $GF(2)$. In: Arora, S., Jansen, K., Rolim, J.D.P., Sahai, A. (eds.) RANDOM 2003 and APPROX 2003. LNCS, vol. 2764, pp. 188–199. Springer, Heidelberg (2003)
2. Ambainis, A.: An upper bound on the communication complexity of private information retrieval. In: Degano, P., Gorrieri, R., Marchetti-Spaccamela, A. (eds.) ICALP 1997. LNCS, vol. 1256, pp. 401–407. Springer, Heidelberg (1997)
3. Arora, S.: Probabilistic checking of proofs and the hardness of approximation problems. PhD thesis, UC Berkeley (1994)
4. Arora, S., Lund, C., Motwani, R., Sudan, M., Szegedy, M.: Proof verification and the hardness of approximation problems. JACM 45,3, 501–555 (1998); Preliminary Version in 33rd FOCS, 1992
5. Arora, S., Safra, S.: Probabilistic checking of proofs: A new characterization of NP. J. ACM 45(1), 70–122 (1998); Preliminary Version in 33rd FOCS 1992

---

[42] We stress that it is not required that $\Pr[M^{C(x)}(1^k, i) = \perp] > 2/3$ for $i \in [k] \setminus I_w$. Adding this requirement collapses the notion of relaxed-decodability to ordinary decodability (cf. [23]).

6. Babai, L., Fortnow, L., Lund, C.: Non-deterministic exponential time has two-prover interactive protocols. Computational Complexity 1(1), 3–40 (1991)
7. Babai, L., Fortnow, L., Levin, L.A., Szegedy, M.: Checking computations in polylogarithmic time. In: Proc. 23rd ACM Symposium on the Theory of Computing, pp. 21–31 (May 1991)
8. Barak, B.: How to go beyond the black-box simulation barrier. In: Proc. 42nd IEEE Symposium on Foundations of Computer Science, pp. 106–115 (October 2001)
9. Beimel, A., Ishai, Y., Kushilevitz, E., Raymond, J.F.: Breaking the $O(n^{1/(2k-1)})$ barrier for information-theoretic private information retrieval. In: Proc. 43rd FOCS, pp. 261–270 (November 2002)
10. Bellare, M., Coppersmith, D., Håstad, J., Kiwi, M., Sudan, M.: Linearity testing in characteristic two. In: Proceedings of the 36th FOCS, pp. 432–441 (1995)
11. Bellare, M., Goldreich, O., Sudan, M.: Free bits, PCPs, and nonapproximability—towards tight results. In: SICOMP, vol. 27, 3, pp. 804–915 (1998); Preliminary Version in 36th FOCS 1995
12. Bellare, M., Goldwasser, S., Lund, C., Russell, A.: Efficient probabilistically checkable proofs and applications to approximation. In: Proc. 25th STOC, pp. 294–304 (May 1993)
13. Bellare, M., Sudan, M.: Improved non-approximability results. In: Proceedings of the 26th Annual ACM Symposium on the Theory of Computing, pp. 184–193 (1994)
14. Ben-Sasson, E., Goldreich, O., Sudan, M.: Bounds on 2-Query Codeword Testing. In: Arora, S., Jansen, K., Rolim, J.D.P., Sahai, A. (eds.) RANDOM 2003 and APPROX 2003. LNCS, vol. 2764, pp. 216–227. Springer, Heidelberg (2003)
15. Ben-Sasson, E., Goldreich, O., Harsha, P., Sudan, M., Vadhan, S.: Robust PCPs of proximity, shorter PCPs and applications to coding. In: Proc. 36th STOC, pp. 1–10 (June 2004); See ECCC Technical Report TR04-021, March 2004
16. Ben-Sasson, E., Goldreich, O., Harsha, P., Sudan, M., Vadhan, S.: Short PCPs verifiable in polylogarithmic time. In: 20th IEEE Conference on Computational Complexity, pp. 120–134 (2005)
17. Ben-Sasson, E., Guruswami, V., Kaufman, T., Sudan, M., Viderman, M.: Locally testable codes require redundant testers. In: IEEE Conference on Computational Complexity, pp. 52–61 (2009)
18. Ben-Sasson, E., Harsha, P., Raskhodnikova, S.: Some 3CNF properties are hard to test. In: Proc. 35th STOC, pp. 345–354. s, f (2003)
19. Ben-Sasson, E., Sudan, M.: Robust Locally Testable Codes and Products of Codes. In: Jansen, K., Khanna, S., Rolim, J.D.P., Ron, D. (eds.) RANDOM 2004 and APPROX 2004. LNCS, vol. 3122, pp. 286–297. Springer, Heidelberg (2004)
20. Ben-Sasson, E., Sudan, M.: Short PCPs with polylog query complexity. In: SICOMP, vol. 38(2), pp. 551–607 (2008); Preliminary Version in 37th STOC 2005)
21. Ben-Sasson, E., Sudan, M., Vadhan, S., Wigderson, A.: Randomness-efficient low degree tests and short PCPs via epsilon-biased sets. In: Proc. 35th STOC, pp. 612–621 (June 2003)
22. M. Blum, M. Luby, and R. Rubinfeld. Self-testing/correcting with applications to numerical problems. *JCSS 47*, 3 (December 1993), 549–595. (Preliminary Version in *22nd STOC*, 1990).
23. Buhrman, H., de Wolf, R.: On relaxed locally decodable codes(July 2004) Unpublished manuscript
24. Canetti, R., Goldreich, O., Halevi, S.: The random oracle methodology. In: Proc. 30th STOC, pp. 209–218 (May 1998) (revisited)

25. Chor, B., Goldreich, O., Kushilevitz, E., Sudan, M.: Private Information Retrieval. Journal of the ACM 45(6), 965–982 (1998)
26. Dinur, I.: The PCP theorem by gap amplification. J. ACM 54(3) Art. 12 (2007); Extended abstract in 38th STOC 2006
27. Dinur, I., Harsha, P.: Composition of low-error 2-query PCPs using decodable PCPs. In: 50th FOCS, pp. 472–481 (2009)
28. Dinur, I., Reingold, O.: Assignment-testers: Towards a combinatorial proof of the PCP-Theorem. SICOMP 36(4), 975–1024 (2006); Extended abstract in 45th FOCS 2004
29. Efremenko, K.: 3-query locally decodable codes of subexponential length. In: 41st STOC, pp. 39–44 (2009)
30. Ergün, F., Kumar, R., Rubinfeld, R.: Fast approximate PCPs. In: Proc. 31st STOC, pp. 41–50 (May 1999)
31. Feige, U., Goldwasser, S., Lovász, L., Safra, S., Szegedy, M.: Interactive proofs and the hardness of approximating cliques. J. ACM 43(2), 268–292 (1996); Preliminary version in 32nd FOCS 1991
32. Forney, G.D.: Concatenated Codes. MIT Press, Cambridge (1966)
33. Fortnow, L., Rompel, J., Sipser, M.: On the power of multi-prover interactive protocols. Theoretical Computer Science 134, 2, 545–557 (November 1994)
34. Friedl, K., Sudan, M.: Some improvements to total degree tests. In: Proc. 3rd Israel Symposium on Theoretical and Computing Systems, Tel Aviv, Israel, January 4–6, pp. 190–198 (1995)
35. Gemmell, P., Lipton, R., Rubinfeld, R., Sudan, M., Wigderson, A.: Self-testing/correcting for polynomials and for approximate functions. In: Proc. 23rd STOC, pp. 32–42 (1991)
36. Goldreich, O.: Short locally testable codes and proofs (survey). ECCC Technical Report TR05-014 (January 2005)
37. Goldreich, O.: Computational Complexity: A Conceptual Perspective. Cambridge University Press, Cambridge (2008)
38. Newman, I.: Property testing of massively parametrized problems - A survey. In: Goldreich, O. (ed.) Property Testing. LNCS, vol. 6390, pp. 142–157. Springer, Heidelberg (2010)
39. Goldreich, O., Goldwasser, S., Ron, D.: Property testing and its connection to learning and approximation. J. ACM 45(4), 653–750 (1998); Preliminary Version in 37th FOCS 1996
40. Goldreich, O., Ron, D.: On proximity oblivious testing. ECCC, TR08-041, 2008. Also in the proceedings of the 41st STOC (2009)
41. Goldreich, O., Karloff, H., Schulman, L., Trevisan, L.: Lower bounds for linear locally decodable codes and private information retrieval. In: Proc. 17th Conference on Computational Complexity, Montréal, Québec, Canada, May 21–24, pp. 175–183 (2002)
42. Goldreich, O., Sudan, M.: Locally testable codes and PCPs of almost linear length. In: Proc.43rd FOCS, pp. 13–22 (November 2002); See ECCC Report TR02-050 2002
43. Harsha, P., Sudan, M.: Small PCPs with low query complexity. Computational Complexity 9, 3–4, 157–201 (2001); Preliminary Version in 18th STACS, 2001
44. Håstad, J.: Clique is hard to approximate within $n^{1-\epsilon}$. Acta Mathematica 182, 105–142 (1999); Preliminary Versions in 28th STOC 1996 and 37th FOCS 1997
45. Håstad, J.: Some optimal inapproximability results. Journal of the ACM 48(4), 798–859 (2001); Preliminary Version in 29th STOC 1997

46. Katz, J., Trevisan, L.: On the efficiency of local decoding procedures for error-correcting codes. In: Proc. 32nd STOC, pp. 80–86 (2000)
47. Kaufman, T., Litsyn, S., Xie, N.: Breaking the ϵ-soundness bound of the linearity test over GF(2). SICOMP 39(5), 1988–2003 (2009/2010)
48. Kerenidis, I., de Wolf, R.: Exponential lower bound for 2-query locally decodable codes via a quantum argument. In: Proc. 35th ACM Symposium on the Theory of Computing, pp. 106–115 (June 2003)
49. Kilian, J.: A note on efficient zero-knowledge proofs and arguments (extended abstract). In: Proc. 24th ACM Symposium on the Theory of Computing, pp. 723–732 (May 1992)
50. Lapidot, D., Shamir, A.: Fully parallelized multi prover protocols for NEXP-time (extended abstract). In: Proc. 32nd IEEE Symposium on Foundations of Computer Science, pp. 13–18 (October 1991)
51. Lund, C., Fortnow, L., Karloff, H., Nisan, N.: Algebraic methods for interactive proof systems. JACM 39(4), 859–868 (1992)
52. Meir, O.: Combinatorial construction of locally testable codes. SICOMP 39(2), 491–544 (2009); Extended abstrat in 40th STOC (2008)
53. Meir, O.: Combinatorial PCPs with efficient verifiers. In: 50th FOCS, pp. 463–471 (2009)
54. Micali, S.: Computationally sound proofs. SICOMP 30(4), 1253–1298 (2000); Preliminary Version in 35th FOCS (1994)
55. Moshkovitz, D., Raz, R.: Two query PCP with sub-constant error. In: 49th FOCS, pp. 314–323 (2008)
56. Polishchuk, A., Spielman, D.A.: Nearly-linear size holographic proofs. In: Proc. 26th STOC, pp. 194–203 (May 1994)
57. Raz, R.: A parallel repetition theorem. SIAM Journal of Computing 27(3), 763–803 (1998); Preliminary Version in 27th STOC (1995)
58. Rubinfeld, R., Sudan, M.: Robust characterizations of polynomials with applications to program testing. SICOMP 25(2), 252–271 (1996); Preliminary Version in 3rd SODA (1992)
59. Spielman, D.: Computationally efficient error-correcting codes and holographic proofs. PhD thesis, Massachusetts Institute of Technology (June 1995)
60. Sudan, M.: Efficient Checking of Polynomials and Proofs and the Hardness of Approximation Problems. LNCS, vol. 1001. Springer, Heidelberg (1995)
61. Szegedy, M.: Many-valued logics and holographic proofs. In: Wiedermann, J., Van Emde Boas, P., Nielsen, M. (eds.) ICALP 1999. LNCS, vol. 1644, pp. 676–686. Springer, Heidelberg (1999)
62. Yekhanin, S.: Towards 3-Query locally decodable codes of subexponential length. In: 39th STOC, pp. 266–274 (2007)

Bravely, Moderately: A Common Theme in Four Recent Works

Oded Goldreich

Abstract. We highlight a common theme in four relatively recent works that establish remarkable results by an iterative approach. Starting from a trivial construct, each of these works applies an ingeniously designed sequence of iterations that yields the desired result, which is highly non-trivial. Furthermore, in each iteration, the construct is modified in a relatively moderate manner. The four works we refer to are

1. the polynomial-time approximation of the permanent of non-negative matrices (by Jerrum, Sinclair, and Vigoda, *33rd STOC*, 2001);
2. the iterative (Zig-Zag) construction of expander graphs (by Reingold, Vadhan, and Wigderson, *41st FOCS*, 2000);
3. the log-space algorithm for undirected connectivity (by Reingold, *37th STOC*, 2005);
4. and, the alternative proof of the PCP Theorem (by Dinur, *38th STOC*, 2006).

Keywords: Approximation, Expander Graphs, Log-Space, Markov Chains, NP, Permanent, PCP, Space Complexity, Undirected Connectivity.

An early version of this survey appeared as TR05-098 of *ECCC*.

1 Introduction

Speude bradeos.[1]

The title of this essay employs more non-technical terms than one is accustomed to encounter in the title of a technical survey, let alone that some are rarely used in a technical context. Indeed, this is an unusual survey, written in an attempt to communicate a feeling that cannot be placed on sound grounds. The feeling is that there is a common theme among the works to be reviewed here, and that this common theme is intriguing and may lead to yet additional important discoveries. We hope that also readers that disagree with the foregoing feeling may benefit from the perspective offered by lumping the said works together and highlighting a common theme.

---

[1] This Ancient Greek proverb, reading *hasten slowly*, is attributed to Augustus; see C. Suetonius Tranquillus, *D. Octavius Caesar Augustus*, paragraph XXV. The intention seems to be a calling for action that is marked by determination and thoroughness, which characterizes the "moderate revolution" of Rome under Augustus.

O. Goldreich et al.: Studies in Complexity and Cryptography, LNCS 6650, pp. 373–389, 2011.

We are going to review four celebrated works, each either resolving a central open problem or providing an alternative proof for such a central result. The common theme that we highlight is the (utmost abstract) attitude of these works towards solving the problem that they address. Rather than trying to solve the problem by one strong blow, each of these works goes through a long sequence of iterations, gradually transforming the original problem into a trivial one. (At times, it is more convenient to view the process as proceeding in the opposite direction; that is, gradually transforming a solution to the trivial problem into a solution to the original problem.) Anyhow, each step in this process is relatively simple (in comparison to an attempt to solve the original problem at one shot), and it is the multitude of iterated steps that does the job. Let us try to clarify the foregoing description by providing a bird's eye view of each of these works.

1.1   A Bird's Eye View of the Four Works

Following are very high level outlines of the aforementioned works. At this point we avoid almost all details (including crucial ones), and refrain from describing the context of these works (i.e., the history of the problems that they address). Instead, we focus on the iterative processes eluded to above. More detailed descriptions as well as comments about the history of the problems are to be found in corresponding sections of this essay.

Approximating the permanent of non-negative matrices. The probabilistic polynomial-time approximation algorithm of Jerrum, Sinclair, and Vigoda [18] is based on the following observation: Knowing (approximately) certain parameters of a non-negative matrix M allows to approximate the same parameters for a matrix M', provided that M and M' are sufficiently similar. Specifically, M and M' may differ only on a single entry, and the ratio of the corresponding values must be sufficiently close to one. Needless to say, the actual observation (is not generic but rather) refers to specific parameters of the matrix, which include its permanent. Thus, given a matrix M for which we need to approximate the permanent, we consider a sequence of matrices $M_0, ..., M_t \approx M$ such that M_0 is the all 1's matrix (for which it is easy to evaluate the said parameters), and each M_{i+1} is obtained from M_i by reducing some adequate entry by a factor sufficiently close to one. This process of (polynomially many) gradual changes, allows to transform the dummy matrix M_0 into a matrix M_t that is very close to M (and hence has a permanent that is very close to the permanent of M). Thus, approximately obtaining the parameters of M_t allows to approximate the permanent of M.

The iterative (Zig-Zag) construction of expander graphs. The construction of constant-degree expander graphs by Reingold, Vadhan, and Wigderson [26] proceeds in iterations. Its starting point is a very good expander G of constant size, which may be found by exhaustive search. The construction of a large expander graph proceeds in iterations, where in the ith iteration the current graph G_i and the fixed graph G are combined (via a so-called Zig-Zag product) to obtain the larger graph G_{i+1}. The combination step guarantees that the expansion property of G_{i+1} is at least as good as the expansion of G_i, while G_{i+1} maintains the degree of G_i and is a constant times larger than G_i. The process is initiated with $G_1 = G^2$, and terminates when we obtain a graph of approximately the desired size (which requires a logarithmic number of iterations). Thus, the last graph is a constant-degree expander of the desired size.

The log-space algorithm for undirected connectivity. The aim of Reingold's algorithm [25] is to (deterministically) traverse an arbitrary graph using logarithmic amount of space. Its starting point is the fact that any expander is easy to traverse in deterministic logarithmic-space, and thus the algorithm gradually transforms any graph into an expander, while maintaining the ability to map a traversal of the latter into a traversal of the former. Thus, the algorithm traverses a virtual graph, which being an expander is easy to traverse in deterministic logarithmic-space, and maps the virtual traversal of the virtual graph to a real traversal of the actual input graph. The virtual graph is constructed in (logarithmically many) iterations, where in each iteration the graph becomes easier to traverse. Specifically, in each iteration the expansion property of the graph improves by a constant factor, while the graph itself only grows by a constant factor, and each iteration can be performed (or rather emulated) in constant space. Since each graph has some noticeable expansion (i.e., expansion inversely related to the size of the graph), after logarithmically many steps this process yields a good expander (i.e., constant expansion).

The alternative proof of the PCP Theorem. Dinur's novel approach [12] to the proof of the PCP Theorem is based on gradually improving the performance of PCP-like systems. The starting point is a trivial PCP-like system that detects error with very small but noticeable probability. Each iterative step increases the detection probability of the system by a constant factor, while incurring only a small overhead in other parameters (i.e., the randomness complexity increases by a constant term). Thus, the PCP Theorem (asserting constant detection probability for $\mathcal{NP}$) is obtained after logarithmically many such iterative steps. Indeed, the heart of this approach is the detection amplification step, which may be viewed as simple only in comparison to the original proof of the PCP Theorem.

1.2   An Attempt to Articulate the Thesis

The current subsection will contain an attempt to articulate the thesis that there is a common theme among these works. Readers who do not care about philosophical discussions (and other attempts to say what cannot be said) are encouraged to skip this subsection. In order to emphasize the subjective nature of this section, it is written in first person singular.

I will start by saying a few works about bravery and moderation. I consider as brave the attempt to resolved famous open problems or provide alternative proofs for central celebrated results.[2] To try a totally different approach is also brave, and so is *realizing one's limitations* and trying a moderate approach: Rather than trying to resolve the problem in a single blow, one wisely designs a clever scheme that gradually progresses towards the desired goal. Indeed, this is the victory of moderation.

Getting to the main thesis of this essay (i.e., the existence of a common theme among the reviewed works), I believe that I have already supported a minimalistic interpretation of this thesis by the foregoing bird's eye view of the four works. That is, there is an obvious similarity among these bird's eye views. However, some researchers may claim (and indeed have claimed) that this similarity extends also to numerous other works and to various other types of iterative procedures. This is the claim I wish to oppose here: I believe that the type of iterative *input-modification* process that underlies the aforementioned works is essentially novel and amounts to a new algorithmic paradigm.

Let me first give a voice to the skeptic. For example, Amnon Ta-Shma, playing the Devil's advocate, claims that many standard iterative procedures (e.g., repeated squaring) may be viewed as "iteratively modifying the input" (rather than iteratively computing an auxiliary function of it, as I view it). Indeed, the separation line between input-modification and arbitrary computation is highly subjective, and I don't believe that one can rigorously define it. Nevertheless, rejecting Wittgenstein's advice [29, §7], I will try to speak about it.

My claim is that (with the exception of the iterative expander construction of [26]) the reviewed works do not output the modified input, but rather a function of it, and they modify the input in order to ease the computation of the said function. That is, whereas the goal was to compute a function of the original input, they compute a function of the final modified input, and obtain the originally desired value (of the function evaluated at the original input) by a process that relies on the relatively simplicity of the intermediate modifications. The line that I wish to draw is between *iteratively producing modified inputs* (while maintaining a relation between the corresponding outputs) and *iteratively producing better refinements of the desired output while keeping the original input intact*. Indeed, I identify the latter with standard iterative processes (and the former with the common theme of the four reviewed works).

My view is that in each of these works, *the input itself undergoes a gradual transformation* in order to ease some later process. This is obvious in the case of approximating the permanent [18] and in the case of traversing a graph in log-space [25], but it is also true with respect to the other two cases: In Dinur's

---

[2] Consider the problems addressed by the four reviewed works: The problem of approximating the permanent was open since Valiant's seminal work [28] and has received considerable attention since Broder's celebrated work [11]. Constructions of expander graphs were the focus of much research since the 1970's, and were typically based on non-elementary mathematics (cf. [21,16,20]). The existence of deterministic log-space algorithms for undirected connectivity has been in the focus of our community since the publication of the celebrated randomized log-space algorithm of Aleliunas *et. al.* [1]. The PCP Theorem, proved in the early 1990's [5,6], is closely related (via [14,5]) to the study of approximation problems (which dates to the early 1970's).

proof [12] of the PCP Theorem the actual iterative process consists of a sequence of Karp-reductions (which ends with a modified instance that has a simple PCP system), and in the iterative construction of expanders [26] the size of the desired expander increases gradually. In contrast, in typical proofs by induction, it is the problem itself that gets modified, whereas standard iterative procedures refer to sub-problems that relate to auxiliary constructs. Indeed, the separation line between the iterative construction of expanders and standard iterative analysis is the thinnest, but the similarity between it and the results of Reingold [25] and Dinur [12] may appeal to the skeptic.

I wish to stress that the aforementioned iterative process that gradually transforms the input is marked by the relative simplicity of each iteration, especially in comparison to the full-fledged task being undertaken. In the case of Reingold's logspace algorithm [25], each iteration needs to be implemented in constant amount of space, which is indeed a good indication to its simplicity. In the case of the approximation of the permanent [18], each iteration is performed by a modification of a known algorithm (i.e., of [17]). In the iterative construction of expanders [26], a graph powering and a new type of graph product are used and analyzed, where the analysis is simple in comparison to either of [21,16,20]. Lastly, in Dinur's proof [12] of the PCP Theorem, each iteration is admittedly quite complex, but not when compared to the original proof of the PCP Theorem [6,5].

The similarity among the iterated Zig-Zag construction of [26], the log-space algorithm for undirected connectivity of [25], and the new approach to the PCP Theorem of [12] has been noted by many researchers (see, e.g., [25,12] themselves). However, I think that the noted similarity was more technical in nature, and was based on the role of expanders and "Zig-Zag like" operations in these works. In contrast, my emphasis is on the sequence of gradual modifications, and thus I view the permanent approximator of [18] just as close in spirit to these works. In fact, as is hinted in the foregoing discussion, I view [25,12] as closer in spirit to [18] than to [26].

2 Approximating the Permanent of Non-negative Matrices

The permanent of a n-by-n matrix $(a_{i,j})$ is the sum, taken over all permutations $\pi : [n] \to [n]$, of $\prod_{i=1}^{n} a_{i,\pi(i)}$. Although defined very similarly to the determinant (i.e., just missing the minus sign in half of the terms), the permanent seems to have a totally different complexity than the determinant. In particular, in a seminal work [28], Valiant showed that the permanent is $\#\mathcal{P}$-complete; that is, counting the number of solutions to any NP-problem is polynomial-time reducible to *computing* the permanent of 0-1 matrices, which in turn counts the number of perfect matchings in the corresponding bipartite graph. Furthermore, the reduction of NP-counting problems to the permanent of *integer* matrices preserves the (exact) number of solutions (when normalized by an easy to compute factor), and hence *approximating* the permanent of such matrices seems infeasible (because it will imply $\mathcal{P} = \mathcal{NP}$). It was noted that the same does not hold for 0-1 matrices (or even non-negative matrices). In fact, Broder's celebrated work [11] introduced an approach having the potential to yield efficient algorithms for approximating the permanent of non-negative matrices. Fifteen years later, this potential was fulfilled by Jerrum, Sinclair, and Vigoda, in a work [18] to be reviewed here.

The algorithm of Jerrum, Sinclair, and Vigoda [18] follows the general paradigm of Broder's work (which was followed by all subsequent works in the area): The approach is based on the relation between *approximating* the ratio of the numbers of perfect and nearly perfect matchings of a graph and *sampling* uniformly a perfect or nearly perfect matching of a graph, where a nearly perfect matching is a matching that leave unmatched a single pair of vertices. In order to perform the aforementioned sampling, one sets-up a huge Markov Chain with states corresponding to the set of perfect and nearly perfect matchings of the graph. The transition probability of the Markov Chain maps each perfect matching to a nearly perfect matching obtained by omitting a uniformly selected edge (in the perfect matching). The transition from a nearly perfect matching that misses the vertex pair (u, v) is determined by selecting a random vertex z, adding (u, v) to the matching if $z \in \{u, v\}$ and (u, v) is an edge of the graph, and adding (u, z) to the matching and omitting (x, z) from it if $z \notin \{u, v\}$ and (u, z) is an edge of the graph. By suitable modification, the stationary distribution of the chain equals the uniform distribution over the set of perfect and nearly perfect matchings of the graph. The stationary distribution of the chain is approximately sampled by starting from an arbitrary state (e.g., any perfect matching) and taking a sufficiently long walk on the chain.

This approach depends on the mixing time of the chain (i.e., the number of steps needed to get approximately close to its stationary distribution), which in turn is linearly related to the ratio of the numbers of nearly perfect and perfect matchings in the underlying graph (see [17]). (We mention that the later ratio also determines the complexity of the reduction from approximating this ratio to sampling the stationary distribution of the chain.) When the latter ratio is polynomial, this approach yields a polynomial-time algorithms, but it is easy to see that there are graphs for which the said ratio is exponential.

One key observation of [18] is that the latter problem can be fixed by introducing auxiliary weights that when applied (as normalizing factors) to all nearly perfect matchings yield a situation in which the set of perfect matchings has approximately the same probability mass (under the stationary distribution) as the set of nearly perfect matchings. Specifically, for each pair (u, v), we consider a weight $w(u, v)$ such that the probability mass assigned to perfect matchings approximately equals $w(u, v)$ times the probability mass assigned to nearly perfect matchings that leaves the vertices u and v unmatched. Needless to say, in order to determine the suitable weights, one needs to know the corresponding ratios, which seems to lead to a vicious cycle.

Here is where the main idea of [18] kicks in: *Knowing the approximate sizes of the sets of perfect and nearly perfect matchings in a graph G allows to efficiently approximate these parameters for a related graph G' that is closed to G, by running* *the Markov Chain that corresponds to G' under weights as determined for G.* This observation is the basis of the iterative process outlined in the Introduction: We start with a trivial graph G_0 for which the said quantities are easy to determine, and consider a sequence of graphs $G_1, ..., G_t$ such that G_{i+1} is sufficiently close to G_i, and G_t is sufficiently close to the input graph. We approximate the said quantities for G_{i+1} using the estimated quantities for G_i, and finally obtain an approximation of the number of perfect matchings in the input graph.

The algorithm actually works with weighted graphs, where the weight of a matching is the product of the weights of the edges in the matching. We start with G_0 that is a complete graph (i.e., all edges are present, each at weight 1), and let G_{i+1} be a graph obtained from G_i by reducing the weight of one of the non-edges of the input graph by a factor of $\rho = 9/8$. Using such a sequence, for $t = \widetilde{O}(n^3)$, we can obtain a graph G_t in which the edges of the input graph have weight 1 while non-edges of the input graph have weight lower than $1/(n!)$. Approximating the total weight of the weighted perfect matchings in G_t provides the desired approximation to the input graph.

Digest. The algorithm of Jerrum, Sinclair, and Vigoda [18] proceeds in iterations, using a sequence of weighted graphs $G_0, ..., G_t$ such that G_0 is the complete (unweighted) graph, G_{i+1} is a sufficiently close approximation of G_i, and G_t is a sufficiently close approximation to the input graph. We start knowing the numbers of perfect and nearly perfect matchings in G_0 (which is easily determined by the number of vertices). In the ith iteration, using approximations for the numbers of perfect and nearly perfect matchings in G_i, we compute such approximations for G_{i+1}. These approximations are obtained by running an adequate Markov Chain (which refers to G_{i+1}), and the fact that we only have (approximations for) the quantities of G_i merely effects the mixing time of the chain (in a non-significant way). Thus, gradually transforming a dummy graph G_0 into the input graph, we obtain approximations to relevant parameters of all the graphs, where the approximated parameters of G_i allow us to obtain the approximated parameters of G_{i+1}, and the approximated parameters of G_t include an approximation of the number of perfect matchings in the input graph.

Comment. We mention that a different iterative process related to the approximation of the permanent was previously studied in [19]. In that work, an input matrix is transformed to an approximately Doubly Stochastic (aDS) matrix, by iteratively applying row and column scaling operations, whereas for any aDS n-by-n matrix the permanent is at least $\Omega(\exp(-n))$ and at most 1.

3   The Iterative (Zig-Zag) Construction of Expander Graphs

By **expander graphs** (or expanders) of degree d and eigenvalue bound $\lambda < d$, we mean an infinite family of d-regular graphs, $\{G_n\}_{n \in S}$ ($S \subseteq \mathbb{N}$), such that G_n is a d-regular graph with n vertices and the absolute value of all eigenvalues, except the biggest one, of the adjacency matrix of G_n is upper-bounded by λ. This algebraic definition is related to the combinatorial definition of expansion in which it is required that any (not too big) set of vertices in the graph must have a relatively large set of strict neighbors (i.e., is "expanding"); see [3] and [2]. It is often more convenient to refer to the relative eigenvalue bound defined as λ/d.

We are interested in *explicit constructions* of expander graphs, where the minimal notion of explicitness requires that the graph be constructed in time that is polynomial in its size (i.e., there exists a polynomial time algorithm that, on input 1^n, outputs G_n).[3] A *stronger notion of explicitness* requires that there exists a polynomial-time algorithm that on input n (in binary), a vertex $v \in G_n$ and an index $i \in [d] \overset{\text{def}}{=} \{1, ..., d\}$, returns the ith neighbor of v. Many explicit constructions of expanders were given, starting in [21] (where S is the set of all quadratic integers), and culminating in the optimal construction of [20] (where $\lambda = 2\sqrt{d-1}$ and S is somewhat complex). These constructions are quite simple to present, but their analysis is based on non-elementary results from various branches of mathematics. In contrast, the following construction of Reingold, Vadhan, and Wigderson [26] is based on an iterative process, and its analysis is based on a relatively simple algebraic fact regarding the eigenvalues of matrices.

The starting point of the construction (i.e., the base of the iterative process) is a very good expander G of *constant size*, which may be found by an exhaustive search. The construction of a large expander graph proceeds in iterations, where in the ith iteration the graphs G_i and G are combined to obtain the larger graph G_{i+1}. The combination step guarantees that the expansion property of G_{i+1} is at least as good as the expansion of G_i, while G_{i+1} maintains the degree of G_i and is a constant times larger than G_i. The process is initiated with $G_1 = G^2$ and terminates when we obtain a graph G_t of approximately the desired size (which requires a logarithmic number of iterations).

The heart of the combination step is a new type of "graph product" called *Zig-Zag product*. This operation is applicable to any pair of graphs $G = ([D], E)$ and $G' = ([N], E')$, provided that G' (which is typically larger than G) is D-regular. For simplicity, we assume that G is d-regular (where typically $d \ll D$). The Zig-Zag product of G' and G, denoted $G' ⓩ G$, is defined as a graph with vertex set $[N] \times [D]$ and an edge set that includes an edge between $\langle u, i \rangle \in [N] \times [D]$ and $\langle v, j \rangle$ if and only if $(i, k), (\ell, j) \in E$ and the kth edge incident at u equals the ℓth edge incident at v.

It will be convenient to represent graphs like G' by their edge rotation function[4], denoted $R' : [N] \times [D] \to [N] \times [D]$, such that $R'(u, i) = (v, j)$ if (u, v) is the ith edge incident at u as well as the jth edge incident at v. That is, applying R' to (u, i) "rotates" the ith edge incident at vertex u, yielding its representation

---

[3] We also require that the set S for which G_n's exist is sufficiently "tractable": say, that given any $n \in \mathbb{N}$ one may efficiently find $s \in S$ so that $n \le s < 2n$.

[4] In [26] (and [25]) these functions are called rotation maps. As these functions are actually involutions (i.e., $R(R(x)) = x$ for every $x \in [N] \times [D]$), one may prefer terms as "edge rotation permutations" or "edge rotation involutions".

from its other endpoint view (i.e., as the jth edge incident at vertex v, assuming $R'(u,i) = (v,j)$). For simplicity, we assume that G is edge-colorable with d colors, which in turn yields a natural edge rotation function (i.e., $R(i,\alpha) = (j,\alpha)$ if the edge (i,j) is colored α). We will denote by $E_\alpha(i)$ the vertex reached from $i \in [D]$ by following the edge colored α (i.e., $E_\alpha(i) = j$ iff $R(i,\alpha) = (j,\alpha)$). The Zig-Zag product of G' and G, denoted $G' \textcircled{z} G$, is then defined as a graph with the vertex set $[N] \times [D]$ and the edge rotation function

$$(\langle u,i\rangle, \langle \alpha,\beta\rangle) \mapsto (\langle v,j\rangle, \langle \beta,\alpha\rangle) \quad \text{if } R'(u, E_\alpha(i)) = (v, E_\beta(j)). \tag{1}$$

That is, edges are labeled by pairs over $[d]$, and the $\langle \alpha,\beta\rangle^{\text{th}}$ edge out of vertex $\langle u,i\rangle \in [N] \times [D]$ is incident at the vertex $\langle v,j\rangle$ (as its $\langle \beta,\alpha\rangle^{\text{th}}$ edge) if $R(u, E_\alpha(i)) = (v, E_\beta(j))$. (Pictorially, based on the $G' \textcircled{z} G$-label $\langle \alpha,\beta\rangle$, we take a G-step from $\langle u,i\rangle$ to $\langle u, E_\alpha(i)\rangle$, then viewing $\langle u, E_\alpha(i)\rangle \equiv (u, E_\alpha(i))$ as an edge of G' we rotate it to obtain $(v,j') \stackrel{\text{def}}{=} R'(u, E_\alpha(i))$, and finally take a G-step from $\langle v,j'\rangle$ to $\langle v, E_\beta(j')\rangle$, while defining $j = E_\beta(j')$ and using $j' = E_\beta(E_\beta(j')) = E_\beta(j)$.)

Clearly, the graph $G' \textcircled{z} G$ is d^2-regular and has $D \cdot N$ vertices. The key fact, proved in [26], is that the relative eigenvalue of the zig-zag produce is upper-bounded by the sum of the relative eigenvalues of the two graphs (i.e., $\lambda(G' \textcircled{z} G) \leq \lambda(G') + \lambda(G)$, where $\lambda(\cdot)$ denotes the relative eigenvalue of the relevant graph).[5]

The iterated expander construction uses the aforementioned zig-zag product as well as graph squaring. Specifically, the construction starts with a d-regular graph $G = ([D], E)$ such that $D = d^4$ and $\lambda(G) < 1/4$. Letting $G_1 = G^2 = ([D], E^2)$, the construction proceeds in iterations such that $G_{i+1} = G_i^2 \textcircled{z} G$ for $i = 1, 2, ..., t-1$. That is, in each iteration, the current graph is first squared and then composed with the fixed graph G via the zig-zag product. This process maintains the following two invariants:

1. The graph G_i is d^2-regular and has D^i vertices.
   This holds for $G_1 = G^2$ (since G is d-regular with D vertices), and is maintained for the other G_i's because a zig-zag product (of a D-regular N'-vertex graph) with a d-regular (D-vertex) graph yields a d^2-regular graph (with $D \cdot N'$ vertices).

2. The relative eigenvalue of G_i is smaller than one half.
   Here we use the fact that $\lambda(G_{i-1}^2 \textcircled{z} G) \leq \lambda(G_{i-1}^2) + \lambda(G)$, which in turn equals $\lambda(G_{i-1})^2 + \lambda(G) < (1/2)^2 + 1/4$. (Note that graph squaring is used to reduce the relative eigenvalue of G_i before allowing its moderate increase by the zig-zag product with G.)

To ensure that we can construct G_i, we should show that we can actually construct the edge rotation function that correspond to its edge set. This boils down to showing that, given the edge rotation function of G_{i-1}, we can compute the

---

[5] In fact, the upper-bound proved in [26] is stronger. In particular, it also implies that $1 - \lambda(G' \textcircled{z} G) \geq (1 - \lambda(G)^2) \cdot (1 - \lambda(G'))/2$.

edge rotation function of G_{i-1}^2 as well as of its zig-zag product with G. Note that this computation amounts to two recursive calls to computations regarding G_{i-1} (and two computations that correspond to the constant graph G). But since the recursion is logarithmic in the size of the final graph, the time spend in the recursive computation is polynomial in the size of the final graph. This suffices for the minimal notion of explicitness, but not for the stronger one.

To achieve a *strongly explicit construction*, we slightly modify the iterative construction. Rather than letting $G_{i+1} = G_i^2 \textcircled{z} G$, we let $G_{i+1} = (G_i \times G_i)^2 \textcircled{z} G$, where $G' \times G'$ denotes the tensor product of G' with itself (i.e., if $G' = (V', E')$ then $G' \times G' = (V' \times V', E'')$, where $E'' = \{(\langle u_1, u_2\rangle, \langle v_1, v_2\rangle) : (u_1, v_1), (u_2, v_2) \in E'\}$ with an edge rotation function $R''(\langle u_1, u_2\rangle, \langle i_1, i_2\rangle) = (\langle v_1, v_2\rangle, \langle j_1, j_2\rangle)$ where $R'(u_1, i_1) = (v_1, j_1)$ and $R'(u_2, i_2) = (v_2, j_2)$). (We still use $G_1 = G^2$.) Using the fact that tensor product preserves the relative eigenvalue and using a d-regular graph $G = ([D], E)$ with $D = d^8$, we note that the modified $G_i = (G_{i-1} \times G_{i-1})^2 \textcircled{z} G$ is a d^2-regular graph with $(D^{2^{i-1}-1})^2 \cdot D = D^{2^i-1}$ vertices, and $\lambda(G_i) < 1/2$ (because $\lambda((G_{i-1} \times G_{i-1})^2 \textcircled{z} G) \leq \lambda(G_{i-1})^2 + \lambda(G)$). Computing the neighbor of a vertex in G_i boils down to a constant number of such computations regarding G_{i-1}, but due to the tensor product operation the depth of the recursion is only double-logarithmic in the size of the final graph (and hence logarithmic in the length of the description of vertices in it).

Digest. In the first construction, the zig-zag product was used both in order to increase the size of the graph and to reduce its degree. However, as indicated by the second construction (where the tensor product of graphs is the main vehicle for increasing the size of the graph), the primary effect of the zig-zag product is to reduce the degree, and the increase in the size of the graph is merely a side-effect (which is actually undesired in Section 4). In both cases, graph squaring is used in order to compensate for the modest increase in the relative eigenvalue caused by the zig-zag product. In retrospect, the second construction is the "correct" one, because it decouples three different effects, and uses a natural operation to obtain each of them: Increasing the size of the graph is obtained by tensor product of graphs (which in turn increases the degree), a degree reduction is obtained by the zig-zag product (which in turn increases the relative eigenvalue), and graph squaring is used in order to reduce the relative eigenvalue.

A second theme. In continuation to the previous comment, we note that the successive application of several operations, each improving a different parameter (while not harming too much the others), reappears in the works of Reingold [25] and Dinur [12]. This theme has also appeared before in several other works (including [6,5,13]).[6]

---

[6] We are aware of half a dozen of other works, but guess that they are many more. We choose to cite here only works that were placed in the reference list for other reasons. Indeed, this second theme appears very clearly in PCP constructions (e.g., first optimizing randomness at the expense of number of queries and then reducing the latter at the expense of a bigger alphabet (not to mention the very elaborate combination in [13])).

4 The Log-Space Algorithm for Undirected Connectivity

For more than two decades, undirected connectivity was one of the most appealing examples of the computational power of randomness. Whereas every graph can be efficiently traversed by a deterministic algorithm, the classical (deterministic) linear-time algorithms (e.g., BFS and DFS) require an extensive use of (extra) memory (i.e., space linear in the size of the graph). On the other hand, in 1979 Aleliunas et. al. [1] showed that, with high probability, a random walk of polynomial length visits all vertices (in the corresponding connected component). Thus, the randomized algorithm requires a minimal amount of auxiliary memory (i.e., logarithmic in the size of the graph). In the early 1990's, Nisan [22,23] showed that any graph can be traversed in polynomial-time and poly-logarithmic space, but despite more than a decade of research attempts (see, e.g., [4]), a significant gap remained between the space complexity of randomized and deterministic polynomial-time algorithms for this natural and ubiquitous problem. This gap was recently closed by Reingold, in a work [25] reviewed next.

Reingold presented a deterministic polynomial-time algorithm that traverses any graph while using a logarithmic amount of auxiliary memory. His algorithm is based on a novel approach that departs from previous attempts, which tried to derandomize the random-walk algorithm. Instead, Reingold's algorithm traverses a virtual graph, which (being an expander) is easy to traverse (in deterministic logarithmic-space), and maps the virtual traversal of the virtual graph to a real traversal of the actual input graph. The virtual graph is constructed in (logarithmically many) iterations, where in each iteration the graph becomes easier to traverse. Specifically, in each iteration, each connected component of the graph becomes closer to a constant-degree expander in the sense that (the graph has constant degree and) the gap between its relative eigenvalue and 1 doubles.[7] Hence, after logarithmically many iterations, each connected component becomes a constant-degree expander, and thus has logarithmic diameter. Such a graph is easy to traverse deterministically using logarithmic space (e.g., by scanning all paths of logarithmic length going out of a given vertex, while noting that each such path can be represented by a binary string of logarithmic length).

The key point is to maintain the connected components of the graph while making each of them closer to an expander. Towards this goal, Reingold applies a variant of the iterated zig-zag construction (presented in Section 3), starting with the input graph, and iteratively composing the current graph with a constant-size expander. Details follow.

For adequate positive integers d and c, we first transform the actual input graph into a d^2-regular graph (e.g., by replacing each vertex v with a (multi-edge) cycle C_v and using each vertex on C_v to take care of an edge incident to v). Denoting the resulting graph by $G_1 = (V_1, E_1)$, we go through a logarithmic number of iterations letting $G_{i+1} = G_i^c \textcircled{z} G$ for $i = 1, ..., t-1$, where G is a fixed d-regular graph with d^{2c} vertices. Thus, G_i is a d^2-regular graph

---

[7] See Section 3 for definition of expander and its relative eigenvalue.

with $d^{2c \cdot i} \cdot |V_1|$ vertices, and $1 - \lambda(G_i) > \max(2(1 - \lambda(G_{i-1})), 1/6)$, where the latter upper-bound on $\lambda(G_i)$ relies on a result of [26] (see Footnote 5). We infer that $1 - \lambda(G_i) > \max(2^i \cdot (1 - \lambda(G_1)), 1/6)$, and using the fact that $\lambda(G_1) < 1 - (1/\text{poly}(|V_1|))$, which holds for any connected and non-bipartite graph, it follows that $\lambda(G_t) < 5/6$ for $t = O(\log |V_1|)$. (Indeed, it is instructive to assume throughout the analysis that (the original input and thus) G_1 is connected, and to guaranteed that it is non-bipartite (e.g., by adding self-loops).)

One detail of crucial importance is the ability to transform G_1 into G_t via a log-space computation. It is easy to see that the transformation of G_i to G_{i+1} can be performed in constant-space (with an extra pointer), but the standard composition lemma for space-bounded complexity incurs a logarithmic space overhead per each composition (and thus cannot be applied here). Still, taking a closer look at the transformation of G_i to G_{i+1}, one may note that it is highly structured and supports a stronger composition result that incurs only a constant space overhead per composition. An alternative implementation, outlined in [25], is obtained by unraveling the composition. The details of these alternative implementations are beyond the scope of the current essay.[8]

A minor variant. It is simpler to present a direct implementation of a minor variant of the foregoing process. Specifically, rather than using the zig-zag product $G' \circledz G$ (of Section 3), one may use the replacement product $G' \circledr G$ defined as follows for a D-regular graph $G' = (V', E')$ and a d-regular graph $G = ([D], E)$:[9] The resulting $2d$-regular graph has vertex set $V' \times [D]$ and the following edge rotation function (which actually induces an edge coloring)

$$(\langle u, i \rangle, \langle 0, \alpha \rangle) \mapsto (\langle u, E_\alpha(i) \rangle, \langle 0, \alpha \rangle)$$
$$\text{and} \quad\quad\quad\quad\quad\quad\quad\quad\quad\quad\quad\quad\quad\quad\quad (2)$$
$$(\langle u, i \rangle, \langle 1, \alpha \rangle) \mapsto (R'(u, i), (1, \alpha)),$$

where E_α is as in Section 3. That is, every $\langle u, i \rangle \in V' \times [D]$ has d incident edges that correspond to the edges incident at i in G, and d parallel copies of the ith edge of u in G'. It can be shown that, in the relevant range of parameters, the replacement product effect the eigenvalues in a way that is similar to the affect of the zig-zag product (because the two resulting graphs are sufficiently related).

---

[8] We cannot refrain from saying that we prefer an implementation based on composition, and provide a few hints regarding such an implementation (detailed in [15, Sec. 5.2.4]). Firstly, we suggest to consider the task of computing the neighbor of a given vertex in G_i, where the original graph is viewed as an oracle and the actual input is the aforementioned vertex. This computation can be performed by a constant-space oracle machine provided that its queries are answered by a similar machine regarding G_{i-1}. Second, the overhead involved in standard composition can be avoided by using a model of "shared memory for procedural calls" and noting that the aforementioned reduction requires only constant-space in addition to the log-space shared memory. The key point is that the latter need only be charged once.

[9] Since this product yields a $2d$-regular graph, in the context of the log-space algorithm one should set $D = (2d)^c$ (rather than $D = d^{2c}$).

Another variant. A more significant variant on the construction was subsequently presented in [27]. As a basic composition, they utilize a derandomized graph squaring of a large D-regular graph $G' = (V', E')$ using a d-regular (expander) graph $G = ([D], E)$: Unlike the previous composition operations, the resulting graph, which is a subgraph of the square of G', has V' itself as the vertex set but the (vertex) degree of the resulting graph is larger than that of G'. Specifically, the edge rotation function is

$$(u, \langle i, \alpha \rangle) \mapsto (v, \langle j, \alpha \rangle) \quad \text{if } R'(u, i) = (w, k) \text{ and } R'(w, E_\alpha(k)) = (v, j). \quad (3)$$

where E_α is as in Section 3. That is, the edge set contains a subset of the edges of the standard graph square, where this subset corresponds to the edges of the small (expander) graph G. It can be shown that the derandomized graph squaring effect the eigenvalues in a way that is similar to the combination of squaring and zig-zag product, but the problem is that the (vertex) degree does not remain constant through the iterated procedure. Nevertheless, two alternatives ways of obtaining a log-space algorithm are known, one of which is presented in [27].

5 The Alternative Proof of the PCP Theorem

The PCP Theorem [5,6] is one of the most influential and impressive results of complexity theory. Proven in the early 1990's, the theorem asserts that membership in any NP-set can be verified, with constant error probability (say 1%), by a verifier that probes a polynomially long (redundant) proof at a constant number of randomly selected locations. The PCP Theorem led to a breakthrough in the study of the complexity of combinatorial approximation problems (see, e.g., [14,5]). Its original proof is very complex and involves the composition of two highly non-trivial proof systems, each minimizing a different parameter of the PCP system (i.e., proof length and number of probed locations). An alternative approach to the proof of the PCP Theorem was recently presented by Dinur [12], and is reviewed below. In addition to yielding a simpler proof of the PCP Theorem, Dinur's approach resolves an important open problem regarding PCP systems (i.e., constructing a PCP system having proofs of almost-linear rather than polynomial length).

The original proof of the PCP Theorem focuses on the construction of two PCP systems that are highly non-trivial and interesting by themselves, and combines them in a natural manner. Loosely speaking, this combination (via proof composition) preserves the good features of each of the two systems; that is, it yields a PCP system that inherits the (logarithmic) randomness complexity of one system and the (constant) query complexity of the other. In contrast, Dinur's approach is focused at the "amplification" of PCP systems, via a gradual process of logarithmically many steps. It start from a trivial "PCP" system that rejects false assertions with probability inversely proportional to their length, and double the rejection probability in each step. In each step, the constant query complexity is preserved and the length of the PCP oracle is increased only by a constant factor. Thus, the process gradually transforms a very weak PCP system into a remarkable PCP system as postulated in the PCP Theorem.

In order to describe the aforementioned process we need to redefine PCP systems so to allow arbitrary soundness error. In fact, for technical reasons it is more convenient to describe the process as an iterated reduction of a "constraint satisfaction" problem to itself. Specifically, we refer to systems of 2-variable constraints, which are readily represented by (labeled) graphs.

Definition 5.1 (CSP with 2-variable constraints): *For a fixed finite set Σ, an instance of* CSP *consists of a graph $G = (V, E)$ (which may have parallel edges and self-loops) and a sequence of 2-variable constraints $\Phi = (\phi_e)_{e \in E}$ associated with the edges, where each constraint has the form $\phi_e : \Sigma^2 \to \{0, 1\}$. The value of an assignment $\alpha : V \to \Sigma$ is the number of constraints satisfied by α; that is, the value of α is $|\{(u, v) \in E : \phi_{(u,v)}(\alpha(u), \alpha(v)) = 1\}|$. We denote by* $\mathtt{vlt}(G, \Phi)$ *the fraction of unsatisfied constraints under the best possible assignment; that is,*

$$\mathtt{vlt}(G, \Phi) = \min_{\alpha:V \to \Sigma} \{|\{(u, v) \in E : \phi_{(u,v)}(\alpha(u), \alpha(v)) = 0\}|/|E|\} \qquad (4)$$

For various functions $t : \mathbb{N} \to [0, 1]$, we will consider the promise problem $\mathtt{gapCSP}_t^{\Sigma}$, *having instances as above, such that the* YES-*instances are fully satisfiable instances* (i.e., $\mathtt{vlt} = 0$) *and the* NO-*instances are pairs (G, Φ) satisfying* $\mathtt{vlt}(G, \Phi) > t(|G|)$, *where $|G|$ denotes the number of edges in G.*

Note that 3SAT (and thus any other set in $\mathcal{NP}$) is reducible to $\mathtt{gapCSP}_t^{\{1,\dots,7\}}$ for $t(m) = 1/m$. Our goal is to reduce 3SAT (or rather $\mathtt{gapCSP}_t^{\{1,\dots,7\}}$) to $\mathtt{gapCSP}_c^{\Sigma}$, for some fixed finite Σ and constant $c > 0$. The PCP Theorem follows by showing a simple PCP system for $\mathtt{gapCSP}_c^{\Sigma}$ (e.g., consider an alleged proof that encodes an assignment $\alpha : V \to \Sigma$, and a verifier that inspects the values of a uniformly selected constraint). The desired reduction is obtained by iteratively applying the following reduction logarithmically many times.

Lemma 5.2 (amplifying reduction of gapCSP to itself): *For some finite Σ and constant $c > 0$, there exists a polynomial-time reduction of $\mathtt{gapCSP}^{\Sigma}$ to itself such that the following conditions hold with respect to the mapping of any instance (G, Φ) to the instance (G', Φ').*

1. *If $\mathtt{vlt}(G, \Phi) = 0$, then $\mathtt{vlt}(G', \Phi') = 0$.*
2. $\mathtt{vlt}(G', \Phi') \geq \min(2 \cdot \mathtt{vlt}(G, \Phi), c)$.
3. $|G'| = O(|G|)$.

Proof outline: The reduction consists of three steps. We first apply a preprocessing step that makes the underlying graph suitable for further analysis. The value of $\mathtt{vlt}$ may decrease during this step by a constant factor. The heart of the reduction is the second step in which we may increase $\mathtt{vlt}$ by any desired constant factor. The latter step also increases the alphabet Σ, and thus a postprocessing step is employed to regain the original alphabet (by using any inner PCP systems; e.g., the Hadamard-based one presented in [5]). Details follow.

We first note that the aforementioned Σ and c, as well as the auxiliary parameters d and t, are fixed constants that will be determined to satisfy various conditions that arise in the course of our argument.

We start with the pre-processing step. Our aim in this step is to reduce the input (G, Φ) of $\mathtt{gapCSP}^\Sigma$ to an instance (G_1, Φ_1) such that G_1 is a d-regular expander graph. Furthermore, each vertex in G_1 will have at least $d/2$ self-loops, $|G_1| = O(|G|)$, and $\mathtt{vlt}(G_1, \Phi_1) = \Theta(\mathtt{vlt}(G, \Phi))$. This step is quite simple: Essentially, the original vertices are replaced by expanders of size proportional to their degree, and a big (dummy) expander is superimposed on the resulting graph.

The main step is aimed at increasing the fraction of violated constraints by a sufficiently large constant factor. This is done by reducing the instance (G_1, Φ_2) of $\mathtt{gapCSP}^\Sigma$ to an instance (G_2, Φ_2) of $\mathtt{gapCSP}^{\Sigma'}$ such that $\Sigma' = \Sigma^{d^t}$. Specifically, the vertex set of G_2 is identical to the vertex set of G_1, and each t-edge long path in G_1 is replaced by a corresponding edge in G_2, which is thus a d^t-regular graph. The constraints in Φ_2 are the natural ones, viewing each element of Σ' as a Σ-labeling of the ("distance $\leq t$") neighborhood of a vertex, and checking that two such labelings are consistent and satisfy Φ_1. That is, suppose that there is a path of length at most t in G_1 going from vertex u to vertex v and passing through vertex w. Then, there is an edge in G_2 between vertices u and v, and the constraint associated with it mandates that the entries corresponding to vertex w in the Σ'-labeling of vertices u and v are identical. In addition, if the G_1-edge (w, w') is on a path of length at most t starting at v, then the corresponding edge in G_2 is associated with a constraint that enforces the constraint that is associated to (w, w') in Φ_1.

Clearly, if $\mathtt{vlt}(G_1, \Phi_1) = 0$, then $\mathtt{vlt}(G_2, \Phi_2) = 0$. The interesting fact is that the fraction of violated constraints increases by a factor of $\Omega(\sqrt{t})$; that is, $\mathtt{vlt}(G_2, \Phi_2) \geq \min(\Omega(\sqrt{t} \cdot \mathtt{vlt}(G_1, \Phi_1)), c)$. The intuition is that any Σ'-labeling to the vertices of G_2 must either be consistent with some Σ-labeling of G_1 or violate many edges in G_2 (due to the equality conditions that were inserted to all new constraints). Focusing on the first case and relying on the hypothesis that G_1 is an expander, it follows that the set of violated edge-constraints (of Φ_1) with respect to the aforementioned Σ-labeling causes many more edge-constraints of Φ_2 to be violated. The point is that a set F of edges of G_1 is likely to appear on a $\min(\Omega(t) \cdot |F|/|G_1|, \Omega(1))$ fraction of the edges of G_2 (i.e., t-paths of G_1). (Note that the claim is obvious if G_1 were a complete graph, but it also holds for an expander.)[10]

For a suitable choice of the constant t, the factor of $\Omega(\sqrt{t})$ gained in the second step, makes up for the constant factor lost in the first step (as well as the constant factor to be lost in the last step), while leaving us with a net amplification by a constant factor. However, we obtained an instance of $\mathtt{gapCSP}^{\Sigma'}$ rather than an instance of $\mathtt{gapCSP}^\Sigma$, where $\Sigma' = \Sigma^{d^t}$. The purpose of the last step is to reduce the latter instance to an instance of $\mathtt{gapCSP}^\Sigma$. This is done by viewing the instance of $\mathtt{gapCSP}^{\Sigma'}$ as a (weak) PCP system and composing it with an inner-verifier, using the proof composition paradigm (of [9,13], which in turn follow [6]). We stress that the inner-verifier used here needs only handle instances

---

[10] We also note that due to a technical difficulty it is easier to establish the claimed bound of $\Omega(\sqrt{t} \cdot \mathtt{vlt}(G_1, \Phi_1))$ rather than $\Omega(t \cdot \mathtt{vlt}(G_1, \Phi_1))$.

O. Goldreich

of constant size (i.e., having description length $O(d^t \log |\Sigma|)$), and so the one presented in [5] (or [8]) will do. The resulting PCP-system uses randomness $r \stackrel{\text{def}}{=}$ $\log_2 |G_2| + (d^t \log |\Sigma|)^2$ and a constant number of binary queries, and has rejection probability $\Omega(\text{vlt}(G_2, \Phi_2))$, which is independent of the choice of the constant t. For $\Sigma = \{0,1\}^{O(1)}$, we obtain an instance of $\texttt{gapCSP}^\Sigma$ that has a $\Omega(\text{vlt}(G_2, \Phi_2))$ fraction of violated constraints. Furthermore, the size of the resulting instance is $O(2^r) = O(|G_2|)$, because d and t are constants. This completes the description of the last step as well as the proof of the entire lemma.   □

Application to short PCPs. Recall that the PCP Theorem asserts that membership in any NP-set can be verified, with constant error probability, by a verifier that probes a polynomially long (redundant) proof at a constant number of randomly selected locations. Denoting by N the length of the standard proof, the length of the redundant proof was reduced in [9] to $\exp((\log N)^\epsilon) \cdot N$, for any $\epsilon > 0$. An open problem, explicitly posed in [9], is whether the length of the redundant proof can be reduced to $\text{poly}(\log N) \cdot N$. Building on prior work of [10], this seemingly difficult open problem was resolved by Dinur [12]: Specifically, viewing the system of [10] (which makes $\text{poly}(\log N)$ queries into a proof of length $\text{poly}(\log N) \cdot N$) as a PCP system with rejection probability $1/\text{poly}(\log N)$, Dinur applies the foregoing amplification step for a double-logarithmic number of times, thus deriving the desired PCP system.

Acknowledgments. First of all, I wish to thank Irit Dinur, Mark Jerrum, Omer Reingold, Alistair Sinclair, Salil Vadhan, Eric Vigoda, and Avi Wigderson (i.e., the authors of [12,18,25,26]) for their fascinating works. Next, I wish to thank Amnon Ta-Shma and Avi Wigderson for playing the Devil's advocate, thus forcing me to better articulate my feelings. Finally, I am grateful to Omer Reingold, Salil Vadhan, and Avi Wigderson for critical comments regarding an early version of this essay.

References

1. Aleliunas, R., Karp, R.M., Lipton, R.J., Lovász, L., Rackoff, C.: Random walks, universal traversal sequences, and the complexity of maze problems. In: 20th FOCS, pp. 218–223 (1979)
2. Alon, N.: Eigenvalues and expanders. Combinatorica 6, 83–96 (1986)
3. Alon, N., Milman, V.D.: λ_1, Isoperimetric Inequalities for Graphs and Superconcentrators. J. Combinatorial Theory, Ser. B 38, 73–88 (1985)
4. Armoni, R., Ta-Shma, A., Wigderson, A., Zhou, S.: $SL \subseteq L^{4/3}$. In: 29th STOC, pp. 230–239 (1997)
5. Arora, S., Lund, C., Motwani, R., Sudan, M., Szegedy, M.: Proof Verification and Intractability of Approximation Problems. JACM 45, 501–555 (1998); Preliminary version in 33rd FOCS (1992)
6. Arora, S., Safra, S.: Probabilistic Checkable Proofs: A New Characterization of NP. JACM 45, 70–122 (1998); Preliminary version in 33rd FOCS (1992)
7. Babai, L., Fortnow, L., Levin, L., Szegedy, M.: Checking Computations in Polylogarithmic Time. In: 23rd STOC, pp. 21–31 (1991)

8. Bellare, M., Goldreich, O., Sudan, M.: Free Bits, PCPs and Non-Approximability
   – Towards Tight Results. SICOMP 27(3), 804–915 (1998); Extended abstract in
   36th FOCS (1995)
9. Ben-Sasson, E., Goldreich, O., Harsha, P., Sudan, M., Vadhan, S.: Robust PCPs
   of proximity, Shorter PCPs and Applications to Coding. In: 36th STOC, pp. 1–10
   (2004); Full version in ECCC, TR04-021 (2004)
10. Ben-Sasson, E., Sudan, M.: Simple PCPs with Poly-log Rate and Query Complex-
    ity. In: ECCC, TR04-060 (2004)
11. Broder, A.: How hard is it to marry at random (On the approximation of the
    Permanent). In: 18th STOC, pp. 50–58 (1986)
12. Dinur, I.: The PCP Theorem by Gap Amplification. In: 38th STOC, pp. 241–250
    (2006)
13. Dinur, I., Reingold, O.: Assignment-testers: Towards a combinatorial proof of the
    PCP-Theorem. In: 45th FOCS, pp. 155–164 (2004)
14. Feige, U., Goldwasser, S., Lovász, L., Safra, S., Szegedy, M.: Approximating Clique
    is almost NP-complete. JACM 43, 268–292 (1996); Preliminary version in 32nd
    FOCS (1991)
15. Goldreich, O.: Computational Complexity: A Conceptual Perspective. Cambridge
    University Press, Cambridge (2008)
16. Gaber, O., Galil, Z.: Explicit Constructions of Linear Size Superconcentrators.
    JCSS 22, 407–420 (1981)
17. Jerrum, M., Sinclair, A.: Approximating the Permanent. SICOMP 18, 1149–1178
    (1989)
18. Jerrum, M., Sinclair, A., Vigoda, E.: A Polynomial-Time Approximation Algorithm
    for the Permanent of a Matrix with Non-Negative Entries. JACM 51(4), 671–697
    (2004); Preliminary version in 33rd STOC pp. 712–721 (2001)
19. Linial, N., Samorodnitsky, A., Wigderson, A.: A Deterministic Strongly Polynomial
    Algorithm for Matrix Scaling and Approximate Permanents. Combinatorica 20(4),
    545–568 (2000)
20. Lubotzky, A., Phillips, R., Sarnak, P.: Ramanujan Graphs. Combinatorica 8, 261–
    277 (1988)
21. Margulis, G.A.: Explicit Construction of Concentrators. Prob. Per. Infor. 9(4),
    71–80 (1973) (in Russian); English translation in Problems of Infor. Trans., pp.
    325–332 (1975)
22. Nisan, N.: Pseudorandom Generators for Space Bounded Computation. Combina-
    torica 12(4), 449–461 (1992)
23. Nisan, N.: $\mathcal{RL} \subseteq \mathcal{SC}$. Journal of Computational Complexity 4, 1–11 (1994)
24. Nisan, N., Zuckerman, D.: Randomness is Linear in Space. JCSS 52(1), 43–52
    (1996)
25. Reingold, O.: Undirected ST-Connectivity in Log-Space. In: 37th STOC (2005)
26. Reingold, O., Vadhan, S., Wigderson, A.: Entropy Waves, the Zig-Zag Graph Prod-
    uct, and New Constant-Degree Expanders and Extractors. Annals of Mathemat-
    ics 155(1), 157–187 (2001); Preliminary version in 41st FOCS, pp. 3–13 (2000)
27. Rozenman, E., Vadhan, S.: Derandomized Squaring of Graphs. In: Chekuri, C.,
    Jansen, K., Rolim, J.D.P., Trevisan, L. (eds.) APPROX 2005 and RANDOM 2005.
    LNCS, vol. 3624, pp. 436–447. Springer, Heidelberg (2005)
28. Valiant, L.G.: The Complexity of Computing the Permanent. Theoretical Com-
    puter Science 8, 189–201 (1979)
29. Wittgenstein, L.: Tractatus Logico-Philosophicus (1922)

On the Complexity of Computational Problems Regarding Distributions

Oded Goldreich and Salil Vadhan

Abstract. We consider two basic computational problems regarding discrete probability distributions: (1) approximating the statistical difference (aka variation distance) between two given distributions, and (2) approximating the entropy of a given distribution. Both problems are considered in two different settings. In the first setting the approximation algorithm is only given samples from the distributions in question, whereas in the second setting the algorithm is given the "code" of a sampling device (for the distributions in question).

We survey the know results regarding both settings, noting that they are fundamentally different: The first setting is concerned with the number of samples required for determining the quantity in question, and is thus essentially information theoretic. In the second setting the quantities in question are determined by the input, and the question is merely one of computational complexity. The focus of this survey is actually on the latter setting. In particular, the survey includes proof sketches of three central results regarding the latter setting, where one of these proofs has only appeared before in the second author's PhD Thesis.

Keywords: Approximation, Reductions, Entropy, Statistical Difference, Variation Distance, Sampleable Distributions, Zero-Knowledge, and Promise Problems.

This survey was first drafted in 2003, and was recently posted as ECCC TR11-004.

1 Introduction

We consider two basic computational problems regarding discrete probability distributions:

1. Computing (or rather approximating) the statistical difference (aka variation distance) between two given distributions.
2. Computing (or rather approximating) the entropy of a given distribution.

The foregoing informal phrases avoid the question of representation; that is, how are the distributions given to the algorithms. Both computational problems are quite trivial in the case that the distributions are explicitly given to the algorithm (i.e., by a list of all elements in the support of the distribution coupled with the probability mass assigned to them). Very good additive approximations can be

O. Goldreich et al.: Studies in Complexity and Cryptography, LNCS 6650, pp. 390–405, 2011.
© Springer-Verlag Berlin Heidelberg 2011

obtained also in the case that the algorithm is given sufficiently many samples (drawn independently) from the distribution, where "sufficiently many" means linear in the size of the distribution's support. For example, given $N/\text{poly}(\epsilon)$ samples from a distribution that has support size (at most) N, one can estimate the distribution's entropy up-to an additive deviation of ϵ (w.v.h.p.). The same number of samples suffices for approximating the statistical distance between two such distributions (again, up to an additive deviation of ϵ, w.v.h.p.).

The question is whether such approximations (or even weaker ones) can be obtained based on significantly less samples. At the very least, we are interested in algorithms that take $o(N)$ samples (i.e., a "sub-linear" (in the support size) number of samples). In Section 3, we survey what is known regarding this question. The bottom-line is that weak approximations of both quantities can be obtained using N^e samples, for some $e < 1$, but nothing significant can be achieved with $N^{o(1)}$ samples.

We note that the foregoing question is essentially an information-theoretical one; that is, the question refers to the number of samples required to make some estimations regarding the distribution(s). In contrast, in Section 4, we consider a purely computational-complexity problem: We consider algorithms that are given the "code" of a sampling device (for the distributions in question). We stress that such a device fully determines the distribution (from an information-theoretic point of view), and the issue is what quantities can be *efficiently* computed based on this description of the distribution. Note that the algorithm may, of course, use the sampling device in order to generate samples. However, the algorithm is not confined to this usage of the sampling device and may try to analyze the device in other ways (e.g., try to "reverse-engineer" it).

To be concrete, the sampling device is represented by a circuit $C : \{0,1\}^m \to \{0,1\}^n$, which can be used to generate samples by feeding it with a uniformly selected m-bit long string. Alternatively, one may say that C is an implicit representation of a distribution over $\{0,1\}^n$, obtained by feeding C with a uniformly selected m-bit long string. Typically, the circuit's size is polynomial in n, whereas the distribution defined by it can have support size 2^n. Thus, when we consider the aforementioned computational problems in terms of the circuit size, polynomial-time algorithms correspond to algorithms that run in time that may be poly-logarithmic in the size of the support. We stress that, in this model, the algorithm has full information regarding the distribution in question, but it does not have enough time to use this information in a straightforward way (i.e., feed the circuit with all possible inputs). The question is whether the algorithm can obtain approximations to the aforementioned quantities within time that is polynomial in the size of the circuit. In Section 4, we survey what is known regarding this question. The bottom-line is that the complexity of approximating each of the foregoing computational problems is complete (under polynomial-time reductions) for the complexity class $\mathcal{SZK} \subseteq \mathcal{AM} \cap \text{co}\mathcal{AM}$, which is conjectured to extend beyond $\mathcal{BPP}$ (i.e., probabilistic polynomial-time). In particular, under the widely believed conjecture that the Discrete Logarithm Problem is intractable, it follows that the approximation versions of each of the foregoing computational problems are intractable. It is also known that the two types of computational problems are actually computationally equivalent; that is, each is efficiently reducible to the other.

Organization: In Section 3 we briefly survey the known results regarding sampling-based algorithms (i.e., algorithms that only get samples from the distributions in question). In Section 4 we survey the known results regarding the second setting; that is, we consider algorithms that are given as input a full description of a sampling device for the distributions in question. In Section 5 we present the main ideas underlying the proofs of the three theorems stated in Section 4. One of these proofs has only appeared before in the second author's PhD Thesis [22]. Sections 4 and 5 are actually the main part of this survey.

2 Preliminaries

Traditionally, (discrete) probability distributions are represented by the list of probabilities assigned to the various elements in their range (or potential support). That is, a distribution is presented by a sequence $(p_1, ..., p_N)$ of non-negative numbers (which sum-up to one) such that p_i represents the probability mass that is assigned to the ith element, denoted e_i. Without loss of generality, we may assume that $\{e_i : i = 1, ..., N\} = [N] \stackrel{\text{def}}{=} \{1, ..., N\}$.

In this survey, we prefer to represent probability distributions by corresponding random variables that represent an element selected according to the distribution in question. That is, for a sequence $(p_1, ..., p_N)$ as above, we consider a random variable $X \in [N]$ such that $p_i = \Pr[X = e_i]$, and identify the random variable X with the probability distribution that assigns to e_i the probability mass $\Pr[X = e_i]$.

The statistical difference (or variation distance) between the distributions (or the random variables) X and Y is defined as

$$\Delta(X, Y) \stackrel{\text{def}}{=} \frac{1}{2} \cdot \sum_e |\Pr[X = e] - \Pr[Y = e]| = \max_S \{\Pr[X \in S] - \Pr[Y \in S]\} \quad (1)$$

We say that X and Y are δ-close if $\Delta(X, Y) \leq \delta$ and that they are δ-far if $\Delta(X, Y) \geq \delta$. Note that X and Y are identical if and only if they are 0-close, and are disjoint (or have disjoint support) if and only if they are 1-far.

The entropy of a distribution (or random variables) X is defined as

$$\mathrm{H}(X) \stackrel{\text{def}}{=} \sum_e \Pr[X = e] \cdot \log_2(1/\Pr[X = e]). \quad (2)$$

The entropy of a distribution is always non-negative and is zero if and only if the distribution is concentrated on a single element. In general, a distribution that has support size N has entropy at most $\log_2 N$.

3   Sampling-Based Algorithms

In this section we consider algorithms that approximate quantities related to distributions solely on the basis of samples of the relevant distributions. We refer to such algorithms as sampling-based algorithms, and consider such algorithms for approximating the distance between pairs of distributions and approximating the entropy of a distribution. We denote by N an upper bound on the size of the support of these distributions, and focus on algorithms that obtain $o(N)$ samples.

We review the known results regarding the relationship between the number of samples and the quality of the approximation. In other words, we consider the sample complexity of these approximation problems.

3.1   Approximating the Distance between Distributions

The study of sampling-based algorithms for approximating the statistical distance between distributions was initiated by Batu et. al. [6]. They show that $\Theta(N^{1/2})$ samples are necessary and sufficient in order to distinguish a pair of identical distributions from a pair of disjoint distributions (i.e., to distinguish the case that the two distributions are 0-close from the case that they are 1-far), where N is an upper bound on the support of the distribution. Regarding the more general problem of distinguishing pairs of identical distributions from pairs of distributions that are δ-far, Batu et. al. [6] showed that $\widetilde{O}(N^{2/3}\delta^{-4})$ samples suffice, and claimed that $\Omega(N^{2/3})$ samples are necessary. The latter claim was proved by P. Valiant [25]. Regarding the even more general problem of approximating the statistical distance between distributions, it was shown by P. Valiant [25] that $N^{1-o(1)}$ samples are required. That is, for every fixed $0 < \delta_1 < \delta_2 < 1$, it is the case that $N^{1-o(1)}$ samples are required in order to distinguish distribution-pairs that are δ_1-close from distribution-pairs that are δ_2-far apart.

Our conclusion is that in order to obtain any meaningful information regarding the distance between two distributions (in this model), one must obtain $\Omega(N^{1/2})$ samples. Furtherthmore, while $O(N^{2/3})$ samples suffice for distinguishing identical distribution-pairs from distribution-pairs that are far apart (say 0.1-far), in the general case $N^{1-o(1)}$ samples are required in order to approximate (up to any constant additive term) the statistical distance between two distributions (of support size N).

3.2   Approximating the Entropy of a Distribution

Batu et. al. [4] considered the problem of approximating the entropy of a distribution based on samples from it; that is, they considered sampling-based algorithms for this task. They presented an algorithm that, for any $\gamma > 1$, using $\widetilde{O}(N^{1/\gamma})$ samples of a distribution that has entropy $\Omega(\gamma)$ provides a γ-factor approximation of its entropy. We comment that some lower-bound on the en-

tropy is necessary for obtaining any approximation-factor based on samples.[1] On
the other hand, also in the case that the entropy is lower-bounded (as in Footnote 1 or even more), a constant factor approximation of the entropy requires
$N^{\Omega(1)}$ samples (i.e., a γ-factor approximation requires $\Omega(N^{(1/\gamma)-o(1)})$ samples;
see [25]).

Our conclusion is that, except in pathological cases (of distributions having
very small entropy), the sample complexity of obtaining a γ-factor approximation
of the entropy of a distribution is $N^{(1/\gamma)\pm o(1)}$, where N is an upper bound on
the support of the distribution.

Additive error approximation. The foregoing discussion refers to multiplicative
error approximation. Recent work by G. Valiant and P. Valiant [23,24] refers to
additive error approximations and shows that $\Theta(n/\log n)$ samples are necessary
and sufficient in such a case.

3.3   Additional Comments

A general framework for analyzing the sample complexity of various computational problems regarding distributions was recently provided by P. Valiant [25].
Indeed, some of the aforementioned lower-bounds are derived using this framework. Furthermore, this framework may be applied to other natural measures of
distance between distributions.

Some of the aforementioned results can be cast naturally within the formalism
of property testing (cf. [20,12,9]). For example, one may consider the property
of two distributions being identical, and the task of accepting pairs having the
property and rejecting pairs that are far from having the property according to
a natural distance measure (cf. [9]).

Related work. Batu *et. al.* [5] have considered the task of approximating the distance between a fixed distribution and a second distribution for which one only
obtains samples.[2] They present an algorithm that, for a parameter δ, determines
whether the two distributions are $\mu(N)\cdot\delta^3$-close or δ-far based on $\widetilde{O}(N^{1/2}\delta^{-O(1)})$
samples, where $\mu(N) = \widetilde{O}(1/\sqrt{N})$. This matches a lower bound of $\Omega(\sqrt{N})$ samples requires to distinguish the case that the distribution is uniform over $[N]$
from the case that it is (say) 0.1-far from being uniform. Batu *et. al.* [4] considered the problem of approximating the entropy of a distribution also in a model
in which the algorithm has access to an "evaluation oracle" instead or in addition to the samples, where the evaluation oracle is defined to answer the query x
with the probability mass assigned to x.

4   Algorithms That Are Given a Sampling Device

In this section we consider algorithms that are given a succinct description of
the distributions in question. That is, the algorithm is given a "sampling device"

---

[1] Consider, for example, the family of distributions (parameterized by $\epsilon > 0$) having
support size 2, assigning probability ϵ to one element and $1 - \epsilon$ to the other.

[2] Alternatively, the first distribution may be given explicitly (as input to the algorithm), which in this case has running time linear in N.

(in the form of a circuit) and is supposed to approximate a quantity that refers to the distribution defined by this sampling device. A sampling device is actually an algorithm, and the distribution defined by it is the output distribution of the device when fed with a random input of adequate length. For concreteness, for a feasibility parameter n, we consider $\mathrm{poly}(n)$-size circuits that map $\mathrm{poly}(n)$-bit long inputs to n-bit long outputs. Note that such circuits define a distribution over $\{0,1\}^n$, which may contain $N = 2^n$ elements. In other words, a distribution over $\{0,1\}^n$ is represented by a corresponding ($\mathrm{poly}(n)$-size) sampling device (or circuit), which typically means that we use a succinct representation of the distribution.

We consider algorithms that are given such a representation (i.e., a circuit) as input, and need to approximate some quantities of the represented distribution. Indeed, one thing that such an algorithm can do is evaluate the circuit on inputs of its choice, and in particular on uniformly selected inputs. Thus, the algorithm can certainly produce samples of the distribution, where these samples are indeed of the type used in Section 3. However, the algorithm is not confined to operating in that way, and it may try to "reverse engineer" the circuit in order to learn more about the distribution (than by merely observing random samples generated according to the distribution). Needless to say, we don't really believe that "reverse engineering" can help to answer the computational problems considered here, still we cannot rule out this possibility.

We stress that unlike in Section 3, the algorithm gets full information of the distribution. That is, from an information theoretic point of view, the sampling device (or circuit) determines the distribution, and thus determines its entropy and its distance from another distribution. The question is how much time is required in order to compute these quantities from the information that fully-determines them. In the rest of this section we associate the sampling circuits with the distributions generated by them. That is, we associate the circuit C with the distribution it outputs when fed with a uniformly selected input.

We study the complexity of approximation problems by defining corresponding promise problems (cf. [7]), where the latter are pairs of disjoint sets (cf. [10]). A promise problem (A, B) consists of distinguishing between inputs in A and inputs in B, where inputs out of $A \cup B$ are ignored (or one is "promised" that the input is in $A \cup B$).

We briefly recall the standard definitions of reductions, when applied to promise problems. The promise problem (A_1, B_1) is Karp-reducible to (A_2, B_2) if there exists a polynomial-time computible function f such that if $x \in A_1$ (resp., $x \in B_1$) then $f(x) \in A_2$ (resp., $f(x) \in B_2$). More generally, (A_1, B_1) is Cook-reducible (or just reducible) to (A_2, B_2) if there exists a polynomial-time oracle machine M that on input $x \in A_1$ (resp., $x \in B_1$) and oracle access to (A_2, B_2), outputs 1 (resp., 0), where query q to the oracle (A_2, B_2) is answered arbitrarily in case $q \notin A_2 \cup B_2$. Two problems are said to be computationally equivalent (resp., computationally equivalent under Karp-reductions) if each is Cook-reducible (resp., Karp-reducible) to the other.

4.1 Approximating the Distance between Distributions

We consider promise problems that take as input a pair of circuits and refer to the statistical difference between the two corresponding distributions
(generated by the two circuits). For (threshold) functions $c, f : \mathsf{N} \to [0, 1]$,
where $c \leq f$, the promise problem $\mathsf{GapSD}^{c,f} = (\mathtt{Close}^c, \mathtt{Far}^f)$ is defined such
that $(C_1, C_2) \in \mathtt{Close}^c$ if $\Delta(C_1, C_2) \leq c(|C_1| + |C_2|)$ and $(C_1, C_2) \in \mathtt{Far}^f$
if $\Delta(C_1, C_2) > f(|C_1| + |C_2|)$. In particular, we focus on promise problem
$\mathsf{GapSD} \overset{\text{def}}{=} \mathsf{GapSD}^{\frac{1}{3}, \frac{2}{3}}$. Interestingly, the complexity of this gap problem, which captures a moderately good approximation requirement, is computationally equiv-
alent to a very crude approximation requirement. That is, the former problem
is Karp-reducible to the latter:

Theorem 1 ([21], see proof sketch in Section 5.1): *There exists a Karp-
reduction of* $\mathsf{GapSD}^{\frac{1}{3}, \frac{2}{3}}$ *to* $\mathsf{GapSD}^{\epsilon, 1-\epsilon}$, *where* $\epsilon(n) = 2^{-n}$. *More generally, for every
polynomial-time computible* $c, f : \mathsf{N} \to [0, 1]$ *such that* $c(n) < f(n)^2 - (1/\text{poly}(n))$
it holds that $\mathsf{GapSD}^{c,f}$ *is Karp-reducible to* $\mathsf{GapSD}^{\epsilon, 1-\epsilon}$.

Using a trivial reduction in the other direction, we conclude that *for every* $c, f :$
$\mathsf{N} \to [0, 1]$ *such that* $c(n) \geq 2^{-n}$, $c(n) < f(n)^2 - (1/\text{poly}(n))$ *and* $f(n) \geq 1 - 2^{-n}$,
the problems $\mathsf{GapSD}^{c,f}$ *and* $\mathsf{GapSD} = \mathsf{GapSD}^{\frac{1}{3}, \frac{2}{3}}$ *are computationally equivalent*
(under Karp reductions). This equivalence is useful in determining the complex-
ity of GapSD (as well as all these $\mathsf{GapSD}^{c,f}$'s). Sahai and Vadhan [21] showed that
any promise problem having a statistical zero-knowledge proof system is Karp-
reducible to $\mathsf{GapSD}^{\frac{1}{2p^2}, \frac{1}{p}}$, for some polynomial p, and that $\mathsf{GapSD}^{\epsilon, 1-\epsilon}$ (where
$\epsilon(n) = 2^{-n}$) has a statistical zero-knowledge proof system. Denoting the class of
promise problem having statistical zero-knowledge proof systems by $\mathcal{SZK}$, we
have:

Theorem 2 [21]: *The promise problem* GapSD *is* $\mathcal{SZK}$-*complete* (under Karp-
reductions).

Recall that $\mathcal{SZK}$ contains some promise problems (e.g., one equivalent to Discrete Logarithm Problems) that are widely believed not to be in $\mathcal{BPP}$ (cf. [13]).
On the other hand, $\mathcal{SZK} \subseteq \mathcal{AM} \cap \text{co}\mathcal{AM}$ (cf. [11,1]), which in turn is quite low
in the Polynomial-Time Hierarchy.
    We comment that $\mathsf{GapSD} = (\mathtt{Close}, \mathtt{Far})$ *is Karp-reducible to its complement*
$(\mathtt{Far}, \mathtt{Close})$ [21]; that is, there is a Karp-reduction that maps pairs (C_1, C_2) to
pairs (C_1', C_2') such that if $\Delta(C_1, C_2) \leq 1/3$ then $\Delta(C_1', C_2') > 2/3$ whereas if
$\Delta(C_1, C_2) > 2/3$ then $\Delta(C_1', C_2') \leq 1/3$.

4.2 Approximating the Entropy of a Distribution

We consider two computational problems related to approximating the entropy
of a distribution. The first problem is captured by promise problems that take as
input a circuit and a value and refers to the relation between the entropy of (the

distribution generated by) the circuit and the given value. For a (slackness) function $s : \mathsf{N} \to \mathsf{R}$, where $s > 0$, the promise problem $\mathtt{GapEnt}^s = (\mathtt{Smaller}^s, \mathtt{Larger})$ is defined such that $(C, v) \in \mathtt{Smaller}^s$ if $\mathrm{H}(C) \leq v - s(|C|)$ and $(C, v) \in \mathtt{Larger}$ if $\mathrm{H}(C) \geq v$. In particular, we focus on promise problem $\mathtt{GapEnt} \overset{\text{def}}{=} \mathtt{GapEnt}^1$ (which refers to approximating the entropy up to an additive error of 1). It is easy to see that, for every polynomial p and for every $\epsilon > 0$ and $\ell(n) = n^{1-\epsilon}(n)$, the problems $\mathtt{GapEnt}^{1/p}$, $\mathtt{GapEnt}^1$ and $\mathtt{GapEnt}^\ell$ are computational equivalent (under Karp reductions).[3]

We also consider promise problems that take as input a pair of circuits and refer to the relation between the entropies of the corresponding distributions (generated by the two circuits). For a (slackness) function $s : \mathsf{N} \to \mathsf{R}$, where $s > 0$, the promise problem $\mathtt{GapCmprEnt}^s = (\mathtt{Smaller}^s, \mathtt{Larger}^s)$ is defined such that $(C_1, C_2) \in \mathtt{Smaller}^s$ if $\mathrm{H}(C_1) \leq \mathrm{H}(C_2) - s(|C_1| + |C_2|)$ and $(C_1, C_2) \in \mathtt{Larger}^s$ if $\mathrm{H}(C_1) \geq \mathrm{H}(C_2) + s(|C_1| + |C_2|)$. In particular, we focus on promise problem $\mathtt{GapCmprEnt} \overset{\text{def}}{=} \mathtt{GapCmprEnt}^1$, and note that it is computationally equivalent (under Karp reductions) to $\mathtt{GapCmprEnt}^{1/p}$ and $\mathtt{GapCmprEnt}^\ell$ (where p and ℓ are as above). Two easy observations follow:

Observation 1: *The problems* $\mathtt{GapEnt}$ *and* $\mathtt{GapCmprEnt}$ *are computationally equivalent* (under Cook reductions). Specifically, $\mathtt{GapEnt}$ is Karp-reducible to $\mathtt{GapCmprEnt}$, whereas $\mathtt{GapCmprEnt}$ is Cook-reducible to $\mathtt{GapEnt}$.
For example, one may use a Karp-reduction that maps an instance (C, v) of $\mathtt{GapEnt}$ to the intance $(C, C_{v-0.5})$ of $\mathtt{GapCmprEnt}^{1/3}$ such that $C_{v-0.5}$ is a standard circuit that generates some distribution of entropy (approximately) $v - 0.5$. For the other direction, consider an oracle machine that decides intances of $\mathtt{GapCmprEnt}$ by using queries to $\mathtt{GapEnt}^{1/3}$ in order to determine the entropy of each of the two input distributions (up to an additive error of $1/3$).

Observation 2: *The problem* $\mathtt{GapCmprEnt} = (\mathtt{Smaller}, \mathtt{Larger})$ *is Karp-reducible to its complement* $(\mathtt{Larger}, \mathtt{Smaller})$; e.g., by the reduction that maps
(C_1, C_2) to (C_2, C_1).

It is not know whether or not $\mathtt{GapCmprEnt}$ is Karp-reducible to $\mathtt{GapEnt}$ and whether or not $\mathtt{GapEnt}$ is Karp-reducible to its complement. In fact, both questions are equivalent (cf. [16]), and we conjecture that the answer (to both of them) is negative. It turns out that all these computational problems (regarding entropy) are computationally equivalent to the computational problems regarding statistical distance:

Theorem 3 ([17], see proof sketch in Section 5.3): *The promise problems* $\mathtt{GapCmprEnt}$ *and* $\mathtt{GapSD}$ *are computationally equivalent under Karp reductions.*

Combining Theorem 3 and Observation 2, it follows that $\mathtt{GapSD} = (\mathtt{Close}, \mathtt{Far})$ *is Karp-reducible to its complement* $(\mathtt{Far}, \mathtt{Close})$. We comment that this result

---

[3] The tighter (additive) approximation is reduced to the looser one by combining sufficiently many copies of the circuit.

(which was already stated at the end of Subsection 4.1) was originally proved in [21] without using the equivalence of GapSD and GapCmprEnt (i.e., without using Theorem 3).

4.3   Additional Comments

We comment that the promise problems GapSD, GapEnt and GapCmprEnt were originally introduced as tools in the study of statistical zero-knowledge.[4] Consequently, the original presentations (cf. [21,17,16]) focus on the derivation and presentation of results regarding statistical zero-knowledge, and the relation between the promise problems themselves is sometimes only implicit (and is typically not at the main focus). In fact, redeeming this state of affairs has been our initial motivation for writing the current survey.

The bottom-line of the foregoing results is that many of the approximation versions of the two problems (i.e., approximating the distance between distributions and approximating their entropy) are computationally-equivalent. The exceptional versions that are not known to be equivalent to the other versions refer to too small gaps (which may yield even harder versions). Whereas in the case of approximating the entropy the definition of "too small gaps" is a natural one, it is somewhat artificial in the case of $GapSD^{c,f}$ where we require $c < f^2$. An interesting open problem is to determine the complexity of $GapSD^{c,f}$ in the case that $c > f^2$ (but $c < f$, of course)[5]; that is, is this problem computationally equivalent to GapSD or is it strictly harder?

An alternative perspective on the current section is that it concerns only probability distributions that have a succinct representation, where such a representation is one allowing to efficiently obtain samples from the distribution. Specifically, for a feasibility parameter n, we consider probability distributions over $\{0,1\}^n$. The support of such a distribution may contain 2^n elements, while we consider algorithms operating in poly(n)-time. Thus, such algorithms cannot read an explicit representation of the distribution (in the form of a sequence of length 2^n), and hence the distribution is given to it in a succinct representation. Specifically, we have considered algorithms that are given a sampling device, which is a poly(n)-size circuit that when feed with a random input output a sample that is distributed according to the distribution. We have considered the complexity of estimating various quantities of distributions given by such a succinct representation.

5   Proof Sketches for the Three Theorems

In this section we outline the main ideas used in the proofs of the three theorems stated in Section 4. Theorem 2 is the only one that refers to statistical zero-knowledge and its proof is the only one that assumes any familiarity with zero-

---

[4] For more details regarding statistical zero-knowledge see either [21,15,17,16] or [22].

[5] The above formulation refers to constant c and f. For $c, f : \mathbb{N} \to [0, 1]$, we have to require that $c(n) < f(n) - (1/p(n))$ for some polynomial p.

knowledge. The other two proofs are based merely on elementary results from probability theory and probabilistic analysis.

As in Section 4, we associate the sampling circuits with the distributions generated by them. That is, we associate the circuit C with the distribution it outputs when fed with a uniformly selected input.

5.1 Proof Sketch for Theorem 1

Theorem 1 was proven by Sahai and Vadhan [21], and here we provide an outline of their proof. Recall that the theorem claims a Karp-reduction of $\mathsf{GapSD}^{\frac{1}{3},\frac{2}{3}}$ (or any adequate $\mathsf{GapSD}^{c,f}$) to $\mathsf{GapSD}^{\epsilon,1-\epsilon}$, where $\epsilon(n) = 2^{-n}$. This reduction (called the *Polarization Lemma* in [21]) has the interesting effect of "polarizing the situation": pairs of distributions that are somewhat close (e.g., are at most at distance $1/3$ apart) are mapped to pairs of almost identical distributions (i.e., having negligible distance between them), whereas pairs of distributions that are somewhat far apart (e.g., at distance at least $2/3$) are mapped to pairs of distributions that are very different (e.g., have distance negligiblly close to 1). The "polarizing" reduction is obtained by composing three Karp-reductions, which in turn are of two types. These two types of Karp-reductions (among $\mathsf{GapSD}^{c,f}$ problems) are described next, starting with the simpler one.

The Direct Product reduction: This reduction increases both bounds in the definition of $\mathsf{GapSD}^{c,f}$ (but not in a tight manner). For any (polynomial) t, we reduce $\mathsf{GapSD}^{c,f}$ to $\mathsf{GapSD}^{t\cdot c,1-2\exp(-t\cdot f^2/2)}$ by constructing circuits that generate t samples of each of the corresponding input distributions. That is, we map the circuit pair (C_1, C_2) to (C_1', C_2'), where $C_i'(r_1,...,r_t) \stackrel{\text{def}}{=} (C_i(r_1),...,C_i(r_t))$. Clearly, the statistical distance between the distributions grows by at most a factor of t. On the other hand, it can be shown that if two distributions are at distance δ then the statistical difference between their t-products is at least $1 - 2\exp(-t\cdot\delta^2/2)$. (Indeed, it is not true that the statistical difference between the t-products is exactly $t\cdot\delta$, the latter is merely an upper bound on the former.)[6]

The XOR reduction: This reduction decreases both bounds (in a tight manner). For any (polynomial) t, we reduce $\mathsf{GapSD}^{c,f}$ to GapSD^{c^t,f^t} by mapping the circuit pair (C_0, C_1) to (C_0', C_1'), where

$$C_i'(b_1,...,b_{t-1},r_1,...,r_{t-1},r_t) \stackrel{\text{def}}{=} \left(C_{b_1}(r_1),...,C_{b_{t-1}}(r_{t-1}), C_{i+\sum_{j=1}^{t-1} b_j \bmod 2}(r_t)\right).$$

---

[6] The lower bound of $1 - 2\exp(-t\cdot\delta^2/2)$ can be proved by referring to the second definition in Eq. (1). Specifically, for an adequate set S, it holds that $p \stackrel{\text{def}}{=} \Pr_r[C_1(r) \in S] = \Pr_r[C_2(r) \in S] - \delta$. Thus, C_1' (resp., C_2') is expected to have $t\cdot p$ (resp., $t\cdot(p+\delta)$) elements in S. By applying a Chernoff Bound, we note that with probability at least $1 - \exp(-t\cdot\delta^2/2)$, the output of C_1' (resp., C_2') will have less than $t\cdot(p+\frac{\delta}{2})$ (resp., more than $t\cdot(p+\frac{\delta}{2})$) elements in S. This yields a set S' that demonstrates the claimed lower bound on the statistical difference between C_1' and C_2'.

That is, the two output circuits (i.e., the C_i''s) select samples from the two input distributions (respresented by the C_i's), and differ only in the parity of the number samples taken from the (say) first input distribution. Specifically, C_0' (resp., C_1') takes an even (resp., odd) number of samples from C_1. It can be shown that if two input distributions are at distance δ then the statistical difference between the constructed (output) distributions is exactly δ^t. (Intuitively, a single sample drawn for one of the two input distributions corresponds to a "weak" encryption of a bit, whereas a sample drawn from one of the output circuits corresponds to encrypting a bit by applying "weak" encryptions to a random sequence of bits that have the desired parity. The "weakness" of the resulting encryption decays exponentially with t; cf. [26].)[7]

We now turn to the actual reduction of $\mathsf{GapSD}^{\frac{1}{3},\frac{2}{3}}$ (or any adequate $\mathsf{GapSD}^{c,f}$) to $\mathsf{GapSD}^{\epsilon,1-\epsilon}$, where $\epsilon(n) = 2^{-n}$. This reduction is composed of the following three reductions:

1. A Karp-reduction of $\mathsf{GapSD}^{\frac{1}{3},\frac{2}{3}}$ (or any $\mathsf{GapSD}^{c,f}$ such that $c(n) < f(n)^2 - \frac{1}{\mathrm{poly}(n)}$) to some $\mathsf{GapSD}^{c',f'}$ such that $f'(n) > \sqrt{8n \cdot c'(n)}$.

   Specifically, for an adequate parameter t, we use the XOR reduction and get $c' = c^t$ and $f' = f^t$, which satisfies the desired condition (regarding c' and f') provided that $c < f^2$ (or actually $c(n) < (8n)^{-t/2} \cdot f(n)^2$). In particular, for $c = 1/3$ and $f = 2/3$, we set $t = O(\log n)$ and reduce $\mathsf{GapSD}^{c,f}$ to $\mathsf{GapSD}^{c',f'}$, where $c'(n) \stackrel{\mathrm{def}}{=} c^t = 1/\mathrm{poly}(n)$ and $f'(n) \stackrel{\mathrm{def}}{=} f^t = (f^2/c)^{t/2} \cdot c^{t/2} > \sqrt{8n \cdot c'(n)}$. In general, we set $t = \mathrm{poly}(n)$ such that $(f(n)^2/c(n))^{t/2} \geq 8n$, which is possible because $\frac{f(n)^2}{c(n)} > 1 + \frac{1}{p(n)}$ for some positive polynomial p.

2. A Karp-reduction of a $\mathsf{GapSD}^{c',f'}$ (with c' and f' as obtained in Step 1) to $\mathsf{GapSD}^{c'',f''}$, where $c''(n) = 1/4$ and $f''(n) \geq 1 - 2\exp(-n)$.

   Specifically, for an adequate parameter t (i.e., $t = 1/4c'(n)$), we use the Direct Product reduction and get $c''(n) \stackrel{\mathrm{def}}{=} t \cdot c'(n) = 1/4$ and $f''(n) \stackrel{\mathrm{def}}{=} 1 - 2\exp(-t \cdot f'(n)^2/2)$. Using the hypothesis $f'(n) \geq \sqrt{8n \cdot c'(n)}$, it follows that $f''(n) = 1 - 2\exp(-f'(n)^2/8c'(n)) \geq 1 - 2\exp(-n)$.

3. A Karp-reduction of a $\mathsf{GapSD}^{c'',f''}$ (with c'' and f'' as obtained in Step 2) to $\mathsf{GapSD}^{\epsilon,1-\epsilon}$, where $\epsilon(n) = 2^{-n}$.

   Specifically, we apply the XOR reduction again, but this time with $t = n/2$, and use $(1/4)^t = 2^{-n} = \epsilon(n)$ and $(1 - 2\exp(-n))^t > 1 - 2^{-n} = 1 - \epsilon(n)$.

Combining the above three reductions, we obtain a Karp-reduction of $\mathsf{GapSD}^{\frac{1}{3},\frac{2}{3}}$ (or any $\mathsf{GapSD}^{c,f}$ such that $c(n) < f(n)^2 - \frac{1}{\mathrm{poly}(n)}$) to $\mathsf{GapSD}^{\epsilon,1-\epsilon}$, where $\epsilon(n) = 2^{-n}$.

---

[7] Alternatively, consider the following problem. For pairs of random variables, (X_0, X_1) and (Y_0, Y_1), we define a new pair of random variables, (Z_0, Z_1), such that $Z_i = (X_b, Y_{i \oplus b})$, where $b \in \{0, 1\}$ is uniformly distributed. Using the first definition in Eq. (1) and expanding the expression for $\Delta(Z_0, Z_1)$, one can show that $\Delta(Z_0, Z_1) = \Delta(X_0, X_1) \cdot \Delta(Y_0, Y_1)$. The general claim (stated above) follows by induction on t.

On the use of the condition $c < f^2$ *in the current reduction:* Note that in Step 2 we have assumed that $f'(n) \geq \sqrt{8n \cdot c'}$, where (by Step 1) $f' = f^t$ and $c' = c^t$. It follows that we must have $f(n)^t \geq (8n)^{t/2} \cdot (\sqrt{c(n)})^t$, and in particular $f(n)^2 > c(n)$. As discussed in Section 4.3, it is an open problem whether or not there exists an alternative reduction that uses a more relaxed condition (regarding c and f).

5.2 Proof Sketch for Theorem 2

Theorem 2 was also proven by Sahai and Vadhan [21], and here we sketch the ideas underlying their proof. The proof consists of two parts: (1) showing that GapSD has a statistical zero-knowledge proof system, and (2) showing that any problem in $\mathcal{SZK}$ is Karp-reducible to GapSD. We try to present the proof ideas while assuming only a superficial familiarity with the notion of statistical zero-knowledge proof systems. A reader that does not feel comfortable with this assumption is invited to skip the current subsection.

The problem GapSD *has a statistical zero-knowledge proof system:* Using Theorem 1, it suffices to show such a proof system for $\mathrm{GapSD}^{\epsilon, 1-\epsilon}$, where $\epsilon(n) = 2^{-n}$. Actually, we present such a proof system for the complement problem (i.e., $(\mathrm{Far}^{1-\epsilon}, \mathrm{Close}^{\epsilon})$), and rely on the (highly non-trivial) fact that GapSD is reducible to its complement.[8] Employing the same idea as in [18,14], the verifier selects one of the input distributions at random and presents the prover with a random sample generated according to this distribution. The verifier accepts if and only if the prover correctly identifies the distribution from which the sample was taken. Observe that if the input distributions are far apart then the prover can answer correctly with very high probability. On the other hand, if the input distributions are very close then the prover cannot guess the correct answer with probability significantly larger than $1/2$. This establishes that the protocol is an interactive proof (and thus that GapSD is in $\mathrm{co}\mathcal{AM}$). It can be shown that this protocol is actually statistical zero-knowledge, intuitively because the verifier learns nothing from the prover's correct answer which is a priori known to to the verifier (in case the two distributions are far apart).

Any problem in $\mathcal{SZK}$ *is Karp-reducible to* GapSD: We rely on Okamoto's Theorem by which any problem in $\mathcal{SZK}$ has a *public-coin* statistical zero-knowledge proof system. (We comment that an alternative proof of that theorem has subsequently appeared in [17].) We consider an arbitrary (*public-coin*) statistical zero-knowledge proof system. Following Fortnow [11], we observe a discrepency between the behavior of the simulator on YES-instances versus NO-instances:

– In case the input is a YES-instance, the simulator outputs transcripts that are very similar to those in the real interaction. In particular, these trascripts

---

[8] As mentioned in Section 4, this fact follows by combining Theorem 3 with Observation 2. An alternative proof of the fact that GapSD is reducible to its complement was given in [21]. (Actually this alternative proof was discovered before Theorem 3.)

are accepting and the verifier's behavior in them is as in a real interaction. In particular, resorting to the public-coin condition, this means that the verifier's messages in the simulation are (almost) uniformly distributed independently of prior messages.

- In case the input is a NO-instance, the simulator must output either rejecting transcripts or transcripts in which the verifier's behavior is significantly different from the verifier's behavior in a real interaction. In particular, the only way the simulator can produce accepting transcripts is by producing transcripts in which the verifier's messages are not "random enough" (i.e., they depend, in a noticeable way, on previous messages).

Thus assuming, without loss of generality, that the simulator only produces accepting transcripts, we consider two types of distributions. The first type of the distributions is obtained by truncating a random simulator-produced transcript at a random "location" (after some verifier message), whereas the second type is obtained by doing the same while replacing the last verifier message by a random one. Note that both distributions can be implemented by polynomial-size circuits that depend on the input to the proof system being analyzed (and that these two circuits can be constructed in polynomial-time given the said input). The key observation is that if the input is a YES-instance then the two corresponding distributions will be very close, whereas if the input is a NO-instance then there will be a noticeable distance between the two corresponding distributions. Thus, we reduced any problem having a (public-coin) statistical zero-knowledge proof system to $\mathsf{GapSD}^{\mu,\nu}$, where μ is a negligible function and $\nu(n)$ is a noticeable function.[9] The proof is completed by using Theorem 1 (while noting that $\mu(n) < \nu(n)^2 - (1/\mathrm{poly}(n))$).

5.3   Proof Sketch for Theorem 3

Theorem 3 was proven by Goldreich and Vadhan [21], by showing that GapCmprEnt is $\mathcal{SZK}$-complete (under Karp-reductions) and invoking Theorem 2 (which shows the same for GapSD). Here we follow a more direct proof, which has appeared in Vadhan's PhD Thesis [22]. The proof consists of two parts: (1) showing that GapSD is Karp-reducible to GapCmprEnt, and (2) showing that GapCmprEnt is Karp-reducible to GapSD.

Reducing GapSD *to* GapCmprEnt: Using Theorem 1, it suffices to reduce $\mathsf{GapSD}^{\epsilon,1-\epsilon}$ to GapCmprEnt, for $\epsilon(n) = 2^{-n}$. Actually, we will reduce $\mathsf{GapSD}^{\epsilon,1-\epsilon}$ to a related problem, denoted GapCmprEnt$'$, that refers to distinguishing pairs of distributions that have approximately the same entropy from pairs in which the first distribution has (say half a unit of) more entropy.[10] We reduce $\mathsf{GapSD}^{\epsilon,1-\epsilon}$ to

---

[9] A function $\mu : \mathsf{N} \to [0,1]$ is called negligible if $\mu(n) < 1/p(n)$ for every positive polynomial p and all sufficiently large n. A function $\nu : \mathsf{N} \to [0,1]$ is called noticeable if $\nu(n) > 1/p(n)$ for some positive polynomial p and all sufficiently large n.

[10] Indeed, the reduction from GapCmprEnt$'$ to GapCmprEnt is easy: we just increase the gap in entropy (by repeated sampling), and move the gap location (by augmenting the second distribution with a few random bits).

GapCmprEnt$'$ by mapping the circuit pair (C_0, C_1) to (C_1', C_2'), where $C_1'(r, s, b) \stackrel{\text{def}}{=} (C_s(r), b)$ and $C_2'(r, s, b) \stackrel{\text{def}}{=} (C_s(r), s)$. That is, C_2 outputs a sample of one of the input distributions along with the "selection bit" s used to determine the input distribution being sampled, whereas C_1 outputs such a sample along with an independently distributed random bit (denoted b). Clearly, the entropy of C_1' is always $v + 1$, where $v \stackrel{\text{def}}{=} \frac{H(C_0) + H(C_1)}{2}$. Now, if the two input distributions are very far apart then the selection bit s will be determined by the sample and so the entropy of C_2' will be approximately v, which is significantly smaller than $H(C_1')$. On the other hand, if the two input distributions are very close then (even conditioned on the sampled selected) the selection bit s will be almost random and so $H(C_2') \approx v + 1$, which is approximately the same as $H(C_1')$.

A warm-up: reducing GapEnt *to* GapSD. We first reduce GapEnt to GapEnt$^\ell$, where $\ell(n) = \sqrt{n}$, by using sufficiently many samples (of the input distribution): for example, we may map (C, v) to (C', v'), where $C'(r_1, ..., r_n) = (C(r_1), ..., C(r_n))$ and $v' = n \cdot v$. Next, we assume that the input distribution is "flat", where a distribution is called flat if it is uniform over some set (i.e., if all elements in its support are assigned the same probability mass). We note that by taking sufficiently many samples, we can transform each distribution to one that is "almost flat" (in a sense that is sufficient for the rest of the proof), while maintaining its "relative entropy" (i.e., the average entropy per output bit). Now, suppose that we are given a pair (C, v) such that $C : \{0, 1\}^m \to \{0, 1\}^n$ is flat and $|H(C) - v| \geq \sqrt{n}$, and we are interested in the relation between $H(C)$ and v. Suppose that h is a random hash function[11] mapping m-bit strings to $(m-v-\log_2^2 n)$-bit long string. Now, consider the distributions $(h, C(r), h(r))$ and $(h, C(r), h(r'))$, where $r, r' \in \{0, 1\}^m$ and h are uniformly selected. By the property of the hashing function, the third part of the distribution $(h, C(r), h(r'))$ is almost uniform over $\{0, 1\}^{m-v-\log_2^2 n}$, even when conditioning on the first parts (specifically on h). On the other hand, the third part of the distribution $(h, C(r), h(r))$ is distributed as $h(r)$ conditioned on $C(r)$ (i.e., $h(r)|C(r)$). We note that $H(r|C(r)) = m - H(C)$, and that the distribution $r|C(r)$ is flat. Furthermore, if $H(C) \leq v$ then $H(r|C(r)) \geq m - v$ and the distribution $h(r)|C(r)$ is almost uniform over $\{0, 1\}^{m-v-\log_2^2 n}$, whereas if $H(C) \geq v + 2\log_2^2 n$ then $H(r|C(r)) \leq m - v - 2\log_2^2 n$ and the distribution $h(r)|C(r)$ is very far from being uniform over $\{0, 1\}^{m-v-\log_2^2 n}$. Now, recall that $|H(C) - v| \geq \sqrt{n}$, and observe that if $H(C) < v$ then the distribution $(h, C(r), h(r))$ is almost identical to the distribution $(h, C(r), h(r'))$, whereas if $H(C) > v$ then $(h, C(r), h(r))$ is very far from $(h, C(r), h(r'))$. Thus, we have reduced GapEnt to GapSD.

Reducing GapCmprEnt *to* GapSD: As in the warm-up, we first reduce GapCmprEnt to GapCmprEnt$^\ell$, where $\ell(n) = \sqrt{n}$, such that each of the two distributions is almost flat. Suppose that we are given a pair of circuits (C_1, C_2) such that both are

---

[11] Formally speaking, we mean a uniformly selected function in a collection of universal2 hashing functions [8]. For example, we may select h uniformly among all affine mappings of $GF(2^m)$ to $GF(2^k)$, for $k = m - v - \log_2^2 n$.

(almost) flat and $|H(C_1) - H(C_2)| \geq \sqrt{n}$, and we are interested in the question of which circuit (or distribution represented by it) has higher entropy. Further suppose that $C_1, C_2 : \{0,1\}^m \to \{0,1\}^n$. Suppose that h is a random hash function mapping $(n + m)$-bit strings to $(m - \log_2^2 n)$-bit long string. Now, consider the distributions $(h, C_1(r_1), h(C_2(r_2), r_1))$ and $(h, C_1(r_1), h(0^n, r_2))$, where $r_1, r_2 \in \{0,1\}^m$ and h are uniformly selected. By the property of the hashing function, the third part of the distribution $(h, C_1(r_1), h(0^n, r_2))$ is almost uniform over $\{0,1\}^{m-\log_2^2 n}$, even when conditioning on the first parts. On the other hand, the third part of the distribution $(h, C_1(r_1), h(C_2(r_2), r_1))$ is distributed as $h(C_2(r_2), r_1)|C_1(r_1))$. We note that $u \stackrel{\text{def}}{=} H(C_2(r_2), r_1|C_1(r_1)) = H(C_2) + (m - H(C_1))$, and that the distribution $(C_2(r_2), r_1)|C_1(r_1)$ is flat. Furthermore, if $u \geq m$ then the distribution $h(C_2(r_2), r_1)|C_1(r_1))$ is almost uniform over $\{0,1\}^{m-\log_2^2 n}$, whereas if $u \leq m - 2\log_2^2 n$ then the distribution $h(C_2(r_2), r_1)|C_1(r_1))$ is very far from being uniform over $\{0,1\}^{m-\log_2^2 n}$. Now, recall that $|H(C_1) - H(C_2)| \geq \sqrt{n}$, and observe that if $H(C_2) > H(C_1)$ then $u = m + (H(C_2) - H(C_1)) > m$, whereas if $H(C_2) < H(C_1)$ then $u \leq m - \sqrt{m}$. We conclose that in the first case the distribution $(h, C_1(r_1), h(C_2(r_2), r_1))$ is almost identical to the distribution $(h, C_1(r_1), h(0^n, r_2))$, whereas in the second case $(h, C_1(r_1), h(C_2(r_2), r_1))$ is very far from $(h, C_1(r_1), h(0^n, r_2))$. Thus, we have reduced GapCmprEnt to GapSD.

6 Conclusions

In Section 4 we considered the complexity of approximating the entropy of a distribution when given the full description of a sampling device for the distribution. In contrast, the results of Section 3 can be viewed as referring to the case that we are only given "black-box" access to such a sampling device. Thus, the results surveys in these sections represent a potential gap between black-box and "non-black-box" access to sampling devices. This gap may become a real separation if $\mathcal{SZK}$ is contained in sub-exponential time (i.e., $\mathcal{SZK} \subseteq \text{Dtime}(f)$ for some $f(n) = 2^{o(n)}$). On the other hand, the hypothetical existence of "sampling obfuscators" (see [3, Def. 6.2]), which means that non-black-box access to sampling devices does not actually help, implies that $\mathcal{SZK} \neq \mathcal{BPP}$ (see [3, Prop. 6.4]).

We comment that the general study of the relation between black-box and non-black-box algorithms has received considerable attention lately. The interested reader is referred to Barak's PhD Thesis [2].

References

1. Aiello, W., Håstad, J.: Perfect Zero-Knowledge Languages can be Recognized in Two Rounds. In: 28th FOCS, pp. 439–448 (1987)
2. Barak, B.: Non-Black-Box Techniques in Cryptography. Ph.D. Thesis, Weizmann Institute of Science (January 2004)
3. Barak, B., Goldreich, O., Impagliazzo, R., Rudich, S., Sahai, A., Vadhan, S.P., Yang, K.: On the (Im)possibility of obfuscating programs. In: Kilian, J. (ed.) CRYPTO 2001. LNCS, vol. 2139, pp. 1–18. Springer, Heidelberg (2001)

4. Batu, T., Dasgupta, S., Kumar, R., Rubinfeld, R.: The Complexity of Approximating the Entropy. In: 34th STOC (2002)
5. Batu, T., Fischer, E., Fortnow, L., Kumar, R., Rubinfeld, R., White, P.: Testing random variables for independence and identity. In: 42nd FOCS (2001)
6. Batu, T., Fortnow, L., Rubinfeld, R., Smith, W.D., White, P.: Testing that distributions are close. In: 41st FOCS, pp. 259–269 (2000)
7. Bellare, M., Goldreich, O., Sudan, M.: Free Bits, PCPs and Non-Approximability – Towards Tight Results. SICOMP 27(3), 804–915 (1998)
8. Carter, L., Wegman, M.: Universal Hash Functions. JCSS 18, 143–154 (1979)
9. Ergun, F., Kannan, S., Kumar, S.R., Rubinfeld, R., Viswanathan, M.: Spot-checkers. JCSS 60(3), 717–751 (2000)
10. Even, S., Selman, A.L., Yacobi, Y.: The Complexity of Promise Problems with Applications to Public-Key Cryptography. Inform. and Control 61, 159–173 (1984)
11. Fortnow, L.: The Complexity of Perfect Zero-Knowledge. In: 19th STOC, pp. 204–209 (1987)
12. Goldreich, O., Goldwasser, S., Ron, D.: Property testing and its connection to learning and approximation. JACM, 653–750 (July 1998)
13. Goldreich, O.V., Kushilevitz, E.V.: A Perfect Zero-Knowledge Proof for a Decision Problem Equivalent to Discrete Logarithm. JofC 6(2), 97–116 (1993)
14. Goldreich, O., Micali, S., Wigderson, A.: Proofs that Yield Nothing but their Validity or All Languages in NP Have Zero-Knowledge Proof Systems. JACM 38(1), 691–729 (1991); Preliminary version in 27th FOCS (1986)
15. Goldreich, O., Sahai, A., Vadhan, S.: Honest-Verifier Statistical Zero-Knowledge equals general Statistical Zero-Knowledge. In: 30th STOC, pp. 399–408 (1998)
16. Goldreich, O., Sahai, A., Vadhan, S.P.: Can Statistical Zero-Knowledge be Made Non-Interactive? or On the Relationship of SZK and NISZK. In: Wiener, M. (ed.) CRYPTO 1999. LNCS, vol. 1666, pp. 467–484. Springer, Heidelberg (1999)
17. Goldreich, O., Vadhan, S.: Comparing Entropies in Statistical Zero-Knowledge with Applications to the Structure of SZK. In: 14th IEEE Conference on Computational Complexity, pp. 54–73 (1999)
18. Goldwasser, S., Micali, S., Rackoff, C.: The Knowledge Complexity of Interactive Proof Systems. SICOMP 18, 186–208 (1989); Preliminary version in 17th STOC (1985); Earlier versions date to (1982)
19. Okamoto, T.: On relationships between statistical zero-knowledge proofs. In: 28th STOC, pp. 649–658 (1996)
20. Rubinfeld, R., Sudan, M.: Robust Characterizations of Polynomials with Applications to Program Checking. SICOMP 25(2), 252–271 (1996); Preliminary version in 3rd SODA (1992)
21. Sahai, A., Vadhan, S.: A Complete Promise Problem for Statistical Zero-Knowledge. In: 38th FOCS, pp. 448–457 (1997)
22. Vadhan, S.: A Study of Statistical Zero-Knowledge Proofs. PhD Thesis, Department of Mathematics, MIT (1999)
23. Valiant, G., Valiant, P.: A CLT and tight lower bounds for estimating entropy. In: ECCC, TR10-179 (2010)
24. Valiant, G., Valiant, P.: Estimating the unseen: A sublinear-sample canonical estimator of distributions.In: ECCC, TR10-180 (2010)
25. Valiant, P.: Testing symmetric properties of distributions. In: ECCC, TR07-135 (2007)
26. Wyner, A.D.: The wire-tap channel. Bell System Technical Journal 54(8), 1355–1387 (1975)

Basing Non-Interactive Zero-Knowledge on (Enhanced) Trapdoor Permutations: The State of the Art

Oded Goldreich

Abstract. The purpose of this article is to correct the inaccurate account of this subject that is provided in our two-volume work *Foundation of Cryptography*. Specifically, as pointed out by Jonathan Katz, it seems that the construction of Non-Interactive Zero-Knowledge proofs for $\mathcal{NP}$ requires the existence of a doubly-enhanced collection of trapdoor permutations (to be defined below). We stress that the popular candidate collections of trapdoor permutations do satisfy this doubly-enhanced condition. In fact, any collection of trapdoor permutations that has dense and easily recognizable domain satisfies this condition.

Keywords: Non-Interactive Zero-Knowledge, Trapdoor Permutations.

This article was completed in Nov. 2008, and appeared on the author's webpage.

1 Introduction

The purpose of this article is to correct the inaccurate account of the construction of Non-Interactive Zero-knowledge proofs (NIZK) for $\mathcal{NP}$ that is provided in [G1, Sec. 4.10.2] and modified in [G2, Apdx. C.4.1]. We briefly recall the relevant facts.

In [G1, Rem. 4.10.6], a construction of NIZK for $\mathcal{NP}$ is sketched based on a collection of trapdoor permutations in which each permutation f_α has domain $\{0, 1\}^{|\alpha|}$. This description is correct, but the problem is with the unsupported claim (at the end of [G1, Rem. 4.10.6]) by which the construction can be extended to arbitrary collections of trapdoor permutations (in which the domain of the permutation f_α may be a sparse subset of $\{0, 1\}^{|\alpha|}$ and may not be easy to recognize (although it is easy to sample from)).

In [G2, Apdx. C.4.1] it was claimed that such a construction (of NIZK for $\mathcal{NP}$) can be obtained based on any *enhanced* collections of trapdoor permutations, where the enhancement is as defined in [G2, Apdx. C.1]. But again, this claim was not fully supported. Furthermore, as pointed out by Jonathan Katz, it seems that this construction requires an additional enhancement. In this article we define the resulting notion of a *doubly-enhanced* collection of trapdoor permutations, and provide full details to the claim that using such permutations one can construct NIZK for $\mathcal{NP}$. We stress that the popular candidate collections of trapdoor permutations do satisfy this doubly-enhanced condition. In fact, any collection of trapdoor permutations that has dense and easily recognizable domain satisfies

O. Goldreich et al.: Studies in Complexity and Cryptography, LNCS 6650, pp. 406–421, 2011.

this condition. More generally, if the domain-sampler S' of an enhanced collection of trapdoor permutations has a "reversed sampler" (which given α, y generates a random r such that $S'(\alpha, r) = y$), then this collection is doubly-enhanced.

On the non-technical level, we believe that this unfortunate line of events demonstrates the importance of not being tempted by hand-waving arguments and working out detailed proofs. Indeed, we believe that the source of trouble is that the basic idea is presented in [G1, Rem. 4.10.6] as a patch, and further modifications are also presented as patches (see [G2, Apdx. C.4.1]). These patches are replaced by the detailed description provided in Section 3, which is the core of the current article.

2   Background

In this section we recall the standard definition of non-interactive zero-knowledge proof systems as well as the construction of such systems based on proof systems in the *hidden-bits model*. Since proof systems for $\mathcal{NP}$ in the hidden-bits model are known to exists (unconditionally, see [G1, Sec. 4.10.2]), our focus in this article is on transforming such systems into standard NIZK systems. We stress that intractability assumptions are used in the latter transformation.

The rest of this section is essentially reproduced from [G1, Sec. 4.10.1&4.10.2], and its first subsection (i.e., Section 2.1) can be skipped by readers who are familiar with the standard definition of non-interactive zero-knowledge proof systems.

2.1   The Basic Definition

Recall that the model of non-interactive (zero-knowledge) proof systems consists of three entities: a prover, a verifier and a uniformly selected sequence of bits (which can be thought of as being selected by a trusted third party). Both verifier and prover can read the random sequence, and each can toss additional coins. The interaction consists of a single message sent from the prover to the verifier, who is then left with the decision (whether to accept or not). Here we present only the basic definition that supports the case of proving a single assertion of a-priori bounded length. Various extensions are presented in [G1, Sec. 4.10.3] and in [G2, Sec. 5.4.4.4]; we recall that the construction of such stronger NIZKs can be reduced to the construction of basic NIZKs (as defined below).

The model of non-interactive proofs seems closer in spirit to the model of NP-proofs than to general interactive proofs. In a sense, the NP-proof model is extended by allowing the prover and verifier to refer to a common random string, as well as toss coins by themselves. Otherwise, as in case of NP-proofs, the interaction is minimal (i.e., it is unidirectional (from the prover to the verifier)). Thus, in the definition below both the prover and verifier are ordinary probabilistic machines that, in addition to the common-input, also get a uniformly distributed (common) *reference-string*. We stress that, in addition to the aforementioned common input and common reference-string, both the prover and verifier may toss coins and get auxiliary inputs. However, for sake of simplicity, we present a definition for the case in which none of these machines gets an auxiliary input (yet, they may both toss additional coins). Finally, note that the verifier also gets as input the output produced by the prover.

Definition 1 (non-interactive proof system): *A pair of probabilistic machines,* (P, V), *is called a* non-interactive proof system for a language L *if* V *is polynomial-time and the following two conditions hold:*

- Completeness: *For every* $x \in L$, *it holds that*

$$\Pr\left[V(x, R, P(x, R)) = 1\right] \geq \frac{2}{3}$$

  where R is a random variable uniformly distributed in $\{0, 1\}^{\text{poly}(|x|)}$.
- Soundness: *For every* $x \notin L$ *and every algorithm* B, *it holds that*

$$\Pr\left[V(x, R, B(x, R)) = 1\right] \leq \frac{1}{3}$$

  where R is a random variable uniformly distributed in $\{0, 1\}^{\text{poly}(|x|)}$.

The uniformly chosen string R is called the common reference-string.

As usual, the error probability in both conditions can be reduced (from $\frac{1}{3}$) up to $2^{-\text{poly}(|x|)}$, by repeating the process sufficiently many times (using a sequence of many independently chosen reference-strings). In stating the soundness condition, we have deviated from the standard formulation that allows $x \notin L$ to be adversarially selected after R is fixed; the latter "adaptive" formulation of soundness is used in [G1, Sec. 4.10.3], and it is easy to transform a system satisfying the above ("non-adaptive") soundness condition into one satisfying the adaptive soundness condition (see [G1, Sec. 4.10.3]).

Every language in $\mathcal{NP}$ has a non-interactive proof system (in which no randomness is used). However, this NP-proof system is unlikely to be zero-knowledge (as defined next). The definition of zero-knowledge for the non-interactive model is simplified by the fact that, since the verifier cannot affect the prover's actions, it suffices to consider the simulatability of the view of a single verifier (i.e., the prescribed one). Actually, we can avoid considering the verifier at all (since its view can be generated from the common reference-string and the message sent by the prover).

Definition 2 (non-interactive zero-knowledge): *A non-interactive proof system,* (P, V), *for a language L is* zero-knowledge *if there exist a polynomial p and a probabilistic polynomial-time algorithm M such that the probability ensembles $\{(x, U_{p(|x|)}, P(x, U_{p(|x|)}))\}_{x \in L}$ and $\{M(x)\}_{x \in L}$ are computationally indistinguishable, where U_m is a random variable uniformly distributed over $\{0, 1\}^m$.*

This definition too is "non-adaptive" (i.e., the common input may not depend on the common reference-string). An adaptive formulation of zero-knowledge is presented and discussed in [G1, Sec. 4.10.3]. Note that zero-knowledge is actually a property of the perscribed prover P, and so we may say that P is zero-knowledge.

2.2   The Hidden-Bits Model

A fictitious abstraction, which is nevertheless very helpful for the design of non-interactive zero-knowledge proof systems, is the *hidden bits model*. In this model the common reference-string is uniformly selected as before, but only the prover can see all of it. The 'proof' that the prover sends to the verifier consists of two parts; a 'certificate' and the specification of some bit positions in the common reference-string. The verifier may only inspect the bits of the common reference-string that reside in the locations that have been specified by the prover. Needless to say, in addition, the verifier inspects the common input and the 'certificate'.

Definition 3 (proof systems in the Hidden Bits Model): *A pair of probabilistic machines, (P, V), is called a* hidden-bits proof *system for L if V is polynomial-time and the following two conditions hold:*

- Completeness: *For every $x \in L$, it holds that*

$$\Pr\left[V(x, R_I, I, \pi) = 1\right] \geq \frac{2}{3}$$

where $(I, \pi) \overset{\text{def}}{=} P(x, R)$, R is a random variable uniformly distributed in $\{0, 1\}^{\text{poly}(|x|)}$ and R_I is the substring of R at positions $I \subseteq \{1, 2, ..., \text{poly}(|x|)\}$. That is, $R_I = r_{i_1} \cdots r_{i_t}$, where $R = r_1 \cdots r_t$ and $I = (i_1, ..., i_t)$.
- Soundness: *For every $x \notin L$ and every algorithm B, it holds that*

$$\Pr\left[V(x, R_I, I, \pi) = 1\right] \leq \frac{1}{3}$$

where $(I, \pi) \overset{\text{def}}{=} B(x, R)$, R is a random variable uniformly distributed in $\{0, 1\}^{\text{poly}(|x|)}$ and R_I is the substring of R at positions $I \subseteq \{1, 2, ..., \text{poly}(|x|)\}$.

In both cases, I is called the set of revealed bits *and π is called the* certificate. *Zero-knowledge is defined as in Def. 2, except that here we need to simulate $(x, R_I, P(x, R)) = (x, R_I, I, \pi)$ rather than $(x, R, P(x, R))$.*

As stated above, we do not suggest the Hidden-Bits Model as a realistic model. The importance of the model stems from two facts. Firstly, it is a 'clean' model that facilitates the design of proof systems (in it), and secondly proof systems in the Hidden-Bits Model can be easily transformed into non-interactive proof systems (i.e., the realistic model). The transformation (which utilizes a one-way permutation f with hard-core b) follows.

Construction 4 (from Hidden Bits proof systems to non-interactive ones): *Let (P, V) be a hidden-bits proof system for L, and suppose that $f : \{0, 1\}^* \to \{0, 1\}^*$ and $b : \{0, 1\}^* \to \{0, 1\}$ are polynomial-time computable. Furthermore, let $m = \text{poly}(n)$ denote the length of the common reference-string for common inputs of length n, and suppose that f is 1-1 and length preserving. Following is a specification of a non-interactive system, denoted (P', V'):*

- Common Input: $x \in \{0,1\}^n$.
- Common Reference-String: $s = (s_1, ..., s_m)$, where each s_i is in $\{0,1\}^n$.
- Prover (denoted P'):
    1. Computes $r_i = b(f^{-1}(s_i))$, for $i = 1, 2, ..., m$.
    2. Invokes P to obtain $(I, \pi) = P(x, r_1 \cdots r_m)$.

    The prover P' outputs (I, π, p_I), where $p_I \stackrel{\text{def}}{=} (f^{-1}(s_{i_1}) \cdots f^{-1}(s_{i_t}))$ for $I = (i_1, ..., i_t)$.

    That is, P' augments the proof (I, π), obtained from P, with the f-preimages of the blocks in the reference-string that have indices in I. These preimages reveal the values of the corresponding "revealed" bits in the hidden-bits model, while the values of the other bits remain essentially hidden.
- Verifier (denoted V'), given prover's output $(I, \pi, (p_1 \cdots p_t))$:
    1. Checks that $s_{i_j} = f(p_j)$, for each $i_j \in I$.

    In case a mismatch is found, V' halts and rejects.
    2. Computes $r_i = b(p_i)$, for $i = 1, ..., t$. Let $r = r_1, ..., r_t$.
    3. Invokes V on (x, r, I, π), and accepts if and only if V accepts.

    That is, using the p_j's, the verifier V' reconstructs the the values of the corresponding "revealed" bits in the hidden-bits model, and invokes V on these values.

We comment that P' is not perfect (or statistical) zero-knowledge even in case P is. Furthermore (and more central to this article), the prover P' may not be implemented in polynomial-time even if P is (and even with the help of auxiliary inputs). See further discussion in the next section.

Proposition 5 (analysis of Construction 4): Let (P,V), L, f, b and (P',V') be as in Construction 4. Then, (P',V') is a non-interactive proof system for L, provided that $\Pr[b(U_n) = 1] = \frac{1}{2}$. Furthermore, if P is zero-knowledge and b is a hard-core of f, then P' is zero-knowledge too.

Proof: To see that (P',V') is a non-interactive proof system for L we note that uniformly chosen strings $s_i \in \{0,1\}^n$ induce uniformly distributed bits $r_i \in \{0,1\}$. This follows by $r_i = b(f^{-1}(s_i))$, the fact that f is one-to-one, and the fact that $b(f^{-1}(U_n)) \equiv b(U_n)$ is unbiased. Thus, the actions of the parties in the real model (i.e., in Construction 4) perfectly emulate the actions of the parties in the hidden bits model.

Note that if b is a hard-core of f, then b is almost unbiased (i.e., $\Pr[b(U_n) = 1] = \frac{1}{2} \pm \mu(n)$, where μ is a negligible function), and the said emulation is only guaranteed to be almost-perfect (i.e., deviates negligibly from the original). Thus, saying that b is a hard-core for f essentially suffices for concluding that (P',V') is a non-interactive proof system for L.

To see that P' is zero-knowledge note that we can convert an efficient simulator for P into an efficient simulator for P'. Specifically, we first invoke the P-simulator and obtain a triple (α, I, π), where α denotes the (simulated) sequence of revealed bits, I denotes their positions in the common reference-string, and π denotes the simulated certificate. Next, for each revealed bit of value σ, we uniformly select a string $r \in \{0,1\}^n$ such that $b(r) = \sigma$ and place $f(r)$ in the corresponding position in the common reference-string (being simulated for P'). That is, if the said bit corresponds to position $i \in I$, then we place $f(r)$ in the ith block of the reference-string. For each *unrevealed* bit (i.e., bit position $i \notin I$), we uniformly select a string $s \in \{0,1\}^n$ and place it in the corresponding position in the common reference-string (i.e., place s in the ith block of the reference-string). The output of the P'-simulator consists of the common reference-string generated as above, the sequence of all r's generated by the P'-simulator for bits revealed by the P-simulator (i.e., bit in I), and the pair (I, π) as in the output of the P-simulator. Following is a rigorous description of the P'-simulator, when invoked on input $x \in \{0,1\}^n$ and using the P-simulator, denoted M.

1. Obtain $(x, (\sigma_1, ..., \sigma_t), (i_1, ..., i_t), \pi) \leftarrow M(x)$.
2. For every $j = 1, .., t$, select uniformly $p_j \in \{0,1\}^n$ such that $b(p_j) = \sigma_j$ and set $s_{i_j} = f(p_j)$.
3. For every $i \in [m] \setminus \{i_j : j = 1, .., t\}$, select s_i uniformly in $\{0,1\}^n$.
4. Output $(x, (s_1, ..., s_m), ((i_1, ..., i_t), \pi, (p_1, ..., p_t)))$.
   That is the sequence $(s_1, ..., s_m)$ is the simulated "common reference-string" whereas the triple $((i_1, ..., i_t), \pi, (p_1, ..., p_t))$ is the simulated proof.

Using the hypothesis that b is a hard-core of f, it follows that the output of the P'-simulator is computationally indistinguishable from the verifier's view (when receiving a proof from P'). Note that the only difference between the simulation output and the real execution is that in the real execution the blocks of the (actual) reference-string match the values of the bits of the (imaginary) reference-string that is given to P (and only partially revealed to V). In contrast, in the simulation, the blocks that correspond to *unrevealed* bits (in the hidden bits model) do not necessarily match the values of these (imaginary) unrevealed bits.[1] However, this difference is computationally indistinguishable (by the hypothesis that b is a hard-core of f). ∎

3   Efficient Implementations of the Prover of Construction 4

As hinted in Section 2.2, in general, the strategy P' (described in Construction 4) may not be efficiently implemented given black-box access to P. What is needed for such an efficient implementation is the ability (of P') to invert f. On the other

---

[1] To illustrate the issue, consider a strategy P (for the hidden bits model) that just reveals $m/3$ bits in the m-bit long reference-string such that each revealed bit holds the value 1. Then, the corresponding P' reveals the corresponding f-preimages of $m/3$ blocks in the m-block long reference-string (i.e., the f-preimage of a block is sent only if the value of this preimage under b equals 1). However, the simulator constructed for P' generates a simulated m-block long reference-string in which the f-preimages that are not revealed are random (rather than being suitablly biased towards evaluating to 0 under b).

hand, for P' to be zero-knowledge f must be one-way. The obvious solution is to use a collection of trapdoor permutations and let the prover know the trapdoor.

The basic construction is presented based on a collection of trapdoor permutations that have simple domains (i.e., the domain of each permutation is the set of all strings of some fixed string). Furthermore, the collection should have the property that its members can be efficiently recognized (i.e., given a description of a function one can efficiently decide whether it is in the collection).

3.1   The Basic Construction

Using such a collection of trapdoor permutations, P' starts by selecting a permutation f over $\{0,1\}^n$ such that it knows its trapdoor, and proceeds as in Construction 4, except that it also appends the description of f to the 'proof'. Indeed, the knowledge of the trapdoor allows P' to invert f on any element in f's domain. The verifier acts as in Construction 4 with respect to the function f specified in the proof. In addition the verifier also checks that f is indeed in the collection.

Both the completeness and the zero-knowledge conditions follow exactly as in the proof of Proposition 5. For the soundness condition we need to consider all possible members of the collection (w.l.o.g., there are at most 2^n such permutations). For each such permutation, the argument is as before, and our soundness claim thus follows by a counting argument (as applied in [G1, Sec. 4.10.3]). Actually, we also need to repeat the (P, V) system for $O(n)$ times, so to first reduce the soundness error to $\frac{1}{3} \cdot 2^{-n}$.

The foregoing text is reproduced from [G1, Rem. 4.10.6] and is indeed valid. The only problem is that it refers to a restricted notion of a collection of trapdoor permutations. Specifically, when compared with the general definition of such collections (as provided in [G1, Def. 2.4.5]), the foregoing description corresponds to the special case in which for every index α the domain of the permutation f_α (i.e., the permutation described by α) equals $\{0,1\}^{|\alpha|}$. In contrast, in general, the domain of f_α may be an arbitrary subset of $\{0,1\}^{|\alpha|}$ (as long as this subset is easy to sample from). The focus of this article is on trying to extend the foregoing construction by using more general forms of trapdoor permutations.

3.2   Extending the Basic Construction

We start by recalling the (general) definition of a collection of trapdoor permutations, and considering a couple of enhancements.

Enhanced collections of trapdoor permutations. Recall that a collection of trapdoor permutations, as defined in [G1, Def. 2.4.5], is a collection of finite permutations, denoted $\{f_\alpha : D_\alpha \to D_\alpha\}$, accompanied by four probabilistic polynomial-time algorithms, denoted G, S, F and B (for *generator*, *sample*, *forward* and *backward*), such that the following (syntactic) conditions hold:

1. On input 1^n, algorithm G selects a random n-bit long index α of a permutation f_α, along with a corresponding trapdoor τ;

2. On input α, algorithm S *samples* the domain of f_α, returning an almost uniformly distributed element in it;
3. For any x in the domain of f_α, given α and x, algorithm F returns $f_\alpha(x)$ (i.e., $F(\alpha, x) = f_\alpha(x)$);
4. For any y in the range of f_α if (α, τ) is a possible output of $G(1^n)$, then, given τ and y, algorithm B returns $f_\alpha^{-1}(y)$ (i.e., $B(\tau, y) = f_\alpha^{-1}(y)$).

The standard hardness condition (as in [G1, Def. 2.4.5]) refers to the difficulty of inverting f_α on a uniformly distributed element of its range, when given only the range-element and the index α. That is, letting $G_1(1^n)$ denote the first element in the output of $G(1^n)$ (i.e., the index), it is required that, for every probabilistic polynomial-time algorithm A (resp., every non-uniform family of polynomial-size circuit $A = \{A_n\}_n$), it holds that

$$\Pr[A(G_1(1^n), f_{G_1(1^n)}(S(G_1(1^n)))) = S(G_1(1^n))] = \mu(n), \qquad (1)$$

where μ denotes a generic negligible function. Namely, A (resp., A_n) fails to invert f_α on $f_\alpha(x)$, where α and x are selected by G and S as above. An equivalent way of writing Eq. (1) is

$$\Pr[A(G_1(1^n), S'(G_1(1^n), R_n)) = f_{G_1(1^n)}^{-1}(S'(G_1(1^n), R_n))] = \mu(n), \qquad (2)$$

where S' is the residual two-input (deterministic) algorithm obtained from S when treating the coins of the latter as an auxiliary input, and R_n denote the distribution of the coins of S on n-bit long inputs. That is, A fails to invert f_α on x, where α and x are selected as above.

Enhanced trapdoor permutations. Although the foregoing definition suffices for many applications, in some cases we will need an enhanced hardness condition. Specifically, we will require that it is hard to invert f_α on a random input x (in the domain of f_α) *even when given the coins used by S in the generation of x.* (Note that, given these coins (and the index α), the resulting domain element x is easily determined, and so we may omit it from the input given to the potential inverter.)

Definition 6 (enhanced trapdoor permutations [G2, Def. C.1.1]): *Let $\{f_\alpha : D_\alpha \to D_\alpha\}$ be a collection of trapdoor permutations as in [G1, Def. 2.4.5]. We say that this collection is* enhanced *(and call it an* enhanced collection of trapdoor permutations) *if, for every probabilistic polynomial-time algorithm A, it holds that*

$$\Pr[A(G_1(1^n), R_n) = f_{G_1(1^n)}^{-1}(S'(G_1(1^n)), R_n))] = \mu(n), \qquad (3)$$

where S' and μ are as above. The non-uniform version is defined analogously.

Note that the special case of [G1, Def. 2.4.5] in which the domain of f_α equals $\{0, 1\}^{|\alpha|}$ satisfies Definition 6 (because, without loss of generality, the sampling algorithm S' may satisfy $S'(\alpha, r) = r$). This implies that modified versions of the RSA and Rabin collections satisfy Definition 6. More natural versions of both collections can also be shown to satisfy Definition 6. For further discussion see the Appendix.

Doubly-enhanced trapdoor permutations. Although collection of enhanced trap-door permutations suffice for the construction of Oblivious Transfer (see [G2, Sec. 7.3.2]), it seems that they do not suffice for our current purpose of provid-ing an efficient implementation of the prover of Construction 4.[2] Thus, we further enhance Definition 6 so to provide for such an implementation. Specifically, we will require that, given α, it is feasible to generate a random pair (x, r) such that r is uniformly distributed in $\{0, 1\}^{\text{poly}(|\alpha|)}$ and x is a preimage of $S'(\alpha, r)$ under f_α; that is, we should generate a random $x \in D_\alpha$ along with coins that fit the generation of $f_\alpha(x)$ (rather than coins that fit the generation of x).

Definition 7 (doubly-enhanced trapdoor permutations): *Let $\{f_\alpha : D_\alpha \to D_\alpha\}$ be an enhanced collection of trapdoor permutations (as in Def. 6). We say that this collection is* doubly-enhanced *(and call it a* doubly-enhanced collection of trapdoor permutations*) if there exists a probabilistic polynomial-time algorithm that on input α outputs a pair (x, r) such that r is distributed identically to $R_{|\alpha|}$ and $f_\alpha(x) = S'(\alpha, r)$.*

We note that Definition 7 is satisfied by any collection of trapdoor permuta-tions that has a **reversed domain-sampler** (i.e., a probabilistic polynomial-time algorithm that on input (α, y) outputs a string that is uniformly distributed in $\{r : S'(\alpha, r) = y\}$).

A useful relaxation of Definition 7 allows r to be distributed almost-identically (rather than identically) to $R_{|\alpha|}$, where by almost-identical distributions we mean that the corresponding variation distance is negligible (i.e., the distributions are statistically close). Needless to say, in this case the definition of a reversed domain-sampler should be relaxed accordingly.

We stress that suitable implementations of the popular candidate collections of trapdoor permutations (e.g., the RSA and Rabin collections) do satisfy the foregoing doubly-enhanced condition. In fact, any collection of trapdoor permu-tations that has dense and easily recognizable domains satisfies this condition. For further details see the Appendix.

Actually implementing the prover. Recall that the basic construction pre-sented in Section 3.1 relies on two extra properties of the collection of trapdoor permutations.

1. It was assumed that the set of possible descriptions of the possible per-mutations, denoted $\bar{I}$, is easily recognizable (i.e., the support of $G(1^n)$ is recognizable in poly(n)-time).
2. It was assumed that the domain of every permutation f_α equals $\{0, 1\}^{|\alpha|}$.

The first assumption was waived by Bellare and Yung [BY], and we briefly sketch their underlying idea first. This relaxation is crucial, since no candidate

---

[2] We mention that the enhancement of Definition 6 was intended to suffice for both purposes. Indeed, in [G2, Apdx. C.4] it was claimed that enhanced trapdoor per-mutations do suffice for providing an efficient implementation of the prover of Con-struction 4. Needless to say, we retract this claim here. Further historical comments appear in Section 4.

collection of trapdoor permutations that satisfies this assumption is known (i.e., for all popular candidates, the corresponding index set $\bar{I}$ is not known to be efficiently recognizable).

The problem that arises is that the prover may select (and send) a function that is not in the collection (i.e., an index α that is not in $\bar{I}$). In such a case, the function is not necessarily 1-1, and, consequently, the soundness property may be violated. This concern can be addressed by using a (simple) non-interactive (zero-knowledge) proof for convincing the verifier that the function is "typically 1-1" (or, equivalently, is "almost onto the designated range"). The proof proceeds by presenting preimages (under the function) of random elements that are specified in the reference string. Note that, for any fixed polynomial p, we can only prove that the function is 1-1 on at least a $1 - (1/p(n))$ fraction of the designated range (i.e., $\{0,1\}^n$), yet this suffices for moderate soundness of the entire proof system (which in turn can be amplified by repetitions). For further details, consult [BY].

Note that this solution extends to the case that the collection of permutations $\{f_\alpha : D_\alpha \rightarrow D_\alpha\}_{\alpha \in \bar{I}}$ does not satisfy $D_\alpha = \{0,1\}^{|\alpha|}$, but is rather an arbitrary collection of doubly-enhanced trapdoor permutations. In this case the reference string will contain a sequence of coin-sequences to be used by the domain-sampling algorithm (rather than consisting of elements of the function's domain). By virtue of the extra condition in Definition 7, we can simulate the inverting of each domain element by generating a pair (x, r), placing r on the reference string, and providing x as the inverse of $S'(\alpha, r)$ under f_α. (See an analogous discussion in next paragraph.)

We now turn to the second aforementioned assumption; that is, the assumption that the domain of f_α equals $\{0,1\}^{|\alpha|}$ (i.e., $D_\alpha = \{0,1\}^{|\alpha|}$). We would have liked to waive this assumption completely, but are only able to do so in the case that the collection of trapdoor permutations is *doubly-enhanced*. The basic idea is letting the reference string consist of coin-sequences to be used by the domain-sampling algorithm (rather than of elements of the function's domain). The corresponding domain elements, which depend on the choice of the index α, are then obtained by applying the domain-sampling algorithm to these coin-sequences. The enhanced hardness property (stated in Def. 6) is used in order to note that the corresponding preimages under f_α are not revealed by these coin-sequences, whereas the additional enhancement (stated in Def. 7) is used for arguing that revealing such preimages does not reveal additional knowledge. That is, the two additional properties (stated in Def. 6 and Def. 7) are used in the (analysis of the) simulation and not in the proof system itself. For sake of simplicity, in the following exposition, we again use the (problematic) assumption by which $\bar{I}$ is efficiently recognizable.

Construction 8 (Construction 4, revised): *Let (P, V) be a zero-knowledge hidden-bits proof system for L with exponentially vanishing soundness error (i.e., soundness error at most 2^{-n-2}), and let $m = \text{poly}(n)$ denote the length of the common reference-string for common inputs of length n. Suppose that $\{f_\alpha : D_\alpha \rightarrow D_\alpha\}_{\alpha \in \bar{I}}$ is a doubly-enhanced collection of trapdoor permutations, where $\bar{I}$ is efficiently recognizable, and $b: \{0,1\}^* \rightarrow \{0,1\}$ is a corresponding hard-*

core predicate (i.e., $b(f_\alpha^{-1}(S'(\alpha, U_\ell)))$ is infeasible to predict when given $(\alpha, U_\ell))$.[3]
Following is a specification of a non-interactive system, denoted (P', V'):

- Common Input: $x \in \{0,1\}^n$.
- Prover's auxiliary input: w.
- Common Reference-String: $s = (s_1, ..., s_m)$, where each s_i is in $\{0,1\}^\ell$ and ℓ is the number of coins used by the domain-sampler when given an n-bit long index of a permutation.
- Prover (denoted P'):
  1. Select at random an n-bit long index α and a corresponding trapdoor τ; i.e., $(\alpha, \tau) \leftarrow G(1^n)$.
  2. Using the trapdoor τ, compute $r_i = b(f_\alpha^{-1}(S'(\alpha, s_i)))$, for $i = 1, 2, ..., m$.
  3. Invokes P to obtain $(I, \pi) = P(x, w, r_1 \cdots r_m)$.

     The prover P' outputs (α, I, π, p_I), where $p_I \overset{\text{def}}{=} (f_\alpha^{-1}(S'(\alpha, s_{i_1})) \cdots f_\alpha^{-1}(S'(\alpha, s_{i_t})))$ for $I = (i_1, ..., i_t)$.
- Verifier (denoted V'), given prover's output $(\alpha, I, \pi, (p_1 \cdots p_t))$:
  1. Check if $\alpha \in \bar{I}$, otherwise halts and rejects.
  2. Check that $S'(\alpha, s_{i_j}) = f_\alpha(p_j)$, for each $i_j \in I$.
     In case a mismatch is found, V' halts and rejects.
  3. Compute $r_i = b(p_i)$, for $i = 1, ..., t$. Let $r = r_1, ..., r_t$.
  4. Invoke V on (x, r, I, π), and accepts if and only if V accepts.

Clearly, the foregoing strategy P' is efficient, provided that so is P.

Proposition 9 (Proposition 5, revised): *Let (P, V), L, f, b and (P', V') be as in Construction 8. Then, (P', V') is a zero-knowledge non-interactive proof system for L.*

Proof: Following the proof of Proposition 5, we note that for any fixed choice $\alpha \in \bar{I} \cap \{0,1\}^n$ the soundness error is at most 2^{-n-2}. Taking a union bound over all possible $\alpha \in \bar{I} \cap \{0,1\}^n$ and discarding all $\alpha \notin \bar{I}$ (by virtue of the explicit check), we establish that (P', V') is a non-interactive proof system for L.

To show that P' is zero-knowledge we convert any (efficient) simulator for P into an (efficient) simulator for P'. First, the new simulator selects at random an index α (of a permutation) just as P' does. We stress that although the P'-simulator obtains the corresponding trapdoor (just as P' does), we will not use this fact in the simulation. Next, we proceed as in the proof of Proposition 5, modulo adequate adaptations that address the crucial difference between Construction 4 and Construction 8. Recall that the difference is that in Construction 4 the reference string is viewed as a sequence of images of the permutation, whereas in Construction 8 the reference string is viewed as a sequence of ℓ-bit long random-sequences that may be used to generate such images. Following is a rigorous description of the current P'-simulator, when invoked on input $x \in \{0,1\}^n$ and using the P-simulator, denoted M.

---

[3] Such a hard-core predicate is obtained by applying the techniques of [GL] (see [G1, Sec. 2.5.2] or better [G3, Sec. 7.1.3]) to any (doubly-)enhanced collection of trapdoor permutations.

1. Obtain $(\alpha, \tau) \leftarrow G(1^n)$.
2. Obtain $((\sigma_1, ..., \sigma_t), (i_1, ..., i_t), \pi) \leftarrow M(x)$.
3. For every $j = 1, .., t$, generate a random pair $(p_j, s_{i_j}) \in D_\alpha \times \{0, 1\}^\ell$ such that $f_\alpha(p_j) = S'(\alpha, s_{i_j})$ and $b(p_j) = \sigma_j$.
   Note that this operation can be efficiently implemented by either relying on the additional enhancement introduced in Def. 7 or by merely relying on the fact that the simulator knows the trapdoor τ and can thus invert f_α. (The "forced" use of the additional enhancement of Def. 7 arises in the proof of indistinguishabilitry provided below.)
4. For every $i \in [m] \setminus \{i_j : j = 1, .., t\}$, select s_i uniformly in $\{0, 1\}^\ell$.
5. Output $(x, (s_1, ..., s_m), (\alpha, (i_1, ..., i_t), \pi, (p_1, ..., p_t)))$.

Using the hypothesis that b is a hard-core of the collection $\{f_\alpha\}$ and the doubly-enhanced hardness of this collection, we will show that the output of the P'-simulator is computationally indistinguishable from the verifier's view (when receiving a proof from P'). Again, the only difference between the simulation and the real execution is that in the simulation the blocks of the (actual) reference strings do not necessarily match the b-values of the corresponding hidden bits seen by P. Intuitively, this difference is computationally indistinguishable by the hypothesis that $b(f_\alpha^{-1}(S'(\alpha, U_\ell)))$ is infeasible to predict when given (α, U_ℓ), which is guaranteed by the enhanced hardness assumption (of Def. 6). However, we need to show that, for $H \stackrel{\text{def}}{=} [m] \setminus \{i_j : j = 1, .., t\}$, it is infeasible to distinguish a sequence of $|H|$ uniformly selected n-bit strings (representing the sequence $(s_i)_{i \in H}$ produced in the simulation) from a corresponding sequence of s_i's that fits a (partially) given sequence of $b(f_\alpha^{-1}(S'(\alpha, s_i)))$ values (as in the real interaction). At this point, we encounter a difficulty that seems to require the doubly-enhanced hypothesis (of Def. 7).

The point is that the indistinguishability of the two sequences is demonstrated by showing that, given a prefix of the second sequence, it is infeasible to predict the $b(f_\alpha^{-1}(S'(\alpha, \cdot)))$-value of the next element. That is, we wish to show that, for every i, given a randomly selected α and a uniformly selected sequence $s_1, ..., s_{i-1}, s_i$ along with the values $b(f_\alpha^{-1}(S'(\alpha, s_1))), ..., b(f_\alpha^{-1}(S'(\alpha, s_{i-1})))$, it is infeasible to predict the value of $b(f_\alpha^{-1}(S'(\alpha, s_i)))$. Recall that the standard approach toward this task is to use a reducibility argument in order to derive a contradiction to the hard-core hypothesis (which refers to a single $s = s_i$ for which $b(f_\alpha^{-1}(S'(\alpha, s)))$ is unpredictable), by generating the auxiliary prefix $s_1, ..., s_{i-1}$ along with $b(f_\alpha^{-1}(S'(\alpha, s_1))), ..., b(f_\alpha^{-1}(S'(\alpha, s_{i-1})))$. Thus, given only α (and $s = s_i$), we need to be able to generate a random sequence $s_1, ..., s_{i-1}$ along with the corresponding $b(f_\alpha^{-1}(S'(\alpha, s_j)))$'s. But this is easy to do given the doubly-enhanced hypothesis (of Def. 7), and once this is done we just rely on the infeasiblity of predicting $b(f_\alpha^{-1}(S'(\alpha, s)))$ based on s and α (which is guaranteed by the enhanced hardness assumption of Def. 6). ∎

Subsequent work: A closer look at the use of the doubly-enhanced hypothsis (in the foregoing construction as well as in other settings) led Rothblum to introduce and study a taxonomy of enhanced trapdoor permutations. Indeed, one of the primitives in his taxonomy fits the aforemention argument perfectly. The interested reader is referred to [R].

Open Problem: *Under what intractability assumptions is it possible to construct non-interactive zero-knowledge proofs* (NIZKs) *with efficient prover strategies for any set in* $\mathcal{NP}$*? In particular, does the existence of arbitrary collections of trapdoor permutations suffice?* We mention that the assumption used in constructing such NIZKs effects the assumption used in (general) constructions of public-key encryption schemes that are secure under chosen ciphertext attacks (see, e.g., [G2, Thm. 5.4.31]).

4   The Story

The story begins with the fact that, while the notion of trapdoor permutations was widely referred to in the 1980's, the exact structural requirements from it were not cmmonly agreed upon at the time. Here we refer to secondary issues regarding the structure of the index set as well as the domains of the various permutations. Bellare and Yung seem to have been the first who explicitly addressed this type of issues, but their focus was on the fact that the index set cannot be assumed to be efficiently recognizable. As for the domains of the permutations, they just assumed that the domain of f_α is $\{0,1\}^{|\alpha|}$, which is indeed the case for minor modifications of all popular trapdoor permutations. In general, it seems that most researchers had in mind dense and efficiently recognizable domains, but these additional requirements were not needed in the main classical applications of trapdoor permutations (e.g., constructions of secure public-key encryption schemes).

When writing [G1], we decided to use the most liberal definition of trapdoor permutations that agrees with the basic intuitions regarding this notion. This led to [G1, Def. 2.4.5], which is the definition that is the starting point of Section 3.2. While this definition suffices for the constructions of passively-secure public-key encryption schemes, we failed to notice at the time that it does not suffice for two less traditional but quite important applications: (1) the construction of Oblivious Transfer, and (2) the construction of NIZKs with efficient provers for $\mathcal{NP}$.

We missed the first opportunity to detect the problem, when addressing the second application in [G1, Sec. 4.10.2]. As stated at the end of the Introduction, we believe that the source of evil is the careless presentation of this topic as a laconic comment (i.e., [G1, Rem. 4.10.6]) that focuses on a simplified setting (i.e., the one discussed in Section 3.1).

When writing [G2, Sec. 7.3.2], we discovered that the known of construction of Oblivious Transfer based on trapdoor permutations [EGL] may be insecure, in general, and that standard proof of security seems to require the enhancement of Definition 6 (which was introduced in [G2, Apdx. C.1] for that purpose).[4] It was evident that this enhancement is also needed for the argument in [G1, Sec. 4.10.2]. At this point, we missed our second opportunity to detect the problem; using some hand-waving, we argued in [G2, Apdx. C.4.1] that enhanced trapdoor permuations (as defined in [G2, Apdx. C.1]) suffice for the construction of NIZKs with efficient provers for $\mathcal{NP}$. Needless to say, we retract this claim here.

The flaw was eventually discovered by others: Specifically, Jonathan Katz called out attention to the flaw in [G2, Apdx. C.4.1], and suggested the notion of doubly-enhanced trapdoor permutations (as in Definition 7).

Acknowledgments. We are grateful to Jonathan Katz for pointing out the gap in [G2, Apdx. C.4.1]. While being embarrassed about such flaws, we feel deeply indebted to those discovering them and bringing them to our attention.

We thank Ron Rothblum for pointing out that a previous version of this write-up failed to deliver the crucial point, which is currently spelled out at the end of the proof of Proposition 9.

References

[ACGS] Alexi, W., Chor, B., Goldreich, O., Schnorr, C.P.: RSA/Rabin Functions: Certain Parts are As Hard As the Whole. SIAM Jour. on Comput. 17, 194–209 (1988)

[BY] Bellare, M., Yung, M.: Certifying Permutations: Noninteractive Zero-Knowledge Based on Any Trapdoor Permutation. Journal of Cryptology 9, 149–166 (1996)

[EGL] Even, S., Goldreich, O., Lempel, A.: A Randomized Protocol for Signing Contracts. CACM 28(6), 637–647 (1985)

[G1] Goldreich, O.: Foundation of Cryptography: Basic Tools. Cambridge University Press, Cambridge (2001)

[G2] Goldreich, O.: Foundation of Cryptography: Basic Applications. Cambridge University Press, Cambridge (2004)

[G3] Goldreich, O.: Computational Complexity: A Conceptual Perspective. Cambridge University Press, Cambridge (2008)

[GL] Goldreich, O., Levin, L.A.: Hard-core Predicates for any One-Way Function. In: 21st STOC, pp. 25–32 (1989)

[H] Haitner, I.: Implementing oblivious transfer using collection of dense trapdoor permutations. In: Naor, M. (ed.) TCC 2004. LNCS, vol. 2951, pp. 394–409. Springer, Heidelberg (2004)

[R] Rothblum, R.: A Taxonomy of Enhanced Trapdoor Permutations. In: ECCC, TR10-145 (September 2010)

---

[4] Indeed, Oblivious Transfer can be based on any enhanced trapdoor permutations [G2, Sec. 7.3.2]. We mention that an alternative construction of Oblivious Transfer was obtained based on an alternative restriction of the notion of trapdoor permutations: Specifically, it was proved that trapdoor permutations with dense domains suffice [H].

Appendix: On the RSA and Rabin Collections

In this appendix we show that suitable versions of the RSA and Rabin collections satisfy the two aforementioned enhancements (presented in Definitions 6 and 7, respectively). Establishing this claim is quite straightforward for the RSA collection, whereas for the Rabin collection some modifications (of the straightforward version) seem necessary. In order to establish this claim we will consider a variant of the Rabin collection in which the corresponding domains are dense and easy to recognize, and will show that having such domains suffices for establishing the claim.

A.1   The RSA Collection Satisfies Both Enhancements

We start our treatment by considering the RSA collection (as presented in [G1, Sec. 2.4.3.1] and further discussed in [G1, Sec. 2.4.3.2]). Note that in order to discuss the enhanced hardness condition (of Def. 6) it is necessary to specify the domain sampler, which is not entirely trivial (since sampling Z_N^* (or even Z_N) by using a sequence of unbiased coins is not that trivial).

A natural sampler for Z_N^* (or Z_N) generates random elements in the domain by using a regular mapping from a set of sufficiently long strings to Z_N^* (or to Z_N). Specifically, the sampler uses $\ell \stackrel{\text{def}}{=} 2\lfloor \log_2 N \rfloor$ random bits, views them as an integer in $i \in \{0, 1, ..., 2^\ell - 1\}$, and outputs $i \bmod N$. This yields an almost uniform sample in Z_N, and an almost uniform sample in Z_N^* can be obtained by discarding the few elements in $Z_N \setminus Z_N^*$.

The fact that the foregoing implementation of the RSA collection satisfies Definition 6 (as well as Definition 7) follows from the fact that it has an efficient reversed-sample (which eliminates the potential gap between having a domain element and having a random sequence of coins that makes the domain-sample output this element). Specifically, given an element $e \in Z_N$, the reversed-sampler outputs an almost uniformly distributed element of $\{i \in \{0, 1, ..., 2^\ell - 1\} : i \equiv e \pmod{N}\}$ by selecting uniformly $j \in \{0, 1, ..., \lfloor 2^\ell/N \rfloor - 1\}$ and outputting $i \leftarrow j \cdot N + e$.

A.2   Versions of the Rabin Collection that Satisfy both Enhancements

In contrast to the case of the RSA, the Rabin Collection (as defined in [G1, Sec. 2.4.3.3]), does not satisfy Definition 6 (because the coins of the sampling algorithm give away a modular square root of the domain element). Still, the Rabin Collection can be easily modify to yield an *doubly-enhanced* collection of trapdoor permutations, provided that factoring is hard (in the same sense as assumed in [G1, Sec. 2.4.3]).

The modification is based on modifying the domain of these permutations (following [ACGS]). Specifically, rather than considering the permutation induced (by the modular squaring function) on the set Q_N of the quadratic residues modulo N, we consider the permutations induced on the set M_N, where M_N

contains all integers in $\{1, ..., N/2\}$ that have Jacobi symbol modulo N that equals 1. Note that, as in case of Q_N, each quadratic residue has a unique square root in M_N (because exactly two square roots have Jacobi symbol that equals 1 and their sum equals N; indeed, as in case of Q_N, we use the fact that -1 has Jacobi symbol 1). However, unlike Q_N, membership in M_N can be determined in polynomial-time (when given N without its factorization). Lastly, note that squaring modulo N is a 1-1 mapping of M_N to Q_N. In order to obtain a permutation over M_N, we modify the function a little such that if the result of modular squaring is bigger than $N/2$, then we use its additive inverse (i.e., rather than outputting $y > N/2$, we output $N - y$).

Using the fact that M_N is dense (w.r.t $\{0,1\}^{\lfloor \log_2 N \rfloor + 1}$) and easy to recognize, we may proceed in one of two ways, which are actually generic. Thus, let us assume that we are given an arbitrary collection of trapdoor permutations, denoted $\{f_\alpha : D_\alpha \to D_\alpha\}_{\alpha \in \bar{I}}$, such that $D_\alpha \subseteq \{0,1\}^{|\alpha|}$ is dense (i.e., $|D_\alpha| > 2^{|\alpha|}/\text{poly}(|\alpha|)$)[5] and easy to recognize (i.e., there exists an efficient algorithm that given (α, x) decides whether or not $x \in D_\alpha$).

1. The most natural way to proceed is showing that the collection $\{f_\alpha\}$ itself is *doubly-enhanced*. This is shown by presenting a rather straightforward domain-sampler that satisfies the enhanced hardness condition (of Def. 6), and noting that this sampler has an efficient reversed sampler (which implies that Def. 7 is satisfied).
    The domain-sampler that we have in mind repeatedly selects random (i.e., uniformly distributed) $|\alpha|$-bit long strings and output the first such string that resides in D_α (and a special failure symbols if $|\alpha| \cdot 2^{|\alpha|}/|D_\alpha|$ attempts have failed). This sampler has an efficient reversed-sampler that, given $x \in D_\alpha$, generates a random sequence of $|\alpha|$-bit long strings and replaces the first string that resides in D_α by the string x.
2. An alternative way of obtaining a doubly-enhanced collection is to first define a (rather artificial) collection of *weak* trapdoor permutations, $\{f'_\alpha : \{0,1\}^{|\alpha|} \to \{0,1\}^{|\alpha|}\}_{\alpha \in \bar{I}}$, such that $f'_\alpha(x) = f_\alpha(x)$ if $x \in D_\alpha$ and $f'_\alpha(x) = x$ otherwise. Using the amplification of a weak one-way property to a standard one-way property (as in [G1, Sec. 2.3&2.6]), we are done.

Indeed, in the first alternative we amplified the trivial domain-sampler that succeeds with noticeable probability, whereas in the second alternative we amplified the one-way property of the trivial extension of f_α to the domain $\{0,1\}^{|\alpha|}$. Either way we obtain a *doubly-enhanced* collection of trapdoor permutations, provided that $\{f_\alpha\}$ is an ordinary collection of trapdoor permutations.

We mention that the foregoing modifications of the Rabin collection follows the outline of the second modification that is presented in [G2, Apdx. C.1]. In contrast, as pointed out by Jonathan Katz, the first implementation (of an enhanced trapdoor permutation based on factoring) that is presented in [G2, Apdx. C.1] is not doubly-enhanced.

---

[5] Actually, a more general case, which is used for the Rabin collection, is one in which $D_\alpha \subseteq \{0,1\}^{\ell(|\alpha|)}$ satisfies $|D_\alpha| > 2^{\ell(|\alpha|)}/\text{poly}(|\alpha|)$, where $\ell : \mathbb{N} \to \mathbb{N}$ is a fixed function.

Average Case Complexity, Revisited

Oded Goldreich

Abstract. More than two decades elapsed since Levin set forth a theory of average-case complexity. In this survey we present the basic aspects of this theory as well as some of the main results regarding it. The current presentation deviates from our old "Notes on Levin's Theory of Average-Case Complexity" (*ECCC*, TR97-058, 1997) in several aspects. In particular:

- We currently view average-case complexity as referring to the performance on "average" (or rather typical) instances, and not as the average performance on random instances. (Thus, it may be more justified to refer to this theory by the name typical-case complexity, but we retain the name average-case for historical reasons.)
- We include a treatment of search problems, and a presentation of the reduction of "NP with sampleable distributions" to "NP with P-computable distributions" (due to Impagliazzo and Levin, *31st FOCS*, 1990).
- We include Livne's result (*ECCC*, TR06-122, 2006) by which all natural NPC-problems have average-case complete versions. This result seems to shed doubt on the association of P-computable distributions with natural distributions.

Keywords: Average-Case Complexity.

This text has been revised based on [6, Sec. 10.2].

1 Introduction

In light of the apparent infeasibility of solving numerous useful computational problems, it is natural to ask whether these problems can be relaxed such that the relaxation is both useful and allows for feasible solving procedures. We stress two aspects about the foregoing question: on one hand, the relaxation should be sufficiently good for the intended applications; but, on the other hand, it should be significantly different from the original formulation of the problem so to escape the infeasibility of the latter. We note that whether a relaxation is adequate for an intended application depends on the application, and thus much of the material in this chapter is less robust (or generic) than the treatment of the non-relaxed computational problems.

One commonly considered type of relaxation refers to the computational problems themselves; that is, for each problem instance we *extend the set of admissible solutions*. In the context of search problems this means settling for solutions that have a value that is "sufficiently close" to the value of the optimal solution

O. Goldreich et al.: Studies in Complexity and Cryptography, LNCS 6650, pp. 422–450, 2011.
© Springer-Verlag Berlin Heidelberg 2011

(with respect to some value function). Needless to say, the specific meaning of 'sufficiently close' is part of the definition of the relaxed problem. In the context of decision problems this means that for some instances both answers are considered valid; specifically, we shall consider promise problems in which the no-instances are "far" from the yes-instances in some adequate sense (which is part of the definition of the relaxed problem).

In this survey, we consider a different type of relaxation. We do not relax the computational problems themselves, but rather the notion of solving them efficiently. Specifcally, this type of relaxation deviates from the requirement that the solver provides an adequate answer on each valid instance. Instead, the behavior of the solver is analyzed with respect to a predetermined input distribution (or a class of such distributions), and bad behavior may occur with negligible probability where the probability is taken over this input distribution. That is, we replace worst-case analysis by *average-case* (or rather *typical-case*) *analysis*. Needless to say, a major component in this approach is limiting the class of distributions in a way that, on one hand, allows for various types of natural distributions and, on the other hand, prevents the collapse of the corresponding notion of average-case hardness to the standard notion of worst-case hardness.

1.1   The Basic Mindframe of Average-Case Complexity

The common approach of complexity theory is termed worst-case complexity, because it refers to the performance of potential algorithms on each legitimate instance (and hence to the performance on the worst possible instance). That is, computational problems were defined as referring to a set of instances and performance guarantees were required to hold for each instance in this set. In contrast, average-case complexity allows ignoring a negligible measure of the possible instances, where *the identity of the ignored instances is determined by the analysis of potential solvers and not by the problem's statement.*

A few comments are in place. Firstly, as just hinted, the standard statement of the worst-case complexity of a computational problem (especially one having a promise) may also ignores some instances (i.e., those considered inadmissible or violating the promise), but these instances are determined by the problem's statement. In contrast, the inputs ignored in average-case complexity are not inadmissible in any inherent sense (and are certainly not identified as such by the problem's statement). It is just that they are viewed as exceptional when claiming that a specific algorithm solve the problem; that is, these exceptional instances are determined by the analysis of that algorithm. Needless to say, these exceptional instances ought to be rare (i.e., occur with negligible probability).

The last sentence raises a couple of issues. Most importantly, a distribution on the set of admissible instances has to be specified. In fact, we shall consider a new type of computational problems, each consisting of a standard computational problem coupled with a probability distribution on instances. Consequently, the question of which distributions should be considered in a theory of average-case complexity arises. This question and numerous other definitional issues will be addressed in Section 2.1.

Before proceeding, let us spell out the rather straightforward motivation to the study of the average-case complexity of computational problems: It is that, in real-life applications, one may be perfectly happy with an algorithm that solves the problem fast on almost all instances that arise in the relevant application. That is, one may be willing to tolerate error provided that it occurs with negligible probability, where the probability is taken over the distribution of instances encountered in the application. The study of average-case complexity is aimed at exploring the possible benefit of such a relaxation, distinguishing cases in which a benefit exists from cases in which it does not exist. A key aspect in such a study is a good modeling of the type of distributions (of instances) that are encountered in natural algorithmic applications.

Let us consider the foregoing motivation from a slightly different perspective: The conjecture that $\mathcal{P} \neq \mathcal{NP}$ (or rather $\mathcal{NP} \not\subseteq \mathcal{BPP}$) only asserts that intractability is a feature of some instances of some problems in $\mathcal{NP}$. These intractable instances may be very rare and pathological. The theory of average-case complexity addresses the question of whether intractability can also be a feature of "typical" instances (i.e., whether intractable instances may occur with noticeable probability with respect to some simple distributions). Needless to say, the meaningfulness of the latter question depends on restricting the class of distributions such that only simple (rather than pathological) distributions are allowed. We shall consider two such classes of distributions (see Section 2.1 and Section 3.2, respectively) and show that if intractability occurs with respect to the wider class then it occurs also with respect to the more restricted class (see Theorem 14).

An Average-Case version of the $\mathcal{P} \neq \mathcal{NP}$ Question. Indeed, a fundamental question that arises is *whether every natural computational problem can be solved efficiently when restricting attention to typical instances?* The conjecture that underlies this section is that, for a well-motivated choice of definitions, the answer is negative; that is, our conjecture is that the "distributional version" of NP is not contained in the average-case (or typical-case) version of P. This means that some NP problems are not merely hard in the worst-case, but are rather "typically hard" (i.e., hard on typical instances drawn from some simple distribution). This suggests that *hard instances may occur in natural algorithmic applications* (and not only in cryptographic (or other "adversarial") applications that are design on purpose to produce hard instances).[1]

The foregoing conjecture motivates the development of an average-case analogue of NP-completeness, which will be presented in this survey. In particular, this (average-case) theory identifies distributional problems that are "typically

---

[1] We highlight two differences between the current context (of natural algorithmic applications) and the context of cryptography. Firstly, in the current context and when referring to problems that are typically hard, the simplicity of the underlying input distribution is of great concern: the simpler this distribution, the more appealing the hardness assertion becomes. This concern is irrelevant in the context of cryptography. On the other hand (see, e.g., [5]), cryptographic applications require the ability to efficiently generate hard instances *together with corresponding solutions*.

hard" provided that distributional problems that are "typically hard" exist at all. If one believes the foregoing conjecture then, for such complete (distributional) problems, one should not seek algorithms that solve these problems efficiently on typical instances.

1.2 Organization

A significant part of our exposition is devoted to the definitional issues that arise when developing a general theory of average-case complexity. These issues are discussed in Section 2.1. In Section 2.2 we prove the existence of distributional problems that are "NP-complete" in the corresponding average-case complexity sense. Furthermore, we show how to obtain such a distributional version for any natural NP-complete decision problem. In Section 2.3 we extend the treatment to randomized algorithms. Additional ramifications are presented in Section 3.

2 The Basic Theory

In this section we provide a basic treatment of the theory of average-case complexity, while postponing important ramifications to Section 3. The basic treatment consists of the preferred definitional choices for the main concepts as well as the identification of complete problems for a natural class of average-case computational problems.

2.1 Definitional Issues

The theory of average-case complexity is more subtle than may appear at first thought. In addition to the generic conceptual difficulty involved in defining relaxations, difficulties arise from the "interface" between standard probabilistic analysis and the conventions of complexity theory. This is most striking in the definition of the class of feasible average-case computations. Referring to the theory of worst-case complexity as a guideline, we shall address the following aspects of the analogous theory of average-case complexity.

1. *Setting the general framework.* We shall consider distributional problems, which are standard computational problems coupled with distributions on the relevant instances.
2. *Identifying the class of feasible* (distributional) *problems.* Seeking an average-case analogue of classes such as $\mathcal{P}$, we shall reject the first definition that comes to mind (i.e., the naive notion of "average polynomial-time"), briefly discuss several related alternatives, and adopt one of them for the main treatment.
3. *Identifying the class of interesting* (distributional) *problems.* Seeking an average-case analogue of the class $\mathcal{NP}$, we shall avoid both the extreme of allowing arbitrary distributions (which collapses average-case hardness to worst-case hardness) and the opposite extreme of confining the treatment to a single distribution such as the uniform distribution.

4. *Developing an adequate notion of reduction among* (distributional) *problems.*
   As in the theory of worst-case complexity, this notion should preserve feasible
   solveability (in the current distributional context).

We now turn to the actual treatment of each of the aforementioned aspects.

Step 1: Defining Distributional Problems. Focusing on decision problems,
we define distributional problems as pairs consisting of a decision problem and a
probability ensemble.[2] For simplicity, here a probability ensemble $\{X_n\}_{n\in\mathbb{N}}$ is a sequence of random variables such that X_n ranges over $\{0,1\}^n$. Thus, $(S, \{X_n\}_{n\in\mathbb{N}})$
is the distributional problem consisting of the problem of deciding membership
in the set S with respect to the probability ensemble $\{X_n\}_{n\in\mathbb{N}}$. (The treatment
of search problem is similar; see Section 3.1.) We denote the uniform probability
ensemble by $U = \{U_n\}_{n\in\mathbb{N}}$; that is, U_n is uniform over $\{0,1\}^n$.

Step 2: Identifying the Class of Feasible Problems. The first idea that
comes to mind is defining the problem $(S, \{X_n\}_{n\in\mathbb{N}})$ as feasible (on the average)
if there exists an algorithm A that solves S such that the *average running time*
of A on X_n is bounded by a polynomial in n (i.e., there exists a polynomial p
such that $\mathsf{E}[t_A(X_n)] \leq p(n)$, where $t_A(x)$ denotes the running-time of A on input
x). The problem with this definition is that it is very sensitive to the model of
computation and is not closed under algorithmic composition. Both deficiencies
are a consequence of the fact that t_A may be polynomial on the average with
respect to $\{X_n\}_{n\in\mathbb{N}}$ but t_A^2 may fail to be so (e.g., consider $t_A(x'x'') = 2^{|x'|}$ if
$x' = x''$ and $t_A(x'x'') = |x'x''|^2$ otherwise, coupled with the uniform distribution
over $\{0,1\}^n$). We conclude that the *average running-time* of algorithms is not
a robust notion. We also doubt the soundness of the appeal of this notion, and
view the *typical running time* of algorithms (as defined next) as a more natural
notion. Thus, we shall consider an algorithm as feasible if its running-time is
typically polynomial.[3]

We say that A is typically polynomial-time on $X = \{X_n\}_{n\in\mathbb{N}}$ if there exists a
polynomial p such that the probability that A runs more that $p(n)$ steps on X_n

---

[2] We mention that even this choice is not evident. Specifically, Levin [10] (see discussion in [4]) advocates the use of a single probability distribution defined over the
set of all strings. His argument is that this makes the theory less representation-
dependent. At the time we were convinced of his argument (see [4]), but currently
we feel that the representation-dependent effects discussed in [4] are legitimate. Furthermore, the alternative formulation of [10,4] comes across as unnatural and tends
to confuse some readers.

[3] An alternative choice, taken by Levin [10] (see discussion in [4]), is considering as
feasible (w.r.t $X = \{X_n\}_{n\in\mathbb{N}}$) any algorithm that runs in time that is polynomial
in a function that is linear on the average (w.r.t X); that is, requiring that there
exists a polynomial p and a function $\ell : \{0,1\}^* \rightarrow \mathbb{N}$ such that $t(x) \leq p(\ell(x))$ for
every x and $\mathsf{E}[\ell(X_n)] = O(n)$. This definition is robust (i.e., it does not suffer from
the aforementioned deficiencies) and is arguably as "natural" as the naive definition
(i.e., $\mathsf{E}[t_A(X_n)] \leq \text{poly}(n)$).

is *negligible* (i.e., for every polynomial q and all sufficiently large n it holds that $\Pr[t_A(X_n) > p(n)] < 1/q(n)$). The question is what is required in the "untypical" cases, and two possible definitions follow.

1. The simpler option is saying that $(S, \{X_n\}_{n \in \mathbb{N}})$ is (typically) feasible if there exists an algorithm A that solves S such that A is typically polynomial-time on $X = \{X_n\}_{n \in \mathbb{N}}$. This effectively requires A to *correctly solve S on each instance*, which is more than was required in the motivational discussion. (Indeed, if the underlying motivation is ignoring rare cases, then we should ignore them altogether rather than ignoring them in a partial manner (i.e., only ignore their affect on the running-time).)

2. The alternative, which fits the motivational discussion, is saying that (S, X) is (typically) feasible if there exists an algorithm A such that A typically solves S on X in polynomial-time; that is, there exists a polynomial p such that *the probability that on input X_n algorithm A either errs or runs more that $p(n)$ steps is negligible*. This formulation totally ignores the untypical instances. Indeed, in this case we may assume, without loss of generality, that A always runs in polynomial-time, but we shall not do so here (in order to facilitate viewing the first option as a special case of the current option).

We stress that both alternatives actually define *typical* feasibility and not *average-case* feasibility. To illustrate the difference between the two options, consider the distributional problem of deciding whether a uniformly selected (n-vertex) graph is 3-colorable. Intuitively, this problem is "typically trivial" (with respect to the uniform distribution),[4] because the algorithm may always say **no** and be wrong with exponentially vanishing probability. Indeed, this trivial algorithm is admissible by the second approach, but not by the first approach. In light of the foregoing discussions, we adopt the second approach.

Definition 1 (the class tpc$\mathcal{P}$): *We say that A typically solves $(S, \{X_n\}_{n \in \mathbb{N}})$ in polynomial-time if there exists a polynomial p such that the probability that on input X_n algorithm A either errs or runs more that $p(n)$ steps is negligible.*[5] *We denote by* tpc$\mathcal{P}$ *the class of distributional problems that are typically solvable in polynomial-time.*

Clearly, for every $S \in \mathcal{P}$ and every probability ensemble X, it holds that $(S, X) \in$ tpc$\mathcal{P}$. However, tpc$\mathcal{P}$ contains also distributional problems (S, X) with $S \notin \mathcal{P}$ (albeit this assertion refers to unnatural distributional versions of problems not in $\mathcal{P}$). The big question, which underlies the theory of average-case complexity, is whether all *natural distributional versions* of $\mathcal{NP}$ are in tpc$\mathcal{P}$. Thus, we turn to identify such versions.

---

[4] In contrast, testing whether a given graph is 3-colorable seems "typically hard" for other distributions (see, e.g., Theorem 7). Needless to say, in the latter distributions both yes-instances and no-instances appear with noticeable probability.

[5] Recall that a function $\mu : \mathbb{N} \to \mathbb{N}$ is negligible if for every positive polynomial q and all sufficiently large n it holds that $\mu(n) < 1/q(n)$. We say that A errs on x if $A(x)$ differs from the indicator value of the predicate $x \in S$.

Step 3: Identifying the Class of Interesting Problems. Seeking to identify reasonable distributional versions of $\mathcal{NP}$, we note that two extreme choices should be avoided. On the one hand, we must limit the class of admissible distributions so as to prevent the collapse of average-case hardness to worst-case hardness (by a selection of a pathological distribution that resides on the "worst case" instances). On the other hand, we should allow for various types of natural distributions rather than confining attention merely to the uniform distribution.[6] Recall that our aim is addressing all possible input distributions that may occur in applications, and thus there is no justification for confining attention to the uniform distribution. Still, arguably, the distributions occuring in applications are "relatively simple" and so we seek to identify a class of simple distributions. One such notion (of simple distributions) underlies the following definition, while a more liberal notion will be presented in Section 3.2.

Definition 2 (the class dist$\mathcal{NP}$): *We say that a probability ensemble $X = \{X_n\}_{n \in \mathbb{N}}$ is simple if there exists a polynomial time algorithm that, on any input $x \in \{0,1\}^*$, outputs $\Pr[X_{|x|} \leq x]$, where the inequality refers to the standard lexicographic order of strings. We denote by dist$\mathcal{NP}$ the class of distributional problems consisting of decision problems in $\mathcal{NP}$ coupled with simple probability ensembles.*

Note that the uniform probability ensemble is simple, but so are many other "simple" probability ensembles. Actually, it makes sense to relax the definition such that the algorithm is only required to output an approximation of $\Pr[X_{|x|} \leq x]$, say, to within a factor of $1 \pm 2^{-2|x|}$. We note that Definition 2 interprets simplicity in computational terms; specifically, as the feasibility of answering very basic questions regarding the probability distribution (i.e., determining the probability mass assigned to a single (n-bit long) string and even to an interval of such strings).

Doudts Regarding Definition 2. We admit that the identification of simple distributions as the class of interesting distribution is significantly more questionable than any other identification advocated in this book. Nevertheless, we believe that we were fully justified in rejecting both the aforementioned extremes (i.e., of either allowing all distributions or allowing only the uniform distribution). Yet, the reader may wonder whether or not we have struck the right balance between "generality" and "simplicity" (in the intuitive sense). One specific concern is that we might have restricted the class of distributions too much. We briefly address this concern next.

A more intuitive and very robust class of distributions, which seems to contain all distributions that may occur in applications, is the class of polynomial-time sampleable probability ensembles (treated in Section 3.2). Fortunately, the

---

[6] Confining attention to the uniform distribution seems misguided by the naive belief according to which this distribution is the *only* one relevant to applications. In contrast, we believe that, for most natural applications, the uniform distribution over instances is not relevant at all.

combination of the results presented in Section 2.2 and Section 3.2 seems to retrospectively endorse the choice underlying Definition 2. Specifically, we note that enlarging the class of distributions weakens the *conjecture* that the corresponding class of distributional NP problems contains infeasible problems. On the other hand, the *conclusion* that a specific distributional problem is not feasible becomes more appealing when the problem belongs to a smaller class that corresponds to a restricted definition of admissible distributions. Now, the combined results of Section 2.2 and Section 3.2 assert that a conjecture that refers to the larger class of polynomial-time sampleable ensembles implies a conclusion that refers to a (very) simple probability ensemble (which resides in the smaller class). Thus, the current setting in which both the conjecture and the conclusion refer to simple probability ensembles may be viewed as just an intermediate step.

Does dist$\mathcal{NP}$ *Contain Only Feasible Problems?* Indeed, the big question in the current context is whether dist$\mathcal{NP}$ is contained in tpc$\mathcal{P}$. A positive answer (especially if extended to sampleable ensembles) would deem the P-vs-NP Question to be of little practical significant. However, our daily experience as well as much research effort indicate that some NP problems are not merely hard in the worst-case, but rather "typically hard". This leads to the *conjecture that* dist$\mathcal{NP}$ *is not contained in* tpc$\mathcal{P}$.

Needless to say, the latter conjecture implies $\mathcal{P} \neq \mathcal{NP}$, and thus we should not expect to see a proof of it. In particular, we should not expect to see a proof that some specific problem in dist$\mathcal{NP}$ is not in tpc$\mathcal{P}$. What we may hope to see is "dist$\mathcal{NP}$-complete" problems; that is, problems in dist$\mathcal{NP}$ that are not in tpc$\mathcal{P}$ unless the entire class dist$\mathcal{NP}$ is contained in tpc$\mathcal{P}$. An adequate notion of a reduction is used towards formulating this possibility.

Step 4: Defining Reductions among (Distributional) Problems. Intuitively, such reductions must preserve average-case feasibility. Thus, in addition to the standard conditions (i.e., that the reduction be efficiently computable and yield a correct result), we require that the reduction "respects" the probability distribution of the corresponding distributional problems. Specifically, the reduction should not map very likely instances of the first ("starting") problem to rare instances of the second ("target") problem. Otherwise, having a typically polynomial-time algorithm for the second distributional problem does not necessarily yield such an algorithm for the first distributional problem. Following is the adequate analogue of a Cook reduction (i.e., general polynomial-time reduction), and the analogue of a Karp-reduction (many-to-one reduction) can be easily derived as a special case.[7]

Definition 3 (reductions among distributional problems): *We say that the oracle machine M* reduces *the distributional problem (S, X) to the distributional problem (T, Y) if the following three conditions hold.*

---

[7] See Footnote 9. We mention that the special case of many-to-one reductions, which suffices for the dist$\mathcal{NP}$-completeness results (e.g., Theorem 5).

1. Efficiency: *The machine M runs in polynomial-time.*[8]
2. Validity: *For every $x \in \{0,1\}^*$, it holds that $M^T(x) = 1$ if an only if $x \in S$, where $M^T(x)$ denotes the output of the oracle machine M on input x and access to an oracle for T.*
3. Domination:[9] *The probability that, on input X_n and oracle access to T, machine M makes the query y is upper-bounded by $\mathrm{poly}(|y|) \cdot \Pr[Y_{|y|} = y]$. That is, there exists a polynomial p such that, for every $y \in \{0,1\}^*$ and every $n \in \mathbb{N}$, it holds that*

$$\Pr[Q(X_n) \ni y] \leq p(|y|) \cdot \Pr[Y_{|y|} = y], \tag{1}$$

where $Q(x)$ denotes the set of queries made by M on input x and oracle access to T.

In addition, we require that the reduction does not make too short queries; that is, there exists a polynomial p' such that if $y \in Q(x)$ then $p'(|y|) \geq |x|$.

In this case we say that the distributional problem (S, X) is reducible *to the distributional problem (T, Y).*

The l.h.s. of Eq. (1) refers to the probability that, on input distributed as X_n, the reduction makes the query y. This probability is required not to exceed the probability that y occurs in the distribution $Y_{|y|}$ by more than a polynomial factor in $|y|$. In this case we say that the l.h.s. of Eq. (1) is dominated by $\Pr[Y_{|y|} = y]$.

Indeed, the domination condition is the only aspect of Definition 3 that extends beyond the worst-case treatment of reductions and refers to the distributional setting. The domination condition does not insist that the distribution induced by $Q(X)$ equals Y, but rather allows some slackness that, in turn, is bounded so to guarantee preservation of typical feasibility. [10]

Proposition 4 (typical feasibility is preserved by reduction): *Suppose that the distributional problem (S, X) is reducible to the distributional problem (T, Y), and that $(T, Y) \in \mathrm{tpc}\mathcal{P}$. Then, $(S, X) \in \mathrm{tpc}\mathcal{P}$.*

---

[8] In fact, one may relax the requirement and only require that M is typically polynomial-time with respect to X. The validity condition may also be relaxed similarly.

[9] Let us spell out the meaning of Eq. (1) in the special case of many-to-one reductions (i.e., $M^T(x) = 1$ if and only if $f(x) \in T$, where f is a polynomial-time computable function): in this case $\Pr[Q(X_n) \ni y]$ is replaced by $\Pr[f(X_n) = y]$. That is, Eq. (1) simplifies to $\Pr[f(X_n) = y] \leq p(|y|) \cdot \Pr[Y_{|y|} = y]$. Indeed, this condition holds vacuously for any y that is not in the image of f.

[10] We stress that the notion of domination is incomparable to the notion of statistical (resp., computational) indistinguishability. On one hand, domination is a local requirement (i.e., it compares the two distribution on a point-by-point basis), whereas indistinguishability is a global requirement (which allows rare exceptions). On the other hand, domination does not require approximately equal values, but rather a ratio that is bounded in one direction. Indeed, domination is not symmetric. We comment that a more relaxed notion of domination that allows rare violations (as in Footnote 8) suffices for the preservation of typical feasibility.

Proof Sketch: Let M, Q, p, and p' be as in Definition 3, and suppose that A is an algorithm that typically solves (T, Y) in polynomial-time. Let B denote the set of instances on which A errs (or runs more than polynomial time), and $B_m \stackrel{\text{def}}{=} B \cap \{0, 1\}^m$. By the domination condition, for every n, it holds that

$$\Pr[M^A(X_n) \text{ errs}] \leq \Pr[Q(X_n) \cap B \neq \emptyset]$$

$$\leq \sum_{m:p'(m) \geq n} \sum_{y \in B_m} \Pr[Q(X_n) \ni y]$$

$$\leq \sum_{m:p'(m) \geq n} \sum_{y \in B_m} p(m) \cdot \Pr[Y_m = y]$$

$$\leq \sum_{m:p'(m) \geq n} p(m) \cdot \Pr[Y_m \in B_m]$$

where the second (resp., third) inequality uses the additional (resp., main) guarantee in the domination condition. It follows that the probability that M^A errs on X_n is negligible (as a function of n). $\qquad\square$

Perspective. We note that the reducibility arguments that are extensively used in cryptopgraphy (see, e.g., [5, Chap. 2]) are actually reductions in the spirit of Definition 3 (except that they refer to different types of computational tasks).

2.2 Complete Problems

Recall that our conjecture is that dist$\mathcal{NP}$ is not contained in tpc$\mathcal{P}$, which in turn strengthens the conjecture $\mathcal{P} \neq \mathcal{NP}$ (making infeasibility a typical phenomenon rather than a worst-case one). Having no hope of proving that dist$\mathcal{NP}$ is not contained in tpc$\mathcal{P}$, we turn to the study of complete problems with respect to that conjecture. Specifically, we say that a distributional problem (S, X) is dist$\mathcal{NP}$-complete if $(S, X) \in$ dist$\mathcal{NP}$ and every $(S', X') \in$ dist$\mathcal{NP}$ is reducible to (S, X) (under Definition 3).

Distributional Bounded Halting. Recall that it is quite easy to prove the mere existence of NP-complete problems and that many natural problems are NP-complete. In contrast, in the current context, establishing completeness results is quite hard. This should not be surprising in light of the restricted type of reductions allowed in the current context. The restriction (captured by the domination condition) requires that "typical" instances of one problem should not be mapped to "untypical" instances of the other problem. In contrast, it is fair to say that standard Karp-reductions (used in establishing NP-completeness results) map "typical" instances of one problem to somewhat "bizarre" instances of the second problem. Thus, the current section may be viewed as a study of reductions that do not commit this sin.[11]

---

[11] The latter assertion is somewhat controversial. While this assertion seems totally justified with respect to the proof of Theorem 5, opinions regarding the proof of Theorem 7 may differ.

Theorem 5 (dist$\mathcal{NP}$-completeness): dist$\mathcal{NP}$ *contains a distributional problem* (T, Y) *such that each distributional problem in* dist$\mathcal{NP}$ *is reducible* (per Definition 3) *to* (T, Y). *Furthermore, the reductions are via many-to-one mappings.*

Proof: We start by introducing such a (distributional) problem, which is a natural distributional version of the "universal decision problem", denoted S_{u}, and often referred to as *Bounded Halting*. Specifically, we define S_{u} such that the instance $\langle M, x, 1^t \rangle$ is in S_{u} if there exists $y \in \cup_{i \leq t} \{0, 1\}^i$ such that machine M accepts the input pair (x, y) within t steps. We couple S_{u} with the "quasi-uniform" probability ensemble U' that assigns to the instance $\langle M, x, 1^t \rangle$ a probability mass proportional to $2^{-(|M|+|x|)}$. Specifically, for every $\langle M, x, 1^t \rangle$ it holds that

$$\Pr[U'_n = \langle M, x, 1^t \rangle] = \frac{2^{-(|M|+|x|)}}{\binom{n}{2}} \tag{2}$$

where $n \stackrel{\text{def}}{=} |\langle M, x, 1^t \rangle| \stackrel{\text{def}}{=} |M| + |x| + t$. Note that, under a suitable natural encoding, the ensemble U' is indeed simple.[12]

The reader can easily verify that the generic reduction used when reducing any set in $\mathcal{NP}$ to S_{u} (see the proof of [6, Thm. 2.19]), fails to reduce dist$\mathcal{NP}$ to (S_{u}, U'). Specifically, in some cases (see next paragraph), these reductions do not satisfy the domination condition. Indeed, the difficulty is that we have to reduce all dist$\mathcal{NP}$ problems (i.e., pairs consisting of decision problems and simple distributions) to one single distributional problem (i.e., (S_{u}, U')). In contrast, considering the distributions induced by the aforementioned reductions, we end up with many distributional versions of S_{u}, and furthermore the corresponding distributions are very different (and are not necessarily dominated by a single distribution).

Let us take a closer look at the aforementioned generic reduction (of S to S_{u}), when applied to an arbitrary $(S, X) \in$ dist$\mathcal{NP}$. This reduction maps an instance x to a triple $(M_S, x, 1^{p_S(|x|)})$, where M_S is a machine verifying membership in S (while using adequate NP-witnesses) and p_S is an adequate polynomial. The problem is that x may have relatively large probability mass (i.e., it may be that $\Pr[X_{|x|} = x] \gg 2^{-|x|}$) while $(M_S, x, 1^{p_S(|x|)})$ has "uniform" probability mass (i.e., $\langle M_S, x, 1^{p_S(|x|)} \rangle$ has probability mass smaller than $2^{-|x|}$ in U'). This violates the domination condition, and thus an alternative reduction is required.

The key to the alternative reduction is an (efficiently computable) encoding of strings taken from an arbitrary *simple* distribution by strings that have a similar probability mass under the uniform distribution. This means that the encoding should shrink strings that have relatively large probability mass under the original distribution. Specifically, this encoding will map x (taken from the ensemble

---

[12] For example, we may encode $\langle M, x, 1^t \rangle$, where $M = \sigma_1 \cdots \sigma_k \in \{0, 1\}^k$ and $x = \tau_1 \cdots \tau_\ell \in \{0, 1\}^\ell$, by the string $\sigma_1 \sigma_1 \cdots \sigma_k \sigma_k 01 \tau_1 \tau_1 \cdots \tau_\ell \tau_\ell 01^t$. Then $\binom{n}{2} \cdot \Pr[U'_n \leq \langle M, x, 1^t \rangle]$ equals $(i_{|M|,|x|,t} - 1) + 2^{-|M|} \cdot |\{M' \in \{0, 1\}^{|M|} : M' < M\}| + 2^{-(|M|+|x|)} \cdot |\{x' \in \{0, 1\}^{|x|} : x' \leq x\}|$, where $i_{k,\ell,t}$ is the ranking of $\{k, k+\ell\}$ among all 2-subsets of $[k + \ell + t]$.

$\{X_n\}_{n\in\mathbb{N}})$ to a codeword x' of length that is upper-bounded by the logarithm of $1/\Pr[X_{|x|}=x]$, ensuring that $\Pr[X_{|x|}=x] = O(2^{-|x'|})$. Accordingly, the reduction will map x to a triple $(M_{S,X}, x', 1^{p'(|x|)})$, where $|x'| < O(1) + \log_2(1/\Pr[X_{|x|}=x])$ and $M_{S,X}$ is an algorithm that (given x' and x) first verifies that x' is a proper encoding of x and next applies the standard verification (i.e., M_S) of the problem S. Such a reduction will be shown to satisfy all three conditions (i.e., efficiency, validity, and domination). Thus, instead of forcing the structure of the original distribution X on the target distribution U', the reduction will incorporate the structure of X in the reduced instance. A key ingredient in making this possible is the fact that X is simple (as per Definition 2).

With the foregoing motivation in mind, we now turn to the actual proof; that is, proving that any $(S, X) \in \text{dist}\mathcal{NP}$ is reducible to (S_u, U'). The following technical lemma is the basis of the reduction. In this lemma as well as in the sequel, it will be convenient to consider the (accumulative) distribution function of the probability ensemble X. That is, we consider $\mu(x) \stackrel{\text{def}}{=} \Pr[X_{|x|} \leq x]$, and note that $\mu : \{0,1\}^* \to [0,1]$ is polynomial-time computable (because X satisfies Definition 2).

Coding Lemma:[13] Let $\mu : \{0,1\}^* \to [0,1]$ be a polynomial-time computable function that is monotonically non-decreasing over $\{0,1\}^n$ for every n (i.e., $\mu(x') \leq \mu(x'')$ for any $x' < x'' \in \{0,1\}^{|x'|}$). For $x \in \{0,1\}^n \setminus \{0^n\}$, let $x - 1$ denote the string preceding x in the lexicographic order of n-bit long strings. Then there exist an encoding function C_μ that satisfies the following three conditions.

1. **Compression:** For every x it holds that $|C_\mu(x)| \leq 1 + \min\{|x|, \log_2(1/\mu'(x))\}$, where $\mu'(x) \stackrel{\text{def}}{=} \mu(x) - \mu(x-1)$ if $x \notin \{0\}^*$ and $\mu'(0^n) \stackrel{\text{def}}{=} \mu(0^n)$ otherwise.
2. **Efficient Encoding:** The function C_μ is computable in polynomial-time.
3. **Unique Decoding:** For every $n \in \mathbb{N}$, when restricted to $\{0,1\}^n$, the function C_μ is one-to-one (i.e., if $C_\mu(x) = C_\mu(x')$ and $|x| = |x'|$ then $x = x'$).

Proof: The function C_μ is defined as follows. If $\mu'(x) \leq 2^{-|x|}$ then $C_\mu(x) = 0x$ (i.e., in this case x serves as its own encoding). Otherwise (i.e., $\mu'(x) > 2^{-|x|}$) then $C_\mu(x) = 1z$, where z is chosen such that $|z| \leq \log_2(1/\mu'(x))$ and the mapping of n-bit strings to their encoding is one-to-one. Loosely speaking, z is selected to equal the shortest binary expansion of a number in the interval $(\mu(x) - \mu'(x), \mu(x)]$. Bearing in mind that this interval has length $\mu'(x)$ and that the different intervals are disjoint, we obtain the desired encoding. Details follows.

We focus on the case that $\mu'(x) > 2^{-|x|}$, and detail the way that z is selected (for the encoding $C_\mu(x) = 1z$). If $x > 0^{|x|}$ and $\mu(x) < 1$, then we let z be the longest common prefix of the binary expansions of $\mu(x-1)$ and $\mu(x)$; for example, if $\mu(1010) = 0.10010$ and $\mu(1011) = 0.10101111$ then $C_\mu(1011) = 1z$

---

[13] The lemma actually refers to $\{0,1\}^n$, for any fixed value of n, but the efficiency condition is stated more easily when allowing n to vary (and using the standard asymptotic analysis of algorithms). Actually, the lemma is somewhat easier to state and establish for polynomial-time computable functions that are monotonically non-decreasing over $\{0,1\}^*$ (rather than over $\{0,1\}^n$); see [4, Sec. 3].

with $z = 10$. Thus, in this case $0.z1$ is in the interval $(\mu(x-1), \mu(x)]$ (i.e., $\mu(x-1) < 0.z1 \leq \mu(x)$). For $x = 0^{|x|}$, we let z be the longest common prefix of the binary expansions of 0 and $\mu(x)$ and again $0.z1$ is in the relevant interval (i.e., $(0, \mu(x)]$). Finally, for x such that $\mu(x) = 1$ and $\mu(x-1) < 1$, we let z be the longest common prefix of the binary expansions of $\mu(x-1)$ and $1 - 2^{-|x|-1}$, and again $0.z1$ is in $(\mu(x-1), \mu(x)]$ (because $\mu'(x) > 2^{-|x|}$ and $\mu(x-1) < \mu(x) = 1$ imply that $\mu(x-1) < 1 - 2^{-|x|} < \mu(x)$). Note that if $\mu(x) = \mu(x-1) = 1$ then $\mu'(x) = 0 < 2^{-|x|}$.

We now verify that the foregoing C_μ satisfies the conditions of the lemma. We start with the compression condition. Clearly, if $\mu'(x) \leq 2^{-|x|}$ then $|C_\mu(x)| = 1 + |x| \leq 1 + \log_2(1/\mu'(x))$. On the other hand, suppose that $\mu'(x) > 2^{-|x|}$ and let us focus on the sub-case that $x > 0^{|x|}$ and $\mu(x) < 1$. Let $z = z_1 \cdots z_\ell$ be the longest common prefix of the binary expansions of $\mu(x-1)$ and $\mu(x)$. Then, $\mu(x-1) = 0.z0u$ and $\mu(x) = 0.z1v$, where $u, v \in \{0,1\}^*$. We infer that

$$\mu'(x) = \mu(x) - \mu(x-1) \leq \left(\sum_{i=1}^{\ell} 2^{-i} z_i + \sum_{i=\ell+1}^{\text{poly}(|x|)} 2^{-i}\right) - \sum_{i=1}^{\ell} 2^{-i} z_i < 2^{-|z|},$$

and $|z| < \log_2(1/\mu'(x)) \leq |x|$ follows. Thus, $|C_\mu(x)| \leq 1 + \min(|x|, \log_2(1/\mu'(x)))$ holds in both cases. Clearly, C_μ can be computed in polynomial-time by computing $\mu(x-1)$ and $\mu(x)$. Finally, note that C_μ satisfies the unique decoding condition, by separately considering the two aforementioned cases (i.e., $C_\mu(x) = 0x$ and $C_\mu(x) = 1z$). Specifically, in the second case (i.e., $C_\mu(x) = 1z$), use the fact that $\mu(x-1) < 0.z1 \leq \mu(x)$.   □

In order to obtain an encoding that is one-to-one when applied to strings of different lengths, we augment C_μ in the obvious manner; that is, we consider $C'_\mu(x) \stackrel{\text{def}}{=} (|x|, C_\mu(x))$, which may be implemented as $C'_\mu(x) = \sigma_1 \sigma_1 \cdots \sigma_\ell \sigma_\ell 01 C_\mu(x)$ where $\sigma_1 \cdots \sigma_\ell$ is the binary expansion of $|x|$. Note that $|C'_\mu(x)| = O(\log|x|) + |C_\mu(x)|$ and that C'_μ is one-to-one (over $\{0,1\}^*$).

The machine associated with (S, X). Let μ be the accumulative probability function associated with the probability ensemble X, and M_S be the polynomial-time machine that verifies membership in S while using adequate NP-witnesses (i.e., $x \in S$ if and only if there exists $y \in \{0,1\}^{\text{poly}(|x|)}$ such that $M(x, y) = 1$). Using the encoding function C'_μ, we introduce an algorithm $M_{S,\mu}$ with the intension of reducing the distributional problem (S, X) to (S_u, U') such that all instances (of S) are mapped to triples in which the first element equals $M_{S,\mu}$. Machine $M_{S,\mu}$ is given an alleged encoding (under C'_μ) of an instance to S along with an alleged proof that the corresponding instance is in S, and verifies these claims in the obvious manner. That is, on input x' and $\langle x, y \rangle$, machine $M_{S,\mu}$ first verifies that $x' = C'_\mu(x)$, and next verifiers that $x \in S$ by running $M_S(x, y)$. Thus, $M_{S,\mu}$ verifies membership in the set $S' = \{C'_\mu(x) : x \in S\}$, while using proofs of the form $\langle x, y \rangle$ such that $M_S(x, y) = 1$ (for the instance $C'_\mu(x)$).[14]

---

[14] Note that $|y| = \text{poly}(|x|)$, but $|x| = \text{poly}(|C'_\mu(x)|)$ does not necessarily hold (and so S' is not necessarily in $\mathcal{NP}$). As we shall see, the latter point is immaterial.

The reduction. We maps an instance x (of S) to the triple $(M_{S,\mu}, C'_\mu(x), 1^{p(|x|)})$, where $p(n) \stackrel{\text{def}}{=} p_S(n) + p_C(n)$ such that p_S is a polynomial representing the running-time of M_S and p_C is a polynomial representing the running-time of the encoding algorithm.

Analyzing the reduction. Our goal is proving that *the foregoing mapping constitutes a reduction of (S, X) to (S_u, U').* We verify the corresponding three requirements (of Definition 3).

1. Using the fact that C'_μ is polynomial-time computable (and noting that p is a polynomial), it follows that the foregoing mapping can be computed in polynomial-time.
2. Recall that, on input $(x', \langle x, y \rangle)$, machine $M_{S,\mu}$ accepts if and only if $x' = C'_\mu(x)$ and M_S accepts (x, y) within $p_S(|x|)$ steps. Using the fact that $C'_\mu(x)$ uniquely determines x, it follows that $x \in S$ if and only if $C'_\mu(x) \in S'$, which in turn holds if and only if there exists a string y such that $M_{S,\mu}$ accepts $(C'_\mu(x), \langle x, y \rangle)$ in at most $p(|x|)$ steps. Thus, $x \in S$ if and only if $(M_{S,\mu}, C'_\mu(x), 1^{p(|x|)}) \in S_u$, and the validity condition follows.
3. In order to verify the domination condition, we first note that the foregoing mapping is one-to-one (because the transformation $x \to C'_\mu(x)$ is one-to-one). Next, we note that it suffices to consider instances of S_u that have a preimage under the foregoing mapping (since instances with no preimage trivially satisfy the domination condition). Each of these instances (i.e., each image of this mapping) is a triple with the first element equal to $M_{S,\mu}$ and the second element being an encoding under C'_μ. By the definition of U', for every such image $\langle M_{S,\mu}, C'_\mu(x), 1^{p(|x|)} \rangle \in \{0, 1\}^n$, it holds that

$$\Pr[U'_n = \langle M_{S,\mu}, C'_\mu(x), 1^{p(|x|)} \rangle] = \binom{n}{2}^{-1} \cdot 2^{-(|M_{S,\mu}| + |C'_\mu(x)|)}$$
$$> c \cdot n^{-2} \cdot 2^{-(|C_\mu(x)| + O(\log |x|))},$$

where $c = 2^{-|M_{S,\mu}|-1}$ is a constant depending only on S and μ (i.e., on the distributional problem (S, X)). Thus, for some positive polynomial q, we have

$$\Pr[U'_n = \langle M_{S,\mu}, C'_\mu(x), 1^{p(|x|)} \rangle] > 2^{-|C_\mu(x)|}/q(n). \tag{3}$$

By virtue of the compression condition (of the Coding Lemma), we have $2^{-|C_\mu(x)|} \geq 2^{-1-\min(|x|, \log_2(1/\mu'(x)))}$. It follows that

$$2^{-|C_\mu(x)|} \geq \Pr[X_{|x|} = x]/2. \tag{4}$$

Recalling that x is the only preimage that is mapped to $\langle M_{S,\mu}, C'_\mu(x), 1^{p(|x|)} \rangle$ and combining Eq. (3) & (4), we establish the domination condition.

The theorem follows. ∎

Reflections: The proof of Theorem 5 highlights the fact that the reduction used in establishing the NP-completeness of S_u does not introduce much structure in the reduced instances (i.e., does not reduce the original problem to a "highly structured special case" of the target problem). Put in other words, unlike more advanced worst-case reductions, this reduction does not map "random" (i.e., uniformly distributed) instances to highly structured instances (which occur with negligible probability under the uniform distribution). Thus, the reduction used in establishing the NP-completeness of S_u suffices for reducing any distributional problem in dist$\mathcal{NP}$ to a distributional problem consisting of S_u coupled with *some* simple probability ensemble.[15]

However, Theorem 5 states more than the latter assertion. That is, it states that any distributional problem in dist$\mathcal{NP}$ is reducible to the *same* distributional version of S_u. Indeed, the effort involved in proving Theorem 5 was due to the need for mapping instances taken from any simple probability ensemble (which may not be the uniform ensemble) to instances distributed in a manner that is dominated by a single probability ensemble (i.e., the quasi-uniform ensemble U').

Other dist$\mathcal{NP}$-complete Problems. Once we have established the existence of one dist$\mathcal{NP}$-complete problem, we may establish the dist$\mathcal{NP}$-completeness of other problems (in dist$\mathcal{NP}$) by reducing some dist$\mathcal{NP}$-complete problem to them (and relying on the transitivity of reductions).[16] Thus, the difficulties encountered in the proof of Theorem 5 are no longer relevant. Unfortunately, a seemingly more severe difficulty arises: almost all known reductions in the theory of NP-completeness work by introducing much structure in the reduced instances (i.e., they actually reduce to highly structured special cases). Furthermore, this structure is too complex in the sense that the distribution of reduced instances does not seem simple (in the sense of Definition 2). Actually, as demonstrated next, the problem is not the existence of a structure in the reduced instances but rather the complexity of this structure. In particular, if the aforementioned reduction is "monotone" and "length regular" then the distribution of the reduced instances is simple enough (i.e., is simple in the sense of Definition 2):

Proposition 6 (sufficient condition for dist$\mathcal{NP}$-completeness): *Suppose that f is a Karp-reduction of the set S to the set T such that, for every $x', x'' \in \{0, 1\}^*$, the following two conditions hold:*

1. *(f is monotone): If $x' < x''$ then $f(x') < f(x'')$, where the inequalities refer to the standard lexicographic order of strings.[17]*
2. *(f is length-regular): $|x'| = |x''|$ if and only if $|f(x')| = |f(x'')|$.*

---

[15] Note that this cannot be said of most known Karp-reductions, which do map random instances to highly structured ones.

[16] When establishing the transitivity of reductions, it is again essential to use the additional guarantee in the domination condition. Compare Proposition 4.

[17] In particular, if $|z'| < |z''|$ then $z' < z''$. Recall that for $|z'| = |z''|$ it holds that $z' < z''$ if and only if there exists $w, u', u'' \in \{0, 1\}^*$ such that $z' = w0u'$ and $z'' = w1u''$.

Then, if there exists an ensemble X such that (S, X) is dist$\mathcal{NP}$-complete, then there exists an ensemble Y such that (T, Y) is dist$\mathcal{NP}$-complete.

Proof Sketch: Note that the monotonicity of f implies that f is one-to-one and that for every x it holds that $f(x) \geq x$. Furthermore, as shown next, f is polynomial-time invertible. Intuitively, the fact that f is both monotone and polynomial-time computable implies that a preimage can be found by a binary search. Specifically, given $y = f(x)$, we search for x by iteratively halving the interval of potential solutions, which is initialized to $[0, y]$ (since $x \leq f(x)$). Note that if this search is invoked on a string y that is not in the image of f, then it terminates while detecting this fact.

Relying on the fact that f is one-to-one (and length-regular), we define the probability ensemble $Y = \{Y_n\}_{n \in \mathbb{N}}$ such that for every x it holds that $\Pr[Y_{|f(x)|} = f(x)] = \Pr[X_{|x|} = x]$. Specifically, letting $\ell(m) = |f(1^m)|$ and noting that ℓ is one-to-one and monotonically non-decreasing, we define

$$\Pr[Y_{|y|} = y] = \begin{cases} \Pr[X_{|x|} = x] & \text{if } x = f^{-1}(y) \\ 0 & \text{if } \exists m \text{ s.t. } y \in \{0,1\}^{\ell(m)} \setminus \{f(x) : x \in \{0,1\}^m\} \\ 2^{-|y|} & \text{otherwise (i.e., if } |y| \notin \{\ell(m) : m \in \mathbb{N}\})^{18}. \end{cases}$$

Clearly, (S, X) is reducible to (T, Y) (via the Karp-reduction f, which, due to our construction of Y, also satisfies the domination condition). Thus, using the hypothesis that dist$\mathcal{NP}$ is reducible to (S, X) and the transitivity of reductions, it follows that every problem in dist$\mathcal{NP}$ is reducible to (T, Y). The key observation, to be established next, is that Y is a simple probability ensemble, and it follows that (T, Y) is in dist$\mathcal{NP}$.

Loosely speaking, the simplicity of Y follows by combining the simplicity of X and the properties of f (i.e., the fact that f is monotone, length-regular, and polynomial-time invertible). The monotonicity and length-regularity of f implies that $\Pr[Y_{|f(x)|} \leq f(x)] = \Pr[X_{|x|} \leq x]$. More generally, for any $y \in \{0,1\}^{\ell(m)}$, it holds that $\Pr[Y_{\ell(m)} \leq y] = \Pr[X_m \leq x]$, where x is the lexicographicly largest string such that $f(x) \leq y$ (and, indeed, if $|x| < m$ then $\Pr[Y_{\ell(m)} \leq y] = \Pr[X_m \leq x] = 0$).[19] Note that this x can be found in polynomial-time by the inverting algorithm sketched in the first paragraph of the proof. Thus, we may compute $\Pr[Y_{|y|} \leq y]$ by finding the adequate x and computing $\Pr[X_{|x|} \leq x]$. Using the hypothesis that X is simple, it follows that Y is simple (and the proposition follows). $\square$

On the Existence of Adequate Karp-Reductions. Proposition 6 implies that a sufficient condition for the dist$\mathcal{NP}$-completeness of a distributional version of a (NP-complete) set T is the existence of an adequate Karp-reduction from the set S_u to the set T; that is, this Karp-reduction should be monotone and length-regular. While the length-regularity condition seems easy to impose (by

---

[18] Having Y_n be uniform in this case is a rather arbitrary choice, which is merely aimed at guaranteeing a "simple" distribution on n-bit strings (also in this case).

[19] We also note that the case in which $|y|$ is not in the image of ℓ can be easily detected and taken care off accordingly.

using adequate padding), the monotonicity condition seems more problematic. Fortunately, it turns out that the monotonicity condition can also be imposed by using adequate padding (or rather an adequate "marking" – see [6, Exer. 2.30] and [6, Exer. 10.21]. We highlight the fact that the existence of an adequate padding (or "marking") is a property of the set T itself, and mention that all popular NP-complete sets satisfy it. Observing that any Karp-reduction to a "monotonically markable" set T can be transformed into a Karp-reduction (to T) that is monotone and length-regular, we conclude that *any natural NP-complete decision problem can be coupled with a simple probability ensemble such that the resulting distributional problem is* dist$\mathcal{NP}$-complete. As a concrete illustration of this thesis, we state the corresponding (formal) result for the twenty-one NP-complete problems treated in Karp's paper on NP-completeness [9].

Theorem 7 (a modest version of a general thesis): *For each of the twenty-one NP-complete problems treated in [9] there exists a simple probability ensemble such that the combined distributional problem is* dist$\mathcal{NP}$-complete.

The said list of problems includes SAT, Clique, and 3-Colorability.

2.3   Probabilistic Versions

The definitions in Section 2.1 can be extended so to account for randomized computations. For example, extending Definition 1, we have:

Definition 8 (the class tpc$\mathcal{BPP}$): *For a probabilistic algorithm A, a Boolean function f, and a time-bound function $t : \mathbb{N} \to \mathbb{N}$, we say that the string x is t-bad for A with respect to f if with probability exceeding $1/3$, on input x, either $A(x) \neq f(x)$ or A runs more that $t(|x|)$ steps. We say that A typically solves $(S, \{X_n\}_{n \in \mathbb{N}})$ in probabilistic polynomial-time if there exists a polynomial p such that the probability that X_n is p-bad for A with respect to the characteristic function of S is negligible. We denote by* tpc$\mathcal{BPP}$ *the class of distributional problems that are typically solvable in probabilistic polynomial-time.*

The definition of reductions can be similarly extended. This means that in Definition 3, both $M^T(x)$ and $Q(x)$ (mentioned in Items 2 and 3, respectively) are random variables rather than fixed objects. Furthermore, validity is required to hold (for every input) only with probability $2/3$, where the probability space refers only to the internal coin tosses of the reduction. Randomized reductions are closed under composition and preserve typical feasibility.

Randomized reductions allow the presentation of a dist$\mathcal{NP}$-complete problem that refers to the (perfectly) uniform ensemble. Recall that Theorem 5 establishes the dist$\mathcal{NP}$-completeness of (S_u, U'), where U' is a quasi-uniform ensemble (i.e., $\Pr[U'_n = \langle M, x, 1^t \rangle] = 2^{-(|M|+|x|)}/\binom{n}{2}$, where $n = |\langle M, x, 1^t \rangle|$). We first note that (S_u, U') can be randomly reduced to (S'_u, U''), where $S'_u = \{\langle M, x, z \rangle : \langle M, x, 1^{|z|} \rangle \in S_u\}$ and $\Pr[U''_n = \langle M, x, z \rangle] = 2^{-(|M|+|x|+|z|)}/\binom{n}{2}$ for every $\langle M, x, z \rangle \in \{0,1\}^n$. The randomized reduction consists of mapping $\langle M, x, 1^t \rangle$ to $\langle M, x, z \rangle$, where z is uniformly selected in $\{0,1\}^t$. Recalling that

$U = \{U_n\}_{n\in\mathbb{N}}$ denotes the uniform probability ensemble (i.e., U_n is uniformly distributed on strings of length n) and using a suitable encoding we get.

Proposition 9 (dist$\mathcal{NP}$-completeness w.r.t the uniform distribition): *There exists $S \in \mathcal{NP}$ such that every $(S', X') \in$ dist$\mathcal{NP}$ is randomly reducible to (S, U).*

Proof Sketch: By the forgoing discussion, every $(S', X') \in$ dist$\mathcal{NP}$ is randomly reducible to (S'_u, U''), where the reduction goes through (S_u, U'). Thus, we focus on reducing (S'_u, U'') to (S''_u, U), where $S''_\mathrm{u} \in \mathcal{NP}$ is defined as follows. The string $\mathrm{bin}_\ell(|u|) \cdot \mathrm{bin}_\ell(|v|) \cdot u \cdot v \cdot w$ is in S''_u if and only if $\langle u, v, w \rangle \in S'_\mathrm{u}$ and $\ell = \lceil \log_2 |uvw| \rceil + 1$, where $\mathrm{bin}_\ell(i)$ denotes the ℓ-bit long binary encoding of the integer $i \in [2^{\ell-1}]$ (i.e., the encoding is padded with zeros to a total length of ℓ). The reduction maps $\langle M, x, z \rangle$ to the string $\mathrm{bin}_\ell(|x|) \cdot \mathrm{bin}_\ell(|M|) \cdot M \cdot x \cdot z$, where $\ell = \lceil \log_2(|M| + |x| + |z|) \rceil + 1$. Noting that this reduction satisfies all conditions of Definition 3, the proposition follows. $\qquad\square$

3 Ramifications

In our opinion, the most problematic aspect of the theory described in Section 2 is the choice to focus on simple probability ensembles, which in turn restricts "distributional versions of NP" to the class dist$\mathcal{NP}$ (Definition 2). As indicated Section 2.1, this restriction raises two opposite concerns (i.e., that dist$\mathcal{NP}$ is either too wide or too narrow).[20] Here we address the concern that the class of simple probability ensembles is too restricted, and consequently that the conjecture dist$\mathcal{NP} \not\subseteq$ tpc$\mathcal{BPP}$ is too strong (which would mean that dist$\mathcal{NP}$-completeness is a weak evidence for typical-case hardness). An appealing extension of the class of simple probability ensembles is presented in Section 3.2, yielding an corresponding extension of dist$\mathcal{NP}$, and it is shown that *if this extension of* dist$\mathcal{NP}$ *is not contained in* tpc$\mathcal{BPP}$, *then* dist$\mathcal{NP}$ *itself is not contained in* tpc$\mathcal{BPP}$. Consequently, dist$\mathcal{NP}$-complete problems enjoy the benefit of both being in the more restricted class (i.e., dist$\mathcal{NP}$) and being hard as long as some problems in the extended class is hard.

A different extension appears in Section 3.1, where we extend the treatment from decision problems to search problems. This extension is motivated by the realization that search problem are actually of greater importance to real-life applications (see, e.g., discussions in [6, Sec. 2.1.1]) and hence a theory motivated by real-life applications must address such problems, as we do next.

Prerequisites: For the technical development of Section 3.1, we assume familiarity with the notion of unique solution and results regarding it (see, e.g., [6, Sec. 6.2.3]). For the technical development of Section 3.2, we assume familiarity with hashing functions (see, e.g., [6, Apdx. D.2]). In addition, the technical development of Section 3.2 relies on Section 3.1.

---

[20] On one hand, if the definition of dist$\mathcal{NP}$ were too liberal, then membership in dist$\mathcal{NP}$ would mean less than one may desire. On the other hand, if dist$\mathcal{NP}$ were too restricted, then the conjecture that dist$\mathcal{NP}$ contains hard problems would have been very questionable.

3.1   Search versus Decision

Indeed, as in the case of worst-case complexity, search problems are at least as important as decision problems. Thus, an average-case treatment of search problems is indeed called for. We first present distributional versions of the search problem classes $\mathcal{PF}$ and $\mathcal{PC}$ (which correspond to $\mathcal{P}$ and $\mathcal{NP}$, resp.),[21] following the underlying principles of the definitions of tpc$\mathcal{P}$ and dist$\mathcal{NP}$.

Definition 10 (the classes tpc$\mathcal{PF}$ and dist$\mathcal{PC}$): *We consider only polynomially bounded search problems; that is, binary relations $R \subseteq \{0,1\}^* \times \{0,1\}^*$ such that for some polynomial q it holds that $(x,y) \in R$ implies $|y| \leq q(|x|)$. We use the notation $R(x) \overset{\text{def}}{=} \{y : (x,y) \in R\}$ and $S_R \overset{\text{def}}{=} \{x : R(x) \neq \emptyset\}$.*

- *A distributional search problem consists of a polynomially bounded search problem coupled with a probability ensemble.*
- *The class tpc$\mathcal{PF}$ consists of all distributional search problems that are typically solvable in polynomial-time. That is, $(R, \{X_n\}_{n \in \mathbb{N}}) \in$ tpc$\mathcal{PF}$ if there exists an algorithm A and a polynomial p such that the probability that on input X_n algorithm A either errs or runs more that $p(n)$ steps is negligible, where A errs on $x \in S_R$ if $A(x) \notin R(x)$ and errs on $x \notin S_R$ if $A(x) \neq \perp$.*
- *A distributional search problem (R, X) is in dist$\mathcal{PC}$ if $R \in \mathcal{PC}$ and X is simple (as in Definition 2).*

Likewise, the class tpc$\mathcal{BPPF}$ consists of all distributional search problems that are typically solvable in *probabilistic* polynomial-time (cf., Definition 8). The definitions of *reductions among distributional problems*, presented in the context of decision problem, extend to search problems.

Fortunately, as in the context of worst-case complexity, the study of distributional search problems "reduces" to the study of distributional decision problems.

Theorem 11 (reducing search to decision): dist$\mathcal{PC} \subseteq$ tpc$\mathcal{BPPF}$ *if and only if* dist$\mathcal{NP} \subseteq$ tpc$\mathcal{BPP}$. *Furthermore, every problem in* dist$\mathcal{NP}$ *is reducible to some problem in* dist$\mathcal{PC}$, *and every problem in* dist$\mathcal{PC}$ *is randomly reducible to some problem in* dist$\mathcal{NP}$.

Proof Sketch: The furthermore part is analogous to the actual contents of the proof that the devision and search versions of the P-vs-NP question are equivalent (see, e.g., [6, Thm. 2.6] and [6, Thm. 2.16]). Indeed the standard reduction of $\mathcal{NP}$ to $\mathcal{PC}$ extends to the current context. Specifically, for any

---

[21] Specifically $\mathcal{PF}$ (standing for Polynomial-time Find) is the class of efficiently solvable search problems; that is, $R \in \mathcal{PF}$ if there exists a polynomial-time algorithm that on input x replies with $y \in R(x) \overset{\text{def}}{=} \{z : (x,z) \in R\}$ (and with $\perp$ if $R(x) = \emptyset$). The class $\mathcal{PC}$ (standing for Polynomial-time Check) is the class of search problems having efficiently checkable solutions; that is, the search problem of a polynomially bounded relation $R \subseteq \{0,1\}^* \times \{0,1\}^*$ is in $\mathcal{PC}$ if there exists a polynomial time algorithm A such that, for every x and y, it holds that $A(x,y) = 1$ if and only if $(x,y) \in R$. For more deatils, see [6, Sec. 2.1.1].

$S \in \mathcal{NP}$, we consider a relation $R \in \mathcal{PC}$ such that $S = \{x : R(x) \neq \emptyset\}$, and note that, for any probability ensemble X, the identity transformation reduces (S, X) to (R, X).

A difficulty arises in the opposite direction. Recall that in the standard reduction of $\mathcal{PC}$ to $\mathcal{NP}$, one reduces the search problem of $R \in \mathcal{PC}$ to deciding membership in $S'_R \stackrel{\text{def}}{=} \{\langle x, y' \rangle : \exists y'' \text{ s.t. } (x, y'y'') \in R\} \in \mathcal{NP}$. The difficulty encountered here is that, on input x, this reduction makes queries of the form $\langle x, y' \rangle$, where y' is a prefix of some string in $R(x)$. These queries may induce a distribution that is not dominated by any simple distribution. Thus, we seek an alternative reduction.

As a warm-up, let us assume for a moment that R has unique solutions; that is, for every x it holds that $|R(x)| \leq 1$. In this case we may easily reduce the search problem of $R \in \mathcal{PC}$ to deciding membership in $S''_R \in \mathcal{NP}$, *where* $\langle x, i, \sigma \rangle \in S''_R$ *if and only if $R(x)$ contains a string in which the ith bit equals σ.* Specifically, on input x, the reduction issues the queries $\langle x, i, \sigma \rangle$, where $i \in [\ell]$ (with $\ell = \text{poly}(|x|)$) and $\sigma \in \{0, 1\}$, which allows for determining the single string in the set $R(x) \subseteq \{0, 1\}^\ell$ (whenever $|R(x)| = 1$). The point is that this reduction can be used to reduce any $(R, X) \in \text{dist}\mathcal{PC}$ (having unique solutions) to $(S''_R, X'') \in \text{dist}\mathcal{NP}$, *where X'' equally distributes the probability mass of x* (under X) *to all the tuples $\langle x, i, \sigma \rangle$; that is, for every $i \in [\ell]$ and $\sigma \in \{0, 1\}$, it holds that* $\Pr[X''_{|\langle x,i,\sigma \rangle|} = \langle x, i, \sigma \rangle]$ *equals* $\Pr[X_{|x|} = x]/2\ell$.

Unfortunately, in the general case, R may not have unique solutions. Nevertheless, applying the main idea that underlies the reduction of NP to "unique-NP" (cf. [6, Thm. 6.29]), this difficulty can be overcome. We first note that the foregoing mapping of instances of the distributional problem $(R, X) \in \text{dist}\mathcal{PC}$ to instances of $(S''_R, X'') \in \text{dist}\mathcal{NP}$ satisfies the efficiency and domination conditions even in the case that R does not have unique solutions. What may possibly fail (in the general case) is the validity condition (i.e., if $|R(x)| > 1$ then we may fail to recover any element of $R(x)$).

Recall that the core of the reduction of NP to unique-NP is a randomized mapping of instances x (of any $R \in \mathcal{PC}$) to triples of the form (x, m, h) such that m is uniformly distributed in $[\ell]$ and h is uniformly distributed in a family of hashing function H^m_ℓ, where $\ell = \text{poly}(|x|)$ and H^m_ℓ is a family of pairwise indepence hashing functions. Furthermore, if $R(x) \neq \emptyset$ then, with probability $\Omega(1/\ell)$ over the choices of $m \in [\ell]$ and $h \in H^m_\ell$, there exists a unique $y \in R(x)$ such that $h(y) = 0^m$. Defining $R'(x, m, h) \stackrel{\text{def}}{=} \{y \in R(x) : h(y) = 0^m\}$, this yields a randomized reduction of the search problem of R to the search problem of R' such that with noticeable probability[22] the reduction maps instances that have solutions to instances having a unique solution. Furthermore, this reduction can be used to reduce any $(R, X) \in \text{dist}\mathcal{PC}$ to $(R', X') \in \text{dist}\mathcal{PC}$, *where X' distributes the probability mass of x* (under X) *to all the triples (x, m, h) such*

---

[22] Recall that the probability of an event is said to be noticeable (in a relevant parameter) if it is greater than the reciprocal of some positive polynomial. In the context of randomized reductions, the relevant parameter is the length of the input to the reduction.

that for every $m \in [\ell]$ and $h \in H_\ell^m$ it holds that $\Pr[X'_{|(x,m,h)|} = (x, m, h)]$ equals
$\Pr[X_{|x|} = x]/(\ell \cdot |H_\ell^m|)$. (Note that with a suitable encoding, X' is indeed simple.)

The theorem follows by combining the two aforementioned reductions. That is, we first apply the randomized reduction of (R, X) to (R', X'), and next reduce the resulting instance to an instance of the corresponding decision problem $(S_{R'}'', X'')$, where X'' is obtained by modifying X' (rather than X). The combined randomized mapping satisfies the efficiency and domination conditions, and is valid with noticeable probability. The error probability can be made negligible by straightforward amplification.   □

3.2   Simple versus Sampleable Distributions

Recall that the definition of simple probability ensembles (underlying Definition 2) requires that the accumulating distribution function is polynomial-time computable. Recall that $\mu : \{0, 1\}^* \to [0, 1]$ is called the **accumulating distribution function** of $X = \{X_n\}_{n \in \mathbb{N}}$ if for every $n \in \mathbb{N}$ and $x \in \{0, 1\}^n$ it holds that $\mu(x) \stackrel{\text{def}}{=} \Pr[X_n \leq x]$, where the inequality refers to the standard lexicographic order of n-bit strings.

As argued in Section 2.1, the requirement that the accumulating distribution function is polynomial-time computable imposes severe restrictions on the set of admissible ensembles. Furthermore, it seems that these simple ensembles are indeed "simple" in some intuitive sense, and that they represent a reasonable (alas disputable) model of distributions that may occur in practice. Still, in light of the fear that this model is too restrictive (and consequently that dist$\mathcal{NP}$-hardness is weak evidence for typical-case hardness), we seek a maximalistic model of distributions that may occur in practice. Such a model is provided by the notion of polynomial-time sampleable ensembles (underlying Definition 12). Our maximality thesis is based on the belief that the real world should be modeled as a *feasible* randomized process (rather than as an arbitrary process). This belief implies that all objects encountered in the world may be viewed as samples generated by a feasible randomized process.

Definition 12 (sampleable ensembles and the class samp$\mathcal{NP}$): *We say that a probability ensemble $X = \{X_n\}_{n \in \mathbb{N}}$ is* (polynomial-time) **sampleable** *if there exists a probabilistic polynomial-time algorithm A such that for every $x \in \{0, 1\}^*$ it holds that $\Pr[A(1^{|x|}) = x] = \Pr[X_{|x|} = x]$. We denote by* samp$\mathcal{NP}$ *the class of distributional problems consisting of decision problems in $\mathcal{NP}$ coupled with sampleable probability ensembles.*

We first note that all simple probability ensembles are indeed sampleable, and thus dist$\mathcal{NP} \subseteq$ samp$\mathcal{NP}$. On the other hand, there exist sampleable probability ensembles that do not seem simple (and so it seems that dist$\mathcal{NP} \subset$ samp$\mathcal{NP}$).

Extending the scope of distributional problems (from dist$\mathcal{NP}$ to samp$\mathcal{NP}$) facilitates the presentation of complete distributional problems. We first note that it is easy to prove that every natural NP-complete problem has a distributional version in samp$\mathcal{NP}$ that is dist$\mathcal{NP}$-hard. Furthermore, it is possible to prove that all natural NP-complete problem have distributional versions that are samp$\mathcal{NP}$-complete. (In both cases, "natural" means that the corresponding Karp-reductions do not shrink the input, which is a weaker condition than the one in Proposition 6.)

Theorem 13 (samp$\mathcal{NP}$-completeness): *Suppose that $S \in \mathcal{NP}$ and that every set in $\mathcal{NP}$ is reducible to S by a Karp-reduction that does not shrink the input. Then, there exists a polynomial-time sampleable ensemble X such that any problem in* samp$\mathcal{NP}$ *is reducible to* (S, X).

The proof of Theorem 13 is based on the observation that *there exists a polynomial-time sampleable ensemble that dominates all polynomial-time sampleable ensembles.* The existence of this ensemble is based on the notion of a universal (sampling) machine.

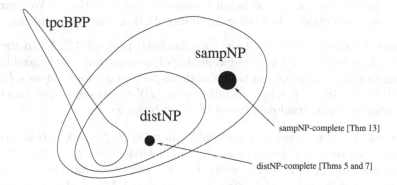

Fig. 1. Two types of average-case completeness

Theorem 13 establishes a rich theory of samp$\mathcal{NP}$-completeness, but does not relate this theory to the previously presented theory of dist$\mathcal{NP}$-completeness (see Figure 1). This is essentially done in the next theorem, which asserts that the existence of typically hard problems in samp$\mathcal{NP}$ implies their existence in dist$\mathcal{NP}$.

Theorem 14 (samp$\mathcal{NP}$-completeness   versus   dist$\mathcal{NP}$-completeness): *If* samp$\mathcal{NP}$ *is not contained in* tpc$\mathcal{BPP}$ *then* dist$\mathcal{NP}$ *is not contained in* tpc$\mathcal{BPP}$.

Thus, the two "typical-case complexity" versions of the P-vs-NP Question are equivalent. That is, if some "sampleable distribution" versions of NP are not typically feasible then some "simple distribution" versions of NP are not typically feasible. In particular, if samp$\mathcal{NP}$-complete problems are not in tpc$\mathcal{BPP}$ then dist$\mathcal{NP}$-complete problems are not in tpc$\mathcal{BPP}$.

The foregoing assertions would all follow if samp$\mathcal{NP}$ were (randomly) reducible to dist$\mathcal{NP}$ (i.e., if every problem in samp$\mathcal{NP}$ were reducible (under a randomized version of Definition 3) to some problem in dist$\mathcal{NP}$); but, unfortunately, we do not know whether such reductions exist. Yet, underlying the proof of Theorem 14 is a more liberal notion of a reduction among distributional problems.

Proof Sketch: We shall prove that if dist$\mathcal{NP}$ is contained in tpc$\mathcal{BPP}$ then the same holds for samp$\mathcal{NP}$ (i.e., samp$\mathcal{NP}$ is contained in tpc$\mathcal{BPP}$). Relying on Theorem 11 and an analogous result for the sampleable classes, it suffices to show that if dist$\mathcal{PC}$ is contained in tpc$\mathcal{BPPF}$, then the sampleable version of dist$\mathcal{PC}$, denoted samp$\mathcal{PC}$, is contained in tpc$\mathcal{BPPF}$. This will be shown by showing that, under a relaxed notion of a randomized reduction, every problem in samp$\mathcal{PC}$ is reduced to some problem in dist$\mathcal{PC}$. Loosely speaking, this relaxed notion (of a randomized reduction) only requires that the validity and domination conditions (of Definition 3 (when adapted to randomized reductions)) hold with respect to a noticeable fraction of the probability space of the reduction.[23] We start by formulating this notion, when referring to distributional *search* problems.

Definition: A relaxed reduction of the distributional problem (R, X) to the distributional problem (T, Y) is a probabilistic polynomial-time oracle machine M that satisfies the following conditions with respect to a family of sets $\{\Omega_x \subseteq \{0,1\}^{m(|x|)} : x \in \{0,1\}^*\}$, where $m(|x|) = \text{poly}(|x|)$ denotes an upper-bound on the number of the internal coin tosses of M on input x:

Density (of Ω_x): There exists a noticeable function $\rho : \mathbb{N} \to [0,1]$ (i.e., $\rho(n) > 1/\text{poly}(n)$) such that, for every $x \in \{0,1\}^*$, it holds that $|\Omega_x| \geq \rho(|x|) \cdot 2^{m(|x|)}$.

Validity (with respect to Ω_x): For every $r \in \Omega_x$ the reduction yields a correct answer; that is, $M^T(x, r) \in R(x)$ if $R(x) \neq \emptyset$ and $M^T(x, r) = \bot$ otherwise, where $M^T(x, r)$ denotes the execution of M on input x, internal coins r, and oracle access to T.

Domination (with respect to Ω_x): There exists a positive polynomial p such that, for every $y \in \{0,1\}^*$ and every $n \in \mathbb{N}$, it holds that

$$\Pr[Q'(X_n) \ni y] \leq p(|y|) \cdot \Pr[Y_{|y|} = y], \tag{5}$$

where $Q'(x)$ is a random variable, defined over the set Ω_x, representing the set of queries made by M on input x, coins in Ω_x, and oracle access to T. That is, $Q'(x)$ is defined by uniformly selecting $r \in \Omega_x$ and considering the set of queries made by M on input x, internal coins r, and oracle access to T. (In addition, as in Definition 3, we also require that the reduction does not make too short queries.)

---

[23] We warn that the existence of such a relaxed reduction between two specific distributional problems does not necessarily imply the existence of a corresponding (standard average-case) reduction. Specifically, although standard validity can be guaranteed (for problems in $\mathcal{PC}$) by repeated invocations of the reduction, such a process will *not* redeem the violation of the standard domination condition.

The reader may verify that this relaxed notion of a reduction preserves typical feasibility; that is, for $R \in \mathcal{PC}$, if there exists a relaxed reduction of (R, X) to (T, Y) and (T, Y) is in tpc$\mathcal{BPPF}$ then (R, X) is in tpc$\mathcal{BPPF}$. The key observation is that the analysis may discard the case that, on input x, the reduction selects coins not in Ω_x. Indeed, the queries made in that case may be untypical and the answers received may be wrong, but this is immaterial. What matter is that, on input x, with noticeable probability the reduction selects coins in Ω_x, and produces "typical with respect to Y" queries (by virtue of the relaxed domination condition). Such typical queries are answered correctly by the algorithm that typically solves (T, Y), and if x has a solution then these answers yield a correct solution to x (by virtue of the relaxed validity condition). Thus, if x has a solution then with noticeable probability the reduction outputs a correct solution. On the other hand, the reduction never outputs a wrong solution (even when using coins not in Ω_x), because incorrect solutions are detected by relying on $R \in \mathcal{PC}$.

Our goal is presenting, for every $(R, X) \in$ samp$\mathcal{PC}$, a relaxed reduction of (R, X) to a related problem $(R', X') \in$ dist$\mathcal{PC}$. (We use the standard notation $X = \{X_n\}_{n \in \mathbb{N}}$ and $X' = \{X'_n\}_{n \in \mathbb{N}}$.)

An oversimplified case: For starters, *suppose that X_n is uniformly distributed on some set $S_n \subseteq \{0,1\}^n$ and that there is a polynomial-time computable and invertible mapping μ of S_n to $\{0,1\}^{\ell(n)}$, where $\ell(n) = \log_2 |S_n|$. Then, mapping x to $1^{|x|-\ell(|x|)}0\mu(x)$, we obtain a reduction of (R, X) to (R', X'), where X'_{n+1} is uniform over $\{1^{n-\ell(n)}0v : v \in \{0,1\}^{\ell(n)}\}$ and $R'(1^{n-\ell(n)}0v) = R(\mu^{-1}(v))$ (or, equivalently, $R(x) = R'(1^{|x|-\ell(|x|)}0\mu(x))$). Note that X' is a simple ensemble and $R' \in \mathcal{PC}$; hence, $(R', X') \in$ dist$\mathcal{PC}$. Also note that the foregoing mapping is indeed a valid reduction (i.e., it satisfies the efficiency, validity, and domination conditions). Thus, (R, X) is reduced to a problem in dist$\mathcal{PC}$ (and indeed the relaxation was not used here).*

A simple but more instructive case: Next, we drop the assumption that there is a polynomial-time computable and invertible mapping μ of S_n to $\{0,1\}^{\ell(n)}$, but maintain the assumption that X_n is uniform on some set $S_n \subseteq \{0,1\}^n$ and assume that $|S_n| = 2^{\ell(n)}$ is easily computable (from n). In this case, we may map $x \in \{0,1\}^n$ to its image under a suitable randomly chosen hashing function h, which in particular maps n-bit strings to $\ell(n)$-bit strings. That is, we randomly map x to $(h, 1^{n-\ell(n)}0h(x))$, where h is uniformly selected in a set $H_n^{\ell(n)}$ of suitable hash functions (i.e., pairwise independent ones). This calls for redefining R' such that $R'(h, 1^{n-\ell(n)}0v)$ corresponds to the preimages of v under h that are in S_n. Assuming that h is a 1-1 mapping of S_n to $\{0,1\}^{\ell(n)}$, we may define $R'(h, 1^{n-\ell(n)}0v) = R(x)$ such that x is the unique string satisfying $x \in S_n$ and $h(x) = v$, where the condition $x \in S_n$ may be *verified by providing the internal coins of the sampling procedure that generate x*. Denoting the sampling procedure of X by S, and letting $S(1^n, r)$ denote the output of S on input 1^n and internal coins r, we actually redefine R' as

$$R'(h, 1^{n-\ell(n)}0v) = \{\langle r, y \rangle : h(S(1^n, r)) = v \wedge y \in R(S(1^n, r))\}. \tag{6}$$

We note that $\langle r, y \rangle \in R'(h, 1^{|x|-\ell(|x|)}0h(x))$ yields a desired solution $y \in R(x)$ if $S(1^{|x|}, r) = x$, but otherwise "all bets are off" (since y will be a solution for $S(1^{|x|}, r) \neq x$). Now, although typically h will not be a 1-1 mapping of S_n to $\{0, 1\}^{\ell(n)}$, it is the case that *for each $x \in S_n$, with constant probability over the choice of h, it holds that $h(x)$ has a unique preimage in S_n under h.* In this case $\langle r, y \rangle \in R'(h, 1^{|x|-\ell(|x|)}0h(x))$ implies $S(1^{|x|}, r) = x$ (which, in turn, implies $y \in R(x)$). We claim that *the randomized mapping of x to $(h, 1^{n-\ell(n)}0h(x))$, where h is uniformly selected in $H_{|x|}^{\ell(|x|)}$, yields a relaxed reduction of (R, X) to (R', X'), where $X'_{n'}$ is uniform over $H_n^{\ell(n)} \times \{1^{n-\ell(n)}0v : v \in \{0, 1\}^{\ell(n)}\}$.* Needless to say, the claim refers to the reduction that (on input x, makes the query $(h, 1^{n-\ell(n)}0h(x))$, and) returns y if the oracle answer equals $\langle r, y \rangle$ and $y \in R(x)$.

The claim is proved by considering the set Ω_x of choices of $h \in H_{|x|}^{\ell(|x|)}$ for which $x \in S_n$ is the only preimage of $h(x)$ under h that resides in S_n (i.e., $|\{x' \in S_n : h(x') = h(x)\}| = 1$). In this case (i.e., $h \in \Omega_x$) it holds that $\langle r, y \rangle \in R'(h, 1^{|x|-\ell(|x|)}0h(x))$ implies that $S(1^{|x|}, r) = x$ and $y \in R(x)$, and the (relaxed) validity condition follows. The (relaxed) domination condition follows by noting that $\Pr[X_n = x] \approx 2^{-\ell(|x|)}$, that x is mapped to $(h, 1^{|x|-\ell(|x|)}0h(x))$ with probability $1/|H_{|x|}^{\ell(|x|)}|$, and that x is the only preimage of $(h, 1^{|x|-\ell(|x|)}0h(x))$ under the mapping (among $x' \in S_n$ such that $\Omega_{x'} \ni h$).

Before going any further, let us highlight the importance of hashing X_n to $\ell(n)$-bit strings. On one hand, this mapping is "sufficiently" one-to-one, and thus (with constant probability) the solution provided for the hashed instance (i.e., $h(x)$) yield a solution for the original instance (i.e., x). This guarantees the validity of the reduction. On the other hand, for a typical h, the mapping of X_n to $h(X_n)$ covers the relevant range almost uniformly. This guarantees that the reduction satisfies the domination condition. Note that these two phenomena impose conflicting requirements that are both met at the correct value of ℓ; that is, the one-to-one condition requires $\ell(n) \geq \log_2 |S_n|$, whereas an almost uniform cover requires $\ell(n) \leq \log_2 |S_n|$. Also note that $\ell(n) = \log_2(1/\Pr[X_n = x])$ for every x in the support of X_n; the latter quantity will be in our focus in the general case.

The general case: Finally, we get rid of the assumption that X_n is *uniformly distributed* over some subset of $\{0, 1\}^n$. All that we know is that there exists a probabilistic polynomial-time ("sampling") algorithm S such that $S(1^n)$ is distributed identically to X_n. In this (general) case, we map instances of (R, X) according to their probability mass such that x is mapped to an instance (of R') that consists of $(h, h(x))$ and additional information, where h is a random hash function mapping n-bit long strings to ℓ_x-bit long strings such that

$$\ell_x \overset{\text{def}}{=} \lceil \log_2(1/\Pr[X_{|x|} = x]) \rceil. \tag{7}$$

Since (in the general case) there may be more than 2^{ℓ_x} strings in the support of X_n, we need to augment the reduced instance in order to ensure that it is uniquely associated with x. The basic idea is augmenting the mapping of x to $(h, h(x))$ with additional information that restricts X_n to strings that occur with probability at least $2^{-\ell_x}$. Indeed, when X_n is restricted in this way, the value of $h(X_n)$ uniquely determines X_n.

Let $q(n)$ denote the randomness complexity of S and $S(1^n, r)$ denote the output of S on input 1^n and internal coin tosses $r \in \{0, 1\}^{q(n)}$. Then, we randomly map x to $(h, h(x), h', v')$, where $h : \{0, 1\}^{|x|} \to \{0, 1\}^{\ell_x}$ and $h' : \{0, 1\}^{q(|x|)} \to \{0, 1\}^{q(|x|)-\ell_x}$ are random hash functions and $v' \in \{0, 1\}^{q(|x|)-\ell_x}$ is uniformly distributed. The instance (h, v, h', v') of the redefined search problem R' has solutions that consists of pairs $\langle r, y \rangle$ such that $h(S(1^n, r)) = v \wedge h'(r) = v'$ and $y \in R(S(1^n, r))$. As we shall see, this augmentation guarantees that, with constant probability (over the choice of h, h', v'), the solutions to the reduced instance $(h, h(x), h', v')$ correspond to the solutions to the original instance x.

The foregoing description assumes that, on input x, we can efficiently determine ℓ_x, which is an assumption that cannot be justified. Instead, we select ℓ uniformly in $\{0, 1, ..., q(|x|)\}$, and so with noticeable probability we do select the correct value (i.e., $\Pr[\ell = \ell_x] = 1/(q(|x|)+1) = 1/\text{poly}(|x|)$). For clarity, we make n and ℓ explicit in the reduced instance. Thus, we randomly map $x \in \{0, 1\}^n$ to $(1^n, 1^\ell, h, h(x), h', v') \in \{0, 1\}^{n'}$, where $\ell \in \{0, 1, ..., q(n)\}$, $h \in H_n^\ell$, $h' \in H_{q(n)}^{q(n)-\ell}$, and $v' \in \{0, 1\}^{q(n)-\ell}$ are uniformly distributed in the corresponding sets.[24] This mapping will be used to reduce (R, X) to (R', X'), where R' and $X' = \{X'_{n'}\}_{n' \in \mathbb{N}}$ are redefined (yet again). Specifically, we let

$$R'(1^n, 1^\ell, h, v, h', v') = \{\langle r, y \rangle : h(S(1^n, r)) = v \wedge h'(r) = v' \wedge y \in R(S(1^n, r))\} \quad (8)$$

and $X'_{n'}$ assigns equal probability to each $X_{n', \ell}$ (for $\ell \in \{0, 1, ..., n\}$), where each $X_{n', \ell}$ is isomorphic to the uniform distribution over $H_n^\ell \times \{0, 1\}^\ell \times H_{q(n)}^{q(n)-\ell} \times \{0, 1\}^{q(n)-\ell}$. Note that indeed $(R', X') \in \text{dist}\mathcal{PC}$.

The foregoing randomized mapping is analyzed by considering the correct choice for ℓ; that is, on input x, we focus on the choice $\ell = \ell_x$. Under this conditioning (as we shall show), *with constant probability over the choice of h, h' and v', the instance x is the only value in the support of X_n that is mapped to* $(1^n, 1^{\ell_x}, h, h(x), h', v')$ *and satisfies* $\{r : h(S(1^n, r)) = h(x) \wedge h'(r) = v'\} \neq \emptyset$. It follows that (for such h, h' and v') any solution $\langle r, y \rangle \in R'(1^n, 1^{\ell_x}, h, h(x), h', v')$ satisfies $S(1^n, r) = x$ and thus $y \in R(x)$, which means that the (relaxed) validity condition is satisfied. The (relaxed) domination condition is satisfied too, because (conditioned on $\ell = \ell_x$ and for such h, h', v') the probability that X_n is mapped to $(1^n, 1^{\ell_x}, h, h(x), h', v')$ approximately equals $\Pr[X'_{n', \ell_x} = (1^n, 1^{\ell_x}, h, h(x), h', v')]$.

We now turn to analyze the probability, over the choice of h, h' and v', that the instance x is the only value in the support of X_n that is mapped to

---

[24] As in other places, a suitable encoding will be used such that the reduction maps strings of the same length to strings of the same length (i.e., n-bit string are mapped to n'-bit strings, for $n' = \text{poly}(n)$). For example, we may encode $\langle 1^n, 1^\ell, h, h(x), h', v' \rangle$ as $1^n 01^\ell 01^{q(n)-\ell} 0 \langle h \rangle \langle h(x) \rangle \langle h' \rangle \langle v' \rangle$, where each $\langle w \rangle$ denotes an encoding of w by a string of length $(n' - (n + q(n) + 3))/4$.

to $(1^n, 1^{\ell_x}, h, h(x), h', v')$ and satisfies $\{r : h(S(1^n, r)) = h(x) \wedge h'(r) = v'\} \neq \emptyset$. Firstly, we note that $|\{r : S(1^n, r) = x\}| \geq 2^{q(n)-\ell_x}$, and thus, with constant probability over the choice of $h' \in H_{q(n)}^{q(n)-\ell_x}$ and $v' \in \{0,1\}^{q(n)-\ell_x}$, there exists r that satisfies $S(1^n, r) = x$ and $h'(r) = v'$. Furthermore, with constant probability over the choice of $h' \in H_{q(n)}^{q(n)-\ell_x}$ and $v' \in \{0,1\}^{q(n)-\ell_x}$, it also holds that there are at most $O(2^{\ell_x})$ strings r such that $h'(r) = v'$. Fixing such h' and v', we let $S_{h', v'} = \{S(1^n, r) : h'(r) = v'\}$ and we note that, with constant probability over the choice of $h \in H_n^{\ell_x}$, it holds that x is the only string in $S_{h', v'}$ that is mapped to $h(x)$ under h. Thus, with constant probability over the choice of h, h' and v', the instance x is the only value in the support of X_n that is mapped to $(1^n, 1^{\ell_x}, h, h(x), h', v')$ and satisfies $\{r : h(S(1^n, r)) = h(x) \wedge h'(r) = v'\} \neq \emptyset$. The theorem follows. □

Reflection: Theorem 14 implies that if samp$\mathcal{NP}$ is not contained in tpc$\mathcal{BPP}$ then every dist$\mathcal{NP}$-complete problem is not in tpc$\mathcal{BPP}$. This means that the hardness of some distributional problems that refer to sampleable distributions implies the hardness of some distributional problems that refer to simple distributions. Furthermore, by Proposition 9, this implies the hardness of distributional problems that refer to the uniform distribution. Thus, hardness with respect to some distribution in an utmost wide class (which arguably captures all distributions that may occur in practice) implies hardness with respect to a single simple distribution (which arguably is the simplest one).

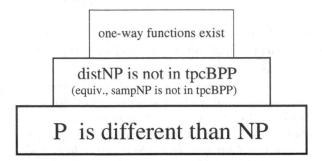

Fig. 2. Worst-case vs average-case assumptions

Relation to One-Way Functions. We note that the existence of one-way functions (see, e.g., [5, Chap. 2]) implies the existence of problems in samp$\mathcal{PC}$ that are not in tpc$\mathcal{BPPF}$ (which in turn implies the existence of such problems in dist$\mathcal{PC}$). Specifically, for a length-preserving one-way function f, consider the distributional search problem $(R_f, \{f(U_n)\}_{n \in \mathbb{N}})$, where $R_f = \{(f(r), r) : r \in \{0,1\}^*\}$.[25] On the other hand, it is not known whether the existence of a problem in samp$\mathcal{PC}$\tpc$\mathcal{BPPF}$ implies the existence of one-way functions. In particular, the

―――――――――
[25] Note that the distribution $f(U_n)$ is uniform in the special case that f is a permutation over $\{0,1\}^n$.

existence of a problem (R, X) in samp$\mathcal{PC} \setminus$ tpc$\mathcal{BPPF}$ represents the feasibility of generating hard instances for the search problem R, whereas the existence of one-way function represents the feasibility of generating instance-solution pairs such that the instances are hard to solve. Indeed, the gap refers to whether or not *hard instances can be efficiently generated together with corresponding solutions*. Our world view is thus depicted in Figure 2, where lower levels indicate seemingly weaker assumptions.

Bibliographic Notes

The theory of average-case complexity was initiated by Levin [10], who in particular proved Theorem 5. In light of the laconic nature of the original text [10], we refer the interested reader to a survey [4], which provides a more detailed exposition of the definitions suggested by Levin as well as a discussion of the considerations underlying these suggestions.

As noted in Section 2.1, the current text uses a variant of the original definitions. In particular, our definition of "typical-case feasibility" differs from the original definition of "average-case feasibility" in totally discarding exceptional instances and in even allowing the algorithm to fail on them (and not merely run for an excessive amount of time). The alternative definition was suggested by several researchers, and appears as a special case of the general treatment provided in [2].

Turning to Section 2.2, we note that while the existence of dist$\mathcal{NP}$-complete problems (cf. Theorem 5) was established in Levin's original paper [10], the existence of dist$\mathcal{NP}$-complete versions of all natural NP-complete decision problems (cf. Theorem 7) was established more than two decades later in [11].

Section 3 is based on [1,8]. Specifically, Theorem 11 (or rather the reduction of search to decision) is due to [1] and so is the introduction of the class samp$\mathcal{NP}$. A version of Theorem 14 was proven in [8], and our proof follows their ideas, which in turn are closely related to the ideas underlying the construction of pseudorandom generators based on any one-way function [7].

Recall that we know of the existence of problems in dist$\mathcal{NP}$ that are hard provided samp$\mathcal{NP}$ contains hard problems. However, these distributional problems do not seem very natural (i.e., they either refer to somewhat generic decision problems such as S_u or to somewhat contrived probability ensembles (cf. Theorem 7)). The presentation of dist$\mathcal{NP}$-complete problems that combine a more natural decision problem (like SAT or Clique) with a more natural probability ensemble is an open problem.

A natural question at this point is what have we gained by relaxing the requirements. In the context of average-case complexity, the negative side seems more prevailing (at least in the sense of being more systematic). In particular, assuming the existence of one-way functions, every natural NP-complete problem has a distributional version that is (typical-case) hard, where this version refers to a sampleable ensemble (and, in fact, even to a simple ensemble). Furthermore, in this case, some problems in NP have hard distributional versions that refer to the uniform distribution.

References

1. Ben-David, S., Chor, B., Goldreich, O., Luby, M.: On the Theory of Average Case Complexity. JCSS 44(2), 193–219 (1992)
2. Bogdanov, A., Trevisan, L.: Average-case complexity. Foundations and Trends in Theoretical Computer Science 2(1) (2006)
3. Cook, S.A.: The Complexity of Theorem Proving Procedures. In: 3rd STOC, pp. 151–158 (1971)
4. Goldreich, O.: Notes on Levin's Theory of Average-Case Complexity. In: Goldreich, O., et al.: Studies in Complexity and Cryptography. LNCS, vol. 6650, pp. 424–454. Springer, Heidelberg (2011); See also ECCC, TR97-058 (December 1997)
5. Goldreich, O.: Foundation of Cryptography: Basic Tools. Cambridge University Press, Cambridge (2001)
6. Goldreich, O.: Computational Complexity: A Conceptual Perspective. Cambridge University Press, Cambridge (2008)
7. Håstad, J., Impagliazzo, R., Levin, L.A., Luby, M.: A Pseudorandom Generator from any One-way Function. SICOMP 28(4), 1364–1396 (1999); Combines papers of Impagliazzo et al., 21st STOC (1989) and Håstad 22nd STOC (1990)
8. Impagliazzo, R., Levin, L.A.: No Better Ways to Generate Hard NP Instances than Picking Uniformly at Random. In: 31st FOCS, pp. 812–821 (1990)
9. Karp, R.M.: Reducibility among Combinatorial Problems. In: Miller, R.E., Thatcher, J.W. (eds.) Complexity of Computer Computations, pp. 85–103. Plenum Press, New York (1972)
10. Levin, L.A.: Average Case Complete Problems. SICOMP 15, 285–286 (1986)
11. Livne, N.: All Natural NPC Problems Have Average-Case Complete Versions. In: ECCC, TR06-122 (2006)

Basic Facts about Expander Graphs

Oded Goldreich

Abstract. In this survey we review basic facts regarding expander graphs that are most relevant to the theory of computation.

Keywords: Expander Graphs, Random Walks on Graphs.

This text has been revised based on [8, Apdx. E.2].

1 Introduction

Expander graphs found numerous applications in the theory of computation, ranging from the design of sorting networks [1] to the proof that undirected connectivity is decidable in determinstic log-space [15]. In this survey we review basic facts regarding expander graphs that are most relevant to the theory of computation. For a wider perspective, the interested reader is referred to [10].

Loosely speaking, expander graphs are regular graphs of small degree that exhibit various properties of cliques.[1] In particular, we refer to properties such as the relative sizes of cuts in the graph (i.e., relative to the number of edges), and the rate at which a random walk converges to the uniform distribution (relative to the logarithm of the graph size to the base of its degree).

Some Technicalities. Typical presentations of expander graphs refer to one of several variants. For example, in some sources, expanders are presented as bipartite graphs, whereas in others they are presented as ordinary graphs (and are in fact very far from being bipartite). We shall follow the latter convention. Furthermore, at times we implicitly consider an augmentation of these graphs where self-loops are added to each vertex. For simplicity, we also allow parallel edges.

We often talk of expander graphs while we actually mean an infinite collection of graphs such that each graph in this collection satisfies the same property (which is informally attributed to the collection). For example, when talking of a d-regular expander (graph) we actually refer to an infinite collection of graphs such that each of these graphs is d-regular. Typically, such a collection (or family) contains a single N-vertex graph for every $N \in \mathbb{S}$, where $\mathbb{S}$ is an infinite subset of $\mathbb{N}$. Throughout this section, we denote such a collection by $\{G_N\}_{N \in \mathbb{S}}$, with the understanding that G_N is a graph with N vertices and $\mathbb{S}$ is an infinite set of natural numbers.

---

[1] Another useful intuition is that expander graphs exhibit various properties of random regular graphs of the same degree.

O. Goldreich et al.: Studies in Complexity and Cryptography, LNCS 6650, pp. 451–464, 2011.

2   Definitions and Properties

We consider two definitions of expander graphs, two different notions of explicit constructions, and two useful properties of expanders.

2.1   Two Mathematical Definitions

We start with two different definitions of expander graphs. These definitions are qualitatively equivalent and even quantitatively related. We start with an algebraic definition, which seems technical in nature but is actually the definition typically used in complexity theoretic applications, since it directly implies various "mixing properties" (see §2.3). We later present a very natural combinatorial definition (which is the source of the term "expander").

The Algebraic Definition (Eigenvalue Gap). Identifying graphs with their adjacency matrix, we consider the eigenvalues (and eigenvectors) of a graph (or rather of its adjacency matrix). Any d-regular graph $G = (V, E)$ has the uniform vector as an eigenvector corresponding to the eigenvalue d, and if G is connected and non-bipartite then the absolute values of all other eigenvalues are strictly smaller than d. The eigenvalue bound, denoted $\lambda(G) < d$, of such a graph G is defined as a tight upper-bound on the *absolute value* of all the other eigenvalues. (In fact, in this case it holds that $\lambda(G) < d - \Omega(1/d|V|^2)$.)[2] The algebraic definition of expanders refers to an infinite family of d-regular graphs and requires the existence of a *constant* eigenvalue bound that holds for all the graphs in the family.

Definition 1 (eigenvalue gap): *An infinite family of d-regular graphs, $\{G_N\}_{N \in \mathbb{S}}$, where $\mathbb{S} \subseteq \mathbb{N}$, satisfies the eigenvalue bound β if for every $N \in \mathbb{S}$ it holds that $\lambda(G_N) \leq \beta$. In such a case, we say that $\{G_N\}_{N \in \mathbb{S}}$ is a family of (d, β)-expanders, and call $d - \beta$ its eigenvalue gap.*

It will be often convenient to consider relative (or normalized) versions of the foregoing quantities, obtained by division by d.

The Combinatorial Definition (Expansion). Loosely speaking, expansion requires that any (not too big) set of vertices of the graph has a relatively large set of neighbors. Specifically, a graph $G = (V, E)$ is c-expanding if, for every set $S \subset V$ of cardinality at most $|V|/2$, it holds that

$$\Gamma_G(S) \overset{\text{def}}{=} \{v : \exists u \in S \text{ s.t. } \{u, v\} \in E\} \tag{1}$$

has cardinality at least $(1 + c) \cdot |S|$. Assuming the existence of self-loops on all vertices, the foregoing requirement is equivalent to requiring that $|\Gamma_G(S) \setminus S| \geq$

---

[2] This follows from the connection to the combinatorial definition (see Theorem 3). Specifically, the square of this graph, denoted G^2, is $|V|^{-1}$-expanding and thus it holds that $\lambda(G)^2 = \lambda(G^2) < d^2 - \Omega(|V|^{-2})$.

$c \cdot |S|$. In this case, every connected graph $G = (V, E)$ is $(1/|V|)$-expanding.[3] The combinatorial definition of expanders refers to an infinite family of d-regular graphs and requires the existence of a *constant* expansion bound that holds for all the graphs in the family.

Definition 2 (expansion): *An infinite family of d-regular graphs, $\{G_N\}_{N \in \mathbb{S}}$ is c-expanding if for every $N \in \mathbb{S}$ it holds that G_N is c-expanding.*

The two definitions of expander graphs are related (see [6, Sec. 9.2] or [10, Sec. 4.5]). Specifically, the "expansion bound" and the "eigenvalue bound" are related as follows.

Theorem 3 (equivalence of the two definitions [3,5]): *Let G be a d-regular graph having a self-loop on each vertex.*[4]

1. *The graph G is c-expanding for $c \geq (d - \lambda(G))/2d$.*
2. *If G is c-expanding then $d - \lambda(G) \geq c^2/(4 + 2c^2)$.*

Thus, any non-zero bound on the combinatorial expansion of a family of d-regular graphs yields a non-zero bound on its eigenvalue gap, and vice versa. Note, however, that the back-and-forth translation between these measures is not tight. We note that most applications in complexity theory refer to the algebraic definition, and that the loss incurred in Theorem 3 is immaterial for them.

Amplification. The "quality of expander graphs improves" by raising these graphs to any power $t > 1$ (i.e., raising their adjacency matrix to the tth power), where this operation corresponds to replacing t-paths (in the original graphs) by edges (in the resulting graphs). Specifically, when considering the algebraic definition, it holds that $\lambda(G^t) = \lambda(G)^t$, but indeed the degree also gets raised to the power t. Still, the ratio $\lambda(G^t)/d^t$ deceases with t. An analogous phenomenon occurs also under the combinatorial definition, provided that some suitable modifications are applied. For example, if for every $S \subseteq V$ it holds that $|\Gamma_G(S)| \geq \min((1 + c) \cdot |S|, |V|/2)$, then for every $S \subseteq V$ it holds that $|\Gamma_{G^t}(S)| \geq \min((1 + c)^t \cdot |S|, |V|/2)$.

---

[3] In contrast, a bipartite graph $G = (V, E)$ is not expanding, because it always contains a set S of size at most $|V|/2$ such that $|\Gamma_G(S)| \leq |S|$ (although it may hold that $|\Gamma_G(S) \setminus S| \geq |S|$).

[4] Recall that in such a graph $G = (V, E)$ it holds that $\Gamma_G(S) \supseteq S$ for every $S \subseteq V$, and thus $|\Gamma_G(S)| = |\Gamma_G(S) \setminus S| + |S|$. Furthermore, in such a graph all eigenvalues are greater than or equal to $-d+1$, and thus if $d - \lambda(G) < 1$ then this is due to a positive eigenvalue of G. These facts are used for bridging the gap between Theorem 3 and the more standard versions (see, e.g., [6, Sec. 9.2]) that refer to variants of both definitions. Specifically, [6, Sec. 9.2] refers to $\Gamma_G^+(S) = \Gamma_G(S) \setminus S$ and $\lambda_2(G)$, where $\lambda_2(G)$ is the second largest eigenvalue of G, rather than referring to $\Gamma_G(S)$ and $\lambda(G)$. Note that, in general, $\Gamma_G(S)$ may be attained by the difference between the smallest eigenvalue of G (which may be negative) and $-d$.

The optimal eigenvalue bound. For every d-regular graph $G = (V, E)$, it holds that $\lambda(G) \geq 2\gamma_G \cdot \sqrt{d-1}$, where $\gamma_G = 1 - O(1/\log_d |V|)$. Thus, for any infinite family of (d, λ)-expanders, it must holds that $\lambda \geq 2\sqrt{d-1}$.

2.2   Two Levels of Explicitness

Towards discussing various notions of explicit constructions of graphs, we need to fix a representation of such graphs. Specifically, throughout this section, when referring to an infinite family of graphs $\{G_N\}_{N \in \mathbb{S}}$, we shall assume that the vertex set of G_N equals $[N]$. Indeed, at times, we shall consider vertex sets having a different structure (e.g., $[m] \times [m]$ for some $m \in \mathbb{N}$), but in all these cases there exists a simple isomorphism of these sets to the canonical representation (i.e., there exists an efficiently computable and invertible mapping of the vertex set of G_N to $[N]$).

Recall that a mild notion of explicit constructiveness refers to the *complexity of constructing the entire object* (i.e., the graph). Applying this notion to our setting, we say that an infinite family of graphs $\{G_N\}_{N \in \mathbb{S}}$ is explicitly constructible if there exists a *polynomial-time algorithm that, on input 1^N (where $N \in \mathbb{S}$), outputs the list of the edges in the N-vertex graph G_N*. That is, the entire graph is constructed in time that is polynomial in its size (i.e., in poly(N)-time).

The foregoing (mild) level of explicitness suffices when the application requires holding the entire graph and/or when the running-time of the application is lower-bounded by the size of the graph. In contrast, other applications refer to a huge virtual graph (which is much bigger than their running time), and only require the computation of the neighborhood relation in such a graph. In this case, the following stronger level of explicitness is relevant.

A strongly explicit construction of an infinite family of (d-regular) graphs $\{G_N\}_{N \in \mathbb{S}}$ is a *polynomial-time algorithm that on input $N \in \mathbb{S}$ (in binary), a vertex v in the N-vertex graph G_N (i.e., $v \in [N]$), and an index $i \in [d]$, returns the ith neighbor of v*. That is, the "neighbor query" is answered in time that is polylogarithmic in the size of the graph. Needless to say, this strong level of explicitness implies the basic (mild) level.

An additional requirement, which is often forgotten but is very important, refers to the "tractability" of the set $\mathbb{S}$. Specifically, we require the existence of an *efficient algorithm that given any $n \in \mathbb{N}$ finds an $s \in \mathbb{S}$ such that $n \leq s < 2n$*. Corresponding to the two foregoing levels of explicitness, "efficient" may mean either running in time poly(n) or running in time poly($\log n$). The requirement that $n \leq s < 2n$ suffices in most applications, but in some cases a smaller interval (e.g., $n \leq s < n + \sqrt{n}$) is required, whereas in other cases a larger interval (e.g., $n \leq s < \text{poly}(n)$) suffices.

Greater Flexibility. In continuation to the foregoing paragraph, we comment that expanders can be combined in order to obtain expanders for a wider range of graph sizes. For example, given two d-regular c-expanding graphs, $G_1 = (V_1, E_1)$ and $G_2 = (V_2, E_2)$ where $|V_1| \leq |V_2|$ and $c \leq 1$, we can obtain a $(d+1)$-regular $c/2$-expanding graph on $|V_1| + |V_2|$ vertices by connecting the two graphs using a perfect matching of V_1 and $|V_1|$ of the vertices of V_2 (and adding self-loops to the remaining vertices of V_2). More generally, combining the d-regular c-expanding graphs $G_1 = (V_1, E_1)$ through $G_t = (V_t, E_t)$, where $N' \stackrel{\text{def}}{=} \sum_{i=1}^{t-1} |V_i| \le |V_t|$, yields a $(d + 1)$-regular $c/2$-expanding graph on $\sum_{i=1}^{t} |V_i|$ vertices (by using a perfect matching of $\cup_{i=1}^{t-1} V_i$ and N' of the vertices of V_t).

2.3   Two Properties

The following two properties provide a quantitative interpretation to the statement that expanders approximate the complete graph (or behave approximately like a complete graph). When referring to (d, λ)-expanders, the deviation from the behavior of a complete graph is represented by an error term that is linear in λ/d.

The Mixing Lemma. Loosely speaking, the following (folklore) lemma asserts that in expander graphs (for which $\lambda \ll d$) the fraction of edges connecting two large sets of vertices approximately equals the product of the densities of these sets. This property is called *mixing*.

Lemma 4 (Expander Mixing Lemma): *For every d-regular graph $G = (V, E)$ and for every two subsets $A, B \subseteq V$ it holds that*

$$\left| \frac{|(A \times B) \cap \vec{E}|}{|\vec{E}|} - \frac{|A|}{|V|} \cdot \frac{|B|}{|V|} \right| \le \frac{\lambda(G)\sqrt{|A| \cdot |B|}}{d \cdot |V|} \le \frac{\lambda(G)}{d} \tag{2}$$

where $\vec{E}$ denotes the set of directed edges (i.e., vertex pairs) *that correspond to the undirected edges of G* (i.e., $\vec{E} = \{(u, v) : \{u, v\} \in E\}$ *and* $|\vec{E}| = d|V|$).

In particular, $|(A \times A) \cap \vec{E}| = (\rho(A) \cdot d \pm \lambda(G)) \cdot |A|$, where $\rho(A) = |A|/|V|$. It follows that $|(A \times (V \setminus A)) \cap \vec{E}| = ((1 - \rho(A)) \cdot d \pm \lambda(G)) \cdot |A|$.

Proof: Let $N \stackrel{\text{def}}{=} |V|$ and $\lambda \stackrel{\text{def}}{=} \lambda(G)$. For any subset of the vertices $S \subseteq V$, we denote its density in V by $\rho(S) \stackrel{\text{def}}{=} |S|/N$. Hence, Eq. (2) is restated as

$$\left| \frac{|(A \times B) \cap \vec{E}|}{d \cdot N} - \rho(A) \cdot \rho(B) \right| \le \frac{\lambda\sqrt{\rho(A) \cdot \rho(B)}}{d} .$$

We proceed by providing bounds on the value of $|(A \times B) \cap \vec{E}|$. To this end we let $\bar{a}$ denote the N-dimensional Boolean vector having 1 in the ith component if and only if $i \in A$. The vector $\bar{b}$ is defined similarly. Denoting the adjacency matrix of the graph G by $M = (m_{i,j})$, we note that $|(A \times B) \cap \vec{E}|$ equals $\bar{a}^\top M \bar{b}$ (because $(i, j) \in (A \times B) \cap \vec{E}$ if and only if it holds that $i \in A$, $j \in B$ and $m_{i,j} = 1$). We consider the *orthogonal eigenvector basis*, $\overline{e_1}, ..., \overline{e_N}$, where $\overline{e_1} = (1, ..., 1)^\top$ and $\overline{e_i}^\top \overline{e_i} = N$ for each i, and write each vector as a linear combination of the vectors in this basis. Specifically, we denote by a_i the coefficient of $\bar{a}$ in the direction of $\overline{e_i}$; that is, $a_i = (\overline{a}^\top \overline{e_i})/N$ and $\overline{a} = \sum_i a_i \overline{e_i}$. Note that $a_1 = (\overline{a}^\top \overline{e_1})/N = |A|/N = \rho(A)$ and $\sum_{i=1}^N a_i^2 = (\overline{a}^\top \overline{a})/N = |A|/N = \rho(A)$. Similarly for $\overline{b}$. It now follows that

$$|(A \times B) \cap \vec{E}| = \overline{a}^\top M \sum_{i=1}^N b_i \overline{e_i}$$

$$= \sum_{i=1}^N b_i \lambda_i \cdot \overline{a}^\top \overline{e_i}$$

where λ_i denotes the ith eigenvalue of M. Note that $\lambda_1 = d$ and for every $i \geq 2$ it holds that $|\lambda_i| \leq \lambda$. Thus,

$$\frac{|(A \times B) \cap \vec{E}|}{dN} = \sum_{i=1}^N \frac{b_i \lambda_i \cdot a_i}{d}$$

$$= \rho(A)\rho(B) + \sum_{i=2}^N \frac{\lambda_i a_i b_i}{d}$$

$$\in \left[\rho(A)\rho(B) \pm \frac{\lambda}{d} \cdot \sum_{i=2}^N a_i b_i \right]$$

Using $\sum_{i=1}^N a_i^2 = \rho(A)$ and $\sum_{i=1}^N b_i^2 = \rho(B)$, and applying Cauchy-Schwartz Inequality, we bound $\sum_{i=2}^N a_i b_i$ by $\sqrt{\rho(A)\rho(B)}$. The lemma follows. ∎

The Random Walk Lemma. Loosely speaking, the first part of the following lemma asserts that, as far as remaining "trapped" in some subset of the vertex set is concerned, a random walk on an expander approximates a random walk on the complete graph.

Lemma 5 (Expander Random Walk Lemma): *Let $G = ([N], E)$ be a d-regular graph, and consider walks on G that start from a uniformly chosen vertex and take $\ell - 1$ additional random steps, where in each such step we uniformly selects one out of the d edges incident at the current vertex and traverses it.*

- *Let W be a subset of $[N]$ and $\rho \stackrel{\text{def}}{=} |W|/N$. Then the probability that such a random walk stays in W is at most*

$$\rho \cdot \left(\rho + (1 - \rho) \cdot \frac{\lambda(G)}{d} \right)^{\ell-1} \tag{3}$$

- *For any $W_0, ..., W_{\ell-1} \subseteq [N]$, the probability that a random walk of length ℓ intersects $W_0 \times W_1 \times \cdots \times W_{\ell-1}$ is at most*

$$\sqrt{\rho_0} \cdot \prod_{i=1}^{\ell-1} \sqrt{\rho_i + (\lambda/d)^2}, \tag{4}$$

where $\rho_i \stackrel{\text{def}}{=} |W_i|/N$.

The basic principle underlying Lemma 5 was discovered by Ajtai, Komlos, and Szemerédi [2], who proved a bound as in Eq. (4). The better analysis yielding the first part of Lemma 5 is due to [12, Cor. 6.1]. A more general bound that refer to the probability of visiting W for a number of times that approximates $|W|/N$ is given in [9], which actually considers an even more general problem (i.e., obtaining Chernoff-type bounds for random variables that are generated by a walk on an expander). An alternative approach to obtaining such Chernoff-type bounds has been recently presented in [11].

Proof of Equation (4). The basic idea is viewing events occuring during the random walk as an evolution of a corresponding probability vector under suitable transformations. The transformations correspond to taking a random step in G and to passing through a "sieve" that keeps only the entries that correspond to the current set W_i. The key observation is that the first transformation shrinks the component that is orthogonal to the uniform distribution, whereas the second transformation shrinks the component that is in the direction of the uniform distribution. Details follow.

Let A be a matrix representing the random walk on G (i.e., A is the adjacency matrix of G divided by d), and let $\hat{\lambda} \stackrel{\text{def}}{=} \lambda(G)/d$ (i.e., $\hat{\lambda}$ upper-bounds the absolute value of every eigenvalue of A except the first one). Note that the uniform distribution, represented by the vector $\overline{u} = (N^{-1}, ..., N^{-1})^{\top}$, is the eigenvector of A that is associated with the largest eigenvalue (which is 1). Let P_i be a 0-1 matrix that has 1-entries only on its diagonal such that entry (j,j) is set to 1 if and only if $j \in W_i$. Then, the probability that a random walk of length ℓ intersects $W_0 \times W_1 \times \cdots \times W_{\ell-1}$ is the sum of the entries of the vector

$$\overline{v} \stackrel{\text{def}}{=} P_{\ell-1}A \cdots P_2AP_1AP_0\overline{u}. \qquad (5)$$

We are interested in upper-bounding $\|\overline{v}\|_1$, and use $\|\overline{v}\|_1 \leq \sqrt{N} \cdot \|\overline{v}\|$, where $\|\overline{z}\|_1$ and $\|\overline{z}\|$ denote the L_1-norm and L_2-norm of $\overline{z}$, respectively (e.g., $\|\overline{u}\|_1 = 1$ and $\|\overline{u}\| = N^{-1/2}$). The key observation is that the linear transformation P_iA shrinks every vector.

Main Claim. For every $\overline{z}$, it holds that $\|P_iA\overline{z}\| \leq (\rho_i + \hat{\lambda}^2)^{1/2} \cdot \|\overline{z}\|$.

Proof. Intuitively, A shrinks the component of $\overline{z}$ that is orthogonal to $\overline{u}$, whereas P_i shrinks the component of $\overline{z}$ that is in the direction of $\overline{u}$. Specifically, we decompose $\overline{z} = \overline{z_1} + \overline{z_2}$ such that $\overline{z_1}$ is the projection of $\overline{z}$ on $\overline{u}$ and $\overline{z_2}$ is the component orthogonal to $\overline{u}$. Then, using the triangle inequality and other obvious facts (which imply $\|P_iA\overline{z_1}\| = \|P_i\overline{z_1}\|$ and $\|P_iA\overline{z_2}\| \leq \|A\overline{z_2}\|$), we have

$$\|P_iA\overline{z_1} + P_iA\overline{z_2}\| \leq \|P_iA\overline{z_1}\| + \|P_iA\overline{z_2}\|$$
$$\leq \|P_i\overline{z_1}\| + \|A\overline{z_2}\|$$
$$\leq \sqrt{\rho_i} \cdot \|\overline{z_1}\| + \hat{\lambda} \cdot \|\overline{z_2}\|$$

where the last inequality uses the fact that P_i shrinks any uniform vector by eliminating $1 - \rho_i$ of its elements, whereas A shrinks the length of any eigenvector except $\overline{u}$ by a factor of at least $\hat{\lambda}$. Using the Cauchy-Schwartz inequality[5], we get

$$\|P_i A \overline{z}\| \leq \sqrt{\rho_i + \hat{\lambda}^2} \cdot \sqrt{\|\overline{z_1}\|^2 + \|\overline{z_2}\|^2}$$
$$= \sqrt{\rho_i + \hat{\lambda}^2} \cdot \|\overline{z}\|$$

where the equality is due to the fact that $\overline{z_1}$ is orthogonal to $\overline{z_2}$.   □

Recalling Eq. (5) and using the Main Claim (and $\|\overline{v}\|_1 \leq \sqrt{N} \cdot \|\overline{v}\|$), we get

$$\|\overline{v}\|_1 \leq \sqrt{N} \cdot \|P_{\ell-1} A \cdots P_2 A P_1 A P_0 \overline{u}\|$$
$$\leq \sqrt{N} \cdot \left(\prod_{i=1}^{\ell-1} \sqrt{\rho_i + \hat{\lambda}^2} \right) \cdot \|P_0 \overline{u}\|.$$

Finally, using $\|P_0 \overline{u}\| = \sqrt{\rho_0 N \cdot (1/N)^2} = \sqrt{\rho_0/N}$, we establish Eq. (4).   ■

Rapid Mixing. A property related to Lemma 5 is that a random walk starting at any vertex converges to the uniform distribution on the expander vertices after a logarithmic number of steps. Specifically, we claim that *starting at any distribution $\overline{s}$* (including a distribution that assigns all weight to a single vertex) *after ℓ steps on a (d, λ)-expander $G = ([N], E)$ we reach a distribution that is $\sqrt{N} \cdot (\lambda/d)^\ell$-close to the uniform distribution over $[N]$.* Using notation as in the proof of Eq. (4), the claim asserts that $\|A^\ell \overline{s} - \overline{u}\|_1 \leq \sqrt{N} \cdot \hat{\lambda}^\ell$, which is meaningful only for $\ell > 0.5 \cdot \log_{1/\hat{\lambda}} N$. The claim is proved by recalling that $\|A^\ell \overline{s} - \overline{u}\|_1 \leq \sqrt{N} \cdot \|A^\ell \overline{s} - \overline{u}\|$ and using the fact that $\overline{s} - \overline{u}$ is orthogonal to $\overline{u}$ (because the former is a zero-sum vector). Thus, $\|A^\ell \overline{s} - \overline{u}\| = \|A^\ell (\overline{s} - \overline{u})\| \leq \hat{\lambda}^\ell \|\overline{s} - \overline{u}\|$ and using $\|\overline{s} - \overline{u}\| < 1$ the claim follows.

3   Constructions

Many explicit constructions of (d, λ)-expanders are known. The first such construction was presented in [14] (where $\lambda < d$ was not explicitly bounded), and an optimal construction (i.e., an optimal eigenvalue bound of $\lambda = 2\sqrt{d-1}$) was first provided in [13]. Most of these constructions are quite simple (see, e.g., §3.1), but their analysis is based on non-elementary results from various branches of mathematics. In contrast, the construction of Reingold, Vadhan, and Wigderson [16], presented in §3.2, is based on an iterative process, and its analysis is based on a relatively simple algebraic fact regarding the eigenvalues of matrices.

---

[5] That is, we get $\sqrt{\rho_i}\|z_1\| + \hat{\lambda}\|z_2\| \leq \sqrt{\rho_i + \hat{\lambda}^2} \cdot \sqrt{\|z_1\|^2 + \|z_2\|^2}$, by using $\sum_{i=1}^{n} a_i \cdot b_i \leq \left(\sum_{i=1}^{n} a_i^2 \right)^{1/2} \cdot \left(\sum_{i=1}^{n} b_i^2 \right)^{1/2}$, with $n = 2$, $a_1 = \sqrt{\rho_i}$, $b_1 = \|z_1\|$, etc.

Before turning to these explicit constructions we note that it is relatively easy to prove the existence of 3-regular expanders, by using the Probabilistic Method (cf. [6]) and referring to the combinatorial definition of expansion.[6]

3.1   The Margulis–Gabber–Galil Expander

For every natural number m, consider the graph with vertex set $\mathbb{Z}_m \times \mathbb{Z}_m$ and the edge set in which every $\langle x, y \rangle \in \mathbb{Z}_m \times \mathbb{Z}_m$ is connected to the vertices $\langle x \pm y, y \rangle$, $\langle x \pm (y+1), y \rangle$, $\langle x, y \pm x \rangle$, and $\langle x, y \pm (x+1) \rangle$, where the arithmetic is modulo m. This yields an extremely simple 8-regular graph with an eigenvalue bound that is a constant $\lambda < 8$ (which is independent of m). Thus, we get:

Theorem 6 *There exists a strongly explicit construction of a family of $(8, 7.9999)$-expanders for graph sizes $\{m^2 : m \in \mathbb{N}\}$. Furthermore, the neighbors of a vertex in these expanders can be computed in logarithmic-space.*[7]

An appealing property of Theorem 6 is that, for every $n \in \mathbb{N}$, it directly yields expanders with vertex set $\{0, 1\}^n$. This is obvious in case n is even, but can be easily achieved also for odd n (e.g., use two copies of the graph for $n - 1$, and connect the two copies by the obvious perfect matching).

Theorem 6 is due to Gabber and Galil [7], building on the basic approach suggested by Margulis [14]. We mention again that the (strongly explicit) (d, λ)-expanders of [13] achieve the optimal eigenvalue bound (i.e., $\lambda = 2\sqrt{d-1}$), but there are annoying restrictions on the degree d (i.e., $d - 1$ should be a prime congruent to 1 modulo 4) and on the graph sizes for which this construction works.[8]

---

[6] This can be done by considering a 3-regular graph obtained by combining an N-cycle with a random matching of the first $N/2$ vertices and the remaining $N/2$ vertices. It is actually easier to prove the related statement that refers to the alternative definition of combinatorial expansion that refers to the relative size of $\Gamma_G^+(S) = \Gamma_G(S) \setminus S$ (rather than to the relative size of $\Gamma_G(S)$). In this case, for a sufficiently small $\varepsilon > 0$ and all sufficiently large N, a random 3-regular N-vertex graph is "ε-expanding" with overwhelmingly high probability. The proof proceeds by considering a (not necessarily simple) graph G obtained by combining three uniformly chosen perfect matchings of the elements of $[N]$. For every $S \subseteq [N]$ of size at most $N/2$ and for every set T of size $\varepsilon|S|$, we consider the probability that for a random perfect matching M it holds that $\Gamma_M^+(S) \subseteq T$. The argument is concluded by applying a union bound.

[7] In fact, for m that is a power of two (and under a suitable encoding of the vertices), the neighbors can be computed by a on-line algorithm that uses a constant amount of space. The same holds also for a variant in which each vertex $\langle x, y \rangle$ is connected to the vertices $\langle x \pm 2y, y \rangle$, $\langle x \pm (2y + 1), y \rangle$, $\langle x, y \pm 2x \rangle$, and $\langle x, y \pm (2x + 1) \rangle$. This variant yields a better known bound on λ, i.e., $\lambda \leq 5\sqrt{2} \approx 7.071$.

[8] The construction in [13] allows graph sizes of the form $(p^3 - p)/2$, where $p \equiv 1$ (mod 4) is a prime such that $d - 1$ is a quadratic residue modulo p. As stated in [4, Sec. 2], the construction can be extended to graph sizes of the form $(p^{3k} - p^{3k-2})/2$, for any $k \in \mathbb{N}$ and p as in the foregoing.

3.2   The Iterated Zig-Zag Construction

The starting point of the following construction is a very good expander G of *constant size*, which may be found by an exhaustive search. The construction of a large expander graph proceeds in iterations, where in the ith iteration the current graph G_i and the fixed graph G are combined, resulting in a larger graph G_{i+1}. The combination step guarantees that the expansion property of G_{i+1} is at least as good as the expansion of G_i, while G_{i+1} maintains the degree of G_i and is a constant times larger than G_i. The process is initiated with $G_1 = G^2$ and terminates when we obtain a graph G_t of approximately the desired size (which requires a logarithmic number of iterations).

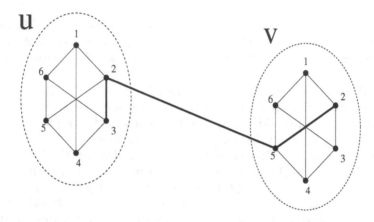

Fig. 1. Detail of the Zig-Zag product of G' and G. In this example G' is 6-regular and G is a 3-regular graph having six vertices. In the graph G' (not shown), the 2nd edge of vertex u is incident at v, as its 5th edge. The wide 3-segment line shows one of the corresponding edges of $G' \textcircled{z} G$, which connects the vertices $\langle u, 3 \rangle$ and $\langle v, 2 \rangle$.

The Zig-Zag Product. The heart of the combination step is a new type of "graph product" called *Zig-Zag product*. This operation is applicable to any pair of graphs $G = ([D], E)$ and $G' = ([N], E')$, provided that G' (which is typically larger than G) is D-regular. For simplicity, we assume that G is d-regular (where typically $d \ll D$). The Zig-Zag product of G' and G, denoted $G' \textcircled{z} G$, is defined as a graph with vertex set $[N] \times [D]$ and an edge set that includes an edge between $\langle u, i \rangle \in [N] \times [D]$ and $\langle v, j \rangle$ if and only if $\{i, k\}, \{\ell, j\} \in E$ and the kth edge incident at u equals the ℓth edge incident at v. That is, $\langle u, i \rangle$ and $\langle v, j \rangle$ are connected in $G' \textcircled{z} G$ if there exists a "three step sequence" consisting of a G-step from $\langle u, i \rangle$ to $\langle u, k \rangle$ (according to the edge $\{i, k\}$ of G), followed by a G'-step from $\langle u, k \rangle$ to $\langle v, \ell \rangle$ (according to the kth edge of u in G' (which is the ℓth edge of v)), and a final G-step from $\langle v, \ell \rangle$ to $\langle v, j \rangle$ (according to the edge $\{\ell, j\}$ of G). See Figure 1 as well as further formalization (which follows).

It will be convenient to represent graphs like G' by their edge-rotation function, denoted $R' : [N] \times [D] \rightarrow [N] \times [D]$, such that $R'(u,i) = (v,j)$ if $\{u,v\}$ is the ith edge incident at u as well as the jth edge incident at v. That is, R' rotates the pair (u,i), which represents one "side" of the edge $\{u,v\}$ (i.e., the side incident at u as its ith edge), resulting in the pair (v,j), which represents the other side of the same edge (which is the jth edge incident at v). For simplicity, we assume that the (constant-size) d-regular graph $G = ([D], E)$ is edge-colorable with d colors, which in turn yields a natural edge-rotation function (i.e., $R(i, \alpha) = (j, \alpha)$ if the edge $\{i,j\}$ is colored α). We will denote by $E_\alpha(i)$ the vertex reached from $i \in [D]$ by following the edge colored α (i.e., $E_\alpha(i) = j$ iff $R(i, \alpha) = (j, \alpha)$). The Zig-Zag product of G' and G, denoted $G' \textcircled{z} G$, is then defined as a graph with the vertex set $[N] \times [D]$ and the edge-rotation function

$$(\langle u,i \rangle, \langle \alpha, \beta \rangle) \mapsto (\langle v,j \rangle, \langle \beta, \alpha \rangle) \quad \text{if } R'(u, E_\alpha(i)) = (v, E_\beta(j)). \tag{6}$$

That is, edges are labeled by pairs over $[d]$, and the $\langle \alpha, \beta \rangle^{\text{th}}$ edge out of vertex $\langle u,i \rangle \in [N] \times [D]$ is incident at the vertex $\langle v,j \rangle$ (as its $\langle \beta, \alpha \rangle^{\text{th}}$ edge) if $R(u, E_\alpha(i)) = (v, E_\beta(j))$, where indeed $E_\beta(E_\beta(j)) = j$. Intuitively, based on $\langle \alpha, \beta \rangle$, we first take a G-step from $\langle u,i \rangle$ to $\langle u, E_\alpha(i) \rangle$, then viewing $\langle u, E_\alpha(i) \rangle \equiv (u, E_\alpha(i))$ as a side of an edge of G' we rotate it (i.e., we effectively take a G'-step) reaching $(v,j') \stackrel{\text{def}}{=} R'(u, E_\alpha(i))$, and finally we take a G-step from $\langle v,j' \rangle$ to $\langle v, E_\beta(j') \rangle$.

Clearly, the graph $G' \textcircled{z} G$ is d^2-regular and has $D \cdot N$ vertices. The key fact, proved in [16] (using techniques as in §2.3), is that the relative eigenvalue-value of the zig-zag product is upper-bounded by the sum of the relative eigenvalue-values of the two graphs; that is, $\bar\lambda(G' \textcircled{z} G) \leq \bar\lambda(G') + \bar\lambda(G)$, where $\bar\lambda(\cdot)$ denotes the relative eigenvalue-bound of the relevant graph. The (qualitative) fact that $G' \textcircled{z} G$ is an expander if both G' and G are expanders is very intuitive (e.g., consider what happens if G' or G is a clique). Things are even more intuitive if one considers the (related) replacement product of G' and G, denoted $G' \textcircled{r} G$, *where there is an edge between $\langle u,i \rangle \in [N] \times [D]$ and $\langle v,j \rangle$ if and only if either* $u = v$ *and* $\{i,j\} \in E$ *or the ith edge incident at u equals the jth edge incident at v.*

The Iterated Construction. The iterated expander construction uses the afore-mentioned zig-zag product as well as graph squaring. Specifically, the construction starts[9] with the d^2-regular graph $G_1 = G^2 = ([D], E^2)$, where $D = d^4$ and $\bar\lambda(G) < 1/4$, and proceeds in iterations such that $G_{i+1} = G_i^2 \textcircled{z} G$ for $i = 1, 2, ..., t-1$, where t is logarithmic in the desired graph size. That is, in each iteration, the current graph is first squared and then composed with the fixed (d-regular D-vertex) graph G via the zig-zag product. This process maintains the following two invariants:

---

[9] Recall that, for a sufficiently large constant d, we first find a d-regular graph $G = ([d^4], E)$ satisfying $\bar\lambda(G) < 1/4$, by exhaustive search.

1. The graph G_i is d^2-regular and has D^i vertices.
   (The degree bound follows from the fact that a zig-zag product with a d-regular graph always yields a d^2-regular graph.)
2. The relative eigenvalue-bound of G_i is smaller than one half (i.e., $\bar{\lambda}(G_i) < 1/2$).
   (Here we use the fact that $\bar{\lambda}(G_{i-1}^2 \textcircled{z} G) \leq \bar{\lambda}(G_{i-1}^2) + \bar{\lambda}(G)$, which in turn equals $\bar{\lambda}(G_{i-1})^2 + \bar{\lambda}(G) < (1/2)^2 + (1/4)$. Note that graph squaring is used to reduce the relative eigenvalue of G_i before increasing it by zig-zag product with G.)

In order to show that we can actually construct G_i, we show that we can compute the edge-rotation function that correspond to its edge set. This boils down to showing that, given the edge-rotation function of G_{i-1}, we can compute the edge-rotation function of G_{i-1}^2 as well as of its zig-zag product with G. Note that this entire computation amounts to two recursive calls to computations regarding G_{i-1} (and two computations that correspond to the constant graph G). But since the recursion depth is logarithmic in the size of the final graph (i.e., $t = \log_D |\text{vertices}(G_t)|$), the total number of recursive calls is polynomial in the size of the final graph (and thus the entire computation is polynomial in the size of the final graph). This suffices for the minimal (i.e., "mild") notion of explicitness, but not for the strong one.

The Strongly Explicit Version. To achieve a *strongly explicit construction*, we slightly modify the iterative construction. Rather than letting $G_{i+1} = G_i^2 \textcircled{z} G$, we let $G_{i+1} = (G_i \times G_i)^2 \textcircled{z} G$, where $G' \times G'$ denotes the *tensor product of G' with itself*; that is, if $G' = (V', E')$ then $G' \times G' = (V' \times V', E'')$, where

$$E'' = \{\{\langle u_1, u_2 \rangle, \langle v_1, v_2 \rangle\} : \{u_1, v_1\}, \{u_2, v_2\} \in E'\}$$

(i.e., $\langle u_1, u_2 \rangle$ and $\langle v_1, v_2 \rangle$ are connected in $G' \times G'$ if for $i = 1, 2$ it holds that u_i is connected to v_i in G'). The corresponding edge-rotation function is

$$R''(\langle u_1, u_2 \rangle, \langle i_1, i_2 \rangle) = (\langle v_1, v_2 \rangle, \langle j_1, j_2 \rangle),$$

where $R'(u_1, i_1) = (v_1, j_1)$ and $R'(u_2, i_2) = (v_2, j_2)$. We still use $G_1 = G^2$, where (as before) G is d-regular and $\bar{\lambda}(G) < 1/4$, but here G has $D = d^8$ vertices.[10] Using the fact that tensor product preserves the relative eigenvalue-bound while squaring the degree (and the number of vertices), we note that the modified iteration $G_{i+1} = (G_i \times G_i)^2 \textcircled{z} G$ yields a d^2-regular graph with $(D^{2^i-1})^2 \cdot D = D^{2^{i+1}-1}$ vertices, and that $\bar{\lambda}(G_{i+1}) < 1/2$ (because $\bar{\lambda}((G_i \times G_i)^2 \textcircled{z} G) \leq \bar{\lambda}(G_i)^2 + \bar{\lambda}(G)$). Computing the neighbor of a vertex in G_{i+1} boils down to a constant number of such computations regarding G_i, but due to the tensor product operation the depth of the recursion is only double-logarithmic in the size of the final graph (and hence logarithmic in the length of the description of vertices in this graph).

---

[10] The reason for the change is that $(G_i \times G_i)^2$ will be d^8-regular, since G_i will be d^2-regular.

Digest. In the first construction, the zig-zag product was used both in order to increase the size of the graph and to reduce its degree. However, as indicated by the second construction (where the tensor product of graphs is the main vehicle for increasing the size of the graph), the primary effect of the zig-zag product is reducing the graph's degree, and the increase in the size of the graph is merely a side-effect.[11] In both cases, graph squaring is used in order to compensate for the modest increase in the relative eigenvalue-bound caused by the zig-zag product. In retrospect, the second construction is the "correct" one, because it decouples three different effects, and uses a natural operation to obtain each of them: Increasing the size of the graph is obtained by tensor product of graphs (which in turn increases the degree), the desired degree reduction is obtained by the zig-zag product (which in turn slightly increases the relative eigenvalue-bound), and graph squaring is used in order to reduce the relative eigenvalue-bound.

Stronger Bound Regarding the Effect of the Zig-Zag Product. In the foregoing description we relied on the fact, proved in [16], that the relative eigenvalue-bound of the zig-zag product is upper-bounded by the sum of the relative eigenvalue-bounds of the two graphs (i.e., $\bar{\lambda}(G' \textcircled{z} G) \leq \bar{\lambda}(G') + \bar{\lambda}(G))$). Actually, a stronger upper-bound is proved in [16]: It holds that $\bar{\lambda}(G' \textcircled{z} G) \leq f(\bar{\lambda}(G'), \bar{\lambda}(G)))$, where

$$f(x,y) \overset{\text{def}}{=} \frac{(1-y^2) \cdot x}{2} + \sqrt{\left(\frac{(1-y^2) \cdot x}{2}\right)^2 + y^2} \tag{7}$$

Indeed, $f(x,y) \leq (1-y^2) \cdot x + y \leq x + y$. On the other hand, for $x \leq 1$, we have $f(x,y) \leq \frac{(1-y^2) \cdot x}{2} + \frac{1+y^2}{2} = 1 - \frac{(1-y^2) \cdot (1-x)}{2}$, which implies

$$\bar{\lambda}(G' \textcircled{z} G) \leq 1 - \frac{(1 - \bar{\lambda}(G)^2) \cdot (1 - \bar{\lambda}(G'))}{2}. \tag{8}$$

Thus, $1 - \bar{\lambda}(G' \textcircled{z} G) \geq (1 - \bar{\lambda}(G)^2) \cdot (1 - \bar{\lambda}(G'))/2$, and it follows that the zig-zag product has a positive eigenvalue-gap if both graphs have positive eigenvalue-gaps (i.e., $\lambda(G' \textcircled{z} G) < 1$ if both $\lambda(G) < 1$ and $\lambda(G') < 1$). Furthermore, if $\bar{\lambda}(G) < 1/\sqrt{3}$ then $1 - \bar{\lambda}(G' \textcircled{z} G) > (1 - \bar{\lambda}(G'))/3$. This fact plays an important role in the celebrated proof that undirected connectivity is decidable in determinstic log-space [15].

References

1. Ajtai, M., Komlos, J., Szemerédi, E.: An $O(n \log n)$ Sorting Network. In: 15th STOC, pp. 1–9 (1983)
2. Ajtai, M., Komlos, J., Szemerédi, E.: Deterministic Simulation in LogSpace. In: 19th STOC, pp. 132–140 (1987)

---

[11] We mention that this side-effect may actually be undesired in some applications. For example, in the proof of [15] we would rather not have the graph grow in size, but we can tolerate the constant size blow-up (caused by zig-zag product with a constant-size graph).

3. Alon, N.: Eigenvalues and expanders. Combinatorica 6, 83–96 (1986)
4. Alon, N., Bruck, J., Naor, J., Naor, M., Roth, R.: Construction of Asymptotically Good, Low-Rate Error-Correcting Codes through Pseudo-Random Graphs. IEEE Transactions on Information Theory 38, 509–516 (1992)
5. Alon, N., Milman, V.D.: λ_1, Isoperimetric Inequalities for Graphs and Superconcentrators, *J. J. Combinatorial Theory, Ser.* B 38, 73–88 (1985)
6. Alon, N., Spencer, J.H.: The Probabilistic Method, 2nd edn. John Wiley & Sons, Chichester (1992, 2000)
7. Gabber, O., Galil, Z.: Explicit Constructions of Linear Size Superconcentrators. JCSS 22, 407–420 (1981)
8. Goldreich, O.: Computational Complexity: A Conceptual Perspective. Cambridge University Press, Cambridge (2008)
9. Healy, A.: Randomness-efficient sampling within *NC1*. In: Díaz, J., Jansen, K., Rolim, J.D.P., Zwick, U. (eds.) APPROX 2006 and RANDOM 2006. LNCS, vol. 4110, pp. 398–409. Springer, Heidelberg (2006)
10. Hoory, S., Linial, N., Wigderson, A.: Expander Graphs and their Applications. Bull. AMS 43(4), 439–561 (2006)
11. Impagliazzo, R., Kabanets, V.: Constructive Proofs of Concentration Bounds. In: ECCC, TR10-072 (2010)
12. Kahale, N.: Eigenvalues and Expansion of Regular Graphs. JACM 42(5), 1091–1106 (1995)
13. Lubotzky, A., Phillips, R., Sarnak, P.: Ramanujan Graphs. Combinatorica 8, 261–277 (1988)
14. Margulis, G.A.: Explicit Construction of Concentrators. Prob. Per. Infor. 9(4), 325–332 (1975) (in Russian); English translation in Problems of Infor. Trans. pp. 325–332 (1975)
15. Reingold, O.: Undirected ST-Connectivity in Log-Space. In: 37th STOC, pp. 376–385 (2005)
16. Reingold, O., Vadhan, S., Wigderson, A.: Entropy Waves, the Zig-Zag Graph Product, and New Constant-Degree Expanders and Extractors. Annals of Mathematics 155(1), 157–187 (2001); Preliminary version in 41st FOCS, pp. 3–13 (2000)

A Brief Introduction to Property Testing

Oded Goldreich

Abstract. This short article provides a brief description of the main issues that underly the study of property testing.

This article was originally written for inclusion in [4], which provides a collection of surveys and extended abstracts that cover various specific subareas and research directions in property testing.

1 Introduction

Property Testing is the study of super-fast (randomized) algorithms for approximate decision making. These algorithms are given direct access to items of a huge data set, and determine whether this data set has some predetermined (global) property or is far from having this property. Remarkably, this approximate decision is made by accessing a small portion of the data set.

Property Testing has been a subject of intensive research in the last couple of decades, with hundreds of studies conducted in it and in closely related areas. Indeed, Property Testing is closely related to Probabilistically Checkable Proofs (PCPs), and is related to Coding Theory, Combinatorics, Statistics, Computational Learning Theory, Computational Geometry, and more.

This brief introduction to the area of Property Testing is confined to conceptual issues; that is, it focuses on the main notions and models being studied, while hardly mentioning the numerous results obtained in the various models. This deficiency of the current article is corrected by the various surveys and extended abstracts presented in [4]. In addition, we refer the interested reader to two recent surveys of Ron [11,12].

2 The Issues

Property testing is a relaxation of decision problems and it focuses on algorithms that can only read parts of the input. Thus, the input is represented as a function (to which the tester has oracle access) and the tester is required to accept functions that have some predetermined property (i.e., reside in some predetermined set) and reject any function that is "far" from the set of functions having the property. Distances between functions are defined as the fraction of the domain on which the functions disagree, and the threshold determining what is considered far is presented as a proximity parameter, which is explicitly given to the tester.

O. Goldreich et al.: Studies in Complexity and Cryptography, LNCS 6650, pp. 465–469, 2011.

An asymptotic analysis is enabled by considering an infinite sequence of domains, functions, and properties. That is, for any n, we consider functions from D_n to R_n, where $|D_n| = n$. (Often, one just assumes that $D_n = [n] \overset{\text{def}}{=} \{1, 2, ..., n\}$.) Thus, in addition to the input oracle, representing a function $f : D_n \to R_n$, the tester is explicitly given two parameters: a size parameter, denoted n, and a proximity parameter, denoted ϵ.

Definition 1. *Let $\Pi = \bigcup_{n \in \mathbb{N}} \Pi_n$, where Π_n contains functions defined over the domain D_n. A* tester *for a* property *Π is a probabilistic oracle machine T that satisfies the following two conditions:*

1. *The tester accepts each $f \in \Pi$ with probability at least $2/3$; that is, for every $n \in \mathbb{N}$ and $f \in \Pi_n$ (and every $\epsilon > 0$), it holds that $\Pr[T^f(n, \epsilon) = 1] \geq 2/3$.*
2. *Given $\epsilon > 0$ and oracle access to any f that is ϵ-far from Π, the tester rejects with probability at least $2/3$; that is, for every $\epsilon > 0$ and $n \in \mathbb{N}$, if $f : D_n \to R_n$ is ϵ-far from Π_n, then $\Pr[T^f(n, \epsilon) = 0] \geq 2/3$, where f is ϵ-far from Π_n if, for every $g \in \Pi_n$, it holds that $|\{e \in D_n : f(e) \neq g(e)\}| > \epsilon \cdot n$.*

If the tester accepts every function in Π with probability 1, then we say that it has one-sided error*; that is, T has one-sided error if for every $f \in \Pi$ and every $\epsilon > 0$, it holds that $\Pr[T^f(n, \epsilon) = 1] = 1$. A tester is called* non-adaptive *if it determines all its queries based solely on its internal coin tosses (and the parameters n and ϵ); otherwise it is called* adaptive.

Definition 1 does not specify the query complexity of the tester, and indeed an oracle machine that queries the entire domain of the function qualifies as a tester (with zero error probability...). Needless to say, we are interested in testers that have significantly lower query complexity.

Research in property testing is often categorized according to the type of functions and properties being considered. In particular, algebraic property testing focuses on the case that the domain and range are associated with some algebraic structures (e.g., groups, fields, and vector spaces) and studies algebraic properties such as being a polynomial of low degree (see, e.g., [3,13]). In the context of testing graph properties (see, e.g., [6,5]), the functions represent graphs or rather allow certain queries to such graphs (e.g., in the adjacency matrix model, graphs are represented by their adjacency relation and queries correspond to pairs of vertices where the answers indicate whether or not the two vertices are adjacent in the graph).[1]

Ramifications. While most research in property testing refers to distances with respect to the uniform distribution on the function's domain, other distributions and even distribution-free models were also considered. That is, for a (known or unknown) distribution μ on the domain, we say that f is ϵ-far from g (w.r.t μ) if $\Pr_{e \sim \mu}[f(e) \neq g(e)] > \epsilon$. Indeed, Definition 1 refers to the case that μ is uniform over the domain (i.e., D_n).

---

[1] In an alternative model, known as the incidence-list model, graphs are represented by functions that assign to the pair (v, i) the ith neighbor of vertex v.

A somewhat related model is one in which the tester obtains random pairs $(e, f(e))$, where each sample e is drawn (independently) from the aforementioned distribution. Such random (f-labeled) example can be either obtained on top of the queries to f or instead of them. This is also the context of testing distributions, where the examples are actually unlabeled and the aim is testing properties of the underlying distribution (rather than properties of the labeling which is null here).

A third ramification refers to the related notions of *tolerant testing* and *distance approximation* (cf. [10]). In the latter, the algorithm is required to estimate the distance of the input (i.e., f) from the predetermined set of instances having the property (i.e., Π). Tolerant testing usually means only a crude distance approximation that guarantees that inputs close to Π (rather than only inputs in Π) are accepted while inputs that are far from Π are rejected (as usual).

On the Current Focus on Query Complexity. Current research in property testing focuses mainly on query (and/or sample) complexity, while either ignoring time complexity or considering it a secondary issue. The current focus on these information theoretic measures is justified by the fact that even the latter are far from being understood. (Indeed, this stands in contrast to the situation in, say, PAC learning.)

On the Importance of Representation. The representation of problems' instances is crucial to any study of computation, since the representation determines the type of information that is explicit in the input. This issue becomes much more acute when one is only allowed partial access to the input (i.e., making a number of queries that result in answers that do not fully determine the input). An additional issue, which is unique to property testing, is that the representation may effect the distance measure (i.e., the definition of distances between inputs). This is crucial because property testing problems are defined in terms of this distance measure.

The importance of representation is forcefully demonstrated in the gap between the complexity of testing numerous natural graph properties in two natural representations: the adjacency matrix representation (cf. [6]) and the incidence lists representation (cf. [7]).

Things get to the extreme in the study of locally testable codes, which may be viewed as evolving around testing whether the input is "well formed" with respect to some fixed error correcting code. Interestingly, the general study of locally testable codes seeks an arbitrary succinct representation (i.e., a code of good rate) such that well-formed inputs (i.e., codewords) are far apart and testing well-formness is easy (i.e., there exists a low complexity codeword test).

3   A Brief Historical Perspective

Property testing first appeared as a tool towards program checking (see the linearity tester of [3]) and the construction of PCPs (see the low-degree tests and their relation to locally testable codes, as discussed in [13]). In these settings it was natural to view the tested object as a function, and this convention continued also in [6], which defined property testing in relation to PAC learning. More importantly, in [6] property testing is promoted as a new type of computational problems, which transcends all its natural applications.

While [3,13] focused on algebraic properties, the focus of [6] was on graph properties. From this perspective the choice of representation became less obvious, and oracle access was viewed as allowing local inspection of the graph rather than being the graph itself.[2] The distinction between objects and their representations became more clear when an alternative representation of graphs was studied in [7,8]. At this point, query complexity that is polynomially related to the size of the object (e.g., its square root) was no longer considered inhibiting. This shift in scale is discussed next.

Recall that initially property testing was viewed as referring to functions that are implicitly defined by some succinct programs (as in the context of program checking) or by "transcendental" entities (as in the context of PAC learning). From this perspective the yardstick for efficiency is being polynomial in the length of the query, which means being polylogarithmic in the size of the object. However, when viewing property testing as being applied to (huge) objects that may exist in explicit form in reality, it is evident that any sub-linear complexity may be beneficial.

The realization that property testing may mean any algorithm that does not inspect its entire input seems crucial to the study of testing distributions, which emerged with [2]. In general, property testing became identified as a study of a special type of sublinear-time algorithms.

Another consequence of the aforementioned shift in scale is the decoupling of the representation from the query types. In the context of graph properties, this culminated in the model of [9].

Nevertheless, the study of testing properties within query complexity that only depends on the proximity parameter (and is thus totally independent of the size of the object) remains an appealing and natural direction. A remarkable result in this direction is the characterization of graph properties that are testable within such complexity in the adjacency matrix model [1].

References

1. Alon, N., Fischer, E., Newman, I., Shapira, A.: A Combinatorial Characterization of the Testable Graph Properties: It's All About Regularity. In: 38th STOC, pp. 251–260 (2006)
2. Batu, T., Fortnow, L., Rubinfeld, R., Smith, W.D., White, P.: Testing that Distributions are Close. In: 41st FOCS, pp. 259–269 (2000)

---

[2] That is, in this case the starting point is the (unlabeled) graph itself, and its representation as a (labeled) graph by either its adjacency matrix or incidence list is an auxiliary conceptual step.

3. Blum, M., Luby, M., Rubinfeld, R.: Self-Testing/Correcting with Applications to Numerical Problems. JCSS 47(3), 549–595 (1990); Extended abstract in 22nd STOC (1990)
4. Goldreich, O. (ed.): Property Testing. LNCS, vol. 6390, pp. 142–157. Springer, Heidelberg (2010)
5. Goldreich, O.: Introduction to Testing Graph Properties. In: Goldreich, O., et al.: Studies in Complexity and Cryptography. LNCS, vol. 6650, pp. 467–471. Springer, Heidelberg (2011)
6. Goldreich, O., Goldwasser, S., Ron, D.: Property testing and its connection to learning and approximation. Journal of the ACM, 653–750 (July 1998); Extended abstract in 37th FOCS (1996)
7. Goldreich, O., Ron, D.: Property Testing in Bounded Degree Graphs. Algorithmica 32(2), 302–343 (2002); Extended abstract in 29th STOC (1997)
8. Goldreich, O., Ron, D.: A Sublinear Bipartitness Tester for Bounded Degree Graphs. Combinatorica 19(3), 335–373 (1999); Extended abstract in 30th STOC (1998)
9. Kaufman, T., Krivelevich, M., Ron, D.: Tight bounds for testing bipartiteness in general graphs. In: Arora, S., Jansen, K., Rolim, J.D.P., Sahai, A. (eds.) RANDOM 2003 and APPROX 2003. LNCS, vol. RANDOM, pp. 341–353. Springer, Heidelberg (2003)
10. Parnas, M., Ron, D., Rubinfeld, R.: Tolerant Property Testing and Distance Approximation. JCSS 72(6), 1012–1042 (2006); Preliminary version in ECCC (2004)
11. Ron, D.: Property Testing: A Learning Theory Perspective. Foundations and Trends in Machine Learning 1(3), 307–402 (2008)
12. Ron, D.: Algorithmic and Analysis Techniques in Property Testing. Foundations and Trends in TCS 5(2), 73–205 (2010)
13. Rubinfeld, R., Sudan, M.: Robust Characterization of Polynomials with Applications to Program Testing. SIAM Journal on Computing 25(2), 252–271 (1996)

Introduction to Testing Graph Properties

Oded Goldreich

Abstract. The aim of this article is to introduce the reader to the study of testing graph properties, while focusing on the main models and issues involved. No attempt is made to provide a comprehensive survey of this study, and specific results are often mentioned merely as illustrations of general themes.

Keywords: Graph Properties, randomized algorithms, approximation problems.

This survey was originally written for inclusion in [32].

1 The General Context

In general, property testing is concerned with super-fast (probabilistic) algorithms for deciding whether a given object has a predetermined property or is *far* from any object having this property. Such algorithms, called testers, obtain local views of the object by making adequate queries; that is, the object is seen as a function and the tester gets oracle access to this function, and thus may be expected to work in time that is sub-linear in the size of the object.

Looking at the foregoing formulation, we first note that property testing is concerned with promise problems (cf. [26,30]), rather than with standard decision problems. Specifically, objects that neither have the property nor are far from having the property are discarded. The exact formulation of these promise problems refers to a *distance measure* defined on the set of all relevant objects (i.e., this distance measure coupled with a distance parameter determine the set of objects that are far from the property). Thus, the choice of natural distance measures is crucial to the study of property testing. Secondly, we note that the requirement that the algorithms operate in sub-linear time (i.e., without reading their entire input) calls for a specification of the *type of queries* that these algorithms can make to their input. Thus, the choice of natural query types is also crucial to the study of property testing. These two general considerations will become concrete once we delve into the actual subject matter (i.e., testing graph properties).

1.1 Why Graphs?

Let us start with an empirical observation, taken from Shimon Even's book *Graph Algorithms* [25] (published in 1979):

O. Goldreich et al.: Studies in Complexity and Cryptography, LNCS 6650, pp. 470–506, 2011.
© Springer-Verlag Berlin Heidelberg 2011

Graph theory has long become recognized as one of the more useful mathematical subjects for the computer science student to master. The approach which is natural in computer science is the algorithmic one; our interest is not so much in existence proofs or enumeration techniques, as it is in finding efficient algorithms for solving relevant problems, or alternatively showing evidence that no such algorithms exist. Although algorithmic graph theory was started by Euler, if not earlier, its development in the last ten years has been dramatic and revolutionary.

Meditating on these facts, one may ask what is the source of this ubiquitous use of graphs in computer science. The most common answer is that graphs arise naturally as a model (or an abstraction) of numerous natural and artificial objects. Another answer is that graphs help visualize binary relations over finite sets. These two different answers correspond to two types of models of testing graph properties that will be discussed below.

1.2   Why Testing?

Suppose we are given a huge graph representing some binary relation over a huge data-set (see below), and we need to determine whether the graph (equivalently, the relation) has some predetermined property. Since the graph is huge, we cannot or do not want to even scan all of it (let alone process all of it). The question is whether it is possible to make meaningful statements about the entire graph based only on a "small portion" of it. Of course, such statements will at best be approximations. But in many settings approximations are good enough.

As a motivation, let us consider a well-known example in which fast approximations are possible and useful. Suppose that some cost function is defined over a huge set, and that one wants to obtain the average cost of an element in the set. To be more specific, let $\mu : S \to [0,1]$ be a cost function, and suppose we want to estimate $\overline{\mu} \stackrel{\text{def}}{=} \frac{1}{|S|} \sum_{x \in S} \mu(x)$. Then, uniformly (and independently) selecting $m \stackrel{\text{def}}{=} O(\epsilon^{-2} \log(1/\delta))$ sample points, $x_1, ..., x_m$, in S we obtain with probability at least $1 - \delta$ an estimate of $\overline{\mu}$ within $\pm\epsilon$. That is,

$$\mathbf{Pr}_{x_1,...,x_m \in S} \left[\left| \frac{1}{m} \sum_{i=1}^{m} \mu(x_i) - \overline{\mu} \right| > \epsilon \right] < \delta . \tag{1}$$

Turning back to graphs, we note that they capture more complex features of data sets; that is, graphs capture relations among pairs of elements (rather then functions of single elements). Specifically, a symmetric binary relation $R \subseteq S \times S$ is represented by a graph $G = (S, R)$, where the elements of S are viewed as vertices and the elements in R are viewed as edges.

The study of testing graph properties reveals that sampling a huge data set may be useful not only towards approximating various statistics regarding a function defined over the set, but also towards approximating various properties regarding a binary relation defined on this set. As we shall see, in many cases, the sampling method used (or at least its analysis) is significantly more sophisticated than the one employed in gathering statistics of the former type. But before doing so, we wish to further discuss the potential benefit in the notion of approximation underlying the definition of property testing.

Firstly, being close to a graph that has the property is a notion of approximation that, in certain applications, may be of direct value. Furthermore, in some cases, being close to a graph having the property translates to a standard notion of approximation (see Section 2.2). In other cases, it translates to a notion of "dual approximation" (see, again, Section 2.2).

Secondly, in some cases, we may be forced to take action without having the time to run a decision procedure, while given the option of modifying the graph in the future, at a cost proportional to the number of added/omitted edges. For example, suppose we are given a graph that represents some suggested design, where bipartite graphs correspond to good designs and changes in the design correspond to edge additions/omissions. Using a `Bipartiteness` tester, we may (with high probability) accept any good design, while rejecting designs that will cost a lot to modify. That is, we may still accept designs that are not good, but only such that are close to being good and thus will not cost too much to modify later.

Thirdly, we may use the property tester as a preliminary stage before running a slower exact decision procedure. In case the graph is far from having the property, with high probability, we obtain an indication towards this fact, and save the time we might have used running the decision procedure. Furthermore, if the tester has one-sided error (i.e., it always accepts a graph having the property) and the tester has rejected, then we have obtained an absolutely correct answer without running the slower decision procedure at all. The saving provided by using a property tester as a preliminary stage may be very substantial in many natural settings where *typical* graphs either have the property or are very far from having the property. Furthermore, *if* it is *guaranteed* that graphs either have the property or are very far from having it *then* we may not even need to run the (exact) decision procedure at all.

1.3   Three Models of Testing Graph Properties

A graph property is a set of graphs closed under graph isomorphism (renaming of vertices).[1] Let Π be such a property. A Π-tester is a *randomized* algorithm that is given oracle access to a graph, $G = (V, E)$, and has to determine whether the graph is in Π or is far from being in Π. The type of oracle (equiv., the type of queries allowed) and distance-measure depend on the model, and we focus on three such models:

1. The adjacency predicate model [33]: Here the Π-tester is given oracle access to a symmetric function $g : V \times V \to \{0, 1\}$ that represents the adjacency

---

[1] That is, Π is a graph property if, for every graph $G = (V, E)$ and every permutation π over V, it holds that $G \in \Pi$ if and only if $\pi(G) \in \Pi$, where $\pi(G) \stackrel{\text{def}}{=} (V, \{\{\pi(u), \pi(v)\} : \{u, v\} \in E\})$.

predicate of the graph G; that is $g(u, v) = 1$ if and only if $(u, v) \in E$. In this model distances between graphs are measured according to their representation; that is, if the graphs G and G' are represented by the functions g and g', then their relative distance is the fraction of pairs (u, v) such that $g(u, v) \neq g'(u, v)$.

Note that saying that $G = ([N], E)$ is ϵ-far from the graph property Π means that for every $G' \in \Pi$ it holds that G is ϵ-far from G'. Since Π is closed under graph isomorphism, this means that G is ϵ-far from any isomorphic copy of $G' = ([N], E')$; that is, for every permutation π over $[N]$, it holds that $|\{(u, v) : g(u, v) \neq g'(\pi(u), \pi(v))\}| > \epsilon N^2$, where g and g' are as above. Finally, note that this notion of distance between graphs is most meaningful in the case that the graph is dense (since in this case fractions of the number of possible vertex pairs are closely related to fractions of the actual number of edges). Thus, this model is often called the **dense graph model**.

2. **The incidence function model** [35]: Here, for some fixed upper bound d (on the degrees of vertices in G), the Π-tester is given oracle access to a function $g : V \times [d] \to V \cup \{\perp\}$ that represents the graph $G = (V, E)$ such that $g(u, i) = v$ if v is the ith vertex incident at u and $g(u, i) = \perp$ if u has less than i neighbors. That is, $E = \{(u, v) : \exists i \; f(u, i) = v\}$, where we always assume that $g(u, i) = v$ if and only if there exists a $j \in [d]$ such that $g(v, j) = u$.

Indeed, only graphs of degree at most d can be represented in this model, which is called the **bounded-degree graph model**.

In this model too, distances between graphs are measured according to their representation, but here the representation is different and so the distances are different. Specifically, if the graphs G and G' are represented by the functions g and g', then their relative distance is the fraction of pairs (u, i) such that $g(u, i) \neq g'(u, i)$. Again, saying that $G = ([N], E)$ is ϵ-far from the graph property Π means that for every $G' \in \Pi$ it holds that G is ϵ-far from G'. Since Π is closed under graph isomorphism (and the ordering of the vertices incident at each vertex is arbitrary), this means that for every permutation π over $[N]$, it holds that

$$\sum_{u \in V} |\{v : \exists i \; g(u, i) = v\} \triangle \{v : \exists i \; g'(\pi(u), i) = \pi(v)\}| > \epsilon d N,$$

where g and g' are as above, and $\triangle$ denotes the symmetric difference (i.e., $A \triangle B = (A \cup B) \setminus (A \cap B)$).

3. **The general graph model** [53,47]: In contrast to the foregoing two models in which the oracle queries and the distances between graphs are linked to the representation of graphs as functions, in the following model the representation is blurred and the query types and distance measure are decoupled.

The relative distance between the graphs $G = ([N], E)$ and $G' = ([N], E)$ is usually defined as $\frac{|E \triangle E'|}{\max(|E|, |E'|)}$; that is, the absolute distance is normalized by the actual number of edges rather than by an absolute upper bound (on the number of edges) such as $N^2/2$ or $dN/2$.

The types of queries typically considered are the two types of queries considered in the previous two models. That is, the algorithm may ask whether two vertices are adjacent in the graph and may also ask for a specific neighbor of a specific vertex.

Needless to say, the general graph model is the most general one, and it is indeed closest to actual algorithmic applications.[2] The fact that this model has so far received relatively little attention merely reflects the fact that its study is overly complex. Given that current studies of the other models still face formidable difficulties (and that these models offer a host of interesting open problems), it is natural that researchers shy away from yet another level of complication.

The Current Focus on Query Complexity. Although property testing is motivated by referring to super-fast algorithms, research in the area tends to focus on the *query complexity* of testing various properties. This focus should be viewed as providing an initial estimate to the actual complexity of the testing problems involved; certainly, query complexity lower bounds imply corresponding bounds on the time complexity, whereas the latter is typically at most exponential in the query complexity. Furthermore, in many cases, the time complexity is polynomial in the query complexity and this fact is typically stated. Thus, we will follow the practice of focusing on the query complexity of testing, but also mention time complexity upper bounds whenever they are of interest.

1.4 Organization

The following three sections are devoted to the three models discussed above: We start with the dense graph model (Section 2), then move to the bounded-degree model (Section 3), and finally get to the general graph model (Section 4). In each model we review the definition of testing (when specialized to that model), provide a taste of the known results, and demonstrate some of the ideas involved (by focusing on testing Bipartiteness, which seems a good benchmark).

We conclude this article with a discussion of a few issues that are relevant to all models; these include the treatment of directed graphs (Section 5.1), the related notions of tolerant testing and distance approximation (Section 5.2), and the notion of proximity oblivious testing (Section 5.3).

The appendix presents three observations that occurred to us in the process of writing this article. These refer to testing (degree) regularity in the dense graph model (Appendix A.1), non-adaptive testers in the bounded-degree graph model (Appendix A.2), and testing strong connectivity of directed graphs by only using forward queries (Appendix A.3).

---

[2] In other words, this model is relevant for most applications, since these seem to refer to general graphs (which model various natural and artificial objects). In contrast, the dense graph model is relevant to applications that refer to (dense) binary relations over finite graphs.

2   The Dense Graph Model

In the adjacency matrix model (a.k.a the dense graph model), an N-vertex graph $G = ([N], E)$ is represented by the Boolean function $g : [N] \times [N] \to \{0, 1\}$ such that $g(u, v) = 1$ if and only if u and v are adjacent in G (i.e., $\{u, v\} \in E$). Distance between graphs is measured in terms of their aforementioned representation (i.e., as the fraction of (the number of) different matrix entries (over N^2)), but occasionally one uses the more intuitive notion of the fraction of (the number of) unordered vertex pairs over $\binom{N}{2}$.[3] Recall that we are interested in *graph properties*, which are sets of graphs that are closed under isomorphism; that is, Π is a **graph property** if for every graph $G = ([N], E)$ and every permutation π of $[N]$ it holds that $G \in \Pi$ if and only if $\pi(G) \in \Pi$, where $\pi(G) \stackrel{\text{def}}{=} ([N], \{\{\pi(u), \pi(v)\} : \{u, v\} \in E\})$. We now spell out the meaning of property testing in this model.

Definition 2.1 (testing graph properties in the adjacency matrix model): *A* **tester** *for a graph property Π is a probabilistic oracle machine that, on input parameters N and ϵ and access to* (the adjacency predicate of) *an N-vertex graph $G = ([N], E)$, outputs a binary verdict that satisfies the following two conditions.*

1. *If $G \in \Pi$ then the tester accepts with probability at least $2/3$.*
2. *If G is ϵ-far from Π then the tester accepts with probability at most $1/3$, where G is ϵ-far from Π if for every N-vertex graph $G' = ([N], E') \in \Pi$ it holds that the symmetric difference between E and E' has cardinality that is greater than $\epsilon \cdot \binom{N}{2}$.*

If the tester accepts every graph in Π with probability 1, then we say that it has **one-sided** *error. A tester is called* **non-adaptive** *if it determines all its queries based solely on its internal coin tosses* (and the parameters N and ϵ); *otherwise it is called* **adaptive***.*

The query complexity of a tester is the number of queries it makes to any N-vertex graph, as a function of the parameters N and ϵ. We say that a tester is efficient if it runs in time that is polynomial in its query complexity, where basic operations on elements of $[N]$ (and in particular, uniformly selecting an element in $[N]$) are counted at unit cost.

We stress that testers are defined as (uniform)[4] algorithms that are given the size parameter N and the distance (or proximity) parameter ϵ as explicit inputs. This uniformity (over the values of the distance parameter) makes the positive results stronger and more appealing (especially in light of a separation result

---

[3] Indeed, there is a tiny discrepancy between these two measures, but it is immaterial in all discussions.

[4] That is, we refer to the standard (uniform) model of computation (cf., e.g., [31, Sec. 1.2.3]), which does not allow for hard-wiring some parameters (e.g., input length) into the computing device (as done in the case of non-uniform circuit families).

shown in [10]). In contrast, negative results typically refer to a fixed value of the distance parameter.

The study of property testing in the dense graph model was initiated by Goldreich, Goldwasser, and Ron [33], as a concrete and yet general framework for the study of property testing at large. From that perspective, it was most natural to represent graphs as Boolean functions, and the adjacency matrix representation was the obvious choice. This dictated the choice of the type of queries as well as the distance measure. In retrospect, the dense graph model seems most natural when graphs are viewed as representing generic (symmetric) binary relations (cf. the second motivation to the study of graphs mentioned in Section 1.1 as well as the discussion of sampling in Section 1.2).

2.1   A Taste of the Known Results

We first mention that graph properties of arbitrary query complexity are known: Specifically, in this model, graph properties (even those in $\mathcal{P}$) may have query complexity ranging from $O(1/\epsilon)$ to $\Omega(N^2)$, and the same holds also for monotone graph properties in $\mathcal{NP}$ (cf. [34]).[5] In this overview, we focus on properties that can be tested within *query complexity that only depends on the proximity parameter* (i.e., ϵ); that is, *the query complexity does not depend on the size of the graph being tested*. Interestingly, there is much to say about this class of properties. Let us start with a brief summary, and provide more details later.

1. A celebrated result of Alon, Fischer, Newman, and Shapira [3] provides a combinatorial characterization of the class of properties that can be tested within query complexity that only depends on the proximity parameter. This class contains natural properties that are not testable in query complexity poly$(1/\epsilon)$; see [1].
2. The prior work of Goldreich, Goldwasser, and Ron [33] provides a natural class of properties that can be tested within query complexity poly$(1/\epsilon)$. This class consists of so-called "partition problems" and includes sets such as k-colorability, for any fixed $k \geq 2$, and graphs containing a clique for density ρ, for any fixed $\rho > 0$.
3. A relatively recent work of Goldreich and Ron [39] initiates a study of the class of properties that can be tested within query complexity $\widetilde{O}(1/\epsilon)$.

Before providing more details on the foregoing results, we mention that, when disregarding a possible quadratic blow-up in the query complexity, we may assume that the tester in canonical in the following sense.

---

[5] We mention that a full query complexity hierarchy is established in [34] by using unnatural graph properties, starting from the $\Omega(N^2)$ lower bound of [33], which also uses an unnatural graph property. In contrast, the $\Omega(N)$ lower bound established in [27] (following [2]) refers to the natural property of testing whether an N-vertex graph consists of two isomorphic copies of some $N/2$-vertex graph.

Theorem 2.2 (canonical testers [41, Thm 2]):[6] *Let Π be any graph property. If there exists a tester with query complexity $q(N, \epsilon)$ for Π, then there exists a tester for Π that uniformly selects a set of $O(q(N, \epsilon))$ vertices and accepts iff the induced subgraph has property Π', where Π' is a graph property that may depend on N as well as on Π. Furthermore, if the original tester has one-sided error, then so does the new tester, and a sample of $2q(N, \epsilon)$ vertices suffices*

Indeed, the resulting tester is called canonical. We warn that Π' need not equal Π (let alone that Π' may depend on N), and that the time complexity of the canonical tester may be significantly larger than the time complexity of the original tester. Still, in many natural cases (e.g., k-colorability), $\Pi' = \Pi$.

2.1.1 Testability in $q(\epsilon)$ Queries, for Any Function q

As stated above, a celebrated result of Alon *et al.* [3] provides a combinatorial characterization of the class of properties that can be tested within query complexity that only depends on the proximity parameter. This characterization refers to the notion of a *regularity instance*, where regularity is in the sense of Szemerédi's Regularity Lemma [58]. The result essentially asserts that a graph property can be tested in query complexity that only depends on ϵ if and only if it can be characterized in terms of a constant number of regularity instances. The lesson from this characterization is that, when ignoring the specific dependency on ϵ, *testing graph properties in query complexity that only depends on ϵ reduces to graph regularity*. This lesson makes more concrete the feeling already raised by Theorem 2.2 that testing in this model reduces to combinatorics.

The downside of the algorithms that emerge from this characterization is that their query complexity is related to the proximity parameter via a function that grows tremendously fast. Specifically, in the general case, the query complexity is only upper bounded by a tower of a tower of exponents (in a monotonically growing function of $1/\epsilon$, which in turn depends on the property at hand).

Interestingly, it is known that a super-polynomial dependence on the proximity parameter is inherent to the foregoing result. Actually, as shown by Alon [1], such a dependence is essential even for testing *triangle freeness*. Indeed, this fact provides a nice demonstration of the non-triviality of testing graph properties. *One might have guessed that $O(1/\epsilon)$ or $O(1/\epsilon^3)$ queries would have sufficed to detect a triangle in any graph that is ϵ-far from being triangle free, but Alon's result asserts that this guess is wrong and that $\mathrm{poly}(1/\epsilon)$ queries do not suffice.* We mention that the best upper bound known for the query complexity of testing triangle freeness is $\mathtt{tf}(\mathrm{poly}(1/\epsilon))$, where $\mathtt{tf}$ is the tower function defined inductively by $\mathtt{tf}(n) = \exp(\mathtt{tf}(n-1))$ with $\mathtt{tf}(1) = 2$ (cf. [1]).

---

[6] As pointed out in [10], the statement of [41, Thm 2] should be corrected such that the auxiliary property Π' may depend on N and not only on Π. Thus, on input N and ϵ (and oracle access to an N-vertex graph G), the canonical tester checks whether a random induced subgraph of size $s = O(q(N, \epsilon))$ has the property Π', where Π' itself (or rather its intersection with the set of s-vertex graphs) may depend on N. In other words, the tester's decision depends only on the induced subgraph that it sees and on the size parameter N.

Perspective. It is indeed an amazing fact that many properties can be tested within (query) complexity that only depends on the proximity parameter (rather than also on the size of the object being tested). This amazing statement seems to shadow the question of the form of the aforementioned dependence, and blurs the difference between a reasonable dependence (e.g., a polynomial relation) and a prohibiting one (e.g., a tower-function relation). We beg to disagree with this sentiment and claim that, as in the context of standard approximation problems (cf. [45]), *the dependence of the complexity on the approximation* (or proximity) *parameter is a key issue.*

We wish to stress that we do value the impressive results of [2,7,8,29] (let alone [3]), which refer to graph property testers having query complexity that is independent of the graph size but depends prohibitively on the proximity parameter. We view such results as an impressive first step, which called for further investigation directed at determining the actual dependency of the complexity on the proximity parameter.

While it is conceivable that there exist (natural) graph properties that can be tested in $\exp(1/\epsilon)$ queries but not in $\text{poly}(1/\epsilon)$ queries, we are not aware of such a property.[7] We thus move directly from complexities of the form $\text{tf}(1/\epsilon)$ (and larger) to complexities of the form $\text{poly}(1/\epsilon)$.

2.1.2 Testability in $\text{poly}(1/\epsilon)$ Queries
Testers of query complexity $\text{poly}(1/\epsilon)$ are known for several natural graph properties [33].

- k-Colorability, for any fixed $k \geq 2$. The query-complexity is $\text{poly}(k/\epsilon)$. For $k = 2$ the running-time is $\tilde{O}(1/\epsilon^3)$, whereas for $k > 2$ the running-time is $\exp(\text{poly}(1/\epsilon))$ (and running-time polynomial in $1/\epsilon$ is unlikely, since k-Colorability is NP-complete, for $k \geq 3$).
  The k-Colorability tester has one-sided error; that is, in case the graph is k-colorable, the tester always accepts. Furthermore, when rejecting a graph, this tester always supplies a small counterexample (i.e., a $\text{poly}(1/\epsilon)$-size subgraph that is not k-colorable).
  The 2-Colorability (equivalently, Bipartiteness) Tester is presented in §2.3. An improved analysis has been obtained by Alon and Krivelevich [4].
- ρ-Clique, for any fixed $\rho > 0$, where ρ-Clique is the set of graphs that have a clique of density ρ (i.e., N-vertex graphs having a clique of size ρN).
- ρ-CUT, for any fixed $\rho > 0$, where ρ-CUT is the set of graphs that have a cut of density at least ρ (compared to N^2).
  A generalization to k-way cuts has query-complexity $\text{poly}((\log k)/\epsilon)$.
- ρ-Bisection, for any fixed $\rho > 0$, where ρ-Bisection is the set of graphs that have a bisection of density at most ρ (i.e., an N-vertex graph is in ρ-Bisection if its vertex set can be partitioned into two equal parts with at most ρN^2 edges going between them).

---

[7] Needless to say, demonstrating the existence of such (natural) properties is an interesting open problem.

Except for k-Colorability, all the other testers have two-sided error, and this is unavoidable for any tester of $o(N)$ query complexity for any of these properties.

All the above property testing problems are special cases of the General Graph Partition Testing Problem, which is parameterized by a set of lower and upper bounds. In this problem one needs to determine whether there exists a k-partition of the vertices so that the number of vertices in each part as well as the number of edges between each pair of parts falls between the corresponding lower and upper bounds (in the set of parameters). For example, ρ-clique is expressible as a 2-partition in which one part has ρN vertices, and the number of edges in this part is $\binom{\rho N}{2}$. A tester for the general problem also appears in [33]: The tester uses $\widetilde{O}(k^2/\epsilon)^{2k+O(1)}$ queries, and runs in time exponential in its query-complexity.

From Testing to Searching. Interestingly, the testers for (all cases of) the General Graph Partition Problem can be modified into algorithms that find an (implicit representation of an) approximately adequate partition whenever it exists. That is, if the graph has the desired (partitioning) property, then the testing algorithm may actually output auxiliary information that allows to reconstruct, in $\text{poly}(1/\epsilon) \cdot N$-time, a partition that approximately obeys the property. For example, for ρ-CUT, we can construct a partition with at least $(\rho - \epsilon) \cdot N^2$ crossing edges. We comment that this notion of an implicit representation of an adequate structure may be relevant for other sets in $\mathcal{NP}$, where this structure corresponds to an NP-witness. (Indeed, an interesting algorithmic application was presented in [28], where an implicit partition of an imaginary hypergraph is used in order to efficiently construct a regular partition (with almost optimal parameters) of a given graph.)

Back to Testing Graph Properties. Although many natural graph properties can be formulated as partition problems, many other properties that can be tested with $\text{poly}(1/\epsilon)$ queries cannot be formulated as such problems. The list include the set of regular graphs, connected graphs, planar graphs, and more. We identify three classes of such natural properties:

1. Properties that only depends on the vertex degree distribution (e.g., degree regularity and average degree). For example, for any fixed $\rho > 0$, the set of N-vertex graphs having ρN^2 edges can be tested using $O(1/\epsilon^2)$ queries, which is the best result possible.[8] The same holds with respect to testing degree regularity, where the $\Omega(1/\epsilon^2)$ queries lower bound follows by reduction to estimating the average value of Boolean functions and a corresponding upper bound can be obtained by building on the $\widetilde{O}(1/\epsilon^3)$-query algorithm presented in the proof of [33, Prop. 10.2.1.3].[9]

---

[8] Both upper and lower bounds can be proved by reduction to the problem of estimating the average value of Boolean functions (cf. [22]).

[9] For the lower bound, consider the problem of distinguishing between a random N-vertex graph in which each vertex has degree either $(0.5 + \epsilon)N$ or $(0.5 - \epsilon)N$ and a random $(N/2)$-regular N-vertex graph. For the upper bound, see Appendix A.1.

2. Properties that are satisfied only by sparse graphs (i.e., N-vertex graphs having $O(N)$ edges)[10] such as `Cycle-freeness` and `Planarity`. These properties can be tested by rejecting any graph that is not sufficiently sparse (see [33, Prop. 10.2.1.2]).

3. Properties that are almost trivial in the sense that, for some constant $c > 0$ and every $\epsilon > N^{-c}$, all N-vertex graphs are ϵ-close to the property. For example, every N-vertex graph is N^{-1}-close to being connected (or being Hamiltonian or Eulerian). These properties can be tested by accepting any N-vertex graph if $\epsilon > N^{-c}$ (without making any query), and inspecting the entire graph otherwise (where, in this case $\binom{N}{2} = \text{poly}(1/\epsilon)$). (See [33, Prop. 10.2.1.1].)

In view of all of the foregoing, we believe that characterizing the class of graph properties that can be tested in $\text{poly}(1/\epsilon)$ queries may be very challenging. We mention that the special case of induced subgraph freeness properties was resolved in [9].

2.1.3   Testability in $\widetilde{O}(1/\epsilon)$ Queries

While Theorem 2.2 may be interpreted as suggesting that testing in the dense graph model leaves no room for algorithmic design, this conclusion is valid only if one ignores a possible quadratic blow-up in the query complexity (and also disregards the time complexity). As advocated by Goldreich and Ron [39], a finer examination of the model, which takes into account the exact query complexity (i.e., cares about a quadratic blow-up), reveals the role of algorithmic design. In particular, the results in [39] distinguish adaptive testers from non-adaptive ones, and distinguish the latter from canonical testers. These results refer to testability in $\widetilde{O}(1/\epsilon)$ queries. In particular, it is shown that:

– Testing every "non-trivial for testing" graph property requires $\Omega(1/\epsilon)$ queries, even when adaptive testers are allowed. Furthermore, any canonical tester for such a property requires $\Omega(1/\epsilon^2)$ queries.

– There exists a natural graph property that can be tested by $\widetilde{O}(1/\epsilon)$ adaptive queries, requires $\Omega(\epsilon^{-4/3})$ non-adaptive queries, and is actually testable by $O(\epsilon^{-4/3})$ non-adaptive queries.

– There exists a natural graph property that can be tested by $\widetilde{O}(1/\epsilon)$ adaptive queries but requires $\Omega(\epsilon^{-3/2})$ non-adaptive queries.

– There exist an infinite class of natural graph properties that can be tested by $\widetilde{O}(1/\epsilon)$ non-adaptive queries.

All the above testers have one-sided error probability and are efficient, whereas the lower bounds hold also for two-sided error testers (regardless of efficiency).

The foregoing results seem to indicate that even at this low complexity level (i.e., testing in $\widetilde{O}(1/\epsilon)$ adaptive queries) there is a lot of structure and much to be understood. In particular, it is conjectured in [39] that, for every $t \geq 4$, there

---

[10] Actually, this class can be extended by considering a more relaxed notion of sparseness that includes N-vertex graphs having $O(N^{2-\Omega(1)})$ edges.

exists graph properties that can be tested by $\widetilde{O}(1/\epsilon)$ adaptive queries and have non-adaptive query complexity $\Theta(\epsilon^{-2+\frac{2}{t}})$.

2.1.4 Reflections
Let us reflect about some issues that arise from the foregoing exposition.

Adaptive Testers versus Non-Adaptive Ones. Recall that Theorem 2.2 asserts that canonical testers (which are in particular non-adaptive) have query complexity that is at most quadratic in the query complexity of general (possibly adaptive) testers. Still the results surveyed in §2.1.3 indicate that such a gap may exist. An interesting question, raised by Michael Krivelevich, is whether such a gap exists also for properties having query complexity that is significantly larger than $\widetilde{O}(1/\epsilon)$. In particular, we mention that testing Bipartiteness, which has non-adaptive query complexity $\widetilde{\Theta}(\epsilon^{-2})$ (cf. [4,21])[11] and requires $\Omega(\epsilon^{-3/2})$ adaptive queries [21], may be testable in $o(\epsilon^{-2})$ adaptive queries (cf. [42]).

One-Sided versus Two-Sided Error Probability. As noted above, for many natural properties there is a significant gap between the complexity of one-sided and two-sided error testers. For example, ρ–CUT has a two-sided error tester of query complexity poly$(1/\epsilon)$, but no one-sided error tester of query complexity $o(N^2)$. In general, the interested reader may contrast the characterization of two-sided error testers in [3] with the results in [8].

A Contrast to Recognizing Graph Properties. The notion of testing a graph property Π is a *relaxation* of the classical notion of *recognizing the graph property* Π, which has received much attention since the early 1970's (cf. [48]). In the classical (recognition) problem there are no margins of error; that is, one is required to accept all graphs having property Π and reject all graphs that lack property Π. In 1975, Rivest and Vuillemin resolved the Aanderaa–Rosenberg Conjecture, showing that any deterministic procedure for deciding any non-trivial monotone N-vertex graph property must examine $\Omega(N^2)$ entries in the adjacency matrix representing the graph. The query complexity of randomized decision procedures was conjectured by Yao to be $\Omega(N^2)$, and the currently best lower bound is $\Omega(N^{4/3})$. This stands in striking contrast to the aforementioned results regarding testing graph properties that establish that many natural (non-trivial) monotone graph properties can be *tested* by examining a constant number of locations in the matrix (where this constant depends on the constant value of the proximity parameter).

Graph Properties Are Poor Codes. We note that with the exception of two properties, which each contains a single N-vertex graph, the adjacency matrix representation of any property Π_N of N-vertex graphs yields a code over $\{0,1\}^{\binom{N}{2}}$

---

[11] The $\widetilde{O}(\epsilon^{-2})$ upper bound is due to [4], improving over [33], whereas the $\Omega(\epsilon^{-2})$ lower bound is due to [21].

with relative distance at most $O(1/N)$. Specifically, if Π_N neither consists of the N-vertex clique nor of the N-vertex independent set, then Π_N contains a graph $G = ([N], E)$ that contains two vertices $u, v \in [N]$ that have different neighborhoods in G. Consider a permutation π that transposes u and v, while leaving the rest of $[N]$ intact, and let $G' = ([N], \{\pi(a), \pi(b) : (a, b) \in E\})$. Then $G' \in \Pi_N$, but G' is $\frac{2N}{\binom{N}{2}}$-close to G.

2.2 Testing versus other Forms of Approximation

We shortly discuss the relation of the notion of approximation underlying the definition of testing graph properties (in the dense graph model)[12] to more traditional notions of approximation. Throughout this section, we refer to randomized algorithms that have a small error probability, which we ignore for simplicity.

Application to the Standard Notion of Approximation: The relation of testing graph properties to standard notions of approximation is best illustrated in the case of Max-CUT. Any tester for the set ρ-CUT, working in time $T(\epsilon, N)$, yields an algorithm for approximating the size of the maximum cut in an N-vertex graph, up to additive error ϵN^2, in time $\frac{1}{\epsilon} \cdot T(\epsilon, N)$. Thus, for any constant $\epsilon > 0$, using the above tester of [33], we can approximate the size of the max-cut to within ϵN^2 in constant time. This yields a constant time approximation scheme (i.e., to within any constant relative error) for dense graphs, which improves over previous work of Arora *et al.* [12] and de la Vega [24] who solved this problem in polynomial-time (i.e., in $O(N^{1/\epsilon^2})$–time and $\exp(\widetilde{O}(1/\epsilon^2)) \cdot N^2$–time, respectively). In the latter works the problem is solved by actually finding approximate max-cuts. Finding an approximate max-cut does not seem to follow from the mere existence of a tester for ρ-cut; yet, as mentioned above, the tester in [33] can be used to find such a cut in time linear in N.

Relation to "Dual Approximation" (cf. [45, Chap. 3]): To illustrate this relation, we consider the aforementioned ρ-Clique Tester. The traditional notion of approximating Max-Clique corresponds to distinguishing the case in which the max-clique has size at least ρN from, say, the case in which the max-clique has size at most $\rho N/2$. On the other hand, when we talk of testing ρ-Clique, the task is to distinguish the case in which an N-vertex graph has a clique of size ρN from the case in which it is ϵ-far from the class of N-vertex graphs having a clique of size ρN. This is equivalent to the "dual approximation" task of distinguishing the case in which an N-vertex graph has a clique of size ρN from the case in which any ρN subset of the vertices misses at least ϵN^2 edges. To demonstrate that these two tasks are vastly different we mention that whereas the former task is NP-Hard, for $\rho < 1/4$ (see [15,43]), the latter task can be solved in $\exp(O(1/\epsilon^2))$-time, for any $\rho, \epsilon > 0$. We believe that there is no absolute sense in which one of these approximation tasks is more important than the other: Each of these tasks may be relevant in some applications and irrelevant in others.

---

[12] Analogous relations hold also in the other models of testing graph properties.

2.3   A Benchmark: Testing Bipartiteness

The Bipartite tester is extremely simple: It selects a tiny, random set of vertices and checks whether the induced subgraph is bipartite.

Algorithm 2.3 (Bipartite Tester in the Dense Graph Model [33]): *On input N, ϵ and oracle access to an adjacency predicate of an N-vertex graph,* G = (V, E):

1. *Uniformly select a subset of $\widetilde{O}(1/\epsilon^2)$ vertices of* V.
2. *Accept if and only if the subgraph induced by this subset is bipartite.*

Step (2) amounts to querying the predicate on all pairs of vertices in the subset selected at Step (1), and testing whether the induced graph is bipartite (e.g., by running BFS). As will become clear from the analysis, it actually suffice to query only $\widetilde{O}(1/\epsilon^3)$ of these pairs. We comment that a more complex analysis due to Alon and Krivelevich [4] implies that the Algorithm 2.3 is a Bipartite Tester even if one selects only $\widetilde{O}(1/\epsilon)$ vertices (rather than $\widetilde{O}(1/\epsilon^2)$) in Step (1)).

Theorem 2.4 [33]: *Algorithm 2.3 is a* Bipartite *Tester (in the dense graph model). Furthermore, the algorithm always accepts a bipartite graph, and in case of rejection it provides a witness of length* poly$(1/\epsilon)$ *(that the graph is not bipartite).*

Proof: Let R be the subset selected in Step (1), and G_R the subgraph of G induced by R. Clearly, if G is bipartite then so is G_R, for any R. The point is to prove that if G is ϵ-far from bipartite then the probability that G_R is bipartite is at most 1/3. Thus, from this point on we assume that at least ϵN^2 edges have to be omitted from G to make it bipartite.

We view R as a union of two disjoint sets U and S, where $t \stackrel{\text{def}}{=} |U| = O(\epsilon^{-1} \cdot \log(1/\epsilon))$ and $m \stackrel{\text{def}}{=} |S| = O(t/\epsilon)$. We will consider all possible partitions of U, and associate a partial partition of V with each such partition of U. The idea is that in order to be consistent with a given partition, (U_1, U_2), of U, all neighbors of U_1 (respectively, U_2) must be placed opposite to U_1 (respectively, U_2). We will show that, with high probability, most high-degree vertices in V do neighbor U and so are forced by its partition. Since there are relatively few edges incident to vertices that do not neighbor U, it follows that, with very high probability, each such partition of U is detected as illegal by G_R. Details follow, but before we proceed let us stress the key observation: *It suffices to rule out relatively few* (partial) *partitions of* V (i.e., these induced by partitions of U), rather than all possible partitions of V.

We use the notations $\Gamma(v) \stackrel{\text{def}}{=} \{u : (u, v) \in E\}$ and $\Gamma(X) \stackrel{\text{def}}{=} \cup_{v \in X} \Gamma(v)$. Given a partition (U_1, U_2) of U, we define a (possibly partial) partition, (V_1, V_2), of V so that $V_1 \stackrel{\text{def}}{=} \Gamma(U_2)$ and $V_2 \stackrel{\text{def}}{=} \Gamma(U_1)$ (assume, for simplicity that $V_1 \cap V_2$ is indeed empty). As suggested above, if one claims that G can be "bi-partitioned" with U_1 and U_2 on different sides, then $V_1 = \Gamma(U_2)$ must be on the opposite side to U_2 (and $\Gamma(U_1)$ opposite to U_1). Note that the partition of U places no restriction on vertices that have no neighbor in U. Thus, we first ensure that *almost all* "influential" (i.e., "high-degree") vertices in V have a neighbor in U.

Technical Definition 2.4.1 (high-degree vertices and good sets): *We say that a vertex $v \in V$ is of* high-degree *if it has degree at least $\frac{\epsilon}{3}N$. We call U* good *if all but at most $\frac{\epsilon}{3}N$ of the high-degree vertices in V have a neighbor in U.*

We comment that NOT insisting that a good set U neighbors *all* high-degree vertices allows us to show that, with high probability, a random U of size unrelated to the size of the graph is good. (In contrast, if we were to insist that a good U neighbors *all* high-degree vertices, then we would have had to use $|U| = \Omega(\log N)$.)

Claim 2.4.2. *With probability at least 5/6, a uniformly chosen set U of size t is good.*

Proof: For any high-degree vertex v, the probability that v does not have any neighbor in a uniformly chosen U is at most $(1 - \epsilon/3)^t < \frac{\epsilon}{18}$ (since $t = \Omega(\epsilon^{-1} \log(1/\epsilon))$). Hence, the expected number of high-degree vertices that do not have a neighbor in a random set U is less than $\frac{\epsilon}{18} \cdot N$, and the claim follows by Markov's Inequality.   □

Technical Definition 2.4.3 (disturbing a partition of U): *We say that an edge* disturbs *a partition* (U_1, U_2) *of U if both its end-points are in the same* $\Gamma(U_i)$, *for some* $i \in \{1, 2\}$.

Claim 2.4.4. *For any good set U and any partition of U, at least $\frac{\epsilon}{3}N^2$ edges disturb the partition.*

Proof: Each partition of V has at least ϵN^2 violating edges (i.e., edges with both end-points on the same side). We upper bound the number of these edges that are not disturbing. Actually, we upper bound the number of edges that have an end-point not in $\Gamma(U)$.

- The number of edges incident to high-degree vertices that do not neighbor U is bounded by $\frac{\epsilon}{3}N \cdot N$ (since there are at most $\frac{\epsilon}{3}N$ such vertices).
- The number of edges incident to vertices that are not of high-degree is bounded by $N \cdot \frac{\epsilon}{3}N$ (since each such vertex has at most $\frac{\epsilon}{3}N$ incident edges).

This leaves us with at least $\frac{\epsilon}{3}N^2$ violating edges connecting vertices in $\Gamma(U)$ (i.e., edges disturbing the partition of U).   □

The theorem follows by observing that G_R is bipartite only if either (1) the set U is not good; or (2) the set U is good and there exists a partition of U so that none of the disturbing edges occurs in G_R. Using Claim 2.4.2 the probability of event (1) is bounded by 1/6, whereas by Claim 2.4.4 the probability of event (2) is bounded by the probability that there exists a partition of U so that none of the corresponding $\geq \frac{\epsilon}{3}N^2$ disturbing edges has both end-points in the second sample S. Actually, we pair the m vertices of S, and consider the probability that none of these pairs is a disturbing edge for a partition of U. Thus the probability of event (2) is bounded by

$$2^{|U|} \cdot \left(1 - \frac{\epsilon}{3}\right)^{m/2} < \frac{1}{6}$$

where the inequality holds since $m = \Omega(t/\epsilon)$. The theorem follows.   ∎

Comment: The procedure employed in the proof yields a randomized poly$(1/\epsilon) \cdot$ N-time algorithm for 2-partitioning a bipartite graph such that (with high probability) at most ϵN^2 edges lie within the same side. This is done by running the tester, determining a partition of U (defined as in the proof) that is consistent with the bipartite partition of R, and partitioning V as done in the proof (with vertices that do not neighbor U, or neighbor both U_1, U_2, placed arbitrarily). Thus, the placement of each vertex is determined by inspecting at most $\widetilde{O}(1/\epsilon)$ entries of the adjacency matrix. Furthermore, the aforementioned partition of U constitutes a succinct representation of the 2-partition of the entire graph. All this is a typical consequence of the fact that the analysis of the tester follows the "enforce-and-test" paradigm (see [56, Sec. 4]).

3   The Bounded-Degree Graph Model

The bounded-degree model refers to a fixed degree bound, denoted $d \geq 2$. An N-vertex graph $G = ([N], E)$ (of maximum degree d) is represented in this model by a function $g : [N] \times [d] \to \{0, 1, ..., N\}$ such that $g(v, i) = u \in [N]$ if u is the ith neighbor of v and $g(v, i) = 0$ if v has less than i neighbors.[13] Distance between graphs is measured in terms of their aforementioned representation (i.e., as the fraction of (the number of) different array entries (over dN)), but occasionally we shall use the more intuitive notion of the fraction of (the number of) edges over $dN/2$. We now spell out the meaning of property testing in this model.

Definition 3.1 (testing graph properties in the bounded-degree model): *For a fixed d, a* tester *for a graph property Π is a probabilistic oracle machine that, on input parameters N and ϵ and access to (the incidence function of) an N-vertex graph $G = ([N], E)$ of maximum degree d, outputs a binary verdict that satisfies the following two conditions.*

1. *If $G \in \Pi$ then the tester accepts with probability at least 2/3.*
2. *If G is ϵ-far from Π then the tester accepts with probability at most 1/3, where G is ϵ-far from Π if for every N-vertex graph $G' = ([N], E') \in \Pi$ of maximum degree d it holds that the symmetric difference between E and E' has cardinality that is greater than $\epsilon \cdot dN/2$.*

One-sided testers and non-adaptive testers are defined as in Definition 2.1.

---

[13] For simplicity, we assume here that the neighbors of v appear in an arbitrary order in the sequence $g(v, 1), ..., g(v, \deg(v))$, where $\deg(v) \overset{\text{def}}{=} |\{i : g(v, i) \neq 0\}|$. Also, we shall always assume that if $g(v, i) = u \in [N]$ then there exists $j \in [d]$ such that $g(u, j) = v$.

The query complexity of a tester is defined as in Section 2; ditto for its efficiency.

The study of property testing in the bounded-degree graph model was initiated by Goldreich and Ron [35], with the aim of allowing the consideration of sparse graphs, which appear in numerous applications (cf. the first motivation to the study of graphs mentioned in Section 1.1). The point was that the dense graph model seems irrelevant to sparse graphs, both because the distance measure that underlies it deems all sparse graphs as close to one another, and because adjacency queries seems unsuitable for sparse graphs. Sticking to the paradigm of representing graphs as functions, where both the distance measure and the type of queries are determined by the representation, the aforementioned representation seemed the most natural choice. Indeed, a conscious decision was (and is) made not to capture, at this point (and in this model), sparse graphs that do not have constant (or low) maximum degree.

3.1   A Taste of the Known Results

We first mention that, also in this model, graph properties of arbitrary query complexity are known: Specifically, in this model, graph properties (in $\mathcal{NP}$) may have query complexity ranging from $O(1/\epsilon)$ to $\Omega(N)$, and furthermore such properties are monotone and natural (cf. [34], which builds over [20]). In particular, testing 3-Colorability requires $\Omega(N)$ queries, whereas testing 2-Colorability (i.e., Bipartiteness) requires $\Omega(\sqrt{N})$ queries [35] and can be done using $\widetilde{O}(\sqrt{N}) \cdot \mathrm{poly}(1/\epsilon)$ queries [36]. We also mention that many natural properties are testable in query complexity that only depends on the proximity parameter (i.e., ϵ). A partial list includes k-edge connectivity, for every fixed k, and Planarity (cf. [35] and [18], respectively). Details follow.

3.1.1   Testability in $q(\epsilon)$ Queries, for Any Function q

We first mention, that with the exception of properties that only depend on the degree distribution, adaptive testers are essential for obtaining query complexity that only depends on ϵ (cf. [55]).[14] Still, as observed in [40], at the cost of an exponentially blow-up in the query complexity, we may assume that the tester's adaptivity is confined to performing (full, BFS-like) searches of a predetermined depth from several randomly selected vertices. However, the best testing results are typically obtained by testers that either perform more adaptive searchers or perform DFS-like rather than BFS-like searchers. A few examples follow, where all testers are efficient (i.e., their running time is polynomial in their query complexity).

Testing Connectivity. Graph connectivity can be tested in $\widetilde{O}(1/\epsilon)$ queries [35]. Essentially, the tester starts a search (e.g., a BFS) from a few randomly selected

---

[14] Actually, the result extends to query complexity of the form $o(\sqrt{N} \cdot q(\epsilon))$, for any function q. In contrast, note that triangle-freeness can be tested by $O(\sqrt{N/\epsilon})$ non-adaptive queries; see Appendix A.2.

vertices, but each such search is terminated after a predetermined number of vertices is encountered (rather than after visiting all vertices that are at a predetermined distance from the start vertex). This tester rejects if and only if it detects a small connected component, and thus it has one-sided error. The result essentially extends to k-edge connectivity, for any $k \geq 2$, but the query complexity is $\widetilde{O}(k^3/\epsilon^c)$, where $c = \min(k-1, 3)$ (cf. [35]).

Testing Cycle-Freeness. Cycle-freeness can be tested in $\widetilde{O}(\epsilon^{-3})$ queries, by a tester having two-sided error [35]. Essentially, the tester compares the number of edges to the number of connected components, while fully exploring any small connected components that it happens to visit. The two-sided error is unavoidable by any tester that has query complexity $o(\sqrt{N})$ (cf. [35, Prop. 4.3]). Viewing cycle-free graphs as graphs that have no K_3-minor, leads us to the following general result of Benjamini, Schramm, and Shapira [18], which refers to graph minors (to be briefly recalled next).

The graph H is a minor of the graph G, if H can be obtained from G by a sequence of edge removal, vertex removal, and edge contraction operations. We say that G is H-minor free if H is not a minor of G. Thus, a graph is cycle-free if and only if it is K_3-minor free, where K_k denotes the k-vertex clique. (The notion of minor freeness extends to sets of graphs; that is, for a set of graphs $\mathcal{H}$, the graph G is $\mathcal{H}$-minor free if no element of $\mathcal{H}$ is a minor of G.) Lastly, a graph property is minor-closed if it is closed under removal of edges, removal of vertices, and edge contraction. Note that, for every finite sets of graphs $\mathcal{H}$, the property of being $\mathcal{H}$-minor free (e.g., Planarity) is minor-closed.

Theorem 3.2 ([44], improving over [18]):[15] *Any minor-closed property can be tested in query complexity* $\exp(\mathrm{poly}(1/\epsilon))$.

We mention that this tester has two-sided error, which is unavoidable for any tester of query complexity $o(\sqrt{N})$, except for the case that the forbidden minors are all cycle-free.

3.1.2   Testability in $\widetilde{O}(N^{1/2}) \cdot \mathrm{poly}(1/\epsilon)$ Queries

The query complexity of testing two natural properties is $\widetilde{\Theta}(N^{1/2}) \cdot \mathrm{poly}(1/\epsilon)$, and in both cases the time complexity has the same form. The properties are Bipartiteness and Expansion. In both cases, the algorithm is based on taking many (i.e., $\widetilde{O}(N^{1/2}) \cdot \mathrm{poly}(1/\epsilon)$) *random walks* from a few randomly selected vertices, where each walk has length $\mathrm{poly}(\epsilon^{-1} \log N)$.

The foregoing algorithmic approach originates in [36], where it was applied to testing Bipartiteness; for further details see §3.2.2. This approach was also suggested for testing Expansion [37], but the analysis was successfully completed only in [46,51]. We mention that the Bipartite tester has one-sided error, and whenever it rejects it may also output a short proof that the graph is not bipartite (i.e., an odd cycle of length $\mathrm{poly}(\epsilon^{-1} \log N)$).

---

[15] The query complexity obtained in [18] is triple-exponential in $1/\epsilon$.

The $\Omega(N^{1/2})$ lower bound on the query complexity of testing each of the aforementioned properties was proved in [35]; for details see §3.2.1. We note that the lower bound for testing Bipartiteness stands in sharp contrast to the situation in the dense graph model, where Bipartite testing is possible in poly$(1/\epsilon)$-time. This discrepancy is due to the difference between the notions of relative distance employed in the two models.

An application to the Study of the Dense Graph Model. We mention that the Bipartiteness tester of the bounded-degree model was used in order to derive an alternative Bipartite tester for the dense graph model [42]. In the case that almost all vertices in the N-vertex graph have degree $O(\epsilon^{0.99}N)$, this tester improves over the ones presented in [33,4]. Essentially, this dense-graph model tester invokes the bounded-degree model tester on the subgraph induced by a sample S of $\widetilde{O}(1/\epsilon)$ random vertices (and emulates neighbor queries regarding a vertex $v \in S$ by making adjacency queries of the form (v, w) for every $w \in S$).

3.1.3   Reflections
The fact that the bounded-degree model is closer (than the dense graph model) to standard algorithmic research offers greater interaction at the technical level. Indeed, techniques such as local search and random walks are quite basic in both domains, and the relationship becomes even tighter when we shall move to the general graph model (in Section 4). At the current point, we mention that the idea underlying the cycle-freeness tester (outlined in §3.1.1) was employed to the design of an algorithm for approximating the minimum spanning tree weight in sub-linear time [23].

We also mention that the idea underlying the expansion tester has become quite pivotal in the contents of testing distributions, which emerged with [13].

3.2   A Benchmark: Testing Bipartiteness
Both the following lower and upper bounds reflect the fact that being far from Bipartiteness does not require having constant size cycles of odd length. We comment that a simplified version of the upper bound implies that odd cycles of logarithmic length must exist (cf. [36, Prop. 1]).

3.2.1   A Lower Bound
In contrast to Theorem 2.4, under the incidence function representation, there exists no Bipartite tester of complexity that is independent of the graph size.

Theorem 3.3 [35]: *Testing* Bipartiteness (with constant ϵ and d) *requires* $\Omega(\sqrt{N})$ *queries* (in the incidence function model).

Proof Idea: For any (even) N, we consider the following two families of graphs:

1. The first family, denoted $\mathcal{G}_1^N$, consists of all degree-3 graphs that are composed of the union of a Hamiltonian cycle and a perfect matching. That is, there are N edges connecting the vertices in a cycle, and the other $N/2$ edges are a perfect matching.

2. The second family, denoted $\mathcal{G}_2^N$, is the same as the first *except* that the perfect matchings allowed are restricted as follows: the distance on the cycle between every two vertices that are connected by a perfect matching edge must be odd.

Clearly, all graphs in $\mathcal{G}_2^N$ are bipartite. It can be shown that almost all graphs in $\mathcal{G}_1^N$ are far from being bipartite. On the other hand, one can prove that a testing algorithm that performs $o(\sqrt{N})$ queries cannot distinguish between a graph chosen randomly from $\mathcal{G}_2^N$ (which is always bipartite) and a graph chosen randomly from $\mathcal{G}_1^N$ (which with high probability is far from bipartite). Loosely speaking, this is the case since in both cases the algorithm is unlikely to encounter a cycle (among the vertices that it has inspected).   ∎

3.2.2   An Algorithm
The lower bound of Theorem 3.3 is essentially tight. Furthermore, the following natural algorithm constitutes a **Bipartite** tester of running time $\mathrm{poly}((\log N)/\epsilon) \cdot \sqrt{N}$.

Algorithm 3.4 (Bipartite Tester in the Bounded-Degree Model [36]): *On input N, d, ϵ and oracle access to an incidence function for an N-vertex graph, $\mathrm{G} = (\mathrm{V}, \mathrm{E})$, of degree bound d, repeat $T \stackrel{\text{def}}{=} \Theta(\frac{1}{\epsilon})$ times:*

1. *Uniformly select s in V.*
2. *(Try to find an odd cycle through vertex s):*
    (a) Perform $K \stackrel{\text{def}}{=} \mathrm{poly}((\log N)/\epsilon) \cdot \sqrt{N}$ random walks starting from s, each of length $L \stackrel{\text{def}}{=} \mathrm{poly}((\log N)/\epsilon)$.
    (b) Let R_0 (respectively, R_1) denote the vertices set reached from s in an even (respectively, odd) number of steps in any of these walks.
    (c) If $R_0 \cap R_1$ is not empty then reject.

If the algorithm did not reject in any of the foregoing T iterations, then it accepts.

Theorem 3.5 [36]: *Algorithm 3.4 is a* **Bipartite** *Tester* (in the incidence function model). *Furthermore, the algorithm always accepts a bipartite graph, and in case of rejection it provides a witness of length $\mathrm{poly}((\log N)/\epsilon)$* (that the graph is not bipartite).

Motivation – the Special Case of Rapid Mixing Graphs. The proof of Theorem 3.5 is quite involved. As a motivation, we consider the special case where the graph has a "rapid mixing" feature. It is convenient to modify the random walks so that at each step each neighbor is selected with probability $1/2d$, and otherwise (with probability at least $1/2$) the walk remains in the present vertex. Furthermore, we will consider a single execution of Step (2) starting from an arbitrary vertex, s, which is fixed in the rest of the discussion. The rapid mixing feature we assume is that, for every vertex v, a (modified) random walk of length L starting at s reaches v with probability approximately $1/N$ (say, up-to a factor of 2). Note that if the graph is an expander then this is certainly the case (since $L = \omega(\log N)$).

The key quantities in the analysis are the following probabilities, referring to the parity *of the length of a path obtained from the random walk by omitting the self-loops* (transitions that remain at current vertex). Let $p^0(v)$ (respectively, $p^1(v)$) denote the probability that a (modified) *random walk of length L, starting at s, reaches v while making an even* (respectively, *odd*) *number of real* (i.e., non-self-loop) *steps*. By the rapid mixing assumption (for every $v \in V$), it holds that

$$\frac{1}{2N} < p^0(v) + p^1(v) < \frac{2}{N}. \tag{2}$$

We consider two cases regarding the sum $\sum_{v \in V} p^0(v)p^1(v)$: If the sum is (relatively) "small", we show that V can be 2-partitioned so that there are relatively few edges between vertices that are placed in the same side, which implies that G is close to being bipartite. Otherwise (i.e., when the sum is not "small"), we show that with significant probability, when Step (2) is started at vertex s it is completed by rejecting G. These two cases are analyzed in the following two (corresponding) claims.

Claim 3.5.1. *Suppose* $\sum_{v \in V} p^0(v)p^1(v) \leq \epsilon/50N$. *Let* $V_1 \stackrel{\text{def}}{=} \{v \in V : p^0(v) < p^1(v)\}$ *and* $V_2 = V \setminus V_1$. *Then, the number of edges with both end-points in the same* V_σ *is bounded above by* ϵdN.

Proof Sketch: Consider an edge (u, v) where, without loss of generality, both u and v are in V_1. Then, both $p^1(v)$ and $p^1(u)$ are greater than $\frac{1}{2} \cdot \frac{1}{2N}$. However, one can show that $p^0(v) > \frac{1}{3d} \cdot p^1(u)$: Observe that an $(L-1)$-step walk of path-parity 1 ending at u is almost as likely as an L-step walk of path-parity 1 ending at u, and that once an $(L-1)$-step walk reaches u, with probability exactly $1/2d$, it continues to v in the next step. Thus, the edge (u, v) contributes at least $\frac{(1/4N)^2}{3d}$ to the sum $\sum_{w \in V} p^0(w)p^1(w)$. It follows that we can have at most $(\epsilon/50N)/(1/48dN^2)$ such edges, and the claim follows.   □

Claim 3.5.2. *Suppose* $\sum_{v \in V} p^0(v)p^1(v) \geq \epsilon/50N$, *and that Step (2) is started with vertex s. Then, with probability at least 2/3, the set* $R_0 \cap R_1$ *is not empty (and rejection follows).*

Proof Sketch: Consider the probability space defined by an execution of Step (2) with start vertex s. For every $i \neq j$ such that $i, j \in [K]$, we define an indicator random variable $\zeta_{i,j}$ representing *the event that the vertex encountered in the* L^{th} *step of the* i^{th} *walk equals the vertex encountered in the* L^{th} *step of the* j^{th} *walk, and that the* i^{th} *walk corresponds to an even-path whereas the* j^{th} *to an odd-path.* (That is, $\zeta_{i,j} = 1$ if the foregoing event holds, and $\zeta_{i,j} = 0$ otherwise.) Then

$$\mathbf{E}[|R_0 \cap R_1|] > \sum_{i \neq j} \mathbf{E}[\zeta_{i,j}]$$

$$= K(K-1) \cdot \sum_{v \in V} p^0(v)p^1(v)$$

$$> \frac{500N}{\epsilon} \cdot \sum_{v \in V} p^0(v)p^1(v)$$

$$\geq 10$$

where the second inequality is due to the setting of K, and the third to the claim's hypothesis. Intuitively, with high probability, it should hold that $|R_0 \cap R_1| > 0$. This is indeed the case, but proving it is less straightforward than it seems; the problem being that the $\zeta_{i,j}$'s are not pairwise independent. Yet, since the sum of the covariances of the dependent $\zeta_{i,j}$'s is quite small, Chebyshev's Inequality is still very useful (cf. [11, Sec. 4.3]). Specifically, letting $\mu \stackrel{\text{def}}{=} \sum_{v \in V} p^0(v)p^1(v)$ $(= \mathbf{E}[\zeta_{i,j}])$, and $\overline{\zeta}_{i,j} \stackrel{\text{def}}{=} \zeta_{i,j} - \mu$, we get:

$$\mathbf{Pr}\left[\sum_{i \neq j} \zeta_{i,j} = 0\right] < \frac{\mathbf{Var}\left[\sum_{i \neq j} \zeta_{i,j}\right]}{(K^2\mu)^2}$$

$$= \frac{1}{K^4\mu^2} \cdot \left(\sum_{i,j} \mathbf{E}\left[\overline{\zeta}_{i,j}^2\right] + 2\sum_{i,j,k} \mathbf{E}\left[\overline{\zeta}_{i,j}\overline{\zeta}_{i,k}\right]\right)$$

$$< \frac{1}{K^2\mu} + \frac{2}{K\mu^2} \cdot \mathbf{E}[\zeta_{1,2}\zeta_{1,3}]$$

For the second term, we observe that $\mathbf{Pr}[\zeta_{1,2} = \zeta_{1,3} = 1]$ is upper bounded by $\mathbf{Pr}[\zeta_{1,2} = 1] = \mu$ times the probability that the Lth vertex of the first walk appears as the Lth vertex of the third path. Using the rapid mixing hypothesis, we upper bound the latter probability by $2/N$, and obtain

$$\mathbf{Pr}[|R_0 \cap R_1| = 0] < \frac{1}{K^2\mu} + \frac{2}{K\mu^2} \cdot \mu \cdot \frac{2}{N}$$

$$< \frac{1}{3}$$

where the last inequality uses $\mu \geq \epsilon/50N$ and $K^2 \geq 6 \cdot 50N/\epsilon$ (along with $\epsilon > 5000/N$). The claim follows.   □

Beyond Rapid Mixing Graphs. The proof in [36] refers to a more general sum of products; that is, $\sum_{u \in U} p_{\text{odd}}(u)p_{\text{even}}(u)$, where $U \subseteq V$ is an appropriate set of vertices, and $p_{\text{odd}}(v)$ (respectively, $p_{\text{even}}(v)$) is essentially the probability that an L-step random walk (starting at s) passes through v after more than $L/2$ steps and the corresponding path to v has odd (respectively, even) parity. Much of the analysis in [36] goes into selecting the appropriate U (and an appropriate starting vertex s), and pasting together many such U's to cover all of V. Loosely speaking, U and s are selected so that there are few edges from U and the rest of the graph, and $p_{\text{odd}}(u) + p_{\text{even}}(u) \approx 1/\sqrt{|V| \cdot |U|}$, for every $u \in U$. The selection is based on the "combinatorial treatment of expansion" of Mihail [50]. Specifically, we use the contrapositive of the standard analysis, which asserts that rapid mixing occurs when all cuts are relatively large, to assert the existence of small cuts which partition the graph so that vertices reached with relatively high probability (in a short random walk) are on one side and the rest of the graph on the other. The first set corresponds to the aforementioned U and the cut is relatively small with respect to U. A start vertex s for which the corresponding sum is big is shown to cause Step (2) to reject (when started with this s), whereas a small corresponding sum enables to 2-partition U while having few violating edges among the vertices in each part of U.

The actual argument of [36] proceeds in iterations. In each iteration a vertex s for which Step (2) accepts with high probability is fixed, and an appropriate set of remaining vertices, U, is found. The set U is then 2-partitioned so that there are few violating edges inside U. Since we want to paste all these partitions together, U may not contain vertices treated in previous iterations. This complicates the analysis, since it must refer to the part of G, denoted H, not treated in previous iterations. We consider walks over an (imaginary) Markov Chain representing the H-part of the walks performed by the algorithm on G. Statements about rapid mixing are made with respect to the Markov Chain, and linked to what happens in random walks performed on G. In particular, a subset U of H is determined so that the vertices in U are reached with probability $\approx 1/\sqrt{|V| \cdot |U|}$ (in the chain) and the cut between U and the rest of H is small. Linking the sum of products defined for the chain with the actual walks performed by the algorithm, we infer that U may be partitioned with few violating edges inside it. Edges to previously treated parts of the graphs are charged to these parts, and edges to the rest of $H \setminus U$ are accounted for by using the fact that this cut is small (relative to the size of U).

4   The General Graph Model

In contrast to the foregoing two models in which the oracle queries and the distances between graphs are linked to the representation of graphs as functions, in the following model the representation is blurred and the query types and distance measure are decoupled. This decoupling makes the current model closer in spirit to standard studies in graph algorithms.

Giving up on the representation as a yardstick for the relative distance between graphs, leaves us with no absolute point of reference. Instead, we just define the relative distance between graphs in relation to the actual number of edges in these graphs; specifically, the relative distance between the graphs $G = ([N], E)$ and $G' = ([N], E)$ may be defined as $\frac{|E \triangle E'|}{\max(|E|, |E'|)}$ (or, alternatively, as $\frac{|E \triangle E'|}{(|E|+|E'|)/2}$).[16]

Turning to the question of query types, we again need to make a choice, which is now free from representation considerations. The most natural choice is to allow both *adjacency queries* and *incidence queries* (i.e., the two types of

---

[16] Needless to say, these two definitions may not yield the same result, but they are related by a factor of at most 2.

queries that were each allowed in one of the previous queries).[17] However, other choices has been considered too (cf. [17]). We note that, typically, adjacency queries become more useful as the graph becomes more dense, whereas incidence queries (a.k.a neighbor queries) become more useful as the graph becomes more sparse (cf. [17]).

Definition 4.1 (testing graph properties in the general model): *A* tester *for a graph property Π is a probabilistic oracle machine that, on input parameters N and ϵ and access to a function answering adjacency queries and incidence queries regarding an N-vertex graph $G = ([N], E)$, outputs a binary verdict that satisfies the following two conditions.*

1. *If $G \in \Pi$ then the tester accepts with probability at least $2/3$.*
2. *If G is ϵ-far from Π then the tester accepts with probability at most $1/3$, where G is ϵ-far from Π if for every N-vertex graph $G' = ([N], E') \in \Pi$ it holds that the symmetric difference between E and E' has cardinality that is greater than $\epsilon \cdot \max(|E|, |E'|)$.*

One-sided testers and non-adaptive testers are defined as in Definition 2.1.

The query complexity of a tester is defined as in Section 2; ditto for its efficiency.

The study of property testing in the general graph model was initiated by Parnas and Ron [53], who only considered incidence queries, and extended by Kaufman, Krivelevich, and Ron [47], who considered both types of queries. Needless to say, the aim of these works was to allow the consideration of arbitrary graphs and so strengthen the relation between property testing and standard algorithmic studies. However, forsaking the paradigm of representing graphs as functions means that the connection to the rest of property testing is a bit weakened (or at least becomes more cumbersome). Still, we believe that the trade-off is worthwhile.

4.1   A Taste of the Known Results

It is natural to attempt to extend testers designed for the bounded-degree model to the general graph model. Such extensions face two potential difficulties, which refer to two ways in which the general graph model extends the bounded-degree model:

1. Firstly, the maximum degree of vertices in the graph may no longer be constant, and the question is how does the performance of the tester depends on the degree bound, d. Formally, one should think of the degree bound d as a variable, and analyze the tester accordingly.

   Note that when d increases, relative distances decrease and so testing may become easier. On the other hand, we can no longer scan all neighbors of a given vertex at constant cost.

---

[17] Recall that the incidence query (u, i) is answered with 0 if u has less than i neighbors. Thus, the incidence queries allow to emulate degree queries at logarithmic cost.

2. Treating the maximum degree as a variable, raises the question of what happens when there is a significant discrepancy among the degrees of the various vertices. Such a situation can break the balance between the afore-mentioned positive and negative effects of increasing the maximum degree. Specifically, the algorithmic operations may becomes more costly when the maximum degree increases, but when using the distance measure of Definition 4.1 the distances no longer vary with the maximum degree (i.e., d) but rather vary with the average degree. Thus, we may be in trouble if the maximum degree is significantly larger than the average degree.

The effect of the foregoing issues is tester-dependent. For example, the operation of the Connectivity tester (outlined in §3.1.1) is not affected by the possible discrepancies in the vertex degrees, and so this tester (as is) applies also to the general graph model (cf. [53]). In contrast, the Bipartiteness tester presented in Algorithm 3.4 should be modified to the current setting. Details follow.

4.2   A Benchmark: Testing Bipartiteness

Firstly, it was shown in [47] that the algorithm's performance does not deteriorate when d increases. Next, an algorithm for the general graph model was obtained by emulating Algorithm 3.4 on an imaginary graph that is obtained by replacing vertices of high degree by adequate gadgets. Specifically, a vertex having degree that is t times larger than the average degree is replaced by a t-by-t bipartite expander graph, while connecting the original neighbors to vertices on one of the sides of the expander (such that no vertex has degree greater than twice the average degree). This replacement preserves the distance to Bipartiteness (up to a constant factor). We warn that implementing the emulation (of Algorithm 3.4 on this imaginary graph) is not straightforward. In particular, it seems to require a procedure for sampling edges in the actual graph such that almost all edges are sampled with probability that is approximately (up to a constant factor) the uniform one.[18] For details, see [47].

As evident from the above description, the extension of a tester from the bounded-degree model to the general graph model may require ideas that are specific to the property at hand. For example, the gadgets used above should preserve Bipartiteness (as well as distance to Bipartiteness).

Another issue that arises is that one may hope to perform better when the degree bound d (whether maximum or average) is large. Indeed, we know that in case of Bipartiteness, dense graphs can be tested with much fewer queries than sparse graphs (recall Algorithm 2.3). Thus, an optimal tester for the general graph model should be able to match the result of the dense graph model whenever the actual graph happens to be dense. Such a result is indeed provided by [47], who show a Bipartiteness tester (for the general graph model) that is optimal for all possible edge densities.

---

[18] A more accurate sampling procedure is implicit in the subsequent work of [38].

Theorem 4.2 (Testing Bipartiteness in the General Graph Model [47]): *Ignoring factors that are polynomial in $\epsilon^{-1} \log N$, the query (and time) complexity of testing* Bipartiteness *is* $\min(\sqrt{N}, N^2/M)$, *where M denotes the number of edges in the input graph.*

Note that dealing with $M \gg N^{3/2}$ requires some deviation from the aforementioned emulation (of Algorithm 3.4). Indeed, in such a case the tester of [47] behaves quite differently. Specifically, it takes $K = \sqrt{N^2/M}$ random walks (rather than N^2/M random walks), from each random start vertex, and checks for collisions among the endpoints these K walks by using $\binom{K}{2}$ adjacency queries. We mention that the use of adjacency queries is necessary for an $o(\sqrt{N})$ query tester of Bipartiteness.

An Opposite Behavior. In contrast to the case of testing Bipartiteness, where the complexity improves with the edge density, in the case of testing triangle-freeness we see the opposite behavior [5].[19] Furthermore, in contrast to testing Bipartiteness, there is a gap between the complexity of testing triangle-freeness in the bounded-degree model and the corresponding complexity in the general graph model even when the graph is sparse (i.e., $M = O(N)$). For example, in the general graph model, the complexity is $\Omega(N^{1/3})$ as long as $M = N^{2-o(1)}$ [5].

4.3 Reflections

The bulk of algorithmic research regarding graphs refers to general graphs. Of special interest are graphs that are neither very dense nor have a bounded degree. In contrast, research in testing properties of graphs started (in [33]) with the study of dense graphs, proceeded to the study of bounded-degree graphs (in [35]), and reached general graphs only in [53,47]. This evolution has historical reasons to be reviewed first.

Testing graph properties was initially conceived (in [33]) as a special case of the framework of testing properties of functions. Thus, graphs had to be represented by functions, and two standard representations of graphs (indeed, the two reviewed in Sections 2 and 3) seemed most fitting in this context. We stress that both models were formulated in a way that identifies the graphs with a specific functional representation, which in turn defines the type of queries allowed to the tester as well as the notion of fractional distance (which underlies the performance guarantee).

The identification of graphs with any specific functional representation was abandoned by Parnas and Ron [53] who developed a more general model by decoupling the type of queries allowed to the tester from the distance measure: Whatever is the mechanism of accessing the graph, the distance between graphs is defined as the number of edges in their symmetric difference (rather

---

[19] This is to be expected in light of the fact that testing triangle-freeness has complexity $O(d/\epsilon)$ in the bounded-degree model [35], whereas in the dense graph model testing triangle-freeness requires more than poly$(1/\epsilon)$ queries [1].

than the number of different entries with respect to some specific functional representation). Furthermore, the relative distance may be defined as the size of the symmetric difference divided by the actual (total) number of edges in both graphs (rather than divided by some (possibly non-tight) upper-bound on the latter quantity). Also, as advocated by Kaufman *et al.* [47], it is reasonable to allow the tester to perform both adjacency and neighbor queries (and indeed each type of query may be useful in a different range of edge densities). Needless to say, this model seems adequate for the study of testing properties of arbitrary graphs, and it strictly generalizes the positive aspects of the two prior models (i.e., the models based on the adjacency matrix and bounded-degree incidence list representations).

We wish to advocate further study of the latter model. We believe that this model, which allows for a meaningful treatment of property testing of general graphs, is the one that is most relevant to computer science applications. Furthermore, it seems that designing testers in this model requires the development of algorithmic techniques that may be applicable also in other areas of algorithmic research. As an example, we mention that techniques in [47] underly the average degree approximation of [38]. (Likewise techniques of [35] underly the minimum spanning tree weight approximation of [23]; indeed, as noted next, the bounded-degree incidence list model is also more algorithmic oriented than the adjacency matrix model.)

Let us focus on the algorithmic contents of property testing of graphs. Recall that, when ignoring a quadratic blow-up in the query complexity, property testing in the adjacency matrix representation reduces to sheer combinatorics (as reflected in the notion of canonical testers, see Theorem 2.2). Indeed, as shown in [39], a finer look (which does not allow for ignoring quadratic blow-ups in complexity) reveals the role of algorithmic design also in this model. But still property testing in the incidence list representation seems to require more sophisticated algorithms. Testers in the general graph models seem to require even more algorithmic ideas (cf. [47]).

To summarize, we advocate further study of the model of [53,47] for two reasons. The first reason is that we believe in the greater relevance of this model to computer science applications. The second reason is that we believe in the greater potential of this model to have cross fertilization with other branches of algorithmic research. Nevertheless, this advocation is not meant to undermine the study of the dense graph and bounded-degree models. The latter have their own merits and also offer a host of interesting open problems, which are potentially relevant to computer science at large.

5 Additional Issues

In this section we discuss three issues that are relevant to each of the three models discussed in the prior corresponding three sections.

5.1   Directed Graphs

So far our discussion was confined to undirected graphs. Nevertheless, the three models extend naturally to the case of directed graphs. Actually, when considering incidence queries, two different sub-models emerge (cf. [16]): In the first model the tester may only query for edges in the forward direction (resp., backward direction), whereas in the second model both forward and backward directions are allowed. That is, in the second model, the directed graph $G = ([N], E)$ is represented by two functions, g_{out} and g_{in}, such that $g_{out}(u, i) = v$ (resp., $g_{in}(u, i) = v$) if the ith out-going edge of u leads to v (resp., the ith in-coming edge of u arrives from v).

The gap between these two query models was demonstrated by Bender and Ron, who initiated the study of testing properties of directed graphs [16]. In particular, they showed that while strong connectivity in bounded-degree directed graphs can be tested by $\tilde{O}(1/\epsilon)$ forward and backward queries [16, Sec. 5.1], when only forward (resp., backward) queries are allowed no tester can work with $o(\sqrt{N})$ queries (even when allowing two-sided error [16, Sec. 5.2]).[20]

Another task studied in [16] is testing whether a given directed graph is acyclic (i.e., has no directed cycles). They presented an `Acyclicity` tester of $poly(1/\epsilon)$ complexity in the adjacency predicate model, and showed that in the incidence list model no `Acyclicity` tester can work with $o(N^{1/3})$ queries (even when both forward and backward queries are allowed). The question of whether `Acyclicity` can be tested with $o(N)$ queries (in the bounded-degree digraph model) remains open. In general, it seems that the study of this model deserves more attention than it has received so far. (We mention that testing directed graphs in the dense digraph model was further studied in [6,52].)

5.2   Tolerant Testing and Distance Approximation

Recall that property testing calls for distinguishing objects having a predetermined property from object that are far from any objects that has this property (i.e., are far from the property). A more "tolerant" notion requires distinguishing objects that are close to having the property from objects that are far from this property. Such a distinguisher is called a tolerant tester, and is a special case of a distance approximator that given any object is required to approximate its distance to the property. The study of these related notions was initiated by Parnas, Ron, and Rubinfeld [54].

Definition 5.1 (sketch for the generic case): *Let Π be a set of functions over a finite set Ω. A distance approximator for Π is a probabilistic oracle machine T that on input an approximation parameter ϵ and access to any function f outputs with probability at least $2/3$ a value that approximates the relative distance of $f*

---

[20] The lower bound can be strengthened to $\Omega(N)$ when considering only one-sided error testers. In the case of two-sided error, some improvements are possible; see Appendix A.3.

to Π up to an additive term of ϵ; that is, $\mathbf{Pr}[|T^f - \delta_\Pi(f)| \leq \epsilon] \geq 2/3$, where $\delta_\Pi(f) \stackrel{\text{def}}{=} \min_{g \in \Pi}\{\delta(f,g)\}$ and $\delta(f,g) \stackrel{\text{def}}{=} \mathbf{Pr}_{x \in \Omega}[f(x) \neq g(x)]$.

A simple observation is that any tester that makes uniformly distributed queries offers some level of tolerance. Specifically, if a tester makes $q(\epsilon)$ queries and each query is uniformly distributed, then this tester distinguishes between objects that are ϵ-far from the property and objects that are $(\epsilon/10q(\epsilon))$-close to the property. Needless to say, the challenge is to provide stronger relations between property testing and distance approximators. Such a result was provided by Fischer and Newman [29]: They showed that, *in the dense graph model, testability in a number of queries that only depends on ϵ implies distance approximator in a number of queries that only depends on ϵ.* In the the bounded-degree model, many of the known testers were extended to yield distance approximators (cf. [49]).

5.3 Proximity Oblivious Testing

Note that in order to satisfy the property testing requirement, any tester (of a reasonable property) must obtain the proximity parameter as auxiliary input and determine its actions accordingly. The question, addressed here, is what does the tester do with this parameter (or how does the parameter affect the actions of the tester). A very minimal effect is exhibited by testers that, based on the value of the proximity parameter, determine the number of times that a basic test is invoked, where the basic test is oblivious of the proximity parameter. For example, the celebrated linearity tester of [19] repeats a basic test that consists of selecting two random points, x and y, and probing the value of the function at the points x, y, and $x + y$. This basic test is repeated for a number of times that is inversely proportional to the proximity parameter.

Our focus here is on such basic tests (i.e., basic tests that are oblivious of the proximity parameter), called proximity oblivious testers. Although proximity oblivious testers were implicit in prior works (see, e.g., [19,2,3]), their general study was initiated by Goldreich and Ron [40].

Definition 5.2 (sketch for the generic case): *Let Π be a set of functions over a finite set Ω. A proximity-oblivious tester for Π is a probabilistic oracle machine T that, when given oracle access to any function f over Ω, satisfies the following two conditions:*

1. *The machine T accepts each function in Π with probability 1.*
2. *For some* (monotone) *function $\rho : (0,1] \to (0,1]$, each function $f \notin \Pi$ is rejected by T with probability at least $\rho(\delta_\Pi(f))$, where $\delta_\Pi(f)$ is as in Definition 5.1.*

The function ρ is called the detection probability *of the tester T.*

Indeed, we require that $\rho(\epsilon) > 0$ for every $\epsilon > 0$, whereas extending Item 2 to $f \in \Pi$ (while avoiding contradiction with Item 1) mandates extending ρ so that $\rho(0) = 0$. The requirement that ρ is monotone (i.e., monotonically increasing) does not rule out cases where the tight lower-bound is non-monotone (e.g., [14]), because ρ is not required to be tight.

Indeed, using a proximity-oblivious tester T, we can obtain a standard (one-sided error) tester (of error probability at most $1/3$). Specifically, given the proximity parameter ϵ, the standard tester invokes T for $\Theta(1/\rho(\epsilon))$ times, and accepts if and only if all these invocations accept. Two natural questions regarding proximity oblivious testers are:

1. Which properties have proximity oblivious tests (of small query complexity)?
2. How does the detection probability of such tests grow as a function of the distance of the object from the property, and how does this relate to the query complexity of the best (standard) tester for the corresponding property.

Goldreich and Ron [40] provide a mix of positive and negative results regarding the foregoing questions. In particular, they provide a characterizations of the graph properties that have constant-query proximity-oblivious testers in the two main models discussed in this article (i.e., the dense graphs model and the bounded-degree graph model). It follows that constant-query proximity-oblivious testers do not exist for many easily testable properties (e.g., Bipartiteness in the dense graph model). Also, even when proximity-oblivious testers exist, repeating them does not necessarily yield the best standard testers for the corresponding property (e.g., Clique Collection in the dense graph model).

Acknowledgments. We are grateful to Tali Kaufman, Michael Krivelevich, Dana Ron, Asaf Shapira, and Omer Tamuz for useful comments and suggestions regarding this article.

References

1. Alon, N.: Testing subgraphs of large graphs. Random Structures and Algorithms 21, 359–370 (2002)
2. Alon, N.V., Fischer, E., Krivelevich, M.V., Szegedy, M.V.: Efficient Testing of Large Graphs. Combinatorica 20, 451–476 (2000)
3. Alon, N., Fischer, E., Newman, I., Shapira, A.: A Combinatorial Characterization of the Testable Graph Properties: It's All About Regularity. In: 38th STOC, pp. 251–260 (2006)
4. Alon, N.V., Krivelevich, M.V.: Testing k-Colorability. SIAM Journal on Disc. Math. 15(2), 211–227 (2002)
5. Alon, N., Kaufman, T., Krivelevich, M., Ron, D.: Testing triangle freeness in general graphs. In: 17th SODA, pp. 279–288 (2006)
6. Alon, N., Shapira, A.: Testing subgraphs in directed graphs. JCSS 69, 354–482 (2004)
7. Alon, N., Shapira, A.: Every Monotone Graph Property is Testable. In: 37th STOC, pp. 128–137 (2005)
8. Alon, N., Shapira, A.: A Characterization of the (natural) Graph Properties Testable with One-Sided. In: 46th FOCS, pp. 429–438 (2005)
9. Alon, N., Shapira, A.: A Characterization of Easily Testable Induced Subgraphs. Combinatorics Probability and Computing 15, 791–805 (2006)

10. Alon, N., Shapira, A.: A Separation Theorem in Property Testing. Combinatorica 28(3), 261–281 (2008)
11. Alon, N., Spencer, J.H.: The Probabilistic Method. John Wiley & Sons, Inc., Chichester (1992)
12. Arora, S., Karger, D., Karpinski, M.: Polynomial time approximation schemes for dense instances of NP-hard problems. JCSS 58(1), 193–210 (1999)
13. Batu, T., Fortnow, L., Rubinfeld, R., Smith, W.D., White, P.: Testing that Distributions are Close. In: 41st FOCS, pp. 259–269 (2000)
14. Bellare, M., Coppersmith, D., Håstad, J., Kiwi, M., Sudan, M.: Linearity testing in characteristic two. In: The 36th FOCS, pp. 432–441 (1995)
15. Bellare, M., Goldreich, O., Sudan, M.: Free Bits, PCPs and Non-approximability – Towards Tight Results. SIAM Journal on Computing 27(3), 804–915 (1998)
16. Bender, M.V., Ron, D.V.: Testing acyclicity of directed graphs in sublinear time. In: Random Structures and Algorithms, pp. 184–205 (2002)
17. Ben-Eliezer, I., Kaufman, T., Krivelevich, M., Ron, D.: Comparing the strength of query types in property testing: the case of testing k-colorability. In: 19th SODA (2008)
18. Benjamini, I.V., Schramm, O., Shapira, A.: Every Minor-Closed Property of Sparse Graphs is Testable. In: 40th STOC, pp. 393–402 (2008)
19. Blum, M., Luby, M., Rubinfeld, R.: Self-Testing/Correcting with Applications to Numerical Problems. JCSS 47(3), 549–595 (1993)
20. Bogdanov, A., Obata, K., Trevisan, L.: A lower bound for testing 3-colorability in bounded-degree graphs. In: 43rd FOCS, pp. 93–102 (2002)
21. Bogdanov, A., Trevisan, L.: Lower Bounds for Testing Bipartiteness in Dense Graphs. In: IEEE Conference on Computational Complexity, pp. 75–81 (2004)
22. Canetti, R., Even, G., Goldreich, O.: Lower Bounds for Sampling Algorithms for Estimating the Average. IPL 53, 17–25 (1995)
23. Chazelle, B., Rubinfeld, R., Trevisan, L.: Approximating the minimum spanning tree weight in sublinear time. In: 19th ICALP, pp. 190–200 (2001)
24. de la Vega, W.F.: MAX-CUT has a randomized approximation scheme in dense graphs. Random Structures and Algorithms 8(3), 187–198 (1996)
25. Even, S.: Graph Algorithms. Computer Science Press (1979)
26. Even, S., Selman, A.L., Yacobi, Y.: The Complexity of Promise Problems with Applications to Public-Key Cryptography. Inform. and Control 61, 159–173 (1984)
27. Fischer, E., Matsliah, A.: Testing graph isomorphism. In: 17th SODA, pp. 299–308 (2006)
28. Fischer, E., Matsliah, A., Shapira, A.: Approximate hypergraph partitioning and applications. In: 48th FOCS, pp. 579–589 (2007)
29. Fischer, E., Newman, I.: Testing versus estimation of graph properties. In: 37th STOC, pp. 138–146 (2005)
30. Goldreich, O.: On promise problems: A survey. In: Goldreich, O., Rosenberg, A.L., Selman, A.L. (eds.) Theoretical Computer Science. LNCS, vol. 3895, pp. 254–290. Springer, Heidelberg (2006)
31. Goldreich, O.: Computational Complexity: A Conceptual Perspective. Cambridge University Press, Cambridge (2008)
32. Goldreich, O. (ed.) Property Testing. LNCS, vol. 6390. Springer, Heidelberg (2010)
33. Goldreich, O., Goldwasser, S., Ron, D.: Property testing and its connection to learning and approximation. Journal of the ACM, 653–750 (July 1998); Extended abstract in 37th FOCS (1996)

34. Goldreich, O., Krivelevich, M., Newman, I., Rozenberg, E.: Hierarchy Theorems for Property Testing. In: ECCC, TR08-097 (2008); Extended abstract in the proceedings of RANDOM 2009
35. Goldreich, O., Ron, D.: Property testing in bounded degree graphs. Algorithmica, 302–343 (2002)
36. Goldreich, O., Ron, D.: A sublinear bipartite tester for bounded degree graphs. Combinatorica 19(3), 335–373 (1999)
37. Goldreich, O., Ron, D.: On Testing Expansion in Bounded-Degree Graphs. In: ECCC, TR00-020 (March 2000)
38. Goldreich, O., Ron, D.: Approximating Average Parameters of Graphs. Random Structures and Algorithms 32(3), 473–493 (2008)
39. Goldreich, O., Ron, D.: Algorithmic Aspects of Property Testing in the Dense Graphs Model. In: ECCC TR08-039 (2008)
40. Goldreich, O., Ron, D.: On Proximity Oblivious Testing. In: ECCC, TR08-041 (2008); Also in the proceedings of the 41st STOC (2009)
41. Goldreich, O., Trevisan, L.: Three theorems regarding testing graph properties. Random Structures and Algorithms 23(1), 23–57 (2003)
42. Gonen, M., Ron, D.: On the Benefits of Adaptivity in Property Testing of Dense Graphs. In: Charikar, M., Jansen, K., Reingold, O., Rolim, J.D.P. (eds.) RANDOM 2007 and APPROX 2007. LNCS, vol. 4627, pp. 525–539. Springer, Heidelberg (2007)
43. Håstad, J.: Clique is hard to approximate within $n^{1-\epsilon}$. Acta Mathematica 182, 105–142 (1999); (Preliminary Version in 28th STOC, 1996 and 37th FOCS (1996)
44. Hassidim, A., Kelner, J., Nguyen, H., Onak, K.: Local Graph Partitions for Approximation and Testing. In: 50th FOCS, pp. 22–31 (2009)
45. Hochbaum, D.: Approximation Algorithms for NP-Hard Problems. PWS (1996)
46. Kale, S., Seshadhri, C.: Testing expansion in bounded degree graphs. In: 35th ICALP, pp. 527–538 (2008); Preliminary version appeared as TR07-076, ECCC (2007)
47. Kaufman, T., Krivelevich, M., Ron, D.: Tight Bounds for Testing Bipartiteness in General Graphs. SIAM Journal on Computing 33(6), 1441–1483 (2004)
48. Lovász, L., Young, N.: Lecture notes on evasiveness of graph properties. Technical Report TR–317–91, Princeton University, Computer Science Department (1991)
49. Marko, S., Ron, D.: Distance approximation in bounded-degree and general sparse graphs. Transactions on Algorithms 5(2) Article number 22 (2009)
50. Mihail, M.: Conductance and convergence of Markov chains– A combinatorial treatment of expanders. In: 30th FOCS, pp. 526–531 (1989)
51. A. Nachmias and A. Shapira. Testing the expansion of a graph. TR07-118. In: ECCC (2007)
52. Orenstein, Y.: Testing properties of directed graphs. Master's thesis, School of Electrical Engineering (2010)
53. Parnas, M., Ron, D.: Testing the diameter of graphs. Random Structures and Algorithms 20(2), 165–183 (2002)
54. Parnas, M., Ron, D., Rubinfeld, R.: Tolerant Property Testing and Distance Approximation. Journal of Computer and System Sciences 72(6), 1012–1042 (2006)
55. Raskhodnikova, S., Smith, A.: A note on adaptivity in testing properties of bounded-degree graphs. In: ECCC TR06-089 (2006)

56. Ron, D.: Algorithmic and Analysis Techniques in Property Testing. Foundations and Trends in TCS 5(2), 73–205 (2010)
57. Rubinfeld, R., Sudan, M.: Robust characterization of polynomials with applications to program testing. SIAM Journal on Computing 25(2), 252–271 (1996)
58. Szemerédi, E.: Regular partitions of graphs. In: Proceedings, Colloque Inter. CNRS, pp. 399–401 (1978)

Appendix: In Passing – Three Unrelated Observations

The following three observations occurred to us in the process of writing this article.

A.1   Testing Degree Regularity in the Dense Graph Model

We improve the $\widetilde{O}(\epsilon^{-3})$ query upper bound of [33, Prop. 10.2.1.3] to an optimal quadratic bound.

Proposition A.1 *In the dense graph model, degree regularity can be tested in $O(\epsilon^{-2})$ non-adaptive queries.*

Proof: We start by reviewing the $\widetilde{O}(\epsilon^{-3})$-query tester presented in the proof of [33, Prop. 10.2.1.3]. This tester selects $O(1/\epsilon)$ random vertices, and estimates the degree of each of them up to $\pm \epsilon N/100$ using a sample of $s = \widetilde{O}(1/\epsilon^2)$ random vertices (and making the corresponding s queries). This tester accepts if and only if all these estimates are at most $\epsilon N/20$ apart. The analysis is based on the observation that if the tester accepts with high probability, then all but $\epsilon' N$ vertices have degree that is within $\pm \epsilon' N$ units of some value, where $\epsilon' = \epsilon/13$. By omitting and adding at most $\epsilon' N^2$ vertices (i.e., from/to the exceptional vertices), we reach a situation in which all vertices have degrees that at most $D \stackrel{\text{def}}{=} 4\epsilon' N$ units apart. At this point, we are done by applying a theorem of Noga Alon (cf. [33, Apdx. D]) that asserts that such a graph is $((3D/N) + o(1))$-close to being regular.

We improve the foregoing upper bound as follows. For a sufficiently large constant c, let $\ell \stackrel{\text{def}}{=} \log_2(c/\epsilon)$, and consider an algorithm that, for every $i \in [\ell]$, proceeds as follows:

1. The algorithm selects uniformly $c \cdot 2^i$ vertices, and estimates the degree of each of these vertices up to $\pm 2^{4i/5}\epsilon \cdot N/c$ units by using a sample of $s_i \stackrel{\text{def}}{=} c^3 \cdot 2^{-3i/2}\epsilon^{-2}$ random vertices.

   Note that with probability at least

   $$1 - c \cdot 2^i \cdot \exp(-2s_i \cdot (2^{4i/5}\epsilon/c)^2) = 1 - c \cdot 2^i \cdot \exp(-2c \cdot 2^{i/10})$$
   $$> 1 - 2^{-i-c}$$

   all these estimates are as desired.

2. If two of these estimates are more than $2^{1+(4i/5)}\epsilon \cdot N$ units apart, then the algorithm rejects.

(The algorithm accepts if and only if it does not reject in any of these ℓ iterations.) The query complexity of this algorithm is $\sum_{i \in [\ell]} c2^i \cdot c^3 2^{-3i/2} \epsilon^{-2} = O(\epsilon^{-2})$, and it accepts each regular graph with high probability (i.e., whenever all the foregoing degree estimates are adequate).

On the other hand, if a graph is accepted with high probability, then, for every $i \in [\ell]$, it holds that all but at most a 2^{-i} fraction of the vertices have degree that is within $2^{1+4i/5}\epsilon \cdot N/c$ of the average degree, denoted ρ. For each value of $i \in [\ell]$, let us denote the set of deviating vertices by B_i; that is, each vertex in $[N] \setminus B_i$ has degree $(\rho \pm 2^{1+4i/5}\epsilon/c) \cdot N$. Thus (dealing separately with each $B_i \setminus B_{i+1}$ as well as with B_ℓ and $[N] \setminus B_1$), we may omit at most $40\epsilon N^2/c$ edges from the graph, and obtain a graph in which every vertex has degree at most $(\rho + 2\epsilon/c)N$. Next, by adding at most $42\epsilon N^2/c$ edges to the graph, we can obtain a graph in which every vertex has degree at least $(\rho - 2\epsilon/c)N$, and if we add these edges uniformly (among the vertices) then each vertex in the resulting graph has degree $(\rho \pm 44\epsilon/c)N$. At this point we can apply the result of aforementioned result of Noga Alon, and be done. ∎

A.2   Non-adaptive Testers in the Bounded-Degree Graph Model

Recall that, for any function q, if a property can be tested in $o(\sqrt{N} \cdot q(\epsilon))$ non-adaptive queries in the bounded-degree graph model, then it depends only on the vertex degree distribution [55]. In contrast, we show that triangle-freeness can be tested by $O(\sqrt{N/\epsilon})$ non-adaptive queries (in the same model).

The tester selects at random $O(\sqrt{N/\epsilon})$ vertices, queries for the neighbors of each of them, and accepts if and only if the subgraph discovered contains no triangles. Note that if the input graph is ϵ-far from triangle-freeness, then it contains $\Omega(\epsilon N)$ triangles, whereas a random sample of $O(\sqrt{N/\epsilon})$ vertices is likely to hit two vertices of such a triangle.

The argument can be extended to testing H-freeness,[21] for any fixed H, with $O((N/\epsilon)^{1-\frac{1}{\beta(H)}})$ non-adaptive queries, where $\beta(H)$ denotes the minimum vertex cover of H. In this case, if the input graph is ϵ-far from being H-free, then a sample of $O((N/\epsilon)^{1-\frac{1}{\beta(H)}})$ random vertices is likely to hit all vertices in a vertex cover of one of the copies of H. A more general statement, with weaker quantitative bounds, follows.

Proposition A.2 *Let Π be a graph property having a q-query proximity-oblivious tester of detection probability ρ, in the bounded-degree model. Then, in this model, Π can be tested by $O(N^{\frac{q-1}{q}}/\rho(\epsilon))$ non-adaptive queries.*

Actually, Proposition A.2 holds also when q is an upper bound on the number of different vertices that appear in the queries of the proximity-oblivious tester.

---

[21] Here, we refer to subgraph freeness.

Proof: The main observation is that a sample of $O(N^{1-(1/q)})$ vertices (along with the neighbor queries that correspond to each vertex) is likely to allow for the emulation of a random execution of the proximity-oblivious tester (POT). Specifically, given a q-query POT, we consider the following non-adaptive POT:

1. Select a random sample of $O(N^{1-(1/q)})$ vertices, denoted S, and query the neighborhood of each vertex in S. For every $(v,i) \in S \times [d]$, denote the oracle answer by $\Gamma_i(v)$.
   These are all the queries made by the new POT, and the following steps only involve computations (and no actual queries).
2. Select and fix random coins for T, deriving a residual deterministic oracle machine T'.
3. Let $S = \{s_1, ..., s_{|S|}\}$, and $\overline{S} \stackrel{\text{def}}{=} \{(s_{(i-1)q+1}, ..., s_{iq}) : i \in [|S|/2q]\}$; that is, $\overline{S}$ consists of q-sequences of elements in S such that no element appears twice. For every $(v_1, ..., v_q) \in \overline{S}$, try to emulate an execution of T using the information obtained in Step 1. For $j = 1, ..., q$, proceed as follows, where initially the permutation $\pi : [N] \to [N]$ is totally undetermined.
   (a) Obtain the jth query of T', denoted (u_j, i_j).
       If π is undetermined on u_j, then determine $\pi(u_j) = v_j$.
       If π is determined on u_j and $\pi(u_j) \notin S$, then this emulation is terminated. Thus, the algorithm proceeds to Step 3b only if $\pi(u_j) \in S$, whereas in this case the value of $\Gamma_{i_j}(\pi(u_j))$ is known.
   (b) Let $a_j = \Gamma_{i_j}(\pi(u_j))$, and suppose that $a_j \in [N]$ (as otherwise we provide a_j as the oracle answer to T', and proceed to the next iteration).[22] If π^{-1} is undetermined on a_j, then select at random a vertex u such that π is undetermined on u, and determine $\pi(u) = a_j$. Provide u as the oracle answer to T', and proceed to the next iteration.
       Note that it is quite likely that $a_j \notin S$, and in this case if T' subsequently issues a query of the form $(u, \cdot)$ then the emulation will be terminated (in the corresponding execution of Step 3a).
   If the current emulation is successfully completed, then halt and output the corresponding verdict of T'. Otherwise, proceed to the next $(v_1, ..., v_q) \in \overline{S}$, while resetting π to be totally undetermined.
4. If no emulation is successfully completed, then halt and output the verdict 1 (i.e., accept).

Each execution of Step 3b may yield a value $a_j \notin S$, with probability at least $1 - (|S|/N)$. However, with probability at least $|S|/2N$, it holds that $a_j \in S$. Thus, for each $(v_1, ..., v_q) \in \overline{S}$, we complete an emulation of T' (in Step 3) with probability at least $(|S|/2N)^{q-1} \gg 1/|\overline{S}|$. Furthermore, such an emulation correspond to the execution of T' on a random isomorphic copy of the input graph.

To see that, with high probability, at least one of the $|\overline{S}|$ emulations is completed, we consider all $|\overline{S}|$ emulations simultaneously. Let $u_1^{(i)}, ..., u_q^{(i)}$ denote the

---

[22] Recall that in this case a_j is a fixed indication that the relevant vertex has less than i_j neighbors.

sequence of vertices that occur in the ith emulation, and let $\pi^{(i)}$ denote the corresponding permutation. We partition the $|S|/2$ samples that do not appear in $\overline{S}$ into q equal sets, denoted $S_1, ..., S_q$, and terminate the ith emulation in iteration $j < q$ if $a_j^{(i)} \notin S_j$. (Indeed, this only makes early termination more likely; cf. Step 3b.) Still, on can show by induction on j, that with high probability the number of emulations that are not terminated by iteration j exceeds $|\overline{S}| \cdot (|S|/4qN)^j$. Furthermore, the queries issued in the $j+1^{\text{st}}$ iteration are mostly different, because they are determined based on different sequences in $\overline{S}$. Using $|\overline{S}| \cdot (|S|/4qN)^{q-1} > 1$, we conclude that, with high probability, there exists an emulation that does not terminate before the last iteration.

It follows that the foregoing non-adaptive POT has detection probability at least $\rho/2$. Applying this POT for $O(1/\rho(\epsilon))$ times, we obtain a non-adaptive tester of query complexity $O(N^{1-(1/q)}/\rho(\epsilon))$. ∎

Conclusion. Recall that all subgraph-freeness properties do have a proximity-oblivious testers of constant-query complexity in the bounded-degree graph model. Our conclusion is that non-adaptive testers are not totally useless in that model.

A.3   Testing Strong Connectivity with Forward Queries only

We show that, for any constant $\epsilon > 0$, strong connectivity in bounded-degree digraphs can be tested by using $N^{1-\Omega(1)}$ forward queries (and no backward queries). Needless to say, the same holds for using only backward queries, and in both cases the tester has two-sided error (which is unavoidable).[23]

Proposition A.3 *In the directed bounded-degree model where only forward queries are allowed, strong connectivity can be tested in query complexity $\exp(1/\epsilon)$ $\cdot N^{1-\frac{1}{t}}$, where $t = \lceil 4/\epsilon d \rceil \cdot d < d + (1/\epsilon)$ and d is the in-degree and out-degree bound.*

Proof Sketch: Our starting point is the observation that if a graph is ϵ-far from being strongly connected, then it contains at least $\epsilon dN/4$ source and sink components each containing at most $\lceil 4/\epsilon d \rceil$ vertices (cf. [16, Cor. 9]).[24] The easy case is when the graph contains at least $\epsilon dN/8$ small sink component, since these are easy to detect by forward queries. The problematic case is the one in which the graph contains $\epsilon dN/8$ source components, and we start by considering the simple case in which each of these source components consists of a single vertex.

In the latter case we can estimate the number of vertices having in-degree zero, by estimating the number of vertices having in-degree d, $d-1$, all through

---

[23] The distributions used in [16, Sec. 5.2] can be used to prove an $\Omega(N)$ query bound for one-sided error. The point is that we can find no direct evidence to the fact that a vertex has in-degree zero.

[24] Throughout this proof, the word component means a strongly connected component, and source (resp., sink) components are components that have no in-coming (resp., out-going) edges.

1. To estimate the number of vertices having in-degree $i > 1$, we estimate the number of i-way collisions at the head of randomly selected[25] directed edges, and use the information we already gathered regarding in-degree j for every $j > i$. The number of vertices having in-degree 1 is estimated by estimating the collisions between a uniformly selected vertex and the vertex at the head of a uniformly selected random edge. Note that, for every $i \geq 2$, the number of i-way collisions can be estimated by a sample of size $O(N^{1-\frac{1}{i}})$.

In the foregoing, we have relied on the fact that a vertex has zero in-degree if and only if it is a source vertex, and on the hypothesis that many source vertices exist. But, in general, we only know that there are many small source components. So the intuitive idea is to "contract" all small components, and consider in-coming edges at the component level. One small difficulty is that we cannot determine the components of the input graph, and so the following modification is used.

For every vertex v, we let C_v denote the set of vertices u such that v and u reside on a directed cycle of size at most $s \overset{\text{def}}{=} \lceil 4/\epsilon d \rceil$. We say that v is good if for every $u \in C_v$ it holds that $C_u = C_v$. Note that, given a vertex v, we can determine C_v as well as whether v is good by using d^s queries. Also note that every vertex that resides in a small source component is good. We now emulate the foregoing procedure on the directed graph in which for every good v the set C_v is contracted to a new vertex, and note that a vertex has in-degree zero in the resulting graph if and only if it represents a small source of G. Noting that the maximum degree in this graph is $s \cdot s$, the claim follows.   ∎

Conclusion. Our lesson is that some non-trivial testing can be carried out also in the model that allows forward queries only.

---

[25] We may select a random directed edge by selecting a vertex uniformly, and selecting each of its out-going edges with probability $1/d$.

Randomness and Computation

Oded Goldreich

Abstract. The interplay of randomness and computation is at the heart of modern Cryptography and plays a fundamental role in the design of algorithms and in the study of computation at large. Specifically, this interplay is pivotal to several intriguing notions of probabilistic proof systems (e.g., interactive proofs, zero-knowledge proofs, and probabilistically checkable proofs), is the focal of the computational approach to randomness, and is essential for some types of sub-linear time algorithms (e.g., property testers). This essay provides a brief outline of these connections.

Keywords: Pseudorandomness, Interactive Proofs, Zero-Knowledge Proofs, Probabilistically Checkable Proofs, Encryption Schemes, Message Authentication Schemes, Property Testing.

This is a revised version of an essay that has appeared in the *Handbook of Probability Theory with Applications*, Sage Publishers, 2008.

Preface. This essay was originally intended for a wide audience of scholars, who may not have any background in computer science. While theoretical computer scientists may find much of the introduction (esp., Sections 1.2 and 1.3) unnecessary, we avoided the temptation to revise and/or omit this part. Our hope is that this part of the text may demonstrate to theoretical computer scientists how one can go about in exposing the field to outsiders. We believe that the rest of this essay may be of more direct interest to many theoretical computer scientists: It contains brief overviews of the theory of pseudorandomness (Section 2), three types of probabilistic proof systems (Section 3), the theoretical foundations of Cryptography (Section 4), and property testing (Section 5). These overviews focus on the clarification of the main issues, while trying to avoid any technical details. Here too, we retained the original style, which attempts to accommodate outsiders, in order to demonstrate to experts the feasibility of communicating the contents of these areas to outsiders.

1 Introduction

While it is safe to assume that any living adult is aware of the revolutionary impact of the computing technology on our society, we fear that few readers have a sense of the theory of computation. This contrast is not so surprising, because people seem so overwhelmed by the wonders of this technology that they do not get to wonder about the theory underlying it. Furthermore, people tend to

O. Goldreich et al.: Studies in Complexity and Cryptography, LNCS 6650, pp. 507–539, 2011.

think of computing in the concrete terms in which they have lastly encountered it rather than in general terms. Consequently, the fascinating intellectual contents of the theory of computation is rarely understood by non-specialists.

One goal of this essay is making a tiny contribution towards a possible change in this sour state of affairs, by discussing one aspect of the theory of computation: Its connection to randomness.

1.1   On the Relation between Computation and Randomness

Our guess is that the suggestion that there is a connection between computation and randomness may meet the skepticism of some readers, because computation seems the ultimate manifestation of determinism.

To address this skepticism, we suggest considering what happens when a deterministic machine (or any deterministic process) is fed with a random input or just with an input that looks random. Indeed, one contribution of the theory of computation (further discussed in Section 2) is a definition of "objects that look random" (a notion which makes sense even if the real world is actually deterministic).

Still one may wonder whether we can obtain or generate objects that look random. For example, can we toss a coin (in the sense that one cannot feasibly predict the answer before seeing it)? Assuming a positive answer, we may also assume that unpredictable values can be obtained by other mechanical and/or electrical processes, which suggest that computers can also obtain such values. The question then is what benefit can be achieved by using such random (or unpredictable) values.

A major application of random (or unpredictable) values is to the area of Cryptography (see Section 4). In fact, the very notion of a *secret* refers to such a random (or unpredictable) value. Furthermore, various natural security concerns (e.g., private communication) can be met by employing procedures that make essential use of such secrets and/or random values.

Another major application of random (or unpredictable) values is to various sampling procedures. In Section 5, we consider a wider perspective on such procedures, viewing them as a special type of super fast procedures called *property testers*. Such a procedure cannot afford to scan the entire input, but rather probes few (randomly) selected locations in it and, based on these few values, attempts to make a meaningful assertion regarding the entire input. Indeed, we assume that the reader is aware of the fact that random sampling allows to approximate the fraction of the population that votes for a particular candidate. Our point is that other global properties of the input, which are not merely averages of various types, can also be approximated by sampling.

Lastly, we mention that randomized verification procedures yield fascinating types of *probabilistic proof systems*, which are discussed in Section 3. In particular, such proof systems demonstrate the advantage of interaction (over one-directional communication) and the possibility of decoupling proving from learning (i.e., the possibility of proving an assertion without yielding anything beyond its validity). Other forms of probabilistic proof systems allow for super fast verification (based on probing few locations in a redundant proof, indeed as in the aforementioned sublinear-time algorithms).

Before discussing the foregoing applications of randomness in greater length, we provide a somewhat wider perspective on the theory of computation as well as present some of its central conventions. We will also clarify what randomness means in that theory (and in this article).

1.2   A Wider Perspective on the Theory of Computation

The theory of computation aims at understanding general properties of computation be it natural, man-made, or imaginary. Most importantly, it aims to understand the nature of efficient computation. We demonstrate these issues by briefly considering a few typical questions.

A key question is *which functions can be efficiently computed?* For example, it is (relatively) easy to multiply integers, but it seems hard to take the product and factor it into its prime components. In general, it seems that there are one-way computations, or put differently *one-way functions*: Such functions are easy to evaluate but hard to invert (even in an average-case sense). But do one-way functions really exist? It is widely believed that the answer is positive, and this question is related to other fundamental questions.

A related question is that of the comparable difficulty of *solving problems versus verifying the correctness of solutions.* Indeed our daily experience is that it is harder to solve a problem than it is to check the correctness of a solution (e.g., think of either a puzzle or a research problem). Is this experience merely a coincidence or does it represent a fundamental fact of life (or a property of the world)? Could you imagine a world in which solving any problem is not significantly harder than checking a solution to it? Would the term "solving a problem" not lose its meaning in such a hypothetical (and impossible in our opinion) world? The denial of the plausibility of such a hypothetical world (in which "solving" is not harder than "checking") is what the celebrated "P different from NP" conjecture means, where P represents tasks that are efficiently solvable and NP represents tasks for which solutions can be efficiently checked for correctness.

The theory of computation is also concerned with finding the most efficient methods for solving specific problems. To demonstrate this line of research we mention that the simple (and standard) method for multiplying numbers that is taught in elementary school is not the most efficient one possible. Multiplying two n-digit long numbers by this method requires n^2 single-digit multiplications (and a similar number of single-digit additions). In contrast, consider writing these numbers as $10^{n/2} \cdot a' + a''$ and $10^{n/2} \cdot b' + b''$, where a', a'', b', b'' are $n/2$-digit long numbers, and note that

$$(10^{n/2} \cdot a' + a'') \times (10^{n/2} \cdot b' + b'') = 10^n \cdot P_1 + 10^{n/2} \cdot (P_2 - P_1 - P_3) + P_3$$
$$\text{where } P_1 = a' \times b', P_2 = (a' + a'') \times (b' + b''), \text{ and } P_3 = a'' \times b''.$$

Thus, multiplying two n-digit long numbers requires only three (rather than four) multiplications of $n/2$-digit long numbers (and a constant number of additions of $n/2$-digit long numbers and "shifts" of n-digit long numbers (indicated by $\cdot$)). Letting $M(n)$ denote the complexity of multiplying two n-digit long numbers, we obtain $M(n) < 3 \cdot M(n/2) + c \cdot n$, where c is some constant (independent of n), which solves to $M(n) < c' \cdot 3^{\log_2 n} = c' \cdot n^{\log_2 3} < n^{1.6}$ (for some constant c'). We mention that this is not the best known algorithm; the latter runs in time $\mathrm{poly}(\log n) \cdot n$.

The theory of computation provides a new viewpoint on old phenomena. We have already mentioned the computational approaches to randomness (see Section 2) and to proofs, interaction, knowledge, and learning (see Section 3). Additional natural concepts given an appealing computational interpretations include the *importance of representation*, the notion of *explicitness*, and the possibility that approximation is easier than optimization (see Section 5). Let us say a few words about representation and explicitness.

The foregoing examples hint to *the importance of representation*, because in all these computational problems the solution is implicit in the problem's statement. That is, the problem contains all necessary information, and one merely needs to process this information in order to supply the answer.[1] Thus, the theory of computation is concerned with the manipulation of information, and its transformation from one representation (in which the information is given) to another representation (which is the one desired). Indeed, a solution to a computational problem is merely a different representation of the information given; that is, a representation in which the answer is explicit rather than implicit. For example, the answer to the question of whether or not a given system of quadratic equations has an integer solution is implicit in the system itself (but the task is to make the answer explicit). Thus, the theory of computation clarifies a central issue regarding representation; that is, the distinction between what is explicit and what is implicit in a representation. Furthermore, it also suggests a quantification of the level of non-explicitness.

1.3   Important Conventions for the Theory of Computation

In light of the foregoing discussion it is important to specify the representation used in computational problems. Actually, a computational problem refer to an infinite set of *finite objects*, called the problem's instances, and specifies the desired solution for each instance. For example, the instances of the multiplication problem are pairs of natural numbers, and the desired solution is the corresponding product. Objects are represented by finite binary sequences, called strings.[2]

---

[1] In contrast, in other disciplines, solving a problem may also require gathering information that is not available in the problem's statement. This information may either be available from auxiliary (past) records or be obtained by conducting new experiments.

[2] Indeed, in the foregoing example, we used the daily representation of numbers as sequences of decimal digits, but in the theory of computation natural numbers are typically represented by their binary expansion.

For a natural number n, we denote by $\{0,1\}^n$ the set of all strings of length n, hereafter referred to as n-bit strings. The set of all strings is denoted $\{0,1\}^*$; that is, $\{0,1\}^* = \cup_{n\in\mathbb{N}}\{0,1\}^n$.

We have already mentioned the notion of an algorithm, which is central to the theory of computation and means an automated procedure designed to solve some computational task. A rigorous definition requires specifying a reasonable model of computation, but the specifics of this model are not important for the current essay. We focus on efficient algorithms, which are commonly defined as making a number of steps that is polynomial in the length of their input.[3] Indeed, asymptotic analysis (or rather a functional treatment of the running time of algorithms in terms of the length of their input) is a central convention in the theory of computation.[4]

Typically, our notion of efficient algorithms will include also *probabilistic* (polynomial-time) algorithms; that is, algorithms that can "toss coins" (i.e., make random choices). For each reasonable model of computation, probabilistic (or randomized) algorithms are defined as standard algorithm augmented with the ability to choose uniformly among a finite number (say two) of predetermined possibilities. That is, at each computation step, such an algorithm makes a move that is chosen uniformly among two predetermined possibilities.

1.4   Randomness in the Context of Computation

Throughout the entire essay we will refer only to *discrete* probability distributions. The support of such distributions will be associated with a set of strings, typically of the same length.

For the purpose of asymptotic analysis, we will often consider probability ensembles, which are sequences of distributions that are indexed either by integers or by strings. For example, throughout the essay, we let $\{U_n\}_{n\in\mathbb{N}}$ denote the uniform ensemble, where U_n is uniform over the set of strings of length n; that is, $\mathrm{Pr}_{z\sim U_n}[z=\alpha]$ equals 2^{-n} if $\alpha \in \{0,1\}^n$ and equals 0 otherwise. More generally, we will typically consider probability ensembles, denoted $\{D_n\}_{n\in\mathbb{N}}$ (or $\{D_s\}_{s\in S}$, where $S \subseteq \{0,1\}^*$), where there exists some function $\ell : \mathbb{N} \to \mathbb{N}$ such that $\mathrm{Pr}_{z\sim D_n}[z\in\{0,1\}^{\ell(n)}] = 1$ (resp., $\mathrm{Pr}_{z\sim D_s}[z\in\{0,1\}^{\ell(n)}] = 1$, where n denotes the length of s). Furthermore, typically, ℓ will be a polynomial.

One important case of probability ensembles is that of ensembles that represent the output of randomized processes (e.g., randomized algorithms). Letting

---

[3] In Section 5 we consider even faster algorithms, which make (significantly) less steps than the length of their input, but such algorithms can only provide approximate solutions.

[4] We stress, however, that asymptotic (or functional) treatment is not essential to this theory, but rather provides a convenient framework. One may develop the entire theory in terms of inputs of fixed lengths and concrete bounds on the number of steps taken by corresponding algorithms. However, such an alternative treatment is more cumbersome.

$A(x)$ denote the output of the probabilistic (or randomized) algorithm A on input x, we may consider the probability ensemble $\{A(x)\}_{x \in \{0,1\}^*}$. Indeed, if A is a probabilistic polynomial-time algorithm then $A(x)$ is distributed over strings of length that is bounded by a polynomial in the length of x.

On the other hand, we say that a probability ensemble $\{D_n\}_{n \in \mathbb{N}}$ (resp., $\{D_s\}_{s \in S}$) is efficiently sampleable if there exists a probabilistic polynomial-time algorithm A such that for every $n \in \mathbb{N}$ it holds that $A(1^n) \equiv D_n$ (resp., for every $s \in S$ it holds that $A(s) \equiv D_s$). That is, algorithm A makes a number of steps that is polynomial in n, and produces a sample distributed according to D_n (resp., D_s, where n denotes the length of s).

We will often talk of "random bits" and mean values selected uniformly and independently in $\{0, 1\}$. In particular, randomized algorithms may be viewed as deterministic algorithms that are given an adequate number of random bits as an auxiliary input. This means that rather than viewing these algorithms as making random choices, we view them as determining these choices according to a sequence of random bits that is generated by some outside process.

1.5   The Rest of This Essay

In the rest of this essay we briefly review the theory of pseudorandomness (Section 2), three types of probabilistic proof systems (Section 3), the theoretical foundations of Cryptography (Section 4), and property testing (Section 5). Needless to say, these overviews are the tip of an iceberg, and the interested reader will be referred to related texts for further information. In general, the most relevant text is [6] (see also [9]), which provides more extensive overviews of the first three areas.

In addition, we recommend textbooks such as [10,23,27] for background on the aspects of the theory of computation that are most relevant for the current essay. We note that randomized algorithms and procedures are valuable also in settings not discussed in the current essay (e.g., for polynomial-time computations as well as in the context of distributed and parallel computation). The interested reader is referred to [22].

An Apology. Our feeling is that in an essay written for a general readership it makes no sense to provide the standard scholarly citations. The most valuable references for such readers are relevant textbooks and expository articles, written with the intension of communicating to non-experts. On the other hand, the general reader may be interested in having some sense of the history of the field, and thus references to few pioneering works seem adequate. We are aware that in trying to accommodate the non-experts, we may annoy the experts, and hence the current apology to all experts who made an indispensable contribution to the development of these areas and who's work was victim to our referencing policy.

2 Pseudorandomness

Indistinguishable things are identical.[5]

G.W. Leibniz (1646–1714)

A fresh view at the *question of randomness* has been taken in the theory of computation: It has been postulated that a distribution is pseudorandom if it cannot be told apart from the uniform distribution by any efficient procedure. The paradigm, originally associating efficient procedures with polynomial-time algorithms, has been applied also with respect to a variety of limited classes of such distinguishing procedures.

At the extreme, this approach says that the question of whether the world is deterministic or allows for some free choice (which may be viewed as sources of randomness) is irrelevant. *What matters is how the world looks to us and to various computationally bounded devices.* That is, if some phenomenon looks random then we may just treat it as if it were random. Likewise, if we can generate sequences that cannot be told apart from the uniform distribution by any efficient procedure, then we can use these sequences in any efficient randomized application instead of the ideal random bits that are postulated in the design of this application.

2.1 A Wider Context and an Illustration

The second half of this century has witnessed the development of three theories of randomness, a notion which has been puzzling thinkers for ages. The first theory (cf., [4]), initiated by Shannon, is rooted in probability theory and is focused at distributions that are not perfectly random (i.e., are not uniform over a set of strings of adequate length). Shannon's Information Theory characterizes perfect randomness as the extreme case in which the *information contents* is maximized (i.e., the strings contain no redundancy at all). Thus, perfect randomness is associated with a unique distribution: the uniform one. In particular, by definition, one cannot (deterministically) generate such perfect random strings from shorter random seeds.

The second theory (cf., [20]), initiated by Solomonov, Kolmogorov, and Chaitin, is rooted in computability theory and specifically in the notion of a universal language (equiv., universal machine or computing device). It measures the complexity of objects in terms of the shortest program (for a fixed universal machine) that generates the object. Like Shannon's theory, Kolmogorov Complexity is quantitative and perfect random objects appear as an extreme

---

[5] This is the *Principle of Identity of Indiscernibles*. Leibniz admits that counterexamples to this principle are conceivable but will not occur in real life because God is much too benevolent. We thus believe that he would have agreed to the theme of this section, which asserts that *indistinguishable things should be considered as identical.*

case. However, in this approach one may say that a single object, rather than a distribution over objects, is perfectly random. Still, Kolmogorov's approach is inherently intractable (i.e., Kolmogorov Complexity is uncomputable), and – by definition – one cannot (deterministically) generate strings of high Kolmogorov Complexity from short random seeds.

The third theory, initiated by Blum, Goldwasser, Micali and Yao [16,3,28], is rooted in the notion of *efficient computations* and is the focus of this section. This approach is explicitly aimed at providing a notion of randomness that nevertheless allows for an efficient generation of random strings from shorter random seeds. The heart of this approach is the suggestion to view objects as equal if they cannot be told apart by any efficient procedure. Consequently, a distribution that cannot be efficiently distinguished from the uniform distribution will be considered as being random (or rather called pseudorandom). Thus, randomness is not an "inherent" property of objects (or distributions) but is rather relative to an observer (and its computational abilities). To demonstrate this approach, let us consider the following mental experiment.

Alice and Bob play "head or tail" in one of the following four ways. In each of them, Alice flips an unbiased coin and Bob is asked to guess its outcome *before* the coin hits the floor. The alternative ways differ by the knowledge Bob has before making his guess.

In the first alternative, Bob has to announce his guess before Alice flips the coin. Clearly, in this case Bob wins with probability 1/2.

In the second alternative, Bob has to announce his guess while the coin is spinning in the air. Although the outcome is *determined in principle* by the motion of the coin, Bob does not have accurate information on the motion and thus we believe that also in this case Bob wins with probability 1/2.

The third alternative is similar to the second, except that Bob has at his disposal sophisticated equipment capable of providing accurate *information* on the coin's motion as well as on the environment effecting the outcome. However, Bob cannot process this information in time to improve his guess.

In the fourth alternative, Bob's recording equipment is directly connected to a *powerful computer* programmed to solve the motion equations and output a prediction. It is conceivable that in such a case Bob can substantially improve his guess of the outcome of the coin.

We conclude that the randomness of an event is relative to the information and computing resources at our disposal. Thus, a natural concept of pseudorandomness arises: a distribution is *pseudorandom* if no efficient procedure can distinguish it from the uniform distribution, where efficient procedures are associated with (probabilistic) polynomial-time algorithms. This notion of pseudorandomness is indeed the most fundamental one, and the current section is focused on it.[6]

The foregoing discussion has focused at one aspect of the pseudorandomness question: the resources or type of the observer (or potential distinguisher). Another important aspect is whether such pseudorandom sequences can be generated from much shorter ones, and at what cost (i.e., at what computational effort). A natural approach is that the generation process has to be at least as efficient as the distinguisher (equiv., that the distinguisher is allowed at least as much resources as the generator). Coupled with the aforementioned strong notion of pseudorandomness, this yields the archetypical notion of pseudorandom generators – these operating in polynomial-time and producing sequences that are indistinguishable from uniform ones by *any* polynomial-time observer. Such (general-purpose) pseudorandom generators enable reducing the randomness complexity of *any efficient application*, and are thus of great relevance to randomized algorithms and Cryptography (see Sections 2.5 and 4). Indeed, these general-purpose pseudorandom generators will be the focus of the current section.[7] Further discussion of the conceptual contents of this approach is provided in Section 2.6.

2.2   The Notion of Pseudorandom Generators

Loosely speaking, a pseudorandom generator is an *efficient* program (or algorithm) that *stretches* short random strings into long *pseudorandom* sequences. We stress that the generator itself is deterministic and that the randomness involved in the generation process is captured by its input. We emphasize three fundamental aspects in the notion of a pseudorandom generator:

1. Efficiency. The generator has to be efficient. Since we associate efficient computations with polynomial-time ones, we postulate that the generator has to be implementable by a deterministic polynomial-time algorithm.

---

[6] We mention that weaker notions of pseudorandomness arise as well; they refer to indistinguishability by weaker procedures such as space-bounded algorithms (see [6, Sec. 3.5] or [9, Sec. 8.4]), constant-depth circuits, etc. Stretching this approach even further one may consider algorithms that are designed on purpose so not to distinguish even weaker forms of "pseudorandom" sequences from random ones (such algorithms arise naturally when trying to convert some natural randomized algorithm into deterministic ones; see [6, Sec. 3.6] or [9, Sec. 8.5]).

[7] We mention that there are important reasons for considering also an alternative that seems less natural; that is, allowing the pseudorandom generator to use more resources (e.g., time or space) than the observer it tries to fool. This alternative is natural in the context of derandomization (i.e., converting randomized algorithms to deterministic ones), where the crucial step is replacing the "random source" of a fixed algorithm by a pseudorandom source, which in turn can be deterministically emulated based on a much shorter random source. For further clarification and demonstration of the usefulness of this approach the interested reader is referred to [6, Sec. 3.4&3.5] (or [9, Chap. 8]).

This algorithm takes as input a string, called its **seed**. The seed captures a bounded amount of randomness used by a device that "generates pseudo-random sequences." The formulation views any such device as consisting of a deterministic procedure applied to a random seed.

2. **Stretching.** The generator is required to stretch its input seed to a longer output sequence. Specifically, it stretches n-bit long seeds into $\ell(n)$-bit long outputs, where $\ell(n) > n$. The function ℓ is called the **stretching measure** (or **stretching function**) of the generator.

3. **Pseudorandomness.** The generator's output has to look random to any efficient observer. That is, any efficient procedure should fail to distinguish the output of a generator (on a random seed) from a truly random bit-sequence of the same length. The formulation of the last sentence refers to a general notion of **computational indistinguishability** that is the heart of the entire approach.

To demonstrate the foregoing, consider the following suggestion for a pseudorandom generator. The seed consists of a pair of 500-bit integers, denoted x and N, and a million-bit long output is obtained by repeatedly squaring the current x modulo N and emitting the least significant bit of each intermediate result (i.e., let $x_i \leftarrow x_{i-1}^2 \bmod N$, for $i = 1, ..., 10^6$, and output $b_1, b_2, ..., b_{10^6}$, where $x_0 \stackrel{\text{def}}{=} x$ and b_i is the least significant bit of x_i). This process may be generalized to seeds of length n (here we used $n = 1000$) and outputs of length $\ell(n)$ (here $\ell(1000) = 10^6$). Such a process certainly satisfies Items (1) and (2) above, whereas the question whether Item (3) holds is debatable (once a rigorous definition is provided). As a special case of Theorem 2.6 (which follows), we mention that, under the assumption that it is difficult to factor large integers, a slight variant of the foregoing process is indeed a pseudorandom generator.

Computational Indistinguishability. Intuitively, two objects are called computationally indistinguishable if no efficient procedure can tell them apart. Here the objects are (fixed) probability distributions (or rather ensembles), and the observer is given a sample drawn from one of the two distributions and is asked to tell from which distribution it was taken (e.g., it is asked to say "1" if the sample is taken from the first distribution). Following the asymptotic framework (see Sections 1.3 and 1.4), the foregoing discussion is formalized as follows.

Definition 2.1 (computational indistinguishability [16,28]). *Two probability ensembles, $\{X_n\}_{n \in \mathbb{N}}$ and $\{Y_n\}_{n \in \mathbb{N}}$, are called* computationally indistinguishable *if for any probabilistic polynomial-time algorithm A, any positive polynomial p, and all sufficiently large n*

$$\left| \Pr_{x \sim X_n}[A(x) = 1] - \Pr_{y \sim Y_n}[A(y) = 1] \right| < \frac{1}{p(n)}. \tag{1}$$

The probability is taken over X_n (resp., Y_n) as well as over the internal coin tosses of algorithm A.

Algorithm A, which is called a potential distinguisher, is given a sample (which is drawn either from X_n or from Y_n) and its output is viewed as an attempt to tell whether this sample was drawn from X_n or from Y_n. Eq. (1) requires that such an attempt is bound to fail; that is, the outcome 1 (possibly representing a verdict that the sample was drawn from X_n) is essentially as likely to occur when the sample is drawn from X_n as when it is drawn from Y_n.

A few comments are in order. Firstly, the distinguisher (i.e., A) is allowed to be probabilistic. This makes the requirement only stronger, and seems essential to the technical development of our approach. Secondly, we view events occurring with probability that is upper bounded by the reciprocal of polynomials as negligible (e.g., $2^{-\sqrt{n}}$ is negligible as a function of n). This is well-coupled with our notion of efficiency (i.e., polynomial-time computations): an event that occurs with negligible probability (as a function of a parameter n), will also occur with negligible probability if the experiment is repeated for poly(n)-many times. Thirdly, for efficiently sampleable ensembles, computational indistinguishability is preserved also when providing the distinguisher with polynomially many samples (of the tested distribution). Lastly we note that computational indistinguishability is a coarsening of statistical indistinguishability; that is, waiving the computational restriction on the distinguisher is equivalent to requiring that the variation distance between X_n and Y_n (i.e., $\sum_z |X_n(z) - Y_n(z)|$) is negligible (in n).

An important case in which computational indistinguishability is strictly more liberal than statistical indistinguishability arises from the notion of a pseudo-random generator.

Definition 2.2 (pseudorandom generators [3,28]). *A deterministic polynomial-time algorithm G is called a* pseudorandom generator *if there exists a* stretching function, *$\ell : \mathsf{N} \rightarrow \mathsf{N}$ (i.e., $\ell(n) > n$), such that the following two probability ensembles, denoted $\{G_n\}_{n \in \mathsf{N}}$ and $\{R_n\}_{n \in \mathsf{N}}$, are computationally indistinguishable.*

1. *Distribution G_n is defined as the output of G on a uniformly selected seed in $\{0,1\}^n$.*
2. *Distribution R_n is defined as the uniform distribution on $\{0,1\}^{\ell(n)}$.*

Note that $G_n \equiv G(U_n)$, whereas $R_n = U_{\ell(n)}$. Requiring that these two ensembles are computationally indistinguishable means that, for any probabilistic polynomial-time algorithm A, the detected (by A) difference between G_n and R_n, denoted

$$d_A(n) \stackrel{\text{def}}{=} \left| \mathsf{Pr}_{s \sim U_n}[A(G(s)) = 1] - \mathsf{Pr}_{r \sim U_{\ell(n)}}[A(r) = 1] \right|$$

is negligible (i.e., $d_A(n)$ vanishes faster than the reciprocal of any polynomial). Thus, pseudorandom generators are efficient (i.e., polynomial-time) deterministic programs that expand short randomly selected seeds into longer pseudorandom bit sequences, where the latter are defined as computationally indistinguishable from truly random bit-sequences. It follows that any efficient randomized algorithm maintains its performance when its internal coin tosses are substituted by a sequence generated by a pseudorandom generator. That is:

Construction 2.3 (typical application of pseudorandom generators). *Let A be a probabilistic polynomial-time algorithm, and $\rho(n)$ denote an upper bound on the number of coins that A tosses on n-bit inputs (e.g., $\rho(n) = n^2$). Let $A(x,r)$ denote the output of A on input x and coin tossing sequence $r \in \{0,1\}^{\rho(n)}$, where n denotes the length of x. Let G be a pseudorandom generator with stretching function $\ell: \mathbb{N} \to \mathbb{N}$ (e.g., $\ell(k) = k^5$). Then A_G is a randomized algorithm that on input $x \in \{0,1\}^n$, proceeds as follows. It sets $k = k(n)$ to be the smallest integer such that $\ell(k) \geq \rho(n)$ (e.g., $k^5 \geq n^2$), uniformly selects $s \in \{0,1\}^k$, and outputs $A(x,r)$, where r is the $\rho(|x|)$-bit long prefix of $G(s)$.*

Thus, using A_G instead of A, the number of random bits used by the algorithm is reduced from ρ to $\ell^{-1} \circ \rho$ (e.g., from n^2 to $k(n) = n^{2/5}$), while it is infeasible to find inputs (i.e., x's) on which the *noticeable behavior* of A_G is different from the one of A. That is, we save randomness while maintaining performance (see Section 2.5).

Amplifying the stretch function. Pseudorandom generators as in Definition 2.2 are only required to stretch their input a bit; for example, stretching n-bit long inputs to $(n+1)$-bit long outputs will do. Clearly, generators with such moderate stretch functions are of little use in practice. In contrast, we want to have pseudorandom generators with an arbitrary long stretch function. By the efficiency requirement, the stretch function can be at most polynomial. It turns out that pseudorandom generators with the smallest possible stretch function can be used to construct pseudorandom generators with any desirable polynomial stretch function. That is:

Theorem 2.4 [7, Sec. 3.3.2]. *Let G be a pseudorandom generator with stretch function $\ell(n) = n + 1$, and ℓ' be any positive polynomial such that $\ell'(n) \geq n + 1$. Then there exists a pseudorandom generator with stretch function ℓ'. Furthermore, the construction of the latter consists of invoking G for ℓ' times.*

Thus, when talking about the existence of pseudorandom generators, we may ignore the specific stretch function.

2.3   How to Construct Pseudorandom Generators

The known constructions of pseudorandomness generators are based on one-way functions. Loosely speaking, a *polynomial-time computable* function is called one-way if any efficient algorithm can invert it only with negligible success probability. For simplicity, we consider only length-preserving one-way functions.

Definition 2.5 (one-way function). *A one-way function, f, is a polynomial-time computable function such that for every probabilistic polynomial-time algorithm A', every positive polynomial $p(\cdot)$, and all sufficiently large n*

$$\Pr_{x \sim U_n}\left[A'(f(x)) \in f^{-1}(f(x))\right] < \frac{1}{p(n)},$$

where $f^{-1}(y) = \{z : f(z) = y\}$.

It is widely believed that one-way functions exists. Popular candidates for one-way functions are based on the conjectured intractability of integer factorization, the discrete logarithm problem, and decoding of random linear code. Assuming that integer factorization is indeed infeasible, one can prove that a minor modification of the construction outlined at the beginning of Section 2.2 constitutes a pseudorandom generator. More generally, it turns out that pseudorandom generators can be constructed based on any one-way function.

Theorem 2.6 (existence of pseudorandom generators [18]). *Pseudorandom generators exist if and only if one-way functions exist.*

To show that the existence of pseudorandom generators implies the existence of one-way functions, consider a pseudorandom generator G with stretch function $\ell(n) = 2n$. For $x, y \in \{0,1\}^n$, define $f(x,y) \stackrel{\text{def}}{=} G(x)$, so that f is polynomial-time computable (and length-preserving). It must be that f is one-way, or else one can distinguish $G(U_n)$ from U_{2n} by trying to invert and checking the result: Inverting f on its range distribution refers to the distribution $G(U_n)$, whereas the probability that U_{2n} has inverse under f is negligible. The interesting direction is the construction of pseudorandom generators based on any one-way function. A treatment of some natural special cases is provided in [7, Sec. 3.4-3.5].

2.4 Pseudorandom Functions

Pseudorandom generators allow one to efficiently generate long pseudorandom sequences from short random seeds (e.g., using n random bits, we can efficiently generate a pseudorandom bit-sequence of length n^2). Pseudorandom functions (defined below) are even more powerful: they allow efficient direct access to a huge pseudorandom sequence (which is infeasible to scan bit-by-bit). For example, based on n random bits, we define a sequence of length 2^n such that we can efficiently retrieve any desired bit in this sequence while the retrieved bits look random. In other words, pseudorandom functions can replace truly random functions in any efficient application (e.g., most notably in Cryptography). That is, pseudorandom functions are indistinguishable from random functions by any efficient procedure that may obtain the function values at arguments of its choice. Such procedures are called oracle machines, and if M is such machine and f is a function, then $M^f(x)$ denotes the computation of M on input x when M's queries are answered by the function f (i.e., during its computation, M generates special strings called queries such that in response to the query q machine M is given the value $f(q)$).

Definition 2.7 (pseudorandom functions [13]). *A* pseudorandom function (ensemble), *with length parameters* $\ell_{\text{D}}, \ell_{\text{R}} : \mathsf{N} \to \mathsf{N}$, *is a collection of functions* $\{F_n\}_{n \in \mathsf{N}}$, *where*

$$F_n \stackrel{\text{def}}{=} \{f_s : \{0,1\}^{\ell_{\text{D}}(n)} \to \{0,1\}^{\ell_{\text{R}}(n)}\}_{s \in \{0,1\}^n},$$

satisfying

- (efficient evaluation). *There exists an efficient* (deterministic) *algorithm that when given a seed, s, and an $\ell_D(n)$-bit argument, x, returns the $\ell_R(n)$-bit long value $f_s(x)$, where n denotes the length of s.*
  (Thus, the seed s is an "effective description" of the function f_s.)
- (pseudorandomness). *For every probabilistic polynomial-time oracle machine M, every positive polynomial p, and all sufficiently large n*

$$\left| \Pr_{s \sim U_n}[M^{f_s}(1^n) = 1] - \Pr_{\rho \sim R_n}[M^\rho(1^n) = 1] \right| < \frac{1}{p(n)},$$

where R_n denotes the uniform distribution over all functions mapping $\{0,1\}^{\ell_D(n)}$ to $\{0,1\}^{\ell_R(n)}$.

Suppose, for simplicity, that $\ell_D(n) = n$ and $\ell_R(n) = 1$. Then a function uniformly selected among 2^n functions (of a pseudorandom ensemble) presents an input-output behavior indistinguishable in poly(n)-time from the one of a function selected at random among all the 2^{2^n} Boolean functions. Contrast this with a distribution over 2^n sequences, produced by a pseudorandom generator applied to a random n-bit seed, which is computationally indistinguishable from the uniform distribution over $\{0,1\}^{\text{poly}(n)}$ (which has a support of size $2^{\text{poly}(n)}$). Still pseudorandom functions can be constructed from any pseudorandom generator.

Theorem 2.8 (how to construct pseudorandom functions [13]). *Let G be a pseudorandom generator with stretching function $\ell(n) = 2n$. For $s \in \{0,1\}^n$, let $G_0(s)$ (resp., $G_1(s)$) denote the first (resp., last) n bits in $G(s)$, and let*

$$G_{\sigma_n \cdots \sigma_2 \sigma_1}(s) \overset{\text{def}}{=} G_{\sigma_n}(\cdots G_{\sigma_2}(G_{\sigma_1}(s)) \cdots).$$

That is, $G_x(s)$ is computed by successive applications of either G_0 or G_1 to the current n-bit long string, where the decision which of the two mappings to apply is determined by the corresponding bit of x. Let $f_s(x) \overset{\text{def}}{=} G_x(s)$ and consider the function ensemble $\{F_n\}_{n\in\mathbb{N}}$, where $F_n = \{f_s : \{0,1\}^n \to \{0,1\}^n\}_{s\in\{0,1\}^n}$. Then this ensemble is pseudorandom (with length parameters $\ell_D(n) = \ell_R(n) = n$).

The foregoing construction can be easily adapted to any (polynomially-bounded) length parameters $\ell_D, \ell_R : \mathbb{N} \to \mathbb{N}$.

2.5 The Applicability of Pseudorandom Generators

Randomness is playing an increasingly important role in computation: it is frequently used in the design of sequential, parallel, and distributed algorithms (see [22]), and is of course central to Cryptography. Whereas it is convenient to design such algorithms making free use of randomness, it is also desirable to minimize the use of randomness in real implementations since generating perfectly random bits via special hardware is quite expensive. Thus, pseudorandom generators (as in Definition 2.2) are a key ingredient in an "algorithmic toolbox": they provide an automatic compiler of programs written with free use of randomness into programs that make an economical use of randomness.

Indeed, "pseudo-random number generators" have appeared with the first computers. However, typical implementations use generators that are not pseudorandom according to Definition 2.2. Instead, at best, these generators are shown to pass *some* ad-hoc statistical test. We warn that the fact that a "pseudorandom number generator" passes some statistical tests does not mean that it will pass a new test and that it is good for a future (untested) application. Furthermore, the approach of subjecting the generator to some ad-hoc tests fails to provide general results of the type stated above (i.e., of the form "for *all* practical purposes using the output of the generator is as good as using truly unbiased coin tosses"). In contrast, the approach encompassed in Definition 2.2 aims at such generality, and in fact is tailored to obtain it: the notion of computational indistinguishability, which underlines Definition 2.2, covers all possible efficient applications postulating that for all of them pseudorandom sequences are as good as truly random ones.

Pseudorandom generators and functions are of key importance in Cryptography. In particular, they are typically used to establish private-key encryption and authentication schemes. For further discussion see Section 4.

2.6 The Intellectual Contents of Pseudorandom Generators

We shortly discuss some intellectual aspects of pseudorandom generators as defined above.

Behavioristic versus Ontological. Our definition of pseudorandom generators is based on the notion of computational indistinguishability. The behavioristic nature of the latter notion is best demonstrated by confronting it with the Kolmogorov-Chaitin approach to randomness. Loosely speaking, a string is *Kolmogorov-random* if its length equals the length of the shortest program producing it. This shortest program may be considered the "true explanation" to the phenomenon described by the string. A Kolmogorov-random string is thus a string that does not have a substantially simpler (i.e., shorter) explanation than itself. Considering the simplest explanation of a phenomenon may be viewed as an ontological approach. In contrast, considering the effect of phenomena (on an observer), as underlying the definition of pseudorandomness, is a behavioristic approach. Furthermore, there exist probability distributions that are not uniform (and are not even statistically close to a uniform distribution) but nevertheless are indistinguishable from a uniform distribution by any efficient procedure. Thus, distributions that are ontologically very different are considered equivalent by the behavioristic point of view taken in the Definition 2.1.

A Relativistic View of Randomness. Pseudorandomness is defined in terms of its observer: It is a distribution that cannot be told apart from a uniform distribution by any efficient (i.e., polynomial-time) observer. However, pseudorandom sequences may be distinguished from random ones by infinitely powerful computers (not at our disposal!). Furthermore, a machine that runs in exponential-time can distinguish the output of a pseudorandom generator from a uniformly selected string of the same length (e.g., just by trying all possible seeds). Thus, pseudorandomness is subjective, dependent on the abilities of the observer.

Randomness and Computational Difficulty. Pseudorandomness and computational difficulty play dual roles: The definition of pseudorandomness relies on the fact that placing computational restrictions on the observer gives rise to distributions that are not uniform and still cannot be distinguished from uniform. Furthermore, the known constructions of pseudorandom generators relies on conjectures regarding computational difficulty (e.g., the existence of one-way functions), and this is inevitable: the existence of pseudorandom generators implies the existence of one-way functions.

Randomness and Predictability. The connection between pseudorandomness and unpredictability (by efficient procedures) plays an important role in the analysis of several constructions of pseudorandom generators (see [7, Sec. 3.3.5&3.5]). We wish to highlight the intuitive appeal of this connection.

2.7 Suggestions for Further Reading

A detailed textbook presentation of the material that is reviewed in this section is provided in [7, Chap. 3]. For a wider perspective, which treats this material as a special case of a general paradigm, the interested reader is referred to [6, Chap. 3] (or [9, Chap. 8]).

3 Probabilistic Proof Systems

The glory attributed to the creativity involved in finding proofs, makes us forget that it is the less glorified procedure of verification which gives proofs their value. Philosophically speaking, proofs are secondary to the verification procedure; whereas technically speaking, proof systems are defined in terms of their verification procedures.

The notion of a verification procedure assumes the notion of computation and furthermore the notion of efficient computation. This implicit assumption is made explicit in the following definition in which efficient computation is associated with deterministic polynomial-time algorithms.

Definition 3.1 (NP-proof systems): *Let $S \subseteq \{0,1\}^*$ and $\nu : \{0,1\}^* \times \{0,1\}^* \to \{0,1\}$ be a function such that $x \in S$ if and only if there exists a $w \in \{0,1\}^*$ that satisfies $\nu(x, w) = 1$. If ν is computable in time bounded by a polynomial in the length of its first argument then we say ν defines an NP-proof system for S and that S is an NP-set. The class of NP-sets is denoted $\mathcal{NP}$.*

Indeed, ν represents a verification procedure for claims of membership in a set S, and a string w satisfying $\nu(x, w) = 1$ is a proof that x belongs to S, whereas $x \notin S$ has no such proofs. For example, consider the set of systems of quadratic equations that have integer solutions, which is a well-known NP-set. Clearly, any integer solution $\bar{v}$ to such a system Q constitutes an "NP-proof" for the assertion **the system Q has an integer solution** (the verification procedure consists of substituting the variables of Q by the values provided in $\bar{v}$ and computing the value of the resulting arithmetic expressions).

We seize the opportunity to note that the celebrated "P different from NP" conjecture asserts that NP-proof systems are useful in the sense that *there are assertions for which obtaining a proof helps to verify the correctness of the assertion.*[8] This conforms with our daily experience by which reading a proof eases the verification of an assertion.

The formulation of NP-proofs restricts the "effective" length of proofs to be polynomial in length of the corresponding assertions (since the running-time of the verification procedure is restricted to be polynomial in the length of the assertion). However, longer proofs may be allowed by padding the assertion with sufficiently many blank symbols. So it seems that NP gives a satisfactory formulation of proof systems (with efficient verification procedures). This is indeed the case if one associates efficient procedures with *deterministic* polynomial-time algorithms. However, we can gain a lot if we are willing to take a somewhat non-traditional step and allow *probabilistic* verification procedures. In particular:

- Randomized and interactive verification procedures, giving rise to *interactive proof systems*, seem much more powerful than their deterministic counterparts (see Section 3.1).
- Such randomized procedures allow the introduction of *zero-knowledge proofs*, which are of great conceptual and practical interest (see Section 3.2).
- NP-proofs can be efficiently transformed into a (redundant) form (called a *probabilistically checkable proof*) that offers a trade-off between the number of bit-locations examined in the NP-proof and the confidence in its validity (see Section 3.3).

In all these types of probabilistic proof systems, explicit bounds are imposed on the computational resources of the verification procedure, which in turn is personified by the notion of a verifier. Furthermore, in all these proof systems, the verifier is allowed to toss coins and rule by statistical evidence. Thus, *all these proof systems carry a probability of error; yet, this probability is explicitly bounded and, furthermore, can be reduced by successive application of the proof system.*

Clarifications. Like the definition of NP-proof systems, the abovementioned types of probabilistic proof systems refer to proving membership in predetermined sets of strings. That is, the assertions are all of the form "the string x

---

[8] NP represents sets of assertions that can be efficiently verified with the help of adequate proofs, whereas P represents sets of assertions that can be efficiently verified from scratch (i.e., without proofs). Thus, "P different from NP" asserts the existence of assertions that are harder to prove than to be convinced of their correctness when presented with a proof. This means that the notion of a proof is meaningful (i.e., that proofs do help when trying to be convinced of the correctness of assertions).

is in a set S", where S is a fixed infinite set and x is a variable input. The definition of an interactive proof system makes explicit reference to a prover, which is only implicit in the definition of an NP-proof system (where the prover is the unmentioned entity providing the proof). We note that, as a first approximation, we are not concerned with the complexity of the prover or the proving task. Our main focus is on the complexity of verification. This is consistent with the intuitive notion of a proof, which refers to the validity of the proof and not to how it was obtained.

3.1   Interactive Proof Systems

In light of the growing acceptability of randomized and distributed computations, it is only natural to associate the notion of efficient computation with probabilistic and interactive polynomial-time computations. This leads naturally to the notion of an interactive proof system in which the verification procedure is interactive and randomized, rather than being non-interactive and deterministic. Thus, a "proof" in this context is not a fixed and static object but rather a randomized (dynamic) process in which the verifier interacts with the prover. Intuitively, one may think of this interaction as consisting of "tricky" questions asked by the verifier, to which the prover has to reply "convincingly". The above discussion, as well as the following definition, makes explicit reference to a prover, whereas a prover is only implicit in the traditional definitions of proof systems (e.g., NP-proofs).

Loosely speaking, an interactive proof is a game between a computationally bounded verifier and a computationally unbounded prover whose goal is to convince the verifier of the validity of some assertion. Specifically, the verifier is probabilistic polynomial-time. It is required that if the assertion holds then the verifier always accepts (i.e., when interacting with an appropriate prover strategy). On the other hand, if the assertion is false then the verifier must reject with probability at least $\frac{1}{2}$, no matter what strategy is being employed by the prover.

Definition 3.2 (Interactive Proofs – IP [17]): *An* interactive proof system *for a set S is a two-party game, between a* verifier *executing a probabilistic polynomial-time strategy* (denoted V) *and a* prover *which executes a computationally unbounded strategy* (denoted P), *satisfying*

- Completeness: *For every $x \in S$ the verifier V always accepts after interacting with the prover P on common input x.*
- Soundness: *For every $x \notin S$ and every possible strategy P^*, the verifier V rejects with probability at least $\frac{1}{2}$, after interacting with P^* on common input x.*

The class of sets having interactive proof systems is denoted by $\mathcal{IP}$.

Recall that the error probability in the soundness condition can be reduced by successive application of the proof system. To clarify the definition and illustrate the power of the underlying concept, we consider the following story.

One day on the Olympus, bright-eyed Athena claimed that Nectar poured out of the new silver-coated jars tastes less good than Nectar poured out of the older gold-decorated jars. Mighty Zeus, who was forced to introduce the new jars by the practically oriented Hera, was annoyed at the claim. He ordered that Athena be served one hundred glasses of Nectar, each poured at random either from an old jar or from a new one, and that she tell the source of the drink in each glass. To everybody's surprise, wise Athena correctly identified the source of each serving, to which the Father of the Gods responded "my child, you are either right or extremely lucky." Since all gods knew that being lucky was not one of the attributes of Pallas-Athena, they all concluded that the impeccable goddess was right in her claim.

Note that the proof system underlying this story establishes the dissimilarity of two objects. This idea can be used to provide an interactive proof system for the set of "pairs of non-isomorphic graphs" [15], which informally refer to the dissimilarity of two given objects.[9] Indeed, typically, proving similarity between objects is easy, because one can present a mapping (of one object to the other) that demonstrates this similarity. In contrast, proving dissimilarity seems harder, because in general there seems to be no succinct proof of dissimilarity. More generally, it is typically easy to prove the existence of an easily verifiable structure in the given object by merely presenting this structure, but proving the non-existence of such a structure seems hard.

Formally speaking, proving the existence of an easily verifiable structure corresponds to NP-proof systems. The forgoing discussion suggests that interactive proof systems can be used to demonstrate the non-existence of such structures. Specifically, the set of pairs of non-isomorphic graphs is *not* known to have an NP-proof system, and does have an interactive proof system. In general, interactive proof systems can be used to prove the non-existence of any easily verifiable structure; that is, for every $S \in \mathcal{NP}$, the set $\{0,1\}^* \setminus S$ has an interactive proof system (i.e., the class $\mathrm{co}\mathcal{NP}$ is contained in $\mathcal{IP}$). We stress that it is widely believed that $\mathrm{co}\mathcal{NP} \stackrel{\mathrm{def}}{=} \{\{0,1\}^* \setminus S : S \in \mathcal{NP}\}$ is *not* contained in $\mathcal{NP}$. For example, the set of systems of quadratic equations that have no integer solutions has an interactive proof system, but is believed not to have an NP-proof system. Furthermore, the class of sets having interactive proof systems coincides with the class $\mathcal{PSPACE}$ containing all sets for which membership is decidable by an algorithm that uses a polynomial amount of work-space.

Theorem 3.3 [21,27]: $\mathcal{IP} = \mathcal{PSPACE}$.

We mention that $\mathcal{NP} \cup \mathrm{co}\mathcal{NP} \subseteq \mathcal{PSPACE}$ and that it is widely believed that $\mathcal{NP}$ contain "little" of $\mathcal{PSPACE}$. Thus, interactive proofs seem to be more powerful

---

[9] A graph $G = (V, E)$ consists of a finite set of vertices V and a finite set of edges E, where each edge is an unordered pair of vertices. Two graphs, $G_1 = (V_1, E_1)$ and $G_2 = (V_2, E_2)$, are called isomorphic if there exists a 1-1 and onto mapping $\phi : V_1 \rightarrow V_2$ such that $\{u, v\} \in E_1$ if and only if $\{\phi(u), \phi(v)\} \in E_1$.

than NP-proofs. This conforms with our daily experience by which interaction facilitates the verification of assertions. As we shall argue next, randomness (and the error probability in the soundness condition) play a key role in this phenomenon.

Interactive proof systems extend NP-proof systems in allowing extensive interaction as well as randomization (and ruling based on statistical evidence). As hinted, extensive interaction by itself does not provide any gain (over NP-proof systems). The reason being that the prover can predict the verifier's part of the interaction and thus it suffices to let the prover send the full transcript of the interaction and let the verifier check that the interaction is indeed valid.[10] The moral is that *there is no point to interact with predictable parties that are also computationally weaker.* This moral represents the prover's point of view (with respect to deterministic verifiers). Certainly, from the verifier's point of view it is beneficial to interact with the prover, since the latter is computationally stronger.

We mention that the power of interactive proof systems remains unchanged under several natural variants. In particular, it turns out that, in this context, *asking clever questions is not more powerful than asking totally random questions.* The reason being that a powerful prover may assist the verifier, which may thus refrain from trying to be clever and focus on checking (by using only random questions) that the help extended to it is indeed valid. Also, the power of interactive proof systems remains unchanged when allowing two-sided error probability (i.e., allowing bounded error probability also in the completeness condition). Recall that, in contrast, one-sided error probability (i.e., error probability in the soundness condition) is essential to the power of interactive proofs.

3.2   Zero-Knowledge Proof Systems

Standard proofs are believed to yield knowledge and not merely establish the validity of the assertion being proven. Indeed, it is commonly believed that (good) proofs provide a deeper understanding of the theorem being proved. At the technical level, assuming that NP-proof are useful at all (i.e., assuming that $\mathcal{P} \neq \mathcal{NP}$), an NP-proof of membership in some sets $S \in \mathcal{NP} \setminus \mathcal{P}$ yields something (i.e., the NP-proof itself) that is typically hard to find (even when assuming that the input is in S). For example, an integer solution to a system of quadratic equations constitutes an NP-proof that this system has an integer solution, but it yields information (i.e., the solution) that is infeasible to find (when given an arbitrary system of quadratic equations that has an integer solution). In contrast to such NP-proofs, which seem to yield a lot of knowledge, zero-knowledge proofs yield no knowledge at all; that is, the latter exhibit an extreme contrast between being convincing (of the validity of a statement) and teaching something on top of the validity of the statement.

Loosely speaking, zero-knowledge proofs are interactive proofs that yield nothing beyond the validity of the assertion. These proofs, introduced in [17], are

---

[10] In case the verifier is not deterministic, the transcript sent by the prover may not match the outcome of the verifier coin tosses.

fascinating and extremely useful constructs. Their fascinating nature is due to their seemingly contradictory definition: zero-knowledge proofs are both convincing and yet yield nothing beyond the validity of the assertion being proven. Their applicability in the domain of Cryptography is vast; they are typically used to force malicious parties to behave according to a predetermined protocol. In addition to their direct applicability in Cryptography, zero-knowledge proofs serve as a good bench-mark for the study of various problems regarding cryptographic protocols.

Zero-knowledge is a property of some interactive proof systems, or more accurately of some prover strategies. Specifically, it is the property of yielding nothing beyond the validity of the assertion; that is, a verifier obtaining a zero-knowledge proof only gains conviction in the validity of the assertion. This is formulated by saying that anything that can be feasibly obtained from a zero-knowledge proof is also feasibly computable from the (valid) assertion itself. Details follow.

The formulation of the zero-knowledge condition refers to two types of probability ensembles, where each ensemble associates a distribution to each valid assertion. The first ensemble represents the output distribution of the verifier after interacting with the specified prover strategy P, where the verifier is not necessarily employing the specified strategy (i.e., V) but rather any efficient strategy. The second ensemble represents the output distribution of some probabilistic polynomial-time algorithm (which does not interact with anyone). The basic paradigm of zero-knowledge asserts that for every ensemble of the first type there exist a "similar" ensemble of the second type. The specific variants differ by the interpretation given to the notion of similarity. The most strict interpretation, leading to *perfect zero-knowledge*, is that similarity means equality.

Definition 3.4 (perfect zero-knowledge, a simplified version[11]): *A prover strategy, P, is said to be* perfect zero-knowledge *over a set S if for every probabilistic polynomial-time verifier strategy, V^*, there exists a probabilistic polynomial-time algorithm, M^*, such that for every $x \in S$ it holds that $(P, V^*)(x) \equiv M^*(x)$, where $(P, V^*)(x)$ denote the distribution that represents the output of verifier V^* after interacting with the prover P on common input x.*[12]

A somewhat more relaxed interpretation of similarity, leading to almost-perfect zero-knowledge, is that similarity means statistical closeness (i.e., negligible difference between the ensembles). The most liberal interpretation, leading to the standard usage of the term zero-knowledge, is that similarity means computational indistinguishability (i.e., failure of any efficient procedure to tell the two ensembles apart). The actual definition is obtained from Definition 2.1, by considering ensembles indexed by strings and providing the distinguisher with the relevant index. That is, *the probability ensembles, $\{Y_x\}_{x \in S}$ and $\{Z_x\}_{x \in S}$, are* indistinguishable *by an algorithm A if*

---

[11] The actual definition allows for a rare event (which occurs with negligible probability) in which M^* halts with no output, and the output of M^* is considered condition on this event not occuring.

[12] As usual, $M^*(x)$ denotes a distribution representing the output of algorithm M^* on input x.

$$d_A(n) \stackrel{\text{def}}{=} \max_{x \in S \cap \{0,1\}^n} \{|prob(A(x, Y_x) = 1) - \Pr(A(x, Z_x) = 1)|\}$$

is a negligible function.[13] The ensembles $\{Y_x\}_{x \in S}$ and $\{Z_x\}_{x \in S}$ are computation-ally indistinguishable if they are indistinguishable by every probabilistic polynomial-time algorithm.

The foregoing discussion refers to simplified versions of the actual definitions. Specifically, in order to guarantee that zero-knowledge is preserved under sequen-tial composition it is necessary to slightly augment the definitions. For details see [7, Sec. 4.3.3-4.3.4].

The Power of Zero-Knowledge. We consider the set of 3-colorable graphs, where a graph[14] $G = (V, E)$ is said to be *3-colorable* if there exists a function $\pi : V \to \{1, 2, 3\}$ (called a *3-coloring*) such that $\pi(v) \neq \pi(u)$ for every $\{u, v\} \in E$. It is easy to prove that a given graph G is 3-colorable by just presenting a 3-coloring of G, but this NP-proof is not a zero-knowledge proof (unless $\mathcal{P} = \mathcal{NP}$). In fact, assuming $\mathcal{P} \neq \mathcal{NP}$, graph 3-colorability has no zero-knowledge NP-proofs, but as we shall see it has zero-knowledge interactive proofs. We first describe these proof systems using (abstract) "boxes" in which information can be hidden and later revealed. Such "boxes" can be implemented using one-way functions.

Construction 3.5 (Zero-knowledge proof of 3-colorability [15]): *On common input, $G = (V, E)$, The following steps are repeated $|V| \cdot |E|$ times.*

- Prover's first step: *Let ψ be a 3-coloring of G. The prover selects a random permutation, π, over $\{1, 2, 3\}$, and sets $\phi(v) \stackrel{\text{def}}{=} \pi(\psi(v))$, for each $v \in V$. Hence, the prover forms a random relabeling of the 3-coloring ψ. The prover sends the verifier a sequence of $|V|$ locked and non-transparent boxes such that the vth box contains the value $\phi(v)$.*
- Verifier's first step: *The verifier uniformly selects an edge $\{u, v\} \in E$, and sends it to the prover. Intuitively, the verifier asks to inspect the colors of vertices u and v.*
- Prover's second step: *The prover sends to the verifier the keys to boxes u and v.*
- Verifier's second step: *The verifier opens boxes u and v, and checks whether or not they contain two different elements in $\{1, 2, 3\}$.*

The verifier accepts if and only if all checks turn out positive.

The foregoing verifier strategy is easily implemented in probabilistic polynomial-time. The same holds with respect to the prover's strategy, provided it is given a 3-coloring of G as auxiliary input. Clearly, if the input graph is 3-colorable then the prover can cause the verifier to accept with probability 1. On the other hand, if the input graph is not 3-colorable then any contents put in the boxes

---

[13] If $S \cap \{0, 1\}^n = \emptyset$ then we define $d_A(n) = 0$.
[14] See Footnote 9.

must be invalid on at least one edge, and consequently each time the foregoing steps are repeated the verifier rejects with probability at least $\frac{1}{|E|}$. Repeating these steps $t \cdot |E|$ times has the effect of reducing the soundness error probability to

$$\left(1 - \frac{1}{|E|}\right)^{t \cdot |E|} \approx e^{-t}.$$

The zero-knowledge property follows easily, in this abstract setting, because one can simulate the real interaction by placing a random pair of different colors in the boxes indicated by the verifier. This indeed demonstrates that the verifier learns nothing from the interaction (since it expects to see a random pair of different colors and indeed this is what it sees). We stress that this simple argument is not possible in the digital implementation because the boxes are not totally unaffected by their contents (but are rather affected, yet in an indistinguishable manner).

As stated, in order to obtain a real interactive proof, the (abstract or physical) "boxes" need to be implemented digitally. This can be done using an adequately defined "commitment scheme" (see [7, Sec. 4.4.1]). Loosely speaking, such a scheme is a two phase game between a sender and a receiver so that after the first phase the sender is "committed" to a value and yet, at this stage, it is infeasible for the receiver to find out the committed value. The committed value will be revealed to the receiver in the second phase and it is guaranteed that the sender cannot reveal a value other than the one committed. Such commitment schemes can be implemented assuming the existence of one-way functions. Thus, the existence of one-way functions implies a zero-knowledge proofs for 3-colorability. In fact, one gets zero-knowledge proofs for any NP-set.

Theorem 3.6 [15]: *Assuming the existence of one-way functions, any NP-proof can be efficiently transformed into a zero-knowledge interactive proof. That is, the prover strategy in the zero-knowledge interactive proof can be implemented in probabilistic polynomial-time provided that it is given an adequate NP-proof as auxiliary input.*

Theorem 3.6 has a dramatic effect on the design of cryptographic protocols (cf., [7,8]). In a different vein and for the sake of elegance, we mention that, using further ideas and under the same assumption, any set having an interactive proof system also has a zero-knowledge interactive proof system.

The Role of Randomness. Again, randomness is essential to all the aforementioned results. Namely, zero-knowledge proof systems in which either the verifier or the prover is deterministic exist only for sets in $\mathcal{BPP}$, where $\mathcal{BPP}$ is the class of sets for which membership is decidable by some probabilistic polynomial-time algorithm. Note that such sets have trivial zero-knowledge proofs in which the prover sends nothing and the verifier just test the validity of the assertion by itself. Thus, randomness is essential to the usefulness of zero-knowledge proofs.

3.3 Probabilistically Checkable Proof Systems

We now return to the non-interactive mode in which the verifier receives a (alleged) written proof. But our focus is on probabilistic verifiers that are capable of evaluating the validity of the assertion by examining few (randomly selected) locations in the alleged proof. Thus, the alleged proof is a string, as in the case of a traditional proof system, but we are interested in probabilistic verification procedures that access only few locations in the proof, and yet are able to make a meaningful probabilistic verdict regarding the validity of the alleged proof. Specifically, the verification procedure should accept any valid proof (with probability 1), but rejects with probability at least 1/2 any alleged proof for a false assertion.

The main complexity measure associated with probabilistically checkable proof (PCP) systems is indeed their query complexity (i.e., the number of bits accessed in the alleged proof). Another complexity measure of natural concern is the length of the proofs being employed, which in turn is related to the randomness complexity of the system. The randomness complexity of PCPs plays a key role in numerous applications (e.g., in composing PCP systems as well as when applying PCP systems to derive non-approximability results), and thus we specify this parameter rather than the proof length.

Loosely speaking, a probabilistically checkable proof system consists of a probabilistic polynomial-time verifier having access to an oracle that represents an alleged proof (in redundant form). Typically, the verifier accesses only few of the oracle bits, and these bit positions are determined by the outcome of the verifier's coin tosses. As in the case of interactive proof systems, it is required that if the assertion holds then the verifier always accepts (i.e., when given access to an adequate oracle); whereas, if the assertion is false then the verifier must reject with probability at least $\frac{1}{2}$, no matter which oracle is used. The basic definition of the PCP setting is given in Item (1) of Definition 3.7. Yet, the complexity measures introduced in Item (2) are of key importance for the subsequent discussions.

Definition 3.7 (Probabilistically Checkable Proofs – PCP):

1. *A* probabilistically checkable proof system (PCP) *for a set S is a probabilistic polynomial-time oracle machine* (called verifier), *denoted V, satisfying*
   - Completeness: *For every $x \in S$ there exists an oracle π_x so that V, on input x and access to π_x, always accepts x.*
   - Soundness: *For every $x \notin S$ and every oracle π, machine V, on input x and access to π, rejects x with probability at least $\frac{1}{2}$.*
2. *Let r and q be integer functions. The complexity class $\mathcal{PCP}(r(\cdot), q(\cdot))$ consists of sets having a probabilistically checkable proof system in which the verifier, on any input of length n, makes at most $r(n)$ coin tosses and at most $q(n)$ oracle queries, where each query is answered by a single bit. For sets of integer functions, R and Q, we let $\mathcal{PCP}(R, Q)$ equal $\cup_{r \in R, q \in Q} \mathcal{PCP}(r(\cdot), q(\cdot))$.*

We stress that the oracle π_x in a PCP system constitutes a proof in the standard mathematical sense. Yet, this oracle has the extra property of enabling a lazy verifier, to toss coins, take its chances and "assess" the validity of the proof without reading all of it (but rather by reading a tiny portion of it).

Letting `poly` denote the set of all polynomials, one may verify that $\mathcal{PCP}(0, \texttt{poly}) = \mathcal{NP}$. Letting `log` denote the set of all logarithmic functions (i.e., $\ell \in \texttt{log}$ if there exists a constant b such that $\ell(n) \leq \log_b n$ for all sufficiently large n), one may also verify that $\mathcal{PCP}(\texttt{log}, \texttt{poly}) \subseteq \mathcal{NP}$ (because the relevant oracles are of polynomial length). It follows that, for every constant c, it holds that $\mathcal{PCP}(\texttt{log}, c) \subseteq \mathcal{NP}$. This upper bound turned out to be tight, but proving this is much more difficult (to say the least). The following result is a culmination of a sequence of great works (see [6, Sec. 2.6.2] for a detailed account).

Theorem 3.8 [2,1]: *There exists a constant c such that* $\mathcal{NP} \subseteq \mathcal{PCP}(\texttt{log}, c)$.

Thus, probabilistically checkable proofs in which the verifier tosses only logarithmically many coins and makes only a constant number of queries exist for every set in the complexity class $\mathcal{NP}$. (Essentially, this constant is three.) Furthermore, NP-proofs can be efficiently transformed into NP-proofs that offer a trade-off between the portion of the proof being read and the confidence it offers. Specifically, if the verifier is willing to tolerate an error probability of ϵ then it suffices to let it examine $c \cdot \log_2(1/\epsilon)$ bits of the (transformed) NP-proof.[15] These bit locations need to be selected at random. We mention that the length of the redundant NP-proofs that provide the aforementioned trade-off can be made almost linear in the length of the standard NP-proofs.

PCP and the Study of Approximation. Following [5] and [1], the characterization of $\mathcal{NP}$ in terms of probabilistically checkable proofs has played a central role in developments concerning the study of approximation problems. For details, see [19, Chap. 10]. We merely mention that Theorem 3.8 implies that, assuming $\mathcal{P} \neq \mathcal{NP}$, there exists a constant $\delta < 1$ such that *given a system of quadratic equations it is infeasible to distinguish the case in which the system has an integer solution from the case that any assignment of integers satisfies at most a δ fraction of the equations.*

The Role of Randomness. The foregoing results rely on the randomness of the verifier and are not possible for deterministic verifiers. Furthermore, $\mathcal{PCP}(0, \texttt{log}) = \mathcal{P}$.

3.4   Suggestions for Further Reading

More detailed overviews of the three types of probabilistically proof systems can be found in [6, Chap. 2] (or [9, Chap. 9]). A detailed textbook treatment of zero-knowledge is provided in [7, Chap. 4].

---

[15] In fact, c can be made arbitrarily close to one, when ϵ is small enough.

4 Cryptography

In this section we focus on the role of randomness in Cryptography. As stated at the beginning of the introduction, the very notion of a secret, which is central to Cryptography, refers to randomness in the sense of unpredictability (i.e., unpredictability of the secret by other parties). Furthermore, the use of randomized algorithms and/or strategies is essential for achieving almost any security goal. We start with the concrete example of providing secret and authenticated communication, and end with a wider perspective.

4.1 Secret and Authenticated Communication

The problem of providing *secret communication over insecure media* is the traditional and most basic problem of Cryptography. The setting of this problem consists of two parties communicating through a channel that is possibly tapped by an adversary. The parties wish to exchange information with each other, but keep the "wire-tapper" as ignorant as possible regarding the contents of this information. The canonical solution to the above problem is obtained by the use of encryption schemes.

Loosely speaking, an encryption scheme is a protocol allowing these parties to communicate *secretly* with each other. Typically, the encryption scheme consists of a pair of algorithms. One algorithm, called encryption, is applied by the sender (i.e., the party sending a message), while the other algorithm, called decryption, is applied by the receiver. Hence, in order to send a message, the sender first applies the encryption algorithm to the message, and sends the result, called the ciphertext, over the channel. Upon receiving a ciphertext, the other party (i.e., the receiver) applies the decryption algorithm to it, and retrieves the original message (called the plaintext).

In order for the foregoing scheme to provide secret communication, the communicating parties (at least the receiver) must know something that is not known to the wire-tapper. (Otherwise, the wire-tapper can decrypt the ciphertext exactly as done by the receiver.) This extra knowledge may take the form of the decryption algorithm itself, or some parameters and/or auxiliary inputs used by the decryption algorithm. We call this extra knowledge the decryption-key. Note that, without loss of generality, we may assume that the decryption algorithm is known to the wire-tapper, and that the decryption algorithm operates on two inputs: a ciphertext and a decryption-key. (The encryption algorithm also takes two inputs: a corresponding encryption-key and a plaintext.) We stress that the existence of a decryption-key, not known to the wire-tapper, is merely a necessary condition for secret communication.

The point we wish to make is that the decryption-key must be generated by a randomized algorithm. Suppose, in contrary, that the decryption-key is a predetermined function of publicly available data (i.e., the key is generated by employing an efficient deterministic algorithm to this data). Then, the wire-tapper can just obtain the key in exactly the same manner (i.e., invoking the same algorithm on the said data). We stress that saying that the wire-tapper does not know which algorithm to employ or does not have the data on which the algorithm is employed just shifts the problem elsewhere; that is, the question remains as to *how do the legitimate parties select this algorithm and/or the data to which it is applied?* Again, deterministically selecting these objects based on publicly available data will not do. At some point, *the legitimate parties must obtain some object that is unpredictable by the wire-tapper*, and such unpredictability refers to randomness (or pseudorandomness).

However, the role of randomness in allowing for secret communication is not confined to the generation of secret keys. To see why this is the case, we need to understand what is "secrecy" (i.e., to properly define what is meant by this intuitive term). Loosely speaking, we say that an encryption scheme is secure if it is *infeasible for the wire-tapper to obtain from the ciphertexts any additional information about the corresponding plaintexts*. In other words, whatever can be efficiently computed based on the ciphertexts can be efficiently computed from scratch (or rather from the a priori known data). Now, assuming that the encryption algorithm is deterministic, encrypting the same plaintext twice (using the same encryption-key) results in two identical ciphertexts, which are easily distinguishable from any pair of different ciphertexts resulting from the encryption of two different plaintexts. This problem does not arise when employing a randomized encryption algorithm (as presented next).

As hinted, an encryption scheme must specify also a method for selecting keys. In the following encryption scheme, the key is a uniformly selected n-bit string, denoted s. The parties use this key to determine a pseudorandom function f_s (as in Definition 2.7). A plaintext $x \in \{0,1\}^n$ is encrypted (using the key s) by uniformly selecting $r \in \{0,1\}^n$ and producing the ciphertext $(r, f_s(r) \oplus x)$, where $\alpha \oplus \beta$ denotes the bit-by-bit exclusive-or of the strings α and β. A ciphertext (r, y) is decrypted (using the key s) by computing $f_s(r) \oplus y$. The security of this scheme follows from the security of an imaginary (ideal) scheme in which f_s is replaced by a totally random function $F : \{0,1\}^n \to \{0,1\}^n$.

Public-Key Encryption Schemes. The foregoing description corresponds to the so called model of a *private-key encryption scheme*, and requires the communicating parties to agree beforehand on a corresponding pair of encryption/decryption keys. This need is removed in *public-key encryption schemes*, envisioned by Diffie and Hellman (and materialized by the RSA scheme of Rivest, Shamir, and Adleman). In a public-key encryption scheme, the encryption-key can be publicized without harming the security of the plaintexts encrypted using it, allowing anybody to send encrypted messages to Party X by using the encryption-key publicized by Party X. But in such a case, the need for randomized encryption is even more clear. Indeed, if a deterministic encryption algorithm is employed and the wire-tapper knows the encryption-key, then it can identity of the plaintext in the case that the number of possibilities is small. In contrast, using a randomized encryption algorithm, the encryption of plaintext **yes** under a known encryption-key may be computationally indistinguishable from the encryption of the plaintext **no** under the say encryption-key. For further discussion of the security and construction of encryption schemes, the interested reader is referred to [8, Chap. 5].

Authenticated Communication. Message authentication is a task related to the setting considered for private-key encryption schemes. Again, there are two designated parties that wish to communicate over an insecure channel. This time, we consider an active adversary that is monitoring the channel and may alter the messages sent on it. The parties communicating through this insecure channel wish to authenticate the messages they send such that their counterpart can tell an original message (sent by the sender) from a modified one (i.e., modified by the adversary). Loosely speaking, a scheme for message authentication should satisfy the following:

- each of the communicating parties can *efficiently produce an authentication tag* to any message of its choice;
- each of the communicating parties can *efficiently verify* whether a given string is an authentication tag of a given message; but
- *it is infeasible for an external adversary* (i.e., a party other than the communicating parties) *to produce authentication tags* to messages not sent by the communicating parties.

Again, such a scheme consists of a randomized algorithm for selecting keys as well as algorithms for tagging messages and verifying the validity of tags. In the following message authentication scheme, a uniformly chosen n-bit key, s, is used for specifying a pseudorandom function (as in Definition 2.7). Using the key s, a plaintext $x \in \{0,1\}^n$ is authenticated by the tag $f_s(x)$, and verification of (x, y) with respect to the key s amounts to checking whether y equals $f_s(x)$. For further discussion of message authentication schemes and the related notion of signature schemes, the interested reader is referred to [8, Chap. 6].

4.2   A Wider Perspective

Modern Cryptography is concerned with the construction of information systems that are robust against malicious attempts to make these systems deviate from their prescribed functionality. The prescribed functionality may be the private and authenticated communication of information through the Internet, the holding of incoercible and secret electronic voting, or conducting any "fault-resilient" multi-party computation. Indeed, the scope of modern Cryptography is very broad, and it stands in contrast to "classical" Cryptography (which has focused on the single problem of enabling secret communication over insecure communication media).

The design of cryptographic systems is a very difficult task. One cannot rely on intuitions regarding the "typical" state of the environment in which the system operates. For sure, the adversary attacking the system will try to manipulate the environment into "untypical" states. Nor can one be content with countermeasures designed to withstand specific attacks, since the adversary (which acts after the design of the system is completed) will try to attack the schemes in ways that are different from the ones the designer had envisioned. The validity of the above assertions seems self-evident, still some people hope that in practice ignoring these tautologies will not result in actual damage. Experience shows that these hopes rarely come true; cryptographic schemes based on make-believe are broken, typically sooner than later.

In view of the foregoing, we believe that it makes little sense to make assumptions regarding the specific *strategy* that the adversary may use. The only assumptions that can be justified refer to the computational *abilities* of the adversary. Furthermore, the design of cryptographic systems has to be based on *firm foundations*; whereas ad-hoc approaches and heuristics are a very dangerous way to go. A heuristic may make sense when the designer has a very good idea regarding the environment in which a scheme is to operate, yet a cryptographic scheme has to operate in a maliciously selected environment which typically transcends the designer's view.

The foundations of Cryptography are the paradigms, approaches and techniques used to conceptualize, define and provide solutions to natural "security concerns". For a presentation of these foundations, the interested reader is referred to [7,8]. Here we merely note that randomness plays a central role in each definition and technique presented there. In almost every case, the inputs of the legitimate parties are assumed to be unpredictable by the adversary, and the task is performing some manipulation (of the inputs) while preserving or creating some unpredictability. In all cases, this is obtained by using randomized algorithms.

Suggestions for Further Reading. As stated above, a (comprehensive) exposition of the foundations of modern Cryptography can be found in the two-volume work [7,8].

5   Property Testing

For starters, let us consider a well-known example in which fast approximations are possible and useful. Suppose that some cost function is defined over a huge data-set, and that one wants to approximate the average cost of an element in the set. To be more specific, let $\mu : S \to [0, 1]$ be a cost function, and suppose we want to estimate $\overline{\mu} \stackrel{\text{def}}{=} \frac{1}{|S|} \sum_{e \in S} \mu(e)$. Then, for some constant c, uniformly (and independently) selecting $m \stackrel{\text{def}}{=} c \cdot \varepsilon^{-2} \log_2(1/\delta)$ sample points, $s_1, ..., s_m$, in S we obtain with probability at least $1 - \delta$ an estimate of $\overline{\mu}$ within $\pm\varepsilon$:

$$\Pr_{s_1,...,s_m \in S} \left[\left| \frac{1}{m} \sum_{i=1}^{m} \mu(s_i) - \overline{\mu} \right| > \varepsilon \right] < \delta.$$

We stress the fact that the number of samples *only depends on the desired level of approximation* (and is independent of the size of S). In this section we discuss analogous phenomena that occur with respect to objectives that are beyond gathering statistics of individual values. We focus on more complex features of a data-set; specifically, relations among pairs of elements rather than values of single elements. Such binary relations are captured by graphs (as defined in

Footnote 9); that is, a symmetric binary relation $R \subseteq S \times S$ is represented by a graph $G = (S, R)$, where the elements of S are called vertices and the elements of R are called edges. Each edge consists of a pair of vertices, called its end-points.

One natural computational question regarding graphs is whether or not they are bi-partite; that is, whether there exists a partition of S into two subsets S_1 and S_2 such that each edge has one end-point in S_1 and the other endpoint in S_2. For example, the graph consisting of a cycle of four vertices is bi-partite, whereas a triangle is not bi-partite. We mention that there exists an efficient algorithm that given a graph G determines whether or not G is bi-partite. Needless to say, this algorithm must inspect all edges of G, whereas we seek sub-linear time algorithms (i.e., algorithms operating in time smaller than the size of the input). In particular, sub-linear time algorithms cannot afford reading the entire input graph. Instead, these algorithm can inspect portions of the input graph by querying for the existence of specific edges (i.e., query whether there is an edge between a specific pair of vertices). It turns out that, by making a number of queries that is independent of the size of the graph, one may obtain meaningful information regarding its "distance" to being bi-partite. Specifically:

Theorem 5.1 [14]: *There exists a randomized algorithm that, on input a parameter ε and access to a graph $G = (S, R)$, makes* $\mathrm{poly}(1/\varepsilon)$ *queries to G and satisfies the following two conditions:*

1. *If G is bi-partite, then the algorithm accepts with probability 1.*
2. *If any partition of S into two subsets S_1 and S_2 has at least $\varepsilon|S|^2$ edges with both end-points in the same S_i, then the algorithm rejects with probability at least 99%.*

The algorithm underlying Theorem 5.1 uniformly selects $m = \mathrm{poly}(1/\varepsilon)$ vertices, and checks whether the induced graph is bi-partite; that is, for a sample of vertices $v_1, ..., v_m$, it checks whether there exists a partition of $\{v_1, ..., v_m\}$ into two subsets V_1 and V_2 such that for every $i \in \{1, 2\}$ and every $u, v \in V_i$ it holds that $(u, v) \notin R$.

We stress that the said algorithm does not solve the question of whether or not the graph is bi-partite, but rather a relaxed (or approximated) version of this question in which one needs to *distinguish graphs that are bi-partite from graphs that a very far from being bi-partite*. This phenomenon is analogous to the case of approximating the average value of $\mu : S \to [0, 1]$. Also, as in the case of approximating the average value of $\mu : S \to [0, 1]$, it is essential that the approximation algorithm be randomized. A similar phenomena occurs with respect to several other natural properties of graphs, but is not generic. That is, there exist graph properties for which even inspecting a constant fraction of the graph does not allow for an approximate decision regarding satisfiability of the property. For details, the interested reader is directed to [12,24].

We note that the notion of approximation underlying Theorem 5.1 refers to disregarding $\varepsilon|S|^2$ edges, where $|S|^2$ is the maximum possible number of edges over S. This notion of approximation is appealing in the case that R is dense (i.e., contains a constant fraction of all possible edges). Going to the other extreme, we may consider the case that R contains only a linear (in $|S|$) number of edge, or even the case that each vertex participates only in a constant number of edges. In this case, we may want to distinguish the case that the graph is bi-partite from the case that any partition of S into two subsets S_1 and S_2 has at least $\varepsilon|S|$ edges with both end-points in the same S_i. It turns out that this problem can be solved by an algorithm that makes $\text{poly}((\log|S|)/\varepsilon) \cdot \sqrt{|S|}$ queries (to the incidence lists of the graph), and that these many queries are essentially necessary. We note that this sub-linear time algorithm operates by inspecting a graph induced by $\text{poly}((\log|S|)/\varepsilon) \cdot \sqrt{|S|}$ vertices that are selected by taking many (relatively short) random walks from few randomly selected starting vertices. For details, the interested reader is directed to [12, Sec. 3.2].

The aforementioned type of approximation is known by the name *property testing*, and was initiated and developed in [25,14]. One archetypical problem, which played a central role in the construction of PCP systems (see Section 3.3), is distinguishing low-degree polynomials from functions that are far from any such polynomial. Specifically, let F be a finite field and m, d be integers. Given access to a function $f : F^m \to F$, we wish to make few queries and distinguish the case that f is am m-variate polynomial of total degree d from the case it disagrees with any such polynomial on at least 1% of the domain. It turns out that making $\text{poly}(d)$ random (but dependent) queries to f suffices for making a decision that is correct with high probability.

In general, property testing problems refer to objects that are represented by functions, where these functions determine both the type of queries that can be made to the objects and the distance between objects. The tester is required to accept functions that have some predetermined property (i.e., reside in some predetermined set) and reject any function that is "far" from the set of functions having the property. Distances between functions are defined as the fraction of the domain on which the functions disagree, and the threshold determining what is considered far is presented as a proximity parameter, which is explicitly given to the tester. Thus, property testing is a relaxation of standard decision problems (and it focuses on algorithms that can only read parts of the input).

Definition 5.2 (property testers) *Let Π be a class of functions defined over the domain D. A tester for a property Π is a probabilistic oracle machine T that satisfies the following two conditions:*

1. *The tester accepts each $f \in \Pi$ with probability at least $2/3$; that is, for every $f \in \Pi$ (and every $\varepsilon > 0$), it holds that $\Pr[T^f(\varepsilon) = 1] \geq 2/3$, where $T^f(\varepsilon)$ denotes the decision of T (on input ε) when given oracle access to f.*
2. *Given $\varepsilon > 0$ and oracle access to any f that is ε-far from Π, the tester rejects with probability at least $2/3$; that is, for every $\varepsilon > 0$, if $f : D \to \{0,1\}^*$ is ε-far from Π, then $\Pr[T^f(\varepsilon) = 0] \geq 2/3$, where f is ε-far from Π if, for every $g \in \Pi$, it holds that $|\{e \in D : f(e) \neq g(e)\}| > \varepsilon \cdot |D|$.*

If the tester accepts every function in Π with probability 1, then we say that it has one-sided error; that is, T has one-sided error if for every $f \in \Pi$ and every $\varepsilon > 0$, it holds that $\Pr[T^f(\varepsilon) = 1] = 1$.

Definition 5.2 does not specify the query complexity of the tester, and indeed an oracle machine that queries the entire domain of the function qualifies as a tester (with zero error probability...). Needless to say, we are interested in testers that have significantly lower query complexity. Indeed, Theorem 5.1 asserts the existence of a (one-sided error) tester of complexity $\text{poly}(1/\varepsilon)$ for the set of bi-partite graphs, where graphs are represented by their adjacency matrices.

On the Importance of Representation. The representation of problems' instances is crucial to any study of computation, since the representation determines the type of information that is explicit in the input. This issue becomes much more acute when one is only allowed partial access to the input (i.e., making a number of queries that result in answers that do not fully determine the input). An additional issue, which is unique to property testing, is that the representation may effect the distance measure (i.e., the definition of distances between inputs). This is crucial because property testing problems are defined in terms of this distance measure.

The importance of representation is forcefully demonstrated in the gap between the complexity of testing numerous natural graph properties in two natural representations: the adjacency matrix representation and the incidence lists representation; for details see [12].

Suggestions for Further Reading. A brief introduction to property testing can be found in [11]. For a more comprehensive treatment, the interested reader is directed to [24]. For the special case of testing graph properties, the interested reader is directed to [12].

Acknowledgments. I am grateful to Tamas Rudas for his comments on early versions of this essay.

References

1. Arora, S., Lund, C., Motwani, R., Sudan, M., Szegedy, M.: Proof Verification and Intractability of Approximation Problems. Journal of the ACM 45, 501–555 (1998)
2. Arora, S., Safra, S.: Probabilistic Checkable Proofs: A New Characterization of NP. Journal of the ACM 45, 70–122 (1998)
3. Blum, M., Micali, S.: How to Generate Cryptographically Strong Sequences of Pseudo-Random Bits. SIAM Journal of Computing 13, 850–864 (1984)
4. Cover, T.M., Thomas, G.A.: Elements of Information Theory. John Wiley & Sons, Inc., New-York (1991)
5. Feige, U., Goldwasser, S., Lovász, L., Safra, S., Szegedy, M.: Approximating Clique is almost NP-complete. Journal of the ACM 43, 268–292 (1996)
6. Goldreich, O.: Algorithms and Combinatorics Series. Algorithms and Combinatorics Series, vol. 17. Springer, Heidelberg (1999)
7. Goldreich, O.: Foundation of Cryptography – Basic Tools. Cambridge University Press, Cambridge (2001)
8. Goldreich, O.: Foundation of Cryptography: Basic Applications. Cambridge University Press, Cambridge (2004)

9. Goldreich, O.: Computational Complexity: A Conceptual Perspective. Cambridge University Press, Cambridge (2008)
10. Goldreich, O.: P, NP, and NP-Completeness: The Basics of Computational Complexity. Cambridge University Press, Cambridge (2010)
11. Goldreich, O.: A Brief Introduction to Property Testing. In: Goldreich, O., et al.: Studies in Complexity and Cryptography. LNCS, vol. 6650, pp. 471–475. Springer, Heidelberg (2011)
12. Goldreich, O.: Introduction to Testing Graph Properties. In: Goldreich, O., et al.: Studies in Complexity and Cryptography. LNCS, vol. 6650, pp. 476–512. Springer, Heidelberg (2011)
13. Goldreich, O., Goldwasser, S., Micali, S.: How to Construct Random Functions. Journal of the ACM 33(4), 792–807 (1986)
14. Goldreich, O., Goldwasser, S., Ron, D.: Property testing and its connection to learning and approximation. Journal of the ACM, 653–750 (July 1998)
15. Goldreich, O., Micali, S., Wigderson, A.: Proofs that Yield Nothing but their Validity or All Languages in NP Have Zero-Knowledge Proof Systems. Journal of the ACM 38(3), 691–729 (1991)
16. Goldwasser, S., Micali, S.: Probabilistic Encryption. Journal of Computer and System Science 28(2), 270–299 (1984)
17. Goldwasser, S., Micali, S., Rackoff, C.: The Knowledge Complexity of Interactive Proof Systems. SIAM Journal on Computing 18, 186–208 (1989)
18. Håstad, J., Impagliazzo, R., Levin, L.A., Luby, M.: A Pseudorandom Generator from any One-way Function. SIAM Journal on Computing 28(4), 1364–1396 (1999)
19. Hochbaum, D. (ed.): Approximation Algorithms for NP-hard Problems. PWS (1996)
20. Li, M., Vitanyi, P.: An Introduction to Kolmogorov Complexity and its Applications. Springer, Heidelberg (1993)
21. Lund, C., Fortnow, L., Karloff, H., Nisan, N.: Algebraic Methods for Interactive Proof Systems. Journal of the ACM 39(4), 859–868 (1992)
22. Motwani, R., Raghavan, P.: Randomized Algorithms. Cambridge University Press, Cambridge (1995)
23. Papadimitriou, C.H.: Computational Complexity. Addison Wesley, Reading (1994)
24. Ron, D.: Algorithmic and Analysis Techniques in Property Testing. Foundations and Trends in TCS 5(2), 73–205 (2010)
25. Rubinfeld, R., Sudan, M.: Robust characterization of polynomials with applications to program testing. SIAM Journal on Computing 25(2), 252–271 (1996)
26. Shamir, A.: IP = PSPACE. Journal of the ACM 39(4), 869–877 (1992)
27. Sipser, M.: Introduction to the Theory of Computation. PWS Publishing Company (1997)
28. Yao, A.C.: Theory and Application of Trapdoor Functions. In: 23rd IEEE Symposium on Foundations of Computer Science, pp. 80–91 (1982)

On Security Preserving Reductions – Revised Terminology

Oded Goldreich

Abstract. Many of the results in Modern Cryptography are actually transformations of a basic computational phenomenon (i.e., a basic primitive, tool or assumption) to a more complex phenomenon (i.e., a higher level primitive or application). The transformation is explicit and is always accompanied by an explicit reduction of the violation of the security of the complex phenomenon to the violation of the simpler one. A key aspect is the efficiency of the reduction. We discuss and slightly modify the hierarchy of reductions originally suggested by Leonid Levin.

Keywords: Foundations of Cryptography, Complexity, Reductions.

An early version of this article appeared as *ePrint* Report 2000/001. The current revision is quite minimal.

1 Introduction

Modern Cryptography is concerned with the construction of *efficient* schemes for which it is *infeasible* to violate the security feature. Thus, we need a notion of efficient computations as well as a notion of infeasible ones. The computations of the legitimate users of the scheme ought to be efficient, whereas violating the security features (via an adversary) ought to be infeasible. Our notions of efficient and infeasible computations are "asymptotic" (or rather functional):[1] They refer to the running time as a function of the security parameter. This is done in order to avoid cumbersome formulations that refer to the actual running-time on a specific model for specific values of the security parameter. Still, one can easily derive such specific statements from the asymptotic treatment.

Efficient computations are commonly modeled by computations that are polynomial-time in the security parameter. The polynomial bounding the running-time of the legitimate user's strategy is fixed and typically explicit and small (still in some cases it is indeed a valuable goal to make it even smaller). Here (i.e., when referring to the complexity of the legitimate user) we are in the same situation as in any algorithmic research. Things are different when referring to our assumptions regarding the computational resources of the adversary. A common approach is to postulate that the latter are polynomial-time too, where

---

[1] Actually, the term "asymptotic" is misleading, since from the functional treatment of the running-time (as a function of the security parameter), one can derive statements for *any* value of the security parameter.

O. Goldreich et al.: Studies in Complexity and Cryptography, LNCS 6650, pp. 540–546, 2011.

the polynomial is *not* a-priori specified. In other words, the adversary is restricted to the class of efficient computations and anything beyond this is considered to be infeasible. Although many definitions explicitly refer to this convention, this convention is *inessential* to all known results (in the area). In all cases, a more general (and yet more cumbersome) statement can be made by referring to adversaries of running-time bounded by any function (or class of functions). For example, for any function $T : \mathbb{N} \to \mathbb{N}$ (e.g., $T(n) = 2^{\sqrt[3]{n}}$), we may consider adversaries that on security parameter n run for at most $T(n)$ steps. Doing so we (implicitly) define as infeasible any computation that (on security parameter n) requires more than $T(n)$ steps.

The results obtained in Modern Cryptography are in most cases conditional ones. That is, based on some relatively simple intractability assumptions (e.g., the existence of one-way functions [3, Chap. 2]) one constructs and establishes the security of more complex applications (e.g., unforgeable signature schemes [3, Chap. 6]). In many cases these results are stated in an oversimplified form, where a typical form reads *if the function f cannot be inverted in polynomial-time, then the scheme S_f (which utilizes f) cannot be broken in polynomial-time*. However, what is actually proved in such works is stronger. Typically, the proof of security of S_f specifies, for any function $T : \mathbb{N} \to \mathbb{N}$, a function $T' : \mathbb{N} \to \mathbb{N}$ such that *if f cannot be inverted on n-bit images in time $T(n)$, then S_f cannot be broken on inputs of length m in time $T'(m)$*. Furthermore, typically, the relation between T' and T takes the form

$$T'(m) = \frac{p_2^{-1}(T(p_1^{-1}(m)))}{p_3(m)} , \qquad (1)$$

where p_1, p_2, p_3 are some fixed polynomials. Such a relation results from the fact that the proof utilizes a reduction of the task of inverting f on strings of length n to the task of breaking S_f on strings of length $p_1(n)$. Thus, assuming on the contrary to the security claim that S_f can be broken in time $T'(m)$ on inputs of length $m = p_1(n)$, one obtains an algorithm inverting f on inputs of length n in time $T(n) \leq p_3(p_1(n)) \cdot p_2(T'(p_1(n)))$.

It should be clear (and it is indeed well-known) that the aforementioned relation between T and T' determines the strength of the theoretical result as well as its potential practical applicability. Specifically, in almost all the cases the relation takes the form of Eq. (1), and in these cases one is interested in the specific polynomials p_1, p_2, p_3.

The purpose of this note is to discuss a popular classification of such reductions, attributed to Leonid Levin and presented in [12]. We suggest to modify this classification a little.

2   Preliminaries

Actually, the foregoing discussion is over-simplified, because it refers only to the running-time of the violating algorithms (and implicitly suggesting that we talk of algorithms that succeed always or almost always). In many cases, the statements are more complex, referring both to the running-time of algorithms and to a (probabilistic) measure of success. Two such common measures are

1. The success probability of easily verified events. For example, the success probability of an inverting algorithm (for a specific one-way function), or the success probability of a forging algorithm (for a signature scheme).
2. The gap in probability between two experiments. An archetypical example is the notion of computational indistinguishability. Here, for two distributions ensembles, $\{X_n\}$ and $\{Y_n\}$, we consider the gap between the probability that an algorithm A outputs 1 on input X_n and the probability A does so on input Y_n. Thus, definitions such as security of encryption schemes [7], pseudorandomness [1,13,4], and (computational) zero-knowledge [8] fall into this category.

The distinction between the foregoing two types is crucial for Levin's suggestion to incorporate the running-time and the success measure into a single measure (see Section 2.2). Note that in order to succeed with probability at least 2/3 in an attempt of the first type one has to repeat trying for $\Theta(1/\epsilon(n))$ times, where $\epsilon(n)$ is the success probability in a single attempt. On the other hand, in order to amplify a distinguishing gap of $\epsilon(n)$ into a gap of 2/3 we need to repeat the experiment(s) for $\Theta(1/\epsilon(n)^2)$ times.[2]

2.1   The General Form of Security Reductions

Before presenting Levin's approach, let us present the general form that most results take. Typically, one starts with a basic primitive, denoted f (for sake of uniformity with the Introduction), and constructs a scheme S_f. (Each of the two is coupled with its own notion of violation, determining the measure of success.) The proof of security of S_f is by a reduction to violation of security of f. That is, such a proof shows, for any $t' : \mathbb{N} \to \mathbb{N}$ and $e' : \mathbb{N} \to \mathbb{R}$, how to convert an algorithm violating S_f with time complexity t' and success measure e' into an algorithm for violating f with time complexity t and success measure e. Calling the former an S_f-violator and the latter an f-violator, the conversion is by a reduction that typically specifies polynomials $p_1, p_2, ..., p_7$ such that on input of length n the f-violator invokes the S_f-violator on inputs of length $m = p_1(n)$, and satisfies $t(n) = p_2(t'(m)) \cdot p_3(1/e'(m)) \cdot p_4(m)$ as well as $e(n) = p_5(e'(m)) \cdot p_6(1/t(m)) \cdot p_7(1/m)$. It follows that, for any function $T:\mathbb{N}\to\mathbb{N}$ and $\epsilon : \mathbb{N} \to \mathbb{R}$, if f cannot be violated on n-bit inputs in time $T(n)$ with success measure $\epsilon(n)$, then S_f cannot be violated on m-bit inputs in time $T'(m)$ with success measure $\epsilon'(m)$, where T' and ϵ' may be any pair of functions satisfying

$$T(p_1^{-1}(m)) = p_2(T'(m)) \cdot p_3(1/\epsilon'(m)) \cdot p_4(m) \tag{2}$$

$$\epsilon(p_1^{-1}(m)) = \frac{p_5(\epsilon'(m))}{p_6(T(m)) \cdot p_7(m)} \tag{3}$$

---

[2] The above discussion refers to an abstract experiment (or pair of experiments). When applied to the examples given above, repeating the experiment means things like inverting a one-way function on one of several independently selected images, or distinguishing between multiple samples of two ensembles.

where $p_1, p_2, ..., p_7$ are the polynomials specified above. (Assuming, on the contrary, that S_f can be violated on m-bit inputs in time $T'(m)$ with success measure $\epsilon'(m)$, implies – via the reduction – violation of f on n-bit inputs in time $T(n)$ with success measure $\epsilon(n)$.)

2.2 Levin's Notion of Work

In order to simplify treatments as above, Levin suggested to incorporate the running-time and the success-measure of each violating algorithm into a single measure called work. The foregoing distinction between easily verifiable and non-verifiable success measures is crucial to his suggestion. For a verifiable success measure, the work of an algorithm A with running-time $t_A : \mathbb{N} \to \mathbb{N}$ and success measure $\epsilon_A : \mathbb{N} \to \mathbb{R}$ is defined as $w_A(n) \stackrel{\text{def}}{=} t_A(n)/\epsilon_A(n)$. For a (non-verifiable) success measure of the gap type, the work of an algorithm A with running-time $t_A : \mathbb{N} \to \mathbb{N}$ and success measure $\epsilon_A : \mathbb{N} \to \mathbb{R}$ is defined as $w_A(n) \stackrel{\text{def}}{=} t_A(n)/\epsilon_A^2(n)$. (We stress that the definition of work is problem specific and ad-hoc in nature.)[3]

In the sequel, we shall adopt Levin's simplification. A reader feeling uncomfortable with this choice, may consider only algorithms with constant success measure (in which case work is identical to time (up-to a constant factor)). Security will be defined as a (possibly postulated) lower bound on the work of violating algorithms. For example, one may assume that the security of factoring is $\exp(n^{1/4})$, and infer (based on this assumption)[4] that pseudorandom generators of security $\exp(n^{1/4})$ exist.

Definition (security): *Let Π be some primitive with an associated notion of violation that specifies a notion of success measure and induces a notion of work of violating algorithms. We say that Π has security $S : \mathbb{N} \to \mathbb{N}$ if any algorithm A violating Π has work function that grows faster than S.*

3 Levin's Hierarchy of Reductions (Revisited)

In order to demonstrate the different quality of certain reductions, Levin has suggested three types of reductions, which were later canonized in Luby's book [12]. Letting $S : \mathbb{N} \to \mathbb{N}$ denote the security of the basic primitive, and $S' : \mathbb{N} \to \mathbb{N}$ the security of the complex primitive constructed from the former, the three types of reductions are:

(L1) A reduction is linearly preserving if it guarantees $S'(n) \geq S(n)/\text{poly}(n)$.
(L2) A reduction is polynomially-preserving if it guarantees $S'(n) \geq S(n)^e/\text{poly}(n)$, for some constant $e > 0$.

---

[3] The abstract discussion above does not fully justify the definition (see Footnote 2). Furthermore, other functionalities of running-time and success-measure may make sense too.

[4] See [2, Sec. 3.4].

(L3) A reduction is weakly-preserving if it guarantees $S'(n) \geq S(n^d)^e/\text{poly}(n)$, for some constants $d, e > 0$.

Levin has noted that, for nicely-behaved security measures, a reduction that guarantees $S'(n) \geq S(n/d)^e/\text{poly}(n)$, for some constants $d, e > 0$, is also polynomially-preserving. The argument is based on the fact that in our context all primitives are breakable within exponential time (i.e., time 2^n on input length n), and so one may assume without loss of generality that $S(n) \leq 2^n$. Furthermore, for "nicely-behaved" functions S, which are exponentially bounds, and for $c > 1$ one may expect that $S(cm) \leq S(m)^c$ holds. Thus, $S'(n) \geq S(n/d)^e/\text{poly}(n) \geq S(n)^{ed}/\text{poly}(n)$. Still, it seems inappropriate to identify the effect of e and d in a guarantee such as the foregoing (L2). Furthermore, when doing so, we lose an important distinction, which is represented in the gap between the following Types (T2) and (T3).

3.1   The Revised Hierarchy

(T1) A reduction is strongly preserving if it guarantees $S'(n) \geq S(n)/\text{poly}(n)$.
(This is identical to (L1) above.)

(T2) A reduction is linearly-preserving if, for some constant $c \geq 1$, it guarantees

$$S'(n) \geq \frac{S(n/c)}{\text{poly}(n)}$$

(This extends (T1), where $c = 1$, in an important way.)

(T3) A reduction is polynomially-preserving if, for some constants $c \geq 1$ and $e > 0$, it guarantees

$$S'(n) \geq \frac{S(n/c)^e}{\text{poly}(n)}$$

(Formally, (T3) extends (L2), where $c = 1$; but, for "nicely behaved security measures" (see the foregoing discussion), type (T3) is equivalent to type (L2).)

(T4) A reduction is weakly-preserving if, for some constants $c, d, e > 0$, it guarantees

$$S'(n) \geq \frac{S(cn^d)^e}{\text{poly}(n)}$$

(This is equivalent to (L3) above.)

Thus, we replace (L2) by the two distinct categories (T2) and (T3).

3.2   Discussion

On the relation between (T2), (T3) and (L2). Levin's category (L2) is a special case of our (T3). In light of the discussion about, we believe that Levin himself would not care much about the extension of (L2) to (T3). In contrast, we believe that the distinction between Types (T2) and (T3) is very important.

We note that many claims made by Luby [12] regarding (L2) actually refer to either (T2) or (T3), and are valid for (L2) only under the above assumption (i.e., $S(cn) \leq S(n)^c$, for every constant $c > 1$) which collapses (T3) into (L2). Furthermore, when referring to (L2) one losses the important distinction between Types (T2) and (T3). These considerations are examplified by considering the following results.

- *A hard-core predicate for any one-way function* [6]: The original reduction of [6] (as well as the better known alternative reduction (as presented in [2, §2.5.2.1–3])) is of Type (T3).[5] In contrast, the improved reduction of Levin [11] (see also [2, §2.5.2.4]) is of Type (T2).
- *Security-preserving amplification of one-way function* [5]: The reduction demonstrating this result for the case of one-way permutations is of Type (T2). In contrast, the known reduction (of [5]) for the case of regular one-way functions is only of Type (T3), for some range of parameters.[6]

Thus, the distinctions between the strengths of the aforementioned pairs of results are reflected in the distinction between (T2) and (T3), but are not reflected by Levin's Hierarchy (since these results are all of type (L2)). We chose these examples because they are famous cases in which the entire point of the corresponding work is obtaining an improvement in quality of reductions among the studied primitives. Thus, the distinction between (T2) and (T3) is essential for making the point (as demonstrated above).

Beyond (T4). With the exception of a single case, all results we are aware of (in the field) are proven by a reduction of Type (T4), or lower. The only exception is Levin's observation regarding the existence of a *universal one-way function* (cf., [10] and [2, Sec. 2.4.1]).

A final Warning. It should be clear that the above classification (as well as the one suggested in [12]) is ad-hoc in nature. Namely, it only represents our knowledge of the current reductions, and an attempt to classify them in a way that reflects their theoretical strength and practical applicability. Each type may be further refined according to the constants (and/or polynomials) appearing in its definition. Furthermore, in some cases (depending on such refinements), a reduction with higher type may be preferable (in practice) to one with lower type (e.g.,. $2^{\sqrt{n}} < n^{100}$ for $n < 10^6$).

---

[5] The claim in [12] by which the reduction is of type (L2) is correct only for "nicely behaved security measures" (see foregoing discussion).

[6] Actually, in the regular case, the construction in [5] depends on the security of the basic (weak) one-way function, and so we have a family of reductions one per each security function S (which needs to be efficiently computable). These reductions are of Type (T3), provided that, for some $d < 1$, $S(n) < 2^{n^d}$. Otherwise they are only of Type (T4).

An Out of Scope Comment: As discussed in Footnote 6, some results are proven by a construction that depend on the security of the basic scheme; that is, for every security function S, a different construction of a complex primitive is presented (assuming that the basic one has security S). One should prefer results proven via a single construction, which is oblivious of the security of the basic scheme. The security of the resulting construct will depend on the security of the basic one, but the latter need not be known a-priori. In practical terms this means that one may make a weak assumption regarding the basic scheme such that this assumption guarantees sufficient security for the construct. If the basic scheme turns out to be more secure than originally assumed then the resulting construct will benefit in security (as per the security guarantee given with the reduction). In contrast, when the construction depends on the assumed security, better than postulated security of the basic scheme may not translate to better security of the construct.

Acknowledgments. We are grateful to Mihir Bellare for helpful comments.

References

1. Blum, M., Micali, S.: How to Generate Cryptographically Strong Sequences of Pseudo-Random Bits. SICOMP 13, 850–864 (1982); Preliminary version in 23rd FOCS (1982)
2. Goldreich, O.: Foundation of Cryptography: Basic Tools. Cambridge University Press, Cambridge (2001)
3. Goldreich, O.: Foundation of Cryptography: Basic Applications. Cambridge University Press, Cambridge (2004)
4. Goldreich, O., Goldwasser, S., Micali, S.: How to Construct Random Functions. JACM 33(4), 792–807 (1986)
5. Goldreich, O., Impagliazzo, R., Levin, L.A., Venkatesan, R., Zuckerman, D.: Security Preserving Amplification of Hardness. In: 31st FOCS, pp. 318–326 (1990)
6. Goldreich, O., Levin, L.A.: Hard-core Predicates for any One-Way Function. In: 21st STOC, pp. 25–32 (1989)
7. Goldwasser, S., Micali, S.: Probabilistic Encryption. JCSS 28(2), 270–299 (1984); Preliminary version in 14th STOC (1982)
8. Goldwasser, S., Micali, S., Rackoff, C.: The Knowledge Complexity of Interactive Proof Systems. SICOMP 18, 186–208 (1989); Preliminary version in 17th STOC (1985) Earlier versions date to 1982
9. Goldwasser, S., Micali, S., Rivest, R.L.: A Digital Signature Scheme Secure Against Adaptive Chosen-Message Attacks. In: SICOMP, pp. 281–308 (April 1988)
10. Levin, L.A.: One-Way Function and Pseudorandom Generators. Combinatorica 7, 357–363 (1987)
11. Levin, L.A.: Randomness and Non-determinism. J. Symb. Logic 58(3), 1102–1103 (1993)
12. Luby, M.: Pseudorandomness and Cryptographic Applications. Princeton University Press, Princeton (1996)
13. Yao, A.C.: Theory and Application of Trapdoor Functions. In: 23rd FOCS, pp. 80–91 (1982)

Contemplations on Testing Graph Properties

Oded Goldreich

Abstract. This article documents two programmatic comments regarding testing graph properties, which I made during the Dagstuhl workshop on Sublinear-Time Algorithms (July 2005). The first comment advocates paying more attention to the dependence of the tester's complexity on the proximity parameter. The second comment advocates paying more attention to the question of testing general graphs (rather than dense or bounded-degree ones). In addition, this article includes a suggestion to view and discuss property testing within the framework of promise problems.

Keywords: Property testing, graph properties.

An early version of this memo appeared on the author's webpage (in July 2005). The current revision is intentionally minimal. For a review of some subsequent developements, the interested reader is directed to [15].

Preliminaries. We assume that the reader is familiar with the basic models and results regarding testing graph properties (see surveys [15, 28]).

1   Complexity as a Function of the Proximity Parameter

It is indeed an amazing fact that many properties can be tested within (query) complexity that only depends on the proximity parameter (rather than also on the size of the object being tested). This amazing statement seems to put in shadow the question of what is the form of the aforementioned dependence, and blurs the difference between a reasonable dependence (e.g., a polynomial relation) and prohibiting one (e.g., a tower-function relation). We claim that, as in the context of standard approximation problems (cf. [25]), the dependence of the complexity on the approximation (or proximity) parameter is a key issue.

For the sake of simplicity we focus on the *query complexity* of testers, and assume that it only depends on the proximity parameter ϵ. We highlight the difference between the following types of dependencies, where ϵ-testing refers to distinguishing between objects that have the property and objects that are ϵ-far from having the property:

1. The query complexity is linearly related to the proximity parameter; that is, ϵ-testing can be achieved by using $O(1/\epsilon)$ queries.
   We note that, for any non-trivial graph property, the query complexity of ϵ-testing in the adjacency matrix model is $\Omega(1/\epsilon)$ (see [22, Prop. 6.1]).[1] The same lower-bound holds in the bounded-degree incidence list model.

---

[1] This result was conjectured in the earlier version, which mentioned the weaker lower bound of $\Omega(\sqrt{1/\epsilon})$, which is easier to establish.

O. Goldreich et al.: Studies in Complexity and Cryptography, LNCS 6650, pp. 547–554, 2011.
© Springer-Verlag Berlin Heidelberg 2011

2. The query complexity is polynomially related to the proximity parameter; that is, ϵ-testing can be achieved by using poly$(1/\epsilon)$ queries.

For example, all graph property testers in [17] have query complexity poly$(1/\epsilon)$. We note, however, that some of these testers (e.g., the one for 3-colorability) have time complexity that is exponential in the proximity parameter ϵ, and this seems unavoidable assuming that $\mathcal{NP}$ does not have sub-exponential time algorithms. We wish to praise [4, 10] for further studying the query complexity of testing k-colorability, and in particular for determining the query complexity of non-adaptively testing bipartiteness (up to a polylogarithmic factor).[2]

A natural problem that is ϵ-testable within query complexity that only depends on ϵ, but requires a super-polynomial dependency on $1/\epsilon$ was pointed out by Alon [1]. He proved that ϵ-testing triangle-freeness requires at least $(1/\epsilon)^{\Omega(\log(1/\epsilon))}$ queries. We comment that this is quite far from the known upper-bound (mentioned in Item 4).

3. The query complexity is exponentially related to the proximity parameter; that is, ϵ-testing can be achieved by using $\exp(1/\epsilon)$ queries.

We are not aware of any natural testing problem having this query complexity.

4. The query complexity is related to the proximity parameter via a function that grows tremendously fast. A notorious example is the tower function tf defined inductively by $\mathrm{tf}(n) = \exp(\mathrm{tf}(n-1))$ with $\mathrm{tf}(1) = 2$. (Indeed, tf is the inverse of the $\log^*$ function.)

Starting in [2], many positive results regarding testing graph properties in the adjacency matrix model establish such a query complexity; that is, they establish ϵ-testers of query complexity $\mathrm{tf}(\mathrm{poly}(1/\epsilon))$ (and sometimes even a tower of towers). In particular, ϵ-testing triangle freeness is known only when using $\mathrm{tf}(\mathrm{poly}(1/\epsilon))$ queries. This dependence is an artifact of these results' application of the Regularity Lemma (or stronger variants of it).

We wish to stress that we do value the impressive results of [2, 5, 6, 14], which refer to graph testers having query complexity that is independent of the graph size but depend prohibitively on the proximity parameter. We view such results as an impressive first step, which called for further investigation directed at determining the actual dependency of the complexity on the proximity parameter.

Addendum (2006): Recently, Alon *et. al.* [3] established a combinatorial characterization of graph properties that are testable using a number of queries that is independent of the graph size. This characterization provides a seemingly ultimate answer to the qualitative question of which graph properties are testable within query complexity that is independent of the graph size and sets the stage for a quantitative study of the said query complexity.

---

[2] Alon and Krivelevich [4] presented an ϵ-tester that inspects the subgraph induced by $\widetilde{O}(1/\epsilon)$ randomly chosen vertices (thus making $\widetilde{O}(1/\epsilon^2)$ non-adaptive queries), whereas Bogdanov and Trevisan [10] prove a lower-bound of $\Omega(1/\epsilon^2)$ non-adaptive queries (and $\Omega(1/\epsilon^{3/2})$ adaptive queries).

2 Models of Testing Graph Properties

The bulk of algorithmic research regarding graphs refers to general graphs. Of special interest are graphs that are neither very dense nor have a bounded-degree. In contrast, research in testing properties of graphs started (in [17]) with the study of dense graphs, next (starting in [18]) bounded-degree graphs were considered, and general graphs were considered only in [26, 27]. This evolution has historical reasons to be reviewed first.

Testing graph properties was initially conceived (in [17]) as a special case of the general framework of testing properties of functions (cf. [29]). Thus, graphs had to be represented by functions, and two standard representations of graphs seemed most fitting in this context:

1. The *adjacency matrix representation* [17]: That is, a graph $G = (V, E)$ is represented by a function $g : V \times V \to \{0,1\}$ such that $g(u,v) = 1$ if and only if $\{u,v\} \in E$. This representation corresponds to the so-called *adjacency queries*, and suggests that the (relative) distance between graphs be measured as the fraction of vertex-pairs on which the corresponding adjacency matrices differ.

   Needless to say, when considering ϵ-testing, this model is interesting mostly for ϵ-dense graphs (i.e., graphs $G = (V, E)$ such that $2|E| > \epsilon|V|^2$).

   (A partial list of the works done in this model includes [1, 2, 4–7, 10, 14, 17, 23].)

2. The (bounded-degree) *incidence list representation* [18]: Specifically, for a fixed integer d, the graph $G = (V, E)$ is represented by a function $g : V \times [d] \to V \cup \{\bot\}$ such that $g(u,i) = v$ if v is the ith neighbor of u and $g(u,i) = \bot$ if u has less than i neighbors. This representation corresponds to the so-called *neighbor queries*, and suggests that the (relative) distance between graphs be measured as the fraction of vertex-index pairs on which the corresponding incidence lists differ.

   Needless to say, this model can only be applied to graphs of *maximum* degree d. Indeed, we may take d to be arbitrary large, but in this case the model is interesting mostly for ϵ-testing graphs having *average degree* at least ϵd.

   (Work done in this model includes [9, 18–20].)

The reader may note that both models were formulated in a way that identifies the graphs with a specific functional representation, which in turn defines the type of queries allowed to the tester as well as the notion of fractional distance (which underlies the performance guarantee).

The identification of graphs with any specific functional representation was abandoned by Parnas and Ron [27] who developed a more general model by decoupling the type of queries allowed to the tester from the distance measure: Whatever is the mechanism of accessing the graph, the distance between graphs is defined as the number of edges in their symmetric difference (rather than the number of different entries with respect to some specific functional representation). Furthermore, the relative distance may be defined as the size of the symmetric difference divided by the actual (total) number of edges in both graphs (rather than divided by some (possibly non-tight) upper-bound on the latter quantity). As advocated by Kaufman *et. al.* [26], it may be reasonable to allow the tester to perform both adjacency and neighbor queries (and indeed each type of query may be useful in a different range of edge densities). Needless to say, this model seems adequate for the study of testing properties of arbitrary graphs, and it strictly generalizes the positive aspects of the two prior models (i.e., the models based on the adjacency matrix and bounded-degree incidence list representations).

We wish to advocate further study of the latter model. We believe that this model, which allows for a meaningful treatment of property testing of general graphs, is the one that is most relevant to computer science applications. Furthermore, it seems that designing testers in this model requires the development of algorithmic techniques that may be applicable also in other areas of algorithmic research. As an example, we mention that techniques in [26] that underly the average degree approximation of [21]. (Likewise techniques of [18] underly the minimum spanning tree weight approximation of [11]; indeed, as noted next, the bonuded-degree incidence list model is also more algorithmic oriented than the adjacency matrix model.)

Let us focus on the algorithmic contents of property testing of graphs. We first note that, ignoring a quadratic blow-up in the query complexity, property testing in the adjacency matrix representation reduces to sheer combinatorics: To ϵ-test if G has property P, it suffices to check whether a random induced graph (of adequate size) of G has some "related" property P' (see [23]).[3] In contrast, property testing in the incidence list representation employs some non-trivial algorithmic techniques such local search (cf. [18]) and random walks (cf. [19]). Testers in the general ("flexible") graph models seem to require even more algorithmic ideas (cf. [26]).

To summarize, we advocate further study of the model of [26, 27] for two reasons. The first reason is that we believe in the greater relevance of this model to computer science applications. The second reason is that we believe in the

---

[3] In the original version of this article we wrote:

> We mention that the transformation of [23] may increase the query complexity in a quadratic manner. It is conceivable that an adaptive tester (which is less dull from an algorithmic perspective) may perform better than the canonical tester of [23] (which merely examines a random induced subgraph).

In fact, this possibility has subsequently materialized in [24] and [22]. In view of these works, we feel that the foregoing sentiments (i.e., "reduction to sheer combinatorics") were somewhat overstated. As argued in Section 1, one should not dismiss the exact dependence of the complexity on ϵ, and it seems that algorithmic ideas can contribute to reducing the complexity by a constant power. We do not recall our thoughts regarding this matter in July 2005, but it seems that we did not really believe in the possibility that we mentioned as conceivable (i.e., that an adaptive tester may perform better than a canonical tester), let alone when natural graph properties were concerned.

greater potential of this model to have cross fertilization with other branches of algorithmic research.

A Parenthetical Comment: We seize the opportunity to call attention also to the study of testing properties of directed graphs, initiated in [8].

3   Property Testing as a Type of a Promise Problem

We advocate viewing property testing within the framework of promise problems, a framework introduced in [12] (see relatively recent survey [16]). Formally, a promise problem is a partition of the set of all strings into three subsets:

1. The set of strings representing YES-instances.
2. The set of strings representing NO-instances.
3. The set of disallowed strings (which represent neither YES-instances nor NO-instances).

The algorithm (or process) that is supposed to solve the promise problem is required to distinguish YES-instances from NO-instances, and is allowed arbitrary behavior on inputs that are neither YES-instances nor NO-instances. Intuitively, this algorithm (or rather its designer) is "promised" that the input is either a YES-instance or a NO-instance, and is only required to distinguish these two cases. This generalizes the standard notion of a decision problem, in which each string is either a YES-instances or a NO-instance.

Gap problems constitute a special type of promise problems in which instances are partitioned according to some metric leaving a "gap" between YES-instances and NO-instances. Standard approximation problems refer to one such type of metric in which instances are positioned according to the value of the best corresponding "solution" (with respect to some predetermined objective function). Property testing refer to a second type of metric in which instances are positioned according to their distance from the set of objects that satisfy some predetermined property.

Indeed, property testing is a relaxation of decision problems, where this relaxation leaves a gap between instances that should be accepted (with high probability) and instances that should be rejected (with high probability). The former contain all instances that have the predetermined property, whereas the latter contain all instances that are "far from having the property" (i.e., being at large distance from any instance in the former set). Typically, the distance (or proximity) parameter is given as input to the tester, which makes the positive results stronger and more appealing (especially in light of a separation showed in [7]). In contrast, negative results typically refer to a fixed value of the distance parameter.

Thus, for any property P and any *distance function* (e.g., Hamming distance between bit strings), two natural types of promise problems emerge:

1. *Testing w.r.t variable distance*: Here instances are pairs (x, δ), where x is a description of an object and δ is a distance parameter. The YES-instances are pairs (x, δ) such that x has property P, whereas (x, δ) is a NO-instance if x is δ-far from any x' that has property P.

2. *Testing w.r.t a fixed distance*: Here we fix the distance parameter δ, and so the instances are merely descriptions of objects, and the partition to YES and NO instances is as above.

For example, for some fixed integer d, consider the following promise problem, denoted BPG_d, regarding bipartiteness of bounded-degree graphs. The YES-instance are pairs (G, δ) such that G is a bipartite graph of maximum degree d, whereas (G, δ) is a NO-instance if G is an N-vertex graph of maximum degree d such that more than $\delta \cdot dN/2$ edges must be omitted from G in order to obtain a bipartite graph. Similarly, for fixed integer d and $\delta > 0$, the promise problem $\text{BPG}_{d,\delta}$ has YES-instances that are bipartite graphs of maximum degree d and NO-instances that are N-vertex graphs of maximum degree d such that more than $\delta \cdot dN/2$ edges must be omitted from the graph in order to obtain a bipartite graph. In [18] it was shown that any tester for $\text{BPG}_{3,0.01}$ must make $\Omega(\sqrt{N})$ neighbor queries. In contrast, for every d and δ, the tester presented in [19] decides $\text{BPG}_{d,\delta}$ in time $\widetilde{O}(\sqrt{N}/\text{poly}(\delta))$. In fact, this algorithm decides BPG_d in time $\widetilde{O}(\sqrt{N}/\text{poly}(\delta))$, where N and δ are explicitly given parameters.

The Formulation Typically Used in the Literature. Indeed, all research on property testing refers to the two aforementioned types of promise problems, where typically positive results refer to the first type and negative results refer to the second type. However, most works do not provide a strictly formal statement of their results (see further discussion next), because the formulation is rather cumbersome and straightforward. Furthermore, in light of the greater focus on positive results (and in accordance with the traditions of algorithmic research), such a formal statement is believed to be unnecessary.[4] Let us consider what is required for a formal statement of property testing results.

- The starting point is a specification of a property and a distance function, *the combination of which yields a promise problem* (of the first type). [Needless to say, this starting point is common to all property testing work, but the fact that it constitutes a promise problem is rarely stated.]
- The first step is to postulate that the potential "solvers" (i.e., property testers) are probabilistic oracle machines that are given oracle access to the "primary" input (i.e., the object in the aforementioned problem types). [Indeed, this step need to be taken and is taken in all works in the area.]

---

[4] Needless to say, a higher level of rigor is typically required in negative statements. Indeed, property testing is positioned between algorithmic research and complexity theory, and seems to be more influenced by the mind-frame of algorithmic research. (We comment that the positioning of a discipline is determined both by its contents and by sociology-of-science factors.).

- Secondly, for a formal asymptotic complexity statement, one needs to specify the "secondary" (explicit) inputs, which consist of various problem-dependent parameters (e.g., N and d in the foregoing examples) and the distance parameter δ (in case of BPG$_d$ and any other problem of the first aforementioned type).
  [This step is rarely done explicitly in the literature. The importance of this step is highlighted in [6, 7], which explicitly distinguish testers that decide obliviously of N from general testers the decision of which may depend on N even in case their query complexity is independent of N.]
- Finally, one should state the complexity of the tester in terms of these explicit inputs.
  [Needless to say, this is always done...]

Thus, the only step that is acutely missing in typical works is a rigorous definition of the relevant explicit inputs (especially, the various problem-dependent parameters). Regardless of whether or not one explicitly uses the promise problem formulation, we suggest to be more careful about specifying the (problem-dependent) explicit inputs given to the tester.

References

1. Alon, N.: Testing subgraphs of large graphs. Random Structures and Algorithms 21, 359–370 (2002)
2. Alon, N.V., Fischer, E.V., Krivelevich, M.V., Szegedy, M.V.: Efficient Testing of Large Graphs. Combinatorica 20, 451–476 (2000)
3. Alon, N.V., Fischer, E.V., Newman, I., Shapira, A.: A Combinatorial Characterization of the Testable Graph Properties: It's All About Regularity. In: 38th STOC, pp. 251–260 (2006)
4. Alon, N.V., Krivelevich, M.V.: Testing k-Colorability. SIAM Journal on Disc. Math. 15(2), 211–227 (2002)
5. Alon, N., Shapira, A.: Every Monotone Graph Property is Testable. In: 37th STOC, pp. 128–137 (2005)
6. Alon, N., Shapira, A.: A Characterization of the (natural) Graph Properties Testable with One-Sided. In: 46th FOCS (2005) (to appear)
7. Alon, N., Shapira, A.: A Separation Theorem in Property Testing (2004) (unpublished manuscript)
8. Bender, M.V., Ron, D.V.: Testing acyclicity of directed graphs in sublinear time. Random Structures and Algorithms, 184–205 (2002)
9. Bogdanov, A., Obata, K., Trevisan, L.: A lower bound for testing 3-colorability in bounded-degree graphs. In: 43rd FOCS, pp. 93–102 (2002)
10. Bogdanov, A., Trevisan, L.: Lower Bounds for Testing Bipartiteness in Dense Graphs. In: IEEE Conference on Computational Complexity, pp. 75–81 (2004)
11. Chazelle, B., Rubinfeld, R., Trevisan, L.: Approximating the minimum spanning tree weight in sublinear time. In: Yu, Y., Spirakis, P.G., van Leeuwen, J. (eds.) ICALP 2001. LNCS, vol. 2076, pp. 190–200. Springer, Heidelberg (2001)
12. Even, S., Selman, A.L., Yacobi, Y.: The Complexity of Promise Problems with Applications to Public-Key Cryptography. Inform. and Control 61, 159–173 (1984)

13. Fischer, E.: The art of uninformed decisions: A primer to property testing. Bulletin of the European Association for Theoretical Computer Science 75, 97–126 (2001)
14. Fischer, E., Newman, I.: Testing versus estimation of graph properties. In: 37th STOC, pp. 138–146 (2005)
15. Goldreich, O.: Introduction to Testing Graph Properties. In: Goldreich, O., et al.: Studies in Complexity and Cryptography. LNCS, vol. 6650, pp. 549–556. Springer, Heidelberg (2011)
16. Goldreich, O.: On Promise Problems. In: memory of Shimon Even (1935–2004). ECCC, TR05-018 (January 2005)
17. Goldreich, O., Goldwasser, S., Ron, D.: Property testing and its connection to learning and approximation. Journal of the ACM, 653–750 (July 1998)
18. Goldreich, O., Ron, D.: Property testing in bounded degree graphs. Algorithmica, 302–343 (2002)
19. Goldreich, O., Ron, D.: A sublinear bipartite tester for bounded degree graphs. Combinatorica 19(3), 335–373 (1999)
20. Goldreich, O., Ron, D.: On Testing Expansion in Bounded-Degree Graphs. In: ECCC, TR00-020 (March 2000)
21. Goldreich, O., Ron, D.: Approximating Average Parameters of Graphs. In: ECCC, TR05-073 (July 2005)
22. Goldreich, O., Ron, D.: Algorithmic Aspects of Property Testing in the Dense Graphs Model. In: ECCC, TR08-039 (2008)
23. Goldreich, O., Trevisan, L.: Three theorems regarding testing graph properties. Random Structures and Algorithms 23(1), 23–57 (2003)
24. Gonen, M., Ron, D.: On the Benefits of Adaptivity in Property Testing of Dense Graphs. In: Charikar, M., Jansen, K., Reingold, O., Rolim, J.D.P. (eds.) RANDOM 2007 and APPROX 2007. LNCS, vol. 4627, pp. 525–539. Springer, Heidelberg (2007)
25. Hochbaum, D. (ed.): Approximation Algorithms for NP-Hard Problems. PWS (1996)
26. Kaufman, T., Krivelevich, M., Ron, D.: Tight Bounds for Testing Bipartiteness in General Graphs. SIAM Journal on Computing 33(6), 1441–1483 (2004)
27. Parnas, M., Ron, D.: Testing the diameter of graphs. Random Structures and Algorithms 20(2), 165–183 (2002)
28. Ron, D.: Algorithmic and Analysis Techniques in Property Testing. Foundations and Trends in TCS 5(2), 73–205 (2010)
29. Rubinfeld, R., Sudan, M.: Robust characterization of polynomials with applications to program testing. SIAM Journal on Computing 25(2), 252–271 (1996)

Another Motivation for Reducing the Randomness Complexity of Algorithms

Oded Goldreich

Abstract. We observe that the randomness-complexity of an algorithm effects the time-complexity of implementing a version of it that utilizes a weak source of randomness (through a randomness-extractor). This provides an additional motivation for the study of the randomness complexity of randomized algorithms. We note that this motivation applies especially in the case that derandomization is prohibitively costly.

Keywords: andomness Complexity, Weak Sources of Randomness, Randomness Extractors, Pseudorandom Generators, Sampling, Property Testing.

This article was completed in Nov. 2006, and appeared on the author's webpage.

1 Introduction: The Standard Motivations

The randomness-complexity of a randomized algorithm is a natural complexity measure associated with such algorithms. Furthermore, randomness is a "real" resource, and so trying to minimize the use of it falls within the standard paradigms of algorithmic design.

In addition to the aforementioned generic motivation (which was highlighted in [2]), there is a more concrete motivation (which was highlighted in [16]): If we manage to reduce the randomness-complexity to become sufficiently low, then this opens the door to a relatively efficient derandomization. Specifically, a randomized algorithm having time-complexity t and randomness-complexity r yields a functionally equivalent deterministic algorithm of time-complexity $2^r \cdot t$.

2 The Main Message: Another Motivation

In this article we highlight another concrete motivation for the study of the randomness-complexity of randomized algorithms. We refer to the effect of the randomness-complexity on the overhead involved in implementing the algorithm when using only weak sources of randomness (rather than perfect ones). Specifically, we refer to the paradigm of implementing randomized algorithms by using a single sample from such a weak source, and trying all possible seeds to a suitable randomness extractor (see next). We will show that the overhead created by this method is determined by the randomness-complexity of the original algorithm.

O. Goldreich et al.: Studies in Complexity and Cryptography, LNCS 6650, pp. 555–560, 2011.
© Springer-Verlag Berlin Heidelberg 2011

Randomness Extractors. Recall that a randomness extractor is a function E : $\{0,1\}^n \times \{0,1\}^s \to \{0,1\}^r$ that uses an s-bit long random seed in order to transform an n-bit long (outcome of a) weak source of randomness into an r-bit long string that is almost uniformly distributed in $\{0,1\}^r$. Specifically, we consider arbitrary weak sources that are restricted (only) in the sense that, for a parameter k, no string appears as the source outcome with probability that exceeds 2^{-k}. Such sources are called (n,k)-sources (and k is called the min-entropy). Now, E is called a (k,ϵ)-extractor if for any (n,k)-source X it holds that $E(X, U_s)$ is ϵ-close to U_r, where U_m denotes the uniform distribution over m-bit strings (and the term 'close' refers to the statistical distance between the two distributions). For further details about (k,ϵ)-extractors, the interested reader is referred to Shaltiel's survey [13].

Next we recall the standard paradigm of implementing randomized algorithms while using a source of weak randomness. Suppose that the algorithm A has time-complexity t and randomness-complexity $r \leq t$. Recall that, typically, the analysis of algorithm A refers to what happens when A obtains its randomness from a perfect random source (i.e., for each possible input α, we consider the behavior of $A(\alpha, U_r)$, where $A(\alpha, \omega)$ denotes the output of A on input α when given randomness ω). Now, suppose that we have at our disposal only a weak source of randomness; specifically, a (n,k)-source for $n \gg k \gg r$ (e.g., $n = 10k$ and $k = 2r$). Then, using a (k,ϵ)-extractor $E : \{0,1\}^n \times \{0,1\}^s \to \{0,1\}^r$, we can transform the n-bit long outcome of this source into 2^s strings, each of length r, and use the resulting 2^s strings (which are "random on the average") in 2^s corresponding invocations of the algorithm A. That is, upon obtaining the outcome $x \in \{0,1\}^n$ from the source, we invoke the algorithm A for 2^s times such that in the ith invocation we provide A with randomness $E(x, i)$. The results of these 2^s invocations are processed in the natural manner. For example, if A is a decision procedure, then we output the majority vote obtained in the 2^s invocations (i.e., when given the input α, we output the majority vote of the sequence $\langle A(\alpha, E(x, i)) \rangle_{i=1,\ldots,2^s}$).[1]

The analysis of the foregoing implementation is based on the fact that "on the average" the 2^s strings extracted from the source approximate a perfect r-bit long source (i.e., a random setting of the s-bit seed yields an almost uniformly distributed r-bit string). In the case of decision procedures this means that if A has error probability p and X is a (n,k)-source, then the number of values in $\langle E(X, i) \rangle_{i=1,\ldots,2^s}$ that fail $A(\alpha, \cdot)$ is at most $(p + \epsilon) \cdot 2^s$, where the expectation is taken over the distribution of X. It follows that the implementation (which rules by majority) errs with probability at most $2(p + \epsilon)$. This means that we should start with $p < 1/4$. (A similar analysis can be applied to the randomized search procedures discussed in Footnote 1.)

---

[1] For search problems in NP, we output any valid solution that is obtained in the relevant 2^s invocations. For general search problems (i.e., outside NP), some extra condition regarding the original randomized algorithm is required (e.g., either that it never outputs a wrong solution or that it outputs some specific correct solution with probability that exceeds 1/2 by a noticeable amount).

The Cost of the Use of a Weak Source of Randomness. Let us consider the cost of the foregoing implementation. We assume, for simplicity, that the running-time of the randomness extractor is dominated by the running-time of A. Then, algorithm A can be implemented using a weak source, while incurring an overhead factor of 2^s. Recalling that $s > \log_2 n$ (see [13]) and that $n > k > r$, it follows that the aforementioned overhead is at least linear in r. On the other hand, if $s = (1 + o(1)) \log_2 n$ and $n = O(r)$ (resp., $s = O(\log n)$ and $n = \text{poly}(r)$) then the aforementioned overhead is in fact linear in r (resp., polynomial in r). This establishes our claim that the time-complexity of implementing randomized algorithms when using weak sources is related to the randomness-complexity of these algorithms. Let us take a closer look at this relationship.

A Focus on Two Common Setting of n w.r.t r. We shall consider two types of (n, k)-sources, which are most appealing. Indeed, these type of sources have received a lot of attention in the literature. Furthermore, it suffices to set the deviation parameter of the extractor (i.e., ϵ) to a small constant (e.g., $\epsilon = 1/10$ will do). Recall that r denotes the number of bits that we need to extract for such a source (in order to feed it to our algorithm). The two cases we consider are:

1. *Linearly related n, k and r*; that is, for some constants $c > c' > 1$, it holds that $n = c \cdot r$ and $k = c' \cdot r$. In other words, we refer to sources having a constant rate of min-entropy.
   In this case, efficient randomness extractors that use $s = \log n + O(\log \log n) = \log_2 \tilde{O}(n)$ are known (cf. [15,13]). Using these extractors, we obtain an implementation of A (using such weak sources) with overhead factor $\tilde{O}(r)$.
2. *Polynomially related n, k and r*; that is, for some $c > c' > 1$, it holds that $n = r^c$ and $k = r^{c'}$. In other words, we refer to a source having min-entropy that is polynomially related to its length.
   In this case, efficient randomness extractors that use $s = \log \tilde{O}(n) = c \log_2 \tilde{O}(r)$ are known (cf. [14,13]). Using these extractors, we obtain an implementation of A (using such weak sources) with overhead factor $\tilde{O}(r^c)$.

In both cases, the overhead factor is approximately linear in the length of the source's outcome (which, in turn, is linearly or polynomially related to r).

   We wish to stress that the implementation paradigm considered above is most relevant in the case that a full derandomization (incurring an overhead factor of 2^r) is prohibitingly costly. Two settings in which this is inherently the case are considered next.

3   Two Settings: Sampling and Property Testing

Derandomization is not a viable option in the setting of sampling and property testing, and thus these settings provide a good demonstration of the importance of the new motivation. We start with the setting of sampling, although the more dramatic results are obtained in the context of property testing, which may be viewed as a generalization of the context of sampling.

3.1   Sampling

In many settings repeated sampling is used to estimate the average of a huge
set of values. Namely, given a "value" function $\nu : \{0,1\}^n \to \mathbb{R}$, one wishes
to approximate $\bar{\nu} \stackrel{\text{def}}{=} \frac{1}{2^n} \sum_{x \in \{0,1\}^n} \nu(x)$ without having to inspect the value
of ν at each point of the domain. The obvious thing to do is sampling the
domain at random, and obtaining an approximation to $\bar{\nu}$ by taking the average
of the values of ν on the sample points. It is essential to have the range of ν
be bounded (or else no reasonable approximation is possible). For simplicity, we
adopt the convention of having $[0,1]$ be the range of ν, and the problem for other
(predetermined) ranges can be treated analogously. Our notion of approximation
depends on two parameters: accuracy (denoted ϵ) and error probability (denoted
δ). We wish to have an algorithm that, with probability at least $1-\delta$, gets within
ϵ of the correct value. That is, a sampler is a randomized oracle machine that
on input parameters n, ϵ, δ and oracle access to *any* function $\nu : \{0,1\}^n \to [0,1]$,
outputs, with probability at least $1 - \delta$, a value that is at most ϵ away from
$\bar{\nu} \stackrel{\text{def}}{=} \frac{1}{2^n} \sum_{x \in \{0,1\}^n} \nu(x)$.

We are interested in "the complexity of sampling" quantified as a function
of the parameters n, ϵ and δ. Specifically, we will consider three complexity
measures: The sample-complexity (i.e., the number of oracle queries made by the
sampler); the randomness-complexity (i.e., the length of the random seed used
by the sampler); and the computational-complexity (i.e., the running-time of the
sampler). We say that a sampler is efficient if its running-time is polynomial in
the total length of its queries (i.e., polynomial in both its sample-complexity and
in n). It is easy to see that a deterministic sampler must have sample-complexity
close to 2^n, and thus derandomization is not an option here.

While efficient samplers that have optimal (up-to a constant factor) sample-
complexity are of natural interest, the motivation for the study of the randomness-
complexity of such samplers is less evident. Indeed, one may offer the generic
answer (i.e., that randomness as any other resource need to be minimized), but
the previous section shows that (in a very natural setting) there is a very concrete
reason to care about the randomness-complexity of the sampler: The randomness-
complexity of the sampler effects the sample-complexity of an implementation
that uses a weak random source.

Recall that the naive sampler uses sample-complexity $s \stackrel{\text{def}}{=} O(\epsilon^{-2} \log(1/\delta))$
and randomness-complexity $r = s \cdot n$. Using sources of constant min-entropy
rate, this yields an implementation of sample-complexity $s' \approx r \cdot s = n \cdot s^2$. However using a better sampler that has sample-complexity $O(s)$ but randomness-
complexity $r'' = 2n + O(\log(1/\delta))$ (cf. [1]), we obtain an implementation of
sample-complexity $s'' \approx r'' \cdot s = 2n \cdot s + O(\epsilon^2 s^2)$. This is a significant saving,
whenever good accuracy is required (i.e., ϵ is small).

3.2   Property Testing

This notion refers to a relaxation of decision problems, where it suffices to distinguish inputs having a (fixed) property from objects that are far from satisfying

the property (cf., e.g., original papers [11,6] or surveys [4,10]). Typically, one seeks sublinear-time algorithms that query the object at few randomly selected locations. In the natural cases, derandomization is not a viable option, because a deterministic algorithm must inspect at least a constant fraction of the object (and thus cannot run in sublinear-time). Let us clarify this discussion by looking at the specific example of testing the bipartiteness property for graphs of bounded-degree.

Fixing a degree bound d, the task is to distinguish (N-vertex) bipartite graphs of maximum degree d from (N-vertex) graphs of maximum degree d that are ϵ-far from bipartite (for some parameter ϵ), where ϵ-far means that $\epsilon \cdot dN$ edges have to be omitted from the graph in order to yield a bipartite graph. It is easy to see that no deterministic algorithm of $o(N)$ time-complexity can solve this problem. Yet, there exists a probabilistic algorithm of time-complexity $\tilde{O}(\sqrt{N}\mathrm{poly}(1/\epsilon))$ that solves this problem correctly (with probability 2/3). This algorithm makes $q \stackrel{\mathrm{def}}{=} \tilde{O}(\sqrt{N}\mathrm{poly}(1/\epsilon))$ incidence-queries to the graph, and (as described in the original work [8]) has randomness-complexity $r > q > \sqrt{N}$ (yet $r < q \cdot \log_2 N$).[2]

Let us now turn to the question of implementing the foregoing tester in a setting where we have access only to a weak source of randomness. In this case, the implementation calls for invoking the original tester $\tilde{O}(r)$ times, which yields a total running time of $\tilde{O}(r) \cdot \tilde{O}(\sqrt{N}\mathrm{poly}(1/\epsilon)) > N$. But in such a case we better use the standard (deterministic) decision procedure for bipartiteness!

Fortunately, a randomness-efficient implementation of the original tester of [8] is possible. This implementation (presented in [9,12]) has randomness-complexity $r' = \mathrm{poly}(\epsilon^{-1} \log N)$ (rather than $r = \mathrm{poly}(\epsilon^{-1} \log N) \cdot \sqrt{N}$). Thus, the cost of the implementation that uses a weak source of randomness is related to $r' \cdot s = \tilde{O}(\sqrt{N}\mathrm{poly}(1/\epsilon))$, which matches the original bound (up to differences hidden in the $\tilde{O}()$ and poly() notation). Needless to say, this is merely an example, and randomness-efficient implementations of other property testers are also presented in [9,12].

4   Conclusions

This essay articulates an additional motivation for studying the randomness-complexity of algorithms. This motivation is relevant even if one believes that generating a random bit is not more expensive than performing a standard machine instruction. It refers to the fact that the randomness-complexity of a randomized algorithm effects the running-time of an implementation that may only utilize a weak source of randomness (and does so by using a randomness-extractor). Specifically, such an implementation incurs a multiplicative overhead factor that is (at least) linearly related to the randomness-complexity of the original algorithm. This fact motivates the attempt to present randomness-efficient

---

[2] We comment that $\Omega(\sqrt{N})$ is a lower-bound on the query-complexity of any property tester of bipartiteness (in the bounded-degree model; see [7]).

versions of randomized algorithms and even justifies the use of pseudorandom generators for this purpose.[3]

Acknowledgments. We are grateful to Ronen Shaltiel for an update regarding the state-of-art of the problem of randomness extraction. We also wish to thank Adi Akavia and Shafi Goldwasser for a discussion that led to the current essay.

References

1. Bellare, M., Goldreich, O., Goldwasser, S.: Randomness in Interactive Proofs. Computational Complexity 4(4), 319–354 (1993)
2. Blum, M., Micali, S.: How to Generate Cryptographically Strong Sequences of Pseudo-Random Bits. SIAM Journal on Computing 13, 850–864 (1984); Preliminary Version in 23rd FOCS (1982)
3. Goldreich, O.: Foundation of Cryptography – Basic Tools. Cambridge University Press, Cambridge (2001)
4. Goldreich, O.: A Brief Introduction to Property Testing. In: Goldreich, O., et al.: Studies in Complexity and Cryptography. LNCS, vol. 6650, pp. 557–562. Springer, Heidelberg (2011)
5. Goldreich, O., Goldwasser, S., Micali, S.: How to Construct Random Functions. Journal of the ACM 33(4), 792–807 (1986)
6. Goldreich, O., Goldwasser, S., Ron, D.: Property testing and its connection to learning and approximation. Journal of the ACM, 653–750 (July 1998)
7. Goldreich, O., Ron, D.: Property testing in bounded degree graphs. Algorithmica, 302–343 (2002)
8. Goldreich, O., Ron, D.: A sublinear bipartite tester for bounded degree graphs. Combinatorica 19(3), 335–373 (1999)
9. Goldreich, O., Sheffet, O.: On the randomness complexity of property testing. In: Charikar, M., Jansen, K., Reingold, O., Rolim, J.D.P. (eds.) RANDOM 2007 and APPROX 2007. LNCS, vol. 4627, pp. 509–524. Springer, Heidelberg (2007)
10. Ron, D.: Algorithmic and Analysis Techniques in Property Testing. Foundations and Trends in TCS 5(2), 73–205 (2010)
11. Rubinfeld, R., Sudan, M.: Robust characterization of polynomials with applications to program testing. SIAM Journal on Computing 25(2), 252–271 (1996)
12. Sheffet, O.: M.Sc.Thesis, Weizmann Institute of Science (in preparation), http://www.wisdom.weizmann.ac.il/~oded/msc-os.html
13. Shaltiel, R.: Recent Developments in Explicit Constructions of Extractors. Bulletin of the EATCS 77, 67–95 (2002)
14. Shaltiel, R., Umans, C.: Simple Extractors for All Min-Entropies and a New Pseudo-Random Generator. In: 32nd IEEE Symposium on Foundations of Computer Science, pp. 648–657 (2001)
15. Ta-Shma, A., Zuckerman, D., Safra, S.: Extractors from Reed-Muller Codes. In: 32nd IEEE Symposium on Foundations of Computer Science, pp. 638–647 (2001)
16. Yao, A.C.: Theory and Application of Trapdoor Functions. In: 23rd IEEE Symposium on Foundations of Computer Science, pp. 80–91 (1982)

---

[3] The running-time of the generator itself can be ignored, because one may use a pseudorandom generator that runs in time that is almost-linear in the length of its output. Such a generator can be constructed by using a pseudorandom function (cf. [5]). Alternatively, such a generator is implied by the notion of an on-line pseudorandom generator (cf. [3, Chap. 3, Exer. 21]).

About the Authors

Oded Goldreich (oded@wisdom.weizmann.ac.il) is a Meyer W. Weisgal Professor at the Faculty of Mathematics and Computer Science of the Weizmann Institute of Science, Israel. Oded was born in 1957, and completed his graduate studies in 1983 under the supervision of Shimon Even. He was a post-doctoral fellow at MIT (1983-1986), a faculty member at the Technion (1986-1994), a visiting scientist at MIT (1995-1998), and a Radcliffe fellow at Harvard (2003-2004). Since 1995, he has been a member of the Computer Science and Applied Mathematics Department of the Weizmann Institute. He is the author of the books *Modern Cryptography, Probabilistic Proofs and Pseudorandomness* (Springer, 1998), *Computational Complexity: A Conceptual Perspective* (Cambridge University Press, 2008), and the two-volume work *Foundations of Cryptography* (Cambridge University Press, 2001 and 2004).

Lidor Avigad (avigadl@gmail.com) works in HP software. He completed his master thesis at the Weizmann Institute of Science in 2009.

Mihir Bellare (mihir@cs.ucsd.edu) is a Professor at the Department of Computer Science and Engineering of the University of California San Diego, USA. Mihir got a BSc from Caltech in 1986 and a PhD from MIT in 1991 under the supervision of Silvio Micali. He was a research staff member at IBM Research in New York from 1991 to 1995. He is a co-designer of the HMAC authentication algorithm (used to secure credit card numbers in Internet shopping) and the OAEP encryption scheme (used for RSA-based encryption). He has received the 2009 ACM Paris Kanellakis Theory and Practice Award, the 2003 RSA Conference Award in Mathematics, and a 1996 David and Lucille Packard Fellowship in Science and Engineering.

Zvika Brakerski (zvika.brakerski@weizmann.ac.il) is currently a PhD student at the Weizmann Institute of Science, Israel.

Shafi Goldwasser (shafi.goldwasser@gmail.com) is a Professor at the Computer Science Department of the Weizmann Institute of Science in Israel, and the Electrical Engineering and Computer Science Department at MIT. Shafi was born in 1958 in New York city, received her BSc in applied mathematics from Carnegie Mellon University in 1979, and completed her graduate studies in 1983 at the computer science department of UC Berkeley under the supervision of Manuel Blum. She joined the MIT faculty in 1984 and Weizmann faculty in 1994. Her research interests include cryptography, complexity theory, randomized algorithms, and computational number theory She is a member of the American Academy of Arts and Science, the US National Academy of Science and the US National Academy of Engineering.

Shai Halevi (`shaih@alum.mit.edu`) is a Research Staff Member in IBM T.J. Watson Research Center. Shai Halevi has a PhD in Computer Science from MIT (1997). His research covers many aspects of cryptography, he is a board member of the International Association for Cryptologic Research and an editor in ACM TISSEC. Shai served as a Program Chair for CRYPTO 2009, a Co-chair for TCC 2006, and as Program Committee member for many other conferences in cryptography.

Tali Kaufman (`kaufmant@mit.edu`) is a member of the Computer Science Department of Bar-Ilan University, Israel. She completed her graduate studies in 2006 under the supervision of Noga Alon and Michael Krivelevich and was a post-doctoral fellow at MIT and IAS (2006-2009) and in the Weizmann Institute (2010).

Leonid Levin (`http://www.cs.bu.edu/~lnd/`) is a Professor of Computer Science at Boston University. He obtained his master's degree in 1970 and a PhD equivalent in 1972 at Moscow University, where he studied under Andrey Kolmogorov. After emigration to USA in 1978, he also received a PhD at MIT in 1979.

Noam Nisan (`noam.nisan@gmail.com`) is a Professor of Computer Science and a member of the Rationality Center at the Hebrew University of Jerusalem and a part-time research scientist in Google, Tel Aviv. He got his PhD in Computer Science from UC Berkeley in 1988. During the last ten years his research has focused on the border of computer science, game theory, and economic theory. He has worked on computational complexity, and in particular on randomness in computation.

Dana Ron (`danar@eng.tau.ac.il`) is a Professor at the School of Electrical Engineering in Tel Aviv University, Israel. Dana was born in 1964, and completed her graduate studies in 1995 under the supervision of Naftali Tishby. She was an NSF post-doctoral fellow at MIT (1995-97) and a science scholar at the Bunting Institute, Radcliffe (1997-98). She was also a Radcliffe fellow at Harvard (2003/04). Since 1998 she has been a faculty member at Tel Aviv University.

Madhu Sudan (`madhu@mit.edu`) is a Principal Researcher at Microsoft Research, Cambridge, MA, USA and the Fujitsu Professor of Electrical Engineering and Computer Science at the Massachusetts Institute of Technology. Madhu received his BSc in Computer Science from the Indian Institute of Technology in 1987 and his PhD from the University of California at Berkeley, supervised by Umesh Vazirani, in 1992. From 1992 to 1997 he was a Research Staff Member at IBM's Thomas J. Watson Research Center. From 1997 he has been a member of the EECS Department at MIT. In 2009 he took a leave of absence to join Microsoft Research at their New England Research Center. Madhu was also Adjunct Associate Professor at the Tata Institute of Fundamental Research in Mumbai, India, from 1999 to 2002, and a Radcliffe fellow during 2003-2004. Madhu Sudan's principal interests are in computational complexity and communication. He is best known for his works on probabilistic checking of proofs and list decoding. His current interests include property testing and semantic communication.

Luca Trevisan (`trevisan@stanford.edu`) is a Professor of Computer Science at Stanford University. Luca received his Dottorato (PhD) in 1997, from the Sapienza University of Rome, working with Pierluigi Crescenzi. After graduating, Luca was a post doctoral fellow at MIT in 1997-1998 and at DIMACS in 1998, he was an Assistant Professor at Columbia University in 1999-2000, and on the faculty of the University of California, Berkeley, from 2000 to 2010. He joined Stanford University in 2010. Luca received the STOC 1997 Danny Lewin (best student paper) Award, the 2000 Oberwolfach Prize, and the 2000 Sloan Fellowship. He was an invited speaker at the 2006 International Congress of Mathematicians in Madrid.

Salil Vadhan (`salil@seas.harvard.edu`) is the Vicky Joseph Professor of Computer Science and Applied Mathematics at the Harvard School of Engineering Applied Sciences. Salil was born in 1973, and completed his graduate studies in 1999 at MIT under the supervision of Shafi Goldwasser. He was a postdoctoral fellow at MIT (1999-2000) and the Institute for Advanced Study (2000-2001), before joining the Harvard faculty in 2001. He was also a fellow at the Radcliffe Institute for Advanced Study at Harvard University (2003-2004) and a Miller Visiting Professor at UC Berkeley (2007-2008). His main research areas are computational complexity and cryptography, with particular interests in zero-knowledge proofs, pseudorandomness, and data privacy.

Avi Wigderson (`avi@ias.edu`) is the Herbert Maass Professor at the School of Mathematics of the Institute for Advanced Study in Princeton. Avi was born in 1956, and completed his graduate studies in 1983 at Princeton University under the supervision of Dick Lipton. After postdoctoral positions at UC Berkeley, IBM labs at San Jose and at MSRI, he became a faculty member at the Hebrew University in 1986. In 1999 he moved to the Institute for Advanced Study, where he is leading a program on theoretical computer science and discrete mathematics.

David Zuckerman (`diz@cs.utexas.edu`) holds an Endowed Professorship in the Computer Science Department at the University of Texas at Austin. David was born in 1965, and completed his graduate studies in 1991 under the supervision of Umesh Vazirani. He was an NSF Postdoctoral Fellow at MIT (1991-93), a Lady Davis Postdoctoral at the Hebrew University of Jerusalem (Fall 1993), a Visiting Scholar and Visiting MacKay Lecturer at UC Berkeley (1999-2000), and a Guggenheim Fellow and Radcliffe Fellow at Harvard (2004-05). Since 1994, he has been on the faculty of the Computer Science Department at the University of Texas at Austin. His research awards include a Guggenheim Fellowship, a David and Lucile Packard Fellowship for Science and Engineering, an Alfred P. Sloan Research Fellowship, and an NSF CAREER Award.

Printed in the United States
By Bookmasters